NONLINEAR OSCILLATIONS

BY

NICHOLAS MINORSKY
Former Professor, Stamford University

ROBERT E. KRIEGER PUBLISHING COMPANY
HUNTINGTON, NEW YORK
1974

Original Edition 1962
Reprint 1974

Printed and Published by
ROBERT E. KRIEGER PUBLISHING CO., INC.
645 NEW YORK AVENUE
HUNTINGTON, NEW YORK 11743

Library of Congress Catalog Card Number 74-8918
ISBN Number 0-88275-186-7

Printed in the United States of America

Library of Congress Cataloging in Publication Data

Minorsky, Nicholas, 1885-
Nonlinear oscillations.

Reprint of the 1962 ed. published by Van Nostrand, Princeton, N. J.
1. Oscillations. 2. Nonlinear theories.
I. Title.
[QA867.5.M56 1974] 531'.322 74-8918
ISBN 0-88275-186-7

TO MADELEINE

Preface

Ce qui nous rend ces solutions périodiques si précieuses, c'est qu'elles sont, pour ainsi dire, la seule brèche par où nous puissions essayer de pénétrer dans une place jusqu'ici reputée inabordable.

—*Henri Poincaré*

This book may be considered as a second edition of the D. Taylor Model Basin Reports issued toward the end of World War II. The intervening years have seen a widespread growth of interest in the theory of oscillations which, during the thirties, attracted little attention except in the USSR.

On major topics the theory of oscillations has gradually become generalized and assumed definitive form. However, new problems have raised new questions and on these the subject is still in a state of evolution.

Considerable interest of mathematicians in the problems of nonlinear oscillations has resulted in important advances in the theory of nonlinear differential equations but, as is to be expected, some of these advances have exceeded the immediate needs of the theory of oscillations and belong rather to the theory of differential equations per se. On the other hand, physicists and engineers continue to supply experimental material, the analysis of which requires special mathematical tools, some of which are not yet available. In view of this it is sometimes difficult to draw a line between what is known definitely and what is known only provisionally and subject to later revisions.

The theory of oscillations has apparently reached the state of its final codification in the domain of small parameters based on the fundamental researches of H. Poincaré, both topological[1] and analytical[2]; here the theory

[1] H. Poincaré, *J. des Math.* (3) 7, 1881; also Oeuvres, T. 1, Gauthier-Villars, Paris, 1928.

[2] H. Poincaré, *Les methodes nouvelles de la mecanique céleste*, Gauthier-Villars, Paris, 1882; also E. Goursat, *Cours d'Analyse*, T. 2, Gauthier-Villars, Paris, 1892.

and the experimental facts are well coordinated; if there are occasionally some difficulties, they are not essential. The first two parts deal with this important field: Part I is devoted to the qualitative (topological) methods, and Part II concerns the quantitative approaches. Both parts follow the aforementioned treatises of Poincaré, although in some of the later methods in Part II appear modifications in details but not in essence, since in all of them solutions are sought in the form of power series of certain small parameters.

In Part III devoted to applications, the aforementioned nonuniformity of our present understanding of problems posed by the experimental evidence is noticeable. Thus, in Chapter 21 one finds oneself on the relatively little explored field of nonlinear difference-differential equations where, besides some existence theorems, the mathematical developments have not progressed enough to offer a practical tool. Something similar appears in Chapter 25 where one feels the presence of certain "hereditary actions" (probably amenable to integro-differential equations) but, again, the problems being nonlinear, practically nothing is available on the mathematical end to enable one to account for the observed facts.

In Part IV dealing with relaxation oscillations, the situation is still further removed from a state of crystallization. Here, moreover, appears a real "parting of the ways" between the efforts of mathematicians on one hand and those of physicists and engineers on the other, the former giving rise to an asymptotic, the latter to a discontinuous, theory. To these two theories has been added recently a third based on the so-called "piecewise linear" idealization.

The difficulty here is that all these problems are characterized by such poor analyticity that it often seems easier to drop entirely the concept of analyticity and sometimes even that of continuity.

Part I of this edition underwent relatively small changes except for the addition of chapters 5 and 6 on stability. The theory of stability has developed considerably in recent years (mostly in the USSR) on the basis of Liapounov's ideas. The so-called *second* (or direct) *method* of Liapounov, in particular, appears now as a cornerstone for investigations of stability in various problems like those arising in connection with nonlinear control problems. Some changes appear in Chapter 7; they reappear later in Chapter 22; these problems of "bifurcations" turned out to be more important than one could think at the time of the first edition.

In Part II, which is devoted to a survey of the various methods of approximations and constitutes the quantitative part of the theory of oscillations, all chapters relate to recent developments except Chapter 9, which deals with the classical perturbation theory. Chapters 10, 11, 12, and 13 are based on the work of Malkin who has adapted the theory of

Poincaré to problems of the theory of oscillations. Chapters 14 and 15 summarize the so-called asymptotic methods of Bogoliubov and Mitropolsky representing a further generalization of the earlier work of Krylov and Bogoliubov; this asymptotic theory is perhaps the most general of all quantitative methods. Chapter 16 concerns recent work carried out by the author in collaboration with M. Schiffer on the so-called stroboscopic method based on the transformation theory; a number of problems are treated by this method in Part III.

In Part III, concerning various nonlinear oscillatory phenomena, the first three chapters follow closely the general theory; we indicate the standard procedure (of Poincaré) as well as that resulting from the use of the stroboscopic method. The subjects of Chapters 21 and 25 transcend the theory of ordinary differential equations and belong to certain functional equations; as this field has not yet been completely explored, some preliminary conclusions are obtained in the first approximation.

Part IV deals with relaxation oscillations; in this field, as we have already mentioned, there are two distinct theories: the discontinuous theory and the asymptotic one; the first is outlined in the first four chapters and the last in Chapter 30. The reason for this imbalance lies not in any preference of the author but in the fact that the present text is intended primarily for physicists and engineers whose interest in a theory lies primarily in the ease with which it explains phenomena; unfortunately developments in the asymptotic theory have not yet reached the stage when the theory can be applied easily to various problems, as can be observed from the last two sections of Chapter 30. Chapter 31 on piecewise linear idealization has been added recently; these developments (in the USSR) became known when this text had been completed. In fact, this method does not belong to either analytical or discontinuous methods and stands by itself because it involves certain idealizations in the differential equations themselves.

In a work of this nature it has been necessary to limit the exposition to certain topics, omitting others not because they are less interesting or less important, but because they are not needed for the program we have selected.

Moreover, we have given preference to the Russian literature as far as recent results are concerned. To some extent this is due to the fact that the Russian literature is likely to be less known to the western readers and because, having established the initial advance in this field (up to 1940), the Russian scientists maintain their leadership and initiative characterized by a remarkable coordination of efforts between the mathematical and the experimental parts of these fundamental researches.

The author wishes to express his gratitude to Professor M. Schiffer of

Stanford University for his many valuable contributions during the initial stages of this work, particularly in connection with the development of the stroboscopic method which turned out to be a useful tool for applied problems treated in Part III. He also wishes to express his gratitude to Professors G. Sansone and R. Conti of the University of Florence for many valuable discussions and to Dr. O. Plaat for his editorial work in Parts I and II. The author is also indebted to the late Dr. B. van der Pol whose brilliant grasp of nonlinear phenomena, particularly those treated in Part IV, led to many interesting discussions.

This work was carried out under the auspices of the Office of Naval Research; the author is grateful for the opportunity to undertake this work as well as for the many facilities offered to him during its progress.

Contents

Chapter *Page*

Preface v

Part I. Qualitative Methods

1 PHASE PLANE; SINGULAR POINTS 3

1. Introductory remarks 3
2. Theorem of Cauchy-Lipschitz; generalities 4
3. Phase plane 8
4. Singular points; elementary singular points 10
5. Examples of singular points of a linear d.e. 14
6. Canonical transformation; abridged equations 19
7. Distribution of singular points; parameter space 26
8. Center 32
9. Certain conclusions 38

2 NONLINEAR CONSERVATIVE SYSTEMS 40

1. Introductory remarks 40
2. Fundamental properties of nonlinear conservative systems 42
3. Motions in the large; separatrix 45
4. Effect of a parameter in a differential equation; bifurcation values 47
5. Problem of the rotating pendulum 48
6. Attraction of current-carrying conductors 51
7. Further properties of conservative systems; Hamiltonian variables; integral invariants 53
8. Oscillating circuit with no resistance but with a nonlinear inductance 62
9. Volterra's problem 65

Chapter		Page
3	LIMIT CYCLES OF POINCARÉ	71
	1. Definitions	71
	2. Examples of limit cycles	72
	3. Physical significance of limit cycles	74
	4. Polycyclic configurations	75
	5. The index of Poincaré	77
	6. Brouwer's fixed point theorem	80
	7. Negative criterion of Bendixson	82
	8. Poincaré-Bendixson theorem	84
	9. Cycles without contact; curve of contacts	85
	10. Complete topological configurations; singular points at infinity	91
	11. Nonanalytic cycles	96
	12. Topological configurations	98
4	GEOMETRICAL ANALYSIS OF PERIODIC SOLUTIONS	101
	1. Introductory remarks	101
	2. Liénard's theory; curves Γ and Δ; criterion of periodicity	103
	3. Liénard's phase plane; graphical construction of curves Γ	108
	4. Asymptotic cases of Liénard's equation	111
	5. Principle of symmetry	113
	6. Energy criterion in a nearly linear case	116
5	STABILITY (VARIATIONAL EQUATIONS, CHARACTERISTIC EXPONENTS)	118
	1. Introductory remarks	118
	2. Definition of stability (Liapounov)	118
	3. Variational equations	121
	4. Variational equations of singular points	123
	5. Variational systems with constant coefficients	124
	6. Linear systems with periodic coefficients	127
	7. Orbital stability	130
6	STABILITY (SECOND METHOD OF LIAPOUNOV)	134
	1. Introductory remarks	134
	2. Theorems of Liapounov (stability)	136
	3. Geometrical interpretation of Liapounov's theorems	138
	4. Certain auxiliary propositions concerning the functions V	140

Chapter		Page
	5. Construction of the function V for a differential system with constant coefficients	143
	6. Stability on the basis of abridged equations	144
	7. Certain generalizations	145
	8. Aiserman's problem	147
	9. Critical cases	150
	10. Functions V containing time explicitly	154
	11. Theorems of Liapounov for functions V containing t explicitly	155
	12. Criteria of stability for equations periodic in t	156
	13. Application to control theory	158
	14. Concluding remarks	161
7	THEORY OF BIFURCATIONS	163
	1. Introductory remarks	163
	2. Successor functions; geometry of bifurcation effects	165
	3. Bifurcation of a cycle from a focus	169
	4. Applications of the bifurcation theory	173
	5. Other bifurcation problems	177
	6. Examples of coalescing limit cycles	178
	7. Algebraic approach to the bifurcation theory	178
	8. Bifurcation diagrams; effect of bifurcation points on structures	182
	9. Noncritical systems	185
	10. Remarks	188
8	CYLINDRICAL AND TOROIDAL PHASE SPACES	190
	1. Introductory remarks	190
	2. Differential equation of an electromechanical system	193
	3. Cylindrical phase trajectories of a conservative system	195
	4. Cylindrical phase trajectories of nonconservative systems	197
	5. Topological configurations on the cylinder	202
	6. Oscillations of a synchronous motor	202
	7. Toroidal phase surface	204
	8. Examples of nonrecurrent trajectories	206
	9. Gliding flight	207

Part II. Quantitative Methods

	INTRODUCTION	211

Chapter		Page
9	PERTURBATION METHOD	217
	1. Secular terms	217
	2. Energy fluctuations in a van der Pol oscillator	219
	3. Lindstedt's method	224
	4. Theorem of Poincaré	228
10	PERIODIC SOLUTIONS (POINCARÉ)	232
	1. Introductory remarks	232
	2. Nonresonance oscillation of an (NA) system	234
	3. Resonance oscillation of an (NA) system	236
	4. Calculation of periodic solutions; example	240
	5. Autonomous systems	243
	6. Calculation of periodic solutions of autonomous systems	246
	7. Nonanalytic case	250
11	OSCILLATIONS IN SYSTEMS WITH SEVERAL DEGREES OF FREEDOM	252
	1. Introductory remarks	252
	2. Periodic solutions of homogeneous linear systems with constant coefficients	254
	3. Nonresonance oscillations of nonautonomous systems	257
	4. Resonance oscillations of nonautonomous systems	260
	5. Periodic resonance solutions of nonautonomous systems with nonanalytic d.e.	263
	6. Oscillations of autonomous systems	267
	7. Self-excited oscillations in coupled circuits	270
	8. Method of averaging	273
12	ALMOST PERIODIC OSCILLATIONS IN NEARLY LINEAR SYSTEMS	282
	1. Introductory remarks	282
	2. Almost periodic solutions	285
	3. Existence of A.P.S. in noncritical cases	289
	4. Transformation of Krylov-Bogoliubov; standard systems	290
	5. Almost periodic solutions of standard systems	291
	6. Almost periodic oscillations when all roots are critical	292
	7. General case: combination of critical and noncritical roots	294
	8. Van der Pol equation with two forcing terms	296

Chapter Page
9. Parameters of the generating system; nonresonance and resonance frequencies 299
10. Forced oscillations of a mono-rail car 301
11. Physical aspects of A.P. oscillations 306

13 DETERMINATION OF CHARACTERISTIC EXPONENTS 308
1. Determination of characteristic exponents on the basis of Poincaré's theory 308
2. Stability of periodic solutions 313
3. Determination of characteristic exponents by approximations 314
4. Second-order systems; invariants A_i 317
5. Zones of stability 319
6. Calculation of zones of stability 322

14 ASYMPTOTIC METHODS OF KRYLOV-BOGOLIUBOV-MITROPOLSKY (AUTONOMOUS SYSTEMS) 329
1. Introductory remarks 329
2. Successive approximations for the autonomous systems 329
3. Differential equations of the first approximation 333
4. Nonlinear damping 338
5. Self-excited systems 341
6. Stationary amplitudes and their stability 343
7. Equivalent linearization 348

15 ASYMPTOTIC METHODS OF KRYLOV-BOGOLIUBOV-MITROPOLSKY (NONAUTONOMOUS SYSTEMS) 356
1. Introductory remarks 356
2. Formulation of the problem 358
3. Successive approximations in nonresonance cases 360
4. Successive approximations for resonance oscillations 367
5. External periodic excitation of a nonlinear oscillator; jumps of amplitude 375
6. Nonstationary processes; slow time 380
7. Successive approximations for nonstationary processes; slow time 383
8. Oscillations of a pendulum with a variable length 387

Chapter Page

16 STROBOSCOPIC METHOD 390

1. Introductory remarks 390
2. Transformation of points and regions; planes (ψ) and (φ); stroboscopic image 390
3. Stroboscopic differential equations 395
4. Application of the stroboscopic method to the Mathieu oscillators 401
5. Application of the method to autonomous systems; second approximation 406
6. Further properties of the stroboscopic transformation 409
7. Existence and stability of the fixed point 411

17 GENERALIZATION OF NYQUIST'S DIAGRAM FOR NONLINEAR SYSTEMS 416

1. Introductory remarks 416
2. Theory of generalized Nyquist diagrams (Theodorchik) 419
3. Stationary state of self-excitation 422
4. Interaction of nonlinear oscillations 425
5. Stability 426
6. Retarded actions 428
7. Concluding remarks 431

Part III. Oscillations of Nearly Linear Systems

INTRODUCTION 433

18 SYNCHRONIZATION 438

1. Introductory remarks 438
2. Theory of van der Pol 439
3. Topological analysis of Andronov and Witt 441
4. Conditions of the stationary state of synchronization 443
5. Theory of synchronization by the stroboscopic method 444
6. Mutual synchronization 448
7. Other forms of synchronization 455

19 NONLINEAR RESONANCE 460

1. Introductory remarks 460
2. Subharmonics 462

Chapter		Page
	3. Theory of L. Mandelstam and N. Papalexi	464
	4. Application to the resonance of the order $\frac{1}{2}$	469
	5. Subharmonic resonance by the stroboscopic method	473
	6. Applications of the stroboscopic method	476
	7. Zones of subharmonic resonance	482
	8. Comparison of different methods	484
20	PARAMETRIC EXCITATION	488
	1. Introductory remarks	488
	2. General form of differential equation of parametric action	489
	3. Special cases—bifurcation surfaces; stability of the state of rest	492
	4. Phenomenon of Bethenod	495
	5. Origin of the parametric action	498
	6. Parametric excitation in electrical circuits	503
	7. Autoparametric excitation	506
	8. Parametric excitation by the asymptotic method	509
21	OSCILLATIONS CAUSED BY RETARDED ACTIONS	514
	1. Introductory remarks	514
	2. Difference-differential equations arising in applications	516
	3. Characteristic equation; neighborhood of a harmonic solution	518
	4. Advanced versus retarded actions	525
	5. Nonlinear problem; stationary state; frequency correction and stability	526
	6. On the physical nature of retarded actions	532
	7. Experimental evidence; electronic analogue	534
	8. Econometric and other problems	537
22	TOPOLOGY OF LIÉNARD'S EQUATION IN A PARAMETER SPACE	541
	1. Introductory remarks	541
	2. Formation of the stroboscopic equation	543
	3. Phase portraits of Liénard's equation	545
	4. Bifurcations of the third kind; more general cases	550
	5. Special cases of Liénard's equation	551
	6. Phase portraits of Rayleigh's and mixed equations	555
	7. Frequency correction	556

Chapter Page

8. Special forms of the van der Pol equation 557
9. Certain physical considerations 559

23 INTERACTION OF NONLINEAR OSCILLATIONS 562

1. Introductory remarks 562
2. The van der Pol theory of interaction 563
3. Interaction of two autoperiodic oscillations 566
4. Analysis of stability of singular points 567
5. Special cases 570
6. Remarks 571

24 ASYNCHRONOUS ACTIONS 572

1. Introductory remarks 572
2. Small external periodic excitation 574
3. Asynchronous quenching 576
4. Finite external periodic excitation 577
5. Asynchronous excitation 580
6. Concluding remarks 582

25 SYSTEMS WITH INERTIAL NONLINEARITIES 585

1. Introductory remarks 585
2. Inertial nonlinearity 587
3. Van der Pol oscillator with a conductor $R(x_0^2)$ 588
4. Oscillations produced by nonlinear conductors; physical considerations 591
5. General problem 592
6. Stability; concluding remarks 595

Part IV. Relaxation Oscillations

INTRODUCTION 599

26 DISCONTINUOUS THEORY OF RELAXATION OSCILLATIONS 605

1. Piecewise analytic phenomena 605
2. Degeneration theory and its physical significance 609
3. Conditions imposed by invariants 614
4. Discontinuous theory of Mandelstam-Chaikin 615

Chapter | Page

27 APPLICATION OF THE DISCONTINUOUS THEORY TO ELECTRICAL PROBLEMS 618
1. Degenerate RC oscillator 618
2. Oscillator with two degrees of freedom with degeneration in each degree 621
3. Connection between critical points and piecewise analytic phenomena 626
4. Symmetrical multivibrator circuits 627
5. Concluding remarks 639

28 APPLICATION OF THE DISCONTINUOUS THEORY TO MECHANICAL PROBLEMS 632
1. Introductory remarks 632
2. Mechanical relaxation oscillations 633
3. Relaxation oscillations of a Prony brake 633
4. Analogy between mechanical and electrical relaxation oscillations 637
5. Clocks 639
6. Froude's pendulum 644

29 DISCONTINUOUS THEORY OF VOGEL 648
1. Introductory remarks 648
2. Fundamentals of the theory 649
3. Analytic formulation; a special case 655
4. Physical interpretation 656
5. A numerical example 657
6. Multivibrator 658
7. A further extension of the theory 661
8. Concluding remarks 662

30 ASYMPTOTIC METHODS 664
1. Introductory remarks 664
2. Asymptotic theory versus discontinuous theory 665
3. Asymptotic theory applied to an asymmetrical multivibrator 670
4. Method of Cartwright-Littlewood 672
5. Asymptotic expansions 676
6. Method of Dorodnitzin 679
7. Concluding remarks 685

Chapter *Page*
31 PIECEWISE LINEAR IDEALIZATION 687
1. Introductory remarks 687
2. Point transformation method 688
3. Calculation of a piecewise linear limit cycle 690
4. Successor function and the fixed point 692
5. Successor function and the fixed point (hard self-excitation) 695
6. Topology of certain relay systems 699
7. Point transformation for $\beta < 1$ 701
8. Nonanalytic cycles and their stability 703
9. Remarks 705
10. Concluding remarks 706

PART I

QUALITATIVE METHODS

Chapter 1

PHASE PLANE; SINGULAR POINTS

1. Introductory remarks

This chapter may be considered as an investigation of various types of equilibria of a physical system with one degree of freedom on the basis of the theory of Poincaré.[1]

The essential aim of this study is the identification of singular points of a differential system with positions of equilibrium. Once this aim is reached, the remaining analysis follows closely the Poincaré theory with slight additions imposed by its physical interpretation. Thus Poincaré stresses more the geometry of curves defined by a differential equation (d.e.) in the neighborhood of singular points, whereas in the theory of

[1] H. Poincaré, *J. des Math.* (3), **7**, 1881; also Œuvres **T.1**, Gauthier-Villars, Paris, 1928.

This subject can also be found in any textbook on the theory of d.e. in the real domain, for instance:

(a) L. Bieberbach, *Differentialgleichungen*, Springer, Berlin, 1923.
(b) L. Cesari, *Asymptotic Behavior and Stability Problems*, Springer, Berlin, 1959.
(c) E. A. Coddington and N. Levinson, *Theory of Ordinary Differential Equations*, New York, 1955.
(d) E. Kamke, *Differentialgleichungen reeler Functionen*, Leipzig, 1930.
(e) S. Lefschetz, *Differential Equations (Geometric Theory)*, Interscience Publishers, New York, 1957.
(f) V. V. Nemitzky and V. V. Stepanov, *Qualitative Theory of Differential Equations*, original text in Russian, Moscow, 1949; English translation, Princeton Mathematics Series, Princeton University Press, Princeton, N.J., 1960.
(g) E. Picard, *Traité d'Analyse* **T.3**, Gauthier-Villars, Paris, 1928.
(h) G. Sansone and R. Conti, *Equazioni differenziali non lineari*, Ed. Cremonese, Roma, 1956.
(i) F. Tricomi, *Equazioni differenziali*, 2nd ed., Einaudi, Torino, 1953.

oscillations it is useful to take also into account the directional and time-dependent element of motion of the "representative point" on these curves, in which case they are considered as *trajectories* rather than as purely geometrical curves. In this manner appear some refinements in the original theory; thus, in addition to the definition of nodal and focal points there appears the property of *stability* according to the direction of motion of the representative point.

These extensions, however, do not change the intrinsic character of the theory but concern rather its *interpretation* in applied problems.

In Section 2 the Cauchy-Lipschitz theorem is stated in its general form, in which it asserts the existence of solutions of a system of differential equations with prescribed initial conditions and brings it into relation with the phase-plane representation which follows from the representation of a d.e. of the second order by an equivalent system of two first-order equations.

2. Theorem of Cauchy-Lipschitz; generalities

The theorem of Cauchy regarding the existence and uniqueness of a solution of a d.e. can be found in any textbook of differential equations and, for that reason, we shall not be concerned with its proof here. Our primary object will be to establish certain geometric consequences of this theorem which will be useful in that which follows.

We first explain what is meant by a "Lipschitz condition." Let f be a function of the m variables: $x_1, x_2, \ldots, x_m$. We say that $f(x_1, x_2, \ldots, x_m)$ satisfies a Lipschitz condition if, given values $x_1^0, x_2^0, \ldots, x_m^0$ of these variables, there are positive numbers k and δ such that the relation

$$|f(x_1^0, x_2^0, \ldots, x_m^0) - f(x_1, x_2, \ldots, x_m)| \leq k \sum_{i=1}^{m} |x_i^0 - x_i|$$

holds, provided $|x_i^0 - x_i| < \delta$; $i = 1, 2, \ldots, m$. The number k depends on $x_1^0, x_2^0, \ldots, x_m^0$ but not on $x_1, x_2, \ldots, x_m$. If the partial derivatives $\partial f/\partial x_i$ exist and are continuous, it can be shown that f satisfies a Lipschitz condition. Let the n functions $f_i(t, x_1, x_2, \ldots, x_n)$; $i = 1, 2, \ldots, n$ of the $n+1$ variables: $t, x_1, x_2, \ldots, x_n$ satisfy Lipschitz conditions, and let $t_0, x_1^0, x_2^0, \ldots, x_n^0$ be prescribed.

The theorem of Cauchy-Lipschitz then states that the system of the n first-order d.e.

$$\frac{dx_i}{dt} = f_i(t, x_1, x_2, \ldots, x_n); \qquad i = 1, 2, \ldots, n \tag{2.1}$$

has a unique solution $x_i = x_i(t)$; $i = 1, 2, \ldots, n$ defined in the neighborhood of $t = t_0$ such that $x_i(t_0) = x_i^0$; $i = 1, 2, \ldots, n$. The theorem thus guarantees not merely the existence of solutions with prescribed initial conditions, but it asserts that the initial conditions determine the solution uniquely. The d.e. we shall consider will be Lipschitzian, unless the contrary is explicitly indicated. To apply this theorem to a single d.e. of order n we solve the d.e. for the highest derivative and write it in the form

$$x^{(n)} = \frac{d^n x}{dt^n} = f(t, x, \dot{x}, \ldots, x^{(n-1)}) \tag{2.2}$$

This d.e. is equivalent to the system of n first-order d.e.

$$\dot{x}_1 = x_2; \quad \dot{x}_2 = x_3; \ldots, \quad x^{(n)} = f(t, x_1, x_2, \ldots, x_{n-1}) \tag{2.3}$$

where we have written x_1 in place of x for the sake of uniformity. This system is a special case of that covered by the Cauchy-Lipschitz theorem, which now states that the solutions of (2.2) are uniquely determined by the values prescribed for x and its first $n-1$ derivatives at $t = t_0$ provided that the function f satisfies a Lipschitz condition.

For later reference we collect at this point a number of theorems concerning the dependence of the solutions of a system of differential equations on the initial conditions and on parameters occurring in the functions f_i. To avoid confusing notation we denote the initial values of the x_i by ξ_i instead of by x_i^0 and we write τ instead of t_0. Let us denote the (unique) solution of (2.1) which at time $t = \tau$ has the value $x_i = \xi_i$ by: $x_i = \varphi_i(t, \tau, \xi_1, \xi_2, \ldots, \xi_n)$. It follows that $\varphi_i(\tau, \tau, \xi_1, \xi_2, \ldots, \xi_n) = \xi_i$; $(i = 1, 2, \ldots, n)$. We state first:

If the functions $f_i(t, x_1, \ldots, x_n)$ satisfy Lipschitz conditions, then the functions $\varphi_i(t, \tau, \xi_1, \ldots, \xi_n)$ are continuous in $\tau, \xi_1, \xi_2, \ldots, \xi_n$. We have furthermore: *If the partial derivatives $\partial f_i/\partial x_j$; $i, j = 1, \ldots, n$ are continuous, then likewise the functions φ_i have continuous partial derivatives with respect to τ and $\xi_1, \ldots, \xi_n$.*

More generally it is true that the solutions φ_i are as differentiable with respect to $\tau, \xi_1, \ldots, \xi_n$ as the f_i are with respect to $t, x_1, \ldots, x_n$. Turning now to analytic systems, let the f_i be analytic at $t = t_0$, $x_i = x_i^0$. The result in this case is:

The functions $\varphi_i(t, \tau, \xi_1, \ldots, \xi_n)$ are analytic in all of their arguments in a neighborhood of $t = t_0$, $\tau = t_0$, $\xi_i = x_i^0$.

It is possible to make estimates concerning the domain of analyticity of the φ_i given that of the f_i, but we shall not need these.

Finally we consider systems of the form (2.1) in which the right-hand sides depend on a parameter μ, thus a system of the form

$$\dot{x}_i = f_i(t, x_1, \ldots, x_n, \mu)$$

We have:

If the f_i are continuous in their arguments and satisfy Lipschitz conditions in $t, x_1, \ldots, x_n$ uniformly in μ in a neighborhood of μ_0, then the solutions $x_i = \varphi_i(t, \tau, \xi_1, \ldots, \xi_n, \mu)$ are unique and continuous in all their arguments for μ in the same neighborhood of μ_0.

The φ_i are as differentiable in their arguments as the f_i are in theirs.

If the f_i are analytic at $t = t_0$, $x_i = x_i^0$, $\mu = \mu_0$, then the φ_i are analytic in all their arguments in a neighborhood of $t = t_0$, $\tau = t_0$, $\xi_i = x_i^0$, $\mu = \mu_0$.

The analyticity in the initial conditions and the parameter will be important to us in Part II.

The results concerning a single parameter can be extended in the obvious way to the case in which the f_i depend on several parameters $\mu_1, \mu_2, \ldots, \mu_k$. If the functions f_i occurring in (2.1) do not depend on the time, the system is said to be *autonomous*, a term whose significance will be more readily understood when we begin considering connections with physical problems.

In this chapter we shall be interested in systems of the form

$$\dot{x} = P(x,y); \qquad \dot{y} = Q(x,y) \tag{2.4}$$

that is, autonomous systems of two first-order equations. By the equivalence mentioned above, these include the special case of one second-order d.e. which is typical of physical systems with one degree of freedom. In the case of (2.4) it is possible to eliminate dt between the equations and write

$$\frac{dy}{dx} = \frac{Q(x,y)}{P(x,y)}; \qquad P(x,y) \neq 0 \tag{2.5}$$

which is a d.e. of integral curves. If $P = 0$, $Q \neq 0$, we may interchange the roles of x and y and consider the d.e.: $dx/dy = P/Q$, in which case the integral curves are given in the form $x = \alpha(y)$.

For the special case of a d.e. of the second order, $\ddot{x} + f(x,\dot{x}) = 0$, one can reduce it to the system (2.4) of the two d.e. of the first order by setting $\dot{x} = y$, which gives the *equivalent system*:

$$\dot{x} = y = P(x,y); \qquad \dot{y} = -f(x,y) = Q(x,y) \tag{2.4a}$$

and the d.e. (2.5) becomes

$$\frac{dy}{dx} = -\frac{f(x,y)}{y} = \frac{Q(x,y)}{P(x,y)}; \qquad P(x,y) \neq 0 \tag{2.5a}$$

It is to be noted that (2.4) and (2.5) are equivalent, that is, have the same integral curves with that difference, however, that (2.5) gives a geometrical curve without any reference to what happens *in time*, whereas (2.4), in addition, tells how this curve is described in time and direction by the representative point $R = R[x(t),y(t)]$ specifying the instantaneous state of the physical system.

This representation of the integral curve in the parametric form

$$x = x(t) = \varphi(t - t_0, x_0, y_0); \qquad y = y(t) = \psi(t - t_0, x_0, y_0) \qquad (2.5b)$$

is called a *trajectory*, a term which will be used later. It is to be noted that this applies only to *autonomous systems* like (2.5); that is, such systems in which the independent variable t does not enter explicitly. In fact, only in this case it is possible in general to eliminate t between equations (2.4). Another remark is noteworthy. If one considers the quantities t_0, x_0, y_0 fixed (that is, given initial conditions), clearly $x(t)$ and $y(t)$ represent the solution of the d.e. having for $t = t_0$ definite initial conditions x_0, y_0 which determines the solution. But from the translation property of autonomous d.e. it is known that if one replaces t by $t + t_0$, where t_0 is an arbitrary constant (the phase), one has still the solution of the same d.e. In physical language the solution $x(t), y(t)$ specifies a certain *motion*; the fact that t is replaced by $t + t_0$ means clearly another motion (with a different phase t_0) and the just mentioned property of translation of autonomous systems can be stated differently, namely:

To a given trajectory corresponds an infinity of motions (solutions) *differing from each other by the phase.*

This property of autonomous systems is very convenient for the geometric study of integral curves and this and the following three chapters are devoted to the geometry (or topology) of integral curves defined by autonomous d.e. which constitutes the fundamental contribution of Poincaré.

In the case when a d.e. contains t explicitly, this topological procedure ceases to hold. Assume, for instance, that instead of (2.4) we have d.e. of the form

$$\dot{x} = P(t,x,y); \qquad \dot{y} = Q(t,x,y) \qquad (2.6)$$

It is clear that we cannot determine the integral curve as we did in the autonomous case by a simple passage from (2.4) to (2.5). In the non-autonomous case the direction field

$$\frac{dy}{dx} = \frac{Q(t,x,y)}{P(t,x,y)} \qquad (2.7)$$

varies in time and it is meaningless even to speak about integral curves in the sense of "trajectories," etc. In fact, if one assumes as previously the phase-plane representation (see next section), one encounters absurd situations; for instance, trajectories may even intersect each other which is contrary to the theorem of Cauchy-Lipschitz and so on. One can obviate this difficulty if one introduces t as a *third dimension*, but the advantage of a planar representation is lost and the procedure becomes impracticable.

Only under very special conditions (which will be encountered later) is it possible to use the phase-plane representation for nonautonomous systems, but this is rather an exception than the rule. In view of this, one has to bear in mind that Part I deals only with autonomous systems.

Summing up, we shall primarily be concerned with the d.e. in the form (2.4) which gives a space-time representation of an oscillatory phenomenon; occasionally, when only the geometry of integral curves is of interest, we shall use the d.e. (2.5). Hence, unless otherwise specified, we shall always deal with the d.e. in the form (2.4), and in this regard the following definitions are important.

Any point (x_0,y_0) for which the two functions $P(x_0,y_0)$ and $Q(x_0,y_0)$ do not vanish *simultaneously* is called *an ordinary point* with respect to the d.e. A point (x_0^*,y_0^*) for which $P(x_0^*,y_0^*) = Q(x_0^*,y_0^*) = 0$ is called a *singular point*.

3. Phase plane

The introduction of the variable $y = \dot{x}$ in (2.4a) suggests the investigation of integral curves in the plane of the variables (x,y), called the *phase plane*. As an example of this representation, consider the d.e. of a harmonic oscillator in its reduced form $\ddot{x} + x = 0$. Written as a system (2.4a) it is:

$$\dot{x} = y; \qquad \dot{y} = -x \tag{3.1}$$

Multiplying the first equations by x and the second by y and adding the two d.e. and then integrating, one has:

$$x^2 + y^2 = r_0^2 = \text{const}$$

which represents a circle in the phase plane. The direction of motion of the point R on the integral curve is obtained from the system (3.1). Since x and $\dot{x}$ are merely the cartesian coordinates in this representation, in the first quadrant, for instance, one has $x > 0$, $y > 0$, and it is seen that $\dot{x} > 0$ and $\dot{y} < 0$, which shows that the positive direction of the integral curve is clockwise as shown by the arrow in Fig. 1.1. Introducing the polar angle φ of the radius vector r, one has $\cos\varphi = x/r_0$, and differentiating with respect to t, one obtains $\dot{\varphi} = -1$. Thus the usual representation (in the x,t plane) of the integral curve of the harmonic oscillator by a sinusoidal curve with t on the abcissa axis is replaced in the phase plane by a polar representation by a circle described by the representative point with a constant angular velocity $\dot{\varphi} = -1$, the minus sign merely showing that the rotation takes place in the direction opposite to that in which the angles are counted as positive—a fact which has been already noted directly from the d.e. (3.1).

The dependence of the solution on two arbitrary constants of integration also appears in the phase-plane representation. One of these constants is obviously r_0^2 (or r_0) since, if one changes the initial conditions x_0, y_0, then r_0 changes and one thus obtains a continuous family of circles depending on the parameter r_0. For a given r_0 there is still another family of motions depending on the initial phase φ_0. One thus has a very simple and compact representation of the totality of solutions of the harmonic oscillator expressed in terms of two constants of integration.

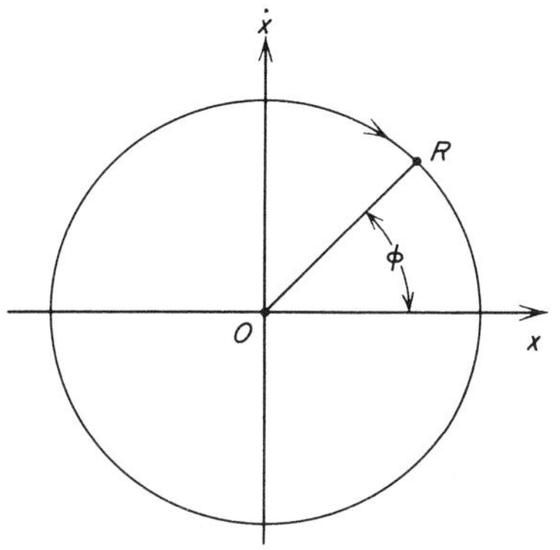

FIGURE 1.1

If one takes the d.e. of the harmonic oscillator in a nonreduced form:

$$\ddot{x} + \omega_0^2 x = 0 \tag{3.2}$$

the result is similar, as one easily verifies by means of a similar argument. One can also obtain the same result by multiplying (3.2) by $\dot{x}$ and integrating, which gives the *first integral*:

$$\dot{x}^2 + \omega_0^2 x^2 = 2C$$

C being an arbitrary constant. If one divides by $2C$ and sets $\dot{x} = y$; $2C = \beta^2$; $2C/\omega_0^2 = \alpha^2$, one obtains the equation of an ellipse with the semi-axes α and β:

$$\frac{x^2}{\alpha^2} + \frac{y^2}{\beta^2} = 1 \tag{3.3}$$

which represents the integral curves in the phase plane (Fig. 1.2). The positive direction on this curve is obtained again from the representation of (3.2) as a system:

$$\dot{x} = y; \qquad \dot{y} = -\omega_0^2 x$$

One has again a continuous family of ellipses corresponding to a continuous variation of the arbitrary constant C. There is still another continuous family of possible motions corresponding to the *same ellipse* but counted with different initial phase angles. At any instant t the projection of the radius vector r on the x axis gives the coordinate x, and the projection on the $y = \dot{x}$ axis gives the velocity $\dot{x}$ corresponding to this instant. In this case, the angular velocity $\dot{\varphi}$ of the radius vector does not remain constant, as in the case of the reduced equation $\ddot{x} + x = 0$, but fluctuates between a maximum and a minimum as the radius vector rotates between the semi-axes of the ellipse.

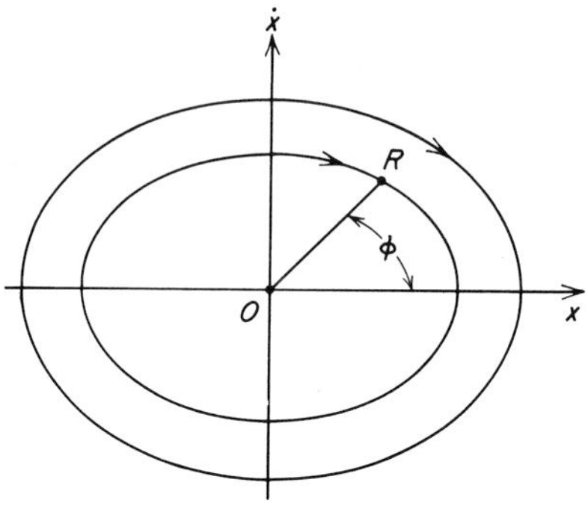

FIGURE 1.2

It is to be noted that the d.e. (3.2) can always be reduced to the form $\ddot{x} + x = 0$ by a change of the independent variable $\tau = \omega_0 t$, which transforms the elliptic integral curves into circles.

4. Singular points; elementary singular points

The Cauchy-Lipschitz theorem applied to the autonomous system

$$\dot{x} = P(x,y); \qquad \dot{y} = Q(x,y) \tag{4.1}$$

has, as a consequence, that *through every point of the plane there passes one and only one integral curve.* We may think of (4.1) as a "flow" in the phase plane defined by the velocity-vector field. The direction of motion at each point being specified by this vector, it follows that the integral curves are tangent at every point to this vector and are, in fact, completely determined by this requirement. If two distinct integral curves were to have a point in common, they would then have to be tangent at this point, a possibility ruled out by the fact that P and Q are Lipschitzian. Thus we arrive at the conclusion that two distinct integral curves have no point in common.

A singular point (x_0,y_0) is a stationary point of the flow, $P(x_0,y_0) = Q(x_0,y_0) = 0$ and the integral curve passing through it consists just of the point itself. This fact can also be deduced directly from (4.1) by noting that $x(t) \equiv x_0$; $y(t) \equiv y_0$ is a solution of (4.1). Now, if $x(t)$, $y(t)$ is an arbitrary solution passing through (x_0,y_0), that is, such that $x(t_0)$

$= x_0$; $y(t_0) = y_0$ for some t_0, then the uniqueness part of the Cauchy-Lipschitz theorem tells us that

$$x(t) \equiv x_0; \qquad y(t) \equiv y_0$$

If (4.1) is obtained from a second-order d.e. describing a dynamical system via the substitution $\dot{x} = y$, then a singular point (x_0,y_0) has a simple interpretation in terms of that dynamical system. In this case $P(x,y) = y$, so that $y_0 = 0$ and, since x is position and y velocity, the fact that $x(t) \equiv x_0$; $y(t) \equiv 0$ is a solution tells us that $x = x_0$ is a position of equilibrium of the dynamical system. We are thus led to identify singular points with positions of equilibrium, and this is of fundamental importance in what follows. In particular, the asymptotic behavior of the trajectories in the neighborhood of a singular point determines the type of equilibrium represented by the singular point.

A further consequence of the Cauchy-Lipschitz theorem follows. If a trajectory passes through an ordinary point, it cannot approach a singular point in finite time; more precisely, if $x = x(t)$; $y = y(t)$ is a nonconstant solution and $x(t) \to x_0$ and $y(t) \to y_0$ as $t \to t_0$, where (x_0,y_0) is a singular point, then $t_0 = \pm\infty$. A singular point is said to be asymptotically *stable* if all trajectories starting sufficiently near it tend to it asymptotically as $t \to \infty$. If there is a trajectory which tends asymptotically to the singular point as $t \to -\infty$, the singular point is said to be asymptotically *unstable*. In the case of an harmonic oscillator the origin is a singular point which is neither stable nor unstable as we have defined these terms, since every trajectory forms a closed curve surrounding the origin (Fig. 1.1 or 1.2). Having in mind the asymptotic character of motion in the neighborhood of a position of equilibrium, one can give a general classification of singular points. We follow closely the basic work of Poincaré. We start with a special case and consider the systems

$$\dot{x} = x; \qquad \dot{y} = ay \tag{4.2}$$

and

$$\dot{x} = -x; \qquad \dot{y} = -ay \tag{4.3}$$

whose solutions are $x = C_1e^t$; $y = C_2e^{at}$ and $x = C_1e^{-t}$; $y = C_2e^{-at}$, respectively. The origin $x = y = 0$ is a singular point for both systems, and the coordinate axes are integral curves. The integral curves on which $x \neq 0$ satisfy the d.e.

$$\frac{dy}{dx} = \frac{ay}{x} \tag{4.4}$$

which has the solution $y = C|x|^a$; this relation can also be deduced directly from the solutions of either system. Thus both systems have the same integral curves. The significance of the signs will appear presently. The integral curves are parabolic if $a > 0$ and hyperbolic if $a < 0$. These two cases give rise to different types of singular points. For reasons of symmetry we may restrict the discussion of the integral curves to positive values of x.

Node. We consider first the case $a > 0$. If $a > 1$, as $dy/dx = Cax^{a-1}$, then $(dy/dx) \to 0$ for $x \to 0$. Hence every integral curve with the exception of the y axis approaches the singular point along the x axis, as is shown in Fig. 1.3a. If $a < 1$, $dy/dx = Ca(1/x^{1-a})$ and in this case every

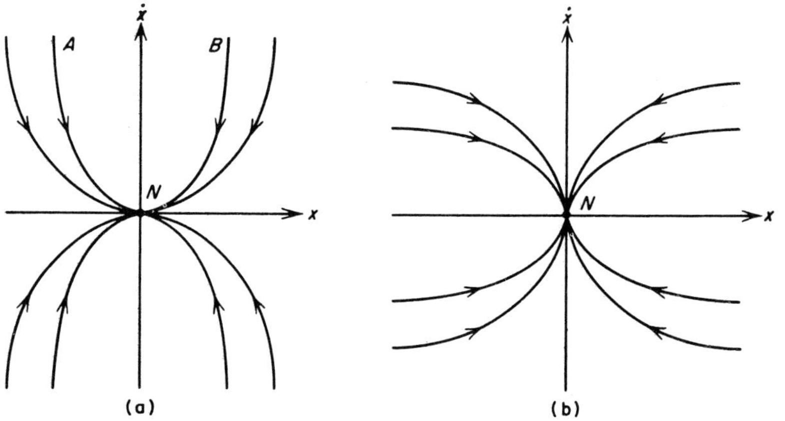

FIGURE 1.3

integral curve with the exception of the x axis approaches the singular point along the y axis (Fig. 1.3b). If $a = 1$, the integral curves are half-lines converging to or radiating from the singular point (Fig. 1.4). Again, *every integral curve has a limiting direction at the singular point*, and this property is taken as the definition of a *node*, or nodal point. If, given any direction, there is an integral curve having this limiting direction at the node, as is the case in the example when $a = 1$, the node is called a proper node, or star. If every integral curve, or every integral curve but one (as in the case $a \neq 1$), has the same limiting direction, the node is called an improper node.

The significance of the signs in (4.2) and (4.3) appears in the solutions. The trajectories of (4.3) actually approach the node (as $t \to \infty$) while the trajectories of (4.2) have the reverse direction. In the first case the node is stable and in the second it is unstable.

We remind the reader that a given parabolic curve such as ANB in

Fig. 1.3a consists, properly speaking, of three different trajectories: AN and BN without N and a singular trajectory consisting of the single point N. All trajectories converge to (or diverge from) the node, but never cross it.

Saddle point. If $a < 0$, the integral curves other than those lying on the coordinate axes are the hyperbolic curves: $y|x|^{|a|} = C$. For $|a| = 1$, they are the ordinary hyperbolas shown in Fig. 1.5. The coordinate axes are the asymptotes of the family. This singular point is called a *saddle point* (in French, *col*). The directions indicated in Fig. 1.5 are those corresponding to (4.2). Only four trajectories tend to the singular point: AS and BS for $t \to \infty$ and DS and CS for $t \to -\infty$. For (4.3) the directions are opposite to those shown in Fig. 1.5. It is clear that a saddle point is always an unstable singularity.

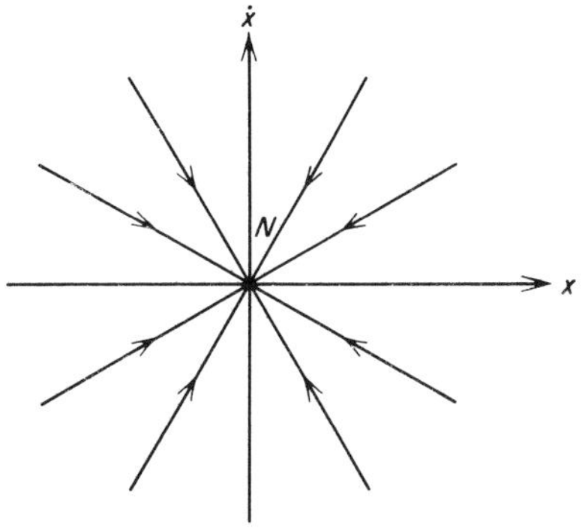

FIGURE 1.4

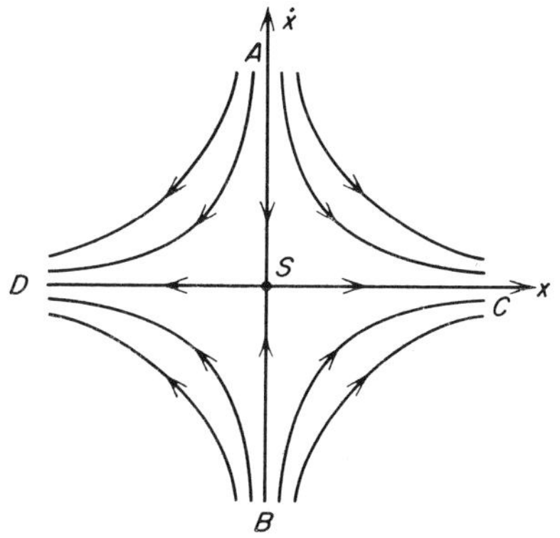

FIGURE 1.5

Focus. To introduce the idea of a focal point we consider the system

$$\frac{dx}{dt} = -ax + y; \qquad \frac{dy}{dt} = -x - ay \tag{4.5}$$

where a is a positive constant. It is convenient to conduct the discussion in polar coordinates by setting $x = r\cos\varphi$; $y = r\sin\varphi$; $r = \sqrt{x^2 + y^2}$; $\varphi = \arctan(y/x)$. In the new variables the d.e. is $\dot{r} = -ar$, $\dot{\varphi} = -1$, which has the solution

$$r = C_1 e^{-at}; \qquad \varphi = -t + C_2 \tag{4.6}$$

The trajectories are thus logarithmic spirals approaching the singular point F at the origin (Fig. 1.6). The rotation of the radius vector is clockwise with a constant angular velocity $\dot{\varphi} = -1$. Every trajectory, in this case, approaches the singular point, which is called a *focal point* (or focus; also, spiral point; in French, *foyer*) *without any definite direction* as the spiral rotates around the singular point an infinite number of turns as it tends to it. If $a < 0$ in (4.5), the preceding conclusions remain except that the motion of R takes place in the opposite direction, so that instead of approaching the singular point F, it *departs* from it. In this way, as in the case of a nodal point, we have to distinguish between the stable focal points ($a > 0$) and the unstable ones ($a < 0$). The whole family of spirals either converges toward, or diverges from, F according to the sign of a.

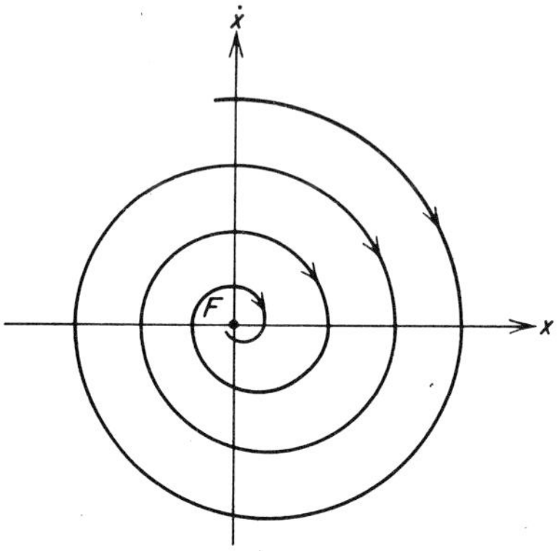

FIGURE 1.6

The nodes, foci, centers, and saddle points constitute the *elementary singularities* of a d.e. in the real domain. We shall be concerned almost exclusively with these in what follows. Besides these, one encounters, occasionally, singularities of higher order, some of which arise from a confluence of two or more simple singular points. In the applications, such cases generally lead to somewhat special conditions of equilibrium as will be mentioned later. Mathematical difficulties here are considerably greater and connections with physical problems are less explored. We do not propose to enter further into the question of these higher-order singularities, and refer the reader to the existing publications.[1]

5. Examples of singular points of a linear d.e.

In the preceding section, we specified the properties of elementary singular points in the cases of particularly simple forms (4.2) and (4.5) of the d.e. Similar conclusions can be obtained for much more general d.e. We shall see later that the nonlinearity of a d.e. generally does not affect the elementary singular points, so that it is useful to consider first the general case of a linear d.e. of the second order with constant coefficients:

$$\ddot{x} + 2b\dot{x} + \omega_0^2 x = 0 \tag{5.1}$$

which represents a damped harmonic motion. It is well known that,

[1] See footnote [1] page 3.

according to whether $b^2 - \omega_0{}^2 > 0$ or $b^2 - \omega_0{}^2 < 0$, one has either an aperiodic or an oscillatory damped motion. In the first case, the singular point of (5.1) is a node; and in the second, a focus. Hence, nothing especially new is gained here as compared to what has already been outlined in the preceding section. The only interest here lies in the manner in which these more complicated cases are reduced to the simpler forms investigated in the preceding section.

Consider first the aperiodically damped case ($b^2 - \omega_0{}^2 > 0$), when the solution of (5.1) is of the form:

$$x(t) = Ae^{-r_1 t} + Be^{-r_2 t} \tag{5.2}$$

A and B being the constants of integration and r_1 and r_2 the negatives of the roots of the characteristic equation, which are *real*. Since our purpose is to investigate the behavior of integral curves in the phase plane, one has to determine $\dot{x}(t) = y(t)$ from (5.2) and then to eliminate the time t between $x(t)$ and $y(t)$. If one carries out this calculation, one obtains the trajectory in the phase plane (or, simply the phase trajectory):

$$|r_1 x + y|^{r_1} = C|r_2 x + y|^{r_2} \tag{5.3}$$

Introducing new variables $v = r_1 x + y$ and $u = r_2 x + y$, (5.3) becomes

$$v = C|u|^a; \qquad a = r_2/r_1 \tag{5.4}$$

and this is already in the form that was used for the investigation of properties of the nodal point. The only point to be noted is the return from the v,u variables to the original ones, x,y. This is a problem of analytic geometry which we do not treat here, giving only the result. The parabolic curves shown in Fig. 1.7a become oblique in the (x,y) plane (Fig. 1.7b). However, the fundamental property of the nodal point N, viz.: tending of trajectories toward N with a *limiting direction*, is invariant as is shown in Fig. 1.7b. Likewise, the approach of parabolic curves to parallelism with one of the axes of coordinates (Fig. 1.3), when the variable along the other axis increases, is also preserved; but, in the "distorted image" of the v,u plane, this parallelism takes place with respect to the "former axis" (Fig. 1.7b), so that nothing is changed in the invariant properties of trajectories in the neighborhood of a nodal point.

If $b^2 - \omega_0{}^2 < 0$, the d.e. (5.1) has a solution of the form $x = x_0 e^{-bt} \cos(\omega_1 t + \alpha)$ where x_0 and α are two integration constants and $\omega_1 = \sqrt{\omega_0{}^2 - b^2}$ is the damped frequency. In order to represent this solution in the phase plane, we consider, as usual, the equivalent system:

$$\dot{x} = y; \qquad \dot{y} = -2by - \omega_0{}^2 x \tag{5.5}$$

which shows that the origin is again a singular point. The elimination of the parameter t results in the d.e.:

$$\frac{dy}{dx} = -\frac{2by + \omega_0^2 x}{y} \tag{5.6}$$

This is a homogeneous d.e. amenable to separation of variables by introducing the auxiliary variable u defined by $y = ux$. Omitting this calculation [2] we indicate the result:

$$y^2 + 2bxy + \omega_0^2 x^2 = C \exp\left[(2b/\omega_1) \arctan (y + bx)/\omega_1 x\right] \tag{5.7}$$

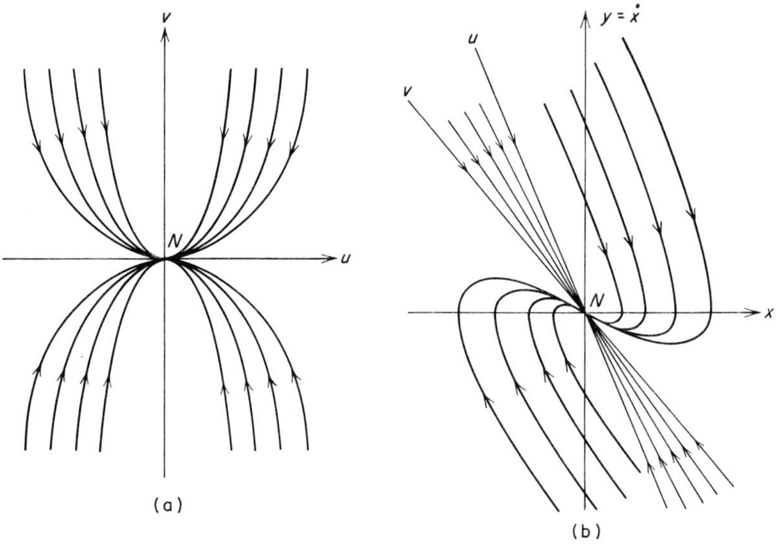

FIGURE 1.7

The left-hand term here reduces to: $(y + bx)^2 + \omega_1^2 x^2$. Introducing first the variables $u = \omega_1 x$; $v = y + bx$, it becomes simply $v^2 + u^2$ so that (5.7) can be written as $v^2 + u^2 = C \exp\left[(2b/\omega_1) \arctan (v/u)\right]$.

Introducing now the polar coordinates $u = r \cos \psi$; $v = r \sin \psi$, the last equation becomes:

$$r = C_1 \exp\left[(b/\omega_1)\psi\right] \tag{5.8}$$

where $C_1 = \sqrt{C}$. The phase trajectory is now a logarithmic spiral and we recognize thus that this singular point is a focal point. This singular point is stable if $b > 0$ and unstable if $b < 0$.

[2] E. Kamke, *Differentialgleichungen reeler Functionen*, Leipzig, 1930.

In electrical engineering the concept of "negative resistance" is frequently used in connection with oscillatory phenomena which *absorb* energy instead of dissipating it. Hence, in such phenomena $b < 0$ and, therefore, the focal point is unstable. The usual representation in such a case is shown in Fig. 1.8, whereas in the phase plane the integral curve (in the u,v variables) is an ordinary logarithmic spiral traversed by R *from* the focal point outward, as shown in Fig. 1.6, but with a changed direction on integral curves. If, however, one considers the x,y variables related to the variables u,v by a linear (affine) transformation, thc spiral is distorted as shown in Fig. 1.9 for dissipative damping ($b > 0$). Thus, the situation is similar to that mentioned in connection with the aperiodic case ($b^2 > \omega_0^2$).

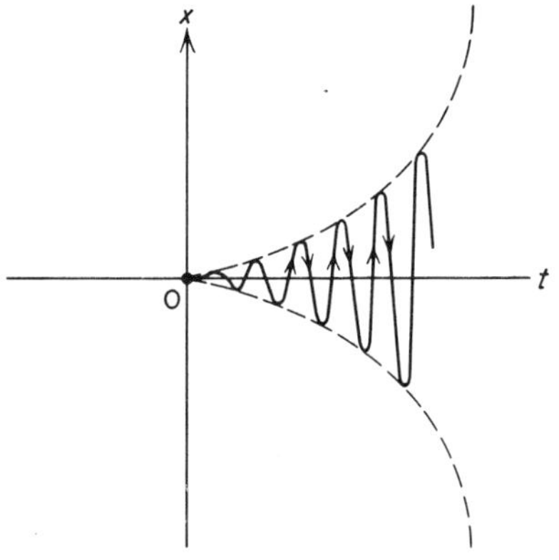

FIGURE 1.8

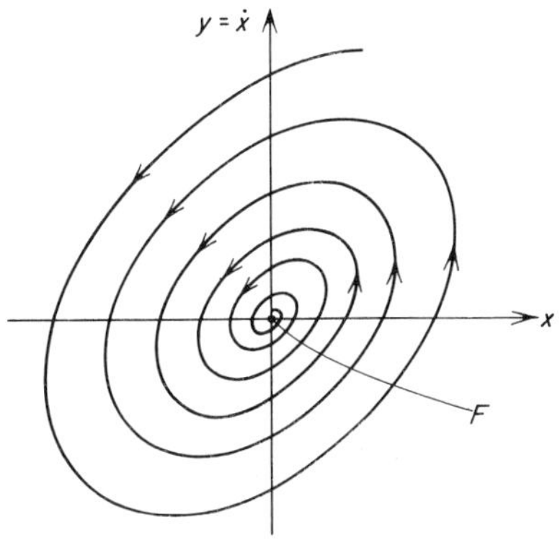

FIGURE 1.9

In this case, the spiral trajectory approaches the singular point F again without any definite direction. The only difference between the form of spirals in Figs. 1.6 and 1.9 is due to the fact that, in the former, one has the logarithmic spiral directly (in u,v variables), whereas in the latter, this spiral is distorted when one passes from the u,v variables back to the x,y variables. In a similar way, the properties of trajectories in the neighborhood of a saddle point can be obtained from the d.e. of the form

$$m\ddot{x} - cx = 0 \tag{5.9}$$

where m and c are some positive constants. This equation has a real exponential solution of the form

$$x = Ae^{rt} + Be^{-rt}$$

where A and B are constants of integration and $r = +\sqrt{c/m}$. This shows that a physical phenomenon represented by (5.9) is unstable. The (x,t)

representation does not yield any further information but, if one translates these relations into the phase-plane representation, some additional conclusions can be obtained as follows.

The equivalent system is:

$$\dot{x} = y; \qquad \dot{y} = (c/m)x \tag{5.10}$$

and the d.e. of the integral curves is

$$\frac{dy}{dx} = \frac{cx}{my} \tag{5.11}$$

which shows that the origin $x = y = 0$ is a singular point. The integral curves are obtained by integrating (5.11) and are

$$y^2 - (c/m)x^2 = C \tag{5.12}$$

C being an integration constant. These curves are thus hyperbolic curves forming a family depending on the parameter C. The asymptotes are obtained by setting $C = 0$, as $y = \pm\sqrt{(c/m)}x$. The positive directions on the curves of the family are obtained by investigating the signs of $\dot{x}$ and $\dot{y}$ in various quadrants, as we did previously. This gives the family of phase trajectories shown in Fig. 1.10. Some conclusions can be derived from this mode of representation.

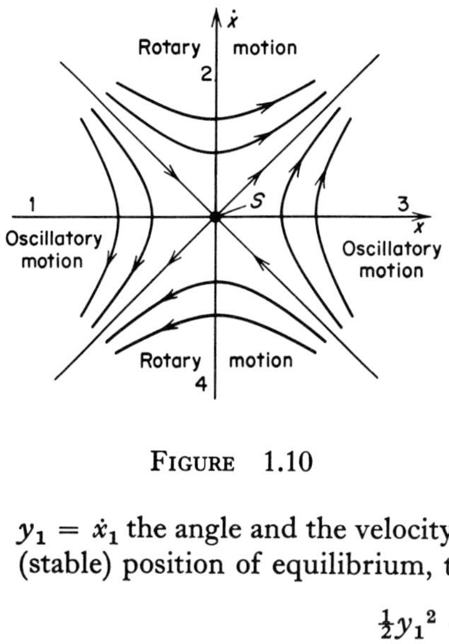

FIGURE 1.10

We consider the simplest possible example of an unstable equilibrium a pendulum in the neighborhood of its upper (unstable) position of equilibrium. If one designates by x_1 and $y_1 = \dot{x}_1$ the angle and the velocity of the pendulum counted from its lower (stable) position of equilibrium, the law of conservation of energy yields

$$\tfrac{1}{2}y_1^2 + V(x_1) = h$$

where $V(x_1)$ is the potential energy, $\tfrac{1}{2}y_1^2$ the kinetic energy (for $m = 1$), and h the total energy, constant since the system is conservative.

In theoretical mechanics one investigates three cases according to $h - V(\pi) \gtreqless 0$. If $h - V(\pi) > 0$, the pendulum rotates all the time in the same direction; its velocity y_1 keeps the same sign but fluctuates in value; its maximum occurs for $x_1 = 0, 2\pi, \ldots$, and minimum for $x_1 = \pi, 3\pi, \ldots$.

We are interested here only in what happens near $x_1 = \pi$ and have plotted in Fig. 1.10 the trajectories near that point. This case corresponds to the quadrants 2 and 4 of Fig. 1.10; the velocity does not change its sign but oscillates in magnitude, being minimum for $x = \pi$. The difference between the quadrants 2 and 4 is only in the direction of the velocity. On the other hand, if $h - V(\pi) < 0$, the pendulum cannot reach the upper position ($x_1 = \pi$) but turns back from some angle $x_1 < \pi$ which is the greater, the greater the total energy. This corresponds to the quadrants 1 and 3 of Fig. 1.10. If, however $h - V(\pi) = 0$, one has the asymptotic case. If the pendulum is projected (by an impulse, for instance) from its lower (stable) position of equilibrium with a kinetic energy just equal to the potential energy in the upper (unstable) position of equilibrium, it tends asymptotically to that position.

6. Canonical transformation; abridged equations

We have given examples of singular points in connection with a few special d.e. (Section 4) and supplemented this, in the preceding section, by more general forms of the linear d.e. of the second order with constant coefficients.

Consider now the general system

$$\dot{x} = P(x,y); \qquad \dot{y} = Q(x,y)$$

where P and Q are real analytic functions of x and y which vanish at the origin. If we develop P and Q into a power series at this point we have

$$\dot{x} = ax + by + P_2(x,y); \qquad \dot{y} = cx + dy + Q_2(x,y) \tag{6.1}$$

where P_2 and Q_2 are power series in x and y beginning with terms of degree at least two. (Notice that we can always make a change of variable that brings a singular point (x_0,y_0) to the origin, and thus gives the d.e. the form (6.1) in the neighborhood of the singular point.) Since we are interested in investigating what happens in the neighborhood of the singular point, x and y may be regarded as small quantities of the first order. In view of this, the terms contained in P_2 and Q_2 are at least of the second order and, thus, in general can be neglected in the neighborhood of the singular point.

This leads to an important simplification of the problem. In fact, if one is interested only in establishing the nature of equilibrium in the neighborhood of the singular point it is frequently sufficient to investigate the linear system

$$\frac{dx}{dt} = ax + by; \qquad \frac{dy}{dt} = cx + dy \tag{6.2}$$

In a great majority of applied problems, the use of the "abridged system" (6.2) instead of the complete system (6.1) is justified. It must be noted, however, that in some special cases, this may not be so (for example, if $a = b = c = d = 0$). In Section 8 we shall investigate a case in which the use of the abridged system does not permit determining the nature of the singular point. Whenever one comes across a situation of this kind the problem becomes more complicated and is to be studied directly.

We shall first investigate the simple abridged equations (6.2) and shall mention later some special cases when the linear terms fail to yield the answer. In the first place we assume that $\begin{vmatrix} a & b \\ c & d \end{vmatrix} \neq 0$, as otherwise the origin is not an isolated singularity. The problem can then be formulated as follows: We introduce a linear transformation of the variables

$$\xi = \alpha x + \beta y; \qquad \eta = \gamma x + \delta y \tag{6.3}$$

with the determinant $\begin{vmatrix} \alpha & \beta \\ \gamma & \delta \end{vmatrix} \neq 0$ so as to reduce (6.2) to the *canonical form*:

$$\dot{\xi} = S_1\xi; \qquad \dot{\eta} = S_2\eta \tag{6.4}$$

where S_1 and S_2 are certain constants real or complex assumed distinct. The d.e. of the integral curves in the new variables is

$$\frac{d\eta}{d\xi} = \left(\frac{S_2}{S_1}\right)\frac{\eta}{\xi}; \qquad \frac{S_2}{S_1} = m \tag{6.5}$$

This can be accomplished by differentiating (6.3) and substituting for $\dot{x}$ and $\dot{y}$ their expressions (6.2), which gives the identities

$$\begin{aligned} S_1(\alpha x + \beta y) &= \alpha(ax + by) + \beta(cx + dy) \\ S_2(\gamma x + \delta y) &= \gamma(ax + by) + \delta(cx + dy) \end{aligned} \tag{6.6}$$

Identifying the coefficients of x and y, one gets two sets of relations

$$\begin{aligned} \alpha(a - S_1) + \beta c = 0; &\qquad \gamma(a - S_2) + \delta c = 0 \\ \alpha b + \beta(d - S_1) = 0; &\qquad \gamma b + \delta(d - S_2) = 0 \end{aligned} \tag{6.7}$$

There are thus two linear systems; one contains the unknowns α and β and the other γ and δ. Nontrivial solutions are possible only when S_1 and S_2 are the roots of the quadratic equation

$$\begin{vmatrix} a - S & c \\ b & d - S \end{vmatrix} = S^2 - (a + d)S + (ad - bc) = 0 \tag{6.8}$$

which is called the *characteristic equation* of (6.2) and which will be of a fundamental importance in what follows. We choose S_1 and S_2 as the two (distinct) roots of (6.8) and proceed with the determination of α, β, γ, and δ. (The case of a double root is discussed in paragraph 5 below.)

Thus $\begin{vmatrix} \alpha & \beta \\ \gamma & \delta \end{vmatrix} \neq 0$ and $\begin{vmatrix} a & b \\ c & d \end{vmatrix} \neq 0$ ensures the existence of two unequal roots S_1 and S_2 and eliminates the possibility of one of them being zero.

Under these restrictions S_1 and S_2 are either real or complex and we shall investigate these various possibilities.

1. If the roots S_1 and S_2 are real and of the same sign, $m > 0$ (from 6.5), the singular point is a node, stable if S_1 and S_2 are negative and unstable if they are positive. The condition for real unequal roots of the same sign is clearly: $0 < ad - bc < [(a + d)/2]^2$. For a stable node $a + d < 0$ and for an unstable one, $a + d > 0$. As to the coefficients α, β, γ, and δ, they are determined by (6.7).

2. If S_1 and S_2 are real but of opposite signs, a similar reduction to the canonical form results in the d.e. (6.5) in which $m < 0$; this, as we saw in Section 4, results in a saddle point. In this case $ad - bc < 0$.

3. If S_1 and S_2 are conjugate complex, we put $\xi = re^{i\varphi}$; $\eta = re^{-i\varphi}$ which gives a system

$$\dot{r} = r(S_1 + S_2)/2; \qquad \dot{\varphi} = (S_1 - S_2)/2i \tag{6.9}$$

The integral curves are given by equations

$$r = C_1 \exp\,[(S_1 + S_2)t/2]; \qquad \varphi = [(S_1 - S_2)(t + C_2)]/2i$$

so that

$$r = C \exp\,[(S_1 + S_2)i\varphi/(S_1 - S_2)] \tag{6.10}$$

where C_1, C_2, and C are the integration constants, which shows that the singular point is a focus provided Re (S), the real part of S_1 and S_2, is not zero. The stability of the singular point is determined by $\mathrm{Re}(S)$. If $\mathrm{Re}(S) < 0$, the focus is stable; if it is positive, the focus is unstable. The condition for a focus is $0 < (a + d)^2 < 4(ad - bc)$.

4. If S_1 and S_2 are purely imaginary, $S_1 + S_2 = a + d = 0$. This requires that $ad - bc = -a^2 - bc > 0$; thus $-bc > a^2$; that is, b and c are of opposite signs and, moreover, $|bc| > |ad|$. We are still in the general situation discussed under 3 (two conjugate roots). Hence (6.10) holds in this case also. Since $S_1 + S_2 = 0$, (6.10) reduces to $r = C$, which shows that the integral curves are circles around the singular point.

In the case of the harmonic oscillator (3.1) we have precisely this situation since $a = d = 0$, $b = 1$, $c = -1$. This special case of $S_1 + S_2 = a + d = 0$ and $-bc > a^2$ characterizes a *center*. Geometrically it is distinguished by a continuous family of closed trajectories around the singularity.

5. If the characteristic equation has a double root $S = S_1 = S_2$, we distinguish two cases. We note first that the condition for a double root is $a = d = S$ and $bc = 0$. If $b = c = 0$, then (6.2) has the form

$$\dot{x} = ax; \qquad \dot{y} = ay$$

which we recognize as a proper node, stable or unstable according as $a < 0$ or $a > 0$. In the other case, that is, not both b and c equal to zero, let us suppose $c = 0$, $b \neq 0$, so that (6.2) has the form

$$\dot{x} = ax + by; \qquad \dot{y} = ay$$

We shall see that the singular point of this system is a type of improper node not previously encountered. The general solution is

$$x(t) = (C_2 + C_1 bt)e^{at}; \quad y(t) = C_1 e^{at}$$

which is found by first integrating the second equation and substituting the solution into the first, which may then be integrated.

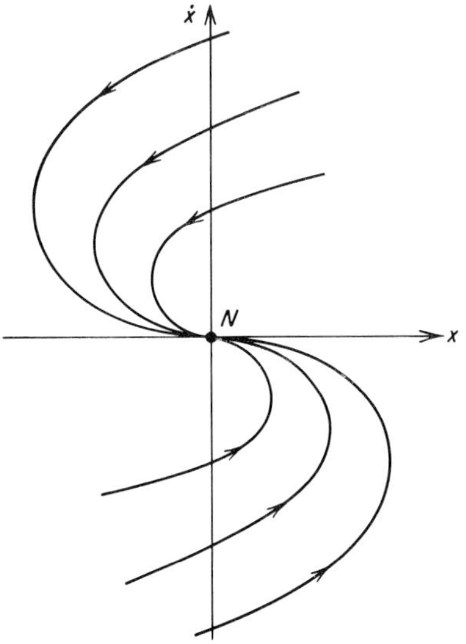

FIGURE 1.11

To describe the behavior of the trajectories near the origin, suppose first that $a < 0$. Then $x(t) \to 0$ and $y(t) \to 0$ as $t \to \infty$. If $C_1 = 0$, we obtain a trajectory on the x axis. If $C_1 \neq 0$, the slope of the trajectory is given by

$$\frac{dy}{dx} = \frac{y'(t)}{x'(t)} = \frac{C_1 a}{a(C_2 + C_1 bt) + bC_1} \to 0 \qquad \text{as } t \to \infty$$

Hence all of the trajectories tend to the origin tangent to the x axis. Thus this node is characterized by the property that all trajectories have the *same* limiting direction at the singular point (see Fig. 1.11). If $a > 0$ the above conclusions apply as $t \to -\infty$, that is, the node is stable or unstable according as $a < 0$ or $a > 0$.

Summing up, we see that the nature of the roots S_1 and S_2 determines the character of the singular point.

(1) If S_1 and S_2 are real and of the same sign, the singular point is a

node. The conditions for a node are: $(ad - bc) > 0$; $(a + d) \neq 0$. If $(a + d) < 0$, the node is stable; if $(a + d) > 0$, it is unstable.

(2) If S_1 and S_2 are real but of opposite signs, the singular point is a *saddle point.* In this case the conditions are: $(ad - bc) < 0$.

(3) If S_1 and S_2 are conjugate complex (not purely imaginary), the singular point is a *focus.* The conditions here are: b and c are of opposite signs and $(a + d) \neq 0$. If $(a + d) < 0$, the focus is stable; if $(a + d) > 0$, it is unstable.

(4) If S_1 and S_2 are purely imaginary, one has a *center* with the conditions: $a + d = 0$ and $|bc| > |ad|$.

Only the linear d.e. of the form (6.2) has been considered so far. It is now necessary to show how the theory is to be applied to the general case of the d.e. (6.1). We shall suppose $S_1 \neq S_2$.

One can apply the same linear transformation between (x,y) and (ξ,η) given by (6.3). However, here it is necessary to express x and y in terms of ξ and η by the inverse transformation, which is possible near the origin since we have assumed that $\Delta = \begin{vmatrix} \alpha & \beta \\ \gamma & \delta \end{vmatrix} \neq 0$. Hence (6.3) can be solved in the form

$$x = (\delta\xi - \beta\eta)/\Delta; \qquad y = -(\gamma\xi - \alpha\eta)/\Delta \tag{6.11}$$

We substitute these values of x and y into the d.e. (6.1). Observe that $P_2(x,y)$ and $Q_2(x,y)$ are transformed into $\pi_2(\xi,\eta)$ and $\kappa_2(\xi,\eta)$ which are at least of degree 2 in the new variables ξ and η. This gives (6.1) in the form

$$\dot{\xi} = S_1\xi + \pi_2(\xi,\eta); \qquad \dot{\eta} = S_2\eta + \kappa_2(\xi,\eta) \tag{6.12}$$

Here S_1 and S_2 may be real numbers, in which case (6.12) is already in the canonical form. If S_1 and S_2 are conjugate complex, we can also set $\xi = u + iv$; $\eta = u - iv$. Comparing the real and the imaginary forms, one has

$$\begin{aligned} \dot{u} &= a_1u - b_1v + p_2(u,v); \qquad & S_1 &= a_1 + ib_1 \\ \dot{v} &= b_1u + a_1v + q_2(u,v); \qquad & S_2 &= a_1 - ib_1 \end{aligned} \tag{6.13}$$

where p_2 and q_2 are at least of the second degree in u and v.

We have to show now that the presence of the nonlinear terms π_2 and κ_2 does not change the general character of the singularity and that the preceding classification obtained on the basis of the linear terms remains valid.

1. We take first the case when S_1 and S_2 are real, unequal, and of the same sign, which in the linear case corresponds to a node. It can be shown that the general properties of the singular point remain substantially the

same in the nonlinear case. We multiply the first equation (6.12) by ξ, the second by η and add, which gives

$$\frac{1}{2}\frac{d}{dt}(\xi^2 + \eta^2) = (S_1\xi^2 + S_2\eta^2) + r_3(\xi,\eta) \tag{6.14}$$

where the last term contains $r = \sqrt{\xi^2 + \eta^2}$ at least in the third power. Hence, if S_1 and S_2 are, for instance, negative and r is small enough, the quantity $\xi^2 + \eta^2 = r^2 = \rho$ decreases continuously, so that the point $R[\xi(t),\eta(t)]$ approaches the singular point at the origin O.

If it is possible to show that this approach to O occurs along a definite direction, the singular point at O is clearly a node. To show this point, we form the second combination

$$\eta\dot{\xi} - \xi\dot{\eta} = (S_1 - S_2)\xi\eta + r_3{}^*(\xi,\eta) \tag{6.15}$$

where $r_3{}^*$ is, at least, of order three. With the variables $r^2 = \xi^2 + \eta^2$, $\psi = \arctan \eta/\xi$ is defined modulo 2π by $\xi = r\cos\psi$, $\eta = r\sin\psi$, and

$$\frac{d\psi}{dt} = -\left(\frac{S_1 - S_2}{2}\right)\sin 2\psi + O(r, \psi) \tag{6.16}$$

Introducing the function

$$h(t) = \cos 2\psi(t) \tag{6.17}$$

and in view of (6.16), we get

$$\frac{dh}{dt} = -(S_2 - S_1)\sin^2 2\psi + O(r, \psi) \tag{6.18}$$

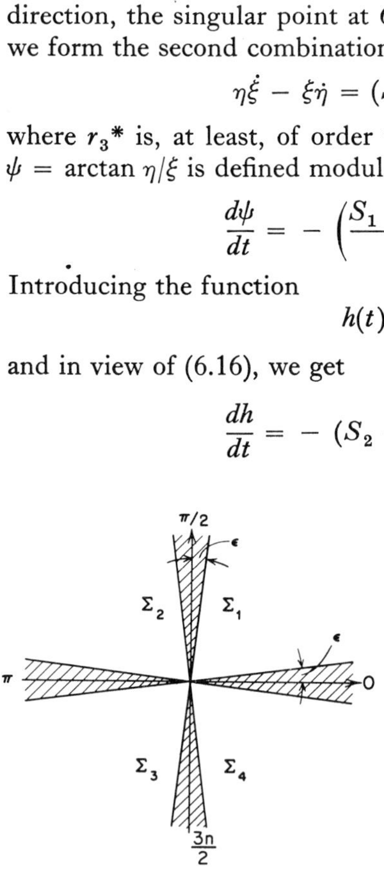

FIGURE 1.12

As $\cos 2\psi = 1$ for $\psi = 0$ and $\psi = \pi$ and $\cos 2\psi = -1$ for $\psi = \pi/2$ and $\psi = 3\pi/2$, we consider the angular regions Σ defined by $\varepsilon < \psi < (\pi/2) - \varepsilon$; $(\pi/2) + \varepsilon < \psi < \pi - \varepsilon$; $\pi + \varepsilon < \psi < (3\pi/2) - \varepsilon$; $(3\pi/2) + \varepsilon < \psi < 2\pi - \varepsilon$; these regions are shown in white in Fig. 1.12, while the remaining angular regions are shown in shading. Observe that, if $\psi \in \Sigma$, there exists a quantity $\delta > 0$ such that

$$-(S_2 - S_1)\sin^2 2\psi > \delta \tag{6.19}$$

assuming that $S_1 > S_2$. Inasmuch as $r \to 0$ with $t \to \infty$, we can assume that our trajectory is so near to O that $|O(r)| < (\delta/2)$. Hence $(dh/dt) > \frac{1}{2}\delta$ if $\psi \in \Sigma$.

Thus, if the point (r,ψ) traveling along a trajectory lies, say, in Σ_1, it will move into the angle $|\psi| < \varepsilon$. Once it has arrived there, it cannot leave this angle since at its boundary the trend of ψ is to move inward into the shaded area. Since ε can be supposed to be as small as we please and the radial motion is inwardly directed, it is clear that the trajectory approaches asymptotically the singular point along a definite direction, that is, O is *a node*. If $S_1 > S_2$, this approach takes place along the horizontal tangent; if $S_1 < S_2$, it takes place along the vertical tangent (Fig. 1.3a and b). We have thus demonstrated this property for the general case of the d.e. (6.12).

In a similar case one can show that the properties of the *star* (equation 4.4, for $a = 1$) are preserved in the general case of equation (6.12) if $S_2 = S_1 = S \neq 0$. The d.e. (6.14) and (6.16) become, in this case,

$$\frac{dr}{dt} = Sr + O(r^2); \qquad \frac{d\psi}{dt} = O(r) \tag{6.20}$$

Thus

$$\frac{d\psi}{dr} = \frac{O(r)}{Sr + O(r^2)} \tag{6.21}$$

or

$$\frac{d\psi}{dr} = \kappa(r,\psi) \tag{6.22}$$

where $\kappa(r,\psi)$ is a continuous differentiable function of r and ψ even for $r = 0$. One can, therefore, integrate (6.22) with an arbitrary initial condition $\psi(0) = \psi_0$ which yields an infinity of curves $\psi(r)$ passing through the origin with definite tangents. For each ψ_0 there is a corresponding trajectory so that we have ascertained the existence of a *star* as the singular point at the origin.

Consider now the case when S_1 and S_2 are real and of opposite signs, say, $S_1 < 0$ and $S_2 > 0$. We have then the equations

$$r\frac{dr}{dt} = -|S_1|\xi^2 + S_2\eta^2 + r_3(\xi,\eta) \tag{6.23}$$

$$\frac{d\psi}{dt} = -\frac{1}{2}(|S_1| + S_2)\sin 2\psi + O(r) \tag{6.24}$$

Observe that ψ tends again to limit values 0, $\pi/2$, π, and $3\pi/2$ if r is small enough. However, it is impossible to deduce from (6.23) that, if $t \to \infty$, r will decrease indefinitely. In fact, in the angular space $\eta^2/\xi^2 < |S_1|/S_2$, r decreases, where for $\eta^2/\xi^2 > |S_1|/S_2$, r increases. In other words, depending on the sign of inequalities

$$(\tan\psi)^2 \gtrless S_2/|S_1| \tag{6.25}$$

the point R will either approach, or recede from, the origin. Since the time variation of ψ (equation 6.24) tends to remove R from the η axis and bring it nearer to the ξ axis, R will end up in the angular space of increasing r. Thus R will not converge into the origin (Fig. 1.13) but, on the contrary, will further depart from it after an initial approach. This shows that the main characteristic of a saddle point is still preserved in the neighborhood of the singularity. For a finer analysis of this case and of the case of conjugate complex roots S_1 and S_2, we refer to the mathematical texts.[1]

It can be shown that the nonlinear terms do not modify in general the form of trajectories defined by the abridged system (6.2). It is to be

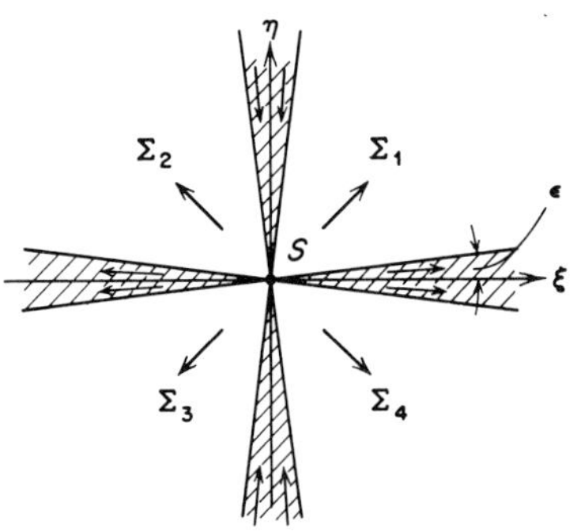

FIGURE 1.13

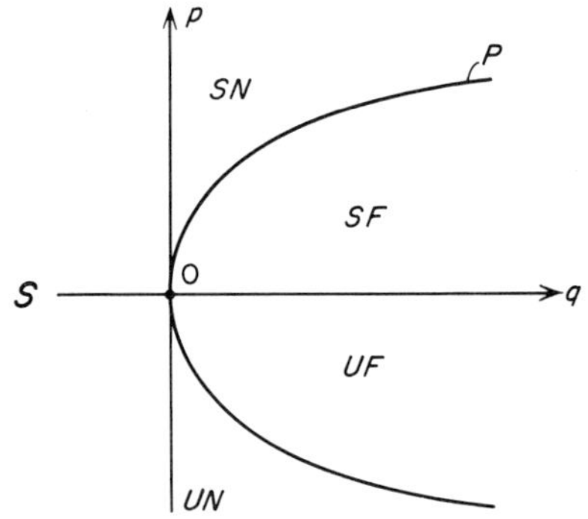

FIGURE 1.14

mentioned that the previous statement does not hold in one special case when the singular point is *a center*, as will be investigated in Section 8.

7. Distribution of singular points; parameter space

A convenient way of investigating the changes in the nature of singular points when the parameters a, b, c, and d vary is to represent graphically the various regions in which the roots S_1 and S_2 have the same form.[3]

[1] See footnote [1], page 3.

[3] (a) A. Andronov and S. Chaikin, *Theory of Oscillations* (original text in Russian), Moscow, 1937.
(b) English translation by S. Lefschetz of A. Andronov and S. Chaikin, *Theory of Oscillations*, Princeton University Press, Princeton, N.J., 1949.
(c) A. Andronov, A. Witt, S. Chaikin, *Theory of Oscillations* (in Russian); this book is the second edition (1959) of A. Andronov and S. Chaikin, *Theory of Oscillations* (original text in Russian), Moscow, 1937.

One can, for instance, set $p = -(a + d)$; $q = \begin{vmatrix} a & b \\ c & d \end{vmatrix}$ and write the characteristic equation (6.8) as

$$S^2 + pS + q = 0 \tag{7.1}$$

A parabola P of equation $p^2 - 4q = 0$ traced in the plane of the variables (p,q) (Fig. 1.14) and the coordinate axes determine five regions: *SN*, *SF*, *UF*, *UN*, and *S* (stable nodal, stable focal, unstable focal, unstable nodal, and saddle points, respectively).

The region of saddle points S is separated from those of other singularities by the p axis for which $\begin{vmatrix} a & b \\ c & d \end{vmatrix} = 0$ and which, as we saw, indicates the presence of at least one zero root.

On the right side of the p axis, there are nodes and foci both stable and unstable. For the former $p > 0$ (that is, $a + d < 0$); for the latter $p < 0$.

It is noted that this applies only to the common frontier between the regions *SF* and *UF* since the regions *SN* and *UN* have no common frontier except at one point. The common frontier between the regions *SN* and *SF*, on one hand, and between *UF* and *UN*, on the other hand, is the parabola P; at this border line one has the relation $(a - d)^2 = -4bc$ which is possible only if b and c are of opposite signs.

The q axis ($p = 0$) corresponds to the relation $a + d = 0$, that is, to the vanishing of the real parts of the roots. This, as we saw, characterizes a degenerate focus when the spiral trajectory from converging becomes diverging or vice versa. The positive q axis is, therefore, a locus of imaginary roots. We shall see later that the q axis plays an important role in the theory of bifurcation (Chapter 7) when the topological configuration (defined in Chapter 3) changes, passing from the *SF* region into the *UF* region or vice versa. It is impossible, however, to pass from the *SN* region to the *UN* region or vice versa, since these two regions have no common border except at one point 0 at which $a + d = 0$ and $ad = bc$ simultaneously, the latter condition meaning that there appears one root equal to zero which excludes the existence of a simple singularity at this point. This representation is less convenient if the parameters a, b, c, and d vary independently. The reason for this is the fact that this representation uses *two combinations* of the four parameters which permit reducing artificially the four-space of parameters to a planar representation.

If, however, one wishes to investigate the stability of a system in terms of its singular points when parameters vary independently, it is necessary to study the problem in the "parameter space." As in this case there are four parameters, a four-space is required; but, in view of the impossibility to give a geometrical interpretation, one has to carry out this study for different "sections" of the four-dimensional space (a,b,c,d). If one takes as "sections" the coordinate planes $a = 0$, $b = 0$,. . ., to each of them corresponds a three-dimensional space (b,c,d), (a,c,d),. . .. One can also take as sections of lower order those which correspond to $a = 0$, $b = 0$, $a = 0$, $c = 0$, in which case one obtains corresponding two-dimensional sections (c,d) (b,d). In this manner it is possible to investigate what happens when some of the parameters are held fixed and the others vary. Often some physical considerations permit reducing the number of "sections" in which the variation of parameters presents a special interest.

As an application of this procedure we consider[3] the conditions of

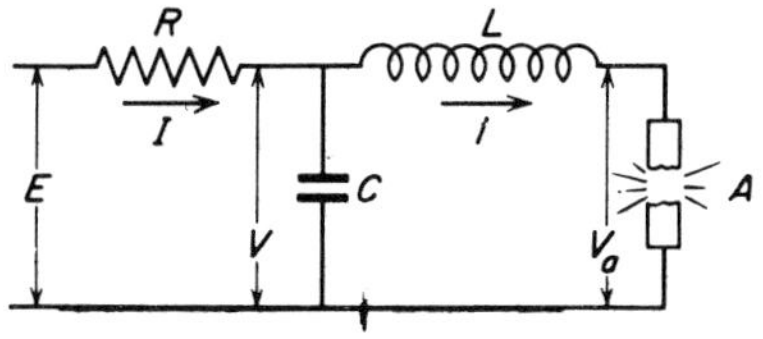

FIGURE 1.15

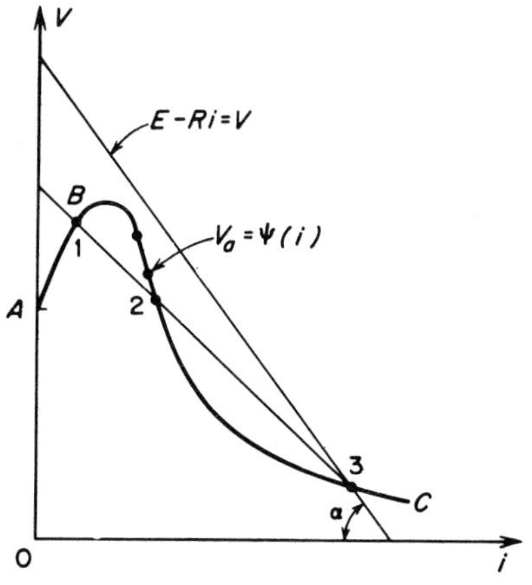

FIGURE 1.16

stability of an electric arc connected to the circuit of a constant voltage E, as shown in Fig. 1.15. As is well known, the arc is a nonlinear conductor of electricity. Its nonlinear characteristic (i, V_a) is shown in Fig. 1.16. The voltage across the arc V_a is thus a certain nonlinear function of the current $i: V_a = \psi(i)$. Kirchhoff's laws yield:

$$V = L\frac{di}{dt} + \psi(i); \qquad E = RI + V; \qquad I = i + C\frac{dV}{dt} \tag{7.2}$$

If one eliminates I between these equations, one gets the d.e.

$$\frac{dV}{dt} = \frac{E - V - Ri}{RC}; \qquad \frac{di}{dt} = \frac{V - \psi(i)}{L} \tag{7.3}$$

[3] See footnote [3], page 26.

For the state of equilibrium ($dV/dt = 0$; $di/dt = 0$), one has, obviously,

$$V_0 = E - Ri_0; \qquad V_0 = \psi(i_0) \tag{7.4}$$

which may be regarded as points of intersection of a straight line $V = E - Ri$ with the characteristic of the arc $V_a = \psi(i)$.

The problem consists in studying the conditions of stability of equilibrium under the various circumstances or, which is the same, the nature of the singular points of the system (7.3). The usual procedure is to examine what happens if the position of equilibrium (V_0, i_0) is disturbed by small perturbations δv, δi so that the disturbed values are now $V = V_0 + \delta v$; $i = i_0 + \delta i$. As the only nonlinear element here is the characteristic $\psi(i)$, we develop it in Taylor's series around the point of equilibrium, limiting the expansion only to the first-order term in δi, which yields

$$\psi(i_0 + \delta i) = \psi(i_0) + \delta i \psi'(i_0) \tag{7.5}$$

If one replaces V and i by $V_0 + \delta v$ and $i_0 + \delta i$ and $\psi(i)$ by $\psi(i_0 + \delta i)$ in (7.3) and cancels out the equilibrium terms according to (7.4), one obtains the so-called *variational* d.e.† of the form

$$\frac{d\delta v}{dt} = \left(-\frac{1}{RC}\right)\delta v + \left(-\frac{1}{C}\right)\delta i; \qquad \frac{d\delta i}{dt} = \left(\frac{1}{L}\right)\delta v + \left(-\frac{\rho}{L}\right)\delta i \tag{7.6}$$

where $\rho = \psi'(i_0)$. The characteristic equation is

$$S^2 + \left(\frac{L + RC\rho}{RCL}\right)S + \frac{\rho + R}{RCL} = 0 \tag{7.7}$$

and its roots are

$$S_{1,2} = \frac{1}{2RCL}\left[-(L + RC\rho) \pm \sqrt{(RC\rho)^2 + L^2 - 2LCR(\rho + 2R)}\right] \tag{7.8}$$

The problem of equilibrium depends thus on four parameters R, L, C, and ρ, of which the first three are positive while ρ may either be positive or negative, depending on the point of the curve $\psi(i)$ at which one wishes to investigate the equilibrium of the circuit. It is seen that the parameter space in this case is four-dimensional, but its "section" corresponding to $\rho = 0$ is of no interest. In view of this, the only two-dimensional sections of the four-space (R,C,L,ρ) which are of interest are (R,ρ), (L,ρ), and (C,ρ). In these sections we consider the indicated symbols as the variables and the nonindicated ones as fixed parameters.

Consider, for instance, the "section" (R,ρ), that is, the plane of the variables R and ρ in which the parameters L and C will be considered as

† This subject is discussed more fully in Chapter 5.

fixed numbers. The condition for the complex roots (for the existence of focal points) is

$$(L - RC\rho)^2 - (2R\sqrt{LC})^2 < 0 \tag{7.9}$$

This region is limited by the two curves:

$$(L - RC\rho) + 2R\sqrt{LC} = 0; \qquad (L - RC\rho) - 2R\sqrt{LC} = 0 \tag{7.10}$$

In the (R,ρ) plane these two curves are hyperbolas having the ρ axis as a common asymptote and the straight lines $\rho_1 = 2\sqrt{L/C}$ and $\rho_2 = -2\sqrt{L/C}$ as other asymptotes, respectively. These curves are shown as 1 and 2 in Fig. 1.17. It is obvious that the hyperbola $R\rho = L/C$ lies between 1 and 2 since it has the axes R, ρ as asymptotes. For the points on this curve, the condition (7.9) is fulfilled, which shows that, in the region limited by the curves 1 and 2, the singular points are foci. Since the stable foci are separated from the unstable ones by the locus of points at which the real parts of the roots vanish, this locus is clearly the hyperbola

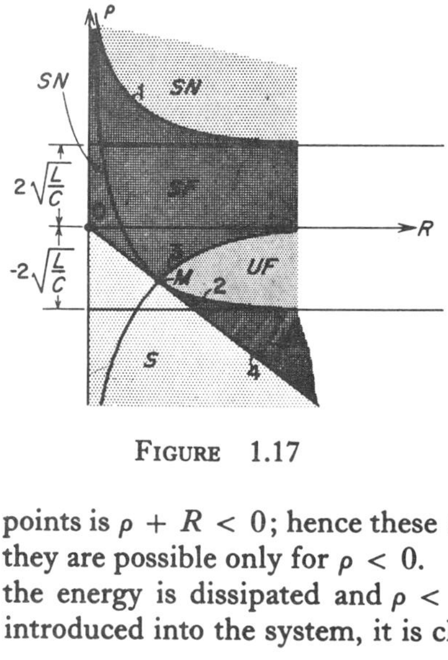

FIGURE 1.17

$$R\rho = -L/C \tag{7.11}$$

indicated on Fig. 1.17 as curve 3 having the positive axis R and the negative axis ρ as asymptotes. The curves 2 and 3 limit a region of unstable foci. The condition for the existence of saddle points is $\rho + R < 0$; hence these singularities lie below the line $|\rho| = R$; they are possible only for $\rho < 0$. Since R is the parameter through which the energy is dissipated and $\rho < 0$ is the parameter through which it is introduced into the system, it is clear that the condition

$$\rho + R < 0 \tag{7.12}$$

means that the saddle points characterize an unstable condition of the system for which it receives more energy than it can dissipate. In the (R,ρ) plane the borderline $\rho + R = 0$ of this region is a bisector of the right angle formed by the positive R axis and the negative ρ axis; it is indicated as line 4 in Fig. 1.17 and the saddle points are situated *below* this line. The line 4 cuts the curve 3 of stability at a point M of coordinates $(+\sqrt{L/C}, -\sqrt{L/C})$. The wedge-shaped area between the curve 2 and the line 4 is the region of unstable nodes; in fact, in this region the roots are real and of the same signs because they become of opposite signs only below the line 4. The remaining regions are to the left of the point M and limited by the curves 2, 3; and the ρ axis and the other above the curve 1 are the regions of stable nodes; they are both stable because they lie above the curve 3 on the one hand, and they characterize real roots because they are outside the zone (between curves 1 and 2) in which these roots are complex.

The point M is thus a *point of bifurcation* (or a branch point), since five different regions meet there. It is noted that if one surrounds the point M by a small closed curve described clockwise, one encounters the different regions in the same order as in Fig. 1.14, in which the choice of parameters is different. The sequence of regions is not affected by the choice of parameters. The region of saddle points (absolute instability) is always separated from the region of focal points (oscillatory process) by intermediate regions of nodal points (aperiodic process). Thus, an oscillatory process cannot degenerate into an unstable one without passing first through an aperiodic form. These conclusions can be given the following physical interpretation. In the regions of stability (stable focal and stable nodal points) the arc exists without any oscillations. If it is disturbed, it returns to its stable equilibrium with damped oscillation in the case of a stable focal point and aperiodically in the case of a stable nodal point. In the case of unstable focal or unstable nodal points, it leaves the position of equilibrium either with gradually increasing oscillations (focal point) or aperiodically (nodal point), but the study of singular points alone is not sufficient here. For this purpose we shall need additional information regarding the existence of certain stationary oscillatory states which will be studied in Chapter 3. In general, such stationary states exist when the singular point is either an unstable nodal or focal point. No such stationary states exist if the singular point is a saddle point, when the physical process either builds up indefinitely until the destruction of the physical system (for example, blowing the fuses), or is prevented from doing so by some other agent not taken into account in the d.e. (for example, insufficient power supply).

Similar graphical investigations can be used in connection with the other two "sections" (L,ρ) and (C,ρ) of the four-dimensional space of parameters. The procedure remains essentially the same, viz.: (1) one establishes the zone of distribution of the complex roots, in which there exist focal points, (2) one determines the limits of stability of foci by equating to zero the real parts of the complex roots which separates the stable focal points from the unstable ones; (3) one determines the limit of existence of saddle points by putting to zero the last term in the characteristic equation (7.7); the saddle points exist in the region in which this term is negative. The remaining regions of the phase plane are attributed to the nodes, and their location with respect to the stability limit indicates the sub-regions in which the nodes are either stable or unstable.

It may be worth mentioning in passing that much confusion existed with regard to investigations of stability of electric arcs as means for producing undamped oscillations prior to the advent of the theory of Poincaré which reduced this question to the analysis of singular points as outlined here. The reason for these difficulties was due to the complexity of stability conditions in the neighborhood of the bifurcation points when a slight change in one of the parameters is already sufficient to swing the process from one form of equilibrium to an entirely different one.

8. Center

In Section 6 it was mentioned that when the roots of the characteristic equation are purely imaginary, the singular point is either a center or a focus, but it is impossible to distinguish between these two singularities on the basis of the linear approximation. This difficulty was noticed from the very beginning by Poincaré to whom is due the first analysis of this problem.[1] We shall see later that the center is a special singularity in the sense that the trajectories around the center can be deformed by a slight perturbation of the equation into convergent or divergent spirals. For the time being we shall limit ourselves to what has been already explained in this chapter, namely, that the center is a singular point possessing the property that all trajectories turn around it without either approaching it or receding indefinitely.

A convenient way of analyzing the situation is to make use of the partial d.e.

$$X\frac{\partial F}{\partial x} + Y\frac{\partial F}{\partial y} = 0 \tag{8.1}$$

There exists a relation between the solution of (8.1) and the system of the d.e.

$$\frac{dx}{dt} = X(x,y); \qquad \frac{dy}{dt} = Y(x,y) \tag{8.2}$$

In fact, each curve $F(x,y) = C$, where $F(x,y)$ is a nonconstant solution of (8.1), is an integral curve of (8.2). This is apparent if one differentiates $F(x,y) = C$ with respect to t and makes use of (8.2), viz.:

$$\frac{\partial F}{\partial x}\cdot\frac{dx}{dt} + \frac{\partial F}{\partial y}\cdot\frac{dy}{dt} = 0 \tag{8.3}$$

[1] See footnote [1], page 3.

Now, by definition, F satisfies (8.1). Hence, either $\begin{vmatrix} \frac{dx}{dt} & \frac{dy}{dt} \\ X & Y \end{vmatrix} = 0$ or $\frac{\partial F}{\partial x} = \frac{\partial F}{\partial y} = 0$. We suppose that, except at isolated points, $\left(\frac{\partial F}{\partial x}\right)^2 + \left(\frac{\partial F}{\partial y}\right)^2 \neq 0$. Hence, along the curve: $F(x,y) = \text{const}$ holds the relations:

$$\frac{dx}{dt} = aX; \qquad \frac{dy}{dt} = aY; \qquad a = a(t); \qquad \text{or} \quad \frac{dy}{dx} = \frac{Y}{X}$$

which is precisely (8.1). Thus, if one knows one solution of (8.1), one has an infinity of integral curves of (8.2). The function $F(x,y)$ enables one, therefore, to make conclusions concerning the general character of the trajectories.

The problem reduces to constructing a solution of (8.1) assuming that X and Y are series of the form

$$X = X_1 + X_2 + X_3 + \ldots; \qquad Y = Y_1 + Y_2 + Y_3 + \ldots \tag{8.4}$$

where X_i and Y_i are homogeneous polynomials of the ith degree in x and y. We shall try to develop likewise $F = F_1 + F_2 + F_3 + \ldots$ into a power series around the origin. Setting

$$X_1 = \alpha x + \beta y; \qquad Y_1 = \gamma x + \delta y \tag{8.5}$$

and assuming, as before, the normal case:

$$\begin{vmatrix} \alpha & \beta \\ \gamma & \delta \end{vmatrix} \neq 0 \tag{8.6}$$

one can show that in this case $F_1 = 0$. In fact, setting $F_1 = mx + ny$, inserting this value into (8.1) and comparing terms with equal degree, one obtains

$$m(\alpha x + \beta y) + n(\gamma x + \delta y) = 0$$

which results in relations

$$\alpha m + \gamma n = 0; \qquad \beta m + \delta n = 0$$

and, in view of (8.6), one has $m = n = 0$, that is, $F_1 = 0$. The development of F begins thus with the term

$$F_2 = ax^2 + 2bxy + cy^2 \tag{8.7}$$

If one substitutes (8.7) into (8.1) and equates again to zero the coefficients of x^2, xy, and y^2, one obtains a homogeneous system of linear equations for a, b, and c which has nontrivial solutions only if

$$\Delta = (\alpha\delta - \beta\gamma)(\alpha + \delta) = 0 \tag{8.7a}$$

Hence, in view of (8.6), F_2 can be different from zero only if

$$\alpha + \delta = 0 \tag{8.8}$$

Under this condition one finds, assuming that $\alpha \neq 0$, $\delta \neq 0$

$$a = -\frac{\gamma}{\alpha} b; \qquad c = -\frac{\beta}{\delta} b = \frac{\beta}{\alpha} b$$

and thus

$$F_2 = \frac{b}{\alpha}(-\gamma x^2 + 2\alpha xy + \beta y^2) \tag{8.9}$$

If the quadratic equation F_2 = const is to represent closed curves surrounding the origin, F_2 has to be definite, that is, must not have any real roots and, for this, one must have $\alpha^2 + \beta\gamma < 0$, which, in view of (8.8) is the condition (8.6) which we shall assume. If the solution F of (8.1) has a convergent series development $F = F_2 + F_3 + \ldots$, the curves $F = \varepsilon$ are closed curves around the origin if ε is small enough. In fact, if we use the above form for F_2 and introduce the polar coordinates: $x = r\cos\varphi$, $y = r\sin\varphi$, one has $F = (b/\alpha)\, r^2(-\gamma\cos^2\varphi + 2\alpha\sin\varphi\cos\varphi + \beta\sin^2\varphi) + O(r^3)$. If r is small enough, the first term on the right side decides the sign of the whole expression. The curves $F = \varepsilon$ and $F_2 = \varepsilon$ become very close if ε is small enough and, since the latter are ellipses around the origin, it can be shown that the trajectories $F = \varepsilon$ have the same general character, that is, are closed curves. The singular point is, thus, obviously a center. It can hardly be expected, however, that a convergent series development of the above type for an arbitrary form of X and Y should exist.

The following analysis of Poincaré will show that in the general case the singular point turns out to be a focus, whenever it is impossible to continue building up $F_3, F_4, \ldots$ indefinitely. In order to see this point we continue building up further approximations involving $F_3, F_4, \ldots$. For the sake of simplicity we assume that the functions X and Y have their linear terms $X_1 = y$; $Y_1 = -x$ of the form typical for a center. Thus

$$X = y + X_2 + X_3 + \ldots; \qquad Y = -x + Y_2 + Y_3 + \ldots \tag{8.10}$$

We start again with F_2 which, by (8.10), has now the form $F_2 = x^2 + y^2$. The equation yields (8.1),

$$y\frac{\partial F}{\partial x} - x\frac{\partial F}{\partial y} + \left(X_2\frac{\partial F}{\partial x} + Y_2\frac{\partial F}{\partial y}\right) + \ldots = 0$$

For the terms of the degree n one has the relation

$$y\frac{\partial F_n}{\partial x} - x\frac{\partial F_n}{\partial y} + \left(X_2\frac{\partial F_{n-1}}{\partial x} + Y_2\frac{\partial F_{n-1}}{\partial y}\right) + \left(X_3\frac{\partial F_{n-2}}{\partial x} + Y_3\frac{\partial F_{n-2}}{\partial y}\right) + \dots = 0$$

This yields the recursive formula for F_n

$$y\frac{\partial F_n}{\partial x} - x\frac{\partial F_n}{\partial y} = H_n(x,y) \tag{8.11}$$

where $H_n(x,y)$ is a homogeneous form of degree n built in a simple way on the known functions $F_{n-1}, F_{n-2},\dots$.

In polar coordinates this expression is

$$\frac{\partial F_n}{\partial \varphi} = H_n(r\cos\varphi, r\sin\varphi) = r^n H_n(\cos\varphi, \sin\varphi) \tag{8.12}$$

This is a simple d.e. for $f_n(\varphi) = F_n(\cos\varphi, \sin\varphi)$. A necessary condition for its solution is obviously

$$\int_0^{2\pi} H_n(\cos\varphi, \sin\varphi)\, d\varphi = 0 \tag{8.13}$$

and it is clear that this condition is also sufficient because, if it holds, we have:

$$\frac{df_n(\varphi)}{d\varphi} = H_n(\varphi) = \sum_{\nu=1}^{n} (a_\nu \cos\nu\varphi + b_\nu \sin\nu\varphi)$$

and, therefore,

$$f_n(\varphi) = \sum_{\nu=1}^{n}\left[\frac{a_\nu}{\nu}\sin\nu\varphi - \frac{b_\nu}{\nu}\cos\nu\varphi\right] + \text{const}$$

If n is odd, clearly $H_n(\varphi) = \sin\varphi P(\cos^2\varphi) + \cos\varphi Q(\cos^2\varphi)$ where P and Q are polynomials; in such a case $\int_0^{2\pi} H_n(\varphi)d\varphi = 0$.

If, however, n is even, it may happen that the H-integral does not vanish and we are thus stopped in our procedure. Assume, therefore, that we have calculated $F_2 = x^2 + y^2, F_3,\dots, F_{2j-1}$, but it is impossible to calculate F_{2j} since $\int_0^{2\pi} H_{2j}d\varphi \neq 0$. In this case we replace the d.e. (8.12) by

$$\frac{dF_{2j}}{d\varphi} = H_{2j}(\varphi) - C_0 \tag{8.14}$$

where $C_0 = \frac{1}{2\pi}\int_0^{2\pi} H_{2j}(\varphi)d\varphi$ and thus achieve the integrability in the $2j$

step. Returning to the cartesian coordinates, we obtain after multiplication by r^{2j}:

$$y \frac{\partial F_{2j}}{\partial x} - x \frac{\partial F_{2j}}{\partial y} = H_{2j}(x,y) - C_0(x^2 + y^2)^j$$

We stop our procedure at this point and consider the cut-off polynomial

$$\Phi = x^2 + y^2 + F_3 + \ldots + F_{2j} \tag{8.15}$$

Consider now $\Phi(x,y)$ as a trajectory: $x(t), y(t)$ of (8.2). Settirg $\Phi[x(t), y(t)] = g(t)$, we have

$$\frac{dg}{dt} = X \frac{\partial \Phi}{\partial x} + Y \frac{\partial \Phi}{\partial y} = (y + X_2 + \ldots) \frac{\partial \Phi}{\partial x} + (-x + Y_2 + \ldots) \frac{\partial \Phi}{\partial y} \tag{8.16}$$

However, all forms of degrees $m < 2j$ cancel out since

$$y \frac{\partial F_m}{\partial x} - x \frac{\partial F_m}{\partial y} + X_2 \frac{\partial F_{m-1}}{\partial x} + Y_2 \frac{\partial F_{m-1}}{\partial y} + \ldots = y \frac{\partial F_m}{\partial x} - x \frac{\partial F_m}{\partial y} - H_m(x,y) = 0$$

There remains thus $(dg(t))/dt = -C_0 r^l +$ terms of order $(l + 1)$ or higher.

If r is small enough, the lowest-order term decides the sign; hence, $dg/dt < 0$ if $C_0 > 0$ so that $\Phi(x,y)$ decreases along the orbit. The function $\Phi(x,y)$ has thus the following structure:

The level lines $\Phi = \varepsilon$ surround the origin and, if ε is small enough, the line of the steepest decrease of Φ approaches the origin. The trajectory in this case may spiral very slowly but finally it ends at a stable focus. In order that a singular point should be a center, it is necessary that the approximations (8.13) should continue indefinitely without ever reaching a step for which $\int_0^{2\pi} H_{2j} d\varphi \neq 0$. This can be obtained only for very special functions X and Y. This question has been studied recently (particularly in the USSR). In one of these papers by Sáharnikov[1] conditions for the existence of a center were studied in connection with the d.e.

$$\dot{x} = y + X(x,y); \qquad \dot{y} = -x + Y(x,y) \tag{8.17}$$

where X and Y are homogeneous polynomials of the third degree:

$$\begin{aligned} X(x,y) &= bx^3 + (c - \beta)x^2y + (3d - \gamma)xy^2 + fy^3 \\ Y(x,y) &= ax^3 + (3b + \alpha)x^2y + (c + \beta)xy + dy^3 \end{aligned} \tag{8.18}$$

[1] See footnote (f), page 3.

Here a, b, c, and d are independent parameters and α, β, and γ are constants. Omitting the somewhat long calculations, we mention only the conclusions. The origin: $x = y = 0$ is a center only in the following four cases:

(1) $\alpha = \beta = \gamma = 0$
(2) $\alpha = \gamma = b = d = 0$
(3) $2(b - d)k^2 + (a + f - 2c)k - 2(b - d) = 0; \quad \alpha + \gamma = 0;$

$$k = \frac{2\beta}{\alpha} = \frac{f - \alpha}{\alpha + 2b + 2d}$$

(4) $\alpha + \gamma = 0; \quad \alpha(a - f) + 2\beta(a + 2b + 2d) = 0;$
$3(a + f) + 2c = 0; \quad c(\alpha^2 - 4\beta^2) = 6\alpha\beta(b - d);$
$a[2\alpha + 5(b + d)] = \beta[2\beta - 5(a - f)]; \quad (\alpha^2 + 4\beta^2)^3$
$= 25[(b - d)(\alpha^2 - 4\beta^2) - 4\alpha\beta(a + f)]^2$

If none of these four conditions is fulfilled, the singular point is a focus. The above conditions can be presented in a more general manner, viz.:

(a) $\partial X/\partial x = \partial Y/\partial y$; (b) $X(y,x) = Y(x,y)$; (c) $X(y,x) = - Y(x,y)$

where $X(x,y) = m(x^3 + 5x^2y - 3xy^2 - y^3)$.

We indicate these conclusions merely in order to emphasize the exclusive character of the center. Moreover, the discussion concerns only a particular case when the functions X and Y are homogeneous polynomials of the third degree; for a more general case these conditions are presumably still more complicated.

Obviously, the existence of a center in the case of d.e. of the type (8.17) must be regarded as a rare exception occurring under very special conditions; whereas the existence of a focus *is the rule* since it depends *on the negation* of these special conditions.

If we now consider physical problems, the chances that a system has a center are extremely remote. In fact, as we have just seen, in order to have a center, a d.e. must have a very special form and, its coefficient must have special values. Very often observation shows that the behavior of trajectories is such that everything indicates the presence of a center; this means that the trajectories are closed and the motion is periodic *as far as this can be ascertained.* The difficulty is in ascertaining by observations whether an orbit corresponds exactly to a center or *to a very weak focus*; the latter appears whenever the criterion: $\int_0^{2\pi} H_{2j} d\varphi = 0$ breaks down for a sufficiently large j.

If one attempts to learn from an observation, one has to wait a sufficient length of time to be sure about the form of the orbit, which is dependent on the scale of time units. In the case of an artificial satellite, for instance, the matter is relatively simple, since the period is of the order of one hour; and, after a reasonably long time (days, weeks, months), one can notice the existence of a convergent spiral trajectory evidencing the presence of a fairly weak focus. On the other hand, on a larger scale (for example, the trajectory of a planet around the sun), the matter is less certain since here it takes a very long time (millenniums or, perhaps, millions of years) before one is able to answer this question with certainty. However, in our terrestrial situations such problems never arise and the singularity of the type *center* has never any *real existence* and appears merely as a convenient mathematical idealization separating convergent and divergent spiral trajectories around weak foci.

9. Certain conclusions

The situation outlined in this chapter reflects to some extent the profound changes in the fundamentals of the theory of oscillations once the theory of Poincaré has been adopted as a mathematical framework for studies of oscillations. In this chapter we have been concerned with the first step in these new developments, namely, the identification of a physical concept, equilibrium; and with a mathematical concept, singular point. It may seem that for the theory of oscillations proper this is not an important point. In fact, such a viewpoint existed in the old theory of oscillations where the problems of equilibrium and of stationary motion were studied more or less independently without trying to relate them to one another.

In Chapter 3 we shall see that in the new theory of oscillations, on the contrary, there exists a close relation between the nature of equilibrium and that of the stationary motion. This appears in a form of certain *topological configurations* and, in this way, the states of equilibrium and of stationary motion become, so to speak, welded together. However, in the new theory the stationary motion is entirely different from that of the old theory where it occurs around a center, since it is only then that the trajectories are closed. On the other hand, we have seen in the preceding section that the center is a very special singularity and that the least change of the form of the d.e. destroys it and converts it into a neighboring weak focus for which the trajectories cease to be closed. In this way a simple harmonic motion which, since the time of Galileo, has been assumed as a pattern of a stationary motion, turned out to be the most difficult to justify mathematically and impossible to produce experimentally. It appears thus that the singular point of the type center does not correspond to any physical reality and is merely a mathematical concept separating the regions of convergent and divergent trajectories. The latter, on the other hand, have a definite physical meaning.

It becomes thus necessary to associate stationary motions with foci (and, as we shall see later, also with nodes). This brings us directly to the most important concept of the new theory, namely, the concept of *the limit cycle* which is outlined in Chapter 3. Its importance lies in the fact that it permits establishing conditions for stationary motions in essentially non-conservative systems; this opens the enormous field of self-sustained oscillations which escaped the old theory. The extremely critical singularity, the center, disappears entirely in this new approach. Before proceeding with this question, we should investigate first the effect of non-linearities in conservative systems in the following chapter. Although this question is somewhat academic, since conservative systems do not appear in applied problems, the matter is still of great importance because it will enable us to introduce certain important definitions that will be useful later in our study of applied problems.

Chapter 2

NONLINEAR CONSERVATIVE SYSTEMS

1. Introductory remarks

Nonlinear conservative systems occupy an important position in the theory of oscillations. The fundamental property of these systems is the existence of a function of the dependent variables which is a constant of the motion and plays the role of the energy.

In a number of applied problems, the conservative character of a given system can be ascertained on physical grounds. When no such a *priori* conclusion can be made, the matter reduces to a search for the existence of a single-valued first integral which, generally, is not a simple problem particularly when the d.e. are not integrable in closed form. In relatively simple problems involving only d.e. with constant coefficients (linear or nonlinear), the presence of a term containing the first derivative (or some power of this derivative) generally indicates either a dissipative (or an energy absorbing) feature of the system and gives, thus, a simple criterion that the system *be not* conservative.

The converse, however, is not true because a lack of such terms does not mean that the system is conservative. Well known examples are the phenomena characterized by the d.e. of Mathieu which, in its standard form, has no term containing the first derivative, and yet the physical systems exhibiting such phenomena are generally not conservative. Likewise, mechanical systems, in which the constants depend explicitly on the time or in which the dynamical parameters (masses, lengths, etc.) are not constant, are not conservative. The pendulum of variable length is a typical example of such a situation.

In this chapter we follow closely the exposition of Andronov[1] who worked out a series of examples, two of which are given in sections 5 and 6.

[1] A. Andronov and S. Chaikin, *Theory of Oscillations* (original text in Russian), Moscow, 1937, and subsequent editions.

All these examples relate to a d.e. of the form $\ddot{x} + f(x) = 0$. The only singular points possible here are either centers, saddle points, or their confluence, and there is no special difficulty in deriving the first integral. Following Poincaré,[2] the discussion of the topology of integral curves, or of the "phase portrait," is easily conducted by introducing a parameter in the d.e. (Section 4). In some cases (Section 3) this can be accomplished by means of a similar parameter in the first integral. In the case of the Volterra problem (Section 9), the conservative character of the system is by no means obvious but, as the exact integration of the d.e. is possible and as the first integral exists, the system is conservative on formal grounds.

Another important feature of conservative systems is their critical character; a slight change in the form of the d.e. generally results in the loss of the conservation of the system. Thus, the d.e. of the harmonic oscillator $\ddot{x} + x = 0$ characterizes a conservative system, but the addition of a term $b\dot{x}$, with b as small as we please, results in the loss of the conservative character.

Before proceeding with the analysis of some special cases of conservative systems, it is useful to define the terms: *first integral* (or, simply, *integral*) and *conservative system*; from these definitions we shall derive some consequences which will be needed in the sequel.

By an integral (or first integral) of a system

$$\dot{x}_i = X_i(x_1, \ldots, x_n); \qquad i = 1, \ldots, n \tag{1.1}$$

we understand a differentiable function $F(x_1, \ldots, x_n)$ defined on a domain D of the phase space, and not constant on any open set, such that $F(u_1(t), \ldots, u_n(t)) \equiv C$; (C) a constant when $x_i = u_i(t)$ is a solution of (1.1). It can be shown that if $x^0 = (x_1^0, \ldots, x_n^0)$ is a regular (that is, nonsingular) point of (1.1), then x^0 has a neighborhood on which the first integral is defined. The existence of an integral, therefore, becomes of interest only in connection with its domain of definition. Our primary interest is in integrals defined on domains containing singular points.

We shall say that a *system is conservative* in a domain D if it has an integral in D and D has the property that *every trajectory having one point in D lies entirely in D* for $t \to +\infty$ or for $t \to -\infty$. In the general case an integral defines a family of $(n-1)$ dimensional surfaces on which the trajectories lie. In the case $n = 2$, the surfaces reduce to trajectories, a fact from which we can immediately deduce an important property of conservative systems.

If $F(x,y)$ is an integral of

$$\dot{x} = X(x,y); \qquad \dot{y} = Y(x,y) \tag{1.2}$$

[2] H. Poincaré, *Acta Math.* **7**, 1885; *Figures d'équilibre d'une masse fluide*, Naud, Paris, 1903.

on the domain D, then the family of curves $F(x,y) = C$ defines the trajectories of (1.2) in D. The trajectories of a conservative system in the plane are, therefore, the level curves of a differentiable surface $z = F(x,y)$. From this it can be seen that a conservative system cannot have any singular points which are stable in the sense in which we defined the term in the previous chapter; for, if all trajectories near the singularity tended to it, the existence of a continuous integral in a neighborhood of the singularity would imply that the value of the integral along any one of integral curves must be equal to its value at the singularity. This, however, implies that $F(x,y)$ must be constant in the whole open neighborhood of the singular point, which contradicts the definition. The same argument holds for trajectories tending toward the singularity as $t \to -\infty$. It follows in particular that a system (1.2) cannot have a node or focus in a region in which it has an integral.

2. Fundamental properties of nonlinear conservative systems

The simplest nonlinear conservative systems encountered in the theory of oscillations with one degree of freedom are those which lead to a d.e. of the form:

$$\ddot{x} + f(x) = 0 \tag{2.1}$$

The d.e. (2.1) may be regarded as an oscillator with a "restoring force," since one can always write $f(x) = F(x)x$ and consider $F(x)$ as a variable spring "constant."

Written as an equivalent system, the d.e. (2.1) is:

$$\dot{x} = y; \qquad \dot{y} = -f(x) \tag{2.2}$$

and the d.e. of integral curves is:

$$\frac{dy}{dx} = -\frac{f(x)}{y} \tag{2.3}$$

which shows that the integral curves have a horizontal tangent at the points x_i which are the roots of $f(x) = 0$, provided $y \neq 0$ at these points. As to the singular points, they require simultaneously $y = 0, f(x) = 0$. In other words, singular points (if they exist) lie necessarily on the x axis.

The energy integral in this case is:

$$\tfrac{1}{2}y^2 + V(x) = h \tag{2.4}$$

where $V(x) = -\int_0^x f(x)dx$. One can consider $\frac{1}{2}y^2 = \frac{1}{2}\dot{x}^2$ as kinetic and $V(x)$ as potential energy; the constant h, the total energy, expresses the fact that the system is conservative.

For a given value of h (2.4) represents a trajectory in the phase plane which exists as long as $h - V(x) > 0$. A position of equilibrium corresponds to the case when $y = 0$ and $V'(x) = \frac{dV(x)}{dx} = 0$. The latter condition implies that the potential energy has an extremum at the equilibrium point.

If one writes (2.4) in the form:

$$\tfrac{1}{2}y^2 = h - V(x) \tag{2.5}$$

a series of obvious conclusions can be obtained by plotting the difference $h - V(x)$ in Fig. 2.1a, and, for the same abscissa x calculating $y/\sqrt{2}$ from the relation $y/\sqrt{2} = \pm\sqrt{h - V(x)}$ in Fig. 2.1b. Given x, the quantity

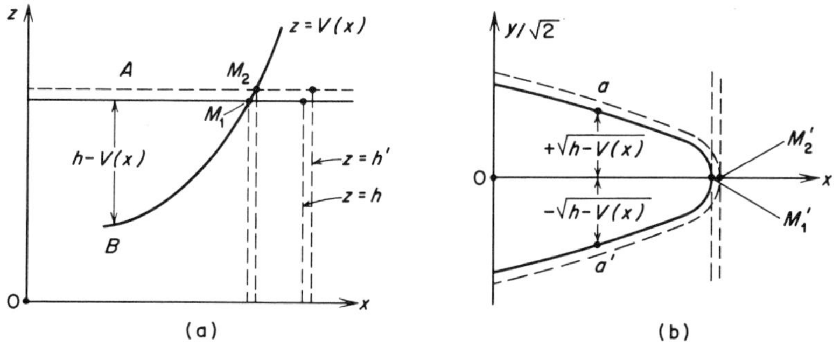

FIGURE 2.1

$h - V(x)$ is measured by the segment AB in (a); by extracting the square root and plotting its values with plus and minus signs, one obtains the points a, a' on the diagram (b); and proceeding thus point by point one obtains the phase trajectory.

If the energy constant changes and becomes h', a similar argument shows that the new trajectory extends further to the right if $h' > h$ (in (a)); it is shown by the broken line in (b). Graphical constructions of this kind permit tracing out changes in more complicated phase trajectories when the constant of energy is varied.

It is useful to discuss briefly the criteria of stability for conservative systems. If one expands the function $f(x)$ in (2.1) in a Taylor's series around the equilibrium point $x = x_1$, one has:

$$f(x) = a_1(x - x_1) + \frac{a_2}{2!}(x - x_1)^2 + \dots \tag{2.6}$$

where $a_1 = f'(x)|_{x=x_1}$; $a_2 = f''(x)|_{x=x_1}$. The potential energy is then:

$$V(x) = h_0 - \frac{a_1}{1 \cdot 2}(x - x_1)^2 - \frac{a_2}{1 \cdot 2 \cdot 3}(x - x_1)^3 - \ldots \quad (2.7)$$

taking into account that

$$f'(x)|_{x=x_1} = -V''(x)|_{x=x_1}; \quad f''(x)|_{x=x_1} = -V'''(x)|_{x=x_1}, \ldots$$

Substituting (2.7) into (2.4) and setting $x - x_1 = \xi$, $y = \eta$, yields

$$\frac{\eta^2}{2} + h_0 - \sum_{n=1}^{\infty} \frac{a_n \xi^{n+1}}{(n+1)!} = h \quad (2.8)$$

As is well known, the stability of equilibrium is associated with the extremum values of the potential energy. If the potential energy is minimum at the equilibrium point, the equilibrium is stable; if it is maximum, the equilibrium is unstable. It is also possible that the potential energy be an extremum without being either maximum or minimum. This intermediate case of "indifferent stability," from a practical point of view, must be considered as unstable.

Equation (2.8) can easily be interpreted in terms of the phase-plane relations. We first consider the case when $a_1 = f'(x)|_{x=x_1} = -V''(x)|_{x=x_1} \neq 0$ and limit the expansion to the term in ξ^2, in which case $z = V(x)$ and the straight line $z = h_0$ have contact of the first order.

If $V(x_1)$ is a minimum, $V'(x_1) = 0$ and $V''(x_1) > 0$, so that $a_1 < 0$ and (2.8) becomes:

$$\frac{\eta^2}{2} + \frac{|a_1|\xi^2}{2} = h - h_0 = \alpha \quad (2.9)$$

which is an ellipse with semi-axes $m = \sqrt{2\alpha}$ and $n = \sqrt{2\alpha/|a_1|}$. Thus the singular point $\eta = \xi = 0$ is a center (and therefore stable), being surrounded by curves which, in the first approximation, are ellipses. (The possibility that the singular point is a focus is ruled out by the fact that the system is conservative.) Retaining now the term in ξ^{k+1} in (2.8), we obtain

$$\frac{\eta^2}{2} + \frac{|a_k|\xi^{k+1}}{(k+1)!} = \alpha \quad (2.10)$$

which is again a closed curve, but not an ellipse.

A similar discussion in the case of a maximum of potential energy $V(x)$ shows that, if $a_1 \neq 0$, the curves in the neighborhood of the point $x = x_1$ at which the maximum occurs are

$$\frac{\eta^2}{2} - \frac{a_1\xi^2}{2} = \alpha \quad (2.11)$$

These are hyperbolas having the lines $\eta = \pm\sqrt{a_1}\,\xi$ as asymptotes. The point $\eta = \xi = 0$ is a saddle point. In a similar manner, if $a_1 = a_2 = \ldots = a_{k-1} = 0$ but $a_k > 0$, one finds that for $a_k > 0$ the trajectories are hyperbolic curves (of order k) of the form:

$$\frac{\eta^2}{2} - \frac{a_k \xi^{k+1}}{(k+1)!} = \alpha \tag{2.12}$$

The motion has features similar to those in the preceding case but the asymptotes are curvilinear, etc., thus even for small values of ξ and η the motion differs quantitatively from the case when $a_1 \neq 0$.

The last case when $V(x)$ is stationary (without being either maximum or minimum) at $x = x_1$ is a critical case and results from the coalescence of a

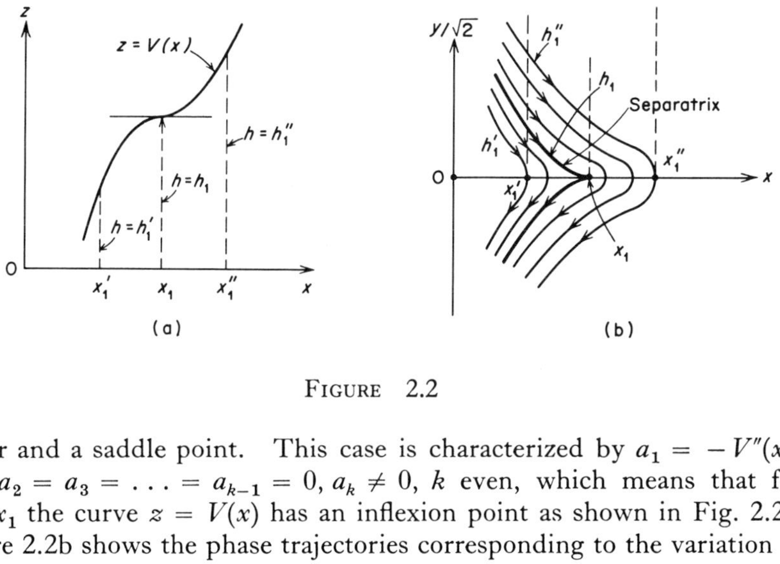

FIGURE 2.2

center and a saddle point. This case is characterized by $a_1 = -V''(x_1) = 0$, $a_2 = a_3 = \ldots = a_{k-1} = 0$, $a_k \neq 0$, k even, which means that for $x = x_1$ the curve $z = V(x)$ has an inflexion point as shown in Fig. 2.2a. Figure 2.2b shows the phase trajectories corresponding to the variation of the constant of energy according to the curve of $V(x)$ shown in Fig. 2.2a. There are exactly two trajectories having the singularity as a limit point. One of these approaches it while the other recedes from it. Together with the singularity these two trajectories form a cusp at the singularity. The singular point is obviously nonelementary and unstable.

3. Motions in the large; separatrix

Although the preceding considerations do not present anything essentially new, their interpretation in the phase plane is of great interest from the standpoint of the topology of phase trajectories and leads to important conclusions regarding the *motions in the large*. One obtains in this manner a mapping of the phase plane into certain domains possessing different oscillatory properties. The boundaries of these domains are certain asymptotic trajectories called *separatrices*. Let us consider, for instance, the potential energy $V(x)$ shown in Fig. 2.3, having two maxima, 3 and 5, and three minima, 2, 4, and 6. To the left of point 1 and to the right of point 7, the function $V(x)$ rises and we propose to investigate what happens only in the region between these two points following the graphical method outlined for Fig. 2.1.

One can consider the curve $V(x)$ in Fig. 2.3a as a profile of the bottom of a sea (or lake) with a level of water h_0, the energy constant, counted from some x axis. This will merely permit using a more condensed language. Assume, to begin with, that the peaks 3 and 5 are at the same level h_0. Starting with this value of $h = h_0$, it gives rise to two saddle points 3 and 5 in the phase-plane diagram shown in Fig. 2.3b. Since no trajectory exists to the left of 1 and to the right of 7, one has a configuration of closed trajectories issuing from the asymptotes of the two saddle points and closed as shown at the points 1 and 7. Since these trajectories tend to, or away from, the saddle points, they are necessarily asymptotic and one has to count them separately, viz.: 3 1 3, 3 5 upper, 5 3 lower, 5 7 5. They are shown in heavy lines and constitute the *separatrices*.

If the level h descends to some lower value, say h_1, "islands" appear in (a), the level being now at points *ab*, *cd*, and *ef*. If one projects points *a*, *b*, *c*, *d*, *e*, and *f* on the lower part of the figure, one obtains the limits between which closed trajectories are possible around the centers 2, 4, and 6 corresponding to points 2, 4, and 6 of the upper part (a) representing the minima of the potential energy. There will appear, thus, corresponding "islands" of periodic motions in (b), but these islands correspond to the submerged parts in (a). If the level h descends further so that the islands in (a) increase, the corresponding islands in (b) shrink. If, for instance, 2 is the point of greatest depth in (a) and if the level reaches this point, in the corresponding island of periodicity (in (b)), the closed trajectory shrinks to one point—the center 2—and disappears thereafter, if the level continues to decrease. In the remaining two islands, the trajectories still exist around the centers 4 and 5.

If, instead of decreasing from $h = h_0$, the level in (a) begins to increase from that point, the peaks 3 and 5 become submerged and this means that in (b) the periodic motion becomes possible on a trajectory enclosing the separatrix in its interior as shown by the broken line in Fig. 2.3b.

It is seen, thus, that the passage of the level through the critical value for which the peaks 3 and 5 become submerged is characterized by a radical change in the configuration of the phase trajectories. Instead of forming three separate islands, the domain of periodicity is now confined to the region *around* the separatrix but excluding the regions inside it.

It will be shown later that these results follow also from a theorem of Poincaré[2] outlined in the next section.

4. Effect of a parameter in a differential equation; bifurcation values

The considerations developed in the preceding sections are derived from the analysis of the influence of the parameter h in the first integral of the d.e. (the energy integral) on the "phase portrait" of integral curves. A still broader approach to this problem can be obtained if a parameter is introduced into the d.e. themselves, as was shown by Poincaré.[2]

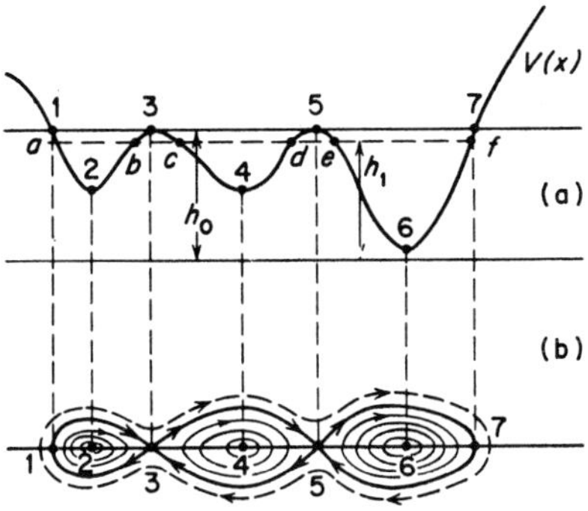

FIGURE 2.3

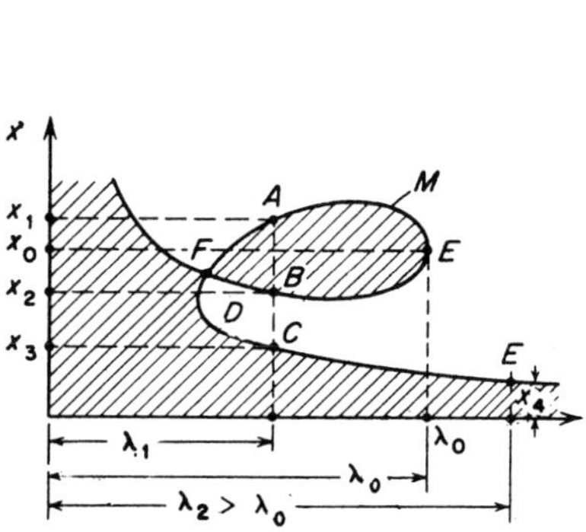

FIGURE 2.4

If a dynamical system (electrical or mechanical) is represented by a d.e. containing a parameter λ, the solution becomes a function of λ. If for some changes of λ the solution varies without undergoing any qualitative changes in its topological structure, such values of λ are called *ordinary* values. If, however, for some special value $\lambda = \lambda_0$ of the parameter, the topological aspect of the phase trajectories undergoes a *qualitative change*, such a special value is called a *critical* or a *bifurcation value.*

In conservative problems the parameter λ enters generally in the expression of the potential energy, which can then be written $V(x,\lambda)$; it appears, therefore, also in the expression of the "restoring force"

$$f(x,\lambda) = -\frac{\partial V(x,\lambda)}{\partial x}.$$

The equation $f(x,\lambda) = 0$ represents a curve in the (λ,x) plane (see, for example, Fig. 2.4). This curve may have a more or less complicated form

[2] See footnote [2], page 41.

depending on each particular problem; we shall need only some simple conclusions for the sake of the examples to follow.

Since the curve $f(x,\lambda) = 0$ represents the positions of equilibrium, it shows how these change when the parameter varies. For example, Fig. 2.4 exhibits, for $\lambda = \lambda_1$, three positions of equilibrium A, B, and C, whereas for $\lambda = \lambda_2$ there is only one, namely G. For $\lambda = \lambda_0$, two positions of equilibrium coalesce at E and disappear thereafter for $\lambda > \lambda_0$. This happens whenever $f_x(x,\lambda) = 0$. This means that the tangent

$$\frac{dx}{d\lambda} = -\frac{f_\lambda(x,\lambda)}{f_x(x,\lambda)} \tag{4.1}$$

is vertical at this point.

Since a point on $f(x,\lambda) = 0$ is stable if $f_x(x,\lambda) < 0$ (that is, the potential energy has a minimum), it follows that, if $f(x,\lambda)$ is positive in the region below the curve $f(x,\lambda) = 0$ and negative in the region above it, the points on $f(x,\lambda)$ are stable. If $f(x,\lambda)$ changes sign in the opposite direction, the points are unstable. Assuming that the shaded area in Fig. 2.4 corresponds to positive values of $f(x,\lambda)$, and the unshaded area to negative values, we conclude that the arc FAE is stable and the arc FBE unstable. Similar considerations apply to other arcs.

5. Problem of the rotating pendulum

Consider a pendulum of mass m and length a constrained to oscillate in a plane P rotating with angular velocity Ω about the vertical line. The moment of the centrifugal force acting on the pendulum (Fig. 2.5) is $m\Omega^2a^2 \sin\theta \cos\theta$ and that of gravity $mga \sin\theta$ so that the d.e. of the rotating pendulum is:

$$I\ddot{\theta} - m\Omega^2a^2(\cos\theta - \lambda)\sin\theta = 0 \tag{5.1}$$

where $I = ma^2$ is the moment of inertia, $\lambda = g/\Omega^2a$ is a parameter, and θ the angular deviation of the pendulum. The equivalent system is:

$$\dot{\theta} = \omega; \qquad \dot{\omega} = \frac{m\Omega^2a^2}{I}(\cos\theta - \lambda)\sin\theta \tag{5.2}$$

The d.e. of the integral curves is:

$$\frac{d\omega}{d\theta} = \frac{m\Omega^2a^2}{I\omega}(\cos\theta - \lambda)\sin\theta \tag{5.3}$$

The singular points of (5.2) are: $\omega_1 = 0, \theta_1 = 0$; $\omega_2 = 0, \theta_2 = \pi$; $\omega_3 = \theta$,

$\theta_3 = \cos^{-1} \lambda$. The last singular point exists only if $\lambda < 1$, that is, if Ω is sufficiently large. As the force $f(x,\lambda)$ of the preceding section is here:

$$f(\theta,\lambda) = m\Omega^2a^2(\cos\theta - \lambda)\sin\theta \tag{5.4}$$

the points of equilibria are clearly $\theta = 0$, $\theta = \pm\pi$, and $\cos\theta = \lambda$.

The corresponding (θ,λ) diagram is shown in Fig. 2.6 with regions in which $f(\theta,\lambda) > 0$ shown in shading. According to the rule of Poincaré, the stable and the unstable branches of the diagram of Fig. 2.6 are shown by black and white points, respectively. The former correspond to the equilibria of the type center and the latter to those of a saddle point.

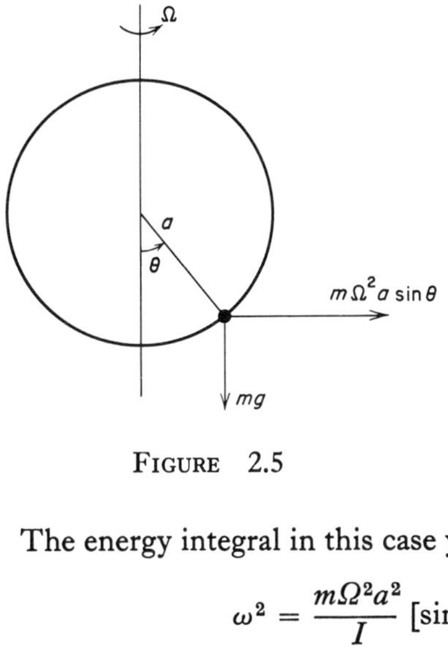

FIGURE 2.5

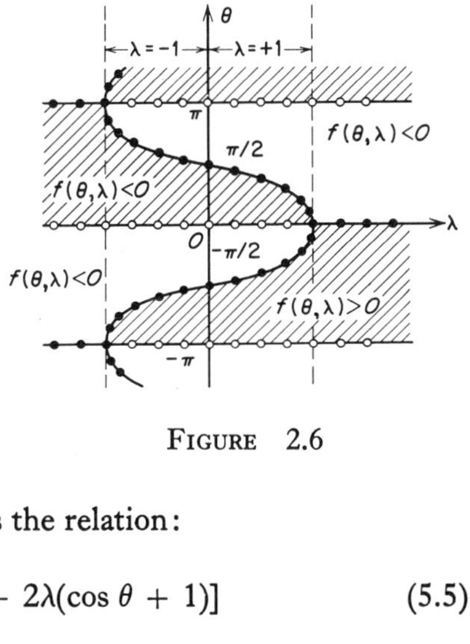

FIGURE 2.6

The energy integral in this case yields the relation:

$$\omega^2 = \frac{m\Omega^2a^2}{I}[\sin^2\theta + 2\lambda(\cos\theta + 1)] \tag{5.5}$$

In this equation, the energy constant has been determined by the condition that the separatrix passes through the saddle point $\theta = \pm\pi$; $\dot{\theta} = 0$. As there also exists a second separatrix corresponding to $\theta = 0$; $\dot{\theta} = 0$ for which $h = -m\lambda\Omega^2a^2$, one also has the relation:

$$\omega^2 = \frac{m\Omega^2a^2}{I}[\sin^2\theta + 2\lambda(\cos\theta - 1)] \tag{5.6}$$

Figure 2.7 shows the phase portrait of the d.e. (5.1) with the separatrices A and B corresponding to (5.5) and (5.6), respectively. It is noted that the center at the origin, if $\Omega = 0$, becomes a saddle point for $\Omega \neq 0$, in which case there appear two centers V_1 and V_2 symmetrically located with respect to the origin. The periodic motions about these centers (within the internal separatrix B) are asymmetrical. When the energy constant h reaches the value corresponding to the separatrix B, the motion changes its character and takes place around two centers V_1, V_2 and the saddle point S at the origin, being still inside the external separatrix A. In this region the motion is still oscillatory with velocity decreasing in the neighborhood of $\theta = 0$. If the energy constant is still further increased and the separatrix A is crossed, the motion becomes rotary (broken line in Fig. 2.7). In the

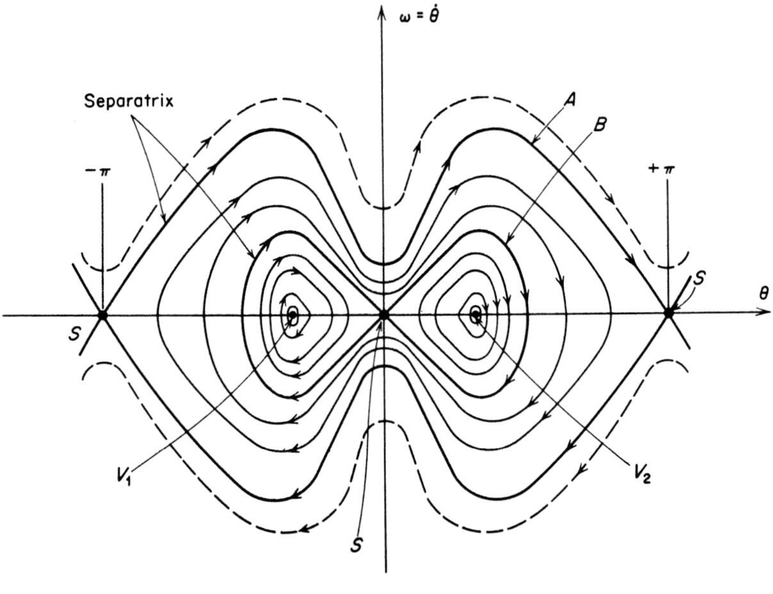

FIGURE 2.7

phase plane this trajectory is not closed; it is, however, closed on the surface of a circular cylinder of radius 1 whose axis is $\dot{\theta}$, since on this surface the point $\theta = +\pi$ is obviously the same as $\theta = -\pi$. If $\lambda \to 0$ (that is, $\Omega \to \infty$), the two separatrices A and B approach each other and the centers V_1 and V_2 approach the points $\theta = \pm\pi/2$, respectively. If $\lambda > 1$, the phase portrait changes again; there appears a center V at the origin ($\theta = 0$) but the intermediate structure of trajectories disappears. The situation is similar to one which has been already encountered (Fig. 2.3).

It is seen that $\lambda = 1, 0, -1$, are the critical or "bifurcation values" of the parameter.

6. Attraction of current-carrying conductors[1]

Another interesting problem investigated also by Andronov concerns the attraction between an elastically constrained current-carrying conductor (of length l, current i) and a fixed conductor (current I, indefinite length, distance a from the wall, Fig. 2.8).

The force acting on the elastically constrained conductor is:

$$f(x,\lambda) = -k\left(x - \frac{\lambda}{a - x}\right) \tag{6.1}$$

where the parameter $\lambda = 2Iil/k$. The term $-kx$ is due to the mechanical

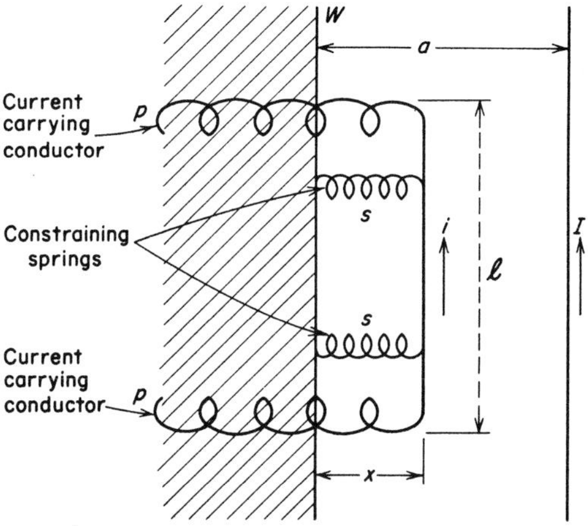

FIGURE 2.8

constraint and $k\lambda/(a - x)$ results from the electrodynamic attraction (Biot and Savart Law).

The value $\lambda_c = a^2/4$ is *critical* because both $f(x,\lambda_c)$ and $f_x(x,\lambda_c)$ vanish.

The d.e. of the system is:

$$m\ddot{x} + k\left(x - \frac{\lambda}{a - x}\right) = 0 \tag{6.2}$$

and the corresponding equivalent system is:

$$\dot{x} = y; \qquad \dot{y} = \frac{k}{m}\,\frac{x^2 - ax + \lambda}{a - x} \tag{6.3}$$

whence:

$$\frac{dy}{dx} = \frac{k}{m}\,\frac{x^2 - ax + \lambda}{y(a - x)} \tag{6.4}$$

[1] See footnote [1], page 40.

The singular points are on the x axis ($y = 0$) of abscissas

$$x_1 = \frac{a}{2} - b; \qquad x_2 = \frac{a}{2} + b; \qquad b = \sqrt{\frac{a^2}{4} - \lambda}$$

If $\lambda < a^2/4$, both x_1 and x_2 are real and positive. If one substitutes the values of x_1 and x_2 into the expression for $f_x(x,\lambda)$, one ascertains that $f_x(x_1,\lambda) < 0$ and $f_x(x_2,\lambda) > 0$. Also, since $f_x(x,\lambda) = -V_{xx}(x,\lambda)$, one concludes that for x_1 the equilibrium is stable (the singular point being a center) and for x_2 it is unstable (the saddle point). Moreover, as $x \to a$,

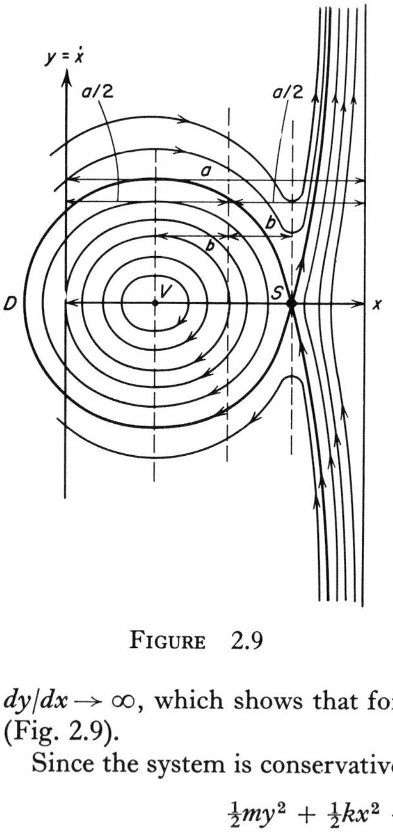

FIGURE 2.9

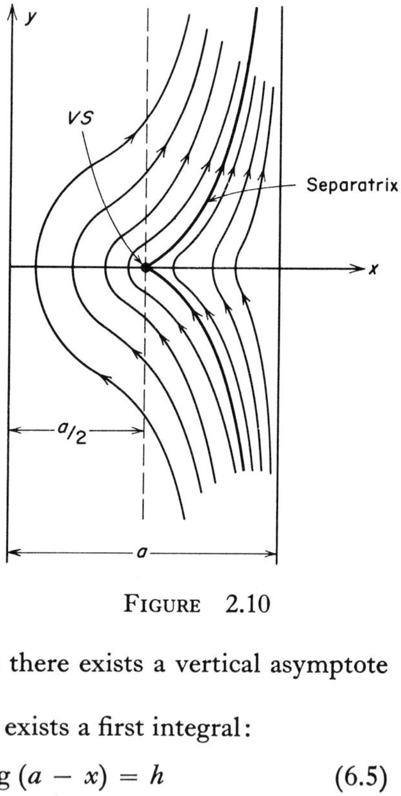

FIGURE 2.10

$dy/dx \to \infty$, which shows that for $x = a$ there exists a vertical asymptote (Fig. 2.9).

Since the system is conservative, there exists a first integral:

$$\tfrac{1}{2}my^2 + \tfrac{1}{2}kx^2 + k\lambda \log(a - x) = h \tag{6.5}$$

One obtains the equation of the separatrix by disposing of the energy constant h so that this particular integral line passes through the saddle point $(x_2,0)$. The energy constant satisfying this condition is:

$$h = \frac{1}{2}k\left(\frac{a}{2} + b\right)^2 + k\lambda \log\left(\frac{a}{2} - b\right) \tag{6.6}$$

thus the equation of the separatrix is:

$$\frac{1}{2}my^2 + \frac{k}{2}\left[x^2 - \left(\frac{a}{2} + b\right)^2\right] + k\lambda \log \frac{a - x}{(a/2) - b} = 0 \qquad (6.7)$$

We obtain again a familiar picture, viz.: inside the separatrix there exists an "island" of periodic trajectories (closed curves around the center $(x_1,0)$). If the value of energy constant (that is, the initial perturbation) is small enough and $\lambda < a^2/4$, the elastically constrained conductor executes a small undamped oscillation. In the phase plane this oscillation is represented by a closed trajectory around a center V. If $\lambda > a^2/4$, equation $x^2 - ax + \lambda = 0$ has no real roots, which shows that there is no singular point (that is, position of equilibrium). This means that the electrodynamic attractive force exceeds everywhere the elastic constraining force and the conductor is ultimately attracted to the fixed conductor which thus ends the motion.

The critical case when $\lambda = a^2/4$ corresponds to the coalescence of the center with the saddle point. This gives rise to a situation analogous to that analyzed previously in connection with the extremum value of the potential energy. The phase-plane diagram of this case is shown in Fig. 2.10.

7. Further properties of conservative systems; Hamiltonian variables; integral invariants

As was mentioned in Section 1, the general criterion of a conservative system is the existence of a *single valued first integral,* which in the previously outlined problems appears as the energy integral. *Single-valuedness* is essential inasmuch as there are systems which have a first integral and still are not conservative because this integral *is not* single valued.

As an example, consider an electric current I flowing in a rectilinear conductor of infinite length. In a plane xy perpendicular to this conductor, the components of the electromagnetic force acting on a magnetic pole m, according to the Biot-Savart Law, are

$$X = -\frac{y}{r^2}Im; \qquad Y = \frac{x}{r^2}Im \qquad (7.1)$$

These forces clearly derive from the potential $\varphi = \arctan(y/x)$ (which is not analytic around the origin) and one has:

$$\frac{\partial\varphi}{\partial x} = -\frac{y/x^2}{1 + (y/x)^2} = -\frac{y}{r^2}$$

and a similar expression for $\partial\varphi/\partial y = x/r^2$.

The equations of motion are thus: $\ddot{x} = X$; $\ddot{y} = Y$. If one multiplies them respectively by $\dot{x}$ and $\dot{y}$ and adds, one has

$$\frac{1}{2}\frac{d}{dt}(\dot{x}^2 + \dot{y}^2) = \frac{\partial\varphi}{\partial x}\dot{x} + \frac{\partial\varphi}{\partial y}\dot{y} = \frac{d\varphi}{dt}$$

This shows that there exists a first integral

$$\tfrac{1}{2}(\dot{x}^2 + \dot{y}^2) - \varphi(x,y) = \text{const} \tag{7.2}$$

However, as in this case the function $\varphi(x,y)$ is not single valued on the whole, the first integral is likewise not a single valued integral and the system is not conservative. In fact, for a rotation of m around I through an angle 2π, work is done.

Let us now return to the general theory. If $F(x_1,\ldots, x_n)$ is an integral of the system

$$\dot{x}_i = X_i(x_1,\ldots, x_n), \qquad i = 1,\ldots, n \tag{7.3}$$

Then differentiating the equation $F(x_1,\ldots, x_n) = C$ with respect to time and using (7.3) yields

$$\sum_{i=1}^{n} \frac{\partial F}{\partial x_i} X_i = 0 \tag{7.4}$$

This partial differential equation for an integral of (7.3) was encountered previously (for the case $n = 2$) in connection with the center problem (Chapter 1, Section 8) where the notion of integral was tacitly introduced.

The d.e. of a dynamical system with n degrees of freedom can be expressed in the Lagrangian form

$$\frac{d}{dt}\left(\frac{\partial T}{\partial \dot{q}_i}\right) - \frac{\partial T}{\partial q_i} = Q_i; \qquad i = 1, 2,\ldots, n \tag{7.5}$$

where T is the kinetic energy of the system, q_i are generalized coordinates (the degrees of freedom), and Q_i are external forces.

As in conservative systems the forces derive from a potential V, they are of the form $Q_i = -\partial V/\partial q_i$, and (7.5) becomes

$$\frac{d}{dt}\left(\frac{\partial L}{\partial \dot{q}_i}\right) - \frac{\partial L}{\partial q_i} = 0; \qquad i = 1, 2,\ldots, n \tag{7.6}$$

where $L = T - V$ is the Lagrangian function.

The differential system (7.6) is of order $2n$ involving the second derivatives of the n unknown functions $q_i(t)$. These higher derivatives appear in an implicit manner and a certain amount of manipulation is necessary to solve for these derivatives; moreover, in this resolution the original symmetry of the Lagrangian equations is lost.

There exists, however, a classical procedure due to Hamilton for transforming the Lagrangian system into a system of $2n$ first order differential equations with a symmetrical character. We shall deal continuously with such systems in what follows.

If we set

$$\frac{\partial L}{\partial \dot{q}_i} = p_i \tag{7.7}$$

from (7.6) one has also

$$\dot{p}_i = \frac{\partial L}{\partial q_i} \tag{7.8}$$

The system (7.7) may be regarded as a system of n equations for n unknowns $\dot{q}_i$ which can be expressed generally in terms of the new variables p_i. One has to transform the system (7.6) in terms of the new variables.

We have thus

$$\delta L = \sum_{i=1}^{n} \left(\frac{\partial L}{\partial q_i} \delta q_i + \frac{\partial L}{\partial \dot{q}_i} \delta \dot{q}_i \right) = \sum_{i=1}^{n} (\dot{p}_i \delta q_i + p_i \delta \dot{q}_i)$$

$$= \delta\left(\sum_{i=1}^{n} p_i \dot{q}_i \right) + \sum_{i=1}^{n} (\dot{p}_i \delta q_i - \dot{q}_i \delta p_i)$$

This gives

$$\delta\left(\sum_{i=1}^{n} p_i \dot{q}_i - L \right) = \sum_{i=1}^{n} (\dot{q}_i \delta p_i - \dot{p}_i \delta q_i) \tag{7.9}$$

One defines the function

$$H(p_i, q_i) = \sum_{i=1}^{n} p_i \dot{q}_i - L(q_i, \dot{q}_i) \tag{7.10}$$

assuming that $\dot{q}_i$ is expressed in terms of q_i and p_i by means of (7.7).

One can write

$$\delta H = \sum_{i=1}^{n} \left(\frac{\partial H}{\partial p_i} \delta p_i + \frac{\partial H}{\partial q_i} \delta q_i \right) \tag{7.11}$$

Comparing with (7.9) one gets

$$\frac{\partial H}{\partial p_i} = \dot{q}_i; \qquad \frac{\partial H}{\partial q_i} = -\dot{p}_i \tag{7.12}$$

Equations (7.12) are the classical d.e. of dynamics in the Hamiltonian form; hence, if L does not contain t explicitly, then obviously H does not contain it either. In such a case one has

$$\frac{dH(p_i,q_i)}{dt} = \sum_{i=1}^{n}\left(\frac{\partial H}{\partial p_i}\dot{p}_i + \frac{\partial H}{\partial q_i}\dot{q}_i\right)$$

$$= \sum_{i=1}^{n}\left(-\frac{\partial H}{\partial p_i}\frac{\partial H}{\partial q_i} + \frac{\partial H}{\partial q_i}\frac{\partial H}{\partial p_i}\right) = 0 \qquad (7.13)$$

which shows that $H(p_i,q_i) = \sum_{i=1}^{n} p_i\dot{q}_i - L(q_i,\dot{q}_i)$ is a first integral.

One has thus an important property of conservative systems, viz.:

If the Lagrangian function does not depend on time explicitly, the Hamiltonian function is a first integral of the dynamical system.

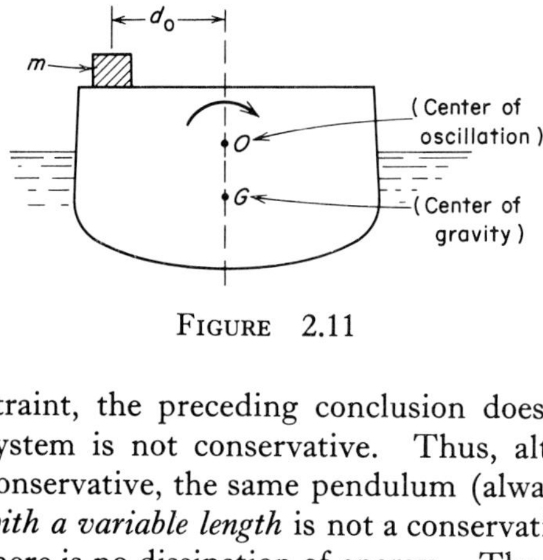

FIGURE 2.11

It can be shown that $H = T + V$ and, since H is an integral, one has

$$H = T + V = h \qquad (7.14)$$

which is the law of conservation of energy.

It is noted that lack of an explicit dependence on time for L or H means that the dynamical parameters and constraints are fixed. If a system has a varying constraint, the preceding conclusion does not hold and in such a case the system is not conservative. Thus, although a frictionless pendulum is conservative, the same pendulum (always in the absence of friction) but *with a variable length* is not a conservative system, in spite of the fact that there is no dissipation of energy. The lack of conservatism in such a case is due to the fact that energy is drained away from (or brought into) the system owing to the work done in the varying constraint.

As an example of a situation in which a dynamical parameter is variable, consider a frictionless pendulum M oscillating with a certain amplitude α. On this pendulum is mounted a moving weight m capable of changing its position across the pendulum M in such a manner that it is always on the *rising* side of M. If the motion of M takes place in accordance with the law $\alpha = \alpha_0 \cos t$, that of the weight m is $d = d_0 \sin t$, so as to fulfill the above condition. Thus, when the angular velocity of M is maximum (the upright position) in the direction shown by the arrow (Fig. 2.11), the weight m is in its extreme position to *the left*.

Such a system is nondissipative and yet it is not conservative. The position of the constraint (the track on which m is moving) is varying in space in such a manner that the weight m is *always rising*, and therefore the force of gravity is continuously doing *negative work*, thus draining energy away from the pendulum. This principle is used in connection with the anti-rolling stabilization of ships by the so-called "moving weight method."

It is clear that in this case the Lagrangian as well as the Hamiltonian functions depend on t explicitly, and the latter does not yield the first integral since the term $\partial H/\partial t$ does not vanish.

Conservative systems possess an important feature associated with the concept of *integral invariants* introduced by Poincaré[3] which, in the case of the Hamiltonian variables, takes a particularly simple form. Poincaré defines as an *integral invariant* a certain integral over a domain depending on t, whereas the value of the integral is independent of t. Intuitively, one can imagine a certain amount of incompressible fluid moving along in space in some manner and undergoing changes in form. The *volume* of the fluid is then an integral invariant.

The theory of integral invariants is connected formally with that of *multipliers* (or integrating factors) in the theory of differential equations and, as the latter is, in turn, connected with the existence of first integrals, it is possible to use the existence of integral invariants for a differential system as a criterion of its conservatism. This criterion is particularly simple if the differential system is expressed in terms of the Hamiltonian variables.

We consider a differential system

$$\dot{x}_i = X_i(x_1, \ldots, x_n); \qquad i = 1, 2, \ldots, n \tag{7.15}$$

and think of it as defining a flow in the phase space. A point which for $t = t_0$ is at some point (x_{i0}) of the n-space, finds itself at the time t at some other point (x_i) of that space. Thus the totality of all initial conditions (for $t = t_0$) forming a certain domain D_{t_0} gives rise to another domain D_t at time $t > t_0$.

If the "fluid" is incompressible, the volume of D_{t_0} is preserved by the flow, that is, $V(D_{t_0}) = V(D_t)$, or

$$\iint_{D_{t_0}} \cdots \int dx_1, dx_2, \ldots, dx_n = \iint_{D_t} \cdots \int dx_1, dx_2, \ldots, dx_n, \qquad t > t_0$$

[3] H. Poincaré, *Les méthodes nouvelles de la mécanique céleste* **T.3**, Gauthier-Villars, Paris, 1892; also E. Goursat, *Cours d'Analyse* **T.2**, Gauthier-Villars, Paris, 1918.

We can express the invariance of this integral by

$$V'(t) = \frac{d}{dt} \iint_{D_t} \cdots \int dx_1, dx_2, \ldots, dx_n = 0 \tag{7.16}$$

We proceed with an elaboration of this idea. Consider more generally scalar functions $M(x_1, \ldots, x_n)$ such that the integral

$$I(t) = \iint_{D_t} \cdots \int M(x_1, \ldots, x_n) dx_1, dx_2, \ldots, dx_n \tag{7.17}$$

over a domain D_t of n-space, does not depend on t. This integral will be invariant if

$$\frac{d}{dt} I(t) = 0 \tag{7.18}$$

A function $M(x_1, \ldots, x_n)$ will be called an *integral invariant* of (7.15) and satisfies (7.18). We wish to find the conditions that must be satisfied by the function $M(x_1, \ldots, x_n)$ in order that (7.18) holds.

Denote the solution of (7.15) whose value at $t = t_0$ is x_i^0 by

$$x_i = g_i(t, x_1^0, \ldots, x_n^0).$$

Let D_{t_0} be an arbitrary domain, and D_t the domain into which D_{t_0} is carried by the transformation $x_i^0 \to x_i$. Let $I(t)$ be defined by (7.17). Since t_0 is arbitrary there is no loss of generality in computing $(d/dt)I(t)$ at $t = t_0$. Denote the Jacobian of the transformation $x_i^0 \to x_i$ by $J(t)$, so

$$J(t) = \begin{vmatrix} \dfrac{\partial g_1}{\partial x_1^0} & \dfrac{\partial g_1}{\partial x_2^0} & \cdots & \dfrac{\partial g_1}{\partial x_n^0} \\ \vdots & & & \vdots \\ \dfrac{\partial g_n}{\partial x_1^0} & \cdots & \cdots & \dfrac{\partial g_n}{\partial x_n^0} \end{vmatrix} \tag{7.19}$$

To compute $I'(t_0)$ we transform $I(t)$ into an integral over D_{t_0}:

$$I(t) = \iint_{D_{t_0}} \cdots \int M(x_1, \ldots, x_n) J(t) dx_1^0, dx_2^0, \ldots, dx_n^0 \tag{7.20}$$

Thus,

$$I(t) - I(t_0)$$
$$= \iint_{D_{t_0}} \cdots \int [M(x_1, \ldots, x_n) J(t) - M(x_1^0, \ldots, x_n^0) J(t_0)] dx_1^0, dx_2^0, \ldots, dx_n^0$$

Setting $t = t_0 + \Delta t$, dividing by Δt, and passing to the limit $\Delta t \to 0$ under the integral sign, we find that

$$I'(t_0) = \iint_{D_{t_0}} \cdots \int \frac{d}{dt}(MJ)dx_1{}^0, dx_2{}^0, \ldots, dx_n{}^0 \tag{7.21}$$

where the integrand is evaluated at $t = t_0$, that is, at $x_i = x_i{}^0$. Now $(d/dt)(MJ) = M\dot{J} + \dot{M}J$ and we proceed to compute $\dot{J}$. Clearly for $t = t_0$, $\partial g_i/\partial x_j{}^0 = \delta_{ij}$ and $J(t_0) = 1$. Also, for $t = t_0 + \Delta t$,

$$\frac{\partial g_i}{\partial x_j{}^0} = \delta_{ij} + \Delta t \left[\frac{d}{dt}\left(\frac{\partial g_i}{\partial x_j{}^0}\right)\right]_{t=t_0} + 0(|\Delta t|^2)$$

Interchanging the order of differentiation,

$$\frac{d}{dt}\left(\frac{\partial g_i}{\partial x_j{}^0}\right) = \frac{\partial}{\partial x_j{}^0}\left(\frac{dg_i}{dt}\right) = \frac{\partial}{\partial x_j{}^0}[X_i(x_1{}^0, \ldots, x_n{}^0)] = \frac{\partial}{\partial x_j}[X_i(x_1, \ldots, x_j)]$$

Thus, $\partial g_i/\partial x_j{}^0 = \delta_{ij} + \Delta t\,(\partial X_i/\partial x_j)_{t=t_0} + 0(|\Delta t|^2)$ for $t = t_0 + \Delta t$, and expanding (7.19) yields

$$J(t_0 + \Delta t) = 1 + \Delta t \left(\sum_{i=1}^{n} \frac{\partial X_i}{\partial x_i}\right)_{t=t_0} + 0(|\Delta t|^2)$$

Therefore,

$$J(t_0 + \Delta t) - J(t_0) = \Delta t \left(\sum_{i=1}^{n} \frac{\partial X_i}{\partial x_i}\right)_{t=t_0} + 0(|\Delta t|^2)$$

and so $\dot{J}(t_0) = \sum_{i=1}^{n} \frac{\partial X_i}{\partial x_i}$. Finally, since $J(t_0) = 1$,

$$\left[\frac{d}{dt}(MJ) = M\sum_{i=1}^{n}\frac{\partial X_i}{\partial x_i} + 1 \cdot \sum_{i=1}^{n}\frac{\partial M}{\partial x_i}\frac{dx_i}{dt} = \sum_{i=1}^{n}\frac{\partial}{\partial x_i}[MX_i]\right]_{t=t_0} \tag{7.22}$$

the partial derivatives being evaluated at $x_i = x_i{}^0$. We can now drop the zero suffixes and write (7.21) in the form

$$\dot{I}(t) = \iint_{D_t} \cdots \int \sum_{i=1}^{n} \frac{\partial(MX_i)}{\partial x_i} dx_1, dx_2, \ldots, dx_n \tag{7.23}$$

If M is to be an integral invariant we must have $\dot{I}(t) = 0$ for *every* choice of the domain D_t. The necessary and sufficient condition that this be the case is that the integrand vanish, that is,

$$\sum_{i=1}^{n} \frac{\partial(MX_i)}{\partial x_i} = 0 \tag{7.24}$$

As a special case we have the result that the motion in the phase space preserves volume if, and only if,

$$\sum_{i=1}^{n} \frac{\partial X_i}{\partial x_i} = 0 \tag{7.25}$$

It is easy to show that this condition is always satisfied by a conservative physical system provided it is expressed in Hamiltonian form

$$\dot{q}_i = \frac{\partial H}{\partial p_i}; \qquad \dot{p}_i = -\frac{\partial H}{\partial q_i}; \qquad i = 1,\ldots, n$$

For this system the condition (7.25) becomes

$$\sum_{i=1}^{n} \frac{\partial}{\partial q_i}\left(\frac{\partial H}{\partial p_i}\right) - \frac{\partial}{\partial p_i}\left(\frac{\partial H}{\partial q_i}\right) = 0$$

which is clearly satisfied, since

$$\frac{\partial}{\partial q_i}\left(\frac{\partial H}{\partial p_i}\right) = \frac{\partial}{\partial p_i}\left(\frac{\partial H}{\partial q_i}\right).$$

Consider now the system

$$\dot{x} = X(x,y); \qquad \dot{y} = Y(x,y) \tag{7.26}$$

The equation

$$\frac{dy}{dx} = \frac{Y(x,y)}{X(x,y)}$$

of the integral curves of (7.26) can be written

$$Y(x,y)dx - X(x,y)dy = 0 \tag{7.26a}$$

The condition that (7.26a) be exact

$$\frac{\partial X}{\partial x} + \frac{\partial Y}{\partial y} = 0$$

is now recognized to be identical with the condition that (7.26) be area-preserving. More generally,

$$\frac{\partial(MX)}{\partial x} + \frac{\partial(MY)}{\partial y} = 0$$

is equally the condition that $M(x,y)$ be an integrating factor of (7.26a) or an integral invariant of (7.26). Thus we see that the existence of an integrating factor is equivalent to the existence of an integral invariant. This result is valid only for the case of the two equations.

To obtain a further connection between conservative and volume-preserving systems, let $g(x_1, \ldots, x_n)$ be an arbitrary differentiable, non-negative, scalar function which vanishes at most at the singular points of (7.15). Then the system

$$\dot{x}_i = g(x_1, \ldots, x_n)X_i(x_1, \ldots, x_n); \qquad i = 1, \ldots, n$$

has the same integral curves as (7.15), since the only effect of the common factor g is to change the *lengths* of the velocity vectors, and thus the speed with which the representative point traverses the trajectories. Now let $M(x_1, \ldots, x_n)$ be an integral invariant of (7.15), and consider the system

$$\dot{x}_i = M(x_1, \ldots, x_n)X_i(x_1, \ldots, x_n); \qquad i = 1, \ldots, n \tag{7.27}$$

By the foregoing argument, the trajectories of (7.27) are the same as those of (7.15). Since (7.27) satisfies (7.24) it is volume-preserving. Hence there is no difference between the phase-space geometry of systems which are merely conservative and those which, in addition, preserve volume.

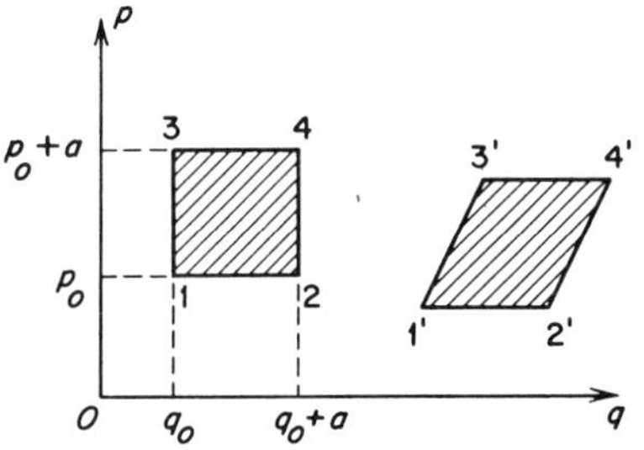

FIGURE 2.12

As an example of the conservation of area, consider the motion of a particle in the field of gravity. In this case we have the relations

$$\frac{dq}{dt} = p; \qquad \frac{dp}{dt} = -g; \qquad q = q_0 + pt - \frac{1}{2}gt^2; \qquad p = p_0 - gt \tag{7.28}$$

For $t = t_0$ we consider four points in the phase plane, namely: (1) q_0, p_0; (2) $(q_0 + a), p_0$; (3) $q_0, p_0 + a$; and (4) $q_0 + a, p_0 + a$, which determine a square shown in shading in Fig. 2.12.

For a later time $t = t_0 + \Delta t$ (Fig. 2.12), these four points become 1′, 2′, 3′, and 4′ calculated by (7.28) which determines the parallelogram 1′2′4′3′ whose area remains the same as that of the original square 1243, as one ascertains easily.

This remarkable property of the Hamiltonian variables to give directly the integral invariant in the form of the conservation of an area of the phase plane does not generally carry over to the Lagrangian variables. In fact, if one wishes to express the integral invariance in these variables, one has to introduce the Jacobian of the transformation $(p,q) \to (\dot{q},q)$; we have thus

$$I(t) = \iint_D dpdq = \iint_{D^*} J\begin{pmatrix} \dfrac{\partial p}{\partial \dot{q}} & \dfrac{\partial p}{\partial q} \\ \dfrac{\partial q}{\partial \dot{q}} & \dfrac{\partial q}{\partial q} \end{pmatrix} dqd\dot{q} = \iint_{D^*} \frac{\partial^2 L}{\partial \dot{q}^2}\, dqd\dot{q} \qquad (7.29)$$

We have now the "density" $\rho = \partial^2 L/\partial \dot{q}^2$ and what is "conserved" (or is invariant) is *not the area* $\iint dqd\dot{q}$ but *the area weighted by the density* ρ. It is clear that the integral invariance subsists if $\partial^2 L/\partial \dot{q}^2$ does not vanish in the domain under consideration.

8. Oscillating circuit with no resistance but with a nonlinear inductance

As an example of application of the general theory just outlined, we consider an oscillating circuit with no resistance but containing a saturated iron core inductance λ; it will be assumed also that the hysteresis loss in iron is negligible. Under these assumptions, the system is obviously conservative and we may expect that the first integral exists.

The d.e. of the circuit is

$$\frac{1}{C}\int_0^t idt + n\frac{d\varphi}{dt} = 0 \qquad (8.1)$$

where C is the capacitance, i the current, φ the magnetic flux through one turn of the inductance coil, and n the number of turns. This d.e. merely means that the electromotive forces across the condenser and that generated in the inductance λ balance each other.

The nonlinearity is $\varphi = f(i)$ since we assume that the iron in λ is saturated.

If one takes as Lagrangian variables: q as the charge in the condenser, $\dot{q} = i$ as the current, the Lagrangian function is

$$L = n\int_{i_0}^{i} \Phi(\dot{q})d\dot{q} - q^2/2C \qquad (8.2)$$

where the first term on the right-hand side is the kinetic energy and the second, the potential energy according to Maxwell's theory. Moreover, L is the Lagrangian function of (8.1) if this d.e. is of the form (7.6) but, since $\varphi = \partial L/\partial \dot{q}$ and $\int idt = q$, this is, indeed, the fact.

In this case there exists the first integral

$$H = \dot{q}\,\frac{\partial L}{\partial \dot{q}} - L = h \tag{8.3}$$

h being a constant; its explicit form is

$$H = n\phi(\dot{q})\dot{q} - n\int_{i_0}^{i}\phi(\dot{q})d\dot{q} + \frac{q^2}{2C} = h \tag{8.4}$$

This, as we saw previously, is the *total energy* $H = T + V$; the potential energy is obviously $q^2/2C$. As to the kinetic energy, it is the work of the electromotive force $nd\varphi(i)/dt$ in producing the current i; thus,

$$T = n\int_{t=0}^{t}\frac{d\varphi(\dot{q})}{dt}\,\dot{q}dt = n\int_{\phi_0}^{\phi}\dot{q}d\varphi(\dot{q}) \tag{8.5}$$

Integrating this expression by parts, one has

$$T = n\varphi(\dot{q})\dot{q} - n\int\varphi(\dot{q})d\dot{q} \tag{8.6}$$

which proves the statement.

It is more convenient to introduce the Hamiltonian variable

$$p = \frac{\partial L}{\partial \dot{q}} = n\varphi(\dot{q}) \tag{8.7}$$

One has the Hamiltonian function

$$H(p,q) = \int\psi(p)dp + \frac{q^2}{2C} \tag{8.8}$$

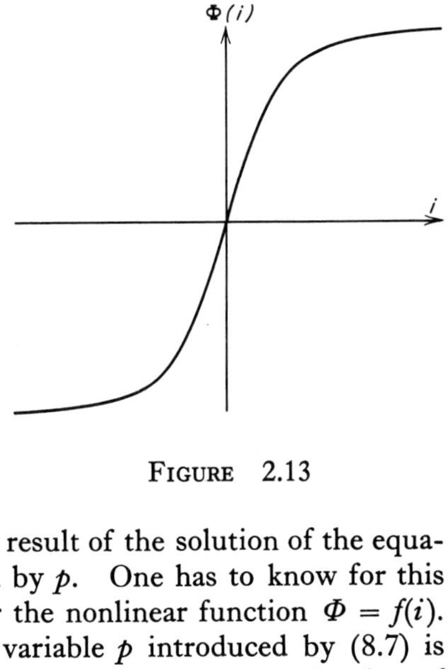

FIGURE 2.13

which gives directly the first integral $H = h$. In this expression $\psi(p)$ is the result of the solution of the equation $p = n\Phi(\dot{q})$ by which $\dot{q}$ is replaced by p. One has to know for this purpose an adequate approximation for the nonlinear function $\Phi = f(i)$. It is to be noted that the Hamiltonian variable p introduced by (8.7) is always related to q in a continuous and single-valued manner in view of the character of the function Φ in this case (Fig. 2.13).

The Hamiltonian equations are here

$$\dot{p} = -\frac{\partial H}{\partial q} = -\frac{q}{C}; \qquad \dot{q} = \frac{\partial H}{\partial p} = \psi(p) \tag{8.9}$$

and the integral invariant is

$$\iint dpdq = \iint\frac{\partial^2 L}{\partial \dot{q}^2}\,dqd\dot{q} = \iint\frac{\partial\varphi(\dot{q})}{\partial \dot{q}}\,dqd\dot{q} \tag{8.10}$$

The quantity $\partial\varphi(\dot{q})/\partial\dot{q}$ plays obviously the role of "density" ρ. The topological configuration describing an oscillatory process of this nature is given directly by the first integral $H = h$ but, in order to have an explicit expression for this integral, one has to determine the function $\psi(p)$.

In electrical engineering there exists a convenient analytical approximation of the function $\Phi(i)$ suggested by Dreyfuss,[4] viz.:

$$\Phi(i) = A \arctan \frac{ni}{S} + B\frac{ni}{S} \tag{8.11}$$

where A, B, and S are certain positive constants by means of which one can "fit" this formula into any particular case of the saturation effect.

If one uses this approximation one obtains the expression for $\partial\varphi/\partial\dot{q}$

$$\frac{\partial\varphi}{\partial\dot{q}} = \frac{An}{S}\frac{1}{1+(n\dot{q}/S)^2} + \frac{Bn}{S} \tag{8.12}$$

and, therefore,

$$n\int\frac{\partial\varphi}{\partial\dot{q}}\,\dot{q}d\dot{q} = \frac{An^2}{S}\int\frac{\dot{q}d\dot{q}}{1+(n\dot{q}/S)^2} + B\frac{n^2}{S}\int\dot{q}d\dot{q}$$

which, upon integration, yields

$$\frac{AS}{2}\log\left(\frac{\dot{q}^2}{2} + \frac{s^2}{2n^2}\right) + \frac{Bn^2}{2S}\dot{q}^2 + \frac{q^2}{2C} = c \tag{8.13}$$

This determines a family of closed curves resembling ellipses with the singular point (center) at the center of the family. Andronov carried out a similar calculation for another type of conservative system in which the nonlinearity is in the capacitance term. It is known that certain minerals used as dielectrics in condensers result in a nonproportionality between the charge and the corresponding voltage across the condenser.

The d.e. of an oscillating circuit in this latter case has the form

$$\lambda_0\ddot{q} + \frac{q}{C(q)} = 0 \tag{8.14}$$

where λ_0 is a linear (that is, constant) coefficient of inductance and $C(q)$ is a nonlinear capacitance. The argument is the same as before, viz.: one has to assume a certain analytical expression for the function $C(q)$ (for example, in a form of a polynomial); one forms the Lagrangian d.e. and passes to the Hamiltonian variables; one obtains finally the first integral by setting $H = h$ where $H = H(q,p)$ is a Hamiltonian function and h is a constant of the family. The remaining thing to be done is to construct the family of the curves $H = h$. One obtains again a family of closed curves around the

[4] L. Dreyfuss, *Electrotechnik und Maschinenbau*, 1911.

center at the origin; these curves look like distorted ellipses (approaching rather a rectangular form with rounded corners).

Summing up, *if one knows that a system is conservative*, it is advantageous to pass to the Hamiltonian variables. It is then sufficient to set $H = h$ and consider this equation as a family of curves, in order to have the corresponding family of closed trajectories around the center.

The essential point in all such procedures is to be *certain that the system is conservative.* This presents no difficulty if the system is either an electrical or mechanical system without dissipation (or, at least, with a negligible dissipation); besides this, the system must be such that its Lagrangian (and, therefore, the Hamiltonian) function does not depend on time explicitly.

If, however, the system is such that the energy consideration does not play any role in its formulation, this ceases to hold and the only criterion of its being conservative is the existence of a single-valued first integral.

In the following section we shall encounter a conservative system in a purely formal sense because no energy considerations are involved.

9. Volterra's problem

The problem of Volterra [5] is interesting as an example of a conservative system in which the question of the energy integral is not involved and the criterion of conservatism is based only on the existence of the single-valued first integral. Volterra formulates his problem in the following manner: in a lake or a closed sea, there exist two species (of fishes). The small species A feeds on vegetation (assumed to be available in an unlimited quantity), whereas the larger species B subsists exclusively by eating the members of the species A. The growth of the two species is governed by two factors:

(1) The natural multiplication of A assumed to occur at a rate proportional to their number N_1 which results in the d.e.:

$$\frac{dN_1}{dt} = \varepsilon_1 N_1 \tag{9.1}$$

and

(2) The dying out of the species B also proportional to their number, that is:

$$\frac{dN_2}{dt} = -\varepsilon_2 N_2 \tag{9.2}$$

[5] V. Volterra, *Théorie Mathématique de la lutte pour la vie*, Gauthier-Villars, Paris, 1931.

where N_1 and N_2 are the numbers of A and B species, respectively, and $\varepsilon_1 > 0$ and $\varepsilon_2 > 0$ are certain coefficients of proportionality.

These somewhat oversimplified hypotheses assimilate the unknown biological probabilities to the simple ones assumed in the kinetic theory of gases or in the theory of ionization. One can presumably argue regarding these assumptions, but the interesting part of the problem is not so much in the answer it yields as in the method followed.

The appearance of the species A or the disappearance of B occurs obviously in unit steps but, as is frequently done in mathematical statistics, one can adopt continuous variables and write the preceding equations as:

$$\frac{dx}{dt} = \varepsilon_1 x \tag{9.3}$$

$$\frac{dy}{dt} = -\varepsilon_2 y \tag{9.4}$$

The significance of these d.e. is sufficiently clear, viz.: if the species A existed alone, its number would increase indefinitely, following the exponential law and, similarly, if B existed alone, it would die out according to the same law.

The coexistence of the two species is to be taken now into account and Volterra assumes that, instead of ε_1 and ε_2, one should have ε_1' and ε_2' (both positive) defined as $\varepsilon_1' = \varepsilon_1 - \gamma_1 y$; $\varepsilon_2' = \varepsilon_2 - \gamma_2 x$. In other words, the coefficient ε_1 of the multiplication process is *decreased* owing to the existence of the extermination process (proportional again to y, the number of the B species) and, likewise, the coefficient ε_2 of the dying-out process for B is decreased owing to the available food offered by the A species.

Under these hypotheses the d.e. are:

$$\frac{dx}{dt} = (\varepsilon_1 - \gamma_1 y)x; \qquad \frac{dy}{dt} = -(\varepsilon_2 - \gamma_2 x)y \tag{9.5}$$

Multiplying the first equation by ε_2/x, the second by ε_1/y, and adding together, one gets:

$$\frac{\varepsilon_2}{x}\frac{dx}{dt} + \frac{\varepsilon_1}{y}\frac{dy}{dt} = -\varepsilon_2\gamma_1 y + \varepsilon_1\gamma_2 x \tag{9.6}$$

Substituting for ε_1 and ε_2, and taking into account (9.3) and (9.4) one has

$$\gamma_2\frac{dx}{dt} + \gamma_1\frac{dy}{dt} - \varepsilon_2\frac{d\log x}{dt} - \varepsilon_1\frac{d\log y}{dt} = 0 \tag{9.7}$$

This d.e. is integrated directly and one has the single-valued first integral:

$$\gamma_2 x + \gamma_1 y - \varepsilon_2 \log x - \varepsilon_1 \log y = h = \text{const}; \qquad x > 0, y > 0 \quad (9.8)$$

which can be written as:

$$F(x,y) = \exp(-\gamma_2 x)\exp(-\gamma_1 y)x^{\varepsilon_2}y^{\varepsilon_1} = H = \text{const} \qquad (9.9)$$

One can ascertain also that the expression $\iint dxdy/xy$ is an integral invariant with the "phase density" $Q = 1/xy$.

These results are already sufficient to guarantee that the system is conservative for $x > 0$, $y > 0$. On a purely intuitive basis, such an assertion may seem to be somewhat paradoxical since one species is continuously destroying the other. This emphasizes once more a lack of any intuitive criteria as soon as one departs from physical systems in which the conservation of energy yields the familiar first integral. One can obtain a somewhat approximate idea in this connection by analyzing the singular points of the system (9.5). Simplifying the notations somewhat and using a, b, c, and d instead of ε_1, γ_1, ε_2, γ_2, respectively, (9.5) is written as:

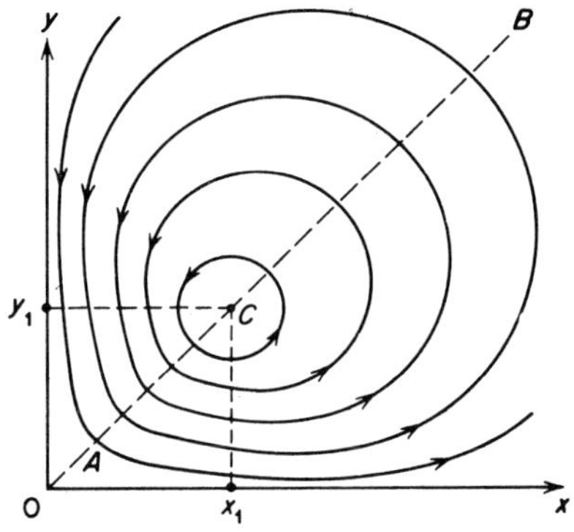

FIGURE 2.14

$$\frac{dx}{dt} = ax - bxy; \qquad \frac{dy}{dt} = -cy + dxy \qquad (9.10)$$

The origin is obviously a singular point. Consider, first, the neighborhood of the origin. In such a case (9.10) becomes approximately:

$$\frac{dx}{dt} \simeq ax; \qquad \frac{dy}{dt} \simeq -cy \qquad (9.11)$$

and it is seen that the origin is a *saddle point*, the axes of coordinates being the asymptotes (Fig. 2.14). If one sets $y = 0$ (the x axis), the first d.e. gives the direction away from the origin (the arrow on the x axis), whereas if $x = 0$, the second equation indicates the direction along the y axis, toward the origin. This merely illustrates (9.1) and (9.2) when one species is absent. Clearly only the first quadrant (for $x > 0$, $y > 0$) is of interest here.

There is a second singular point C of coordinates $x_1 = c/d$ and $y_1 = a/b$.

We transfer the origin to the point C and consider a small neighborhood around this point. Replacing x and y in (9.10) by $x_1 + \xi$ and $y_1 + \eta$, respectively (ξ and η small), the system (9.11) in the neighborhood of C is:

$$\frac{d\xi}{dt} = -\frac{bc}{d}\eta; \qquad \frac{d\eta}{dt} = \frac{ad}{b}\xi \tag{9.12}$$

which shows that the singular point C is a *center*.

Thus, on the basis of Volterra's hypotheses, if one puts into an empty lake a number of the A species equal to $x_1 = c/d$, that of the B species $y_1 = a/b$, there will be no biological fluctuations, the rate of birth of A being exactly compensated for by the rate of devouring A by B. If the initial numbers of each species are not far from x_1 and y_1, respectively, there will be a small fluctuation represented by a small closed curve (practically a circle) around C as center. The point C in this case will be a center. This analysis is not sufficient to give the complete picture of the trajectories in the intermediate region between O and C and one has to investigate a more complicated d.e. (9.10); in fact, the neighborhood of the saddle point at the origin produces a considerable deformation of integral curves which cannot be treated as circles in this region.

The problem is still possible because of the existence of the first integral (9.9) and the problem now reduces to the construction of the family of curves satisfying the equation (9.10). One ascertains that the closed integral curve approaches the origin at its nearest portion A, and spreads away from it on its farthest portion B, as shown in Fig. 2.14.

In the case considered by Volterra the system is conservative only because he assumes the laws (9.1) and (9.2) for the behavior of each species considered by itself. It seems, however, that these more or less a priori laws do not correspond to what may reasonably be expected. In fact, since the system is conservative, the initial conditions (say a point x_0,y_0 in the xy plane) determine a unique curve passing through (x_0,y_0).

Moreover, as the system is conservative, this curve, being closed, determines a periodic phenomenon. Thus, for instance, if (x_0,y_0) is near the origin, the closed curve will accordingly pass through this point. However, the preceding analysis shows that the nearer the closed curve comes to the origin, the further it spreads away from it on its opposite portion B. In other words, in order to produce very large fluctuations in the numbers of both species, it is sufficient to begin the experiment with a few members: x_0 and y_0 of each species.

This does not seem to be in accordance with observations which show that, roughly, the density of distribution of each species remains more or less constant (on the average and in the long run) and merely fluctuates somewhat around these average values. It is clear that the results depend on the choice of the statistical law (equations (9.1) and (9.2)); in the present case it is assumed that this law is the same as that used in the kinetic theory of gases. It is also clear that this is a somewhat doubtful point as there is no certainty whatever that what holds for the behavior of molecules of a gas is applicable also to the reproduction of a species of fishes.

If one assumes another law, results will be different; more specifically, a biological system of this kind may cease to be "conservative" as it now is under the assumption of laws (9.1) and (9.2). In fact, Kolmogorov[6] developed this point of view starting from a more general system of d.e.

$$\frac{dN_1}{dt} = K_1(N_1,N_2)N_1; \qquad \frac{dN_2}{dt} = K_2(N_1,N_2)N_2 \tag{9.13}$$

where K_1 and K_2 are continuous functions of N_1 and N_2 with continuous first derivatives. The quantities N_1 and N_2 are the same as previously and we designate them, as before, x and y.

If P is a point (x,y) in the (x,y) plane, we designate by S the direction OP (from origin to P). It is shown that, by imposing certain conditions one can obtain results different from those of Volterra.

More specifically, if the following conditions are fulfilled:

(1) $\partial K_1/\partial y < 0$; (2) $dK_1/dS < 0$; (3) $K_1(0,0) > 0$; (4) there exists $A > 0$, such that $K_1(0,A) = 0$; (5) there exists $B > 0$ such that $K_1(B,0) = 0$. Likewise, for K_2:

(1′) $\partial K_2/\partial y < 0$; (2′) $dK_2/dS > 0$; and (3′) there exists $C > 0$ such that $K_2(C,0) = 0$; then for $x > C$ and $C < B$ it is possible to have different situations, namely,

(a) Instead of a center, as in Volterra's theory, the point of equilibrium may be either a *stable focus* or a *stable node*.

(b) This point may be an *unstable focus* surrounded by a stable *limit cycle*.†

Apparently the solutions of the type (a) are not encountered, because biological fluctuations have been actually observed. In fact, this observation urged Volterra to undertake his work.

Solution (b) seems to be more appropriate on the basis of observations. In fact, it is utterly improbable that a few members of each species originally placed in a lake would give rise to enormous statistical fluctuations required by Volterra's theory. On the contrary, it seems more probable that by putting a certain number of fishes of each species into an originally empty lake, after a certain time a state of equilibrium will be reached;

[6] A. N. Kolmogorov, *Giorn. dell Istituto Italiano degli Attuari* **14**, 1936.

† This subject is discussed more fully in Chapter 3.

observation adds to this "common sense" picture also the existence of relatively small fluctuations.

Topologically this "common sense" picture (supplemented by fluctuations) is precisely a stable limit cycle in the (x, y) plane onto which wind the spiral trajectories from the outside as well as from the inside (since there is an unstable focus inside the limit cycle). The outside spiral trajectories are those which characterize the establishment of the biological phenomenon and the limit cycle is its representation in a stationary state. As regards the inner unstable focus, its nature is not very clear but it is likely that it means the impossibility of a steady-state *without fluctuations*. In other words, even if one puts into a lake a correct proportion of both species corresponding to this focal point, fluctuations will occur until a stable limit cycle is reached.

As far as is known, no experimental verification of these theoretical results has been made so far. If this is done eventually and the Kolmogorov theory is confirmed, this will give valuable information regarding the actual biological probabilities involved in the coexistence of the two species.

All that can be said at present is that the original hypothesis of Volterra (equations (9.1) and (9.2)) does not seem to be in accordance with the observed facts.

Chapter 3

LIMIT CYCLES OF POINCARÉ

1. Definitions

Poincaré showed[1] that the d.e. of the form

$$\dot{x} = X(x,y); \qquad \dot{y} = Y(x,y) \tag{1.1}$$

admit occasionally special solutions represented by closed curves in the phase plane which he calls *limit cycles*. A limit cycle is a closed trajectory

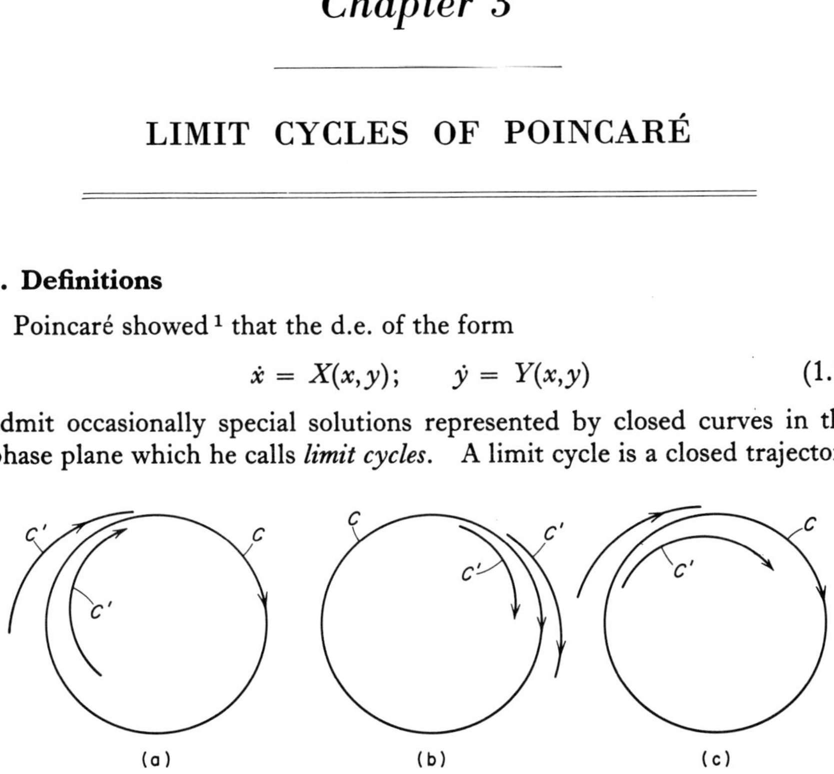

FIGURE 3.1

(hence the trajectory of a periodic solution) such that no trajectory sufficiently near it is also closed. In other words, a limit cycle is an isolated closed trajectory. Every trajectory beginning sufficiently near a limit cycle approaches it either for $t \to \infty$ or for $t \to -\infty$, that is, it either winds itself upon the limit cycle, or unwinds from it. If all nearby trajectories approach a limit cycle C as $t \to \infty$, we say that C is *stable* (Fig. 3.1a); if they approach C as $t \to -\infty$ we say that C is *unstable* (Fig. 3.1b). If the

[1] H. Poincaré, *J. des Math.* (3), **7**, 1881; also Œuvres **T.1**, Gauthier-Villars, Paris, 1928.

trajectories on one side of C approach it while those on the other side depart from it, we sometimes say that C is *semi-stable* (Fig. 3.1c) although from a practical point of view C must be considered unstable.

Limit cycles, and in particular stable limit cycles, are fundamental in the theory of oscillations of nonlinear, nonconservative systems—the only kinds of systems in which they can arise. A stable limit cycle represents a stable stationary oscillation of a physical system in the same way that a stable singular point represents a stable equilibrium.

2. Examples of limit cycles

Consider the following system:

$$\begin{aligned} \dot{x} &= y + \frac{x}{\sqrt{x^2 + y^2}}[1 - (x^2 + y^2)] \\ \dot{y} &= -x + \frac{y}{\sqrt{x^2 + y^2}}[1 - (x^2 + y^2)] \end{aligned} \tag{2.1}$$

In polar coordinates ($x = r\cos\theta$; $y = r\sin\theta$), it becomes:

$$\dot{x} = y + \frac{x}{r}(1 - r^2); \qquad \dot{y} = -x + \frac{y}{r}(1 - r^2)$$

Recalling that $x\dot{x} + y\dot{y} = \frac{1}{2}(dr^2/dt)$ and $y\dot{x} - x\dot{y} = -r^2(d\theta/dt)$ one obtains:

$$\dot{r} = 1 - r^2 \tag{2.2}$$

$$\dot{\theta} = -1 \tag{2.3}$$

The second d.e. merely shows that the radius vector rotates with a constant angular velocity. As to (2.2) it is integrated by the standard procedure which gives:

$$r = \frac{Ae^{2t} - 1}{Ae^{2t} + 1} \tag{2.4}$$

where the constant of integration $A = (1 + r_0)/(1 - r_0)$, r_0 being the initial value of r. It is noted that $r_0 = 1$. The limit cycle in this case is a circle with radius 1. If $r_0 > 1$, the spiral winds itself onto the circle $r_1 = 1$ from the outside; if $r_0 < 1$, it winds itself onto $r_1 = 1$ from the inside.

In a similar manner, one shows that the differential system:

$$\begin{aligned} \dot{x} &= -y + x(x^2 + y^2 - 1) \\ \dot{y} &= x + y(x^2 + y^2 - 1) \end{aligned} \tag{2.5}$$

admits an unstable limit cycle $r_l = 1$ as solution. In polar coordinates (2.5) becomes:

$$\dot{r} = r(r^2 - 1); \qquad \dot{\theta} = 1 \tag{2.6}$$

which is discussed in the same manner.

The differential system:

$$\begin{aligned} \frac{dx}{dt} &= y + x(x^2 + y^2)^{1/2}(x^2 + y^2 - 1)^2 \\ \frac{dy}{dt} &= -x + y(x^2 + y^2)^{1/2}(x^2 + y^2 - 1)^2 \end{aligned} \tag{2.7}$$

gives an example of a semi-stable limit cycle. In polar coordinates this system reduces to the d.e.:

$$\dot{r} = r^2(r^2 - 1)^2; \qquad \dot{\theta} = -1 \tag{2.8}$$

and the integration of the first d.e. permits easily ascertaining this point. Finally, as an example of an exceptional case of accumulation of limit cycles, one can indicate the d.e. of the form:

$$\begin{aligned} \dot{x} &= y + \mu(x^2 + y^2 - 1)x \sin\left(\frac{1}{x^2 + y^2 - 1}\right) \\ \dot{y} &= -x + \mu(x^2 + y^2 - 1)y \sin\left(\frac{1}{x^2 + y^2 - 1}\right) \end{aligned} \tag{2.9}$$

In polar coordinates these d.e. become:

$$\dot{r} = \mu r(r^2 - 1) \sin \frac{1}{r^2 - 1} \quad \text{if } r \neq 1; \quad \frac{dr}{dt} = 0 \quad \text{if } r = 1 \text{ and } \dot{\theta} = -1 \tag{2.10}$$

There exists, obviously, an infinity of circles in the neighborhood of $r = 1$ corresponding to zeros of $\sin [1/(r^2 - 1)]$.†

The simplicity of these examples in polar coordinates is due to the fact that they were originally formulated in polar coordinates and then transformed to cartesian coordinates. In reality, the matter of ascertaining the presence of a limit cycle in a given d.e. is a very difficult problem and one that can be solved by direct methods only in a few isolated cases. In Part II it will be shown that, in contrast with the just-mentioned difficulty of direct methods, the theory of approximation, on the contrary, gives generally simple means of ascertaining the presence of a limit cycle in the theory of the first approximation. Anticipating somewhat the later outline of this subject, it is sufficient to mention that in the theory of the first

† It is to be noted that the d.e. ceases to be analytic for $r = 1$.

approximation the limit cycles appear as circles, the radius of which is determined precisely by this approximation. In this manner, it is possible to determine the essential part of the problem (that is, its topological configuration) at a sacrifice of some secondary facts (for example, the presence of harmonics in the stationary solution).

3. Physical significance of limit cycles

As mentioned previously, limit cycles represent the stationary states of oscillations. The most important application of the theory of limit cycles is in relation to the so-called *self-sustained oscillations* which characterize, for instance, the oscillatory state of an electron-tube oscillator. A clock or a watch may also be regarded as another example of such an oscillator (Section 11 below). A common feature of oscillators of this kind is that their stationary oscillatory state does not depend on the initial conditions (as for conservative systems) but depends uniquely on the parameters of the system, which means that *it is determined by the differential equation itself.* One understands this property easily on the basis of the previous definition of limit cycles by recalling that any initial condition is represented by some point in the phase plane through which passes one and only one trajectory C'. Since, however, the fundamental property of a stable limit cycle is characterized by an approach of any trajectory C' (at least within a certain range) to the closed trajectory C, it is obvious that whatever are the initial conditions (within this range), the ultimate stationary motion will establish itself on C and, in this sense, one can say that this stationary motion is *independent of the initial conditions.* Thus, it is immaterial whether the oscillation of an electron-tube oscillator is started by the closing of a switch or whether at the instant of closing the switch some arbitrary impulse is applied to the system; the ultimate self-sustained oscillation will be exactly the same. The same applies to the mechanism of a clock; if the clock is wound but is initially at a standstill, it is immaterial whether its start is due to a small or to a large impulse as long as it is sufficient for starting.

In addition to the property of being self-sustained, oscillations of the limit cycle type have also another important property—that of self-starting or *self-excitation.* In the above example of an electron-tube oscillator this property manifests itself in that the oscillatory phenomenon develops from the state of rest and reaches its ultimate stationary motion as soon as the switch is closed, that is, as soon as the physical oscillatory system is *completed.* The same applies, generally, to an ordinary watch; as soon as it is wound, it starts going. In both cases the oscillatory phenomenon starts spontaneously from rest and reaches its stationary state on the limit cycle. In applications, conditions of this kind are commonly known under the name of *soft self-excitation.*

In the case of a wound clock that is initially at a standstill, the situation is different, however, in that it is necessary to *start* the clock by some impulse. As was just mentioned, the magnitude of this impulse is immaterial as long as it is sufficient to produce the start. Similar phenomena are observed occasionally in the electron-tube circuits under the appropriate conditions. It is observed in such cases that the circuit is normally in the state of rest, but under the effect of an impulse it starts oscillating. Here again the magnitude of the impulse needed for starting the oscillation is immaterial as long as it is greater than a certain critical or *threshold value.*

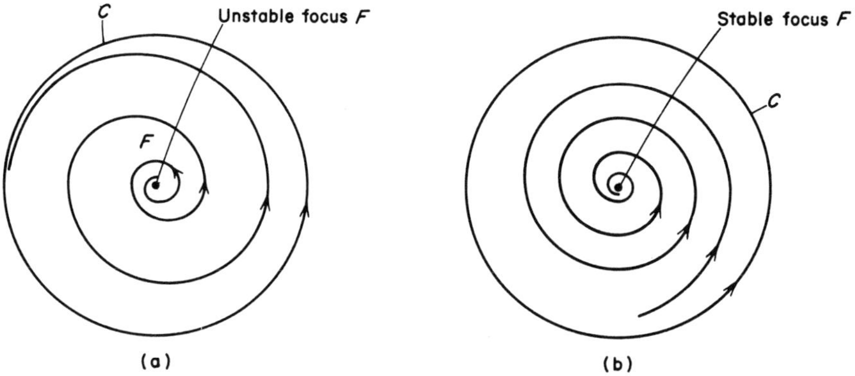

FIGURE 3.2

Phenomena of this nature are usually designated by the term *hard self-excitation.*

As regards the soft self-excitation, it is nothing but the simplest possible topological configuration shown in Fig. 3.2a, according to which the phenomenon is represented by a trajectory unwinding itself from an *unstable* singular point and winding onto the stable limit cycle from the inside.

For a hard self-excitation, the matter is more complicated and it is necessary, for this purpose to investigate additional topological configurations.

4. Polycyclic configurations

In Section 1, a limit cycle C in the phase plane was defined by the property that it is essentially an *isolated closed trajectory* in the sense that all neighboring trajectories C' are spirals winding themselves on C.† This does not exclude, however, the possibility that at some distance there exist other closed trajectories and, in particular, other limit cycles. We shall return to this subject later, but at this time it is useful to mention one configuration which we shall encounter continuously in what follows and which consists of a number of limit cycles enclosed inside each other, with a singular point in the innermost limit cycle. It is intuitively clear that the cycles must be alternately stable or unstable, except that semistable cycles may intervene. For the moment we shall exclude the latter case. If the singular point is unstable (an unstable focus or node), the innermost cycle is stable, the next unstable, etc. (Fig. 3.3a). If the singularity is stable, the first cycle is unstable, the next stable, etc. (Fig. 3.3b).

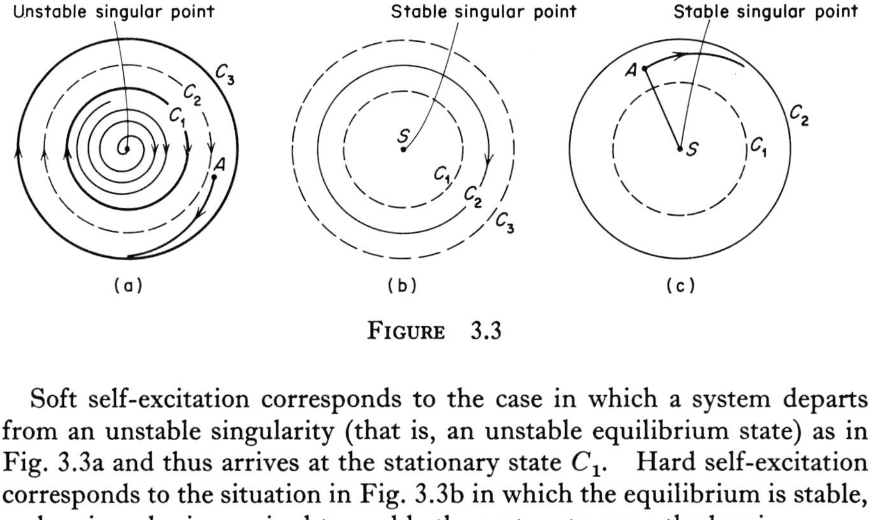

FIGURE 3.3

Soft self-excitation corresponds to the case in which a system departs from an unstable singularity (that is, an unstable equilibrium state) as in Fig. 3.3a and thus arrives at the stationary state C_1. Hard self-excitation corresponds to the situation in Fig. 3.3b in which the equilibrium is stable, and an impulse is required to enable the system to cross the barrier represented by the unstable cycle C_1 and have its initial state correspond to a point in the region between C_1 and C_3. The oscillatory state attained will then be C_2.

There is still another possibility worth mentioning which appears frequently when the d.e. contain a parameter λ. If this parameter varies, the topological configuration like the one shown in Fig. 3.3a varies too, and it may happen that for some critical value $\lambda = \lambda_0$ the stable and the unstable cycles C_1 and C_2 approach each other indefinitely and coalesce at the limit, giving rise to a semi-stable cycle which is, as we mentioned, an essentially

† We exclude from our consideration certain, so to speak, "pathological cases" like the d.e. (2.9) that have no known physical significance.

unstable configuration. Generally, both cycles C_1 and C_2 disappear upon their coalescence. In such a case R finds itself in a region free of limit cycles and is thus within the zone of attraction of the remaining cycle C_3. This circumstance manifests itself in a quasi-discontinuous jump of the amplitude from the cycle C_1 to C_3.

5. The index of Poincaré

The concept of the *index* was introduced by Poincaré for the purpose of establishing a *necessary* criterion for existence of a closed trajectory (limit cycle); this criterion is, however, not sufficient. From the fact that the theorem of the index indicates the possibility of a closed trajectory, one cannot yet conclude that such a trajectory exists. The usefulness of the theorem is that it permits ruling out situations where closed trajectories are impossible. In addition, certain properties of invariance render the theorem useful when the phase portrait undergoes qualitative changes as the result of a parameter variation.

We indicate first a simple (geometrical) definition of the index and outline its analytical definition later.

We consider in the phase plane of the system (1.1) a closed Jordan curve C not traversing any positions of equilibria (singular points). This curve may be regarded as being in the vector field V of trajectories of (1.1). Suppose that a point S moves on C in the positive direction (say, counterclockwise) and that we mark the direction of the tangent to the vector field at S relatively to a fixed system x,y of reference. It is clear that when S turns over an angle 2π on C, the tangent vector will resume its initial position having turned over an angle $2\pi I$, where I is integer.

We consider the rotation of the vector V (and therefore I) as positive if V (with respect to x,y) rotates in the *same direction* as S does on C and as negative in the opposite case.

From this definition it follows that I may be either positive or negative or zero, but it is always an integer; it is also clear that I in a certain sense *does not depend on the form of the curve* C as long as by deforming it we do not change the number of singular points inside.

The integer I so defined is called the index of the curve with respect to the vector field V.

The characteristic cases are obtained if C either does not contain any singular points or contains just one simple singular point.

By tracing the field V of trajectories issuing from the singular point in question and surrounding the latter by closed curve C one can easily verify by the above construction of tangents that

1. For C not containing any singular point, $I = 0$.
2. For C containing either a focus or a node or a center, $I = +1$.
3. For C containing a saddle point, $I = -1$.

We give now a more precise definition of the index. Let $V = V(X,Y)$ be a vector field whose components $X = X(x,y)$ and $Y = Y(x,y)$ are continuously differentiable; other conventions regarding the curve C, the vector field V, and the positive direction on C being the same as first stated. The contour integral:

$$I(C,V) = \frac{1}{2\pi}\int_C d\left(\arctan\frac{Y(x,y)}{X(x,y)}\right) = \frac{1}{2\pi}\int_C \frac{XdY - YdX}{X^2 + Y^2} \tag{5.1}$$

is equal to the number of positive revolutions (counterclockwise − clockwise) of the vector V as the curve C is described once in the positive direction. This number is called the *index* of C with respect to V. Since $I(C,V)$ varies continuously with C so long as no singularity is crossed, while on the other hand $I(C,V)$ is an integer, it follows that $I(C,V)$ is the same for all curves C which can be deformed into each other without any crossing of singular points.

In particular, if p is a regular point, or an isolated singular point of V, then $I(C,V)$ is constant for all C surrounding p and no other singular point. We define $I(p,V)$ the *index* of p with respect to V, to be the common value $I(C,V)$ for all such curves C.

If p is a regular point, it is clear that $I(p,V) = 0$, since by the continuity of V we can find a circle C so small that V lies arbitrarily near a fixed direction for all points on C.

By an argument familiar from the theory of residues of an analytic function, we conclude also that if C surrounds the singular points $p_1, \ldots, p_n$ of V and no others, then

$$I(C,V) = \sum_{i=1}^{n} I(p_i,V) \tag{5.2}$$

Before applying these ideas to differential equations, we state an intuitively obvious but important theorem:

If C has a continuously turning tangent vector T, then $I(C,T) = 1$.

A proof will be found in Coddington and Levinson.[2]

Now we consider the d.e. with the vector field V:

$$\dot{x} = X(x,y); \qquad \dot{y} = Y(x,y) \tag{5.3}$$

[2] E. A. Coddington and N. Levinson, *Theory of Ordinary Differential Equations*, New York, 1955, p. 399.

By the indices of trajectories of singular points of (5.3) we shall mean the indices with respect to the field V. If C is a closed trajectory of (5.3), then V on C is a continuously turning tangent of C and the preceding theorem can be stated thus:

The index of a closed trajectory is 1.

Since the index of a regular point of (5.3) is zero, we have the following important result:

A closed trajectory surrounds at least one singular point.

Let us now compute the index of an elementary singular point of (5.3). We suppose the origin moved to the singular point, so that (5.3) has the form:

$$\dot{x} = X(x,y) = ax + by + F_1(x,y); \qquad \dot{y} = Y(x,y) = cx + dy + F_2(x,y) \tag{5.4}$$

where F_1 and F_2 are of degree two or higher in x,y. The hypothesis that the singular point is elementary means that $D = ad - bc \neq 0$. Since D is the Jacobian

$$\frac{\partial(X,Y)}{\partial(x,y)} = \begin{vmatrix} \dfrac{\partial X}{\partial x} & \dfrac{\partial X}{\partial y} \\ \dfrac{\partial Y}{\partial x} & \dfrac{\partial Y}{\partial y} \end{vmatrix}$$

evaluated at $x = y = 0$, the equations, $u = X(x,y)$ and $v = Y(x,y)$, define a bi-continuous one-to-one transformation T of a neighborhood Ω of $x = y = 0$ onto a neighborhood Ω' of $u = v = 0$. Choose $\alpha > 0$ so small that the circle $u^2 + v^2 = \alpha^2$ lies in Ω'. Its image under T^{-1}, $X^2(x,y) + Y^2(x,y) = \alpha^2$, is thus a simple closed curve in Ω surrounding (0,0).

Denote this curve, with the usual orientation, by C. The index of the singular point is thus:

$$I = \frac{1}{2\pi}\int_C \frac{XdY - YdX}{\alpha^2} = \frac{1}{2\pi}\int_{C'} \frac{udv - vdu}{\alpha^2}; \qquad C' = T(C) \tag{5.5}$$

Parametrizing C' by $u = \alpha \cos \theta$; $v = \alpha \sin \theta$, the last integral becomes $\int_0^{\pm 2\pi} d\theta = \pm 2\pi$, where the plus sign is to be chosen if T is orientation preserving, and the minus sign if T is orientation reversing. Thus $I = \pm 1$. Finally, we observe that T is orientation preserving or reversing according as its Jacobian is positive or negative. Since the sign of the Jacobian in Ω is that of D, we have $I = +1$ if $D > 0$ and $I = -1$ if

$D < 0$.† Referring to Section 6 of Chapter 1, we find that $D > 0$ for nodes, foci, and centers, and $D < 0$ for saddle points. Hence *the index of a node, focus or center is* $+1$; *the index of a saddle point is* -1.

To combine the preceding results, let Γ be a trajectory of (5.3) surrounding only elementary singular points. Then

$$I(\Gamma) = N + F + C - S = 1 \tag{5.6}$$

where N, F, C, and S are, respectively, the number of nodes, foci, centers, and saddles in the interior of Γ.

In the special case of a conservative system this formula becomes

$$C - S = 1 \tag{5.7}$$

which is illustrated by Fig. 2.7 for $C = 2$ and $S = 1$.

6. Brouwer's fixed point theorem

Let R represent a closed segment $a \leq x \leq b$ and let T be a mapping of R into itself, that is, a continuous transformation $x' = f(x)$ defined for all x in R and such that x' also lies in R so that $\alpha \leq f(x) \leq \beta$ where $a \leq \alpha < \beta \leq b$. The existence of fixed points (that is, of points γ such that $\gamma = f(\gamma)$) is known from the elements of calculus.

In fact, $x - f(x)$ is a continuous function defined in $a \leq x \leq b$ and for $x = a$; $x = b$ we have, respectively, $a - f(a) < \alpha - f(a) \leq 0$; $b - f(b) \geq \beta - f(b) \geq 0$, and there is at least one value γ such that $a \leq \gamma \leq b$; $\gamma - f(\gamma) = 0$; that is: $\gamma = f(\gamma)$. This property of continuous transformations admits of extensions for planar or n-dimensional mappings.

† We can obtain this result without assuming the theorem about the sign of the Jacobian. Let us investigate whether the parametrization

$$X(x,y) = \alpha \cos \theta; \qquad Y(x,y) = \alpha \sin \theta$$

of C has the property that the direction of increasing θ corresponds to a positive or negative traversal of C. Let l be a line parallel to the y axis and lying to the right of C. Move l parallel to itself until it just touches C. Let p be a point of contact; then the positive direction on C at p is upward, and it follows that $(\partial y/\partial\theta) > 0$ at p if θ induces the positive orientation on C, and $(\partial y/\partial\theta) < 0$ in the contrary case. A simple calculation shows that $dy/d\theta = (1/2J)(\partial/\partial x)(X^2 + Y^2)$ where J is the Jacobian.

However, at p, $(\partial/\partial x)(X^2 + Y^2) > 0$; hence, the sign of $\partial y/\partial\theta$ is that of J and we have $I = \frac{1}{2\pi}\int_0^{\pm 2\pi} d\theta = \pm 1$, the sign being that of J. In this manner we can perform the integration without an explicit change of variables, although the formal computation is the same as before.

We shall limit ourselves to the planar case, although in what follows we shall need only the above-mentioned unidimensional argument—not the generalization below. Let R be a closed circular disk, that is, the union of a circle and its interior.

By a mapping of R into itself we mean a continuous transformation T: $x' = f(x,y)$; $y' = g(x,y)$ defined for all (x,y) in R and such that (x',y') also lies in R. If p is a point of R, we denote its image under T by $T(p)$. By a fixed point of T we mean a point p such that $T(p) = p$. The theorem of Brouwer states:

Every mapping of R into itself has a fixed point

This theorem has a number of applications in the theory of differential equations. We shall outline a proof based on the idea of the index. First, however, we remark that if, in the theorem, R is replaced by any region S obtainable from it by a bi-continuous one-to-one mapping $\mathfrak{T}$ (that is, any region formed of a simple closed curve and its interior), a corollary theorem is obtained; for if T maps S into itself, $\mathfrak{T}^{-1}T\mathfrak{T}$ is a mapping of R into itself, taking a point p of R first into the point $\mathfrak{T}(p)$ of S, then into the point $T[\mathfrak{T}(p)]$ of S, and finally via the inverse of $\mathfrak{T}$ into the point $\mathfrak{T}^{-1}\{T[\mathfrak{T}(p)]\}$ of R. If, as the theorem asserts $\mathfrak{T}^{-1}T\mathfrak{T}$ has a fixed point, then $\mathfrak{T}^{-1}T\mathfrak{T}(p) = p$ for some p of R so that $T\mathfrak{T}(p) = \mathfrak{T}(p)$. Hence $\mathfrak{T}(p)$ is a fixed point of T.

We define the index $I(C,V)$ of a simple closed curve C relative to a continuous vector field V having no singularities on C to be the number of positive revolutions of V as C is traversed once positively. As was the case for differentiable fields, $I(C,V)$ varies continuously with C so long as no singularities are crossed, and it follows, as before, that $I(C,V) = 0$ if C surrounds no singularity of V.

Now let C be the circumference of R and T a mapping of R into itself. Define the vector field V on R by $V(p) = \overrightarrow{pp'}$ where $p' = T(p)$. V is continuous because T is. If T has no fixed point on C, then V has no singularity on C.

Since V on C always points into the interior of R, it makes exactly one positive revolution as C is traversed once positively. Hence $I(C,V) = 1$, and V has a singularity in R. But a singularity of V is a fixed point of T. This completes the proof.

For a more detailed discussion of the theory of the index and a more rigorous proof of the Brouwer Fixed Point Theorem, the reader is referred to Lefschetz.[3]

[3] S. Lefschetz, *Differential Equations (Geometric Theory)*, Interscience Publishers, New York, 1957.

7. Negative criterion of Bendixson

The theory of indices as we saw gives necessary conditions for the existence of limit cycles; they are not sufficient however.

Poincaré has indicated, and Bendixson has completed, a theorem that gives both necessary and sufficient conditions for the existence of limit cycles. This theorem is known under the name of Poincaré-Bendixson theorem (or positive criterion); it is outlined in the following section. Bendixson,[4] on the other hand, has established a theorem for the *nonexistence* of limit cycles; this theorem is also known under the name of the negative criterion and gives a sufficient condition.

This theorem and an example of its application due to Andronov[5] are indicated in this section.

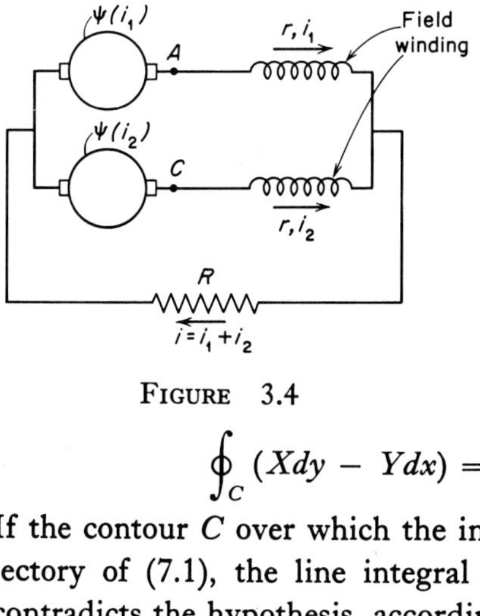

FIGURE 3.4

Given a system of d.e.

$$\dot{x} = X(x,y); \qquad \dot{y} = Y(x,y) \tag{7.1}$$

the negative criterion of Bendixson states:

If the expression $(\partial X/\partial x) + (\partial Y/\partial y)$ does not change its sign (or vanish identically) within a region D of the phase plane, no closed trajectory can exist in D. In fact, by Green's theorem

$$\oint_C (Xdy - Ydx) = \iint_D \left(\frac{\partial X}{\partial x} + \frac{\partial Y}{\partial y}\right) dxdy \tag{7.2}$$

If the contour C over which the integration is performed is a closed trajectory of (7.1), the line integral is $\oint_C (\dot{x}\dot{y} - \dot{y}\dot{x})dt$ and is zero. This contradicts the hypothesis, according to which the double integral cannot vanish. An interesting application of this criterion was indicated by Andronov.[5] It is well known that two series generators cannot work in parallel and fall out of step with each other. (See Fig. 3.4. The $\psi(i)$ are the electromotive forces induced in each generator.)

The application of Kirchhoff's laws yields two equations

$$\begin{aligned} \psi(i_1) - (r + R)i_1 - Ri_2 - L\frac{di_1}{dt} &= 0 \\ \psi(i_2) - (r + R)i_2 - Ri_1 - L\frac{di_2}{dt} &= 0 \end{aligned} \tag{7.3}$$

[4] I. Bendixson, *Acta Math.* **24**, 1901; also H. Poincaré, *J. des Math.* (3), **7**, 1881; also Œuvres **T.1**, Gauthier-Villars, Paris, 1928.

[5] A. Andronov and S. Chaikin, *Theory of Oscillatons* (original text in Russian), Moscow, 1937, and subsequent editions.

If one considers i_1 and i_2 as x and y of the general theory, the expressions $\partial X/\partial x$, $\partial Y/\partial y$ in this case are

$$\frac{\partial X}{\partial i_1} = \frac{\psi'(i_1) - (r + R)}{L}; \qquad \frac{\partial Y}{\partial i_2} = \frac{\psi'(i_2) - (r + R)}{L} \tag{7.4}$$

where $\psi'(i) = [d\psi(i)]/di = \rho$. From this expression it is seen that $\partial X/\partial i_1$ and $\partial Y/\partial i_2$ are never zero as long as the generators remain excited because, in such a case, the electromotive force always outweighs both the ohmic and inductive drops of the voltage in the circuit. One concludes therefore that no closed trajectory (that is, periodic solution) is possible in this case. This can be ascertained by investigating the singular points of the system (7.3) written as

$$\begin{aligned} (\rho - r - R - L\delta)i_1 - Ri_2 &= 0 \\ - Ri_1 + (\rho - r - R - L\delta)i_2 &= 0 \end{aligned}$$

where we set $\delta = d/dt$. The nontrivial solutions are possible here only if the determinant is zero, which gives the condition

$$(\rho - r - R - L\delta) = \pm R \tag{7.5}$$

The roots of the characteristic equation are

$$\delta_1 = \frac{1}{L}[\rho - (r + 2R)]; \qquad \delta_2 = \frac{1}{L}(\rho - r) \tag{7.6}$$

As in the state of self-excitation, $\rho - r - R > 0$, $\delta_2 > 0$. As to δ_1, all depends on the value of R; this root may therefore be either positive or negative. If it is negative, the singular point ($i_1 = i_2 = 0$) is a saddle point; if it is positive, the singular point is an unstable node. In both cases there is instability, which means that any small deviation from the initial condition $i_1 = i_2$ is further emphasized so that, instead of working in parallel, the generators will set themselves into a series operation. If, however, the field connections are reversed (the point A of the first field is connected to the point C of the second generator and vice versa); a similar argument shows that the roots are now

$$\delta_1 = \frac{1}{L}(\rho - r - 2R); \qquad \delta_2 = -\frac{1}{L}(\rho + r) \tag{7.7}$$

It is seen that the root δ_2 is always negative; if δ_1 is also negative, the singularity is a stable node, which shows that the generators work in a stable manner in parallel. If, however, R is not sufficiently large, δ_1 may become positive, in which case the singular point becomes a saddle point, resulting in instability.

This analysis of the nature of singular points does not yet make certain the existence of a stationary oscillation (assuming, of course, that the singularity *is not* a saddle point). What actually rules out the existence of an oscillation is the condition yielded by the negative criterion. It must be noted, however, that the fact that $(\partial X/\partial x) + (\partial Y/\partial y)$ changes sign does not mean that a closed trajectory exists.

8. Poincaré-Bendixson theorem

The Poincaré-Bendixson (P.B.) theorem states:

If a half trajectory C remains in a finite domain D without approaching any singularities, then C is either a closed trajectory or approaches such a trajectory.

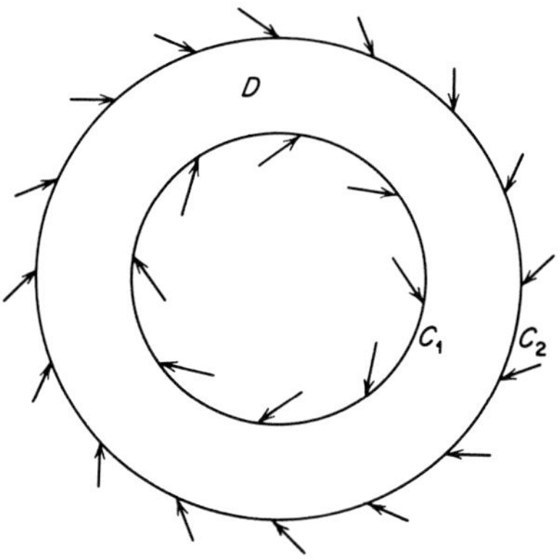

FIGURE 3.5

We refer to the proof of this theorem in Bendixson's treatise.[4] The P.B. theorem gives sufficient conditions for the existence of a closed trajectory. Its principal limitation is the difficulty of determining the domain D satisfying the requirement of the theorem. We shall indicate in the following section a method due to Poincaré which has for its object precisely the determination of the P.B. domain.

If this region D can be determined in one way or another, the theorem gives an immediate answer. A great deal of ingenuity is sometimes required for this determination; we refer to a paper by LaSalle[6] who was able to prove the existence of a periodic solution of the van der Pol equation with a large parameter value by constructing a domain D so as to render the application of the P.B. theorem possible.

In the case of a ring-shaped domain D bounded by two concentric circles C_1 and C_2 (Fig. 3.5), it is sufficient for the existence of at least one closed trajectory that:

(A) Trajectories enter (leave) D *through every point* of C_1 and C_2.
(B) There are no singular points either in D or on C_1 and C_2.

In this form the significance of the theorem is obvious on the basis of the

[4] See footnote [4], page 82.
[6] J. P. LaSalle, *Quart. Appl. Math.* 7, 1949.

hydrodynamical analogy. In fact, if the condition (A) holds, there must be a "sink" either in D or its boundaries; the condition (B) rules out the point-sinks; hence, there must be a curvilinear sink—the limit cycle—in D.

Condition (A) is essential in that in case there are some regions on the bounding curves through which the trajectories change the direction of their entrance into D, the P.B. theorem does not hold. In the following section we shall analyze the fulfillment of condition (A) in greater detail.

9. Cycles without contact; curve of contacts

The difficulty of determining the bounding curves C of the domain D was obviated to some extent by Poincaré in his method of the curves of contact.[1] Let V be the vector with the components X and Y appearing in the system

$$\dot{x} = X(x,y); \qquad \dot{y} = Y(x,y) \tag{9.1}$$

An arc S is called *arc without contact* if V is neither zero nor tangent to S at any of its points. If S is a closed curve, we call it a *cycle without contact.* Given an arbitrary closed curve Γ, the vector V sometimes points into the domain D bounded by Γ, sometimes it is directed outward depending on the sign of the cosine of the angle between V and the inner normal N to Γ at a given point. As cosine of the angle (VN) is a continuous function of the point on Γ, it may vanish sometimes if there is a contact between Γ and V at some point. Clearly, for a given motion on Γ, the vector V, which was pointing *into* the area D limited by Γ, is outwardly directed after a point of contact is traversed. Such a curve Γ *cannot*, therefore, satisfy the condition (A) of the P.B. theorem.

If contacts exist between a family of closed curves Γ depending on a parameter c and the vector field, their locus is a continuous curve called a *curve of contacts.*

Suppose we have a family of curves Γ which Poincaré calls a *topographic system.* This may be, for instance, a family of concentric circles $x^2 + y^2 = c$ centered at a singular point. Our purpose will be to determine such curves of this family which satisfy condition (A) of the preceding section, assuming that condition (B) is fulfilled. It may happen that for some values of the parameter c there are contacts between the curves Γ and vectors V. In such a case, condition (A) is not fulfilled. It may also happen that for some other values of c in the interval, $c_1 < c < c_2$, there are some curves Γ that are cycles without contact and thus fulfill condition (A). Thus if C is a curve of contacts in polar coordinates with r_2 and r_1 being the maximum and minimum radii, one can take two circles of the topographic system with the same radii. In such a case circles with radii $R_2 > r_2$ and $R_1 < r_1$ fulfill condition (A) since they appear as cycles without contact relatively to the vector field V.

If $\Gamma(x,y) = c$ is the equation of the topographic family of curves, Γ_x and Γ_y are components of the vector N normal to any such curve passing through a point x,y. A point of contact occurs whenever this normal vector to Γ is perpendicular to the phase velocity vector V. This condition of orthogonality of Γ and N is thus

$$\Gamma_x X + \Gamma_y Y = 0 \tag{9.2}$$

which gives the equation of the curve of contacts

$$\frac{X}{Y} = -\frac{\Gamma_y}{\Gamma_x} \tag{9.3}$$

The d.e. (9.1) in polar coordinates are

$$\dot{r} = R(r,\varphi); \qquad \dot{\varphi} = \Phi(r,\varphi) \tag{9.4}$$

and the curves of the topographic system satisfy the d.e.

$$\frac{dr}{d\varphi} = F(r,\varphi) \tag{9.5}$$

The points of the curve of contact must satisfy the relation

$$\frac{R(r,\varphi)}{\Phi(r,\varphi)} = F(r,\varphi) \tag{9.6}$$

If the topographic system is a family of circles $F(r,\varphi) = 0$, the equation of the curve of contacts is given by two equations

$$R(r,\varphi) = 0; \qquad \frac{dr}{d\varphi} = 0 \tag{9.7}$$

As an example, consider the d.e. of a pendulum acted on by a constant moment M:

$$\ddot{\varphi} + h\dot{\varphi} + c \sin \varphi = M$$

This d.e. is encountered in the theory of a synchronous motor when M is the driving torque exerted by the rotating magnetic field. As the coordinate φ is cyclic, it is defined only modulo 2π; this, clearly, imposes the use of a cylindrical phase surface. As a topographic system one can take here the lines $\dot{\varphi} = c$ to which correspond circles on the cylinder so that the domain D will be a portion of the cylindrical surface limited by two sections of the cylinder by planes perpendicular to its axis. We have

$$\dot{\varphi} = y = X(\varphi,y); \qquad \dot{y} = \frac{1}{I}(P - hy - mgl \sin \varphi) = Y(\varphi,y)$$

and the equation of the contact curve is $Y = 0$; whence

$$y = \frac{P - mgl \sin \varphi}{h}$$

Hence, a limit cycle, if it exists, is located in the region defined by the inequality

$$(P - mgl)/h < \dot{\varphi} < (P + mgl)/h$$

As another example, consider the van der Pol equation, $\ddot{x} - \mu(1 - x^2)\dot{x} + x = 0$, which, written as a system, is

$$\dot{x} = y = X(x,y); \qquad \dot{y} = \mu(1 - x^2)y - x = Y(x,y)$$

With a circular topographic system the curve of contacts is $(1 - x^2)y^2 = 0$. As the root $y^2 = 0$ is double, the abscissa axis is a locus of even contacts, but these do not violate condition (A). The two branches of the curve of contacts are thus: $x = 1$ and $x = -1$. Hence the smallest circle tangent to the curve of contacts is a circle of radius 1 and there is no largest circle. The limit cycles are thus outside a circle of radius 1, which is obvious from the d.e. which has a "positive friction" if $x > 1$. The problem here is rather indefinite inasmuch as it is impossible to determine the largest circle so as to have the annular region D between the two circles.

There are cases when this can be accomplished as was shown by Andronov and Witt[7] in connection with the d.e. of synchronization. We follow here this calculation in the notations of Stoker.[8] Consider the differential system

$$\begin{aligned} \dot{x} &= -ay + x(1 - r^2) \\ \dot{y} &= ax + y(1 - r^2) + F \end{aligned} \tag{9.8}$$

where $r^2 = x^2 + y^2$ and a and F are positive constants. The d.e. of integral curves is

$$\frac{dy}{dx} = \frac{ax + y(1 - r^2) + F}{-ay + x(1 - r^2)} \tag{9.9}$$

[7] A. Andronov, A. Witt, S. Chaikin, *Theory of Oscillations* (in Russian), 2nd Ed., 1959.

[8] J. J. Stoker, *Nonlinear Vibrations*, Interscience Publishers, New York, 1950.

If x_0, y_0 are coordinates of a singular point of (9.8), it is convenient to introduce the variables $x_1 = x - x_0$, $y_1 = y - y_0$ so that (9.9) becomes

$$\frac{dy_1}{dx_1} = \frac{F + a(x_1 + x_0) + (y_1 + y_0)(1 - r^2)}{-a(y_1 + y_0) + (x_1 + x_0)(1 - r^2)} = \frac{X(x_1,y_1)}{Y(x_1,y_1)} \tag{9.10}$$

where $r^2 = (x_1 + x_0)^2 + (y_1 + y_0)^2$.

The contact curves are obtained by transforming to polar coordinates (r_1,ψ) with the center at $(x_1,y_1) = (0,0)$ which gives

$$\frac{1}{r_1}\frac{dr_1}{d\psi} = \frac{x_1 Y + y_1 X}{x_1 X - y_1 Y} \tag{9.11}$$

where $r_1{}^2 = x_1{}^2 + y_1{}^2$. It is clear that the contact curves correspond to $dr_1/d\psi = 0$, that is, $x_1 Y + y_1 X = 0$; hence

$$x_1[-a(y_1 + y_0) + (x_1 + x_0)(1 - r^2)] + y_1[F + a(x_1 + x_0) + (y_1 + y_0)(1 - r^2)] = 0$$

together with $r_1 = 0$.

As the origin (x_1,y_1) is a singular point, this imposes conditions $X(0,0) = Y(0,0) = 0$ on the determination of F, a, x_0, y_0 which are

$$-ay_0 + x_0(1 - \rho) = 0$$

$$F + ax_0 + y_0(1 - \rho) = 0; \qquad \rho = x_0{}^2 + y_0{}^2 \tag{9.12}$$

If one determines x_0,y_0 in terms of F, a, and ρ, which gives

$$x_0 = -\rho a/F; \qquad y_0 = -\rho(1 - \rho)/F$$

and inserts these values into $x_0{}^2 + y_0{}^2 = \rho$, one has the relation obtained originally by van der Pol by a different argument, viz.:

$$\rho[a^2 + (1 - \rho^2)] = F^2 \tag{9.13}$$

This equation determines the coordinates of the singular point and, hence, the center of the ring domain in which we are interested.

The curve of contacts becomes then

$$r_1{}^2(1 - r^2) - (x_1x_0 + y_1y_0)(r_1{}^2 + 2x_1x_0 + 2y_1y_0) = 0 \tag{9.14}$$

upon the elimination of F and a.

Introducing the quantity k from the relation

$$x_1x_0 + y_1y_0 = r_1(x_0 \cos\psi + y_0 \sin\psi) = r_1 k$$

$$k = x_0 \cos\psi + y_0 \sin\psi$$

where $r^2 = r_1^2 + \rho + 2r_1k$, the curve of contacts becomes

$$r_1^2[r_1^2 + 3kr_1 - (1 - \rho - 2k^2)] = 0 \tag{9.15}$$

Rejecting $r_1 = 0$, the radii of the curve are

$$tr_{1,2} = -\frac{3k}{2} \pm \sqrt{\frac{k^2}{4} - \rho + 1} \tag{9.16}$$

It is noted that the curve has no infinite branches. The radii $r_{1\,max}$ and $r_{1\,min}$ give the circular boundaries of the P.B. ring-shaped domain D. With a few additional relations which we omit, the expression (9.16) can be brought to the form

$$r_{1,2} = -\tfrac{3}{2}\sqrt{\rho} \pm \sqrt{1 - \tfrac{3}{4}\rho}$$

which gives the maximum and the minimum radii r_2 and r, respectively, of the curve of contact.

It is seen that the circles of the topographic system with $r \geq r_2$ and $r \leq r_1$ are curves without contacts and thus satisfy condition (A). Since no singular points exist in D limited by these circles, the P.B. theorem is applicable.

This calculation shows that the determination of boundaries of a P.B. domain D constitutes, in general, a rather complicated problem. In some cases it is possible to obtain simpler results, as was indicated by Nemitzky,[9] owing to the use of the Eulerian derivative for the topographic family of curves. We recall the significance of this derivative. Given a function $\Gamma(x,y)$ and a system (9.1), we have

$$\frac{d\Gamma}{dt} = \frac{\partial\Gamma}{\partial x}\dot{x} + \frac{\partial\Gamma}{\partial y}\dot{y} = \frac{\partial\Gamma}{\partial x}X + \frac{\partial\Gamma}{\partial y}Y \tag{9.17}$$

If the topographic system is a family of circles: $\Gamma(x,y) = x^2 + y^2 = c^2$, by (9.17) one has

$$\frac{d\Gamma}{dt} = 2(xX + yY) \tag{9.18}$$

On the other hand, if one multiplies the first d.e. (9.1) by x, the second by y, adds the two equations, noting that $x\dot{x} + y\dot{y} = \frac{1}{2}(dr^2/dt) = \frac{1}{2}(d\rho/dt)$; $\rho = x^2 + \dot{x}^2 = x^2 + y^2$, one finds that

$$\frac{d\Gamma}{dt} = \frac{d\rho}{dt} \tag{9.19}$$

[9] V. V. Nemitzky and V. V. Stepanov, *Qualitative Theory of Differential Equations*, original text in Russian, Moscow, 1949; English translation, Princeton Mathematics Series, Princeton University Press, Princeton, N.J., 1960.

which means that $\Gamma = \rho + c$, c being some constant. On the other hand, in dynamical problems, ρ by the above definition is the *total energy* stored in the oscillation.

Thus, if one adopts as the topographic system, the energy levels ρ, one has the following theorem due to Nemitzky.[10]

If it is possible to determine two positive constants ρ_2 and ρ_1 ($\rho_2 > \rho_1$) such that for ρ_1 the expression $xX + yY \geq 0$ and for ρ_2, $xX + yY \leq 0$ and if, moreover, the circular ring D between the circles of radii ρ_2 and ρ_1 has no singular points, then there exists a stable limit cycle in D.

If the signs of $xX + yY$ are reversed in this theorem, other conditions being the same, the limit cycle is unstable.

As an example, one can apply this theorem to the system (2.1), in which case $xX + yY = \sqrt{x^2 + y^2}(1 - x^2 - y^2)$. Hence if $x^2 + y^2 = 1 + \varepsilon$, one has $xX + yY < 0$, and for $x^2 + y^2 = 1 - \varepsilon$, $xX + yY > 0$, where ε is as small as we please. By the preceding theorem, it follows that in the ring between the circles of radii $\rho_2 = 1 + \varepsilon$ and $\rho_1 = 1 - \varepsilon$, there exists a stable limit cycle; since ε is arbitrarily small, clearly, the radius of this cycle is *one*, as we found previously. Thus, we obtain the same result as before but without integrating the system (2.1).

This theorem may be applied to d.e. of a somewhat broader type. Thus, it holds for the system

$$\dot{x} = y + \frac{x}{\sqrt{x^2 + y^2}}[1 - (x^2+y^2)] + \frac{x}{r^\alpha}b(x,y), \qquad |b(x,y)| < k$$

$$\dot{y} = -x + \frac{y}{\sqrt{x^2 + y^2}}[1 - (x^2 + y^2)] + \frac{y}{r^\alpha}b(x,y), \qquad 0 < \alpha < 1.$$

In this case for a small circle, one has $[d(r^2)/dt] > 0$; and for large r, $[d(r^2)/dt] < 0$, so that the limit cycle exists if there are no singular points in the ring. The condition for a singular point is clearly

$$x\left(\frac{1 - r^2}{r} + \frac{b}{r^\alpha}\right) + 1 \cdot y = 0$$

$$-1 \cdot x + \left(\frac{1 - r^2}{r} + \frac{b}{r^\alpha}\right)y = 0$$

As the determinant is different from zero, the only solution is $x = y = 0$, which proves that the only singular point is at the origin. The procedure

[10] G. D. Birkhoff, *Dynamical Systems*, Am. Math. Soc., 1927; A. Andronov, A. Witt, S. Chaikin, *Theory of Oscillations* (in Russian); this book is the second edition (1959) of the original Russian text.

just mentioned is not always applicable. This can be seen, for instance, in connection with the van der Pol equation, which, written as a system, is

$$\dot{x} = y; \qquad \dot{y} = -\mu x^2 y + \mu y - x$$

With the use of the circular topographic system and in polar coordinates, $x = r \cos \psi$; $y = r \sin \psi$; $r^2 = \rho$, the Eulerian derivative is

$$\frac{d\rho}{dt} = 2\mu\rho(\sin^2 \psi - \rho \sin^2 \psi \cos^2 \psi) \tag{9.20}$$

In this case, if $\rho < 1$, one has $(d\rho/dt) > 0$, but, if $\rho > 1$, it is impossible to assert that $(d\rho/dt) < 0$, which makes the application of the theorem impossible.

10. Complete topological configurations; singular points at infinity

In the preceding study of singular points and limit cycles there is yet a lack of completeness regarding the behavior of trajectories outside the domain containing singular points and limit cycles. In the definition of a stable limit cycle, it was mentioned that such a cycle is approached (for $t \to \infty$) by nonclosed trajectories C' both from inside and from outside. As regards the latter we still do not know their "sources," if we think in terms of the hydrodynamical analogy.

The answer to this question was given by Poincaré in terms of a *singular point* (or points) *at infinity* for the case when $X(x,y)$ and $Y(x,y)$ are polynomials. These singularities at infinity do not appear directly from *the form* of a given d.e. as do the singular points at a finite distance with which we have been concerned so far. It is convenient, therefore, to transform the d.e. to new variables in which infinity in the original variables is represented by a point with finite coordinates in the new variables.

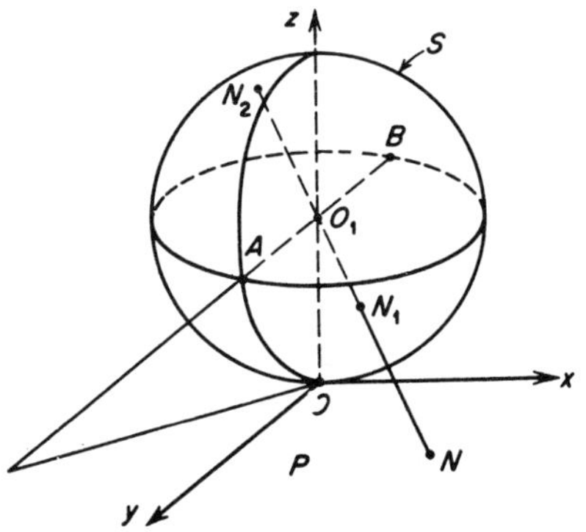

FIGURE 3.6

Consider a sphere S of unit radius touching the phase plane P at the origin O as shown in Fig. 3.6. We take as the center of projection, the center O_1 of the sphere S. To any point N in P correspond two points N_1 and N_2 on the sphere S, as projected by the ray NO_1. We shall consider only the projection on the lower hemisphere. It is clear that any straight line in P is projected as an arc of a great circle on S; more specifically, a straight line through O (in P) is projected as a great circle (the meridian) perpendicular to the equator E in S. The points of the equator correspond to infinitely distant points in P.

This transformation clearly preserves the topology of integral curves in the neighborhood of finite singular points. Thus, focal, nodal, saddle points, and centers in P maintain their characteristic properties on S; the same applies to other elements; thus a closed curve in P remains closed on S, etc.

The transformation from P to S can be effected by selecting cartesian coordinates u, v, z having the origin in O_1, the axes u, v being parallel to the axes x and y in P. To a point (x,y) in P corresponds a point N_1 on S with the coordinates

$$u = \frac{x}{\sqrt{x^2 + y^2 + 1}}; \qquad v = \frac{y}{\sqrt{x^2 + y^2 + 1}}; \qquad z = \frac{1}{\sqrt{x^2 + y^2 + 1}} \tag{10.1}$$

which thus determines the transformation $P \rightarrow S$.

Once this transformation is obtained, it is preferable to return to the planar representation. For this purpose we carry out the second transformation $S \rightarrow P^*$ by projecting the points N_1 on S into another plane. We shall choose either the $v = 1$ plane or the $u = 1$ plane, according to circumstances. Suppose we select the $v = 1$ plane, that is, the plane perpendicular to P touching S at the point $u = 0$, $v = 1$, $z = 0$. If one connects the origin O_1 (the center of S) with some point (u,v,z) on S, this defines a ray (tu,tv,tz) by giving t any value between 0 and ∞. This ray intersects the plane $v = 1$ for $t = 1/v$ and thus gives the coordinates u/v, 1, z/v in this plane. Disregarding the constant coordinate v and using (10.1) we have thus

$$u = \frac{x}{y}; \qquad z = \frac{1}{y} \tag{10.2}$$

which gives the image of the original point N in P after the double projection $P \rightarrow S \rightarrow (v = 1)$.

This double projection can be applied to any point of P except those points lying on the x axis: $y = 0$. For these points, one can use, as a second projection, the projection on the $(u = 1)$ plane defined by $v = y/x$; $z = 1/x$.

Suppose we have a system of d.e.

$$\dot{x} = X(x,y); \qquad \dot{y} = Y(x,y) \tag{10.3}$$

where X and Y are polynomials of degrees p and q, respectively, and use the transformation

$$x = \frac{1}{z}; \qquad y = \frac{v}{z} \tag{10.4}$$

If $z = 0$, one has $x = \infty$, $y = \infty$ but $y/x = v$ is finite. Thus, one moves to infinity along a fixed direction; for a finite v, one has to exclude the direction $x/y = 0$. With the new variables we have $dx = -dz/z^2$ and $dy = (zdv - vdz)/z^2$, which results in the d.e.

$$\dot{z} = -X\left(\frac{1}{z},\frac{v}{z}\right)z^2; \qquad \dot{v} = -X\left(\frac{1}{z},\frac{v}{z}\right)vz + Y\left(\frac{1}{z},\frac{v}{z}\right)z \tag{10.5}$$

As X and Y are of degrees p and q, respectively, one has

$$X\left(\frac{1}{z},\frac{v}{z}\right) = z^{-p}X^*(z,v); \qquad Y\left(\frac{1}{z},\frac{v}{z}\right) = z^{-q}Y^*(z,v) \tag{10.6}$$

Suppose $q \geq p$, then, introducing a new independent variable τ defined by

$$d\tau = x^{q-1}dt = z^{-q+1}dt$$

the system (10.5) becomes

$$\begin{aligned} \frac{dz}{d\tau} &= -z^{q-p+1}X^*(z,v) \\ \frac{dv}{d\tau} &= -vz^{q-p}X^*(z,v) + Y^*(z,v) \end{aligned} \tag{10.7}$$

The singular points at infinity by definition are those for $z = 0$ in (10.7).

One has analogously another system adequate for determination of singular points at infinity except those which lie in the direction of the x axis:

$$\begin{aligned} \frac{dz}{d\tau} &= -zY^{**}(z,u) \\ \frac{du}{d\tau} &= z^{q-p}X^{**}(z,u) - uY^{**}(z,u) \end{aligned} \tag{10.8}$$

If $q = p$, the singular points occur for $z = 0$, except for $z = 0$; $q = 0$.

The d.e. of the harmonic oscillator are $\dot{x} = y$; $\dot{y} = -x$. With new variables $x = 1/z$; $y = v/z$, that is, $z = 1/x$; $v = y/x$, one has the system

$$\dot{z} = -\frac{y}{x^2} = -vz; \qquad \dot{v} = -\frac{x}{x} - \frac{y^2}{x^2} = -1 - v^2 \tag{10.9}$$

At infinity ($z = 0$) there are no singular points, since $\dot{v} \neq 0$. For the d.e. having a singular point of the type *star* at the origin, viz.;

$$\dot{x} = x; \qquad \dot{y} = y$$

the differential system in the new variables is

$$\dot{z} = -\frac{1}{x} = -z; \qquad \dot{v} = \frac{y}{x} - \frac{yx}{x^2} = 0$$

which shows that the axis $z = 0$ is a locus of singular points at infinity.

If one applies these conclusions to the d.e.,

$$\ddot{x} + 2h\dot{x} + x = 0$$

of a linear oscillator with damping, $X(x,y) = y$; $Y(x,y) = -2hy - x$; and the transformation from P to S results in the system

$$dz/d\tau = -zv; \qquad dv/d\tau = -(v^2 + 2hv + 1) \tag{10.10}$$

$$dz/d\tau = 2hz + uz; \qquad du/d\tau = u^2 + 2hu + 1 \tag{10.11}$$

These two systems of d.e. in reality constitute only one system using merely two different projections for different azimuths, as was explained with regard to (10.2) and (10.4).

The infinity $z = 0$ is formed by trajectories. It may be either a limit cycle if there are no singular points at infinity, or it may be formed by separatrices joining two singular points or one to itself.

Let us investigate the system (10.10); its singular point exists if $z = 0$ and $v^2 + 2hv + 1 = 0$; that is, when $z = 0$, $v_1 = -h + \sqrt{h^2 - 1}$ or when $z = 0$, $v_2 = -h - \sqrt{h^2 - 1}$. Likewise, (10.11) has two singular points: $z = 0$, $u_1 = -h + \sqrt{h^2 - 1}$, and $z = 0$, $u_2 = -h - \sqrt{h^2 - 1}$. There is still a fifth singular point at the origin which corresponds to the state of rest.

One determines the stability of these singular points, as usual, by forming the variational equations. Thus, for the singular point $z = 0$, $v = v_1$, these equations are

$$\frac{d\delta z}{d\tau} = -v_1\delta z + 0\cdot\delta v; \qquad \frac{d\delta v}{d\tau} = 0\cdot\delta z + [-2(v_1 + h)]\delta v$$

Hence, the characteristic equation is

$$S^2 + (3v_1 + 2h)S + 2(v_1 + h)v_1 = 0$$

If one replaces v_1 by its value, one has

$$S^2 + (3\sqrt{h^2 - 1} - h)S + 2(h^2 - h\sqrt{h^2 - 1} - 1) = 0 \tag{10.12}$$

In a similar manner for v_2, one gets

$$S^2 + (3\sqrt{h^2 - 1} + h)S + 2(h^2 + h\sqrt{h^2 - 1} - 1) = 0 \quad (10.13)$$

If $h < 1$, there are no singular points on the equator which may be regarded as a kind of unstable limit cycle from which spiral trajectories depart, winding themselves onto a stable focus at the origin as shown in Fig. 3.7. One can say in this case that *the infinity is unstable but the state of rest is stable.*

If $h > 1$, the corresponding topological configuration is different (Fig. 3.8). One ascertains easily that in this case (characteristic equations (10.12) and (10.13)) that the point $z = 0$, $v = v_1$ is an unstable node N_1,

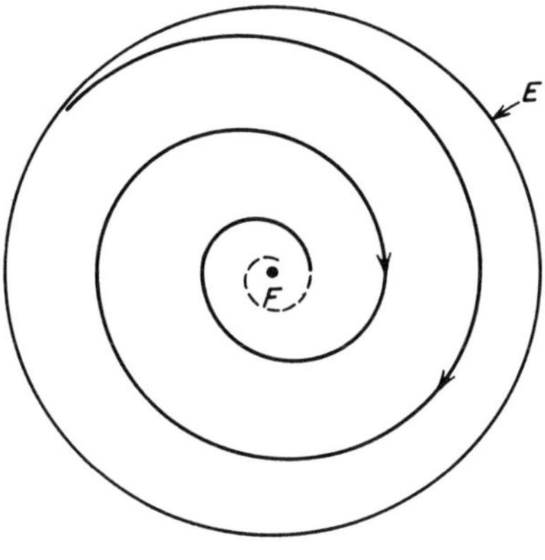

FIGURE 3.7

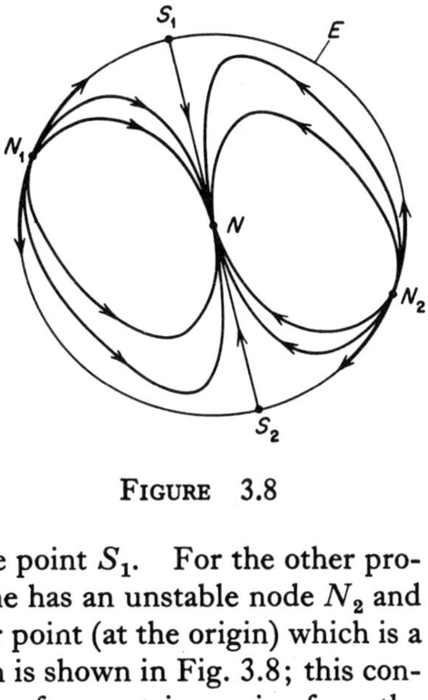

FIGURE 3.8

whereas the point $z = 0$, $v = v_1$ is a saddle point S_1. For the other projection one obtains a similar conclusion: one has an unstable node N_2 and a saddle point S_2. There is a fifth singular point (at the origin) which is a stable node. The topological configuration is shown in Fig. 3.8; this configuration shows that the equator E consists of separatrices going from the unstable nodes and entering saddle points through their stable asymptotes.

Trajectories entering the stable node at the origin come from two kinds of sources, namely: some of them come directly from the unstable nodes N_1 and N_2; there are also separatrices issuing from the unstable asymptotes of saddle points S_1 and S_2 and entering the stable node N.

Andronov[5, 7] applied this procedure to the van der Pol oscillator; in this case there is a stable limit cycle with radius 2. The conclusion is as

[5] See footnote [5], page 82.
[7] See footnote [7], page 87.

follows: the infinity has no singular points; trajectories unwind themselves from the equator (considered as an unstable limit cycle) and wind themselves onto the finite limit cycle ($r_0 = 2$). If, however, the equator has singular points, a similar argument shows that there are two unstable nodes and two saddle points, as in the case of Fig. 3.9. The only difference here is that the trajectories winding onto the finite cycle come from the unstable nodes on the equator (Fig. 3.9); there are also two separatrices issuing from the unstable asymptotes of saddle points and winding themselves also on the van der Pol's stable cycle.

We shall not go beyond these few examples but merely mention that the two aspects of the problem, the theoretical and the applied, do not have the same validity. In fact any d.e. representing a physical phenomenon is valid only in a certain domain. Thus, the representation of operation of an electron-tube oscillator by the van der Pol equation is based on a number of simplifying assumptions, such as, for example, that the grid current and anode reaction are negligible. If one wishes to represent this operation for a larger domain (larger x and $\dot{x}$), these effects cease to be negligible. In other words, if one goes further away from the origin, *the d.e. changes* and, for a sufficiently great distance from the origin, the form of the d.e. changes to such an extent that it becomes physically meaningless to discuss the behavior of trajectories at infinity on the basis of the original d.e.

11. Nonanalytic cycles

So far we have encountered only limit cycles which were analytic curves. Inasmuch as the concept of limit cycles found a widespread application in connection with the description of stationary oscillations, it was soon perceived that on this basis it was possible to account for a number of oscillatory phenomena whose stationary state is not describable in the phase plane by an analytic trajectory. A fuller study of these phenomena is postponed to Part IV, but it is useful to mention here in passing that the definition of limit cycles given in Section 1 does not require the analyticity of trajectories in the phase-plane representation. One of the simplest examples of a nonanalytic limit cycle occurs in the performance of a clock.† A clock is a mechanism consisting essentially of two parts, A and B. Part A is an ordinary torsional pendulum with small damping, and part B, the so-called escapement mechanism, applies impulses to A so as to replenish the energy dissipated by damping.

Consider first the system A; its trajectories are logarithmic spirals converging toward a stable focal point at the origin as shown in Fig. 3.10.

† For a more detailed theory of clocks see Section 5, Chapter 28.

A trajectory starting from a point M after one period of the system A reaches the point N on the y axis; at this moment B delivers an impulse increasing the momentum of A in a quasi-discontinuous manner (we assume it to be discontinuous). This amounts to a discontinuous change of velocity measured by the segment NM' (or NM''). It is clear that, if this segment were exactly NM, the trajectory would be closed and the oscillation would be periodic. It is noted that such trajectory is not analytic at the points of junction of the spiral with the rectilinear stretch NM.

It is easy to establish the periodic condition if one sets: $OM = y_1$, then $ON = y_2 = y_1 e^{-\alpha}$, where α is decrement of the oscillation during one period of A. Immediately after the impulse one has $OM' = y_3 = y_1 + a$,

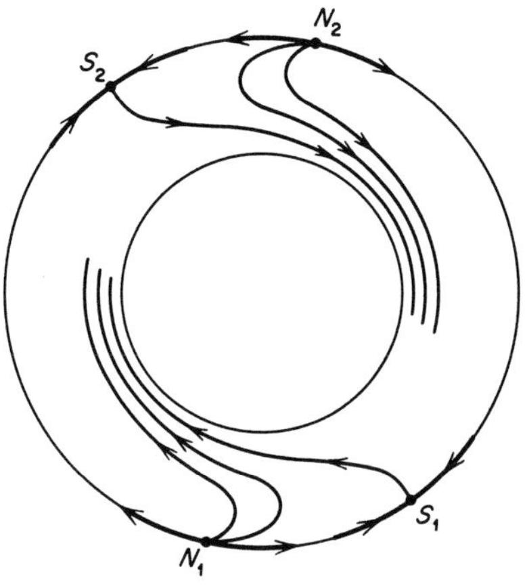

FIGURE 3.9

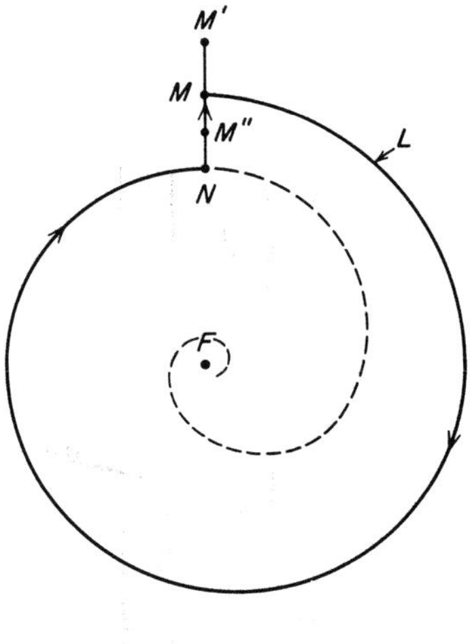

FIGURE 3.10

where a, the impulsive change of velocity, is constant. If the oscillation is stationary, $y_3 = y_1 = y_0$, y_0 being the stationary value of y, and one finds

$$y_0 = a/(1 - e^{-\alpha})$$

One has to investigate also the stability of the stationary y_0. This can be done intuitively as follows: if y_1 is small, the following amplitude y_3 is obviously larger and the amplitudes increase initially; they cannot increase above the value y_0 for which the oscillation becomes stationary. In a similar manner one shows that, if y_1 is large, the amplitudes decrease initially but they cannot decrease below y_0 for the same reason. It is also clear that y_0 depends only on the parameters of the systems A and B and not on the initial conditions.

Thus, the oscillation exhibits in all respects the features of a limit cycle with the only peculiarity that in this case the cycle is made of the two pieces —the spiral and the stretch NM. Very often such cycles are called "piecewise analytic cycles."

Although such cycles are outside the scope of the analytic theory outlined previously, they are of considerable interest in applications if one recalls that the radius vector of a spiral is a measure (in some scale) of the total energy stored in the system A. Hence the decrease of this radius along the spiral characterizes the gradual dissipation of energy and the discontinuous stretch NM which "closes" the curve is the impulsive input of energy into the system A.

The fact that the curve becomes closed in this manner, when the stationary state is reached, means a discontinuous replenishing of energy lost in the process of its continuous dissipation. A system of this kind is again nonconservative in its instantaneous behavior, although it is *conservative on the average*; this means that at the instants separated in time by one period of A the energy content of A is the same. This feature also exists in the case of the analytic cycles, although in these cycles a continuous dissipation of energy is compensated for by a continuous replenishment of it.

On this basis of energy, the stability of oscillation is also obvious. In fact, if one starts with small amplitudes, B delivers impulsively to A more energy than A can dissipate, as the result of which the amplitudes grow initially. Conversely, if one starts with large amplitudes, the dissipation outweighs the impulsive input and the amplitudes decrease. In both cases the stationary state is reached when the continuous dissipation per cycle is just equal to the impulsive input of energy.

It is useful to remark once more that, in spite of the fulfillment of all conditions satisfying the definition of limit cycles, the analytical criteria outlined in the preceding sections do not hold in the case of nonanalytic cycles. Thus a closed stable trajectory does not necessarily need to have in its interior a singular point with index $+1$, and so on. Although the physical nature of these stationary phenomena remains the same as for analytic cycles, the underlying mathematical theory ceases to be applicable here.

12. Topological configurations

In Chapter 1 we investigated the various types of equilibria in terms of the mathematical concept of singular points, and in this chapter we have identified stationary periodic motions of nonconservative systems with the concept of limit cycles. It has been seen that these two concepts are correlated so as to represent certain *topological configurations* connecting singular points with limit cycles. This is illustrated in Figs. 3.2 and 3.3. in which these connections appear in the form of spiral trajectories unwinding themselves from singular points (limit cycles) and winding onto limit cycles (singular points). As the approach (departure) of trajectories to singular points (limit cycles) is asymptotic (either for $t = -\infty$ or for $t = +\infty$), it is clear that the motions on these trajectories occur in the whole interval of t between $t = -\infty$ and $t = +\infty$.

Following this argument it is convenient to consider any such trajectory as a *line of flow* emerging from an unstable element (either an unstable singular point or an unstable limit cycle) for $t = -\infty$ and ending at a stable element (either a stable singular point or a stable cycle) for $t = +\infty$. Such hydrodynamic analogy is useful in analyzing more complicated situations in which the unstable elements appear as *sources* and the stable ones as *sinks*.

It is possible to attack the problem directly in terms of the fundamental properties of trajectories.[10] As such a study is beyond the scope of this text, we indicate merely some principal definitions and conclusions.

Let us consider the system of d.e.

$$\dot{x} = P(x,y); \qquad \dot{y} = Q(x,y) \tag{12.1}$$

where $P(x,y)$ and $Q(x,y)$ are defined and continuous in the whole plane and such that uniqueness of solutions and their continuous dependence on initial data exist for every finite interval of time. The integral curves of (12.1) are:

$$x = \varphi(t - t_0, x_0, y_0) = x(t); \qquad y = \psi(t - t_0, x_0, y_0) = y(t) \tag{12.2}$$

where x_0, y_0 are the initial conditions for $t = t_0$. These equations represent a curve in the (x,y) plane in a parametric form; such a curve is called *trajectory* (or characteristic) and we shall indicate it by L.

The part of the trajectory corresponding to $t \geq t_0$ $(t \leq t_0)$ is called a *positive* (*negative*) *half-trajectory* and will be indicated by $L^+(L^-)$. A point $R = [x(t),y(t)]$ in the (x,y) plane is called according to Birkhoff an ω-limit point (an α-limit point) for a given trajectory L if there exists one sequence at least: $t_1 < t_2 < t_3 < \ldots < t_n$ $(t_1 > t_2 > \ldots > t_n)$ converging to the ω point (or the α point) such that

$$\lim_{n\to\infty} x(t_n) = \xi^*; \qquad \lim_{n\to\infty} y(t_n) = \eta^*$$

The set of the ω-limit points (the α-limit points) for a given trajectory L will be indicated by $\Omega(L^+)$ and $A(L^-)$. When L and $\Omega(L)$ have points in

[10] See footnote [10], page 90.

common, the trajectory L is called stable in the nomenclature of Poisson. It can be proved that every Poisson stable trajectory L is contained in $\Omega(L)$. Analogous definitions and properties hold also for $A(L)$-limit points.

The following properties hold:

(a) Every Poisson stable trajectory is closed (a cycle or a singular point).

(b) If $\Omega(L)$ is a bounded set without singular points, then it is a cycle.

(c) A bounded $\Omega(L)$ set with singular points consists either of a unique singular point or of some singular points plus open trajectories connecting them. In this last case (at least under the assumptions of regularities made at the beginning), such singular points cannot be either foci or nodes and therefore they are saddle points with their separatrices. Summing up, all possible half-trajectories are of the following five types: (1) positions of equilibria; (2) a closed trajectory; (3) a half-trajectory tending to a position of equilibrium; (4) a half-trajectory tending to a closed trajectory; (5) a half-trajectory tending to the limiting structure mentioned in (c).

Case (1) constitutes the object of Chapter 1; (2) was treated in Chapter 2; (3) and (4) were discussed in this chapter; and (5) was encountered in Chapter 2 in regard to the asymptotic trajectories issuing from a saddle point along one asymptote and returning to it along another asymptote. A more general case of a trajectory of this kind was indicated in Fig. 2.3. We note that the center does not enter into this classification and does not result in any definite topological configuration since trajectories around a center appear in continuous families depending only on the initial conditions.

On the other hand, inasmuch as the simple harmonic motion was laid down as a basis of the earlier theory of oscillation, it is obvious that the latter considered only oscillations by themselves without involving any concept of topological configurations. In the new theory the positions of equilibria and of stationary motions are treated together on the basis of the entire trajectory (from $t = -\infty$ to $t = +\infty$).

The foregoing topological concepts can be extended to many other cases; we shall return to these questions later.

Chapter 4

GEOMETRICAL ANALYSIS OF PERIODIC SOLUTIONS

1. Introductory remarks

This chapter deals with the important development of Liénard[1] concerning the geometrical criteria of existence of periodic solutions. His paper appeared in 1928, just one year before the first contact with the theory of Poincaré had been established by Andronov. As the result of this, later developments (mostly in the USSR) followed the analytical methods of Poincaré in connection with the establishment of conditions of periodicity.

The work of Liénard seemed thus to have lost some of its importance, but in some respects this turned out not to be the case. In the first place Liénard's method permits approaching this question in an extremely simple and intuitive manner as was shown by Lefschetz.[2] Also a physical interpretation of Liénard's criterion resulted in the relationship between the periodicity of a solution and a special condition of energy exchanges. In fact, the fundamental point of Liénard's theory reduces the question of periodicity to the vanishing of a certain curvilinear integral. This integral turns out to be the one which specifies the energy exchanges between the oscillating system and the outside sources and, on this basis, Liénard's criterion acquired the very simple interpretation that a stationary state is reached when the energies absorbed and dissipated during one period cancel out. The question of stability acquired also a very elegant interpretation on this basis; thus, if more energy is absorbed than a system can dissipate, the amplitudes grow; if the inverse takes place, they decrease.

[1] A. Liénard, *Rev. Gen. de l'Électricité* **23**, 1928.

[2] S. Lefschetz, *Proc. Fifth Symposium of Appl. Math.* **5**, 1954.

Still another remarkable property of Liénard's method is to be mentioned. His work was inspired by the van der Pol equation

$$\ddot{x} + \mu(x^2 - 1)\dot{x} + x = 0 \tag{1.1}$$

whose form Liénard tried to generalize. In this d.e. appears the parameter μ whose full importance was ascertained later when the analytical methods came into use. A serious limitation of the latter is that they require the smallness of μ since they use solutions in the form of a power series in μ, which converges only if μ is sufficiently small. As Liénard was concerned with a purely geometrical argument, the parameter was not considered and the d.e. appears, therefore, in the form

$$\ddot{x} + f(x)\dot{x} + x = 0 \tag{1.2}$$

which is generally known as *Liénard's equation.* Only at the end of his paper, when the asymptotic cases are considered [(1) $\mu \to 0$ and (2) $\mu \to \infty$], did Liénard introduce the parameter by writing $\mu f(x)$ instead of $f(x)$. In the first case, Liénard found merely what could be found by the classical theory—namely, that the stationary amplitude is $x_0 = 2$—but in the second asymptotic case ($\mu \to \infty$) Liénard made a fundamental discovery which could be appreciated only much later. In fact, a long and persistent search by mathematicians to find an analytic solution (that is, in the form of a power series) of the van der Pol equation when μ is large has not been successful for reasons that will be explained in Part IV.

Since Liénard's argument is purely geometrical, he was able to go to the limit $\mu \to \infty$ with this conclusion†: The oscillation degenerates in this case into four branches; two of these branches are certain analytic curves traversed with a finite velocity and two others are rectilinear stretches traversed by the representative point with a very high (practically infinite) velocity. This fundamental discovery is the basis on which the modern discontinuous theory of relaxation oscillations is built, and we shall return to this subject in Part IV in connection with the so-called piecewise analytic oscillations.

The argument of Liénard has been further generalized and improved in connection with the d.e. of the form

$$\ddot{x} + f(x)\dot{x} + g(x) = 0 \tag{1.3}$$

$$\ddot{x} + f(x,\dot{x})\dot{x} + g(x) = 0 \tag{1.4}$$

A great many mathematicians contributed further to this question.

† It is to be noted that Liénard reached this conclusion in a heuristic manner. The formal proof appeared much later, as will be explained in Part IV.

Among them one should mention: N. Levinson and O. K. Smith,[3] Dragilev,[4] Sansone,[5] Conti,[6] de Castro,[7] Lefschetz[2] (book p. 245) and a number of others.

We note in passing that, as far as electrical applications are concerned (and these are more important than the others), equation (1.2) is more useful than (1.3) or (1.4). This is because the last term is always due to the capacity element of the circuit but, as capacity is always a linear element (at least with the usual circuit elements), the last term always appears as x (with a proper normalization) and not as $g(x)$ in general.

For systems other than electrical, the situation may be different.

2. Liénard's theory; curves Γ and Δ; criterion of periodicity

We reproduce here some theorems by which Liénard's results were brought into a more general form. The essential part of the Liénard argument leading to the establishment of his criterion of re-entrancy of the path remains however the same in all cases.

We mention two theorems, one by Levinson and Smith[3] and the other by Dragilev,[4] which were established independently; as their proofs can easily be found, we omit them here.

Theorem of Levinson and Smith: If $g(x)$ is such that $xg(x) > 0$ for $x > 0$ and $\int_0^\infty g(x)dx = \infty$; if $f(0,0) < 0$ and there exists a number $x_0 > 0$ such that $f(x,\dot{x}) \geq 0$ for $|x| \geq x_0$; moreover $\int_{x_0}^{x_1} f(x,v)dx \geq 10Mx_0$, where $v(x)$ is an arbitrary positive decreasing function of x, then the d.e. (1.4) has at least one periodic solution.

Theorem of Dragilev: The d.e. (1.2) has at least one limit cycle if:

(1) $g(x)$ satisfies a Lipschitz condition and $xg(x) > 0$ for $x \neq 0$ and $g(\infty) = \infty$.

(2) $F(x) = \int_0^x f(x)dx$ is single valued in $(-\infty, +\infty)$, satisfies a Lipschitz condition for any finite interval and, moreover, for small $|x|$, $F(x) < 0$ for $x > 0$ and $F(x) > 0$ for $x < 0$.

(3) There exist numbers M, k, and k', $(k' < k)$ such that $F(x) > k$ for $x > M$, and $F(x) \leq k'$ for $x < -M$.

We shall follow here Bogoliubov and Mitropolsky's exposition[8] which

[2] See footnote [2], page 101.
[3] N. Levinson and O. K. Smith, *Duke Math. J.* **9**, 1942.
[4] A. D. Dragilev, *Prikl. Math. i Mehanika* (Russian) **16**, 1949.
[5] G. Sansone, *Ann. Math. pura e appl.* (4) **28**, 1949.
[6] R. Conti, *Bull. Un. Mat. Italiana* (3) **7**, 1952.
[7] A. de Castro, *Bull. Un. Mat. Italiana* (3) **8**, 1953.
[8] N. Bogoliubov and J. Mitropolsky, *Asymptotic Methods in the Theory of Nonlinear Oscillations* (in Russian), Moscow, 1958.

uses Lefschetz's argument regarding the significance of the variable λ (see below); the assumed conditions are:

(1) $f(x)$ is even, $g(x)$ is odd, $xg(x) > 0$ for all $x \neq 0$, and $f(0) < 0$.
(2) $f(x)$ and $g(x)$ are continuous and $g(x)$ is Lipschitzian.
(3) $F(x) \to \pm\infty$ as $x \to \pm\infty$, where $F(x) = \int_0^x f(x)dx$.
(4) $F(x)$ has one single positive zero $x = a$; for $x \geq a$, the function $F(x)$ increases monotonically with x.

The conclusion is that (1.3) possesses a unique periodic solution, and this solution is stable.

We introduce new variables

$$y = \dot{x} + F(x); \qquad \lambda(x,y) = \frac{y^2}{2} + G(x) \tag{2.1}$$

where $G(x) = \int_0^x g(x)dx$. The term $y^2/2$ may be regarded (with a proper normalization) as kinetic energy, and $G(x)$ as potential energy, so that $\lambda(x,y)$ is the total energy stored in the oscillation, as was pointed out by Lefschetz.[9]

We calculate the rate of change of the energy $d\lambda/dt$ with a view to integrating over one cycle. We have

$$\frac{d\lambda}{dt} = \frac{d}{dt}\left(\frac{y^2}{2} + G(x)\right) = \frac{d}{dt}\left[\frac{1}{2}(\dot{x} + F(x))^2 + G(x)\right]$$

$$= \dot{x}(\ddot{x} + f(x)\dot{x} + g(x)) + F(x)\frac{d}{dt}(\dot{x} + F(x)) \tag{2.2}$$

As the coefficient of $\dot{x}$ vanishes by (1.3), one has

$$d\lambda = F(x)dy \tag{2.3}$$

The energy exchange of the system is $\int F(x)dy$ and, if the system is in a stationary state of oscillation we have the following criterion of Liénard

$$\oint F(x)dy = 0 \tag{2.4}$$

This curvilinear integral is to be taken along a trajectory. The argument of Liénard, as well as of other authors, is that (2.4) is a criterion for the existence of a limit cycle.

We consider the equivalent system in the form:

$$\dot{x} = y - F(x); \qquad \dot{y} = -g(x) \tag{2.5}$$

[9] S. Lefschetz, *Differential Equations (Geometric Theory)*, Interscience Publishers, New York, 1957.

Since $F(x)$ is the integral of a continuous function its derivative is continuous, and hence $F(x)$ satisfies a Lipschitz condition. $g(x)$ satisfies a Lipschitz condition by hypothesis, and thus the fundamental existence and uniqueness theorem applies to the solutions of (2.5).

It is noted that this equivalent system here is different from the usually employed system when one sets $\dot{x} = y$ (the ordinary phase plane). In the *phase plane of Liénard* the velocity $\dot{x}$ is counted not along the ordinate of the usual phase plane $(x,\dot{x})$ down to the abscissa axis but along the ordinate to the "curvilinear abscissa axis" $F(x)$ (Fig. 4.1). This results in a different form of trajectories in the Liénard plane as compared to the same trajectories considered in the ordinary phase plane. Since $F(x)$ and $g(x)$ are odd, it follows that if $x(t)$, $y(t)$ is a solution of (2.5), so is $-x(t)$, $-y(t)$. Thus to every portion of a trajectory corresponds a portion of the same or

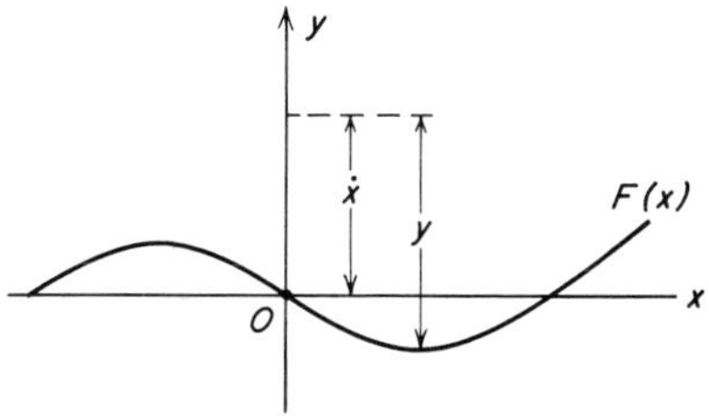

FIGURE 4.1

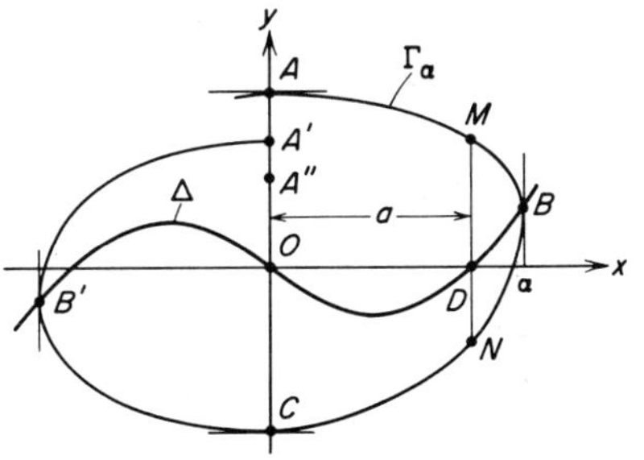

FIGURE 4.2

another trajectory symmetric to it with respect to the origin. Since the origin is the only singular point of (2.5), any closed trajectory must contain the origin in its interior.

The d.e. of integral curves is

$$\frac{dy}{dx} = -\frac{g(x)}{y - F(x)} \tag{2.6}$$

It is seen that for $x = 0$, $y \neq 0$ all integral curves have a horizontal tangent since $g(0) = 0$.

Referring to Fig. 4.2, the curve $\Delta: y - F(x) = 0$ intersects the integral curve Γ at the points B and B' at which the tangent to Γ is vertical since $dy/dx = \infty$. As $xg(x) > 0$, y decreases along Γ to the right of the y axis and increases to the left of Oy. As to x, it increases if Γ is above Δ and decreases in the opposite case. One can assume therefore that the curve has the appearance shown in Fig. 4.2; its actual construction is indicated in the following section. Let us denote the abscissa of the point B by α and the curve Γ by Γ_α; we use this notation in order to attach a definite amplitude α to Γ.

If Γ_α is closed, it is symmetrical with respect to the origin. If it were not symmetrical its image in the origin would be another closed trajectory. This second trajectory would then intersect Γ_α, which is impossible. Hence, if Γ_α is closed, we have $|OA| = |OC|$. Conversely, if $|OA| = |OC|$ the image of the arc $\widehat{AC}$ in the origin forms with $\widehat{AC}$ a closed trajectory and therefore Γ_α is closed. Thus $|OA| = |OC|$ is the necessary and sufficient condition that Γ_α be closed. This condition is also expressed by $\lambda(0,A) = \lambda(0,C)$, which we abbreviate to $\lambda(A) = \lambda(C)$.

In order to prove that this is possible we consider curvilinear integrals along Γ_α setting

$$\varphi(\alpha) = \lambda(C) - \lambda(A) = \int_{ABC} d\lambda = \int_{ABC} F(x)dy \tag{2.7}$$

It is sufficient to study the curvilinear integral along the arc of Γ_α to the right of Oy since everything is symmetrical to the left of it.

If the amplitude $\alpha < a$ (a is the abscissa of the point D at which the curve Δ cuts the x axis), $dy < 0$ (since $\Delta < 0$) but, as $F(x) < 0$, the curvilinear integral $\int_{ABC} F(x)dy > 0$, and therefore $\lambda(C) > \lambda(A)$. Hence *energy is absorbed* and there is no closed trajectory.

We consider now the case when $\alpha > a$ (as indicated in Fig. 4.2). Let MN be a perpendicular to the abscissa axis through D (of abscissa $x = a$) and consider two portions of Γ_α: the first one consisting of the two pieces between the y axis and the line MN and the other one the arc MBN. To simplify, we call the first arcs (I) (consisting of AM and NC); the second (arc MBN) will be called II. We have thus

$$\varphi_{\text{I}}(\alpha) = \int_{AM} d\lambda + \int_{NC} d\lambda; \qquad \varphi_{\text{II}}(\alpha) = \int_{MBN} d\lambda \tag{2.8}$$

Since $d\lambda = F(x)dy$ and $dy/dx = -g(x)/[y - F(x)]$, we have

$$d\lambda = F(x)\frac{dy}{dx}\,dx = -\frac{F(x)g(x)}{y - F(x)}\,dx \tag{2.9}$$

Since $F(x) < 0$ for $x < a$, $d\lambda$ is positive for the piece of trajectory (I) described in the direction $A \to M$; the same holds for the other piece in (I) when it is described in the direction $N \to C$. This shows that $\varphi_{\text{I}}(\alpha) > 0$.

As to $\varphi_{\text{II}}(\alpha)$, one has, on the other hand, $d\lambda < 0$; therefore $\varphi_{\text{II}}(\alpha) < 0$. If α increases, the arc AM goes up and NC goes down; for x fixed, this means that $|y|$ increases, and therefore $\varphi_{\text{I}}(\alpha)$ decreases.

We analyze now the behavior of $\varphi_{\text{II}}(\alpha)$ corresponding to the line integral along the piece of trajectory MBN (Fig. 4.2). If the amplitude increases from α_1 to α_2, this piece becomes $M'B'N'$, Fig. 4.3 . We are going to show that $\varphi_{II}(\alpha_2) < \varphi_{II}(\alpha_1)$.

If one draws parallels to the x axis through the points M and N, one obtains two points P and P' which section the arc $M'B'N'$ into three arcs $M'P$, PP' and $P'N'$; we consider therefore, $J_{M'B'N'} = J_{M'P} + J_{PP'} + J_{P'N}$; where $J_{M'P} = \int_{M'P} F(x)dy$, etc. As $F(x) > 0$ and $dy < 0$, in this region the integrals are negative and one can write

$$J_{M'B'N'} < J_{PP'} \tag{2.10}$$

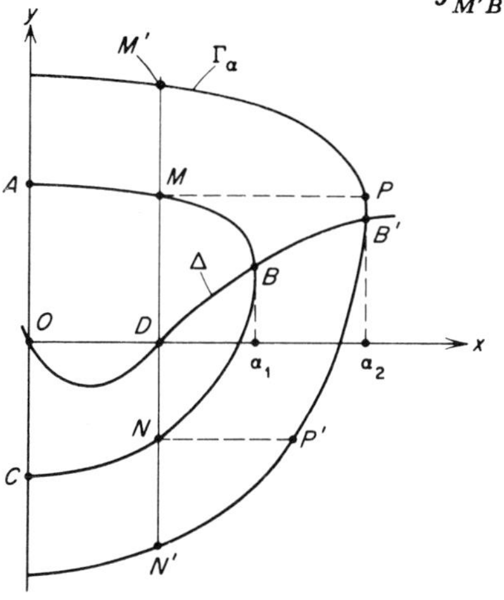

FIGURE 4.3

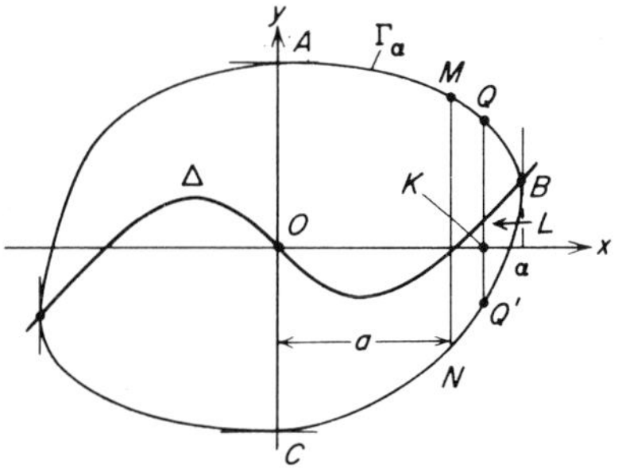

FIGURE 4.4

The limits MP and NP' being the same (for y), it is clear that the integral along MBN is greater than that along PP', inasmuch as for the latter the abscissas are greater than for the former and the integrals are negative; therefore

$$J_{PP'} < J_{MBN} \tag{2.11}$$

It follows that

$$J_{M'B'N'} < J_{MBN} \tag{2.12}$$

and finally that $\varphi_{II}(\alpha_2) < \varphi_{II}(\alpha_1)$ for $\alpha_2 > \alpha_1$; therefore

$$\varphi(\alpha) = \varphi_I(\alpha) + \varphi_{II}(\alpha) \tag{2.13}$$

is a monotonically decreasing function of α. It is noted that, if $\alpha < a$, $\varphi(\alpha) = \varphi_I(\alpha) > 0$.

We now show that $-\varphi_{II}(\alpha) \to \infty$ if $\alpha \to \infty$. We fix some value of x say $a < x_1 < \alpha$ and draw a parallel to the y axis through $x = x_1$ (the line QQ'), Fig. 4.4.

We have $J_{MBN} < J_{QBQ'}$. For the arc QBQ' one has $x > x_1$ so that $F(x) > F(x_1)$, hence $\varphi_{II}(\alpha) = J_{MBN} < F(x_1) \int_{QBQ'} dy = -F(x_1)|\overline{QQ'}|$ and, therefore (calling K the point $(x_1,0)$), we have

$$-\varphi_{II}(\alpha) = -\int_{MBN} d\lambda > \overline{KL} \cdot \overline{QQ'}$$

It is clear that the segments KL and $\overline{QQ'}$ can be as great as we please if α is large enough, which shows that $-\varphi_{II}(\alpha) \to \infty$ for $\alpha \to \infty$. Since for a sufficiently small α, $\varphi(\alpha) > 0$ and $\varphi(\alpha) \to -\infty$ as $\alpha \to \infty$, there exists one and only one value $\alpha = \alpha_0$ for which $\varphi(\alpha_0) = 0$, which shows that there is one and only one closed curve for which $\lambda(A) = \lambda(C)$. The question of stability is studied also geometrically. Thus, if A_0 and C_0 are points of intersection of Γ_{α_0} with the y axis, the point C is nearer to Γ_{α_0} than A, if $\alpha < \alpha_0$; hence A' is nearer to Γ_{α_0} than A. A similar argument is used for $\alpha > \alpha_0$ which gives thus a purely geometrical criterion of stability of the limit cycle.

3. Liénard's phase plane; graphical construction of curves Γ

As was mentioned in the preceding section, the phase plane of Liénard differs from the usual phase plane in that the ordinates in this plane are

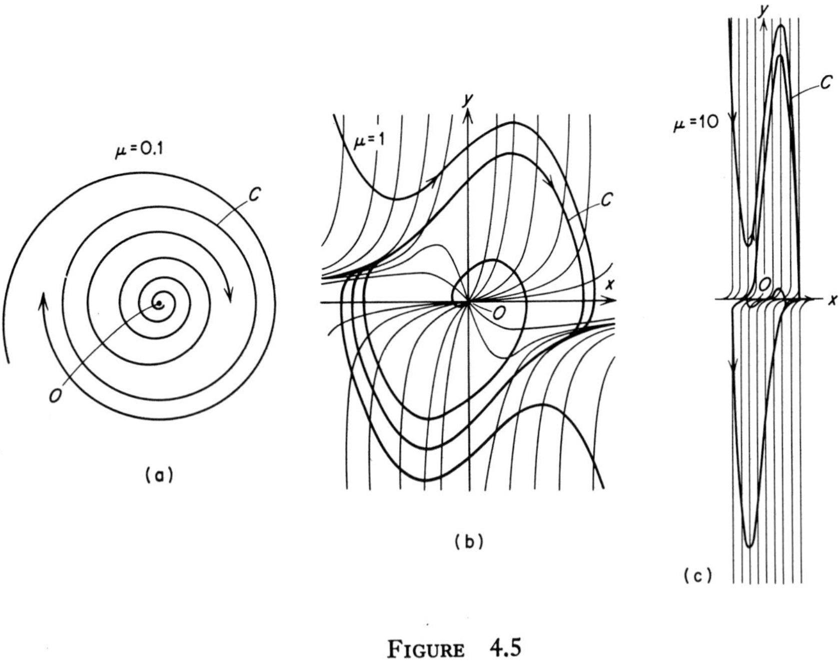

FIGURE 4.5

given by $y = \dot{x} + F(x)$ instead of $y = \dot{x}$. In view of this the same integral curve has different forms in these two planes. Figures 4.5 and 4.6 give a comparison between the integral curves of the van der Pol equation plotted in the phase plane $x,\dot{x}$ (Fig. 4.5) and the same curves traced in Liénard's phase plane (Fig. 4.6). The curves are shown for values of $\mu = 0.1$, 1, and 10. It is observed that for larger μ Liénard's curves approach the form of a rectangle; we shall see in the following section that the horizontal and the vertical sides of this rectangle are traversed with velocities of different order. The curves of Fig. 4.5 were obtained by van der Pol by means of the graphical method of isoclines; those of Fig. 4.6 are taken from a paper by Le Corbeiller.[10]

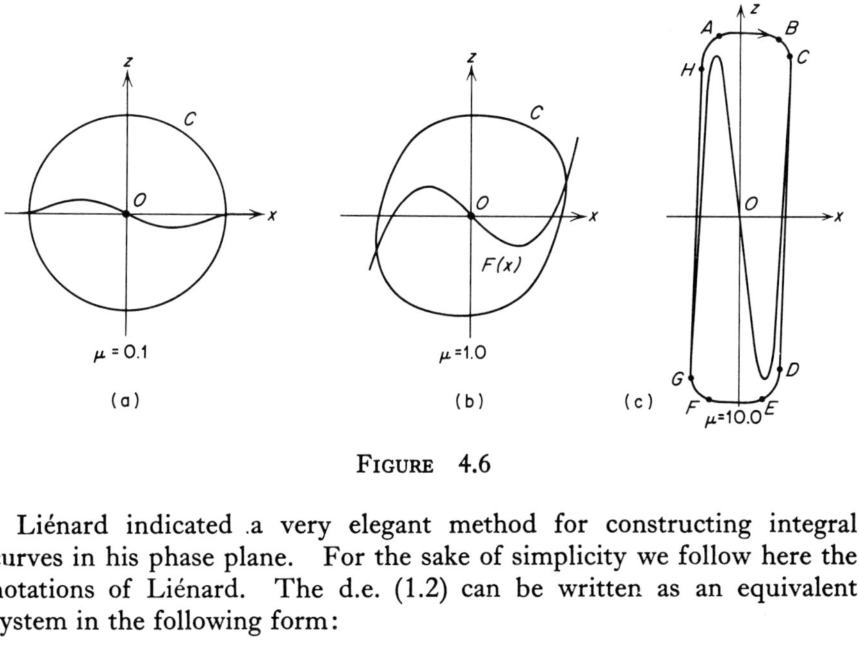

FIGURE 4.6

Liénard indicated a very elegant method for constructing integral curves in his phase plane. For the sake of simplicity we follow here the notations of Liénard. The d.e. (1.2) can be written as an equivalent system in the following form:

$$\dot{x} = v; \qquad \frac{dv}{dx} + f(x) + \frac{x}{v} = 0 \tag{3.1}$$

We introduce here the variable of Liénard as was already mentioned in the preceding section, namely:

$$y = v + F(x), \qquad \text{where } F(x) = \int_0^x f(x)dx$$

[10] Ph. Le Corbeiller, *Systèmes auto-entrenus*, Herman, 1931, Paris.

The d.e. becomes then

$$\frac{dy}{dx} + \frac{x}{y - F(x)} = 0 \tag{3.2}$$

This is, of course, (2.6) with $g(x) = x$; (3.2) can be written also as

$$xdx + (y - F(x))dy = 0 \tag{3.3}$$

This d.e. is that of a normal

$$(x - X)dx + (y - Y)dy = 0 \tag{3.4}$$

which passes through the point $X = 0$; $Y = F(x)$.

Assume, for instance, that we wish to draw Liénard's curves for the van der Pol equation for which $f(x) = x^2 - 1$ and $F(x) = (x^3/3) - x$. This curve $F(x)$ is plotted (Fig. 4.7) and the construction consists in tracing the direction field of lineal elements. We take, for instance, $x = x_1$ and for this point we have the point M_1 on the curve $F(x)$. Let N_1 be the projection of M_1 on the y axis. With N_1 as center one draws small arcs intersecting the line M_1P_1. Then one takes another point $x = x_2$ resulting in N_2 and with N_2 as center one draws again small arcs intersecting the line M_2P_2. Having constructed a sufficient number of ordinates with the corresponding arcs, one draws the Γ curves (Section 2). One of the curves of this family is necessarily closed if the Liénard conditions are fulfilled. It is to be noted that the form of $F(x)$ *before* its intersection with the x axis is immaterial, as Liénard points out; in other words it may have maxima and minima as long as x is in the interval OM (Fig. 4.7); what is essential is that *after* crossing the x axis (at the point M) the curve $F(x)$ should increase monotonically with x. As we saw in the preceding section, this condition amounts to the change of sign of $\int F(x)dy$ along a curve Γ with the corresponding value $\int F(x)dy = 0$ separating the region where $\int F(x)dy > 0$ from that where $\int F(x)dy < 0$ and this, in turn, amounts to the existence of a closed integral curve, that is, a periodic solution when $\int F(x)dy = 0$.

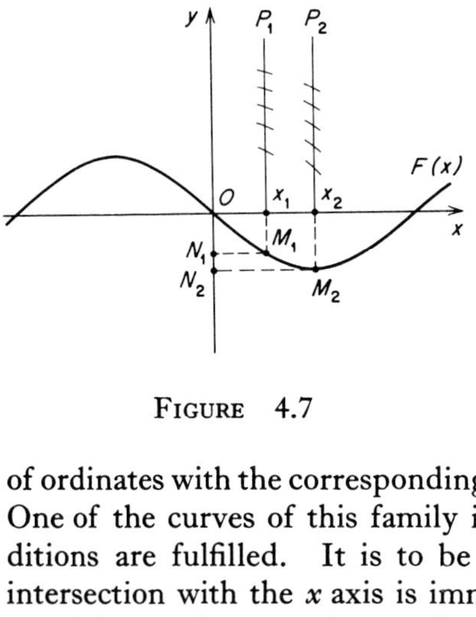

FIGURE 4.7

4. Asymptotic cases of Liénard's equation

Equation (1.2) does not specify the order of magnitude of the term $f(x)\dot{x}$. We write it now with the parameter μ:

$$\ddot{x} + \mu f(x)\dot{x} + x = 0 \tag{4.1}$$

in order to pass to the asymptotic cases $\mu \to 0$ and $\mu \to \infty$.

If one replaces $f(x)$ by $\mu f(x)$ and $F(x)$ by $\mu F(x)$ in the preceding, Liénard's equation becomes

$$\frac{dy}{dx} + \frac{x}{y - \mu F(x)} = 0 \tag{4.2}$$

In order to have the function $F(x)$ as before, one can replace y by μz. The transformed equation takes the form

$$\frac{dz}{dx} = -\frac{x}{\mu^2[z - F(x)]} \tag{4.3}$$

In this form the effect of the parameter can be better ascertained. Thus, if $\mu > 1$, for the same x, the integral curves have smaller slopes than previously ($\mu = 1$). If $\mu \to \infty$, $dz/dx \to 0$. Thus, for increasing μ, the integral curves exhibit flat portions parallel to the x axis.

In the asymptotic case $\mu \to \infty$, (4.3) reduces to

$$[z - F(x)]dz = 0 \tag{4.4}$$

This suggests that the integral curve consists of two branches: on one of them there exists the relation $z = F(x)$ and on the other $dz = 0$, which shows that this branch is merely a straight line parallel to the x axis. (Compare to Fig. 4.6c in Liénard's plane.)

In order to investigate the velocity of the representative point R, it is sufficient to apply the same transformation $y = \mu z$, which gives:

$$dx/dt = \mu(z - F(x)); \qquad dz/dt = -x/\mu \tag{4.5}$$

If R follows the branch $z = F(x)$, in the asymptotic case when μ is large, the velocity dx/dt is finite. For the second branch $z \neq F(x)$, so that dx/dt is large.

Thus the horizontal branches ($z = \text{const}$) are traversed with a very high (practically infinite) velocity, whereas the characteristic $F(x)$ is traversed with a finite velocity. This gives rise to a situation depicted in Fig. 4.8 where $F(x) = -x + (x^3/3)$ corresponding to $f(x) = x^2 - 1$, as in the van der Pol equation. The point R follows $F(x)$ up to the point B where the second branch $dz = 0$ begins. On this branch BC the point R acquires a very high velocity and is traversed practically in no time. At C

begins again the first branch traversed with a finite velocity up to the point D, where another jump DC begins, thus closing the curve $BCDA$ consisting of two pieces of two distinct branches.

This result of Liénard's equation is fundamental for the modern theory of relaxation oscillations, as was mentioned in section 1; we shall return to this question in more detail in Part IV.

The second asymptotic case, when μ is small is less interesting and falls within the scope of the general theory of Poincaré. The argument is as follows: Liénard's equation can be written as $xdx + ydy - \mu F(x)dy = 0$ and, if μ is small, it reduces simply to $xdx + ydy = 0$ which gives the family of concentric (with origin) circles, as may be expected, of course, because in this case Liénard's equation reduces (for $\mu = 0$) to that of the harmonic oscillator.

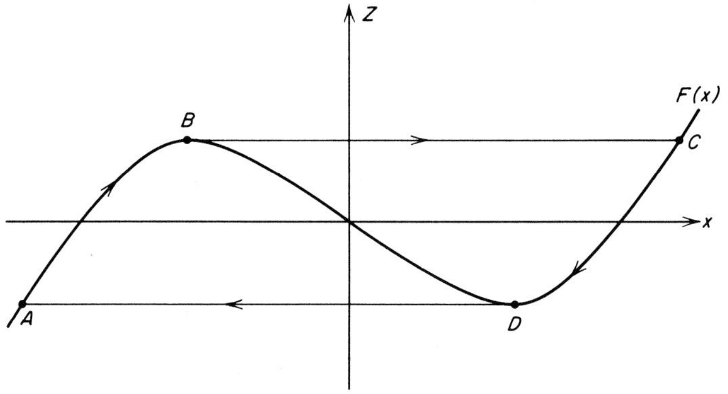

FIGURE 4.8

In polar coordinates the criterion $\int_{AC} F(x)dy = 0$ is

$$\int_0^\pi F(r \cos \varphi)d(r \sin \varphi) = 0$$

and, since in the differentiation $\dot{r} = dr/dt \simeq 0$, $r = \text{const}$, one can write

$$\int_0^\pi F(r \cos \varphi) \cos \varphi d\varphi = 0$$

The integration by parts brings this expression to the form

$$\int_0^\pi f(r \cos \varphi) \sin^2 \varphi d\varphi = 0 \tag{4.6}$$

since $(d/dx)F(x) = f(x)$.

If one replaces $f(x)$ by the van der Pol expression $1 - x^2 = 1 - r^2 \cos^2 \varphi$, one easily obtains the condition $r_0 = 2$. We shall find this value later by a different method.

5. Principle of symmetry

In Section 2 it was shown that, in view of the symmetry with respect to the origin, the proof of existence of a periodic solution reduces to the equality of the intercepts of the integral curve on the y axis and this, in turn, reduces to the Liénard criterion (2.4).

These considerations of symmetry were generalized further by Nemitzky[11] and we shall give a brief account of this work. Consider the differential system

$$\dot{x} = X(x,y); \qquad \dot{y} = Y(x,y) \tag{5.1}$$

and suppose that the origin $x = y = 0$ is a singular point. If $X(x,y)$ is odd with respect to x, that is, $X(-x,y) = -X(x,y)$, and if $Y(x,y)$ is even, that is $Y(-x,y) = Y(x,y)$, the image of an integral curve in the y axis is also an integral curve. (Similar conclusions exist when $X(x,y)$ is even and $Y(x,y)$ is odd in y.) Hence, in order to prove that an integral curve is closed, it is sufficient to show that, starting from a point on the y axis, the curve returns again to this axis.

Consider the d.e.

$$\frac{dy}{dx} = \frac{-x + X_2(x,y)}{y + Y_2(x,y)} \tag{5.2}$$

where X_2 and Y_2 are analytic functions whose series developments begin with terms of, at least, second degree in x and y. From the general theory it is known that the singular point is either a focus or a center.

From the principle of symmetry one can assert that if $X_2(x,y)$ contains only terms of odd degree and $Y_2(x,y)$ only terms of even degree in x, the singular point is a center. Alternatively, if $X_2(x,y)$ contains only terms of even degree and $Y_2(x,y)$ only terms of odd degree in y, then the singular point is a center. Thus, the principle of symmetry in this case gives a very simple criterion for the existence of a center as compared to the classical criterion (Section 8, Chapter 1).

The application of the principle of symmetry is more complicated if the singular point is of a higher order. As an example we give the following case investigated by Philippov.[11]

[11] V. V. Nemitzky and V. V. Stepanov, *Qualitative Theory of Differential Equations*; original text in Russian, Moscow, 1949; English translation, Princeton Mathematics Series, Princeton University Press, Princeton, N.J., 1960.

Consider the d.e.

$$\ddot{x} + f(x)\dot{x} + g(x) = 0 \tag{5.3}$$

We suppose that $f(x)$ and $g(x)$ are odd functions of x with $f(x) > 0$, $g(x) > 0$ for $x > 0$. Then $F(x) = \int_0^x f(x)dx$ and $G(x) = \int_0^x g(x)dx$ are even and monotonically increasing for $x > 0$ and $xG(x) \to \infty$ as $x \to \infty$. We now prove: if there exist constants $x_1 > 0$ and $\varepsilon > 0$ such that $g(x) \geq (\frac{1}{4} + \varepsilon)f(x)F(x)$, then for $0 < x < x_1$ there exists a periodic solution of (5.3) for sufficiently small values of x and $\dot{x}$.

The equivalent system is

$$\dot{x} = y; \qquad \dot{y} = -f(x)y - g(x) \tag{5.4}$$

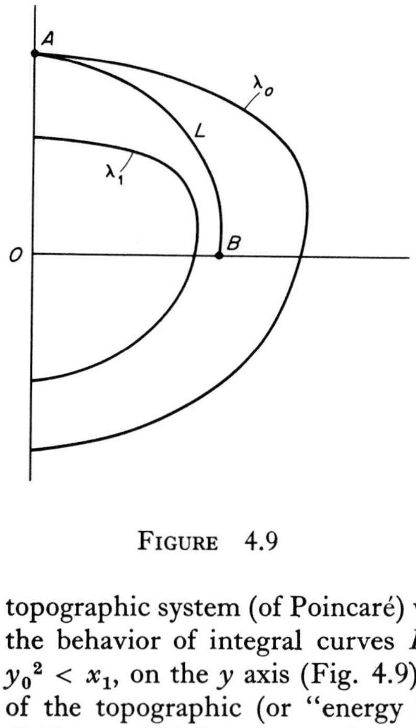

FIGURE 4.9

Multiplying the second d.e. by y, one has

$$g(x)y + y\dot{y} = -f(x)y^2 \tag{5.5}$$

If one sets $\lambda(x,y) = y^2/2 + G(x)$, this d.e. becomes

$$\frac{d\lambda(x,y)}{dt} = -f(x)y^2 \tag{5.6}$$

It is noted that $\lambda(x,y)$ may be regarded as total energy if (5.3) is considered as an oscillator.

By an argument similar to that used in connection with Volterra's problem (Chapter 2), it follows that the curves $\lambda(x,y) = C$ are simple closed curves surrounding the origin. The family $\lambda(x,y) = C$ may play the role of the topographic system (of Poincaré) with respect to which one can investigate the behavior of integral curves L. We start from some point $A(0,y_0)$, $y_0^2 < x_1$, on the y axis (Fig. 4.9) to which is attached the value $\lambda_0(0,y_0)$ of the topographic (or "energy level") curve ω. In the first quadrant $d\lambda/dt = -f(x)y^2 < 0$, so that on a trajectory λ decreases, for $x > 0$, and the representative point R passes from one curve $\lambda = C$ of the family to others nearer to the origin.

The problem is to show that R, following the trajectory L issuing from A on the y axis, returns to the same point on the axis later. Then, in view of the principle of symmetry one can assert that L is closed.

As R remains inside the region limited by $\lambda_0(0,y_0) = \frac{1}{2}y_0^2$, it must intersect the x axis at some point $B(\xi,0)$; at this point $\lambda(\xi,0) = G(\xi) < \frac{1}{2}y_0^2 < \frac{1}{2}x_1$.

In the fourth quadrant ($x \geq 0$; $y \leq 0$) $d\lambda/dt$ continues to be negative, so that R will remain inside the region limited by the topographic curve $\lambda_1(x,y) = G(\xi) < \frac{1}{2}y_0^2$.

To complete the proof it is necessary to show that in the region limited by λ_1, the trajectory L cannot enter the singular point (the origin).

For $y < 0$ and $0 < x < x_1$, one has

$$\frac{dy}{dx} = -f(x) - g(x)/y \geq -[f(x)/y]\left(y + \frac{1}{4}F(x) + \varepsilon F(x)\right)$$

Moreover

$$\frac{d}{dx}\left[\frac{y(x)}{F(x)}\right] = \frac{y'F - yF'}{F^2} \geq \frac{-\dfrac{f}{y}\left(y + \dfrac{1}{4}F + \varepsilon F\right)F - yf}{F^2}$$

$$= -\frac{f}{yF^2}\left(yF + \frac{1}{4}F^2 + \varepsilon F^2 + y^2\right)$$

$$= -\frac{f}{yF^2}\left[\left(y + \frac{F}{2}\right)^2 + \varepsilon F^2\right] > 0$$

since $y < 0$.

For $x \to 0$, y/F decreases; hence there exists

$$\lim_{x\to 0}\left(\frac{y}{F}\right) = -\beta \leq 0$$

We choose x_0 such that for $0 < x < x_0$ one has

$$-\beta < \frac{y}{F} < -\beta + \frac{\varepsilon}{2}$$

On the other hand, in view of the condition: $g(x) \geq (\frac{1}{4} + \varepsilon)f(x)F(x)$ one has the following inequalities.

$$\frac{dy}{dx} = -f(x) - \frac{g(x)}{y(x)} \geq -f(x) + \frac{(\frac{1}{4} + \varepsilon)f(x)F(x)}{\left(\beta - \dfrac{\varepsilon}{2}\right)F(x)} = -f(x)\left[1 - \frac{\frac{1}{4} + \varepsilon}{\beta - \dfrac{\varepsilon}{2}}\right]$$

If one assumes that L goes through the origin, then

$$y(x_0) = \int_0^{x_0}\frac{dy}{dx}dx \geq -\left[1 - \frac{(\frac{1}{4} + \varepsilon)}{\beta - \dfrac{\varepsilon}{2}}\right]\int_0^{x_0} f(x)dx = -\left[1 - \frac{\frac{1}{4} + \varepsilon}{\beta - \dfrac{\varepsilon}{2}}\right]F(x_0)$$

which results in an impossible condition

$$-\left(\beta - \frac{\varepsilon}{2} + \frac{1}{2}\right)^2 + \varepsilon < 0$$

This shows that L does not go to the origin in the fourth quadrant; it must therefore have a negative intercept on the y axis and, by the principle of symmetry, the trajectory L is then a closed curve.

6. Energy criterion in a nearly linear case

The principal conclusion of Liénard's theory is the criterion (2.4) which, as we saw, is equivalent to the statement: out of the whole family of trajectories Γ_α there is one and only one closed trajectory Γ_{α_0} for which the integral of energy exchange over one period vanishes.

It is possible to obtain the same conclusion in a simpler manner when μ is small. Although this anticipates to some extent a later approach through the method of averaging (Part II), the procedure is so simple (and logically associated with the physical significance of Liénard's conclusion) that we indicate it here.

Given an autonomous system

$$\dot{x} = X(x,y); \qquad \dot{y} = Y(x,y) \tag{6.1}$$

We introduce new variables $\rho = r^2 = x^2 + y^2$; $\psi = \arctan (y/x)$ in the ordinary (x,y) plane. This can be done by multiplying the first equation (6.1) by x, the second by y, and adding equations so obtained; this gives

$$x\dot{x} + y\dot{y} = \frac{1}{2}\frac{d\rho}{dt} = xX + yY \tag{6.2}$$

The quantity ρ is obviously the total energy stored in the oscillation; it is, therefore, the same as λ of the preceding theory (Section 2).

As an example we consider a particular case of Liénard's equation when $f(x) = x^2 - 1$, that is, the van der Pol equation. Written in the form of an equivalent system, it is:

$$\dot{x} = y = X(x,y); \qquad \dot{y} = \mu y - \mu x^2 y - x = Y(x,y) \tag{6.3}$$

Replacing $x = r\cos\psi$, $y = r\sin\psi$ (where $r^2 = \rho$) into (6.2), we have

$$\frac{d\rho}{dt} = 2\mu\rho\,(\sin^2\psi - \rho\sin^2\psi\cos^2\psi) \tag{6.4}$$

One can also form a second combination of equations (6.1):

$$x\dot{y} - y\dot{x} = \rho\frac{d\psi}{dt} = xY - yX \tag{6.5}$$

which, in this case, yields after a similar change of variables:

$$\frac{d\psi}{dt} = -1 + \mu\sin\psi\cos\psi - \mu\rho\cos^3\psi\sin\psi \tag{6.6}$$

Since we assume here that μ is very small, we can write approximately: $\dot{\psi} \simeq -1$; $\psi \simeq -t + c$; one can obviously take $c = 0$ so that $\psi = -t$ and (6.4) becomes

$$\frac{d\rho}{dt} \simeq -2\mu\rho(\rho \sin^2 t \cos^2 t - \sin^2 t) \tag{6.7}$$

This equation gives the rate of change of energy or the "energy exchanges" of the system and, following the argument of Liénard, we look for the condition under which these energy exchanges, integrated over the period 2π are zero. It is noted that $d\rho/dt$ is small in view of the small factor μ in (6.7); hence $\rho(t)$ is a slowly varying quantity and in the integration over one period we can replace $\rho(t)$ by its average value ρ_0. We have thus

$$\int_0^{2\pi} \frac{d\rho}{dt} . dt \simeq -2\mu\rho_0\left[\rho_0 \int_0^{2\pi} \sin^2 t \cos^2 t dt - \int_0^{2\pi} \sin^2 t dt\right] = 0 \tag{6.8}$$

Equating to zero the quantity in brackets, one gets $\rho_0 \simeq 4$; that is, $r_0 \simeq 2$, which is the result obtained by Liénard (end of Section 4). This result may be also regarded (geometrically) as the condition for a re-entrant path since $\int_0^{2\pi} \frac{d\rho}{dt} dt = \int_0^{2\pi} d\rho = 0$ is precisely this condition.

The proof of this section is much simpler than that of Liénard (Section 4), but it holds only *when μ is small*, which enabled us to consider ρ as approximately constant $\rho \simeq \rho_0$ in the integration. This method would not hold in the second asymptotic case ($\mu \to \infty$), whereas Liénard's method based on a purely geometrical argument does not depend on the order of magnitude of μ and this, as we saw, constitutes its principal advantage.

For μ small, this method of "averaging" gives the same result as Liénard's method, viz.: an oscillatory phenomenon becomes periodic when the integral of energy exchange over one period vanishes. If the system absorbs more energy than it can dissipate, the amplitude grows; if the contrary takes place it decreases. In Liénard's case this corresponds to $J < 0$ or $J > 0$; in this case, it corresponds to (6.8) being either negative or positive.

Chapter 5

STABILITY (VARIATIONAL EQUATIONS, CHARACTERISTIC EXPONENTS)

1. Introductory remarks

In this and in the following chapter we present a brief discussion of those parts of the theory of stability which will be needed later. We have encountered previously the questions of stability in regard to properties of singular points as well as those of limit cycles. This chapter is devoted to the variational equations of Poincaré, and Chapter 6 to the so-called "second" (or "direct") method of Liapounov.

The subject of stability has grown considerably in recent years, and this discussion attempts little more than an outline, omitting many lengthy proofs that can be found in existing texts on stability.[1, 2, 3] We have defined (Chapter 1) concepts of singular points, trajectories, as well as those of periodic motions, limit cycles (Chapter 3) in relation to one-degree-of-freedom systems. Extensions of these concepts for several degrees of freedom do not present any difficulty except that one cannot use any topological argument and has to use an analytical approach.

2. Definition of stability (Liapounov)

Consider a system of d.e.:

$$\dot{x}_i = f_i(t, x_1, \ldots, x_n); \qquad i = 1, \ldots, n \tag{2.1}$$

[1] H. Poincaré, *Les méthodes nouvelles de la mécanique céleste* **T.1**, Gauthier-Villars, Paris, 1892; also E. Goursat, *Cours d'Analyse* **T.2**, Gauthier-Villars, Paris, 1918.

[2] A. M. Liapounov, *General Problem of Stability of Motion* (in Russian), Charkov, 1892; also L. Cesari, *Asymptotic Behavior and Stability Problems*, Springer, Berlin, 1959.

[3] I. G. Malkin, *Theory of Stability of Motion* (in Russian), Moscow, 1952, English translation.

By its *solution* we mean a set of functions $u_i(t)$; $i = 1, \ldots, n$ that satisfies it. The same definition holds for autonomous systems

$$\dot{x}_i = f_i(x_1, \ldots, x_n) \tag{2.2}$$

We shall be concerned first with autonomous systems and will assume that the solution (or motion) is defined for the interval $(-\infty, +\infty)$ of t. The reader is referred to the discussion in Chapter 1 regarding the difference between the solution (motions) and the trajectory which results from the *translation property* of autonomous systems.

We now define the most important types of stability.

(a) Let $u_i(t)$ be a solution of (2.1) or (2.2). We shall say that $u_i(t)$ is *stable* if, given $\varepsilon > 0$ and t_0, there is $\eta = \eta(\varepsilon, t_0)$ such that any solution $v_i(t)$ for which $|u_i(t_0) - v_i(t_0)| < \eta$ satisfies $|u_i(t) - v_i(t)| < \varepsilon$ for $t \geq t_0$ (if η may be chosen independently of t_0, $u_i(t)$ is said to be *uniformly stable*). If no such η exists, $u_i(t)$ is *unstable.*

(b) *Asymptotic stability:* If $u_i(t)$ is stable, and in addition

$$|u_i(t) - v_i(t)| \to 0$$

as $t \to \infty$, we say that it is *asymptotically stable.*

Both of these definitions concern solutions (or "motions"). We may paraphrase them by saying that a solution (or motion) is stable if all solutions coming near it remain in its neighborhood; it is asymptotically stable if the solutions approach it asymptotically.

The remaining definitions concern closed trajectories (or orbits).

(c) *Orbital stability:* Let C be an orbit of (2.2). We say that C is *orbitally stable* if, given $\varepsilon > 0$, there is $\eta > 0$ such that, if R is a representative point of another trajectory which is within a distance η of C at time τ, then R remains within a distance ε of C for $t > \tau$. If no such η exists, C is *orbitally unstable.*

(d) *Asymptotic orbital stability:* If C is orbitally stable and, in addition, the distance between R and C tends to 0 as $t \to \infty$, it is said to be *asymptotically orbitally stable.*

To illustrate the difference between stability of motion and orbital stability let us consider the system:

$$\dot{\theta} = r; \quad \dot{r} = 0 \tag{2.3}$$

where r, θ are polar coordinates in the plane. In rectangular coordinates (2.3) becomes

$$\dot{x} = -y\sqrt{x^2 + y^2}; \qquad \dot{y} = x\sqrt{x^2 + y^2} \tag{2.3a}$$

The solutions are $r = \rho$ and $\theta = \rho t + \alpha$, or $x = \rho \cos(\rho t + \alpha)$ and

$y = \rho \sin(\rho t + a)$, where ρ and α are the initial values of r and θ and are periodic with period $2\pi/\rho$.

The trajectories are concentric circles centered on the origin. They are obviously all orbitally stable, since to satisfy the condition of orbital stability we may take $\eta = \varepsilon$. On the other hand, all of the solutions (motions), except only the identically zero solution, are unstable. Denote the solution by $x = g(t,\rho,\alpha)$; $y = h(t,\rho,\alpha)$. We investigate the stability of a particular solution, say: $x = g(t,\rho_0 + \eta, \alpha_0)$; $y = h(t,\rho_0 + \eta,\alpha_0)$. At $t = 0$, one has $|g(0,\rho_0,\alpha_0) - g(0,\rho_0 + \eta,\alpha_0)| = |\eta \cos \alpha_0|$; $h(0,\rho_0,\alpha_0) - h(0,\rho_0 - \eta,\alpha_0)| = |\eta \sin \alpha_0|$. By choosing η small enough we can make the initial values of the second solution differ by as little as we please from the initial values $x_0 = \rho_0 \cos \alpha_0$, and $y_0 = \rho_0 \sin \alpha_0$ of the solution being investigated.

Nevertheless we cannot keep the solutions near each other, owing to the circumstance that they have different periods, $2\pi/\rho_0$ and $2\pi/(\rho_0 + \eta)$. In fact, there are arbitrarily small values of η for which the difference $\Delta(t) = g(t,\rho_0,\alpha_0) - g(t,\rho_0 + \eta,\alpha_0)$ assumes the value $2\rho_0 + \eta$ for infinitely many values of t. Without loss of generality we suppose that $\alpha_0 = 0$. We choose $\eta = \rho_0/2N$, where N is an integer, so that $\eta \rightarrow 0$ as $N \rightarrow \infty$.

We evaluate $\Delta(\varepsilon)$ for $t = (2n + 1)N(2\pi/\rho_0)$, n an integer. Now $\Delta(t) = \rho_0 \cos \rho_0 t - (\rho_0 + \eta) \cos(\rho_0 + \eta)t$; and, for certain values of t and η, we obtain $\Delta(t) = 2\rho_0 + \eta$. Thus for fixed N, however large, $\Delta(t) = 2\rho_0 + \eta$ for every value of n. In terms of the phase-space representation this result means that the representative points on the circles $r = \rho_0$ and $r = \rho_0 + \eta$ which have initially the same phase $\theta = \alpha_0$ will be infinitely often $180°$ out of phase and therefore at a distance $2\rho + \eta$ apart.

Summing up, orbital stability requires only that the orbits C and C' (closed trajectories) remain near each other; whereas the stability of motion (or of the solution) requires that, in addition, the representative points R and R' (on C and C', respectively) should remain close to each other if they were close to each other initially.

An analogous distinction holds for asymptotic stability and asymptotic orbital stability. Here there is, however, a further distinction. It is clear that a periodic solution cannot be asymptotically stable if other periodic solutions exist arbitrarily near to it, since one periodic function cannot be asymptotic to another one. But if $u_i(t)$ is a periodic solution of an *autonomous* system, $u_i(t + \tau)$ is another such solution for every value of τ, and therefore there are periodic solutions arbitrarily near to $u_i(t)$. Hence it is *impossible* for a periodic solution of an autonomous system to be asymptotically stable. An orbit which is asymptotically (orbitally) stable is already familiar to us as a stable limit cycle.

One can illustrate the preceding definitions by the following example. Assume that we have two identical physical pendulums (with a negligible friction) provided with an optical attachment capable of registering their motion on a moving photographic film; we shall also assume that the two pendulums have the same period (for a given amplitude of oscillation) and that their records can be made to coincide.

If the pendulums are started at $t = t_0$ from equal initial angles, their records will coincide into one single record.

Assume now that we start the pendulums from slightly different angles which will result in two trajectories (orbits in the phase plane), C and C'; for the sake of argument we may consider C as unperturbed orbit and C' as the perturbed one. As in a physical pendulum the period changes with the amplitude, the periods now will be slightly different, and, although C and C' will remain close to each other, the motion of luminous spots (the representative points R and R') will change; at times they will be in phase, at times out of phase, as was analyzed in connection with (2.3).

In our terminology this will mean that, although there will be the orbital stability (since orbits C and C' remain near to each other), there will be no stability of motion (or of the "solution") since the luminous spots do not remain continuously near to each other. It is noted that the orbital stability is not asymptotic in this case as C' does not have a tendency to approach C.

There will be an entirely different situation if physical pendulums in the foregoing example are replaced by clocks. It is recalled (Chapter 3, Section 11) that, owing to the escapement mechanism, a clock operates on a limit cycle (although the latter is not analytical). If again one considers the same experiment with adjustment of both periods to the same value, a perturbation resulting in the orbit C' (for the second clock) will have a tendency to disappear so that $C' \to C$. In view of this there will be asymptotic orbital stability, as well as stability (of motion) since, during the motion, the two luminous spots will remain always in the neighborhood of each other, if they were near to each other initially.

3. Variational equations

We have encountered the questions of stability with regard to singular points and limit cycles (Chapters 1 and 3). For the former, the stability concerns the equilibrium; for the latter, it relates to a stationary motion on a limit cycle. In both cases we used a geometrical argument regarding the asymptotic behavior of the representative point E either for $t \to \infty$ or for $t \to -\infty$.

The analytic approach to the theory of stability develops from the so-called *variational equations* ("equations aux variations" of Poincaré[1]) which play an important role in the sequel.

Consider a dynamical system whose motion is governed by a system of d.e.:

$$\dot{x}_i = X_i(x_1,\ldots, x_n); \qquad i = 1, 2,\ldots, n \tag{3.1}$$

where the X_i are continuous, twice differentiable functions of the variables x_i which may be regarded as generalized coordinates. Suppose that we know a solution of the system (3.1): $x_{i0} = x_{i0}(t)$ which we may consider as a *nonperturbed solution.*

Let $x_i(t)$ be the solution corresponding to an initial value $x_i(t_0) \neq 0$; it will be called a *perturbed solution.* Between the new and the old solutions there exists a relation

$$x_i(t) = x_{i0}(t) + \xi_i(t) \tag{3.2}$$

where the functions $\xi_i(t)$ are called *the perturbations*: we shall assume that $|\xi_i|$ are sufficiently small to be able to neglect their higher powers.

If one inserts (3.2) into (3.1) and develops the functions X_i around the nonperturbed values $x_{i0}(t)$ to the first order in ξ_i one obtains a system of the *variational equations*

$$\dot{\xi}_i = \sum_{j=1}^{n} \left(\frac{\partial X_i}{\partial x_j}\right)_0 \xi_i \tag{3.3}$$

in which the coefficients of ξ_i are partial derivatives of the functions X_i with respect to the variables x_j into which the nonperturbed values x_{i0} have been replaced after the differentiation.

Since x_{i0} is the known solution and x_i is the perturbed solution, an important case arises when all perturbation functions $\xi_i(t) \to 0$ for $t \to \infty$, in which case $x_i(t) \to x_{i0}(t)$, as follows from (3.2). In this case the stability is called *asymptotic stability.* In what follows we shall be concerned with this form of stability unless specified otherwise.

A remark is noteworthy: in the preceding the perturbed solution (or motion) $x_i(t)$ is compared to the neighboring nonperturbed solution $x_{i0}(t)$.

When $x_{i0}(t) = 0$, we have obviously the position of equilibrium. This is particularly important in what follows and is called the *constant* (or *identically zero*) *solution.* In such an instance the variational equations relate to the behavior of trajectories in the neighborhood of singular points with which we were concerned in Chapter 1. In this manner the stability of equilibrium appears as a particular case of stability of a periodic motion when the trajectory of the latter reduces to a point (which may be the origin).

[1] See footnote [1], page 118.

Here $x_i(t) = \xi_i(t)$, we can replace ξ_i in (3.3) by x_i, inasmuch as the perturbed motion and the perturbations become identical in this case.

4. Variational equations of singular points

Before proceeding with the general theory it is useful to consider first a simple system of d.e. (equations (6.2) Chapter 1):

$$\dot{x} = ax + by; \qquad \dot{y} = cx + dy \tag{4.1}$$

This system can be regarded as the variational system corresponding to the constant solution: $x_0 = y_0 = 0$ (the singular point) if one writes it in the form

$$\begin{aligned} \frac{d(x_0 + \delta x)}{dt} &= a(x_0 + \delta x) + b(y_0 + \delta y); \\ \frac{d(y_0 + \delta y)}{dt} &= c(x_0 + \delta x) + d(y_0 + \delta y) \end{aligned} \tag{4.2}$$

then sets $x_0 = y_0 = 0$ and replaces δx and δy by x and y.

In what follows we shall encounter d.e. of the form

$$\dot{x} = P(x,y); \qquad \dot{y} = Q(x,y) \tag{4.3}$$

where P and Q are generally entire series not containing constant terms, and $P(0,0) = Q(0,0) = 0$. Since in such a case there is again an identically zero solution, the first-order variational system is:

$$\frac{d\delta x}{dt} = P_x{}^0\,\delta x + P_y{}^0\delta y; \qquad \frac{d\delta y}{dt} = Q_x{}^0\delta x + Q_y{}^0\delta y \tag{4.4}$$

where $P_x{}^0$, $P_y{}^0$, $Q_x{}^0$, and $Q_y{}^0$ are partial derivatives of P and Q with respect to x and y at the point: $x = y = 0$.

The characteristic equation (see equation (6.8), Chapter 1) can be written then directly in the form

$$S^2 - (P_x{}^0 + Q_y{}^0)S + (P_x{}^0Q_y{}^0 - P_y{}^0Q_x{}^0) = 0 \tag{4.5}$$

It is seen that the existence of an identically zero solution simplifies the problem since, instead of stability of the periodic motion, one can treat the problem in terms of stability of the state of rest.

Another remark: the limitation of the variational system to the first order limits also the validity of conclusions only to a small neighborhood around the point $x = y = 0$; in view of this, very often stability determined on this basis is called the *infinitesimal stability* in contrast to *stability in the large* yielded by the Second Method of Liapounov to be discussed in Chapter 6.

In the following section we will investigate the general problem of variational equations based on a constant solution corresponding to a particular case studied in this section; in Section 6 the same problem is studied in connection with variational equations based on a periodic solution.

It will be observed that the second problem is more difficult than the first; the difficulty is due to the fact that in order to determine the characteristic exponents, one must know the system of solutions, but the knowledge of the latter requires the determination of the characteristic exponents. One finds oneself thus in a vicious circle from which the only issue is a method of approximations outlined later (Chapter 13).

In view of this, various methods were developed to reduce variational systems based on d.e. with periodic coefficients to those based on the d.e. with constant coefficients or, which is the same, on the constant solution. One such method is outlined in Chapter 16 and most of the problems in Part III are treated on the basis of this reduction. The only case in this text in which the variational problem is based directly on the d.e.'s with periodic coefficients is the method of Mandelstam and Papalexi outlined in Section 3, Chapter 19.

5. Variational systems with constant coefficients

As was pointed out in Section 3, the variational equation of an autonomous system based on a constant solution is of the form

$$\dot{x}_i = \sum_{j=1}^{n} a_{ij}x_j; \qquad i = 1,\ldots, n \tag{5.1}$$

where a_{ij} are real constants.

We shall use the shorthand of vector notation and write:

$$x = (x_1,\ldots, x_n); \quad u(t) = (u_1(t),\ldots, u_n(t)); \quad \dot{u}(t) = (\dot{u}_1(t),\ldots, \dot{u}_n(t))$$

and so on. Subscripts will be reserved for components of a vector, different vectors being distinguished by superscripts. Thus we write, for example, $u'(t) = (u_1'(t),\ldots, u_n'(t)$. From the general theory of linear systems of d.e.'s we know that every solution $u(t)$ of (5.1) can be written as a linear combination

$$u(t) = \sum_{j=1}^{n} k_j w^j(t)$$

of n linearly independent solutions: $w^1(t),\ldots, w^n(t)$, where the constants

k_j are determined by the initial conditions. Furthermore, n solutions $w^j(t)$ are linearly independent if and only if the initial vectors $w^j(0)$ are linearly independent. A set of n linearly independent solutions of (5.1) is called a *basis* for the solutions.

To construct a basis for the solutions of (5.1) we first try to find a solution $w(t)$ not identically zero having the property that $\dot{w}(t) = hw(t)$ for some number h. The solution $w(t)$ must therefore be of the form $w_i(t) = c_i e^{ht}$. Substituting into (5.1) yields

$$hc_i e^{ht} = \sum_{j=1}^{n} a_{ij} c_j e^{ht}$$

Hence $h, c_2, \ldots, c_n$ must satisfy $hc_i = \sum_{j=1}^{n} a_{ij} c_j$.

Writing these n equations out in full, we have

$$\begin{aligned} (a_{11} - h)c_1 + a_{12}c_2 + \ldots + a_{1n}c_n &= 0 \\ a_{21}c_1 + (a_{22} - h)c_2 + \ldots + a_{2n}c_n &= 0 \\ \cdot \quad \cdot \quad \cdot \quad \cdot \quad \cdot \quad \cdot \quad \cdot \quad \cdot \quad \cdot \quad \cdot \quad \cdot \quad \cdot & \\ a_{n1}c_1 + a_{n2}c_2 + \ldots + (a_{nn} - h)c_n &= 0 \end{aligned} \tag{5.2}$$

which can be considered a system of n equations for the n unknowns: $c_1, \ldots, c_n$.

The right-hand sides being zero, a non-zero solution: $c_1, \ldots, c_n$ exists if, and only if, the determinant of the coefficients vanishes. Since the c_i may not all vanish, if $w(t)$ is to be part of a basis, we require

$$\begin{vmatrix} a_{11} - h & a_{12} & \cdots & a_{1n} \\ a_{21} & a_{22} - h & \cdots & a_{2n} \\ \cdots & \cdots & \cdots & \cdots \\ a_{n1} & a_{n2} & \cdots & a_{nn} - h \end{vmatrix} = 0 \tag{5.3}$$

This is an algebraic equation of degree n in the unknown h and is called the *characteristic equation* of (5.1). Its solutions $h_1, \ldots, h_n$ are called the *characteristic exponents* of (5.1). Setting h in (5.2) equal to one of the solutions of (5.3), say $h = h_j$, yields a nonvanishing solution $c_j = (c_1{}^j, c_2{}^j, \ldots, c_n{}^j)$ of (5.2). Thus we finally obtain the solution $w^j(t) = (c_1{}^j \exp h_j t, \ldots, c_n{}^j \exp h_j t)$ of (5.1). We now make the simplifying assumption that the *characteristic exponents are distinct*, that is, $h_i \neq h_j$ for $i \neq j$. It is easily shown that the vectors $c^1, c^2, \ldots, c^n$ are linearly independent. Since c^j is the vector of initial values ($t = 0$) of $w^j(t)$, it follows that the n solutions $w^j(t), j = 1, \ldots, n$ form a basis.

If the characteristic exponents are not distinct, a basis will include functions which, instead of being simply exponentials, are products of exponentials and polynomials.

The exponents are again the characteristic exponents and, since the asymptotic behavior of such products is determined by the exponential factor, no new phenomena are introduced. Thus the conclusions we shall draw concerning the asymptotic behavior of the solutions of (4.1) are completely general.

The h_j are, in general, complex numbers. The corresponding c_i^j are therefore also complex, and we obtain a complex basis $w^j(t)$. We merely note this fact, which has no particular significance. A real basis can always be obtained from a complex basis by suitable linear combinations. For our purposes a real basis has no advantages.

Now suppose that all of the characteristic exponents have negative real parts. Then $\exp(h_j t) \rightarrow 0$ as $t \rightarrow \infty$ for all j so that all solutions of (5.1) are asymptotic to zero. Thus:

If all characteristic exponents of (5.1) *have negative real parts, the identically zero solution of* (5.1) *is asymptotically stable.*

If, on the other hand, at least one characteristic exponent has a positive real part, the corresponding basic solution is unbounded as $t \rightarrow \infty$ and we have:

If at least one characteristic exponent of (5.1) *has a positive real part, the identically zero solution is unstable.*

Considering (5.1) as the variational equation of (2.2) based on a constant solution, the above results combined with the theorem stated in section 3 yield:

If the characteristic exponents of the variational equation based on a constant solution $u_i(t) \equiv \xi_i^0$ *of an autonomous system all have negative real parts, the solution* $u_i(t)$ *is asymptotically stable. If at least one characteristic exponent has a positive real part,* $u_i(t)$ *is unstable.*

Since the signs of the real parts of the characteristic exponent are crucial for the stability problem, we mention the well known criterion of Hurwitz, which gives necessary and sufficient conditions that the roots of a real polynomial have negative real parts.

In order that all solutions of the equation:

$$A_0 h^n + A_1 h^{n-1} + \ldots + A_n = 0; \qquad A_i \text{ real}, A_0 > 0$$

have negative real parts, it is necessary and sufficient that

$$A_1 > 0; \qquad \begin{vmatrix} A_1 & A_3 \\ A_0 & A_2 \end{vmatrix} > 0; \qquad \begin{vmatrix} A_1 & A_3 & A_5 \\ A_0 & A_2 & A_4 \\ 0 & A_1 & A_3 \end{vmatrix} > 0; \quad \ldots$$

$$\begin{vmatrix} A_1 & A_3 & A_5 & \ldots & 0 \\ A_0 & A_2 & A_4 & \ldots & 0 \\ 0 & A_1 & A_3 & \ldots & 0 \\ 0 & A_0 & A_2 & \ldots & 0 \\ \cdot & \cdot & \cdot & \cdot & \cdot \\ \cdot & \cdot & \cdot & \cdot & A_n \end{vmatrix} > 0$$

In order to apply this criterion it is of course necessary to expand the determinant.

For a proof of the Hurwitz criterion and for an extensive discussion of other criteria applicable to the characteristic equation, we refer to the English translation of Gantmacher's treatise.

6. Linear systems with periodic coefficients[4]

The variational equation based on a periodic solution of period ω is a linear system with periodic coefficients:

$$\dot{x}_i = \sum_{j=1}^{n} a_{ij}(t)x_j; \qquad i = 1,\ldots, n \tag{6.1}$$

$$a_{ij}(t + \omega) = a_{ij}(t)$$

whereas the problem of constructing a basis for (5.1) was reducible to the problem of solving an algebraic equation; in this case the problem is more difficult, as will be shown.

There is a theorem due to Floquet which asserts that (6.1) has a basis which is like the basis of a system with constant coefficients, except for the presence of periodic factors of period ω. Let $\varphi^i(t) = [\varphi_1{}^i(t),\ldots, \varphi_n{}^i(t)]$; $i = 1,\ldots, n$ be a basis for the solutions of (6.1), and let $\psi^i(t) = \varphi^i(t + \omega)$. It is readily verified that the functions $\psi^i(t)$ are solutions of (6.1), and in fact form a basis, since the vectors $\psi^i(0) = \varphi^i(\omega)$ are linearly independent. Thus $\psi^i(t) = \varphi^i(t + \omega) = \sum_{j=1}^{n} b_j{}^i\varphi^j(t)$ where the

[4] G. Floquet, *Ann. Ec. Norm. Supér.* (2), **12**, 1883.

vectors b^i are linearly independent. We shall use these vectors b^i in the sequel.

To construct the desired basis from the given basis $\varphi^i(t)$, we try to find a solution $\theta(t) = [\theta_1(t), \ldots, \theta_n(t)]$ with the property that $\theta(t + \omega) = \mu\theta(t)$ for some non-zero number μ. Writing

$$\theta(t) = \sum_{j=1}^{n} c_j\varphi^j(t);$$

$$\theta(t + \omega) = \sum_{j=1}^{n} c_j\varphi^j(t + \omega) = \sum_{j=1}^{n} c_j \sum_{k=1}^{n} b_k{}^j\varphi^k(t)$$

we therefore require

$$\sum_{j=1}^{n} \sum_{k=1}^{n} c_j b_k{}^j\varphi^k(t) = \mu \sum_{j=1}^{n} c_j\varphi^j(t) \tag{6.2}$$

or

$$\sum_{j=1}^{n} \sum_{k=1}^{n} (c_k b_j{}^k - \mu c_j)\varphi^j(t) = 0.$$

Since the $\varphi^j(t)$ are linearly independent, the coefficients in (6.2) must vanish, giving rise to the system

$$\begin{array}{c} (b_1{}^1 - \mu)c_1 + b_1{}^2c_2 + \ldots + b_1{}^nc_n = 0 \\ b_2{}^1c_1 + (b_2{}^2 - \mu)c_2 + \ldots + b_2{}^nc_n = 0 \\ \cdots\cdots\cdots\cdots \\ b_n{}^1c_1 + b_n{}^2c_2 + \ldots + (b_n{}^n - \mu)c_n = 0 \end{array} \tag{6.3}$$

The condition for a nonvanishing solution c is

$$\begin{vmatrix} b_1{}^1 - \mu & b_1{}^2 & b_1{}^3 & \ldots & b_1{}^n \\ b_2{}^1 & b_2{}^2 - \mu & b_2{}^3 & \ldots & b_2{}^n \\ \cdot & \cdot & \cdot & \cdot & \cdot \\ b_n{}^1 & b_n{}^2 & b_n{}^3 & \ldots & b_n{}^n - \mu \end{vmatrix} = 0 \tag{6.4}$$

The solutions $\mu_1, \ldots, \mu_n$ of this characteristic equation are called the *characteristic multipliers* of (6.1). It can be shown that they are independent of the basis $\varphi^i(t)$. Since the column vectors b^i are linearly independent, $\mu = 0$ cannot be a solution of (6.4). Substituting a solution of (6.4) into (6.3) and solving for c therefore yield a desired solution $\theta(t)$ of (6.1). We assume again that the roots μ_i of (6.4) are distinct, and thus obtain a basis $\theta^i(t)$ with the property that $\theta^i(t + \omega) = \mu_i\theta^i(t)$. (The remarks of the preceding section apply again in the case of repeated characteristic multipliers.) The characteristic multipliers are in general complex and thus yield a complex basis $\theta^i(t)$. Again the remarks of the preceding section apply. To show that $\theta^i(t)$ is of the form of an exponential times a function of period ω, let $\log \mu_i$ be any determination of the logarithm of μ_i (the principal value, say) and let

$$f^i(t) = \exp\left(-\frac{t}{\omega}\log \mu_i\right)\theta^i(t)$$

then

$$f^i(t+\omega) = \exp\left(-\frac{t}{\omega}\log \mu_i - \log \mu_i\right)\theta^i(t+\omega)$$

$$= \exp\left(-t\frac{\log \mu_i}{\omega}\right)\frac{1}{\mu_i}\mu_i\theta^i(t) = f^i(t)$$

so that $f^i(t)$ is periodic of period ω. But $\theta^i(t) = \exp[(\log \mu_i)t/\omega]f^i(t)$ and therefore Floquet's theorem is proved. The numbers $h_i = \dfrac{1}{\omega}\log \mu_i$ are called the *characteristic exponents* of (6.1).

Owing to the ambiguity of the logarithm, only the real parts of the h_i are uniquely determined.

Since (6.1) has a basis of the form

$$\theta^i(t) = \exp(h_i t)\cdot f^i(t); \qquad i = 1,\ldots, n$$

where the $f^i(t)$ are periodic, we have the result:

If all characteristic exponents of (6.1) *have negative real parts, the identically zero solution is asymptotically stable. If one characteristic exponent has a positive real part, the identically zero solution is unstable.*

Considering the case where (6.1) is a variational equation based on a periodic solution yields the corollary:

If the characteristic exponents of the variational equation based on a periodic solution have negative real parts, the solution is asymptotically stable. If one characteristic exponent has a positive real part, the solution is unstable.

Instead of the signs of the real parts of the characteristic exponents, we can consider the absolute value of the characteristic multipliers, asymptotic stability resulting when the absolute values are less than unity. It is possible to apply Hurwitz's criterion to the solution of (6.4) by making the change of variable $\mu = (1+\eta)/(1-\eta)$ which maps the interior of the unit circle onto the left half of the complex plane, so the condition $|\mu| < 1$ is equivalent to $\operatorname{Re}\eta < 0$.

The principal difficulty in this analysis is the fact that the basis with which we began, and from which we derived the characteristic equation, is in general unknown. A method for determining the characteristic exponents must therefore be sought elsewhere. We shall return to this problem in Chapter 13 of Part II, which deals with methods of approximation.

7. Orbital stability

Let $u_i(t) = u_i(t + \omega)$ be a periodic solution of the autonomous system (2.2). Differentiating the identity $\dot{u}_i(t) = f_i[u_1(t),\ldots, u_n(t)]$ $i = 1,\ldots, n$ with respect to t shows that $\dot{u}_i(t)$ satisfies the variational equation of (2.2) based on $u_i(t)$. Since $\dot{u}_i(t)$ also has period ω, the variational equation has a periodic solution. Hence one of the characteristic exponents is equal to zero, and the theorem on asymptotic stability stated in the preceding section does not apply because the hypothesis is not satisfied.

In fact, as was shown in Section 2, it is impossible for $u_i(t)$ to be asymptotically stable. There is, however, the following result:

If $n - 1$; $(n > 1)$ of the characteristic exponents associated with a periodic solution $u_i(t)$ of an autonomous system have negative real parts, the trajectory of the solution is asymptotically orbitally stable.

This is another way of saying that for asymptotic orbital stability it is sufficient that $(n - 1)$ solutions of a basis for the variational equation tend to zero. For a proof of this theorem see Cesari[5] or Coddington and Levinson.[6] We now derive a relationship among the characteristic exponents which makes it possible to determine one of them by a quadrature when the remaining ones are known.[1] The following derivation is restricted to the case $n = 2$, but the method generalizes in an obvious way to arbitrary n. Let

$$\begin{aligned} \dot{x} &= a_{11}(t)x + a_{12}(t)y \\ \dot{y} &= a_{21}(t)x + a_{22}(t)y \end{aligned} \tag{7.1}$$

be an arbitrary linear system with differentiable but not necessarily periodic coefficients. Let $u^i(t) = (u_1{}^i(t), u_2{}^i(t))$; $i = 1, 2$ be a basis for (7.1). We form the determinant

$$\Delta(t) = \begin{vmatrix} u_1{}^1(t) & u_1{}^2(t) \\ u_2{}^1(t) & u_2{}^2(t) \end{vmatrix}$$

[5] L. Cesari, *Asymptotic Behavior and Stability Problems*, Springer, Berlin, 1959.

[6] E. A. Coddington and N. Levinson, *Theory of Ordinary Differential Equations*, New York, 1955.

[1] See footnote [1], page 118.

which cannot vanish, since its columns are linearly independent. It is readily verified that

$$\frac{d}{dt}\Delta(t) = \begin{vmatrix} \dot{u}_1{}^1 & \dot{u}_1{}^2 \\ u_2{}^1 & u_2{}^2 \end{vmatrix} + \begin{vmatrix} u_1{}^1 & u_1{}^2 \\ \dot{u}_2{}^1 & \dot{u}_2{}^2 \end{vmatrix} =$$

$$= \begin{vmatrix} a_{11}u_1{}^1 + a_{12}u_2{}^1 & a_{11}u_1{}^2 + a_{12}u_2{}^2 \\ u_2{}^1 & u_2{}^2 \end{vmatrix} + \begin{vmatrix} u_1{}^1 & u_1{}^2 \\ a_{21}u_1{}^1 + a_{22}u_2{}^1 & a_{21}u_1{}^2 + a_{22}u_2{}^2 \end{vmatrix}$$

Subtracting a_{12} times the second row of the first determinant from the first row, and a_{21} times the first row of the second determinant from the second row, we obtain

$$\frac{d}{dt}\Delta(t) = \begin{vmatrix} a_{11}u_1{}^1 & a_{11}u_1{}^2 \\ u_2{}^1 & u_2{}^2 \end{vmatrix} + \begin{vmatrix} u_1{}^1 & u_1{}^2 \\ a_{22}u_2{}^1 & a_{22}u_2{}^2 \end{vmatrix} = [a_{11}(t) + a_{22}(t)]\Delta(t)$$

Integrating this differential equation for $\Delta(t)$ gives the formula

$$\Delta(t) = \Delta(t_0) \exp \int_{t_0}^{t} [a_{11}(s) + a_{22}(s)]ds \tag{7.2}$$

Now let the coefficients a_{ij} have period ω and choose a basis $\varphi^i(t) = u^i(t)$ so that $\varphi_j{}^i(0) = \delta_{ij}$; then the elements $b_j{}^i$ of the determinant (6.4) are given by $b_j{}^i = \varphi_j{}^i(\omega)$. From the theory of linear equations we know that $\Delta(\omega) = \mu_1\mu_2$. Setting $t_0 = 0$, $t = \omega$ in (7.2) yields, since $\Delta(0) = 1$:

$$\Delta(\omega) = \exp \int_0^{\omega} [a_{11}(t) + a_{22}(t)]dt$$

and thus

$$\mu_1\mu_2 = \exp \int_0^{\omega} [a_{11}(t) + a_{22}(t)]dt \tag{7.3}$$

or

$$\log \mu_1 + \log \mu_2 = \int_0^{\omega} [a_{11}(t) + a_{22}(t)]dt$$

Since $h_i = (1/\omega) \log \mu_i$ we obtain the desired formula

$$h_1 + h_2 = \frac{1}{\omega} \int_0^{\omega} [a_{11}(t) + a_{22}(t)]dt \tag{7.4}$$

This formula remains true if h_i is replaced by its real parts, since the sum of the imaginary parts is zero in view of (7.4). Now if (7.1) is a variational equation based on a periodic solution of an autonomous system, one of the h_i's has zero real part, and the real part of the remaining h_i's is given by the right side of (7.4).

In the case of arbitrary n, formula (7.2) becomes

$$\Delta(t) = \Delta(t_0) \exp \int_{t_0}^{t} \sum_{i=1}^{n} a_{ii}(s)ds \tag{7.5}$$

and (7.4) becomes

$$\sum_{i=1}^{n} h_i = \frac{1}{\omega} \int_0^{\omega} \left[\sum_{i=1}^{n} a_{ii}(t) \right] dt \tag{7.6}$$

Let us now consider an example, the system (2.1 of Section 2, Chapter 3):

$$\dot{x} = y + \frac{x}{\sqrt{x^2 + y^2}}(1 - x^2 - y^2) = f(x,y)$$

$$\dot{y} = -x + \frac{y}{\sqrt{x^2 + y^2}}(1 - x^2 - y^2) = g(x,y)$$

which has the periodic solution: $x_0 = \cos t$; $y_0 = -\sin t$. Here $a_{11}(t) = f_x(\cos t, -\sin t) = -2\cos^2 t$ and $a_{22}(t) = g_y(\cos t, -\sin t) = -2\sin^2 t$. Thus the unknown characteristic exponent (strictly speaking, its real part) is given by

$$h = \frac{1}{2\pi} \int_0^{2\pi} (-2\cos^2 t - 2\sin^2 t)dt = -2 < 0$$

and therefore the orbit $x^2 + y^2 = 1$ is asymptotically orbitally stable.

Applying the same procedure to the system (2.7) of Chapter 3 shows that both characteristic exponents have zero real parts, and hence the variational equations yield no information concerning the stability of the orbit $x^2 + y^2 = 1$ which we know to be semi-stable.

In conclusion let us consider the applicability of the foregoing theory to orbits of conservative systems. We pointed out in Chapter 2 that conservative systems cannot have singular points which, in our present terminology, are asymptotically stable. It is easily seen that orbits of systems with integrals or integral invariants cannot be asymptotically (orbitally) stable either. An orbit of a system with an integral has at least two characteristic exponents with zero real parts; if the remaining $n-2$ exponents have negative real parts, the orbit is stable. If $n=2$, the existence of an integral already implies stability. We now show that for a system with an integral invariant the sum of the characteristic exponents vanishes, so that the variational equations can be used at best to establish instability. Let $M(x_1,\ldots, x_n)$ be an integral invariant of (2.2). The stability properties of all singularities and orbits of (2.2) are then identical with those of the system

$$\dot{x}_i = M(x_1,\ldots, x_n)f_i(x_1,\ldots, x_n); \qquad i = 1,\ldots, n \tag{7.7}$$

Since M is an integral invariant,

$$\sum_{i=1}^{n} \partial(Mf_i)/\partial x_i = 0 \tag{7.8}$$

therefore, $\sum_{i=1}^{n} h_i = 0$ for any orbit of (7.7).

Hence, if any h has negative real part, there is at least one h_i with positive real part, and we arrive at the peculiar conclusion that the existence of a negative h_i implies orbital instability. Only if all h_i's have zero real parts is there a possibility of orbital stability, but it is not asymptotic.

An example of this situation is given by the d.e. of the harmonic oscillator: $\ddot{x} + x = 0$, which written as an equivalent system is:

$$\dot{x} = y = 0 \cdot x + 1 \cdot y; \qquad \dot{y} = -x = -1 \cdot x + 0 \cdot y$$

from which one sees that $a_{11}(t) = a_{22}(t) \equiv 0$; then by 7.4 one has

$$h_1 + h_2 = 0 \tag{7.8a}$$

On the other hand, Poincaré has shown[1] that for a periodic solution of an autonomous system one characteristic exponent, say h_1, is always zero, so that from (7.8) it follows that the second exponent h_2 also vanishes. Here we have an autonomous system of the second order in which both characteristic exponents are zero: we also know that the trajectories in this case are orbitally stable but not asymptotically.

One verifies easily this circumstance from the variational equations. In fact, consider a solution $x_0(t)$, $y_0(t)$ as the nonperturbed motion. With the perturbation δx and δy the variational system reduces to

$$\frac{d\delta x}{dt} = \delta y; \qquad \frac{d\delta y}{dt} = -\delta x \tag{7.9}$$

Multiplying the first equation by δx, the second by δy, and adding two equations, one obtains (upon integration)

$$(\delta x)^2 + (\delta y)^2 = \text{const} \tag{7.10}$$

which shows that the perturbation $\sqrt{(\delta x)^2 + (\delta y)^2}$ does not die out in the course of time. There is orbital stability in this case but it is not asymptotic. For further generalizations, see Poincaré.[1]

[1] See footnote [1], page 118.

Chapter 6

STABILITY (SECOND METHOD OF LIAPOUNOV)

1. Introductory remarks

In the preceding chapter we have reviewed the fundamentals of the classical theory of stability based on the use of the variational equations. In his classical work on stability, Liapounov[1] developed a different method which he calls the *second method*; occasionally in modern literature it is also called the *direct method*. The essential feature of the second method is that it gives conditions of stability without any necessity for integrating the system of variational equations which, generally, is a difficult (and often impossible) task.

The method is based on properties of definiteness of certain functions associated with the differential system in such a manner that it is possible to ascertain whether the solution remains in a certain region or not. The criteria of stability (or instability) are derived from this property. The fact that in this theory are involved certain *regions* (and not merely positions of equilibria represented by the singular points) makes it particularly well adapted for the investigation of stability in the large. As the literature of the second method is far beyond what can be condensed in one single chapter of this text, we shall be obliged to touch only the most important points, referring to the existing references.[2] In this outline we shall follow mainly Malkin's presentation,[3] abridging it somewhat, particularly as far as certain special cases are concerned.

We begin with some definitions and consider a function: $V(x_1, \ldots, x_n)$ of n variables—*the function of Liapounov*.

[1] A. M. Liapounov, *General Problem of Stability of Motion*, Charkov, 1892; also L. Cesari, *Asymptotic Behavior and Stability Problems*, Springer, Berlin, 1959.

[2] W. Hahn, *Theorie und Anwendungen der Directen Methode von Liapounov*, Springer, Berlin, 1959.

[3] I. G. Malkin, *Theory of Stability of Motion* (in Russian), Moscow, 1952.

1. The function V is called *definite* (positive or negative): $V = V_d$ in a certain domain D: $|x_i| < h$ ($h > 0$ is a constant), if it has in values of one sign and vanishes only for $x_1 = x_2 = \ldots = x_n = 0$. Thus $V = x_1^2 + x_2^2 + x_3^2 = V_d$.

2. The function V is called *semi-definite*: $V = V_c$ (or, of the same sign) if it has the same sign or is zero in D. Thus: $V = (x_1 + x_2)^2 + x_3^2 = V_c$; in fact, this function vanishes for $x_3 = 0$; $x_1 = -x_2$.

3. The function V is called *indefinite* (or of a variable sign): $V = V_v$ if it changes its sign in any D however small; $x_1 = V_v$; $x_1^2 + x_2^2 - x_3^4 = V_v$. We indicate certain properties of the function V:

(a) If $V(x_1, \ldots x_n)$ is a homogeneous function of the mth degree (a form), that is:

$$V(\lambda x_1, \ldots, \lambda x_n) = \lambda^m V(x_1, \ldots, x_n)$$

and, moreover, m is even, the property V_d is preserved for an unlimited D.

(b) Any odd function: $V(-x_1, \ldots, -x_n) = -V(x_1, \ldots, x_n)$ is always indefinite.

(c) Let V be a homogeneous function of degree m and $V > A(x_1^2 + \ldots + x_m^2)^{m/2}$. If W satisfies the relation $|W| < A(x_1^2 + \ldots + x_m^2)^{m/2}$, then

$$V + W = U_d \tag{1.1}$$

If V is negative semi-definite, then, by adding a small negative function W, one has

$$V + W = U < 0 \tag{1.2}$$

(d) The property $V = V_d$ is preserved for

$$V' = V_d + \lambda \bar{V}_d \tag{1.3}$$

where $\bar{V}_d$ is a homogeneous function of the same degree, and λ is a small coefficient.

(e) One can study the character of definiteness of series developments in powers of x_i. Let $V = V_m + V^*$, where V_m is of the lowest degree and V^* is the rest of higher-order terms. If $V_m > 0$, then, for a sufficiently small Euclidean distance r_i, one has also $V > 0$. In fact, one has:

$$V = V_m + V^* = (a_0 x_1^m + a_1 x_1^{m-1} x_2 + \ldots) + (b_0 x^{m+1} + \ldots)$$
$$= r^m Q_m\left(\frac{x_1}{r}, \frac{x_2}{r}, \ldots, \frac{x_n}{r}\right) + r^{m+1} Q^*\left(\frac{x_1}{r}, \ldots, \frac{x_n}{r}\right)$$

The function Q_m is bounded since $\sum_{i=1}^{n} \left(\frac{x_i}{r}\right)^2 = 1$, as is also Q^*. Moreover

Q_m has a minimum $A > 0$; take M as the maximum of Q; $V = r^m(Q_m + rQ^*)$; $V > r^m(A - rM)$; if $r < A/M$, then $V > 0$, which proves the statement. Thus take:

$$V = x^2 + y^2 + xy^2 + y^3 = r^2\left[1 + \left(\frac{x}{r}\frac{y^2}{r^2} + \frac{y^3}{r^3}\right)r\right]$$
$$= r^2(1 + Q^*); \qquad |Q^*| < 2$$

then, if $r < 1/4$, one has $V > r^2/2$.

The preceding definitions and results can be given a geometrical interpretation. Consider, for instance, a positive V_d for $m = 3$; conclusions remain valid for $m > 3$. The equations

$$V_d(x_1, x_2, x_3) = c \tag{1.4}$$

represent a one parameter family of surfaces, c being the parameter; for $c = 0$ one has $x_1 = x_2 = x_3 = 0$, which is the property: $V = V_d$.

Thus for $c = 0$, the surface $V_d = c$ shrinks to one point, the origin. It is easy to see that for a sufficiently small c the surface V_d is closed, containing O in its interior. In fact, there is a neighborhood of O in which $V = V_d$. Take a sphere ε so small that ε is in this neighborhood. V has a positive minimum on ε, say a. Each level surface $V = c$, $c < a$, cannot intersect ε. Therefore $V = c$ is in ε; it must surround O because otherwise one could connect points on ε with O by a continuous curve C. The function V varies continuously on C, but on ε one has $V \geq a$ and in 0, $V = 0$. Hence one must cross the value $0 < c < a$, which is a contradiction.

2. Theorems of Liapounov (stability)

Given a system of d.e.

$$\dot{x}_i = X_i(x_1, \ldots, x_n) \tag{2.1}$$

We shall assume that the X_i guarantee the existence and unicity of the solution (for example, satisfy the Lipschitz condition) and have a singular point at the origin. Consider a function $V_d(x_1, \ldots, x_n)$. We call Eulerian derivative of this function, the expression:

$$\frac{dV}{dt} = \sum_{i=1}^{n} \frac{\partial V}{\partial x_i}\frac{dx_i}{dt} = \sum_{i=1}^{n} \frac{\partial V}{\partial x_i} X_i = W(x_1, \ldots, x_n) \tag{2.2}$$

so that the Eulerian derivative of V at a point $(x_1, \ldots, x_n)$ is just the time derivative of V along the trajectory through $(x_1, \ldots, x_n)$. It is noted that, in view of (2.1), the derivative dV/dt is a function of x_i vanishing for $x_1 = \ldots = x_n = 0$. Note $dv/dt = \text{grad } V \cdot X$.

The first theorem of Liapounov states:

(I) *Given a differential system* (2.1), *with a singular point* $x_1 = x_2 = \ldots = x_n = 0$, *the equilibrum is stable if it is possible to determine such function* $V = V_d$ *in a certain* D *whose Eulerian derivative* W *is either* W_c *of the sign opposite to* V_d *or which vanishes identically in* D.

PROOF: Without any loss of generality, we assume that V_d is positive in D: $|x_i| \leq h$ except at the origin where it is zero. According to the assumption, $dV/dt = W \leq 0$.

Let $\varepsilon < h$ be a positive number as small as we please and let x be the maximum of $|x_1|, \ldots, |x_n|$ so that $x = \varepsilon$ determines a cube in the n-space with side 2ε. Let L be the minimum of V_d on the boundary of the cube so that

$$V \geq L \qquad \text{for } x = \varepsilon \tag{2.3}$$

The quantity $L > 0$ in view of $V = V_d$ as we assumed.

Consider now a solution $x_i(t)$ of the system (2.1) whose initial conditions satisfy $|x_i^0| \leq \eta$ where $0 < \eta < \varepsilon$ and

$$V(x_1^0, \ldots, x_n^0) < L \tag{2.4}$$

This choice of η is always possible since V is continuous and $V(0, \ldots, 0) = 0$. Substituting $x_i(t)$ into V, one obtains (in view of $W \leq 0$) a non-increasing function as long as $x_i(t)$ remains in D. Thus

$$V(x_1(t), \ldots, x_n(t)) \leq V(x_1^0, \ldots, x_n^0) < L \tag{2.5}$$

Hence, for $t > t_0$ one will have

$$|x_i(t)| < \varepsilon \tag{2.6}$$

In fact, these inequalities would cease to be valid only in the case when at least one of the quantities x_i would reach for $t = T$ the boundary $x = \varepsilon$ where one would have

$$V(x_1(T), \ldots, x_n(T)) \geq L$$

This is impossible, however, since, in view of $\varepsilon < h$, the set of points x_i is still in D where the conditions specified by the theorem hold.

A slight modification of the preceding theorem permits establishing the condition of *asymptotic stability*.

(II) *Given a differential system* (2.1) *with the singular point at the origin, the equilibrium is stable asymptotically if it is possible to determine a function* $V = V_d$ *whose Eulerian derivative* $W = W_d$ *is of the sign opposite to that of* V_d.

Let the hypothesis be satisfied in the region $|x_i| \leq h$. Let $0 < \varepsilon < h$.

Since by the preceding theorem $x = 0$ is at least stable, there exists a positive number $\eta(t)$ such that for every solution $x(t)$ the relation $|x_i(t_0)| \leq \eta$ implies $|x_i(t)| < \varepsilon(t)$ for $t \geq t_0$. We now show that for a solution satisfying this relation, $x_i(t) \to 0$ as $t \to \infty$. Clearly, for such a solution $W(x_1(t), \ldots, x_n(t)) = W(t) < 0$ for $t \geq t_0$, unless $x_i(t) \equiv 0$, since $W(t) = 0$ implies $x(t) = 0$, which, in turn, implies $x(t) \equiv 0$. Hence $V(t)$ is monotonically decreasing for $t \geq t_0$, so that $V(t) \to \alpha$ as $t \to \infty$, for some α, with $V(t) > \alpha$ for all $t \geq t_0$. We show that $\alpha = 0$. The contrary assumption, $\alpha > 0$, entails that $\max_i |x_i(t)|$ is bounded away from zero. Hence $W(t)$ must also be bounded away from zero, say, $W(t) < b < 0$ for $t \geq t_0$. Thus $V(t) = V(t_0) + \int_{t_0}^{t} W(\tau)d\tau < V(t_0) + b(t - t_0)$. But this is impossible, since the right side of the inequality is negative for sufficiently large t, while $V(t) > 0$. Hence, $\alpha = 0$, and thus $x(t) \to 0$ as $t \to 0$, which proves the theorem.

Liapounov indicated also criteria of instability analogous to the theorems (I) and (II) concerning stability. We merely state these theorems, omitting their proofs, as they follow a similar argument.

(III) *Given a differential system* (2.1), *the equilibrium is unstable if it is possible to determine a function V whose Eulerian derivative is $W = W_d$ while V assumes in every neighborhood of O values for which $V \cdot W > 0$.*

(IV) *Given* (2.1) *the equilibrium is unstable if there exists a V such that*

$$dV/dt = W = \lambda V + W^*$$

where λ is a positive constant and either (1) W^* is identically zero, or (2) W^* is W_c^* and every neighborhood of the origin contains points at which $V \cdot W^* > 0$.

3. Geometrical interpretation of Liapounov's theorems

Theorems I and II admit a simple geometrical interpretation. Consider, for instance, $n = 3$ and $V(x_1, x_2, x_3) = V_d > 0$, while $dV/dt = W \leq 0$. The equations $V(x_1, x_2, x_3) = c$ determine, as we saw, a family of closed surfaces, at least for sufficiently small values of c; the surface shrinks to one point for $c = 0$. If $c_1 < c_2$, the surface $V = c_1$ is enclosed inside $V = c_2$.

Consider a trajectory S of the differential system issuing at $t = t_0$ from a point close to the origin. This trajectory S for $t > t_0$ will never intersect a surface $V = c$ *from inside to outside*. In fact, in order to have such an intersection at some point, it is necessary that dV/dt be positive. This, however, is impossible in view of the assumption that $dV/dt = W \leq 0$.

Thus, if R (the representative point) on S was inside some surface $V = c$ initially, it will continue to remain inside that surface. On the other hand, for c sufficiently small the closed surface $V = c$ surrounding O is also small, which indicates the stability of the origin.

In the same manner if $dV/dt = W_d < 0$, the point R on S will cross all surfaces $V = c$ from outside to inside, and R will approach the origin asymptotically; this indicates the existence of asymptotic stability.

The theorems of Liapounov regarding stability amount thus to the establishment of conditions under which all trajectories of a differential system cross the Liapounov surfaces V *inwards*. If, by some means, it is possible to demonstrate this circumstance with weaker conditions than those previously set forth, the validity of the above theorems is still maintained. From this point of view, the condition that all surfaces $V = c$ are free from self-intersections is not necessary.

The preceding conditions hold for a sufficiently small c. In some cases the surface $V = c$ ceases to be closed for $c > c_0$. Thus, for instance, the function

$$V = x_1^2 + [x_2^2/(1 + x_2^2)] = c \quad (3.1)$$

determines a family of closed surfaces $V = c$ only if $c \leq 1$. For $c > 1$ the surface consists of two branches[4] not having any common points (Fig. 6.1). In such a case the function (3.1) can still be used provided $c < 1$. The condition that a surface $V = c$ is a closed surface is guaranteed if V ceases to be bounded when $\sum_{i=1}^{n} x_i^2 \to \infty$. This condition means that, for a large positive number N, one can always find a sufficiently large number L such that for $\sum_{i=1}^{n} x_i^2 > L$, the function V will have values $V > N$.

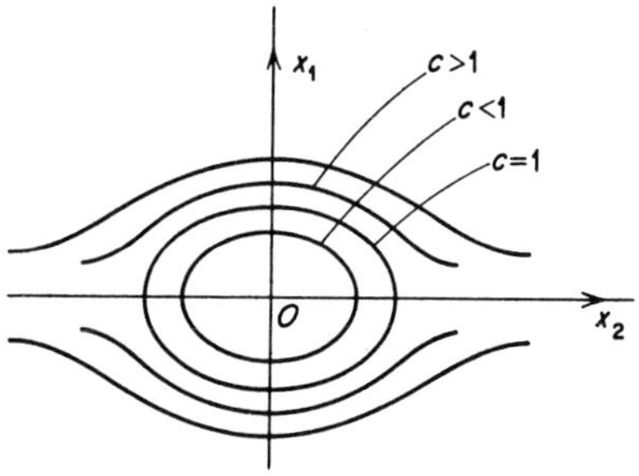

FIGURE 6.1

Liapounov's theorems concerning stability hold always when the function V is not bounded.

In a similar manner one can interpret the theorems on instability. Consider, for example, the case of Theorem III, with $n = 2$ (the planar case) and $W = W_d > 0$. According to the theorem there exists a region near the origin in which $V > 0$. As the curve $V = 0$ separates the region where $V > 0$ from that where $V < 0$ (Fig. 6.2), the shaded area may

[4] M. P. Erougin, *Prikl. Math. i Mehanika* (Russian) **16,** 1952.

represent $V > 0$. A trajectory issuing from some point M of the boundary curve $V = 0$ will necessarily enter into the region $V > 0$ since $(dV/dt) > 0$.

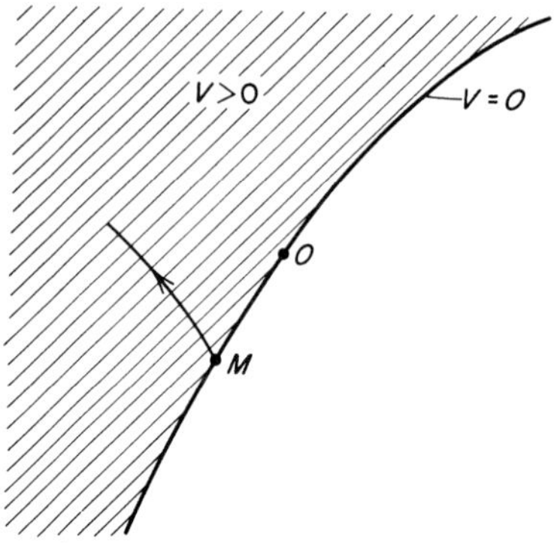

FIGURE 6.2

In fact, if we take a circle around O with radius ε so small that $W > 0$ in this circle, every point M on the line $V = 0$ must move into the region $V > 0$. It cannot reach O without first leaving the circle since, as it moves inside the circle, V increases. It cannot therefore arrive at O where $V = 0$. Hence, O is an unstable point. This permits a completi on of the preceding theorems by a theorem due to Chetaev.[5]

(V) *If for the system* (2.1) *there exists a function V such that every neighborhood of the origin contains a region where $V > 0$ and on whose boundary $V = 0$, and such that at all points of the region $V > 0$, the derivative W is positive, then the origin is unstable.*

4. Certain auxiliary propositions concerning the functions V

We shall narrow the form of the d.e. (2.1) by considering only the linear terms in X_i. The d.e. are thus of the form

$$\dot{x}_i = \sum_{k=1}^{n} p_{ik}x_k \tag{4.1}$$

We shall try to determine a form V, that is, a homogeneous polynomial in the x so as to have

$$\frac{dV}{dt} = \sum_{i=1}^{n} \frac{\partial V}{\partial x_i}(p_{i1}x_1 + \ldots + p_{in}x_n) = W = \lambda V \tag{4.2}$$

where λ is a constant. Assume first that $m = 1$ (m is the degree of V). in which case V is a linear form

$$V = a_1x_1 + \ldots + a_nx_n \tag{4.3}$$

If one substitutes V into (4.2) and equates the coefficients of x_i, one has

$$\begin{aligned} p_{11}a_1 + p_{12}a_2 + \ldots + p_{1n}a_n &= \lambda a_1 \\ \cdot \quad \cdot \quad \cdot \quad \cdot \quad \cdot \quad \cdot &\quad \cdot \quad \cdot \\ p_{n1}a_1 + p_{n2}a_2 + \ldots + p_{nn}a_n &= \lambda a_n \end{aligned} \tag{4.4}$$

[5] N. G. Chetaev, *ibid.* **18**, 1952; **20**, 1955; W. Hahn, *Theorie und Anwendungen der Directen Methode von Liapounov*, Springer, Berlin, 1959.

This system has nontrivial solutions if and only if

$$D(\lambda) = \begin{vmatrix} p_{11} - \lambda & p_{12} & \cdots & p_{1n} \\ \cdot & \cdot & \cdot & \cdot \\ p_{n1} & p_{n2} & \cdots & p_{nn} - \lambda \end{vmatrix} = 0 \tag{4.5}$$

Thus (4.2) can be satisfied by a form V if λ is a characteristic root of (4.1). To each root corresponds a form and, if there are n distinct roots, there are n linear forms (4.3) satisfying (4.2).

In other words, if one has solutions $x_i(t)$ of (4.1), it is possible to find n combinations with constant coefficients $V_i(t) = \sum_{k=1}^{n} a_{ik}x_k(t)$ such that $dV_i/dt = \lambda_i V_i$; that is, $V_i(t) = C_i e^{\lambda_i t}$. Hence expressing x_i as a linear combination of V_i, we find, in fact, $x_i = \sum C_i e^{\lambda_i t}$.

Suppose now that m, the degree of V, is greater than 1. If N is the number of terms in the mth degree, this number is obviously equal to the number of different systems that can be formed by nonnegative integers $m_1, m_2, \ldots, m_n$ under the condition $m_1 + m_2 + \ldots + m_n = m$ and is therefore $N = [n(n + 1) \ldots (n + m - 1)]/m!$.

If all terms of V are arranged in a certain order with coefficients $a_1, a_2, \ldots, a_N$, then, by equating the terms on the right and on the left of (4.2) once V has been substituted, one has a system

$$\sum_{j=1}^{N} A_{ij}a_j = \lambda a_i; \qquad i = 1, 2, \ldots, N \tag{4.6}$$

where A_{ij} are certain constants formed by linear combinations of p_{ij}. This system, again, can have solutions other than the trivial ones if, and only if,

$$D_m(\lambda) = \begin{vmatrix} A_{11} - \lambda & A_{12} & \cdots & A_{1N} \\ \cdot & \cdot & \cdot & \cdot \\ A_{N1} & A_{N2} & \cdots & A_{NN} - \lambda \end{vmatrix} = 0 \tag{4.7}$$

Thus a possibility of satisfying (4.2) by a form of the mth degree depends on the condition that λ be a root of an equation of the Nth degree. Between the roots of (4.7) and those of the characteristic equation of (4.1), that is,

$$D(\lambda) = \begin{vmatrix} p_{11} - \lambda & p_{12} & \cdots & p_{1n} \\ \cdot & \cdot & \cdot & \cdot \\ p_{n1} & p_{n2} & \cdots & p_{nn} - \lambda \end{vmatrix} = 0 \tag{4.8}$$

there exists a simple relation as was pointed out by Liapounov, viz.:

$$\lambda = m_1\lambda_1 + m_2\lambda_2 + \ldots + m_n\lambda_n \qquad (4.9)$$

where $\lambda_1, \lambda_2, \ldots, \lambda_n$ are the roots of (4.8) and $m_1, \ldots, m_n$ are nonnegative integers satisfying the relation

$$m_1 + m_2 + \ldots + m_n = m \qquad (4.10)$$

PROOF: It was shown that to any λ_1 of (4.8) corresponds at least one linear form satisfying (4.2) so that

$$\frac{dV_i}{dt} = \sum_{j=1}^{n} \frac{\partial V_i}{\partial x_j}(p_{j1}x_1 + \ldots + p_{jn}x_n) = \lambda_1 V_1; \qquad i = 1, 2, \ldots, n \quad (4.11)$$

Consider a form of the degree m, viz.:

$$V = V_1^{m_1}, V_2^{m_2}, \ldots, V_n^{m_n} \qquad (4.12)$$

The Eulerian derivative of this form is

$$\frac{dV^{\cdot}}{dt} = m_1 V_1^{m_1-1}, V_2^{m_2}, \ldots, V_n^{m_n}\frac{dV_1}{dt} + \ldots + m_n V_1^{m_1}, V_2^{m_2}, \ldots, V_n^{m_n-1}\frac{dV_n}{dt}$$
$$= (m_1\lambda_1 + m_2\lambda_2 + \ldots + m_n\lambda_n)V \qquad (4.13)$$

which shows that the form of the mth degree satisfies (4.2) with λ given by (4.9). This theorem extends also to cases of multiple roots, but we shall omit these details and pass directly to the following important theorem.

If the roots λ_i of the characteristic equations are such that (4.9) does not vanish for any nonnegative integers m_i related by (4.10), then for any form $U(x_1, \ldots, x_n)$ of the mth degree, there exists one and only one function $V(x_1, \ldots, x_n)$ of the same degree satisfying the equation

$$\frac{dV}{dt} = \sum_{i=1}^{n} (p_{i1}x_1 + \ldots + p_{in}x_n)\frac{\partial V}{\partial x_i} = U \qquad (4.14)$$

PROOF: If one designates by $a_1, \ldots, a_n$ coefficients of V and by $b_1, \ldots, b_n$ those of U, then by equating in (4.14) the coefficients of like powers of x_i, one obtains for a_i equations differing from (4.6) only in the right-hand terms, which are here b_i instead of λa_i.

As the determinant of this system is $\neq 0$ (as λ is assumed to be any root of (4.7) $\neq 0$), this shows that the system

$$A_{i1}a_1 + A_{i2}a_2 + \ldots + A_{iN}a_N = b_N; \qquad i = 1, 2, \ldots, N \quad (4.15)$$

has one and only one solution for a_i. This means that there is only one form V satisfying (4.14).

5. Construction of the function V for a differential system with constant coefficients

We consider again a system

$$\dot{x}_i = \sum_{j=1}^{n} p_{ij} x_j \tag{5.1}$$

with constant coefficients p_{ij} and assume that the solution $x_i = 0$ is asymptotically stable. This, as we saw in Chapter 5, requires that the real parts of all roots of the characteristic equation be negative.

In this case it is possible to invert the first theorem of Liapounov:

If the real parts of the roots of the characteristic equation are negative, there exists one and only one function $V(x_1, \ldots, x_n)$ for any given $U(x_1, \ldots, x_n)$; this function $U = U_d$ satisfies the equation

$$\frac{dV}{dt} = \sum_{i=1}^{n} \frac{\partial V}{\partial x_i} (p_{i1} x_1 + \ldots + p_{in} x_n) = U \tag{5.2}$$

and, moreover, $V = V_d$ of the sign opposite to that of U.

PROOF: As $\mathrm{Re}\,(\lambda_i) < 0$, $\mathrm{Re}\left(\sum_{i=1}^{n} m_i \lambda_i\right) < 0$ and, hence $\sum_{i=1}^{n} m_i \lambda_i \neq 0$ for all nonnegative integers m_i. Hence, there is one and only one V satisfying (5.2).

It remains to be shown that, if $U = U_d$, $V = V_d$ and U_d and V_d are of opposite signs.

Given $U_d < 0$, three cases are possible: (1) V is somewhere negative; (2) $V_c > 0$; (3) $V_d > 0$. In the first case there would be instability in view of Theorem III. This is impossible since $\mathrm{Re}\,(\lambda_i) < 0$. The second case is also impossible; for suppose that at some point $V(x_1^{(0)}, \ldots, x_n^{(0)}) = 0$. Take a trajectory through this point such that $x_i(0) = x_i^{(0)}$. Then $V(t) = V[x_i(t)]$ must decrease with t since its Eulerian derivative is negative. We are thus back in the case (1) which was excluded. Hence, there remains only the possibility (3): $V = V_d > 0$.

Summing up, if $\mathrm{Re}\,(\lambda_i) < 0$ for all roots of the characteristic equation and one selects a form $U = U_d$ of an arbitrary degree, then one can determine a definite form $V = V_d$ of the same degree whose Eulerian derivative dV/dt is equal to U and which is of opposite sign.

This leads to a construction of the V—the forms of Liapounov. One starts with an arbitrary U_d and determines the corresponding V by linear algebra. If $V = V_d$ is of opposite sign, the stability of the system follows from Theorem I (Section 2). If $V \neq V_d$ is of opposite sign, there is instability. It is noted that the asymptotic stability is not covered.

The calculations are simpler if the degree m of V is smaller. It is preferable to take for V a quadratic form, for instance $\sum_{i=1}^{n} x_i^2$.

The criteria of instability are proved in a similar manner (we omit their proofs); they state:

(I) *If, among the roots of the characteristic equation there is at least one* λ *with* Re $(\lambda) > 0$ *and the conditions* (4.10) *and* (4.9) *hold, then for any* $U = U_d$ *there is one and only one* V *satisfying* (5.2), *but in this case* $V \neq V_c$ *or* $V \neq V_d$ *of the sign opposite to* V.

(II) *If among the roots of the characteristic equation there is at least one* λ *with* Re $(\lambda) > 0$, *then for any* $U = U_d$, *there will be always a* V *of the same degree and of such a number* $\alpha > 0$ *that*

$$\frac{dV}{dt} = \alpha V + U$$

but in this case $V \neq V_c$ *of the sign opposite to* U.

6. Stability on the basis of the abridged equations

Consider the nonlinear autonomous system

$$\dot{x}_i = \sum_{j=1}^{n} p_{ij}x_j + X_i(x_1, \ldots, x_n); \qquad i = 1, 2, \ldots, n \tag{6.1}$$

where the p_{ij} are constants and the $X_i(x_1, \ldots, x_n)$ are power series beginning with terms of at least second degree.

The theorems of the preceding section permit establishing criteria of stability on the basis of the variational equation

$$\dot{x}_i = \sum_{j=1}^{n} p_{ij}x_j \tag{6.2}$$

by the following theorems of Liapounov.

(1) *If all roots* λ_i *of the characteristic equation of* (4.8) *have* Re $(\lambda_i) < 0$, *the point of equilibrium* $x_i = 0$ *is asymptotically stable for* (6.1) *whatever the terms* X_i.

In fact consider a quadratic form $V(x_1, \ldots, x_n)$ defined by the equation

$$\sum_{i=1}^{n} \frac{\partial V}{\partial x_i}(p_{i1}x_1 + \ldots + p_{in}x_n) = -(x_1^2 + x_2^2 + \ldots + x_n^2) = U < 0 \tag{6.3}$$

On the basis of the preceding theorem, V exists and is $V_d > 0$. The Eulerian derivative is

$$\frac{dV}{dt} = \sum_{i=1}^{n} \frac{\partial V}{\partial x_i} (p_{i1}x_1 + \ldots + p_{in}x_n + X_i)$$

$$= -(x_1^2 + \ldots + x_n^2) + \sum_{i=1}^{n} X_i \frac{\partial V}{\partial x_i} \qquad (6.4)$$

As the expansion of $\sum X_i(\partial V/\partial x_i)$ begins at least with the terms of the third degree, dV/dt will be definite negative whatever the X_i so that V satisfies the criterion of asymptotic stability with respect to (6.1).

(2) *If among the roots of the characteristic equation* (4.8) *there is at least one* λ *with* Re $(\lambda) > 0$, *the point of equilibrium is unstable whatever the* X_i.

In fact, consider a quadratic form $V(x_1, \ldots, x_n)$ determined by the equation

$$\sum_{i=1}^{n} \frac{\partial V}{\partial x_i} (p_{i1}x_1 + \ldots + p_{in}x_n) = \alpha V + (x_1^2 + \ldots + x_n^2) \qquad (6.5)$$

where $\alpha > 0$. Such a form exists and by Theorem II of Section 5 is $\neq V_c$ with negative sign. Its Eulerian derivative is

$$\frac{dV}{dt} = \alpha V + W(x_1, \ldots, x_n)$$

$$= \alpha V + (x_1^2 + x_2^2 + \ldots + x_n^2) + \sum_{i=1}^{n} X_i \frac{\partial V}{\partial x_i} \qquad (6.6)$$

and $W = W_d$ for any choice of X_i. The form V thus satisfies the conditions of Theorem IV and the equilibrium is unstable.

(3) *If the characteristic equation of the variational equation does not have any roots with positive real parts, but has some roots with zero real parts, then the terms in* X_i *can be chosen as to have either stability or instability.*

The last-mentioned case belongs to the so-called critical cases, which require a special investigation.

7. Certain generalizations

The assumption was made that the perturbing terms X_i are power series which begin with terms of the second order at the origin. One could make weaker assumptions and still have the same conclusions as before.

In fact, we assumed that

$$-(x_1^2 + \dots + x_n^2) + \sum_{i=1}^{n} X_i \frac{\partial V}{\partial x_i} \tag{7.1}$$

is a definite negative expression.

One could assume also that

$$\sum_{i=1}^{n} \left| X_i \frac{\partial V}{\partial x} \right| < A(|x_1| + \dots + |x_n|)^2 \tag{7.2}$$

where A is a sufficiently small positive constant; in fact, it is not necessary at this stage to assume the analyticity of the X_i. Since $\partial V/\partial x_i$ are linear (V is assumed to be quadratic), it is sufficient that X_i satisfy the inequality

$$\sum_{i=1}^{n} |X_i(x_1, \dots, x_n)| < \alpha(|x_1| + \dots + |x_n|) \tag{7.3}$$

where the small constant α depends only on the coefficients of V. It is sufficient, therefore, that X_i be such that (7.2) be fulfilled.

If the X_i are analytic, it is always possible to determine the region, provided that (7.2) is fulfilled.

The problem reduces to the determination of α which, in turn, requires the determination of A, inasmuch as the coefficients of V are known. The only requirement is that the form (7.1) be definite negative. Thus A is to fulfill the inequality

$$A(|x_1| + \dots + |x_n|)^2 < x_1^2 + \dots + x_n^2 \tag{7.4}$$

As both sides of this inequality are quadratic forms, it is fulfilled if one can show that it is fulfilled on a unit sphere $x_1^2 + \dots + x_n^2 = 1$. This gives

$$A < \frac{1}{n^2} < \frac{1}{n} \tag{7.5}$$

where n is the maximum of $(|x_1| + \dots + |x_n|)^2$ on the sphere.

One could also require, for instance, that V satisfy the relation

$$\sum_{i=1}^{n} (p_{i1}x_1 + \dots + p_{in}x_n) \frac{\partial V}{\partial x_i} = U(x_1, \dots, x_n)$$

where U is some negative definite form. Then there is a requirement:

$$A(|x_1| + \dots + |x_n|)^2 < -U(x_1, \dots, x_n) \tag{7.5a}$$

and the condition on A is

$$A < \frac{m}{n^2} \tag{7.6}$$

where m is a minimum of $-U$ on the unit sphere.

Thus, in any case, A depends on the coefficients of the form $-U$. One can determine α directly from the condition that a condition paralleling (7.2) be fulfilled, that is,

$$\sum_{i=1}^{n} X_i \frac{\partial V}{\partial x_i} < -U(x_1, \ldots, x_n) \tag{7.7}$$

In such a case (7.5a) is not necessary. The condition (7.7) is thus a sufficient one.

In general, problems present themselves in three different manners:

(1) Equations of the first approximation are given (with Re $\lambda_i < 0$ for the characteristic exponents); one knows also the forms X_i; the problem consists in determining the region of stability.

(2) Equations of the first approximation (with Re $\lambda_i < 0$) as well as the desired region of stability are specified; the problem is to determine X_i which are allowed under these conditions.

(3) The regions of stability as well as X_i are specified; the problem is to determine the coefficients of linear terms so as to fulfill the prescribed conditions.

8. Aiserman's problem

In investigation of stability of nonlinear control systems (see Section 13 below) one encounters often the case when the nonlinear terms occur only in one equation of the system which thus appears in the form:

$$\dot{x}_1 = \sum_{j=1}^{n} a_{1j}x_j + f(x_k); \qquad \dot{x}_i = \sum_{j=1}^{n} a_{ij}x_j \tag{8.1}$$

where $f(x_k)$ is a nonlinear function.

The following argument due to Aiserman[6] permits analyzing such a case in a simple manner.

Consider the linear system

$$\dot{x}_1 = \sum_{j=1}^{n} a_{1j}x_j + ax_k; \qquad \dot{x}_i = \sum_{j=1}^{n} a_{ij}x_j \tag{8.2}$$

corresponding to (8.1) in which $f(x_k)$ is replaced by the linear term ax_k and assume that the system (8.2) for $a = 0$, namely,

$$\dot{x}_i = \sum_{j=1}^{n} a_{ij}x_j \tag{8.2a}$$

[6] M. A. Aiserman, *Lectures on Theory of Automatic Regulation* (in Russian), Moscow, 1958.

is asymptotically stable (that is, all roots of its characteristic equation have negative real parts).

Using the Hurwitz criterion it is possible then to find an interval (a^*,a^{**}) such that (8.2) is asymptotically stable for $a^* < a < a^{**}$. If a_1 and a_2 are in the interval (a^*,a^{**}), it seems reasonable to assume that, if the linear system (8.2) is stable when a is replaced by a_1 and a_2, the nonlinear system (8.1) will be also stable.

This heuristic assumption has been proved for $n = 2$; for larger n it has not been proved so far, but on physical grounds it seems plausible to expect that its application is more general.

For further details of this question we refer to Aiserman[6] and indicate only the application of this procedure to a control system specified by the d.e.

$$\begin{aligned} \dot{x}_1 &= p_{11}x_1 + \ldots + p_{1n}x_n + f(x_k); \qquad k \leq n \\ \dot{x}_i &= p_{i1}x_1 + \ldots + p_{in}x_n; \qquad i = 2,\ldots, n \end{aligned} \tag{8.3}$$

where the p are constants and $f(x_k)$ is a nonlinear function of one of x_i. By conditions of the problem $f(x_k)$ must be between two limiting straight lines $f_1(x_k) = (a_0 - a_1)x_k$ and $f_2(x_k) = (a_0 + a_2)x_k$ for any x_k where a_1 and a_2 are constants. It is assumed that the linearized problem in which $f(x_k)$ is replaced by a_0x_k has negative characteristic exponents.

The problem consists in determining the numbers a_1 and a_2 for which the state of rest: $x_1 = \ldots = x_n = 0$ is asymptotically stable for any initial conditions; this amounts to the determination of α in (7.3). The problem is solved as follows:

Let $V(x_1,\ldots, x_n) = V_d$ whose Eulerian derivative is given by a prescribed quadratic form $U(x_1,\ldots, x_n) = U_d < 0$, so that

$$\sum_{i=1}^{n} \frac{\partial V}{\partial x_i}(p_{i1}x_1 + \ldots + p_{in}x_n) + a_0x_k\frac{\partial V}{\partial x_1} = U(x_1,\ldots, x_n) \tag{8.4}$$

The quadratic form

$$\begin{aligned} U + ax_k\frac{\partial V}{\partial x_1} &= \sum_{i=1}^{n} \frac{\partial V}{\partial x_1}(p_{i1}x_1 + \ldots + p_{in}x_n) + (a_0 + a)x_k\frac{\partial V}{\partial x_1} \\ &= U_d < 0 \end{aligned} \tag{8.5}$$

if $|a|$ is sufficiently small. Let $-a_1$ and a_2 be the lower and upper limits of a for which the form continues to be $U_d < 0$, which can be determined by any criterion of definiteness of quadratic forms. Thus, if $U_d < 0$ when a is in the interval $-a_1 \leq a \leq a_2$, then the function is

$$\sum_{i=1}^{n} \frac{\partial V}{\partial x_i}(p_{i1}x_1 + \ldots + p_{in}x_n) + f(x_k)\frac{\partial V}{\partial x_1} = U_d^* < 0$$

[6] See footnote [6], page 147.

by the definition of $f(x_k)$. Therefore, for all a in: $-a_1 \leq a \leq a_2$, $f(x_k) = (a_0 + a)x_k$ will make the origin asymptotically stable.

This argument was used by Aiserman in connection with the speed control of a motor. Let x be the departure of speed from its set value; z, the displacement of a control member; s, the displacement of the control element of the servomotor; and y, the displacement of the servomotor. We count $z > 0$ if it corresponds to $\Delta x > 0$; $s > 0$ if it corresponds to $\Delta z > 0$; $y > 0$ if it corresponds to $\Delta s > 0$; likewise $x > 0$ if it corresponds to the increase of speed of the motor.

Disregarding the details of the scheme, we mention only that for the linearized performance the following d.e. exist:

$$\begin{aligned} &\text{Equation of the controlled object: } \dot{x} = -Nax - by \\ &\text{Equation of the servomotor: } \dot{y} = c_3 s \\ &\text{Equation of the control element: } s = c_2 z - d_1 y \\ &\text{Equation of the control member: } z = c_1 x \end{aligned} \tag{8.6}$$

where a, b, c_1, c_2, c_3, and d_1 are positive constants. As to N, it is $+1$ if, in the absence of the control scheme, the speed of the motor is stable; $N = -1$ in the contrary case and $N = 0$ if the speed of the motor does not have any tendency to settle on a definite value.

We consider the case when this natural tendency of the motor to "settle" on a certain speed is a nonlinear function $f(x)$.

One has to replace the linearized term $-Nax$ by $f(x)$. Eliminating s and z one has a system

$$\dot{x} = f(x) - by; \qquad \dot{y} = cx - dy; \qquad c = c_1 c_2 c_3; \qquad d = d_1 c_3 \tag{8.7}$$

We assume $N = -1$. In such a case $f(0) = 0$ has a positive derivative $f'(0) = a_0$.

The system of the first approximation is

$$\dot{x} = a_0 x - by; \qquad \dot{y} = cx - dy$$

The requirement that the roots have negative real parts is here

$$(d - a_0) > 0; \qquad bc - a_0 d > 0 \tag{8.8}$$

we assume that these conditions are fulfilled.

We set:

$$2U = -M(x^2 + y^2); \qquad 2V = Ax^2 + 2Bxy + Cy^2$$

M being a positive constant and determine A, B, and C so as to satisfy the equation

$$\frac{\partial V}{\partial x}(a_0x - by) + \frac{\partial V}{\partial y}(cx - dy) = M(x^2 + y^2) \tag{8.9}$$

Equating the coefficients of the like terms, one gets

$$a_0A + cB = -M; \qquad bB + dC = M; \qquad -bA + (a_0 - d)B + cC = 0 \tag{8.10}$$

Hence

$$\begin{aligned} \Delta A &= M[d(d - a_0) + c(b + c)] \\ \Delta B &= -M(a_0c + bd) \\ \Delta C &= M[b(b + c) - a_0(d - a_0)] \end{aligned} \tag{8.11}$$

where $\Delta = (bc - a_0d)(d - a_0) > 0$ in view of (8.8).

Consider now the form

$$-M(x^2 + y^2) + ax\frac{\partial V}{\partial x} = (-M + aA)x^2 + aBxy - My^2$$

This form is negative definite if

$$B^2a^2 + 4M(-M + aA) < 0$$

This condition is fulfilled if a is in the interval: $-a_1 < a < a_2$, where

$$a_2 = \frac{2M}{B^2}(-A + \sqrt{A^2 + B^2}); \qquad -a_1 = \frac{2M}{B^2}(-A - \sqrt{A^2 + B^2})$$

In view of (8.9), M cancels in these expressions and the roots a_1 and a_2 are

$$a_{1,2} = \frac{2}{B^2}(\pm A^* + \sqrt{A^{*2} + B^{*2}}) \tag{8.12}$$

$$A^* = d(d - a_0) + c(b + c); \qquad B^* = -(a_0c + bd)$$

Thus, if for all values of x, the curve $f = f(x)$ is between the limits $f_1 = (a_0 - a_1)x$; $f_2 = (a_0 + a_2)x$, where a_1 and a_2 are given by (8.12), the equilibrium of the control system is asymptotically stable for any initial conditions.

9. Critical cases

Liapounov investigated also two following *critical cases*; other critical cases were studied later.[1,2]

[1,2] See footnotes [1,2], page 134.

(1) The characteristic equation has one zero root while others have negative real parts.

(2) The characteristic equation has a pair of purely imaginary roots, all others have negative real parts.

In order to give an idea of the difficulty of the problem we outline only case (1). Consider a system of d.e.

$$\dot{y}_i = q_{i1}y_1 + \ldots + q_{i,\,n+1}y_{n+1} + Y_i(y_1, \ldots, y_{n+1}) \tag{9.1}$$

where q_{ij} are constants, and Y_i are analytic in y_i beginning with terms of the second order.

Assume that the characteristic equation of the abridged system ($Y_i = 0$) has one zero root, the others having negative real parts.

We replace one of the y_i by x defined by

$$x = \sum_{i=1}^{n+1} a_i y_i, \qquad a_i \text{ being constants}$$

and shall try to determine a_i so as to have $\dot{x} = 0$. We have an identity

$$\dot{x} = \sum_{i=1}^{n} a_i \dot{y}_i = \sum_{i=1}^{n} a_i(q_{i1}y_1 + \ldots + q_{i,\,n+1}y_{n+1}) = 0 \tag{9.2}$$

Equating to zero the coefficients of y_i, one has a system

$$q_{ik}a_1 + \ldots + q_{n+1,\,k}a_{n+1} = 0; \qquad k = 1, 2, \ldots, n+1 \tag{9.3}$$

As the characteristic equation of (9.1) has one zero root, the determinant of (9.3) vanishes, which means that we can find a_i not all zero. Assume $a_{n+1} \neq 0$. One can take then y_{n+1} as the x just mentioned; we may call x the *critical variable*; all other y_i remain the same, but we shall designate them also as x_i.

This amounts to transforming the abridged system

$$\dot{y}_1 = q_{i1}y_1 + \ldots + q_{i,\,n+1}y_{n+1} \tag{9.4}$$

by means of the transformation

$$\begin{aligned} x &= a_1y_1 + \ldots + a_ny_n + a_{n+1}y_{n+1} \\ x_i &= y_i; \qquad i = 1, 2, \ldots, n \end{aligned} \tag{9.5}$$

which reduces (9.4) to the form

$$\dot{x} = 0; \qquad \dot{x}_i = p_{i1}x_1 + \ldots + p_{in}x_n + p_ix \tag{9.6}$$

For this system the characteristic equation is

$$D(\lambda) = \begin{vmatrix} p_{11} - \lambda & p_{12} & \cdots & p_{1n} & p_1 \\ p_{21} & p_{22} - \lambda & \cdots & p_{2n} & p_2 \\ p_{n1} & p_{n2} & \cdots & p_{nn} - \lambda & p_n \\ 0 & 0 & & 0 & -\lambda \end{vmatrix} = 0 \qquad (9.7)$$

which splits into $\lambda = 0$ and $D_1(\lambda) = 0$, where $D_1(\lambda)$ is shown by a broken line. As the characteristic equation is invariant with respect to linear transformations, all roots of $D_1(\lambda) = 0$ have negative real parts. This transformation reduces (9.1) to a system of the form

$$\begin{aligned} \dot{x} &= X(x,x_1,\ldots, x_n) \\ \dot{x}_i &= p_{i1}x_1 + \ldots + p_{in}x_n + p_i x + X_i(x_3x_1,\ldots, x_n) \end{aligned} \qquad (9.8)$$

Here again, the first equation $\dot{x} = X$ is the *critical* one. We consider first the case: $n = 0$, in which case (9.8) yields

$$\dot{x} = X(x) = gx^m + g_{m+1}x^{m+1} \qquad (9.9)$$

where $m \geq 2$, and g_1 and g_{m+1} are constants. We consider thus only one *critical* equation. If m is even, the motion is clearly unstable; if m is odd, then for $g < 0$, it is asymptotically stable; if $g > 0$, it is unstable.

In fact, if m is even, the right-hand side of (9.8) keeps the same sign, at least in a small neighborhood around the origin, which means that on both sides of the origin the direction of motion is the same; thus, if to the right of O, the motion approaches O, then on the left it will, on the contrary, move away from O, so that the motion is unstable.

If m is odd, the velocity changes when the origin is crossed; for $g > 0$, the motion is always *away* from O; if $g < 0$ it is *toward* O, which proves the statement.

In this case (m odd) by taking the function $V = \frac{1}{2}gx^2$, for dV/dt one has the Eulerian derivative

$$\frac{dV}{dt} = g^2x^{m+1} + gg_{m+1}x^{m+2} + \ldots = W \qquad (9.10)$$

Both functions are $V = V_d$ and $W = W_d$. Moreover, if $g > 0$, V and W are of the same sign; hence one has instability. If $g < 0$, the signs of V and W are opposite and by Theorem I one concludes that the motion is asymptotically stable.

If m is even, we set $V = x$; then dV/dt is definite; as to V, it is of the same sign as dV/dt for $g \gtrless 0$. Thus, according to Theorem III, the motion is unstable.

If $n > 0$ (that is, one has a system (9.8) the investigation of stability is far more complicated, and we refer to Malkin's text,[3] merely mentioning the principal conclusions.

In the first place it is assumed that the right-hand sides of (9.8) are subjected to some further limitations, viz.: denoting by $X^{(0)}(x)$ and $X_i^{(0)}(x)$ terms of X and X_i not containing $x_1, \ldots, x_n$ so that

$$\begin{aligned} X^{(0)}(x) &= X(x, 0, \ldots, 0) = gx^m + g^{(m+1)}x^{m+1} \\ X_i^{(0)}(x) &= X_i(x, 0, \ldots, 0) = g_i x^{m_i} + g^{(m_i+1)}x^{m_i+1} + \ldots \end{aligned} \tag{9.11}$$

where all g are constants. It is assumed that the $X^{(0)}$ are different from zero; $m_i \geq m$; all p_i are zero. In this case the conclusion is: the equilibrium is unstable if m is even; if m is odd, it is stable asymptotically for $g < 0$ and unstable for $g > 0$. In other words, the problem of stability is the same as in the case of one single equation

$$\dot{x} = X^{(0)}(x) = gx^m + \ldots$$

This amounts to the following procedure: one can neglect all noncritical equations and in the critical one neglect all terms not containing the critical variable, which thus reduces the problem to one equation with one unknown function.

In the general case when the right-hand sides of (9.8) are not subject to any additional restriction, the procedure is:

(1) The right-hand sides of noncritical equations are equated to zero, which permits solving them with respect to x_i by means of equations

$$\sum_{j=1}^{n} p_{ij}x_j + p_i x = 0; \qquad i = 1, 2, n \tag{9.12}$$

(2) The variables x_i are replaced by the functions of x (the "critical variable") in the right-hand side of the critical equation. If the result is not zero identically, the stability is obtained from one single equation

$$\dot{x} = X[x, u_1(x), \ldots, u_n(x)] \tag{9.13}$$

In such a case it is sufficient to consider only the lowest degree term in (9.10). If this form is gx^m, we have the previous condition: for $g < 0$, the nonperturbed motion is stable (asymptotically); it is unstable otherwise.

The proof of these propositions is ultimately based on a somewhat delicate analysis of functions V and W and their transitions (V_d, V_c, V_v; W_d, W_c) so as to secure the fulfillment of Liapounov's theorem under the effect of the various terms in the d.e.

[3] See footnote [3], page 134.

The applications of these studies are frequent in the analysis of stability of nonlinear control systems.

We merely indicate examples of the form [7]

$$\dot{x} = ax^2 + bxy + cy^2 = X(x,y); \qquad \dot{y} = -y + kx + lx^2 + mxy + ny^2$$

$$\dot{x} = (3m - 1)x^2 - (m - 1)y^2 - (n - 1)z^2 + (3n - 1)yx - 2mzx - 2nxy$$

$$\dot{y} = -y + x + (x - y + 2z)(y + z - x)$$

$$\dot{z} = -z + x - (x + 2y - z)(y + z - x)$$

in which the problem of stability is completely investigated by Malkin by this method. These critical cases are too complicated to be outlined here but, on the other hand, the full power of the method is felt precisely in their treatment.

10. Functions V containing time explicitly

In Chapter 5 we outlined the question of stability of periodic motions on the basis of the classical theory (variational equations with periodic coefficients). This subject can be also treated on the basis of the second method of Liapounov as explained in Chapter 5 of Malkin's treatise.[3] As it is impossible to discuss this matter in detail here, we shall limit ourselves to the definition of the nature of functions V and W in the case when they contain the variable t explicitly. Once these definitions are made, it is possible to formulate a number of theorems similar to those in Section 2.

We consider a function $V(t, x, \ldots, x_n)$ in the domain

$$t \geq t_0 > 0; \qquad |x_i| \leq h \tag{10.1}$$

where t_0 and h are constants; V has continuous partial derivatives and vanishes for $x_1 = x_2 = \ldots = x_n = 0$.

DEFINITION: *V admits an infinitely small upper limit, if for any $\lambda > 0$, one can find $\mu > 0$, such that for all $t, x_1, \ldots, x_n$ satisfying inequalities*

$$t \geq t_0; \qquad |x_i| \leq \mu$$

the inequality $|V(t, x_1, \ldots, x_n)| \leq \lambda$ is fulfilled.

In other words, V admits (lim 0) if it tends to zero for $x_i^2 \to 0$ uniformly in t.

[7] A. M. Letov, *Stability of Nonlinear Control System* (in Russian), Moscow, 1955; English translation, Princeton University Press, Princeton, N.J., 1961.

[3] See footnote [3], page 134.

Thus $V = (x_1 + \dots + x_n) \sin t$ has (lim 0) uniformly, but $V = \sin [t(x_1 + \dots + x_n)]$ does not have this property. A function V is $V = V_v$ if for t_0 large and h small it cannot have in (10.1) values of a definite sign.

Function V is $V = V_d$ is called positive definite if in (10.1) for t_0 large and h small it satisfies the inequality

$$V(t, x_1, \dots, x_n) \geq V^*(x_1, \dots, x_n) \tag{10.2}$$

where $V^* = V_d^* > 0$. Likewise $V = V_d$ is negative definite if, under the same conditions,

$$V(t, x_1, \dots, x_n) < -V^*(x_1, \dots, x_n)$$

Thus, for instance, the function

$$V = e^{-t}(x_1^2 + \dots + x_n^2)$$

is not $V = V_d$, in spite of the fact that it vanishes for $x_1 = \dots = x_n = 0$, since for the *fixed* x_i it approaches zero for $t \to \infty$. Hence it is not a V_d in the sense of the above definition. On the other hand, the functions

$$V_1 = (2 + \sin t) \sum_i^n x_i^2; \qquad V_2 = (-2 + \sin t) \sum_i^n x_i^2$$

are V_d; namely, $V_1 = V_{1d} > 0$ and $V_2 = V_{2d} < 0$.

One can give a geometrical interpretation of $V_d(t, x_1, \dots, x_n)$ by considering surfaces in n space with t as parameter. Let c_1 be some value of c in $V(t_1, x_1, \dots, x_n) = 0$. Equation $V = c_1$ represents a closed surface surrounding the origin for some t. When t varies, the surface varies also. Consider, on the other hand, a fixed surface $V^*(x_1, \dots, x_n) = c_1$ and assume that $V = V_d > 0$. It can then be shown that $V = c$, $c \leq c_1$ remains inside the fixed surface $V^* = c_1$, as all points of the surface V^* are outside the surface $V = c$ or on it. On the other hand, if $V = V_c$ has an infinitely small upper limit, the surface $V = c$ lies outside a fixed cube $|x| < \mu$. One can, therefore, find μ so small that inside, and on this cube, $|V| < c/2$; thus the surface cannot penetrate into this cube. On the other hand, $V > W$ and, hence, the surface $V = c$ must lie inside the fixed surface $V^* = c_1$.

11. Theorems of Liapounov for functions V containing t explicitly

Consider the d.e.

$$\dot{x}_i = X_i(t, x_1, \dots, x_n); \qquad i = 1, 2, \dots, n \tag{11.1}$$

where X_i are determined for $t > t_0$ in $|x_i| \leq h$. We assume the usual conditions: X_i is continuous and satisfies conditions of uniqueness for given initial data.

We shall state the theorems of Liapounov, omitting their proofs.

(I) *If it is possible to find* $V(t, x_1, \ldots, x_n) = V_d$ *for which the derivative*

$$F = \frac{dV}{dt} = \frac{\partial V}{dt} + W \tag{11.2}$$

W being the Eulerian derivative, is $dV/dt = F_c$ *of the sign opposite to V or is identically zero, the equilibrium is stable.*

(II) *If the conditions of Theorem I hold and besides* $F = F_d$ *and V admits* (lim sup 0) *uniformly, the equilibrium is asymptotically stable.*

For criteria of instability one has analogous theorems.

(III) *If there exists a function* $V(t, x_1, \ldots, x_n)$ *such that* (*a*) *it admits* (lim sup 0) *uniformly*; (*b*) $F = F_d$; (*c*) *for* x_i *arbitrarily small and* $t \to \infty$, *V has the same sign as F, the equilibrium is unstable.*

There exists also a theorem due to Chetaev[5] concerning instability: Given conditions:

(1) For $t \to \infty$ and for a sufficiently small neighborhood around the origin, the function $V > 0$;
(2) In this region V is bounded;
(3) In the same region $F > 0$, if $V \geq \alpha$ and $F \geq L$, α and L being positive numbers; $L = L(\alpha)$;

then the equilibrium is unstable.

12. Criteria of stability for equations periodic in *t*

We shall consider now the d.e. of the form

$$\dot{x}_i = p_{i1}(t)x_1 + \ldots + p_{in}(t)x_n + X_i(t, x_1, \ldots, x_n) \tag{12.1}$$

where p_{ij} are periodic functions with period ω and the X_i are nonlinear in x_i and have period ω in t.

The problem is analogous to that which was studied previously in connection with the d.e. with constant coefficients, namely: What are the necessary and sufficient conditions under which the linear terms in x_i determine the stability of (12.1)?

[5] See footnote [5], page 140.

We assume again that the terms X_i satisfy the following conditions:

(1) There exists a region $t \geq 0$, $|x_s| \leq H$ in which hold the inequalities

$$|X_i(t, x_1, \ldots, x_n)| \leq A[|x_1| + \ldots + |x_n|] \tag{12.2}$$

where A is a constant.

(2) In this region, X_i are continuous and satisfy the usual conditions under which (12.1) has a unique solution for any initial conditions.

We have, moreover, the condition $X_i(t, 0, \ldots, 0) = 0$ (from (12.2)). There are two theorems of Liapounov:

(I) *If the roots* λ_i *of the characteristic equation* (*that is, the characteristic multipliers*), *for the abridged system* 12.1 ($X_i = 0$) *have moduli less than one, the zero solution of the system* (12.1) *is asymptotically stable for any choice of* X_i *provided they satisfy the conditions* (1) *and* (2) *with a sufficiently small* A.

The proof of this theorem depends on another theorem of Liapounov (inasmuch as it can be found in any text on the theory of differential equations, we omit it here) which states that there exists a linear transformation

$$y_j = f_{j1}(t)x_1 + \ldots + f_{jn}(t)x_n; \qquad j = 1, 2, \ldots, n$$

with periodic coefficients which transforms a linear system of d.e. with periodic coefficients into a linear system with constant coefficients. (This theorem is wholly equivalent to the theorem of Floquet, which we proved in Chapter 5, and the method of proof is the same. The characteristic exponents of the resulting system with constant coefficients are precisely the characteristic exponents as defined in Section 5 of Chapter 5.) As the determinant of the transformation never vanishes, the stability with respect to the x variables is the same as that with respect to the y variables.

The original system (12.1) becomes then

$$\dot{y}_i = q_{i1}y_1 + \ldots + q_{in}y_n + Y_i(t, y_1, \ldots, y_n) \tag{12.3}$$

with constant q's where the conditions on Y_i are analogous to those on X_i, viz.:

$$|Y_i(t, y_1, \ldots, y_n)| < B[|y_1| + \ldots + |y_n|] \tag{12.4}$$

B being again a constant, which we assume to be small. The hypothesis guarantees that the characteristic exponents of (12.3) have negative real parts.

Under the stated conditions there exists one and only one quadratic form $V(y_1, \ldots, y_n)$ satisfying the equation

$$\sum_{i=1}^{n} \frac{\partial V}{\partial y_i}(q_{i1}y_1 + \ldots + q_{in}y_n) = -\sum_{i=1}^{n} y_i^2 \tag{12.5}$$

and this $V = V_d > 0$. Its Eulerian derivative is

$$U = \frac{dV}{dt} = -\sum_{i=1}^{n} y_i^2 + \sum_{i=1}^{n} \frac{\partial V}{\partial y_i} Y_i \tag{12.6}$$

If B (in (12.4)) is sufficiently small, $U = U_d < 0$. It is clear that under this condition $V(y_1, \ldots, y_n)$ satisfies Theorem II (Section 2) which thus proves also this theorem.

(II) *If among the roots of the characteristic equation of the abridged system* (12.1), *there is at least one root with a modulus greater than one, the unperturbed motion is unstable for any choice of the functions X_i satisfying the stated conditions if A is sufficiently small.*

The proof is based on the investigation of the transformed system (12.3), which has at least one characteristic root with a positive real part. Hence, one can always find a quadratic form $V(y_1, \ldots, y_n)$ satisfying the condition

$$\sum_{i=1}^{n} \frac{\partial V}{\partial y_i}(q_{i1}y_1 + \ldots + q_{n1}y_n) = \alpha V + \sum_{i=1}^{n} y_i^2 \tag{12.7}$$

where α is a positive constant. There exists a region where $V = 0$. The Eulerian derivative in this case is

$$\frac{dV}{dt} = \alpha V + W(y_1, \ldots, y_n) \tag{12.8}$$

where $W = \sum_{i=1}^{n} y_i^2 + \sum_{i=1}^{n} \frac{\partial V}{\partial y_i} Y_i$. The function is positive definite for any Y_i provided A is small enough. The form V satisfies the conditions of the theorem of Chetaev concerning instability.

13. Application to control theory

In recent years the second method of Liapounov has become a useful tool for investigations of stability of nonlinear control systems and we shall indicate briefly this important development. It can be shown that the d.e. of an uncontrolled motion of a physical system (A) can be specified by a differential system of the form

$$\dot{\eta}_i = \sum_{\alpha=1}^{n} b_{i\alpha}\eta_\alpha; \qquad i = 1, \ldots, n \tag{13.1}$$

in which the quantities η_i are the generalized coordinates of the physical system in question and $b_{i\alpha}$ are certain constants.

The addition of a control action modifies this system and it becomes

$$\dot{\eta}_i = \sum_{\alpha=1}^{n} b_{i\alpha}\eta_\alpha + \eta_i\mu; \qquad i = 1,\ldots, n \tag{13.2}$$

where μ is the coordinate of the control member (servomotor) and η_i are constants.

To this, one has to add the d.e. of the control member itself; namely:

$$V^2\ddot{\mu} + W\dot{\mu} + S\mu = f^*(\sigma) \tag{13.3}$$

where V^2, W, and S are constant parameters (inertia, damping, and restoring force, respectively) and $f^*(\sigma)$ is the generalized force (or moment) acting on the control member; this force is a function (generally nonlinear) of the signal σ.

The latter, in turn, is derived from the dynamical state of the physical system to be controlled and is of the form

$$\sigma = \sum_{\alpha=1}^{m} p_\alpha\eta_\alpha - r\mu \tag{13.4}$$

the quantity r is the so-called "feedback coefficient".

Equations (13.2), (13.3), and (13.4) constitute the differential equations of *the controlled system.* The principal difficulty of this differential system is that it does not yield itself easily to the investigation of stability. Lourje[8] introduced an important transformation of variables by which the abovementioned system can be reduced to the *canonical form* which yields itself easier to the investigation of stability by the second method of Liapounov.

The reduction to the canonical form can be made in different manners but for our purpose it is sufficient to consider the following form

$$\begin{aligned} \dot{x} &= -\rho_k x_k + f(\sigma); \qquad k = 1,\ldots, m+1 \\ \sigma &= \sum_{k=1}^{n+1} \gamma_k x_k \\ \dot{\sigma} &= \sum_{k=1}^{n+1} \beta_k x_k - r'f(\sigma) \end{aligned} \tag{13.5}$$

It is noted that, although the number of degrees of freedom of the uncontrolled system is n, the control action introduces an "additional degree" on a formal basis which is of no further importance here. As to ρ_k, they are roots of the original characteristic equation involving the coefficients

[8] A. I. Lourje, *Nonlinear Problems in the Theory of Automatic Regulation* (in Russian), Moscow, 1951.

b_{ik}; finally γ_k and β_k are some constants. The problem of stability can be formulated now in the form originally given by Lourje.

We define the Liapounov function V by the relation

$$V = \Phi + F + \int_0^\sigma f(\sigma)d\sigma \tag{13.6}$$

with the following definition of functions Φ and F

$$F(a_1x_1,\ldots, a_{n+1}x_{n+1}) = \sum_{k=1}^{n+1}\sum_{i=1}^{n+1} \frac{a_ka_i}{\rho_k + \rho_i} x_kx_i$$

$$\Phi(x_1,\ldots, x_{n+1}) = \tfrac{1}{2}(A_1x_1^2 + \ldots + A_sx_s^2) + C_1x_{s+1}x_{s+2} + \ldots + C_{n-s}x_nx_{n+1} \tag{13.7}$$

In these expressions the ρ's are the roots of the characteristic equation and the constants a, A, and C result from a number of intermediate transformations which are of no interest in this study. The assumption that there are s real roots and $(n - s)$ conjugate complex roots, accounts for the above form of the expression for Φ. It is seen from this definition of F, Φ, and $\int_0^\sigma f(\sigma)d\sigma$ that the function V so constructed can be made positive $V = V_d$ and that one can attempt to determine the regions (in the parameter space) for which the condition of stability is fulfilled.

If one replaces in V the quantities $\dot{x}_i$, etc., from the canonical equations (13.5), one obtains, after somewhat long calculations, the following expression

$$\frac{dV}{dt} = -[+] - [+] + f(\sigma)\bar{(\ )} + f(\sigma)\bar{\bar{(\ )}} \tag{13.8}$$

in which the first two (square) brackets contain only positive terms; those in parentheses $\bar{(\ )}$ and $\bar{\bar{(\ )}}$ may be either positive or negative.

The essence of the method is to set both terms in parentheses to zero, in which case one obtains, clearly, the *sufficient condition* of stability.

The explicit form of this condition is

$$\bar{(\ )} = A_i + \beta_i + 2\sqrt{r}a_i + 2a_i\sum_{k=1}^{n+1}\frac{a_k}{\rho_k + \rho_i} = 0$$

$$\bar{\bar{(\ )}} = C_\alpha + \beta_{s+\alpha} + 2\sqrt{r}a_{s+\alpha} + 2a_{s+\alpha}\sum_{k=1}^{n+1}\frac{a_k}{(\rho_\alpha + \rho_k)} = 0; \tag{13.9}$$

$$\alpha = 1,\ldots, (n + 1 - s)$$

where A_i, C_α, a_i, etc., are certain constants.

It is clear that equations (13.9) define a certain region G in the parameter space and, if the parameters are within this region (that is, when $(\bar{\ }) \leq 0$ and $(\bar{\bar{\ }}) \leq 0$), one is certain that the conditions of the first theorem are fulfilled. The rest of calculation relates to numerical calculations of intervals in which the values of different parameters must be located. In Letov[7] the reader can find numerous applications of the method in connection with various problems of automatic regulation and control.

A remark is noteworthy: the essence of the method is to formulate a *sufficient* condition of stability by annulling the third and the fourth terms in (13.8). It is clear that this condition is by no means *necessary*. One can question as to whether the fulfillment of this sufficient criterion would not impose *practical* difficulties by providing a margin of stability which may be unnecessarily too large and, for that reason, difficult to realize in applied problems.

There is no definite answer to this question at present, but in some cases calculations were made[7] to check the *practicability* of this sufficient criterion by the *necessary and sufficient* criterion yielded by the Hurwitz theorem. Since the latter holds only for linear systems, this requires a preliminary investigation as to the possibility of using conclusions derived for a linearized system as a guide for a comparison, with the results yielded by the sufficient criterion of Liapounov. A few results so obtained seem to show that in normal cases the use of the sufficient criterion does not introduce any practical difficulties. In all cases it guarantees not only the stability on the whole but permits ascertaining the *margin of stability* by a more detailed study of the regions G determined by (13.9).

14. Concluding remarks

The principal advantage of Liapounov's second method is that the difficult (and often impossible) problem of integration of a system of the variational equations is obviated and replaced by a much simpler problem of an algebraic character. The second advantage is that it gives directly *stability in the large* instead of stability in the neighborhood of positions of equilibria as in the classical method. These two basic advantages render the method particularly valuable for applications as has been shown in the preceding section.

The difficult part of the method is in the determination of the function V. It is observed that the method does not give any means for determining such functions, but merely states that if such a function exists (that it satisfies the criterion) the stability condition is fulfilled.

In a special case of systems with one degree of freedom the above criterion

[7] See footnote [7], page 154.

leads to planar regions limited by algebraic curves so that the problem of stability in the large (including the "margin of stability") presents itself in a very simple manner; we refer to Letov[7] where are indicated many examples of application of the second method to the various control problems. It is needless to say that all these problems become more difficult if one attempts to solve them by the classical method (Chapter 5) and, besides this, ascertaining the regions of stability becomes accordingly more difficult.

[7] See footnote [7], page 154.

Chapter 7

THEORY OF BIFURCATIONS

1. Introductory remarks

This chapter is a further extension of the theory of Poincaré concerning the effect of changing a parameter on solutions of a d.e.[1] We discussed this subject briefly in Chapter 2 with regard to conservative systems (Section 4) where it was shown that the passage of the parameter λ through a critical or *bifurcation value* $\lambda = \lambda_0$ causes a qualitative change in the topological structure of the trajectories (Sections 5 and 6). Likewise, in Chapter 3 it was mentioned that a "concentric" pattern of limit cycles may undergo a similar qualitative change, when, for instance, two neighboring cycles (one stable and the other unstable) approach each other as the result of the parameter variation and for $\lambda = \lambda_0$ coalesce, giving rise to a semistable cycle, an essentially unstable structure that disappears if the bifurcation value of the parameter is crossed. We shall now enter into a more detailed study of these effects, which have found important applications in the theory of oscillations.

Two important cases of bifurcations are investigated below. The first case arises in the so-called *phenomena of self-excitation* which we encountered in Chapter 3. In the simplest and, at the same time, most important case of *soft* self-excitation in connection, for instance, with an electron-tube oscillator, the bifurcation effect occurs as follows: if the parameter λ (in this case the parameter is the coefficient M of mutual inductance between the anode and the grid circuits) is sufficiently small ($\lambda < \lambda_0$), the circuit operates as an amplifier and, if there is no signal, one has obviously the state of rest. If λ increases up to the bifurcation value $\lambda = \lambda_0$, the circuit is just on the threshold between the amplification and the generation ranges.

[1] H. Poincaré, *Les méthodes nouvelles de la mécanique céleste* **T.1**, Gauthier-Villars, Paris, 1892; also E. Goursat, *Cours d'Analyse* **T.2**, Gauthier-Villars, Paris, 1918; H. Poincaré, *Acta Math.* **7**, 1885; *Figures d'équilibre d'une masse fluide*, Naud, Paris, 1903.

For $\lambda > \lambda_0$ a self-sustained oscillation appears and its amplitude begins to grow with λ. The passage of λ through its bifurcation value $\lambda = \lambda_0$ can be described by the following scheme.

$$\text{Stable singular point} \xrightarrow{\quad} \begin{cases} \nearrow \text{Stable limit cycle} \\ \searrow \text{Unstable singular point} \end{cases} \tag{I}$$

which thus justifies the term *bifurcation* (that is, "a fork"). The phenomenon is obviously reversible.

The phenomenon of ***hard*** self-excitation is the same except that the limit cycle appears suddenly as soon as $\lambda = \lambda_0$ is crossed; that is, its amplitude does not grow from zero as in the "soft" case. The phenomenon is clearly reversible and can be formulated if one reads scheme I with opposite directions of arrows corresponding to the opposite variation of the parameter λ.

The second important case, also of frequent occurrence in applications, arises when two limit cycles, one stable and the other unstable, coalesce and subsequently vanish. This situation can be described by the scheme:

$$\left.\begin{matrix} \text{Stable limit cycle} \searrow \\ \\ \text{Unstable limit cycle} \nearrow \\ (\lambda < \lambda_0) \end{matrix}\right\rangle \longrightarrow \underset{(\lambda = \lambda_0)}{\text{Semistable limit cycle}} \rightarrow \underset{(\lambda > \lambda_0)}{0} \tag{II}$$

The sign 0 on the right-hand side means the *disappearance* of cycles. There again the scheme is reversible; this means that, under certain conditions, in a region originally free from limit cycles, a semistable cycle appears when a parameter reaches its bifurcation value; if this value is crossed, two cycles, one stable and the other unstable, split from the semistable cycle. The essential point is that the cycles *always appear* (*or disappear*) *in pairs*. As a limit cycle is a periodic solution, scheme (II) is merely a topological representation of a theorem of Poincaré which states:[1]

Periodic solutions disappear (*or appear*) *by couples in the manner of real roots of an algebraic equation.*

Returning to the theory of bifurcations in general, there are two ways to approach this subject: (1) the exact method; and (2) the method resulting from the theory of approximations.

[1] See footnote [1], page 163.

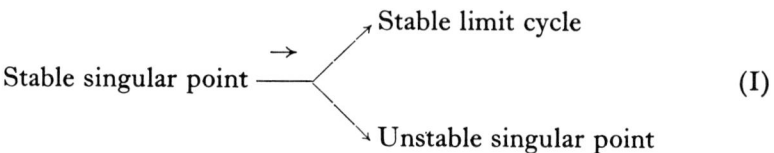

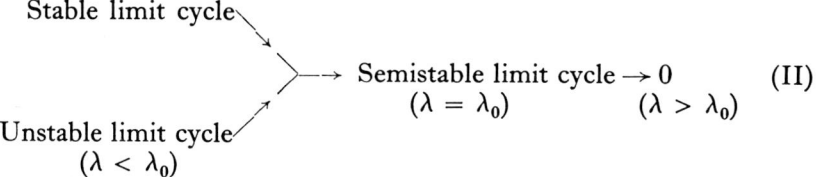

As far as the exact method is concerned, it is based on the use of the so-called "successor function" introduced by Poincaré.† The use of this method is very limited at present and the only case which could be solved completely by this method (Andronov[2]) is the "soft" case of scheme (I). Some attempts have been made to extend the method to other cases, but so far they have not been completed.[3]

With regard to the theory of approximation, the matter is considerably simpler; in particular, in the theory of the first approximation one frequently succeeds in obtaining the d.e. for the amplitude in the form

$$\frac{d\rho}{dt} = \Phi(\rho) \tag{1.1}$$

where $\Phi(\rho)$ is a polynomial. In such a case the problem of circular motion is reduced to the existence of real positive roots of the equation $\Phi(\rho) = 0$, and problem (II) of the bifurcation theory amounts to the condition of existence of one double (or, generally multiple) root of this equation at a point where the roots become complex, so that, from the physical standpoint which admits only real positive roots, such roots cease to exist. Problem (I) also acquires a simple interpretation, as will be shown in Section 7.

2. Successor function; geometry of bifurcation effects

We outline first certain theorems of Poincaré in his "théorie des conséquents."

It is convenient to define the *successor function* for a planar d.e. in the following way. Let AB (Fig. 7.1) be a differentiable arc without contact, and S be a parameter (for example, arc length) on AB. Denote the points of AB by $M = M(S)$. If the trajectory through a point $M_0 = M(S_0)$ of AB has a subsequent intersection with AB at a point $M_1 = M(S_1)$, we call M_1 the *successor* of M_0. The function $S_1 = \varphi(S_0)$ is called the *successor function* and is clearly continuous on any sub-arc on which it is defined. (φ fails to be defined at S_0 if, for example, the trajectory through

† Poincaré uses the term *conséquent*; we use here the term *successor function* suggested by M. Schiffer. The originally used term *function of the sequence* does not seem to translate exactly the meaning of Poincaré's term.

[2] A. Andronov and S. Chaikin, *Theory of Oscillations* (original text in Russian), Moscow, 1937; English translation by S. Lefschetz of A. Andronov and S. Chaikin, *Theory of Oscillations*, Princeton University Press, Princeton, N.J., 1949; A. Andronov, A. Witt, S. Chaikin, *Theory of Oscillations* (in Russian); this book is the second edition (1959) of A. Andronov and S. Chaikin, *Theory of Oscillations* (original text in Russian), Moscow, 1937.

[3] E. A. Leontovich, *Dokl. Ak. Nauk* (USSR) **78**, 1951.

M_0 ends at a singular point, and thus fails to return to AB.) We confine ourselves to the case in which a sub-arc $A'B'$ can be found on which φ is defined and such that the successor of every point of $A'B'$ is also a point of $A'B'$. Thus by the uni-dimensional Brouwer's Fixed Point Theorem (see Chapter 3, beginning of Section 7), the mapping which carries each point into its successor has a fixed point; that is, there is a number S^* such that

$$\varphi(S^*) = S^* \tag{2.1}$$

This is obviously the condition for a closed trajectory.

Poincaré[1] points out that $\varphi(S)$ is continuously differentiable and, furthermore, that $\varphi'(S) \geq 0$. If one takes a point M_0' on AB (Fig. 7.1)

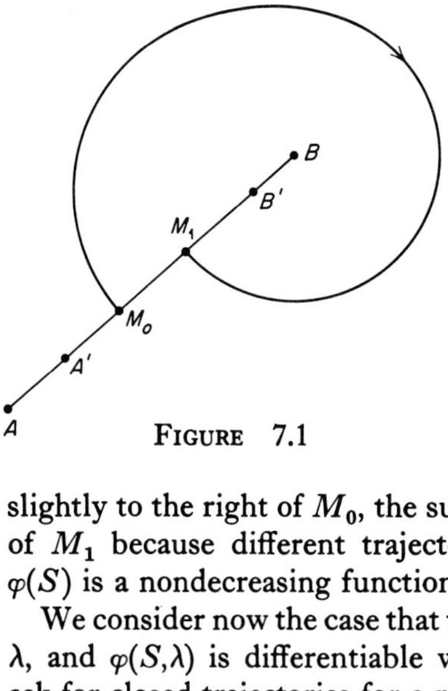

FIGURE 7.1

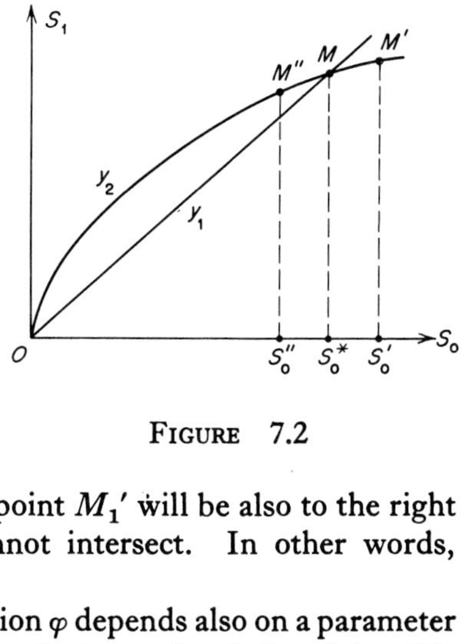

FIGURE 7.2

slightly to the right of M_0, the successor point M_1' will be also to the right of M_1 because different trajectories cannot intersect. In other words, $\varphi(S)$ is a nondecreasing function.

We consider now the case that the function φ depends also on a parameter λ, and $\varphi(S,\lambda)$ is differentiable within certain intervals of S and λ. We ask for closed trajectories for a range of values of λ near λ_0, assuming that for $\lambda = \lambda_0$ a closed trajectory exists:

$$\varphi(S_0{}^*,\lambda_0) = S_0{}^* \tag{2.2}$$

We must therefore find a function

$$S_0 = \tau(\lambda) \tag{2.3}$$

defined for λ near λ_0 such that

$$\varphi(\tau(\lambda),\lambda) = \tau(\lambda) \tag{2.4}$$

[1] See footnote [1], page 163.

The existence of $\tau(\lambda)$ follows from the implicit function theorem provided $\varphi_s(S_0^*,\lambda_0) \neq 1$. We then find that

$$\tau'(\lambda) = \frac{\varphi_\lambda[\tau(\lambda),\lambda]}{1 - \varphi_s[\tau(\lambda),\lambda]} \tag{2.5}$$

Equation (2.2) may be interpreted graphically as the condition for the intersection of the line y_1: $S_1 = S_0$ with the curve y_2: $S_1 = \varphi(S_0,\lambda_0)$ (Fig. 7.2). The abscissa S_0^* of their point of intersection determines the fixed point on the segment AB, that is, the point at which the limit cycle cuts AB in Fig. 7.1. If the trajectory is perturbed so that its intersection with AB occurs at a neighboring point (either at S_0' or S_0'' in Fig. 7.2) there exists stability, which results in the approach to the point S_0^* at which the trajectory becomes closed. The condition of stability results from the elementary considerations regarding limit cycles in Section 1, Chapter 3. In fact, if the disturbance brings the point S_0^* to the point S_0' corresponding to a greater radius vector on the spiral trajectory C', the fact that the value $S_0'M'$ of the successor function at this point is less than the value needed for "closing" the trajectory (since $y_2 < y_1$ in this region) results in the decreasing radii vectors until the point S_0^* is reached when the trajectory is again closed. For a perturbation in the opposite direction ($S_0^* \to S_0''$), the opposite effect occurs; the radii vectors increase (since $y_2 > y_1$ in this region) and ultimately the closing takes place again, since S_0'' is brought to S_0^*.

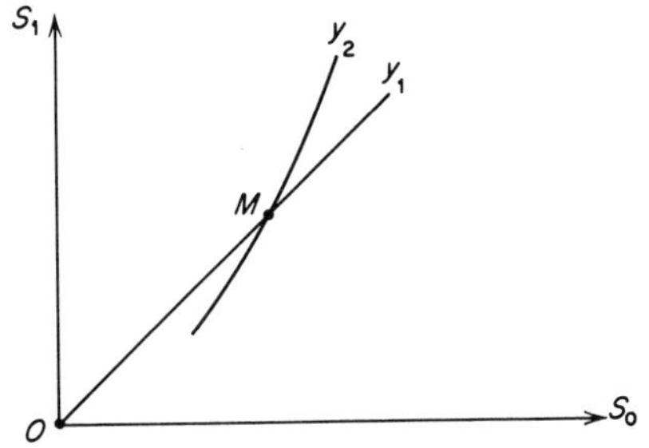

FIGURE 7.3

The configuration shown in Fig. 7.3 corresponds to an unstable cycle, as one ascertains by applying the same argument. It is interesting to note that these graphical criteria resulting from the argument of the successor function have precisely the same form as those used in the stability criteria for electric arcs; stabilized by the insertion of an ohmic resistor. In fact, there exists the so-called Blondel-Kaufman criterion of stability which states: If the nonlinear characteristic of the arc has a smaller slope at the point of its intersection with the straight line (of the ohmic resistance drop), the equilibrium is stable; in the opposite case, it is unstable. It is noted that the graphical argument based on the application of the successor function leads to exactly the same formulation; the successor function y_2 plays the role of the nonlinear characteristic and the straight line y_1, expressing the condition for a closed trajectory, has a role analogous to that of the ohmic resistance line.

The situation is different if one considers a contact between y_2 and y_1, as shown in Fig. 7.4. For the point of contact one has a special situation in that $S_0'' > S_0^*$ is obviously a zone of stability, whereas $S_0' < S_0^*$ is one of instability. One concludes, therefore, that at the point of contact M_0 there appears a semi-stable cycle. If one now changes the parameter λ so that the curve y_2 becomes y_2' (Fig. 7.4), the semistable cycle disappears; but if λ is changed in the opposite direction resulting in the curve y_2'', there appear two cycles at M' (with "radius" S_0') and at M'' (with "radius" S_0'') of which the former is unstable and the latter is stable.

In the same case (that is, in the presence of a contact between y_1 and y_2), but with the concavity of y_2 turned toward the S_1 axis, there is the

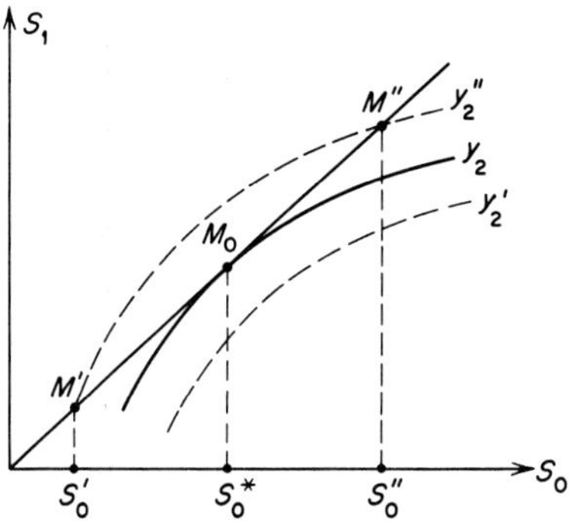

FIGURE 7.4

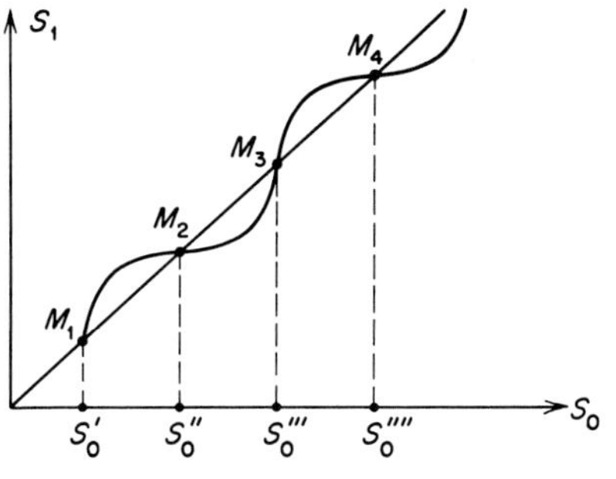

FIGURE 7.5

difference that the zones of stability and instability are interchanged; this applies to the semi-stable cycle as well as to the cycles into which it splits as the result of the parameter variation.

The form of the successor function $S_1 = \Phi(S_0,\lambda)$ depends on the form of the d.e., but the above criteria of stability are quite general. Suppose, for instance, that this function has the form shown in Fig. 7.5. One sees immediately that the limit cycles M_2 and M_4 are stable and that M_1 and M_3 are unstable. If the successor function y_2 either moves as a whole or undergoes a deformation as the result of the parameter variation, one ascertains the appearance or disappearance of cycles, but this happens always in couples of stable and unstable cycles. One can also show that, if the contact is of a higher order, more than one pair of cycles can thus originate in a region which was originally free from limit cycles. Likewise, the parameter variation in the opposite direction produces a mutual annihilation of these cycles in couples.

The geometrical interpretation of the bifurcation theory on the basis of the successor function of Poincaré can be connected with the criterion of stability in terms of the characteristic exponents (Chapter 5), as was also shown by Andronov.[2]

In fact, on the basis of the preceding argument (Figs. 7.2 and 7.3), the limit cycle is stable if $\Phi_S(S_0,\lambda)_{S_0=S_0^*} < 1$ and unstable if it is >1, where the symbol: $\Phi_S(S_0,\lambda)_{S_0=S_0^*}$; means the slope of the successor function at the point $S_0 = S_0^*$.

If one writes

$$\Phi_S(S_0,\lambda)_{S_0=S_0^*} = e^{hT} \tag{2.6}$$

where h is the characteristic exponent and T is the period of one rotation of 2π of the radius vector, it is clear that the above geometrical formulation can be connected with the criterion used in the theory of the characteristic exponents. In fact, if $h < 0$, $\Phi_S(S_0,\lambda)_{S_0=S_0^*} < 1$ which means stability; likewise, $h > 0$ means instability.[1]

3. Bifurcation of a cycle from a focus

The bifurcation effect of type (I) (Section 1) has been solved completely by Andronov on the basis of the general theory outlined in the preceding section, and we shall review this method of attack.

It is recalled (Section 6, Chapter 1) that a linear system corresponding to a focus is

$$\dot{x} = ax - by; \qquad \dot{y} = bx + ay \tag{3.1}$$

One can start therefore with the differential system

$$\dot{x} = ax - by + X_2(x,y); \qquad \dot{y} = bx + ay + Y_2(x,y) \tag{3.2}$$

where X_2 and Y_2 are polynomials of at least the second degree in x and y. The characteristic equation $S^2 + pS + q = 0$ has in this case a pair of complex conjugate roots $S_1 = a + ib$; $S_2 = a - ib$, where $S_1 + S_2 = 2a = -p$; $q = a^2 + b^2$; $S_1 - S_2 = 2ib = i\sqrt{4q - p^2}$; $4q > p^2$; $2b = +\sqrt{4q - p^2}$.

As it is desired to investigate the bifurcation point in terms of a parameter λ, it is convenient to introduce this parameter in the coefficients a

[2] See footnote [2], page 165.
[1] See footnote [1], page 163.

and b as well as in the nonlinear terms X_2 and Y_2. We shall therefore write (3.2) as

$$\dot{x} = a(\lambda)x - b(\lambda)y + X_2(x,y;\lambda); \qquad \dot{y} = b(\lambda)x + a(\lambda)y + Y_2(x,y;\lambda) \tag{3.3}$$

and, likewise, for the roots

$$S_1 = a(\lambda) + ib(\lambda); \qquad S_2 = a(\lambda) - ib(\lambda); \qquad b(\lambda) > 0 \tag{3.4}$$

We wish to investigate the change of stability of the singular point in accordance with the bifurcation scheme (I) (Section 1), and it is logical to expect that this bifurcation point is the root of $a(\lambda_0) = 0$, according to the well known property of the focal point. Written in polar coordinates, the system (3.2) is

$$\begin{aligned} \tfrac{1}{2}dr^2/dt &= a(\lambda)r^2 + X_2(x,y;\lambda)x + Y_2(x,y,\lambda)y \\ d\theta/dt &= (1/r^2)[b(\lambda)r^2 + Y_2(x,y;\lambda)x - X_2(x,y;\lambda)y] \end{aligned} \tag{3.5}$$

The d.e. of the integral curves is

$$\frac{dr}{d\theta} = r\frac{a(\lambda)r + X_2\cos\theta + Y_2\sin\theta}{b(\lambda)r + Y_2\cos\theta - X_2\sin\theta} \tag{3.6}$$

where $X_2 = X_2(r\cos\theta, r\sin\theta,\lambda)$, $Y_2 = Y_2(r\cos\theta, r\sin\theta,\lambda)$. Using the expansion $1/(1 - Z) = 1 + Z + Z^2 + \ldots$, with $Z = (X_2\sin\theta - Y_2 \times \cos\theta)/b(\lambda)r$, equation (3.6) can be written as

$$\frac{dr}{d\theta} = \left(\frac{a(\lambda)}{b(\lambda)}\cdot r + \frac{X_2\cos\theta + Y_2\sin\theta}{b(\lambda)}\right)(1 + Z + Z^2 + \ldots) \tag{3.7}$$

Assuming that the problem is nearly linear, X_2 and Y_2 have a reasonably small upper bound and, as they begin with terms in r^2, $(r < 1)$, $|Z| < 1$, the series in (3.7) converges and, as $b(\lambda) \neq 0$, one can expand the right-hand side in a power series in r, which gives

$$dr/d\theta = rR_1(\theta,\lambda) + r^2R_2(\theta,\lambda) + r^3R_3(\theta,\lambda) + \ldots \tag{3.8}$$

The identification of (3.7) and (3.8) yields

$$R_1(\theta,\lambda) = a(\lambda)/b(\lambda) \tag{3.9}$$

The other coefficients, $R_2, R_3, \ldots$, are polynomials in $\sin\theta$ and $\cos\theta$ and are thus periodic with period 2π.

If $r = r(\theta,r_0,\lambda)$ is a solution of (3.8), one can expand it also in powers of r_0, the initial condition, which gives

$$r = r_0u_1(\theta,\lambda) + r_0^2u_2(\theta,\lambda) + \ldots \tag{3.10}$$

Omitting the intermediate calculations, one obtains finally a recursive system of d.e. for the determination of the functions u_i, viz.:

$$\begin{aligned} du_1/d\theta &= u_1 R_1(\theta,\lambda) \\ du_2/d\theta &= u_2 R_1(\theta,\lambda) + u_1^2 R_2(\theta,\lambda) \\ &\cdots\cdots\cdots\cdots \end{aligned} \tag{3.11}$$

as $r(0,r_0,\lambda) = r_0$, $u_1(0,\lambda) = 1$ and $u_i(0,\lambda) = 0$, $i = 2, 3, \ldots$ with these initial conditions, one determines the functions $u_i(\theta,\lambda)$. One finds that

$$u_1(\theta,\lambda) = \exp\,[a(\lambda)\theta/b(\lambda)] \tag{3.12}$$

We look now for the appearance of a limit cycle at the bifurcation point $\lambda = \lambda_0$, which corresponds to $a(\lambda_0) = 0$. It is seen from (3.5) that $d\theta/dt$ keeps the same sign for small values of r.

The criterion for the existence of a limit cycle (Section 2) is here

$$\psi(r_0,\lambda) = r(2\pi,r_0,\lambda) - r(0,r_0,\lambda) = r(2\pi,r_0,\lambda) - r_0 = 0 \tag{3.13}$$

Since ψ is analytic in r_0, one has

$$\psi(r_0,\lambda) = \alpha_1(\lambda)r_0 + \alpha_2(\lambda)r_0^2 + \alpha_3(\lambda)r_0^3 + \ldots \tag{3.14}$$

where $\alpha_1(\lambda) = u_1(2\pi,\lambda) - 1 = \exp\,[2\pi a(\lambda)/b(\lambda)] - 1$

$$\alpha_i(\lambda) = u_i(2\pi,\lambda); \qquad i = 2, 3, \ldots \tag{3.15}$$

The condition for the existence of a limit cycle is then

$$\psi(r_0, \lambda_0) = r_0[\alpha_1(\lambda_0) + \alpha_2(\lambda_0)r_0 + \alpha_3(\lambda_0)r_0^2 + \ldots] = r_0\varphi(r_0,\lambda_0) = 0 \tag{3.16}$$

Rejecting the trivial solution $r_0 = 0$, this condition is that λ_0 is the root of equation:

$$\varphi(r_0,\lambda) = \alpha_1(\lambda) + \alpha_2(\lambda)r_0 + \alpha_3(\lambda)r_0^2 + \ldots = 0 \tag{3.17}$$

This equation may be regarded as defining a certain curve in the (r_0,λ) plane, and the question arises whether this curve has a branch in the first quadrant, since both r_0 and λ must be positive; moreover, r_0 is supposed to be small, as we are interested here in the problem of a bifurcation only. It is to be noted that, for the bifurcation value $\lambda = \lambda_0$, one has $a(\lambda_0) = 0$.

Differentiating the first equation (3.15), one has for $i = 1$

$$\alpha_1'(\lambda_0) = 2\pi a'(\lambda_0)/b(\lambda_0) \neq 0 \tag{3.18}$$

It can be shown[2] that, if $\alpha_1(\lambda_0) = 0$, one has also $\alpha_2(\lambda_0) = 0$, since $\alpha_1(\lambda_0) = 0$ implies $a(\lambda_0) = 0$ and, therefore $R_1(\theta,\lambda_0) = 0$. Hence, from the first equation (3.11), one has $du_1/d\theta = 0$, so that $u_1(\theta,\lambda_0) = \text{const} = u_1(0,\lambda_0) = 1$. From the second equation (3.11) one has $du_2(\theta,\lambda_0)/d\theta = R_2(\theta,\lambda_0)$. Since $u_2(0,\lambda_0) = 0$, we have also

$$\alpha_2(\lambda_0) = u_2(2\pi,\lambda_0) - u_2(0,\lambda_0) = u_2(2\pi,\lambda_0) = \int_0^{2\pi} R_2(\theta,\lambda_0)d\theta = 0$$

since R_2 is a homogeneous polynomial of the third degree in $\sin\theta$ and $\cos\theta$. More generally, it can be proved that the first nonvanishing term in (3.16) is always odd.

The assumptions that $a(\lambda_0) = \alpha_1(\lambda_0) = 0$; $\alpha_1'(\lambda_0) \neq 0$ show that the point A of the curve $\varphi = 0$ is an ordinary point. If it were a multiple

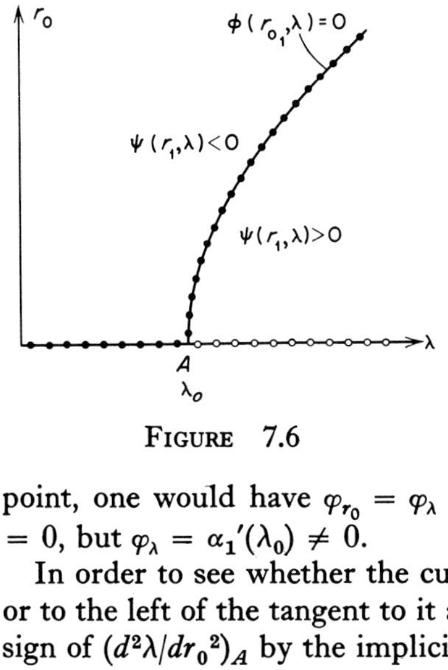

FIGURE 7.6

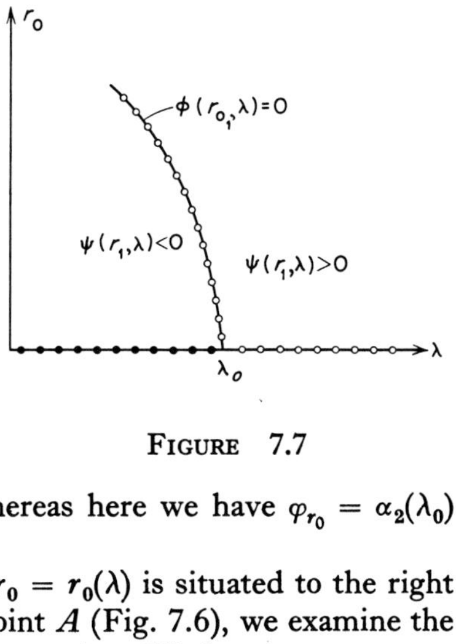

FIGURE 7.7

point, one would have $\varphi_{r_0} = \varphi_\lambda = 0$, whereas here we have $\varphi_{r_0} = \alpha_2(\lambda_0) = 0$, but $\varphi_\lambda = \alpha_1'(\lambda_0) \neq 0$.

In order to see whether the curve C: $r_0 = r_0(\lambda)$ is situated to the right or to the left of the tangent to it at the point A (Fig. 7.6), we examine the sign of $(d^2\lambda/dr_0^2)_A$ by the implicit functions rule. We have:

$$\frac{d\lambda}{dr_0} = -\frac{\varphi_{r_0}}{\varphi_\lambda}; \qquad \frac{d^2\lambda}{dr_0^2} = -\frac{\partial}{\partial\lambda}\left(\frac{\varphi_{r_0}}{\varphi_\lambda}\right)\frac{d\lambda}{dr_0} - \frac{\partial}{\partial r_0}\left(\frac{\varphi_{r_0}}{\varphi_\lambda}\right)$$

Since the derivative is to be taken at the point A: $\lambda = \lambda_0$,

$$\left(\frac{d^2\lambda}{dr_0^2}\right)_{\lambda=\lambda_0} = -\frac{2\alpha_3(\lambda_0)}{\alpha_1'(\lambda_0)} = -\frac{b(\lambda_0)\alpha_3(\lambda_0)}{\pi a'(\lambda_0)} \tag{3.19}$$

As $b(\lambda_0) > 0$, the sign of the second derivative at point A is opposite to that of $\alpha_3(\lambda_0)/a'(\lambda_0)$ which gives the following four cases:

(1) $\qquad a'(\lambda_0) > 0; \qquad \alpha_3(\lambda_0) < 0$

If λ increases monotonically and passes through $\lambda = \lambda_0$, the real part of the roots changes from negative to positive so that the focus changes from stable (for $\lambda < \lambda_0$) to unstable (for $\lambda > \lambda_0$); to this corresponds the positive sign of the second derivative and, therefore, a minimum of φ. In such a case, the curve $\varphi = 0$ exists only for $\lambda > \lambda_0$, as is shown in Fig. 7.6. One verifies easily that the tangent to the curve at the point A is vertical. The solution of $\varphi = 0$ is possible only for $\lambda > \lambda_0$. For some value $\lambda = \lambda_1$ the radius of the limit cycle is obtained by drawing a parallel to the r_0 axis through the point $\lambda = \lambda_1$.

(2) $$a'(\lambda_0) > 0; \qquad \alpha_3(\lambda_0) > 0$$

In this case the sign of the second derivative is negative, which corresponds to the maximum of $\varphi = 0$, Fig. 7.7. As in this region ($\lambda < \lambda_0$) the singular point is stable, the limit cycle is unstable. The bifurcation of the cycle in this case occurs for $\lambda = \lambda_0$.

The two remaining cases:

(3) $$a'(\lambda_0) < 0; \qquad \alpha_3(\lambda_0) > 0$$

and

(4) $$a'(\lambda_0) < 0; \qquad \alpha_3(\lambda_0) < 0$$

are treated in a similar manner, with stabilities and instabilities reversed.

Thus, in all cases a change of stability of the singular point is accompanied by a bifurcation of a limit cycle according to scheme (I) (Section 1).

4. Applications of the bifurcation theory

Numerous applications of the theory have been investigated in the preceding section. Thus, for instance, an amplifier circuit with regeneration begins to work as an oscillator as soon as a certain critical value of the feedback is reached. Likewise, one observes frequently that a control system, operating normally without oscillations, suddenly begins to oscillate or "hunt" after a critical value of a parameter (regulating the intensity of the control action) is reached.

Let us consider a standard electron-tube circuit with inductive coupling (Fig. 7.8). With the usual notations, the d.e. of the oscillating circuit is:

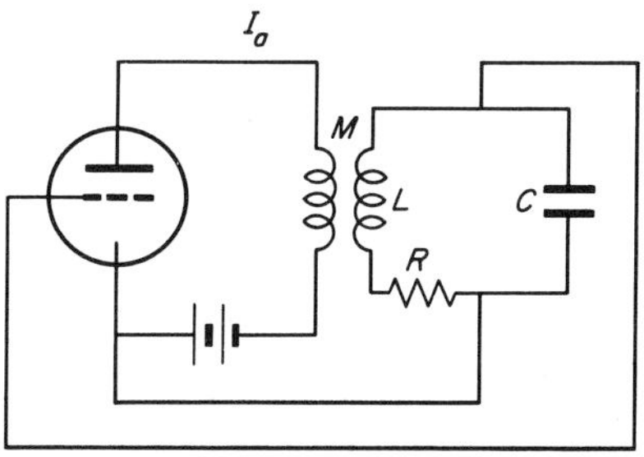

FIGURE 7.8

$$L\frac{di}{dt} + Ri + \frac{1}{C}\int_0^t idt = M\frac{dI_a}{dt} \tag{4.1}$$

The right-hand term in this d.e. indicates the action of the anode current I_a exerted inductively, M being the coefficient of mutual inductance between the plate and the grid circuits. The electron tube may be regarded as a nonlinear conductor defined by a polynomial relation

$$I_a = I_0 + S_1 v + S_2 v^2 + \dots \tag{4.2}$$

where v is the grid voltage and the S_i are certain numerical coefficients by which the polynomial (4.2) is fitted to the experimental curve $I_a = f(v)$. It is customary to introduce the so-called *saturation voltage* V defined as a sufficiently high voltage v for which to a Δv corresponds practically $\Delta I_a \simeq 0$. The expression (4.2) acquires then a more convenient form

$$I_a = V(\beta_1 u + \gamma_1 u^2 - \delta_1 u^3) \tag{4.3}$$

where $u = v/V$, the coefficients β_1, γ_1, δ_1 being positive. The minus sign before the last term results from the usual form of characteristics of electron tubes.†

As $u = (1/CV)\int_0^t i dt$, $\dot{u} = i/CV$, and $\ddot{u} = (di/dt)/CV$, after a differentiation, one has:

$$dI_a/dt = (dI_a/du)(du/dt) = V(\beta_1 + 2\gamma_1 u - 3\delta_1 u^2)\dot{u} \tag{4.4}$$

Substituting (4.4) into the d.e. (4.1) and introducing a new independent variable $\tau = \omega_0 t$, $\omega_0 = \sqrt{1/LC}$, one has

$$\ddot{u} + u = [\beta(M) + 2\gamma(M)u - 3\delta(M)u^2]\dot{u} \tag{4.5}$$

where the differentiations are now with respect to τ and $\beta(M) = (M - RC)\omega_0$

$$\gamma(M) = M\delta_1\omega_0; \qquad \delta(M) = M\delta_1\omega_0 \tag{4.6}$$

Here the coefficient M plays the role of the λ of the preceding theory.

The d.e. (4.5) is of the Liénard type (Chapter 4) with

$$f(u) = -(\beta + 2\gamma u - 3\delta u^2)$$

It is assumed that the coefficients β, γ, and δ are small numbers, as it follows from the assumption that the problem is nearly linear.

The equivalent system is here

$$du/d\tau = w; \qquad dw/d\tau = -u + (\beta + 2\gamma u - 3\delta u^2)w$$

The point $u = w = 0$ is a singular point and, keeping the linear terms only, one finds that it is an unstable focus, the roots being

$$S_{1,2} = (\beta/2) \pm \sqrt{(\beta/2)^2 - 1}$$

† We shall enter more fully into the question of signs in Chapter 22.

It is clear that $\beta(M) = (M_0\beta_1 - RC) = 0$ corresponds to the threshold separating two different phase portraits; for $\beta < 0$ the integral curves are spirals converging to the stable focus; for $\beta > 0$, these spirals diverge from the unstable focus. Hence, $M_0 = RC/\beta_1$ is the bifurcation value of the parameter M.

Setting $u = 2a_1x + 2b_1y$; $w = 2x$ with $a_1(\lambda) = \beta(\lambda)/2$; $b_1(\lambda) = \sqrt{1 + (\beta/2)^2}$ gives the system

$$\begin{aligned} \dot{x} &= a_1x - b_1y + [4\gamma(a_1x + b_1y) - 12\delta(a_1x + b_1y)^2]x \\ \dot{y} &= b_1x + a_1y - \frac{a_1}{b_1}[4\gamma(a_1x + b_1y) - 12\delta(a_1x + b_1y)^2]x \end{aligned} \tag{4.7}$$

As the parameter λ of the general theory in this case is M, the usual procedure of determining the roots of the characteristic equation corresponding to (4.7) shows that the derivative with respect to M of the real part of the roots for $M = M_0$ is positive, which corresponds to the case $a_1'(\lambda_0) > 0$ of the preceding section. In other words, when the parameter M increases passing through the bifurcation point, the singular point changes from stable to unstable; physically this means that the circuit begins to operate as an oscillator for $M > M_0$ after its previous operation ($M < M_0$) as regenerative amplifier.

It remains to be seen that $M = M_0$ is a bifurcation point of the first kind which reduces to showing that $\alpha_3(\lambda_0) < 0$. For the sake of simplicity we shall continue using the letter λ instead of M.

One ascertains that $a_1(\lambda_0) = 0$, $b_1(\lambda_0) = 1$. If one compares (4.7) with (3.11), making use of the polar coordinates, one has

$$\begin{aligned} R_1(\theta,\lambda) &= 0; \\ R_2(\theta,\lambda) &= 4\gamma(\lambda_0)\sin\theta\cos^2\theta; \\ R_3(\theta,\lambda) &= 16\gamma(\lambda_0)\sin^3\theta\cos^3\theta - 12\delta(\lambda_0)\sin^2\theta\cos^2\theta \end{aligned} \tag{4.8}$$

and, for $\lambda = \lambda_0$, the recurrent system has the following form:

$$\frac{du_1}{d\theta} = 0; \qquad \frac{du_2}{d\theta} = 4\gamma(\lambda_0)\sin\theta\cos^2\theta; \qquad \frac{du_3}{d\theta} = 2u_2R_2(\theta,\lambda_0) + R_0(\theta,\lambda_0) \tag{4.9}$$

Upon integration of these d.e., one gets

$$u_1(\theta,\lambda_0) = 1; \qquad u_2(\theta,\lambda) = \tfrac{4}{3}\gamma(\lambda_0)(1 - \cos^3\theta); \qquad u_3(\theta,\lambda) = -3\delta(\lambda_0)\pi$$

Using expressions (4.6), one has

$$\alpha_3(\lambda_0) = -3\pi RC\omega_0(S_{1/\beta_0}) < 0$$

which establishes the existence of a stable cycle.

Similar phenomena are observed in control systems. As an example we consider the problem of anti-rolling stabilization of a ship by the so-called activated tank method. This method consists in impressing on the water column, contained in port and starboard tanks joined by a duct, a control action through an axial pump in the duct. We will not consider here the stabilization problem properly speaking but will investigate only the behavior of the controlled ballast when the ship is on a rigid foundation (for example, in a drydock).

To simplify the problem, we consider the water ballast as a liquid pendulum governed approximately by a linear d.e.

$$J\ddot{\varphi} + b\dot{\varphi} + c\varphi = 0 \tag{4.10}$$

where the significance of the constants J, b, and c is obvious, and φ is the angle measuring the departure of levels in the tanks from their horizontal position. The action of the axial pump on the water ballast may be represented by a function of the form

$$M(\alpha) = a_1\alpha - a_3\alpha^3 \tag{4.11}$$

where α is the angle of the blades which is adjusted continuously by the control system; with a suitable scale, M may represent directly the moment of the water ballast with respect to the center of gravity of the ship.

According to the theory of this method of stabilization, the blade angle α is varied continuously in proportion to the instantaneous angular velocity of the ship, and this, in turn, is proportional to the rate of flow of the ballast in the duct. One can, therefore, replace α by $\dot{\varphi}$ in (4.11) so that the d.e. of the controlled water ballast is

$$J\ddot{\varphi} + (b - a_1)\dot{\varphi} + a_3\dot{\varphi}^3 + c\varphi = 0 \tag{4.12}$$

As the coefficients a_1 and a_3 relative to the control system are generally certain functions of the control parameter (for example, the coefficient of amplification or "gain" in an intermediate control circuit), it is seen that the phase portrait of the system depends on the sign of the coefficient $b - a_1(\lambda) = k$. As long as $k > 0$, the d.e. (4.12) has the state of equilibrium represented by a stable focus; but, if k becomes negative, the focus becomes unstable. Hence, the value $\lambda = \lambda_0$ for which $b = a_1(\lambda_0)$ is again a bifurcation value.

Experiment shows that, if $\lambda < \lambda_0$, the system is at rest, but as soon as $\lambda \geq \lambda_0$, the blades begin to oscillate with corresponding oscillations of the ballast. Here again one is in the presence of the bifurcation scheme (I).

The rest of the problem can be treated in the same manner as in Section 3, but the calculation of the limit cycle is complicated and we do not reproduce it here.

The difficulty is due to the fact that, whereas the determination of the singular point requires only the quantities of the first order, for the determination of limit cycle one has to go to the third-order quantities, as was explained in Section 3. For applications this question is of a somewhat academic interest and, once the existence of the bifurcation point has been ascertained, one is generally certain that oscillations set in, beginning with this point.

5. Other bifurcation problems

Leontovich [3] has investigated the bifurcation of a cycle from a separatrix (Fig. 7.9). The problem may be described as follows. For a range of values of a parameter in the d.e., the separatrix P issuing from a saddle point S is approached by the trajectories issuing from the unstable singular point F in the interior of P. Conditions are then investigated under which, near a certain parameter value, a cycle C appears near P and in its interior, so that the trajectories departing from F now wind themselves onto C.

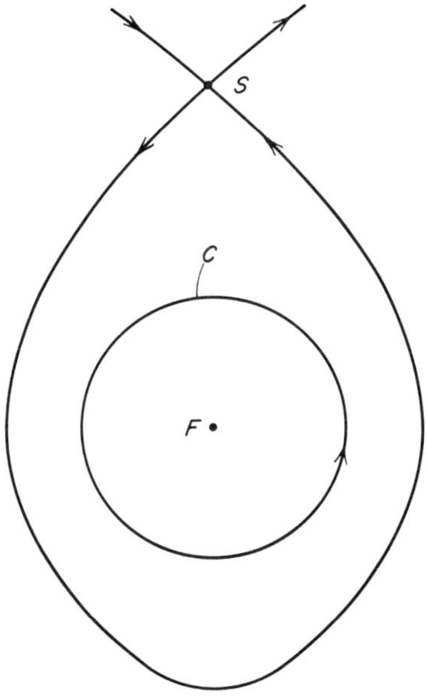

FIGURE 7.9

Another interesting problem of bifurcations has been investigated recently by Sansone and Conti [4] in connection with the d.e.

$$\dot{x} = y^2 - (x + 1)[(x^2 - 1)^2 + \lambda]; \quad \dot{y} = -xy \quad (5.1)$$

indicated by Uno and Yokomi [5]; in this investigation a considerable number of bifurcations of the first kind is indicated when the parameter λ varies.

A few other cases of bifurcations are also indicated in Sansone and Conti's treatise [6] but, aside from the classical case which was investigated in Section 3, in all these cases the connection with the theory of oscillations has not been definitely established so far.

[3] See footnote [3], page 165.

[4] G. Sansone and R. Conti, *Ann. Math. pura e appl.* (4), **27**, 1954; **38**, 1955.

[5] T. Ono and R. Yokomi, *Math. Japonica* **2**, 1952.

[6] G. Sansone and R. Conti, *Equazioni differenziali non lineari*, Ed. Cremonese, Roma, 1956.

Other bifurcation problems are treated in Chapter 22 in connection with Liénard's equation whose parameters may be regarded as bifurcation parameters of the general theory.

6. Examples of coalescing limit cycles

We first give an example of bifurcation from a semi-stable cycle to a stable plus an unstable cycle. The d.e. in polar coordinates is

$$\dot{r} = r(1 - r)(1 + \varepsilon - r); \qquad \dot{\theta} = 1 \tag{6.1}$$

which has the cycles $r = 1$ and $r = 1 + \varepsilon$, $\varepsilon > 0$. For $r < 1$ or $r > 1 + \varepsilon$, $\dot{r} > 0$; for $1 < r < 1 + \varepsilon$, $\dot{r} < 0$. Hence the cycle $r = 1$ is stable and the cycle $r = 1 + \varepsilon$ is unstable. As $\varepsilon \to 0$, the latter cycle converges to the former; and, at $\varepsilon = 0$, it disappears. The cycle $r = 1$ is semi-stable at $\varepsilon = 0$. Thus $\varepsilon = 0$ may be considered as a bifurcation value for (6.1).

To illustrate the fact that the coalescence of two limit cycles may be accompanied by the disappearance of both, we cite the d.e. in polar coordinates

$$\begin{aligned} \dot{r} &= r|1 - r|^{\varepsilon}|1 + \varepsilon - r|^{\varepsilon} \operatorname{sgn}[(1 - r)(1 + \varepsilon - r)] \\ \dot{\theta} &= 1 \end{aligned} \tag{6.2}$$

which also has the cycles $r = 1$ and $r = 1 + \varepsilon$ for $\varepsilon > 0$. As before, the first cycle is stable and the second unstable. For $\varepsilon = 0$ (6.2) becomes

$$\dot{r} = r(1 - r^2); \qquad \dot{\theta} = 1 \tag{6.3}$$

which has no cycles at all, the trajectories being spirals issuing from the origin. In this case we cannot speak of a bifurcation.

7. Algebraic approach to the bifurcation theory

The analytical method (like that of Section 3) based on the use of the successor function in the series solution has not been developed systematically for other problems of bifurcations.

A much simpler approach to problems of this kind appears in the theory of approximations (Part II). It is useful, however, to outline the method here, omitting some details which will be taken up in Chapter 16. We shall encounter numerous applications of this method in applied problems of Part III.

Let the system

$$\dot{x} = X(x,y); \qquad \dot{y} = Y(x,y) \tag{7.1}$$

be transformed to polar coordinates ρ, ψ:

$$\frac{d\rho}{dt} = \Phi(\rho,\psi); \qquad \frac{d\psi}{dt} = \Psi(\rho,\psi) \tag{7.2}$$

where $\rho = x^2 + \dot{x}^2 = x^2 + y^2 = r^2$ and $\psi = \arctan y/x$ where $x = r\cos\psi$, $y = r\sin\psi$.

In a number of problems relative to self-sustained oscillations, equations (7.2) appear in the form

$$\frac{d\rho}{dt} = \Phi(\rho); \qquad \frac{d\psi}{dt} = \text{const} \tag{7.3}$$

D.e. of this kind appear generally in the theory of the first approximation and relate to the van der Pol (or Liénard) equation when $f(x)$ is a polynomial of even degree.

The second equation (7.3) does not generally play any role in the theory which centers only on the first—the amplitude—equation†

$$\frac{d\rho}{dt} = \Phi(\rho) \tag{7.4}$$

Since the circular motion ($d\rho/dt = 0$) corresponds to the roots of the equation

$$\Phi(\rho) = 0 \tag{7.5}$$

and these roots are to be positive, the problem reduces to the determination of real positive roots of $\Phi(\rho)$; if roots are negative or conjugate complex, they are disregarded, which means that no equilibrium exists in such a case.

To each positive root, say ρ_0, in view of a uniform rotation $\dot{\psi} = \text{const}$, corresponds a circle in the phase plane and all such roots determine a polycyclic configuration of a concentric type which we have investigated by the topological argument in Section 5, Chapter 3. The concentric circles corresponding to the positive roots $\rho_0, \rho_0', \ldots$ are thus the limit cycles in the first approximation. If one goes to higher approximations, one finds that there are harmonics superimposed on the fundamental waves with amplitudes $\rho_0, \rho_0', \ldots$ but this is of no importance for our purpose here.

The variational equation of (7.4) corresponding to $\rho = \rho_0$ is

$$\frac{d\delta\rho}{dt} = \Phi_\rho(\rho_0)\delta\rho$$

† The case considered here relates to the d.e. of the form: $\ddot{x} + \mu f(x,\dot{x}) + x = 0$; if, however, one has the d.e. of a more general form: $\ddot{x} + \mu f(x,\dot{x}) + g(x) = 0$ the second equation does not reduce to a constant as will be shown in Chapter 22, but this circumstance does not affect the question of bifurcations investigated here.

and it is seen that a root ρ_0 (and hence the corresponding circular cycle) is stable if

$$\Phi_\rho(\rho_0) < 0 \tag{7.6}$$

Here, as usual, the symbol $\Phi_\rho(\rho_0)$ means the derivative of $\Phi(\rho)$ with respect to ρ in which $\rho = \rho_0$ has been substituted after the differentiation.

In what follows we shall often encounter cases when $\Phi(\rho)$ is of the form

$$\Phi(\rho) = -K(\rho - \rho_1)(\rho - \rho_2)\ldots(\rho - \rho_n)$$

which corresponds to the roots $\rho_1, \rho_2, \ldots$; if these roots are distinct $(\rho_1 \leq \rho_2 \leq \rho_3 \leq \ldots)$ one has a graph of $\Phi(\rho)$ shown in Fig. 7.10.

The alternate stabilities of limit cycles result directly from the signs of the slopes $\Phi_\rho(\rho_i)$ at points $A, B, C, \ldots$ (Fig. 7.10); thus, roots $\rho_1, \rho_3, \ldots$ are

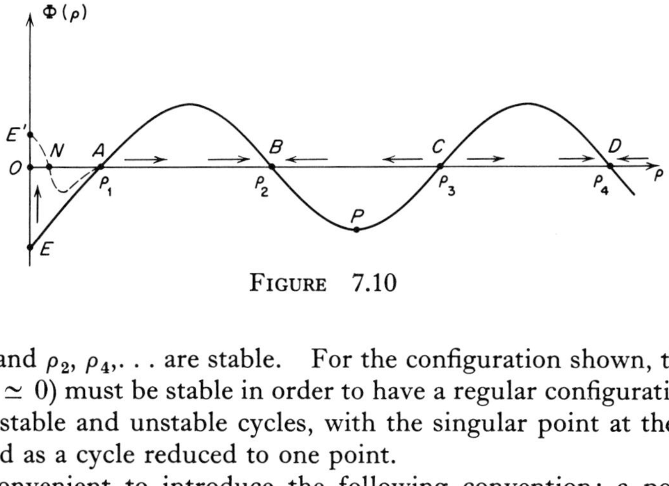

FIGURE 7.10

unstable and $\rho_2, \rho_4, \ldots$ are stable. For the configuration shown, the state of rest $(\rho \simeq 0)$ must be stable in order to have a regular configuration with alternate stable and unstable cycles, with the singular point at the center considered as a cycle reduced to one point.

It is convenient to introduce the following convention: a polycyclic configuration of a "concentric" type will be designated by a series of letters: *S* (stable) and *U* (unstable), the first letter in the sequence relating always to the stability of the singular point, and the subsequent letters relating to the corresponding stabilities of limit cycles from the innermost cycle to the external one. Thus, for instance, the symbol *SUS* is equivalent to the sentence: a stable singular point is surrounded by an unstable cycle which, in turn, is surrounded by a stable cycle; this is the well known configuration of the "hard" self-excitation (Chapter 3). Likewise, *US* means: an unstable singular point is surrounded by a stable cycle which corresponds to the "soft" self-excitation. This convention permits the use of more condensed language when dealing with complicated cases of bifurcations.

Assume that $\Phi(\rho)$ now depends also on a parameter λ in some manner; thus, the parameter λ may modify one or several coefficients of the polynomial $\Phi(\rho,\lambda)$. Suppose, also, that the parameter is chosen so that the curve $\Phi(\rho,\lambda)$ rises as a whole and the roots ρ_2 and ρ_3 approach each other and coalesce into a double root ρ_{23} when the curve touches the axis at point P (Fig. 7.10) corresponding to a bifurcation value $\lambda = \lambda_0$ of the parameter. If λ continues to vary in the same direction, the double root disappears. Thus the bifurcation of type (II) (Section 1) finds its expression in the algebraic formulation of the bifurcation problem in the coalescence of two neighboring positive roots into one double root which disappears when the roots become complex.

A bifurcation of type (I) can be also given a similar algebraic interpretation. Assume, for instance, that the parameter λ is chosen in such a way that its variation causes only a change of stability of the singular point (state of rest) without changing appreciably the values of other roots. This means that the function $\Phi(\rho,\lambda)$ is modified in the neighborhood of its zero root $\rho = 0$. This amounts to raising point E (Fig. 7.10) to point E'. When E coincides with point O, one clearly has a bifurcation point because the function $\Phi(\rho,\lambda)$, as well as its derivative, is zero at this point. As this function is continuous and E varies continuously, it is obvious that for $\lambda > \lambda_0$ there appears an additional small root at point N which did not exist before and, as this root is now stable and the state of rest becomes unstable, one has the situation in agreement with the bifurcation of type (I). In our notations we can also represent the phenomenon by the schemes: before the bifurcation we had $SUSU\ldots$, and after it we have $U_sUSU\ldots$, where s indicates the small stable cycle branching off the state of rest which corresponds to the root at point N of Fig. 7.10. Intuitively one can say that the occurrence of bifurcations is such as to maintain the *regularity* of the structure. An obvious difference arises between the two types of bifurcations. In bifurcations of type (II) the "cyclicity" (number of cycles) of the structure (or configuration) changes always by *two* units and this change is *internal* (within the structure); in bifurcations of type (I), the cyclicity changes always by *one* unit and occurs at the innermost point of the configuration. It is natural to question whether one can have also a kind of a bifurcation (of the "third" type) which would change the cyclicity by one unit by either removing or adding a cycle on the outside of the configuration. In Chapter 22 it will be shown that such is the case and that bifurcations of the *third kind* are also possible.

8. Bifurcation diagrams; effect of bifurcation points on structures

We shall consider now examples of application of the above considerations without entering for the moment into the physical significance of these applications; in Part III we shall have a number of examples which will illustrate these phenomena.

Consider a d.e. of the form

$$\frac{d\rho}{dt} = \rho[(\lambda a - c) - \lambda b\rho] = \Phi(\rho,\lambda) \tag{8.1}$$

where a, b, and c are positive constants, and λ is a variable parameter. In this case, we have

$$\Phi_\rho(\rho) = -2\lambda b\rho + (\lambda a - c) \tag{8.2}$$

The condition of self-excitation from rest is given by (8.1), in which we assume that ρ is very small; this gives: $(d\rho/dt) \simeq (\lambda a - c)\rho$; as in this case one must have $(d\rho/dt) > 0$, the condition of self-excitation is:

$$\lambda > c/a \tag{8.3}$$

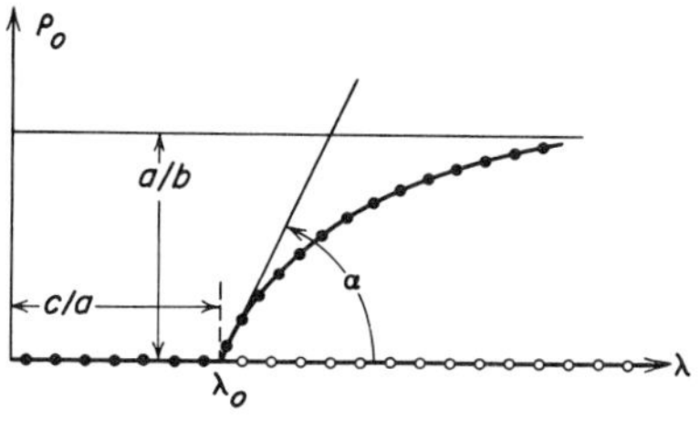

FIGURE 7.11

The stationary amplitude is obtained from (8.1) by equating to zero the bracket which gives

$$\rho_0 = (\lambda a - c)/\lambda b \tag{8.4}$$

As $\rho > 0$, the existence of ρ_0 requires again the condition (8.3). As to the stability of ρ_0, it is obtained by replacing (8.4) in (8.2) which gives:

$$\Phi_\rho(\rho_0) = -2(\lambda a - c) \tag{8.5}$$

Since for stability this expression must be negative, we encounter again condition (8.3) which is thus the necessary and sufficient condition for self-excitation, as well as for the existence of a stable stationary amplitude. In the meantime, it is observed that $\lambda = \lambda_0 = c/a$ is the bifurcation value of the parameter λ for which the bifurcation of the first kind occurs.

Figure 7.11 indicates the corresponding *bifurcation diagram* in the plane of the variables (λ,ρ_0). From (8.4) it is seen that for an increasing λ the curve ρ_0 approaches the straight line $\rho_0 = a/b$ as asymptote. In this diagram the stable singular points are indicated by points on the axis, and the unstable ones by small circles; the same notations are used on the curve ρ_0 to indicate the stability of the stationary oscillation. The slope of the curve at the bifurcation point is

$$\tan\alpha = \left(\frac{d\rho_0}{d\lambda}\right)_{\lambda_0 = c/a} = a^2/cb \tag{8.6}$$

As a second example, we consider the d.e.

$$\frac{d\rho}{d\tau} = -\sigma\rho(\rho^2 - p\rho - q) = \Phi(\rho) \tag{8.7}$$

where σ and p are positive constants; q is a variable parameter (positive or negative). As previously, we have to investigate two questions: the state of rest and the stationary motion. The latter is given by the "radius" (in the first approximation) of the limit cycle; it is determined by the positive root (or roots) of the quadratic equation:

$$\rho_{1,2} = p/2 \pm \sqrt{(p/2)^2 - q} \tag{8.8}$$

If $q > 0$, there are two positive roots which exist for $q < p^2/4$; for $q = p^2/4$ one has one double root $p/2$, and for $q > p^2/4$ the real roots disappear. It is obvious that for $q = p^2/4$ a bifurcation of the second kind takes place.

For the question of stability one has to use the criterion (7.6). In this case we have:

$$\Phi_\rho(\rho) = -\sigma p(2\rho - p)$$

One verifies that for $q < p^2/4$ the smaller root is unstable and the larger one is stable. The state of rest in this case is stable, as this follows from (8.7) in which $\rho \simeq 0$; this gives $\dot{\rho} = -\sigma pq$. Here ($q > 0$ and less than $p^2/4$) one has the configuration: *SUS*. For $q = p^2/4$ the two positive roots coalesce giving rise to the double root $\rho_{12} = p/2$, and for $q > p^2/4$ the real roots disappear so that the configuration becomes, simply, *S*, the stable state of rest. We have thus a bifurcation of the second kind corresponding to the critical value of the parameter $q_0 = p^2/4$. In this algebraic interpretation of the bifurcation effect, the double root $\rho_{12} = p/2$ represents obviously the amplitude of the semi-stable cycle; it separates the region of the *bicyclic* configuration *SUS* from the *acyclic* one, *S*. One ascertains also that the criterion $\Phi_\rho(\rho_0) = \Phi_\rho(p/2) = 0$ as it should be, inasmuch as a semi-stable cycle by its nature is a structure with an *indifferent* stability.

If $q < 0$, the situation is different, since the quadratic equation has only one positive root. This root is stable, as one verifies easily by means of the criterion (7.6); the state of rest is now unstable, giving us the configuration *US* shown in Fig. 7.12 which uses the same notations as previously; this configuration *US* exists for $q < 0$. On the other hand, for $q > 0$, as we saw, the configuration is *SUS*; and when the parameter q increases and passes through the value $q = 0$, one has a bifurcation of the first kind according to the scheme: *US* $\rightarrow$ *SUS*; for $q > 0$ the configuration is bicyclic; and for the second critical value $q = p^2/4$ of the parameter, one has a bifurcation of the second kind because the two cycles, one stable and the other unstable, coalesce at the point M and disappear for $q > p^2/4$.

From Fig. 7.12 one can ascertain the existence of an interesting nonlinear effect which Appleton and van der Pol[7] call "oscillation hysteresis." For the sake of clarity, we reproduce the diagram of Fig. 7.12 in Fig. 7.13, since we are interested only in the variation of the parameter q.

Assume that we investigate the process when the parameter q is negative and has some value $q = q_1$ (point K on the abscissa axis of Fig. 7.13). The oscillation will establish itself with the amplitude KQ and, if q increases from that point, the amplitude will follow the curve QM. The fact that for $q = 0$ the bifurcation point of the first kind will be crossed will have no effect on the oscillation which follows the stable branch QM; at the point

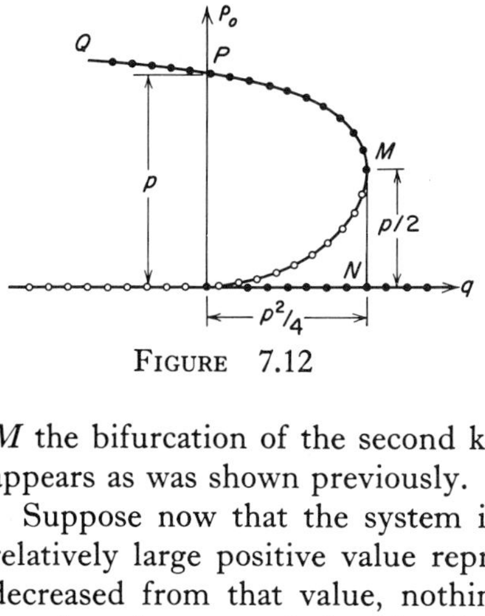

FIGURE 7.12

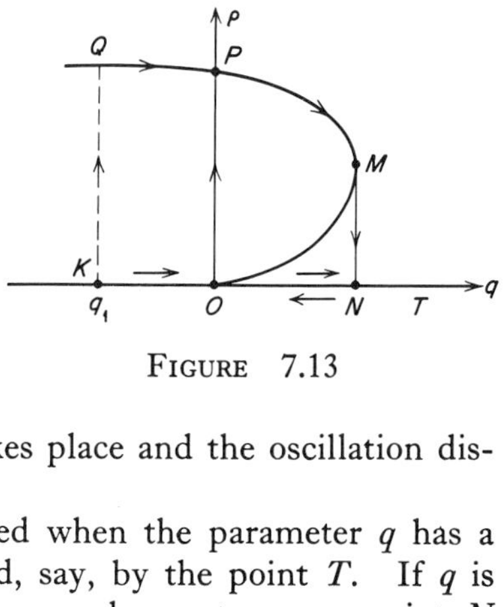

FIGURE 7.13

M the bifurcation of the second kind takes place and the oscillation disappears as was shown previously.

Suppose now that the system is started when the parameter q has a relatively large positive value represented, say, by the point T. If q is decreased from that value, nothing happens when q traverses point N entering the zone of the bicyclic configuration, since the state of rest continues to be stable. Only when q reaches the value $q = 0$, will the amplitude jump abruptly to point P and follow the branch PQ when q decreases further. Thus, the evolution of the oscillatory process for decreasing values of the parameter q is not the same as for increasing ones. For the latter the amplitude follows the path $QPMNT$, and for the former this path is $TNOPQ$. Everything happens as if there were a kind of a hysteresis area $PMNOP$ limited by a stable portion of the curve PM, the abscissa axis, and two jumps OP and MN, of which the former corresponds to a bifurcation point of the first kind and the latter, to that of the second kind. Depending on the form of the d.e., the hysteresis effect may have

[7] E. V. Appleton and B. van der Pol, *Phil. Mag.* **42**, 1921.

also different forms, but in each special case there is no difficulty of ascertaining what happens by a similar argument.

Assume, for instance, that the curve $\Phi(\rho,\lambda)$ is of the form shown in Fig. 7.14. If one increases the parameter from some small values the evolution of the oscillatory phenomenon will follow the path $RUQSM_3T$. If, however, the parameter is decreased from some large values, the path will be TM_3SPUR. In this case there is one bifurcation point of the first kind at the point M_3 on the abscissa axis, and two bifurcation points of the second kind at the points P and Q for $\lambda = \lambda_1$, and $\lambda = \lambda_2$, respectively.

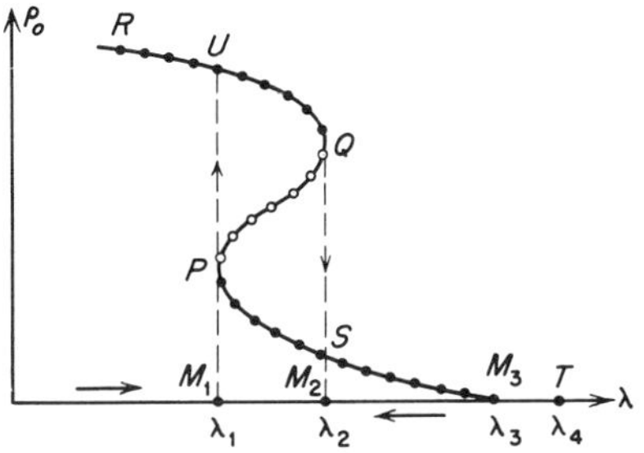

FIGURE 7.14

Summing up, the study of the bifurcation effects in the algebraic approach to these problems reduces to the investigation of the evolution of positive roots of certain polynomials, depending on a parameter. For the quadratic polynomials investigated here the matter is sufficiently simple, but for polynomials of higher degrees one encounters greater difficulties.

9. Noncritical systems

The essential feature of the bifurcation theory is the variation of the qualitative aspect of trajectories (the phase portrait) for small parameter variations around a bifurcation value.

In the preceding sections of this chapter we have given examples of the bifurcation theory. In Chapter 2 similar examples were also given, for instance, in connection with the critical value of the energy constant for which the separatrix is crossed, etc. From the standpoint of applied problems, the bifurcation effects are generally undesirable, and attention has been centered on formulating conditions under which a given physical system is immune against bifurcation effects. This led to the formulation of conditions under which a given system remains *noncritical*, that is, is immune (as far as its qualitative behavior is concerned) against small variations of parameters. These problems received the name of *structural stability*, but the term "stability" is heavily overtaxed and we prefer to use the term "critical" system (or structure) in the above defined sense.

These problems appeared because, in any physical system, parameters are known only approximately and are always subject to fluctuations. If these fluctuations take place when the system is far from its bifurcation thresholds, their effect is negligible. The situation is different if such is not the case. Thus, for instance, an electron-tube system becomes critical in the above defined sense if one tries to receive radio signals with tuning near the critical threshold when the amplifier circuit becomes the generator of oscillations.

The usual approach to this study is to consider the perturbation of the *form* of the original system of d.e.

$$\dot{x} = P(x,y); \qquad \dot{y} = Q(x,y) \tag{9.1}$$

by small perturbing terms $p(x,y)$ and $q(x,y)$; then instead of (9.1), we have:

$$\dot{x} = P(x,y) + p(x,y) = \bar{P}; \qquad \dot{y} = Q(x,y) + q(x,y) = \bar{Q} \tag{9.2}$$

We assume that, in addition to the smallness of $|p(x,y)|$, $|q(x,y)|$, their partial derivatives are also small; we shall also assume that $p(x,y)$ and $q(x,y)$ are analytic, as are $P(x,y)$ and $Q(c,y)$.

Call (9.1) system A and (9.2) $\bar{A}$. If $|p(x,y)|$ and $|q(x,y)|$ are small enough, the cycle without contact for $\bar{A}$ will have the same feature as for A. Under these conditions for any finite time interval, one can always determine $\bar{A}$ sufficiently near to A and take the neighboring initial conditions so that, during a certain time interval, the trajectories (that is, their representative points) remain also near to each other. Of course this does not mean yet that for $t \to \infty$ these features will be preserved.

The condition for a noncritical system can be formulated as follows:

A system A is called noncritical in a domain G if, for any $\varepsilon > 0$, one finds $\delta > 0$ such that for any analytic functions $p(x,y)$ and $q(x,y)$ satisfying (in G) conditions

$$\begin{aligned} &|p(x,y)| < \delta; \qquad |q(x,y)| < \delta; \qquad |p_y(x,y)| < \delta; \\ &|p_y(x,y)| < \delta; \qquad |q_x(x,y)| < \delta; \qquad |q_y(x,y)| < \delta \end{aligned} \tag{9.3}$$

there exists a topological (1:1 *and continuous*) *mapping of G into itself for which any trajectory of A is mapped into the corresponding trajectory of $\bar{A}$ and inversely so that the corresponding points are at distances less than* ε.

In this manner the qualitative behavior of A and $\bar{A}$ will be practically the same.

A number of conditions must be fulfilled in order to secure the noncritical character of a given system. We refer to De Baggis[8] for the theory of these questions and mention only the conclusions.

[8] H. F. De Baggis, *Contributions to the Theory of Nonlinear Oscillations*, Vol. II, Princeton University Press, Princeton, N.J.

1. *First criterion. A noncritical system A cannot have an equilibrium point for which*

$$\Delta = \begin{vmatrix} P_x(x_0,y_0) & P_y(x_0,y_0) \\ Q_x(x_0,y_0) & Q_y(x_0,y_0) \end{vmatrix} = 0 \tag{9.4}$$

In fact, (9.4) means that curves $P(x_0,y_0) = 0$ and $Q(x_0,y_0) = 0$ do not interesect (as they should) but *have a contact.* In such a case (if one admits that for the original system A, condition (9.4) is fulfilled), one shows that a neighboring system $\bar{A}$ (with $\bar{P}(x,y)$, ($\bar{Q}(xy,)$) has several points of intersections of the curves $\bar{P}$ and $\bar{Q}$ in any neighborhood, however small, around the point $O(x_0,y_0)$. The analysis follows closely the argument of Section 3.

2. *Second criterion* (setting $\sigma = P_x + Q_y$). *A noncritical system cannot have positions of equilibria for which*

$$\Delta > 0; \qquad \sigma = P_x + Q_y = 0$$

From these theorems it follows that a noncritical system can have only simple equilibrium points of the following three types:

$$(1)\ \Delta > 0; \quad \sigma^2 - 4\Delta > 0; \qquad (2)\ \Delta < 0; \qquad (3)\ \Delta > 0; \quad \sigma^2 - 4\Delta < 0; \quad \sigma \neq 0$$

Case (1) corresponds to nodes (stable or unstable depending on the sign of σ); case (2) corresponds to saddle points; and case (3) corresponds to foci (also either stable or unstable, depending on the sign of σ).

These positions of equilibria are *noncritical* in the sense that the trajectories of the original system A and of the neighboring (perturbed) system $\bar{A}$ are topologically identical and are merely slightly shifted with respect to each other. In the case of a saddle point, the same situation exists for both systems A and $\bar{A}$.

The argument of Section 3† can be also used for the analysis of noncritical or normal limit cycles. It is recalled that, in general, limit cycles are defined as closed trajectories in the neighborhood of which there are no other closed trajectories, which specifies the noncritical character of *normal* cycles. This results in the theorem:

(2) *Noncritical systems cannot have closed trajectories for which:*

$$\frac{h}{\tau} = \frac{1}{\tau}\int_0^{\tau} [P_x(\varphi,\psi) + Q_y(\varphi,\psi)]dt = 0 \tag{9.5}$$

† If one follows the argument of Section 3, it can be shown in Chapter 1, Section 8, that the position of equilibrium is a weak focus; the order of "weakness" depends on the order of the first nonvanishing coefficient α in the expansion of (3.14). This results in the appearance of a small limit cycle which is the nearer to the point $O(x_0,y_0)$ the smaller is $\bar{a}$ (Section 3).

where φ, ψ, and τ have the same meaning as in Section 2. It is noted that (9.5) expresses the vanishing of the characteristic exponent. It is recalled (Chapter 5) that for the systems of the second order having a periodic solution, one exponent always is zero,[1] but the vanishing of the second exponent (9.5) indicates that the system is orbitally unstable. It is recalled also that this characterizes the trajectories of the harmonic oscillator which exhibit an *indifferent* behavior with respect to incoming perturbations; in other words, such trajectories are determined solely by the initial conditions and not by the parameters of the d.e. itself (like the limit cycles).

We see thus that the condition: $P_x(\varphi,\psi) + Q_y(\varphi,\psi) = 0$ for the systems of the second order amounts to the existence of a *center*; on the other hand, from the criterion mentioned in this section it is shown that this condition accounts for the critical behavior of trajectories. In this manner we were led to identify the critical character of these trajectories with their *indifferent* orbital stability.

We consider an additional case when a separatrix goes from saddle point to either another saddle point or returns to the same saddle point. We saw examples of such situations in Chapter 2. In this connection we had the following important theorem.

(3) *Noncritical systems cannot have separatrices connecting saddle points.*

Summing up: A system: $\dot{x} = P(x,y)$; $\dot{y} = Q(x,y)$ is noncritical in G (limited by the cycle without contact) if:

(1) Its positions of equilibria are such that $\Delta \neq 0$ or, if $\Delta > 0$, $\sigma \neq 0$.
(2) Its limit cycles are such that $h \neq 0$ (being defined by (9.5)).
(3) Its separatrices do not connect saddle points.

If these conditions are not fulfilled, problems become more complicated and special investigations are necessary; such cases, however, are not encountered often in applications; for their study we refer to De Baggis.[8]

10. Remarks

In this chapter we have attempted to indicate the influence of a parameter variation on the modification of the topological structure of solutions of a d.e. As we saw, there are two approaches to this problem. In the analytic approach use is made of the "successor function" of Poincaré, and the aim of the method is to investigate what happens directly from the series expansions satisfying the d.e. This direct method, as we saw, is

[1] See footnote [1], page 163.
[8] See footnote [8], page 186.

generally complicated and could be carried out completely only in one special case (Section 3) which happens to be of a great importance in applications; attempts to extend it to some other cases have not been very conclusive as yet, mainly on account of computational difficulties.

In the algebraic approach (Section 7), the matter is considerably simpler, but here one is concerned only with the theory of the first approximation and, besides, the method is applicable only to d.e. that can be reduced to the form (8.7). It so happens, however, that this form of d.e. is common in applications as the result of the use of the stroboscopic transformation, which will be discussed in Chapter 16. In Chapter 22 we shall return to this subject which is of a great interest in applications, inasmuch as these various bifurcation effects ultimately permit *controlling* the nature of oscillatory processes by an appropriate modification *of the form* of the d.e.

It should be noted, however, that even in the theory of the first approximation in which the algebraic treatment of these bifurcation effects simplifies the problem so much, the difficulties may still be considerable if one encounters problems in which the polynomials are of a degree higher than the second. Thus, for instance, in a problem in which the polynomial is, say, of the third degree, it is necessary to ascertain the existence of only positive roots. If the discriminant of the cubic form passes through zero, one single positive root may give rise either to three or to two positive roots, or to one single root. In general, it is very difficult to establish conclusions algebraically, and it becomes necessary to proceed with long and tedious numerical computations in order to ascertain which of these new roots are stable once the discriminant has changed its sign (see Chapter 18).

Fortunately, for systems with one degree of freedom which play an important role in applications, the problem does not go beyond a very simple discussion of properties of the quadratic polynomials, as we shall see more definitely in Chapter 22.

Chapter 8

CYLINDRICAL AND TOROIDAL PHASE SPACES

1. Introductory remarks

In the preceding chapters the investigation of the topological behavior of trajectories was always conducted in a Euclidean phase space. Often when one or several coordinates are cyclic (for example, an angle determined only modulo 2π), it is more convenient to use appropriate phase spaces (or phase surfaces), rather than a planar representation. Thus, for instance, in Section 5, Chapter 2 it was mentioned that, if the value of the parameter is such that a trajectory appears outside the external separatrix (A in Fig. 2.7), the motion becomes rotary. The periodicity of this motion is not seen directly from the planar representation, although it can still be ascertained somewhat indirectly from the fact that the angular coordinate $+\pi$ denotes the same point as $-\pi$. The matter, on the contrary, becomes obvious if the phase-plane representation of Fig. 2.7 is wrapped on a circular cylinder of radius 1 so that the ω axis becomes parallel to the axis of the cylinder, in which case these two points coincide on the surface of the cylinder.

It is clear that singular points, limit cycles, separatrices, etc., of the phase plane are preserved on the cylindrical phase surface on which they are wrapped. There appears, however, a property inherent to this surface which does not exist in the phase plane. In fact, there may appear now certain closed trajectories going *around* the cylinder and escaping the planar representation. Note that such a closed trajectory does not require the existence of a singular point. One refers generally to this special form of periodicity as *periodicity of the second kind*; it depends on the existence of *one* cyclic coordinate.

With this terminology, a closed integral curve on the surface of the cylinder *which does not go around it* may be called a *periodic trajectory of the first kind*; it is obvious that the latter is just an ordinary planar closed trajectory merely traced on the surface of the cylinder.

If there are two cyclic coordinates, the representation of trajectories may require a toroidal surface. This case was investigated by Poincaré,[1] Denjoy,[2] and, later, by Nemitzky,[3] and others. In this case are possible *periodic trajectories of the third kind.* This happens when a trajectory executes a certain number of complete (2π) turns around the torus in "latitude" θ, while rotating through another number of complete turns in the other cyclic coordinate φ, the "longitude," before closing. In this case trajectories of the second kind are possible when a complete rotation on the surface of the torus takes place only in the θ or in the φ coordinate. Finally, there are possible also trajectories on the surface of the cylinder which do not turn around a cyclic coordinate.

In this chapter we shall discuss some details of the cylindrical phase space with a view toward certain applications, in particular a number of phenomena inherent in the operation of a synchronous motor.

Fewer applications are known at present which require the use of a toroidal phase space; some of these phenomena are indicated in the last section of this chapter.

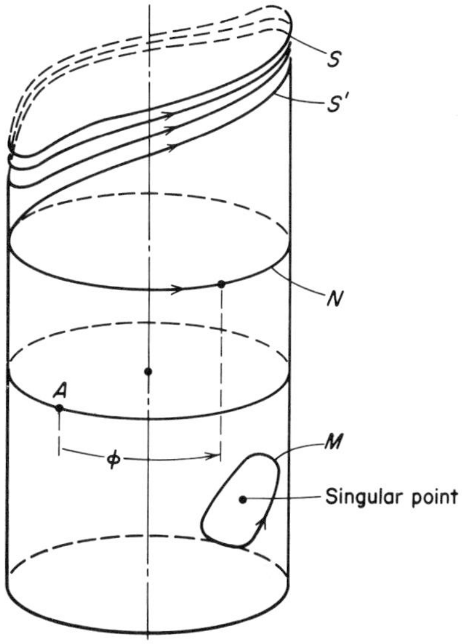

FIGURE 8.1

As regards the cylindrical phase surface, one must mention a certain difference in topological properties between the trajectories of the first and of the second kind. In fact, the trajectories of the first kind always bound off certain areas ("inside" and "outside") on the cylindrical surface on which they are traced, whereas the second do not. This is readily seen from Fig. 8.1, where M shows a periodic trajectory of the first kind and N one of the second kind. From an off-hand consideration it seems im-

[1] H. Poincaré, *J. des Math.* (3), **7**, 1881; also Œuvres **T.1**, Gauthier-Villars Paris, 1928.

[2] A. Denjoy, *J. des Math.* (9), **11**, 1932.

[3] V. V. Nemitzky and V. V. Stepanov, *Qualitative Theory of Differential Equations*, original text in Russian, Moscow, 1949; English translation, Princeton Mathematics Series, Princeton University Press, Princeton, N.J., 1960.

possible to apply to the second kind of trajectories the theorem of Poincaré which asserts that *inside* a closed trajectory there should be at least one singular point whose stability is opposite to that of the trajectory (Section 2, Chapter 3); the reason is that it is impossible to say here where is "the inside" for such a trajectory.

The following argument of Lefschetz clarifies this point. The argument is based on the possibility of representing a cylindrical surface on a plane with the origin left out (Fig. 8.2). It is sufficient to set $\rho = e^z$, where $z = \dot{\varphi}$ in order to see that a point $M(\varphi,z)$ on the cylinder goes into a point $M'(\varphi,z)$ of the plane. As the transformation $M \to M'$ is (1:1) and continuous, it is topological. As z varies between $\pm\infty$, ρ varies between 0 and ∞, the zero being excluded. Thus, the cylinder is transformed into the whole plane, the generating lines of the cylinder appearing as radii of the planar representation; a circle $z = 0$ on the cylinder goes into the circle $\rho = 1$ of the plane. The closed trajectories of the first kind (M in Fig. 8.1) are those which do not go around the origin, whereas those of the second kind enclose the origin in its interior as does the curve N. In this process of "flattening out" the cylinder into a plane, the closed trajectory of the second kind acquires a familiar picture, the origin behaving now in all respects as a singular point.

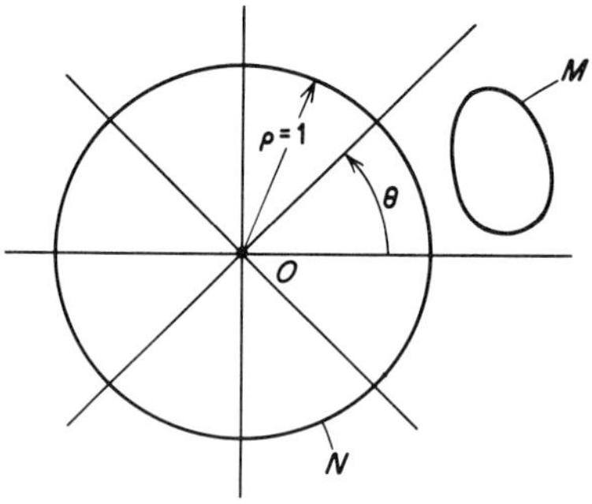

FIGURE 8.2

A most interesting use of the cylindrical phase space was made by Vlasov[4] who discussed the behavior of the synchronous motor on this basis and showed that certain self-excited parasitic oscillations occurring during the synchronous operation appear as limit cycles of the first kind, whereas those of the second kind characterize the abnormal or "asynchronous" operation when the synchronism is lost. A later work of Stoker[5] contributed further to this investigation.

The interesting point of these studies is that a series of complicated physical facts could be thus accounted for on the basis of certain properties of trajectories of the cylindrical phase space.

[4] N. Vlasov, *J. Tech. Phys.* (USSR) **9**, 1939.

[5] J. J. Stoker, *Nonlinear Vibrations*, Interscience Publishers, New York, 1950.

2. Differential equation of an electromechanical system[3, 6]

In this and in the following sections we shall be concerned with a d.e. of the form

$$A\ddot{\theta} + B\dot{\theta} + f(\theta) = M \tag{2.1}$$

where A, B, and M are constants, and $f(\theta)$ is a certain periodic function of θ. In spite of its apparent simplicity, this d.e. presents considerable difficulties which arise from the form of the "restoring force" term $f(\theta)$.

If one takes for $f(\theta)$ the simplest possible form $f(\theta) = C \sin \theta$, it is readily seen that the just mentioned difficulty is in the possibility for this term to change its sign which results in a variable stability of the system, and the question arises whether, under such conditions, a kind of an

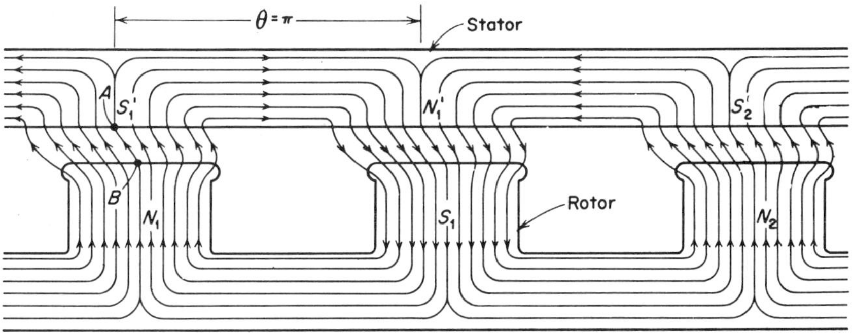

FIGURE 8.3

average stability can still exist so as to enable a trajectory to have a re-entrant path after its complete turn (2π) around the cylinder.

As this d.e. appears in the theory of a synchronous motor, it is useful to outline first this theory without going into too many details. As is well known, a synchronous motor is an electromechanical system consisting of a stationary part, the stator S, and a rotating element, the rotor R, Fig. 8.3. The stator has a polyphase winding which produces a rotating magnetic field. The rotor has a direct-current exciting field. If the rotor runs in synchronism with the rotating field of the stator, the two fields of S and R form in reality one single magnetic field crossing the air-gap, as is shown in Fig. 8.3. The important feature of this situation is that the lines of the

[3] See footnote [3], page 191.

[6] A. Andronov, A. Witt, S. Chaikin, *Theory of Oscillations* (in Russian); this book is the second edition (1959) of A. Andronov and S. Chaikin, *Theory of Oscillations* (original text in Russian), Moscow, 1937.

magnetic force cross the air-gap *obliquely*. It is also known from the Faraday-Maxwell theory that the ponderomotive (mechanical) forces due to the magnetic field act always in the direction of the magnetic lines of force. Everything happens as if these lines were elastic strings. As these lines cross the air-gap obliquely, it is clear that the tangential components of these Maxwellian tensions manifest themselves as a mechanical moment which the stator applies to the rotor enabling it to do the mechanical work. If A and B are the centers of the magnetic field distributions (of opposite polarities) on S and R, it is clear that, on that basis, the angle θ between the radii OA and OB is a measure of the driving moment, O being the center of rotation.

If the resisting moment M, which the rotor has to overcome, increases, the angle θ also increases, but as the driving moment $f(\theta)$ is a nonlinear function of θ such that, beyond a certain value, $df(\theta)/d\theta < 0$, the angle θ increases ultimately more than in proportion to the resisting moment and the value of the driving moment may even change its sign for a sufficiently large value of θ. For a two-pole scheme, this happens when $\theta = \pi$ but, if the motor has several pairs of poles, this happens when the geometrical angle reaches the value π/p, p being the number of pairs of poles. When this occurs, the rotor "drops out of synchronism" with the rotating field of the stator. The synchronism lost at one pair of poles may be still momentarily established at the next pair when the distributions of magnetic force happen to be again of opposite polarities; then it may be lost again, and so on. This abnormal or "asynchronous" performance of a synchronous motor is precisely the one in which appear periodic trajectories of the second kind.

Before proceeding with the analysis of these complicated phenomena, it is useful to simplify first the d.e. (2.1) by assuming $f(\theta) = C \sin \theta$. If one divides the d.e. by C and changes the independent variable from t to τ, where $\tau = \sqrt{(C/A)}t$, the d.e. becomes

$$\ddot{\theta}_\tau + \alpha\dot{\theta}_\tau + \sin \theta - \beta = 0 \tag{2.2}$$

where the subscript τ indicates differentiations with respect to τ. As no confusion is to be feared from now on, we can drop this subscript. The values of the constants are $\alpha = B/\sqrt{AC} \geq 0$; $\beta = M/C \geq 0$. Written as an equivalent system, (2.2) is

$$\dot{\theta} = z; \qquad \dot{z} = -\alpha z - \sin \theta + \beta \tag{2.3}$$

and the d.e. of the integral curves is

$$dz/d\theta = (\beta - \alpha z - \sin \theta)/z \tag{2.4}$$

3. Cylindrical phase trajectories of a conservative system

For a conservative system $\alpha = 0$ and (2.4) becomes

$$dz/d\theta = (\beta - \sin\theta)/z \tag{3.1}$$

This d.e. admits a simple integration which gives

$$z = \pm\sqrt{2(\beta\theta + \cos\theta) + C} \tag{3.2}$$

where C is an integration constant.

The construction of integral curves for the cylindrical surface can be made first on the plane, after which these curves can be wrapped on a circular cylinder of radius 1. For the planar construction of trajectories

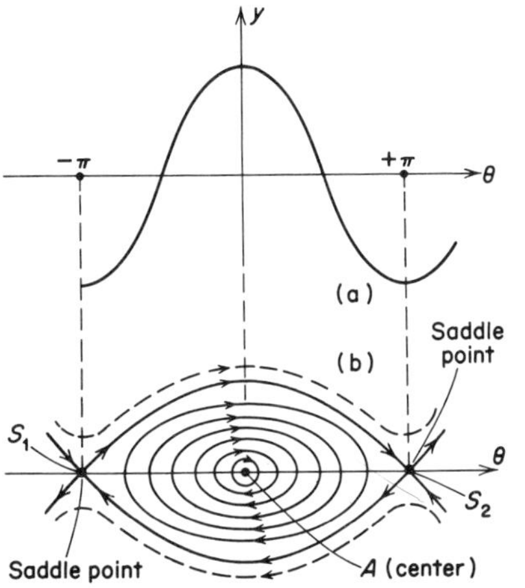

FIGURE 8.4

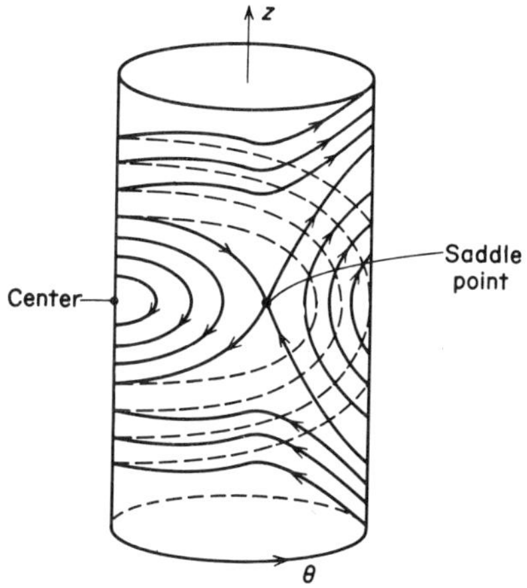

FIGURE 8.5

$z(\theta)$ given by (3.2), we can follow the graphical procedure outlined in the beginning of Chapter 2, namely, we first set $y = 2(\cos\theta + \beta\theta)$ and calculate $z = \pm\sqrt{y + C}$. For any value of C we obtain two values of $z(\theta)$, provided $y + C > 0$. The results will depend also on β.

If $\beta = 0$ (that is, the external moment absent), one obtains the results shown in Fig. 8.4 in which (a) gives the auxiliary curve $y(\theta)$, and (b) gives the integral curves $z(\theta)$ for different values of C. For $C = -2$, one has one point A (Fig. 8.4b) which is a singular point, a center. For $-2 < C < 2$ one has a family of closed integral curves with vertical tangents on the θ axis; for $C = 2$ this gives two curves passing through the points S_1 and S_2. These curves are separatrices of the saddle points; for $C < -2$ there are no integral curves. The same situation was encountered in Chapter 2 (Fig. 2.3), but here the curves $z(\theta)$ are to be wrapped on the surface of a circular cylinder, as shown in Fig. 8.5.

The curves going outside the separatrices (shown in broken lines in Fig. 8.4b) clearly close around the cylinder and thus represent periodic trajectories of the second kind, but they are not limit cycles, since the system is conservative.

If $\beta \neq 0$, the situation differs as $\beta < 1$ or $\beta > 1$. For $\beta < 1$, the curve $y = 2 \cos \theta + 2\beta\theta$ has maxima and minima given by $\theta_1 = \arcsin \beta + 2k\pi$ and $\theta_2 = \arcsin \beta + (2k + 1)\pi$, respectively. For $\beta = 1$, the curve has neither maxima nor minima, but only an inflexion point at $\theta = \pi/2$; at this point the tangent to the curve is horizontal. For $\beta > 1$, the curve increases monotonically. If $\beta < 1$, the graphical construction yields a singular point of the type center and a saddle point, as shown in Fig. 8.6. When these curves are wrapped on the cylinder, one has the situation depicted in Fig. 8.5; in this case there are no trajectories of the second kind, but only those of the first kind, encircling the center inside the separatrix.

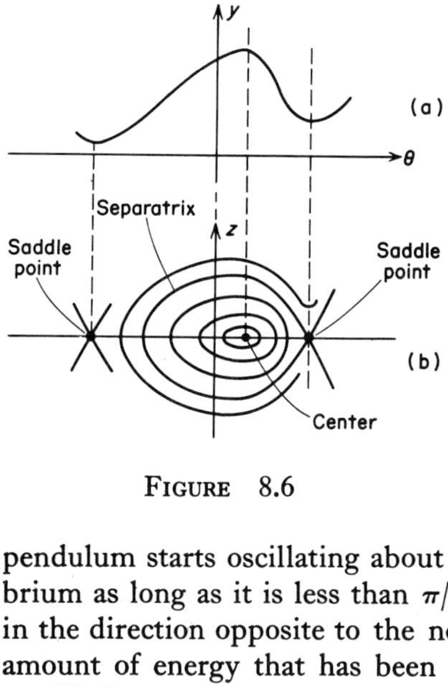

FIGURE 8.6

For $\beta = 1$ one has a singular point of a higher order resulting from the confluence of a center with a saddle point; in this case there are no closed trajectories; the same conclusion holds for $\beta > 1$.

The physical significance of this analysis is sufficiently simple if one considers a physical pendulum acted on by a constant external moment. If this moment is not too large, the pendulum starts oscillating about the new (displaced) position of equilibrium as long as it is less than $\pi/2$. In this case the pendulum moving in the direction opposite to the new equilibrium point releases the same amount of energy that has been stored owing to the displacement of its equilibrium point under the effect of the constant external moment. If, however, the initial deviation is sufficiently great, the effect of the constant external moment may swing the pendulum over its upper (unstable) position of equilibrium which will result in rotary motion increasing in speed after each turn. This occurs when $\beta < 1$ with the initial condition (the initial deviation) large enough to enable the external moment to cause a rotary motion.

Finally, if $\beta > 1$, the external moment exceeds the moment of gravity for any initial position including $\theta = 0$, when the pendulum is at its lowest point. This results in a rotary motion for any initial condition.

All this follows from the consideration of total energy (potential as well as kinetic) stored in the system, inasmuch as we consider here a conservative system ($\alpha = 0$).

Summing up, for conservative systems, the behavior of trajectories on the phase cylinder does not differ much from that of the planar systems, except for the fact that the existence of trajectories of the second kind is more readily seen on the cylinder than in the plane. The whole situation becomes apparent when the planar trajectories are wrapped on the cylinder surface, as was previously mentioned.

4. Cylindrical phase trajectories of nonconservative systems

More interesting and, at the same time, more complicated, is the behavior of nonconservative systems of the type (2.1), which were first investigated by Tricomi, Andronov[6] and others. A brief outline following the presentation of Andronov will suffice here.

We assume now that $\alpha \neq 0$ and $\beta \neq 1$, and consider two cases: $\beta > 1$ and $\beta < 1$.

(1) $\beta > 1$

In this case the system (2.3) has no singular point since $|\sin \theta| \leq 1$. Moreover, the resisting moment M in this case is larger than the maximum value C of the restoring moment $f(\theta) = C \sin \theta$, owing to the electrodynamic couple applied to the rotor R. Intuitively this means that the rotor cannot keep in synchronism with the rotating field, but "drops out of step" or "slips" continuously as previously mentioned. One may expect, therefore, that in this case an *asynchronous* operation will set in, if we are able to show that a periodic solution is still possible.

In order to show this we have to try to find two solutions, $z_1(\theta)$ and $z_2(\theta)$, such that

$$z_1(\theta + 2\pi) \leq z_1(\theta) \qquad \text{and} \qquad z_2(\theta + 2\pi) \geq z_2(\theta) \tag{4.1}$$

If such solutions exist, the existence of the periodic solution $z_0(\theta + 2\pi) = z_0(\theta)$ will follow from the consideration of continuity.

Since $z = \dot{\theta}$, the conditions (4.1) have the obvious meaning: the second condition alone means that, after dropping out of step (that is, loss of velocity $\dot{\theta}$), the rotor is pulled again into synchronism; in such a case we shall not have the asynchronous operation we are looking for. The first condition alone, on the contrary, means that with each "skipping" 2π the velocity $\dot{\theta}$ will decrease, that is, the rotor will finally stop.

[6] See footnote [6], page 193.

If one is able to show that two such solutions exist, one can assert that there must be a periodic solution z_0 somewhere between z_1 and z_2, since no singular points exist in this case that might account for some other situation besides the one just mentioned. The argument is conducted by considering the isocline $dz/d\theta = 0$, which is given by the curve $K: z = (\beta - \sin\theta)/\alpha$. The physical significance of this isocline is obvious:

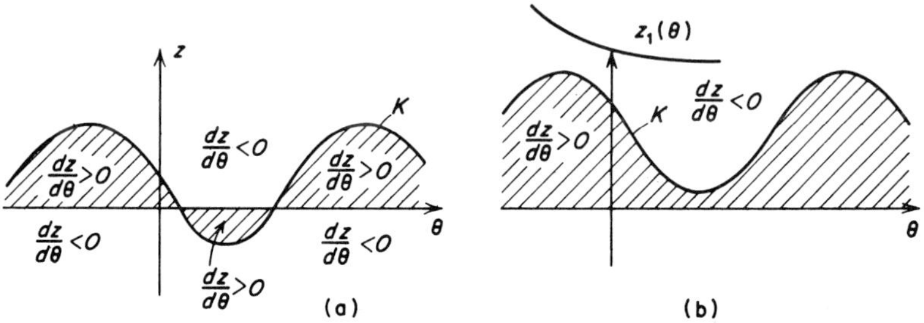

FIGURE 8.7

it characterizes such points in the operation of the motor for which, to a variation $d\theta$ in the angle θ between the stator and the rotor fields, corresponds no change in the velocity $\dot{\theta}$; this is clearly a kind of a threshold separating the regions in which $(dz/d\theta) > 0$ from those in which $(dz/d\theta) < 0$. It is observed that the curve K crosses the θ axis if $\beta < 1$ and is always above that axis if $\beta > 1$; the curve is a sinusoid merely shifted in the direction of the z axis, as shown in Fig. 8.7. The shaded and nonshaded areas of the curve correspond, respectively, to the regions where $(dz/d\theta) > 0$ and $(dz/d\theta) < 0$. As we consider now the case $\beta > 1$, we have to deal with the curve K shown in Fig. 8.7b. For the solution $z_1(\theta)$ we can choose a solution passing through the point $(\beta + 1)/\alpha$, as this point is above K and is thus in the region where $(dz/d\theta) < 0$; this fulfills the condition imposed by the first inequality (4.1).

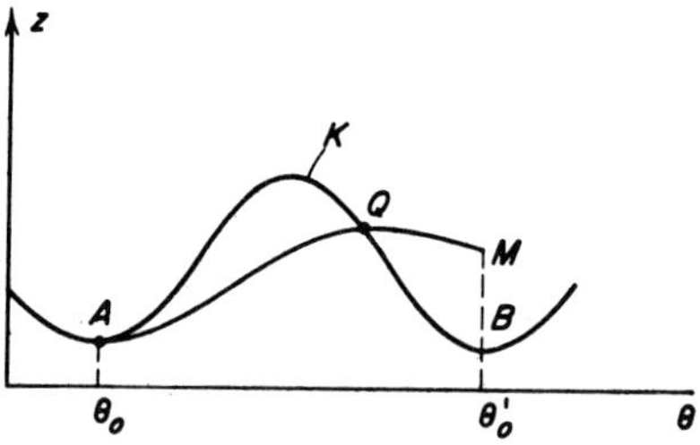

FIGURE 8.8

For the second solution $z_2(\theta)$ (Fig. 8.8), one can take the solution passing through lowest point A of K, with the coordinates $\theta = \pi/2$, $z = (\beta - 1)/\alpha$. The trajectory issuing from this point will remain in the region where $(dz/d\theta) > 0$ until it reaches the point Q on K, at which $(dz/d\theta) = 0$; beyond this point z_2 will be in the region where $(dz/d\theta) < 0$, so that after $\Delta\theta = 2\pi$ (counting from $\theta_0 = \pi/2$), the solution $z_2(\theta)$ will intersect the line $\theta_0' = 5\pi/2$ at a point M which is never lower than B. Thus, for the trajectory $z_2(\theta)$ so selected, we have $z_2(\pi/2 + 2\pi) > z_2(\pi/2)$ which satisfies the second condition (4.1).

It is to be noted that in this argument we have chosen a special initial condition $\theta_0 = \pi/2$, but it is easy to show that one could use any θ in the interval $(0,\pi/2)$. In fact, suppose that the first condition (4.1) holds for some $\theta = \theta_1$; then $\varphi(\theta) = z_1(\theta + 2\pi) - z_1(\theta)$ is a continuous function, negative for $\theta = \theta_1$. Assume that $\varphi(\theta)$ could be made positive for some $\theta = \theta_2$; then one can assert on the basis of continuity that for some $\theta = \theta_3$, $\varphi(\theta_3) = 0$. But this would mean that $z_1(\theta)$ is a periodic solution. Hence, the choice of θ in both conditions (4.1) is immaterial and it is sufficient to prove the statement for some particular θ_0 as we did above for $\theta_0 = \pi/2$.

It remains to be shown that the periodic solution $z_0(\theta)$ whose existence has just been proved is unique. For this we integrate the second d.e. (2.4) between θ_1 and $\theta_1 + 2\pi$, which gives

$$\frac{1}{2}[z^2(\theta_1 + 2\pi) - z^2(\theta_1)] = -\alpha \int_{\theta_1}^{\theta_1+2\pi} z d\theta + 2\pi\beta \tag{4.2}$$

For the periodic solution $z_0(\theta)$ this reduces to

$$\int_{\theta_1}^{\theta_1+2\pi} z_0(\theta) d\theta = 2\pi\beta/\alpha \tag{4.3}$$

If there were two periodic solutions, say z_{01} and z_{02}, one could have either $z_{01} > z_{02}$ or $z_{01} < z_{02}$, since the solutions cannot intersect. In this case the corresponding integrals would be unequal, but in view of (4.3) this is impossible. Hence the periodic solution $z_0(\theta)$ is unique.

Summing up, if $\beta > 1$, an asynchronous periodic solution always exists, and it is necessarily of the *second kind*, that is, the closed trajectory turns around the cylinder. There are no periodic solutions of the first kind since there are no singular points.

(2) $\beta < 1$

In this case singular points exist, namely: $z = 0$; $\beta - \sin\theta = 0$. There are two kinds of points:

(a) Points A_k: $z = 0$; $\quad \theta = 2k\pi + \theta_0$
(b) Points B_k: $z = 0$; $\quad \theta = (2k - 1)\pi - \theta_0$

where $\theta_0 = \arcsin\beta$; k an integer, and we assume $0 < \theta_0 < (\pi/2)$. In order to ascertain the nature of these singular points, one applies the usual variational procedure. Thus, for A_k one sets, $\theta = 2k\pi + \theta_0 + \delta\theta$ and develops $\sin\theta$, keeping only the terms of the first order in $\delta\theta$ which gives the characteristic equation:

$$S^2 + \alpha S + \cos\theta_0 = 0 \tag{4.4}$$

whose roots are

$$S_{1,2} = -\frac{\alpha}{2} \pm \sqrt{\frac{\alpha^2}{4} - \cos\theta_0} \tag{4.5}$$

Hence, as $(\alpha^2/4) > \cos\theta_0$ or $(\alpha^2/4) < \cos\theta_0$, one has either a stable node or a stable focus.

A similar procedure applied to the points B_k leads to the characteristic equation

$$S^2 + \alpha S - \cos\theta_0 = 0 \tag{4.6}$$

whose roots are

$$S_{1,2} = -\frac{\alpha}{2} \pm \sqrt{\frac{\alpha^2}{4} + \cos\theta_0} \tag{4.7}$$

but as $(\cos\theta_0) > 0$, the roots are real and of opposite signs, which indicates a saddle point. The slopes of the separatrices at these saddle points are given by the equation

$$m^2 + \alpha m - \cos\theta_0 = 0 \tag{4.8}$$

so that these slopes are

$$m = -\frac{\alpha}{2} \pm \sqrt{\frac{\alpha^2}{4} + \cos\theta_0} \tag{4.9}$$

If $\alpha = 0$ (conservative system), instead of nodes or foci, one has centers for singular points A_k. As to B_k, they still remain saddle points.

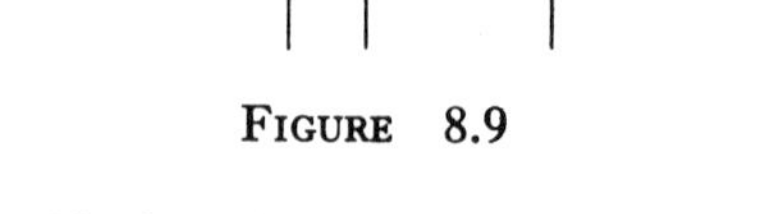

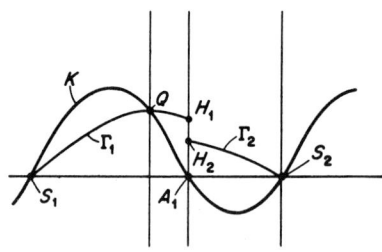

FIGURE 8.9

We can now analyze the operation of the synchronous motor in this case ($\beta < 1$) following an argument similar to that used previously (Fig. 8.9).

The points S_1 and S_2 are two saddle points with the separatrices Γ_1 and Γ_2; the curve K is again the isocline $dz/d\theta = 0$. One ascertains as previously, that there exists a solution $z_1(\theta)$, for which the first condition (4.1) holds. For the second condition we consider the curves Γ_1 and Γ_2 issuing from the separatrices of the saddle points S_1 and S_2. The slope of Γ_1 at the point S_1 is less than that of the sinusoid K at the same point, as is seen from (4.9) and from the expression for the slope of K at the point S_1. The slope of Γ_1 thus decreases, since it becomes zero at the point Q at which Γ_1 intersects the isocline K. Beyond the point Q the curve Γ_1 decreases and intersects the straight line $\theta = \theta_0'$ at the point H_1. The other curve Γ_2 issuing from the saddle point S_2 intersects the same line $\theta = \theta_0'$ at the point H_2. Denote the ordinates of H_1 and H_2 by h_1 and h_2, respectively.

It can be shown by a more detailed analysis[4] that for a sufficiently small α (that is, small damping) $h_2 < h_1$. In such a case the trajectory Γ_1 and other neighboring trajectories above it satisfy the condition

$$z_2(\theta_1) \leq z_2(\theta_1 + 2\pi)$$

Hence, for a sufficiently small α, there exists a periodic solution $z_0(\theta_1) = z_0(\theta_1 + 2\pi)$; and, by the same argument as before, one can show that this solution is unique. It can be shown also that for a sufficiently large α, $h_2 > h_1$. There is thus a certain value $\alpha = \alpha_0$ for which $h_2 = h_1$, so that the curves Γ_1 and Γ_2 form one continuous curve. Moreover, it can be shown that for $\alpha > \alpha_0$ there are no periodic solutions, and for $\alpha < \alpha_0$ there exists only one solution in the upper half plane ($z > 0$). Finally, one can ascertain also that there are no closed solutions of the first kind; this can be shown from a more detailed analysis of the form of the separatrices for $\alpha < \alpha_0$; $\alpha = \alpha_0$; and $\alpha > \alpha_0$. Thus only for $\alpha < \alpha_0$ does there exist a periodic solution of the second kind; for $\alpha > \alpha_0$ there is no such periodic solution.

Summing up, for $\beta > 1$ and any α there exists always a periodic solution of the second kind; for $\beta < 1$ a periodic solution exists provided $\alpha < \alpha_0$; moreover, these periodic solutions are stable. There is an additional circumstance to be noted: for $\beta > 1$ the periodic solution is reached for any initial conditions, whereas for $\beta < 1$ there is a certain region of initial conditions for which the system always falls into step, which manifests itself by a gradual approach to a stable focal point at A_1.

It is seen thus that the case $\beta < 1$ is more complicated than the case $\beta > 1$. The physical significance of this is sufficiently clear; in fact for $\beta > 1$ there are no singular points, and this, as we saw, means that the resisting moment in the synchronous motor always exceeds the synchronizing moment. The rotor "slips" continuously and a stationary condition exists when the work done by the *constant* resisting moment is compensated for by the electrodynamic moment at *a lower* (asynchronous) *mechanical velocity of rotation.*

The case when $\beta > 1$ can be illustrated by the behavior of a physical pendulum to which one applies a sufficiently large moment, as the result of which the pendulum, instead of oscillating, starts rotating in the same

[4] See footnote [4], page 192.

direction. In this case the motion reaches a stationary state when the work done during one revolution by the driving moment is balanced by the dissipation of energy in friction.

5. Topological configurations on the cylinder

Summing up the various cases discussed in the preceding sections, one can state that everything which holds for the phase plane, holds also for the cylindrical phase space, inasmuch as this is merely a matter of "wrapping" the phase-plane diagram on the cylinder, as we have mentioned. This means that for a limit cycle of the first kind there must be a singular point on the surface of the cylinder. The essential feature of the cylindrical phase-surface representation is trajectories of the second kind.

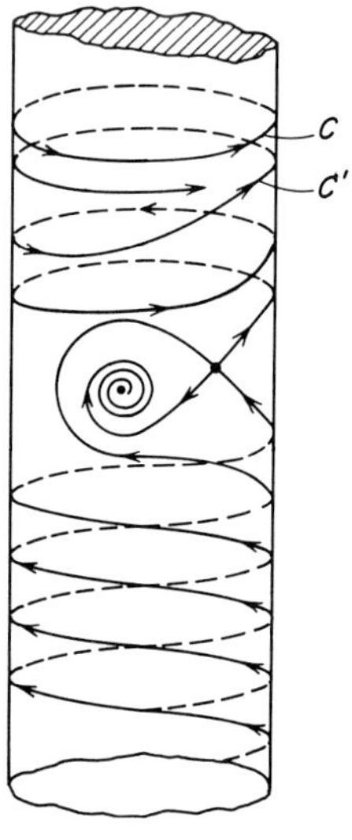

FIGURE 8.10

In the case of conservative systems, the periodic trajectories of the second kind form a family of closed curves depending on a parameter, in the same manner as happens in the phase plane. As regards the limit cycles of the second kind, they are some closed curves C (going around the cylinder) to which approach the helicoidal curves C' (Fig. 8.10) with a gradually decreasing pitch. In the process of "flattening out" the cylinder into the plane, these helicoidal trajectories become the usual spiral curves C'.

6. Oscillations of a synchronous motor

In the preceding section we outlined the physical character of the problem in a rather general manner, but one can elaborate it in more detail if one introduces the theory of the synchronous motor as did Vlasov[4] in his work. We cannot enter here into all details of this interesting investigation, but merely mention its salient points; the d.e. of the synchronous motor is ultimately reduced to the following form:

[4] See footnote [4], page 192.

$$\ddot{\varphi} + \sin\varphi + \left[\frac{c}{k}\cos\rho\sin\varphi(\sin\varphi - 2k\sin\rho) - kb\frac{\sin 4\rho}{\cos\rho}\right]\dot{\varphi} = \frac{\beta}{k\cos\rho} + k\sin\rho \quad (6.1)$$

This d.e. can be obtained from the d.e. of the synchronous motor by introducing notations:

$$k = E_0/U_0; \qquad a^2 = mp^2kU_0\cos\rho/2\theta\omega x; \qquad c = my^2a/2\omega xR_e;$$
$$b = a/4\omega; \qquad \beta = 2M\omega x/mpU_0^2$$

where θ is the moment of inertia of the rotor, p the number of pairs of poles, m the number of phases, E_0 the amplitude of the induced voltage per phase; $y = dE_0/di_e$; i_e is the exciting current; $\omega = 2\pi f$; f is the frequency, U_0 the amplitude of the applied voltage per phase, $\pi - \gamma$ the angle between the vectors E and U; $\rho = \arctan(r/x)$; r is the ohmic resistance (per phase) of the rotor, x the reactance per phase, M the resisting moment, and R_e the resistance of the exciting winding.

If one introduces $\varphi = \gamma + \rho$, at $t = \tau$, one obtains (6.1), where the differentiations are with respect to τ. The equivalent system is obtained, as usual, by setting $\dot{\varphi} = z$, etc.

Summing up, one obtains the familiar system

$$\dot{\varphi} = z; \qquad \dot{z} = \lambda - \sin\varphi + f(\varphi)z \qquad (6.2)$$

The rest reduces to the analysis which has been reproduced in the preceding sections.

The experimental part is particularly interesting and we mention a few principal conclusions, as well as the assumptions under which this investigation was carried out.

(1) It is assumed that the magnetic fields have sinusoidal distributions.
(2) The resistance of the field winding is small.
(3) Magnetic stray fluxes are negligible.
(4) Iron loss is negligible in comparison with ohmic resistance.
(5) The frequency of "hunting" is small compared to the frequency f.
(6) Damping winding removed.

In view of (6.1) it was possible to observe also oscillations of the first kind. The quantity $\rho = \arctan(r/x)$ was used as the parameter, and it was possible to ascertain its bifurcation value $\rho = \rho_0$ for which the limit cycle (of the first kind) disappears in its coalescence with the separatrix.

For sufficiently large z, trajectories become unstable. For other values of ρ, other conditions exist (either a limit cycle or a stable singularity when the motor runs without "hunting", etc.).

Most of these experiments were carried out with knowledge of the numerical values of the various parameters; therefore these results are quantitative. The phase trajectories were plotted on a cylindrical surface; some of these cylindrical diagrams were given in the preceding sections.

So far as is known, no other applications of cylindrical phase space have been made so far, but the matter is still of current interest, especially in the theory of the synchronous motor. Thus, for instance, the assumption of a linear friction is hardly justified when most of this friction consists in "windage," so that the basic d.e. (2.1) is presumably still too oversimplified to account completely for these complicated phenomena.

7. Toroidal phase surface

Let the phase surface with two cyclic coordinates φ and θ be the surface of the torus whose equations are

$$x = (R + r\cos\theta)\cos\varphi;$$
$$y = (R + r\cos\theta)\sin\varphi;$$
$$z = r\sin\theta \qquad (7.1)$$

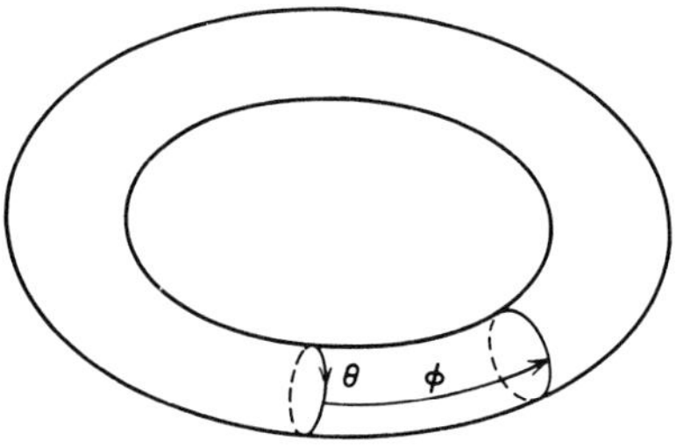

FIGURE 8.11

where $0 \leq \varphi < 2\pi$; $0 \leq \theta < 2\pi$; $0 < r < R$.

One may call the coordinate φ (Fig. 8.11) the "longitude," and θ the "latitude," and any point on the surface of the torus is determined in terms of these two coordinates.

The problem in this case is to find out whether, for a given differential system, there are periodic trajectories "of the third kind" which are characterized by a certain number of complete (2π) rotations in the θ coordinate and some other number of such rotations in the φ coordinate.

There may obviously also be closed trajectories of the second kind; these are of two different types. One of these types is obtained when φ is held constant and θ varies, and the other one when θ is constant and φ varies. There may be also closed trajectories of the first kind, but these require the presence of singular points on the toroidal surface.

Consider a dynamical system of the form

$$d\varphi/dt = \Phi(\varphi,\theta); \qquad d\theta/dt = \Theta(\varphi,\theta) \qquad (7.2)$$

where Φ and Θ have period 2π in φ and θ.

We introduce two restrictions: (a) the system has no singular points on the surface of the torus (which eliminates closed trajectories of the first kind), and (b) the function Φ never vanishes. One can then investigate the behavior of integral curves from the d.e.

$$d\theta/d\varphi = \Theta(\varphi,\theta)/\Phi(\varphi,\theta) = A(\varphi,\theta) \tag{7.3}$$

where $A(\varphi,\theta)$ is a continuous function with period 2π in both arguments and we shall assume that the unicity condition is fulfilled.

It is clear that, instead of studying the motion of the representative point on the surface of the torus, one may use the plane $-\infty < \varphi < +\infty$; $-\infty < \theta < +\infty$ formed by squares of side 2π; that is, $2\pi n \leq \varphi < 2\pi(n + 1)$, $2\pi m \leq \theta < 2\pi(m + 1)$, m and n integers.

Since the function $A(\varphi,\theta)$ is bounded, a solution of (7.3) with arbitrary initial conditions φ_0, θ_0 is defined for all φ. Let

$$\theta = u(\varphi,\varphi_0,\theta_0) \tag{7.4}$$

be the general solution. For the sake of simplicity assume $\varphi_0 = 0$; the solution then depends on only one parameter, θ_0. Out of this family we consider the curve L_0 corresponding to $\theta_0 = 0$, L_0: $\theta = n(\varphi,0)$.

We consider the behavior of L_0 as φ varies. Two cases are possible: (1) L_0 passes through an "integer point" $(2\pi p,2\pi q)$, p and q being integers and $p > 0$. Clearly, in view of the periodicity, L_0 will pass also through the points $2\pi pn$, $2\pi qn$, where $n = \pm 1, \pm 2, \ldots$

On the surface of the torus L_0 corresponds to a closed curve making p rotations in "longitude" and q rotations in "latitude." This is precisely what we have defined as *periodicity of the third kind*. In such a case one has, according to Poincaré,[1] the relation

$$\lim\,[u(\varphi,\theta)/\varphi]_{\varphi\to\infty} = p/q \tag{7.5}$$

for all θ_0. (2) The second possibility arises when L_0 does not pass through an integer point. It can be shown that, if L_0 goes, say, above one point (p,q), it also goes above all points $(2\pi pn, 2\pi qn)$. In the proof this conclusion results from the assumed unicity of solutions.

Each of these integer points is characterized by the ratio $q/p = r$ but, to a given r corresponds an infinity of such points.

If one takes r as a parameter, it is shown that, if a point of the set corresponding to $r = r_1$ lies above L_0 and if $r_2 > r_1$, the points of the set r_2 lie also above those of r_1, etc.

Consider the two classes of integer points for one of which the r numbers correspond to the points above L_0 and, for the other one, below it. Thus, if L_0 cuts $\varphi = 2\pi$ for $\theta = \theta_1$, then the rationals smaller than θ_1 belong to one class and those larger than θ_1 belong to the other class.

These classes of rationals define a real number μ, *the rotation number*, which characterizes the distribution of integral curves on the torus. It is shown that

$$\mu = \lim [u(\varphi,\theta_0)/\varphi]_{\varphi\to\infty} \tag{7.6}$$

The rotation number may be regarded, therefore, as the average rotation of a trajectory on the surface of the torus.

Poincaré establishes the following theorems:

(a) If the rotation number is rational, there exists a closed trajectory on the torus.

(b) If this number is irrational, closed trajectories are impossible; the surface of the torus is densely covered with trajectories which never close; any point on the surface of the torus is approached by every trajectory arbitrarily closely; this is the so-called *ergodic case.*

As an example, consider the d.e.: $d\theta/d\varphi = \sin\theta$. It is obvious that there are two closed trajectories: $\theta = 0$ and $\theta = \pi$. The other trajectories are of the form

$$\theta = 2 \arctan [(\tan \tfrac{1}{2}\theta_0)e^{\varphi}]$$

These trajectories approach the limit cycle $\theta = \pi$ for $\varphi \to \infty$, and $\theta = 0$ for $\varphi \to -\infty$. Hence all trajectories approach the stable orbit $\theta = \pi$ and the rotation number is zero.

8. Examples of nonrecurrent trajectories

The use of the toroidal phase space is helpful each time a problem involves two (or several) oscillations with incommensurate frequencies, as, for instance, in the case of the so-called Lissajous curves.

A similar situation often appears in the use of a cathode-ray oscilloscope. It is well known that a wave of a certain frequency, say f, can be made to "stand still" on the screen by means of a special "sweep circuit" synchronized with that frequency. If the wave in question is represented by a Fourier series, a synchronizing arrangement of this kind "stops" not only the fundamental wave but also the harmonics because the period T of the fundamental harmonic is also the period for all harmonics. In this case the trajectory on the torus has a re-entrant path as, in accordance with the preceding theory, to p rotations in one degree corresponds q rotations in the other degree, p and q being in a rational ratio.

In Chapter 21 (Part III) we shall study a special problem characterized by the existence of a certain transcendental spectrum of frequencies between which no rational ratios exist. If one takes two such frequencies, say p and q, and investigates the resultant oscillation, one finds that, if one makes one of these waves (A in Fig. 8.12) stand still, this does not produce the same effect on the other wave B. This is due to the fact that the switching circuit that starts each electronic beam exactly at the same phase point with respect to the wave A produces the wave B with a gradually shifting phase, since the periods of the two waves do not stand in a rational ratio to each other.

It appears to the eye as if the wave B were "riding" on the wave A with a velocity depending on the difference between the existing (incommensurate) ratio of frequencies and the nearest rational ratio. This situation characterizes precisely the existence of an irrational ratio q/p between the two frequencies.

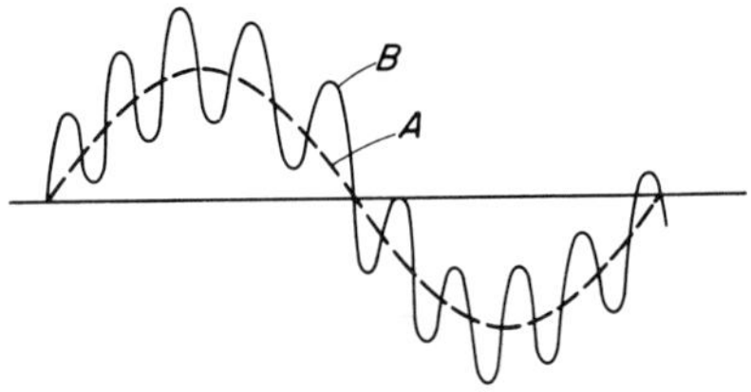

FIGURE 8.12

If a parameter of the system is varied, the whole pattern of the transcendental spectrum generally changes, and it may happen that the incommensurate ratio of frequencies comes nearer to a rational ratio. This manifests itself in a decrease of the speed of "riding" of the wave B on the wave A. It may even happen that, for a suitable value of the parameter, the two frequencies appear, at least momentarily, in a rational ratio, which stops the relative motion of B.

9. Gliding flight

It is possible to reduce the problem of gliding flight to the topological representation on a cylinder.[7] Designating by θ the angle of the tangent to the trajectory, v the speed of the center of gravity of the aircraft, m its mass, F the area of supporting surfaces, g the acceleration of gravity, ρ the air density, and C_x and C_y the aerodynamical coefficients (of drag and lift), the d.e. for the tangential and normal components of acceleration are

$$\begin{aligned} m\dot{v} &= -mg \sin\theta - \tfrac{1}{2}\rho F C_x v^2 \\ mv\dot{\theta} &= -mg\cos\theta + \tfrac{1}{2}\rho F C_y v^2 \end{aligned} \tag{9.1}$$

If one assumes that the moment of inertia of the glider is small in comparison with the aerodynamic forces (which permits neglecting small variations in the angle of attack), one can consider C_x and C_y approximately as constants. Introducing $v = v_0 y$, where $v_0 = \sqrt{2mg/\rho F C_y}$ (that is, the speed of the horizontal flight for which the weight is balanced by the lift)

[7] N. E. Joukovsky, *Collected Works* (in Russian), Moscow, 1959.

and taking as new independent variable t^* given by the formula $t = (v_0/g)t^*$, in the new variables v and t^*, equations (9.1) become

$$\dot{y} = -\sin\theta - ay^2 = F(\theta,y); \qquad \dot{\theta} = -\frac{\cos\theta + y^2}{y} = \Phi(\theta,y) \tag{9.2}$$

where differentiations are with respect to t^* and the constant $a = C_x/C_y$. As conditions $(\theta + 2\pi, y)$ and (θ, y) are physically identical, the right-hand terms in (9.2) are periodic functions of θ with period 2π, and the cylindrical phase space is obviously adequate. One can use a circular cylinder with the y axis along the generating line of the cylinder and the θ axis along the circular cross-section. We consider only trajectories for $y \geq 0$. In such a case the integral curves on the cylinder are given by the d.e.

$$\frac{dy}{d\theta} = \frac{y(\sin\theta + ay^2)}{\cos\theta - y^2} \tag{9.3}$$

It is noted that this d.e. has the solution $y = 0$, which is a special trajectory corresponding to $\theta = -\pi/2$ as soon as the speed v (or y) becomes zero for $y = 0$; $\dot{\theta} = +\infty$ if $-3\pi/2 < \theta < -\pi/2$; and $\dot{\theta} = -\infty$ if $-\pi/2 < \theta < +\pi/2$. This corresponds to an instantaneous rotation of the glider. This absurd conclusion is the result of the hypothesis regarding the constancy of the angle of attack, which does not hold for small speeds.

If one assumes for a moment that $a = 0$, the d.e. (9.3) has the integral:

$$\frac{y^3}{3} - y\cos\theta = C \tag{9.4}$$

It has three singular points of which only one: $\theta = 0, y = +1$ corresponds to the equilibrium of the system (9.2) for $a = 0$; two other singular points are located on a special integral curve: $y = 0$ corresponding to the reversal of the craft for $v = 0$ and which are not, therefore, positions of equilibrium (since $\dot{y} \neq 0$ at these points).

The construction of integral curves is facilitated by the fact that (9.4) can be solved for θ:

$$\theta = \pm \arccos \eta \tag{9.5}$$

This permits tracing the family of the $\eta = \eta(y,C)$ curves; for equilibrium, $\theta = 0$, $y = +1$ degenerates into the singular point of the type center. We refer for the details to Joukovsky [7] but merely mention that according to the value of C, this family is divided into two subfamilies, of which one does not go around the cylinder and the other, on the contrary, goes around it.

[7] See footnote [7], page 207.

The next step is to take into account the aerodynamic forces. The analysis follows the argument analogous to that used previously. The principal conclusion is that, in the presence of air resistance, a periodic solution is impossible and that, after having described a series of closed "loops," the craft approaches a rectilinear descending trajectory with a gradually decreasing swing.

The establishment of d.e. leading to these results is due to Joukovsky (1891) and is based on the observation of the flight of birds.[7]

[7] See footnote [7], page 207.

PART II

QUANTITATIVE METHODS

INTRODUCTION

In Part I we outlined the principal qualitative (or topological) methods and concepts.

In Part II we propose to study the quantitative methods which reduce to the approximations involving solutions in the form of certain power series. This implies that one deals always with approximations, depending on the number of terms used in the series solution. In fact, the cases when the d.e. can be integrated in a closed form are very rare.

The importance of quantitative methods, as the name implies, lies in the fact that they permit obtaining *numerical results*; their drawback is that they do not give any idea regarding the totality of all possible situations as do the topological methods. For that reason it is often useful to start the exploration of a new problem by a qualitative method and, once the character of the solution has been ascertained, to use one of quantitative methods for numerical determination of the solution with a desired degree of approximation.

The origin of quantitative methods lies in astronomical calculations where it received the name of the *perturbation method*. This term acquired later a somewhat different meaning.

There are two different stages in the development of the perturbation theory, one preceding the work of Poincaré and the other, the modern one, following it. We shall be concerned mostly with the latter. It is, however, useful to say a few words about the earlier period limited to astronomical calculations and, as far as known, initiated by Poisson (around 1830). This somewhat broader viewpoint on the situation will be an aid in understanding the important contribution of Poincaré and of later mathematicians, which forms the principal object of Part II.

It is plausible to consider that the motion of a celestial body (for example, the earth) is specified by the d.e. of the form

$$\frac{dx^i}{dt} = X_0^i + \mu X_1^i + \mu^2 X_2^i + \ldots; \qquad i = 1, 2, \ldots, n \qquad \text{(II.1)}$$

where x^i is a coordinate and $X_j^i(x^1,\ldots, x^n)$ are certain functions of coordinates representing the gravitational forces emanating from other bodies arranged according to the order of their smallness ($\mu \ll 1$ is a small factor). On this basis the first (finite) term X_0^i is the gravitational force of the sun and $\mu X_1^i, \mu^2 X_2^i,\ldots$, are much smaller forces (perturbations) from other planets.

This led Poisson to suggest that the solution should be sought in the form of a series:

$$x^i(t) = x_0^i(t) + \mu x_1^i(t) + \mu^2 x_2^i(t) + \ldots + \mu^m x_m^i(t) + \ldots \qquad \text{(II.2)}$$

If one substitutes (II.2) into (II.1) and equates the coefficients with the like powers of μ, one obtains a number of d.e. which can be integrated recursively. We shall go into this matter more in detail later, but at present it is sufficient to mention that this procedure has been adopted later for the quantitative study of nonlinear oscillations. In the simplest case of a system with one degree of freedom with which we shall be concerned in the first place, the d.e. are generally of the form

$$\ddot{x} + x + \mu f(t, x, \dot{x}) = 0 \qquad \text{(II.3)}$$

where $f(t, x, \dot{x})$ is an analytic function of x and $\dot{x}$ and periodic in t with period 2π; μ is a small number (the parameter). The application of the Poisson procedure at this stage is purely formal and it is not certain whether it is legitimate.

In fact, it is not certain as to whether the somewhat heuristic astronomical hypothesis regarding the form of the solution can be transferred into an entirely different problem in which "the perturbation" concerns *the form* of the d.e. and has nothing to do with a more or less intuitive concept of perturbation of the gravitational forces.

A serious difficulty has been encountered in applications of this procedure in the form of the so-called *secular terms*, that is, such terms in (II.2) which grow up indefinitely when $t \to \infty$ and thus destroy the convergence of the series solution. By "secular terms" are meant the terms containing the variable t (the time) as a multiplier of trigonometric functions.

Astronomers were first to encounter these terms in their calculations, and the practice consisted in introducing corrections for the effect of these terms. This did not present any special difficulty as the astronomical tables were usually computed for a certain number of years so that the secular corrections remained always within a reasonable bound.

However, toward the end of the last century there appeared a tendency to determine the constants of integration so as to *eliminate the secular terms*, as the approximations develop, instead of correcting the expansions for these secular terms. The names of Gylden, Lindstedt, Bohlin, and other astronomers of the end of the last century are associated with these developments, but it was especially Poincaré who investigated these new methods in detail.

The secular terms appear whenever one uses a finite number of terms in an infinite series representing a certain function. Thus, for instance, in the case of

$$\sin t = t - \frac{t^3}{3!} + \frac{t^5}{5!} - \dots$$

if one limits the expansion, say, to the first three terms, the approximation of $\sin t$ by these terms is satisfactory up to a certain value of t in $f(t) = t - (t^3/3!) + (t^5/5!)$, after which the two functions $\sin t$ and $f(t)$ will differ from each other by a quantity exceeding the prescribed limit of accuracy. If one takes more terms in $f(t)$ the limit of accuracy can be extended to $t = t_2$, $t_2 > t_1$ but, again, for $t > t_2$ the difference $|\sin t - f(t_2)|$ will exceed the prescribed limit, and so on. In actual calculations one does not know whether the function is periodic or not, and this complicates the situation still further as there is no certainty as to whether the lack of periodicity is in the nature of the unknown function itself or is due to the effect of the secular terms.

Lindstedt suggested (Section 3, Chapter 8) the elimination of secular terms in each step of the approximation procedure owing to the availability of the integration constants, so that the ultimate series solution is free from secular terms. The question of secular terms was of importance to the astronomers when the prediction of the coordinates of celestial bodies was more vital than the question of periodicity.

With Poincaré opens the new period in the perturbation theory in that the periodicity begins to play the primary role, and this question became of fundamental importance in the theory of oscillations. In fact, in astronomy the predictions are made for a relatively small number of periods (years), and the question whether the motion is strictly periodic or is only approximately periodic is not as vital as in the oscillation theory where the departure can be observed in a very short time. For that reason the formulation of the conditions of periodicity is of greater importance for the theory of oscillations than for astronomy. Thus, for instance, an electron-tube oscillator oscillating with a frequency of, say, one megacycle per second in one millisecond passes through the same number of cycles that our planet accomplishes in one thousand years; thus the aims of astronomy and of the theory of oscillations are clearly different.

Instead of looking for the convergence of the series expansion (II.2) for $m \to \infty$, the applicability of the quantitative methods is now sought for the asymptotic behavior for fixed m, for $\mu \to 0$, and for $t \to \infty$, as will be shown in Chapters 14 and 15 concerning the recent developments of Krylov-Bogoliubov-Mitropolsky.

Although the question of secular terms is of a relatively secondary importance for the theory of oscillations, we outline it briefly in Chapter 9 in order to have a better grasp on the situation. The modern aspect of the perturbation theory (in the sense of the perturbation of *the form* of the d.e.) may be considered by itself without any connections with astronomy.

The problem of *periodic solutions* (or the problem of Poincaré) can be formulated with reference to (II.3) as follows: If $\mu = 0$, we have a harmonic oscillator having an infinity of periodic solutions depending on two arbitrary constants of integration. Topologically this amounts to a continuous family of concentric circular trajectories of radius $r = \sqrt{x^2+\dot{x}^2}$. On each of these circles there exists an infinity of "motions" differing from each other by the phase t_0; one has thus a two-parameter family (r_0,t_0) of possible motions.

What will happen if we "perturb" the d.e., $\ddot{x} + x = 0$ by adding a small term, say, $\mu f(t, x, \dot{x})$? It is clear that it is impossible to answer this question on the basis of any offhand consideration. In fact, assuming that $\mu f(t, x, \dot{x}) = \mu b\dot{x}$, it is seen that the perturbation of the d.e. $\ddot{x} + x = 0$ by the term of this nature destroys the original periodicity; the trajectories, instead of being circles, become now logarithmic spirals (convergent if $b > 0$ and divergent for $b < 0$).

To ascertain the conditions under which (II.3) has periodic solutions constitutes the *problem of Poincaré* which we shall study in Chapters 10 (for systems with one degree of freedom) and 11 (several degrees of freedom). It must be noted that this problem is possible only under the assumption that the parameter μ (in II.3) is *sufficiently small* to guarantee the convergence of the series solution. For that reason the method of Poincaré is often called the *small parameter method.* The solution of (II.3) appears thus ultimately in the form

$$\begin{aligned} x(t) &= x_0(t) + \mu x_1(t) + \mu^2 x_2(t) + \dots \\ \dot{x}(t) = y(t) &= y_0(t) + \mu y_1(t) + \mu^2 y_2(t) + \dots \end{aligned} \tag{II.4}$$

For $\mu = 0$ one obtains the simple harmonic solution.

If one goes to the first approximation: $x(t) = x_0(t) + \mu x_1(t)$; $y(t) = y_0(t) + \mu y_1(t)$, the problem is to determine the functions $x_1(t)$ and $y_1(t)$ in such a manner that $x(t)$ and $y(t)$ remain periodic. An interesting feature of this determination is that it is possible only for definite values $x_0^*(t)$ and $y_0^*(t)$ of the "zero order" solution $x_0(t)$ and $y_0(t)$. The particular zero-order solution: $(x_0^*(t), y_0^*(t))$ is called the *generating solution.*

Summing up, for $\mu = 0$ there exists a continuous two-parameter family

$(x_0(t), y_0(t))$ of the "zero order" solution (solution of the harmonic oscillator); if $\mu \neq 0$, but small, and if some special conditions (Chapter 10) are fulfilled, the family $(x_0(t), y_0(t))$ shrinks to *one isolated solution* $(x_0^*(t), y_0^*(t))$, the generating solution (g.s.) and the periodic solutions are possible only in the neighborhood of this g.s. and are given by

$$x(t) = x_0^*(t) + \mu x_1(t); \qquad y(t) = y_0^*(t) + \mu y_1(t) \tag{II.5}$$

where $x_1(t)$ and $y_1(t)$ are yielded by the same procedure.

If one goes to higher-order approximations, this merely adds terms $\mu^2 x_2(t), \mu^3 x_3(t),\ldots; \mu^2 y_2(t), \mu^3 y_3(t),\ldots$, on the right-hand side of (II.5), each subsequent approximation being determined by the data yielded by the preceding one.

There is another fundamental point in the theory of Poincaré, namely: the justification of the heuristic assumption of Poisson regarding the form (II.2) of the series solution. More specifically, Poincaré shows[1] that if a d.e. contains terms with a small parameter (like II.3), the solution is analytic in terms of this parameter (last section of Chapter 9). This theorem has been proved recently by K. O. Friedrichs[2] in a different manner; the proof given in Section 4, Chapter 9, is due to Goursat. In other words, in all small parameter methods it is legitimate to use the power series arranged according to the ascending powers of the parameter as *the solution* giving rise to successive approximations.

In applications one seldom goes beyond the first approximation for the reason that all qualitative features of an oscillatory process generally are completely revealed by the first approximation. The second (or higher) approximation does not add anything new from the *qualitative* point of view, but merely adds a small quantitative correction of the order μ^2 (or higher) which is usually not justified by the amount of the computing work involved. For that reason, although we indicate below the procedure for carrying approximations beyond the first order, in the various applied problems treated in Part III we shall limit ourselves only to the theory of the first approximation.

It must be mentioned that in some (very rare) cases the first approximation may fail to give the answer and a special study is required. In what follows we will not consider such "pathological" cases, because they are not usually encountered in applications.

As regards the remaining chapters of Part II, Chapter 12 gives an

[1] H. Poincaré, *Les méthodes nouvelles de la mécanique céleste* **T.1**, Gauthier-Villars, Paris, 1892.

[2] K. O. Friedrichs, *J. Inst. of Math. and Mechanics*, New York University, New York, 1946.

outline of the theory of almost periodic oscillations; this theory is less developed than that of periodic oscillations and, from the standpoint of applications, these oscillations are less important. We enter into this question in some detail only because almost periodic oscillations appear often spontaneously in oscillatory systems beyond the range of synchronization when autoperiodic and heteroperiodic frequencies separate and an almost periodic process sets in, but this is a somewhat special case.

Chapter 13 concerns the calculation of the characteristic exponents by successive approximations. As is well known, the principal difficulty in calculation of these exponents is the fact that the characteristic equation for d.e. with periodic coefficients depends on the (unknown) solutions and the latter, in turn, depend on these exponents. This results in a vicious circle from which successive approximation is the only issue.

In Chapters 10, 11, 12, and 13 we have followed closely the exposition of Malkin[3] who introduced considerable improvements in the adaptation of the theory of Poincaré to problems of the oscillation theory. Chapters 14 and 15 outline the recent asymptotic theory of Krylov–Bogoliubov–Mitropolsky (KBM) which is a further extension of the earlier (1937) K.B. theory. It is recalled that in the earlier theory, equations of the first approximation were established directly by a procedure reminding of the classical method of Lagrange; but, for approximations of higher orders, it was necessary to use an additional procedure analogous to the Linstedt method. In the new derivation of the theory, approximations proceed symmetrically from a recursive system of d.e. and, owing to an additional condition, the secular terms are absent in all approximations. Chapter 14 deals with autonomous systems and Chapter 15 with nonautonomous ones. The last two sections of Chapter 15 relate to recent (1955) work of Mitropolsky regarding nonstationary processes.

Chapter 16 concerns the stroboscopic method, developed by the author in collaboration with M. Schiffer. It is based on the transformation theory of d.e., and its purpose is to replace the original nonautonomous system by the d.e. of its "stroboscopic image" having the property that the existence of a stable singular point of the stroboscopic system appears as a criterion for the existence of a stable periodic solution of the original system. Owing to this transformation the difficult problem of determination of characteristic exponents is replaced by a much simpler problem of stability of singular point of the stroboscopic system. We shall make extensive use of this method in Part III devoted to the analysis of the various types of nearly linear oscillations. Chapter 17 outlines some recent attempts to extend the Nyquist criterion to nonlinear systems; its connection with the theory of approximations is somewhat indirect.

[3] I. G. Malkin, *Theory of Stability of Motion* (in Russian), Moscow, 1952.

Chapter 9

PERTURBATION METHOD

1. Secular terms

In the Introduction we discussed the basic idea of the perturbation method. It is sufficient, therefore, to give a few examples of the appearance of secular terms in the series solution.

As an example, consider the d.e.

$$\ddot{x} + \mu\dot{x} + x = 0; \qquad |\mu| \ll 1 \tag{1.1}$$

whose exact solution is

$$x = Ae^{-\mu t/2} \sin\left(\sqrt{1 - \mu^2/4}\cdot t + \theta\right) \tag{1.2}$$

where A and θ are the integration constants.

If one attempts to solve this d.e. by the power series solution (II.2), one places it in the d.e. (II.3) and equates the coefficients of the like powers of μ which results in the following recursive system

$$\begin{aligned}
\ddot{x}_0 + x_0 &= 0 \\
\ddot{x}_1 + x_1 &= -f(x_0,\dot{x}_0) \\
\ddot{x}_2 + x_2 &= -[\dot{x}_1 f_x(x_0,\dot{x}_0) + \dot{x}_1 f_{\dot{x}}(x_0,\dot{x}_0)] \\
&\cdots\cdots\cdots\cdots \\
x_i(0) = \dot{x}_i(0) &= 0; \qquad i = 1, 2, \ldots, n
\end{aligned} \tag{1.3}$$

In the particular case of (1.1), $f(x,\dot{x})$ of the general theory is $\dot{x}$. The first d.e. (1.3) yields $x_0 = A \sin(t + \theta)$, A and θ being the integration constants. If one replaces x_0 and $\dot{x}_0$ into the second equation (1.3), one gets

$$\ddot{x}_1 + x_1 = -A\cos(t + \theta)$$

the integration of which yields

$$x_1(t) = -(At/2)\sin(t+\theta) + \frac{A}{2}\sin\theta\sin t \tag{1.4}$$

so that the solution of (1.1) in the first approximation is

$$x(t) = x_0(t) + \mu x_1(t) = A[1 - \mu t/2]\sin(t+\theta) + \frac{\mu A}{2}\sin\theta\sin t \tag{1.5}$$

and it is seen that it has a secular term $(-A\mu t/2)\sin(t+\theta)$ which is at variance with the exact solution (1.2) which is bounded.

As another example, consider the d.e.

$$m\ddot{x} + \alpha x + \gamma x^3 = 0; \qquad \alpha > 0, \qquad \gamma > 0 \tag{1.5a}$$

This d.e. may be regarded as representing a mechanical system with a nonlinear restoring force $p(x) = \alpha x + \gamma x^3$. In order to be within the limits of the nearly linear theory, we assume that $\gamma \ll \alpha$. If one sets $\alpha/m = \omega^2$; $\gamma/m = \mu$ and forms the d.e. of the first approximation, according to (1.3), one has

$$\begin{aligned} \ddot{x}_0 + \omega^2 x_0 &= 0 \\ \ddot{x}_1 + \omega^2 x_1 &= -x_0^3 \end{aligned} \tag{1.6}$$

From the first equation (1.6) one has

$$x_0 = a\cos(\omega t + \theta)$$

The second equation becomes

$$\ddot{x}_1 + \omega^2 x_1 = -\tfrac{3}{4}a^3\cos(\omega t+\theta) - \tfrac{1}{4}a^3\cos 3(\omega t+\theta) \tag{1.7}$$

and its solution is

$$x_1 = -\frac{3}{8\omega}ta^3\sin(\omega t+\theta) + \frac{a^3}{32\omega^2}\cos 3(\omega t+\theta) \tag{1.8}$$

Here again the first term of (1.8) is a secular term.

Although in this case the exact solution is still known in terms of elliptic functions, it is possible to reach the conclusion as to the boundedness of the solution in a simple manner by the consideration of energy since (1.5a) represents obviously a conservative system; that is, has an energy integral. If one multiplies (1.5a) by $\dot{x}$ and integrates, one obtains the first integral

$$\tfrac{1}{2}m\dot{x}^2 + \tfrac{1}{2}\alpha x^2 + \tfrac{1}{4}\gamma x^4 = E \tag{1.9}$$

expressing the law of conservation of energy. From (1.9) it is apparent that for $\alpha > 0$ and $\gamma > 0$, the quantity $x^2 < 2E/\alpha$, which shows that $x(t)$ is bounded so that the result (1.8) cannot be justified.

As was mentioned in the Introduction to Part II, the appearance of the secular terms does not mean that the series solution does actually diverge if one considers *the whole series* just in the same manner in which the whole series $t - (t^3/3!) + (t^5/5!) \ldots$ converges to $\sin t$. As it is impossible to build enough terms by successive approximations to be able to ascertain this fact, the appearance of secular terms renders the method impracticable.

A simple example given by Bogoliubov and Mitropolsky makes this statement particularly convincing. Suppose we have the function: $\sin(\omega + \mu)t$ with period $2\pi/(\omega + \mu)$. If μ is small, one can expand it in the series

$$\sin(\omega + \mu)t = \sin \omega t + \mu t \cos \omega t - (\mu^2 t^2/2!) \sin \omega t - (\mu^3 t^3/3!) \cos \omega t + \ldots \quad (1.10)$$

If one looks only at the right-hand side of this equation, it is impossible to see that it represents a function periodic on the whole axis of real t. The appearance of secular terms, thus, does not mean that the approximation solution is "wrong" but merely that it is presented in such a manner that it is impossible to form an idea as to its periodicity.

As was mentioned in the Introduction to Part II an important practical result was achieved by astronomers (end of the last century) who suggested the elimination of secular terms in each step of the approximation procedure and thus rendered the perturbation method a practical tool in the actual computation of successive approximations. The method of Lindstedt, outlined in Section 3, is particularly suitable for our purpose, but in order to prepare a ground for this procedure we consider first a special case in which the operation of the perturbation method is particularly simple.

2. Energy fluctuations in a van der Pol oscillator

In Chapter 4 it was mentioned that the Liénard criterion means that, after one period, the energy stored in the oscillation is of exactly the same value that it had at the beginning of the period. Aside from this circumstance, Liénard's criterion does not give any information regarding the energy fluctuations *during* the period. On the other hand, since the system is not conservative, the energy cannot remain constant throughout the period.

We propose now to establish what actually happens by means of the perturbation method and consider the d.e. of van der Pol

$$\ddot{x} + \mu(x^2 - 1)\dot{x} + x = 0 \quad (2.1)$$

with the corresponding equivalent system

$$\dot{x} = y; \qquad \dot{y} = \mu(1 - x^2)y - x \quad (2.2)$$

Introducing the polar coordinates $x = r\cos\psi$; $y = r\sin\psi$ and multiplying the first equation (2.2) by x, the second by y, and adding, one has

$$\frac{d\rho}{dt} = 2\mu(1 - \rho\cos^2\psi)\rho\sin^2\psi \tag{2.3}$$

where $\rho = r^2 = x^2 + \dot{x}^2 = x^2 + y^2$; $\psi = \arctan(y/x)$. It is observed that ρ is a measure of total energy stored in the oscillation (up to *a* certain factor of proportionality and with a proper normalization). Likewise, multiplying the first equation by y, the second by x, and subtracting the first equation from the second one, one has

$$\frac{d\psi}{dt} = -1 + \mu(1 - \rho\cos^2\psi)\sin\psi\cos\psi \tag{2.4}$$

If $\mu = 0$, (2.3) and (2.4) give the d.e. of the harmonic oscillator. Equation (2.3) in this case is $d\rho/dt = 0$ and expresses the law of conservation of the total energy; the second one gives $\psi = -t + c$, expressing the rotation of the vector with a constant velocity $\dot{\psi} = -1$ in the clockwise direction, c being an arbitrary constant.

If one wishes to investigate only the *magnitude* of the energy exchanges between the oscillator and the outside source without taking into account the *time element*, it is sufficient to eliminate t between (2.3) and (2.4) and investigate the integral curve given by the equation

$$\frac{d\rho}{d\psi} = \frac{2\mu(1 - \rho\cos^2\psi)\rho\sin^2\psi}{\mu(1 - \rho\cos^2\psi)\sin\psi\cos\psi - 1} \tag{2.5}$$

Assuming $\mu \ll 1$ and expanding the right-hand side into a power series, one gets

$$\frac{d\rho}{d\psi} = -2\rho[\mu(1 - \rho\cos^2\psi)\sin^2\psi + \mu^2(1 - \rho\cos^2\psi)\sin^3\psi\cos\psi$$
$$+ \ldots + \mu^n(1 - \rho\cos^2\psi)^n\sin^{n+1}\psi\cos^{n+1}\psi + \ldots] \tag{2.6}$$

Setting

$$\rho(\psi) = \rho_0(\psi) + \mu\rho_1(\psi) + \mu^2\rho_2(\psi) + \ldots \tag{2.7}$$

and equating the terms with equal powers of μ, one obtains a recursive system of the d.e. yielding successive approximations. Thus, the approximation of the order zero is: $d\rho_0/d\psi = 0$; $\rho_0 = K_0 = \text{const.}$

The first approximation is: $d\rho_1/d\psi = -2(1 - \rho_0\cos^2\psi)\rho_0\sin^2\psi$ which yields

$$\rho_1(\psi) = K_1 - \rho_0(1 - \tfrac{1}{4}\rho_0)\psi + \tfrac{1}{2}\rho_0\sin 2\psi - \tfrac{1}{16}\rho_0{}^2\sin 4\psi \tag{2.8}$$

where K_1 is a constant of integration. It is seen that the second term is a secular one.

Determining the constant K_0 so as to eliminate the secular term, we have

$$K_0 = \rho_0 = 4 \tag{2.9}$$

and one has:

$$\rho_1(\psi) = K_1 + 2 \sin 2\psi - \sin 4\psi$$

the constant K_1 being determined from the second approximation by the same method.

For the second approximation we have:

$$\frac{d\rho_2}{d\psi} = -2\rho_1 \sin^2 \psi + 4\rho_1\rho_0 \sin^2 \psi \cos^2 \psi$$
$$- 2\rho_0 \sin^3 \psi \cos \psi + 4\rho_0{}^2 \sin^3 \psi \cos^3 \psi - 2\rho_0{}^2 \sin^3 \psi \cos^5 \psi$$

Replacing ρ_0 and ρ_1 by their values, integrating, and eliminating again the secular term (which yields $K_1 = 0$), one gets

$$\rho_2(\psi) = K_2 - \tfrac{3}{4} \cos 2\psi + \tfrac{1}{4} \cos 4\psi + \tfrac{5}{12} \cos 6\psi - \tfrac{1}{4} \cos 8\psi \tag{2.10}$$

where K_2 is a constant of integration to be determined in the third approximation by the condition of elimination of the secular term which yields $K_2 = \frac{3}{8}$. Thus,

$$\rho_3(\psi) = -\tfrac{1}{12} \sin 2\psi - \tfrac{5}{192} \sin 4\psi + \tfrac{23}{72} \sin 6\psi$$
$$- \tfrac{37}{192} \sin 8\psi - \tfrac{5}{48} \sin 10\psi + \tfrac{5}{64} \sin 12\psi + K_3 \tag{2.11}$$

the constant K_3 to be determined again in the following approximation.

If one stops at the third approximation which limits its accuracy up to the order μ^4, the integral curve $\rho(\psi)$ is given by the following expression:

$$\begin{aligned}
\rho(\psi) = {} & (4 + \tfrac{3}{8}\mu^2 + \dots) \\
& + [(2\mu - \tfrac{1}{12}\mu^3 + \dots) \sin 2\psi + (-\tfrac{3}{4}\mu^2 + \dots) \cos 2\psi] \\
& + [(-\mu - \tfrac{5}{192}\mu^3 + \dots) \sin 4\psi + (\tfrac{1}{4}\mu^2 + \dots) \cos 4\psi] \\
& + [(\tfrac{23}{72}\mu^3 + \dots) \sin 6\psi + (\tfrac{5}{12}\mu^2 + \dots) \cos 6\psi] \\
& + [(-\tfrac{37}{192}\mu^3 + \dots) \sin 8\psi + (-\tfrac{1}{4}\mu^2 + \dots) \cos 8\psi] \\
& + \dots
\end{aligned} \tag{2.12}$$

where the nonwritten terms in the brackets are of higher degrees in μ; the odd ones for sines, and the even for cosines. This permits ascertaining the order of magnitude of the energy fluctuations for different values of ψ.

These approximations do not as yet give an idea about *the time variations* of these fluctuations inasmuch as we were dealing with the geometrical curve $\rho(\psi)$ and not with trajectories defined by (2.3) and (2.4).

For the latter one has to use the series solutions in the parametric form, viz.:

$$\begin{aligned} \rho(t) &= \rho_0(t) + \mu\rho_1(t) + \mu^2\rho_2(t) + \dots \\ \psi(t) &= \psi_0(t) + \mu\psi_1(t) + \mu^2\psi_2(t) + \dots \end{aligned} \tag{2.13}$$

As the variable ψ enters into the arguments of the trigonometric functions, one has to expand also the latter. Thus, for instance, writing ψ_i for $\psi_i(t)$, we have:

$$\begin{aligned} \cos 2\psi &\simeq \cos 2\psi_0 - 2(\mu\psi_1 + \mu^2\psi_2 + \dots) \sin 2\psi_0 \\ \sin 2\psi &\simeq \sin 2\psi_0 + 2(\mu\psi_1 + \mu^2\psi_2 + \dots) \cos 2\psi_0 \end{aligned} \tag{2.14}$$

The d.e. (2.3) and (2.4) can be written in terms of the multiple arguments, namely:

$$d\rho/dt = \mu\rho(1 - \cos 2\psi) - \tfrac{1}{4}\mu\rho^2(1 - \cos 4\psi) \tag{2.15}$$

$$d\psi/dt = -1 + \tfrac{1}{2}\mu \sin 2\psi - \tfrac{1}{4}\mu\rho \sin 2\psi - \tfrac{1}{8}\mu\rho \sin 4\psi \tag{2.16}$$

We propose to carry out the calculation to the second order; for this purpose it is sufficient to set $\rho \simeq \rho_0 + \mu\rho_1$; $\psi \simeq \psi_0 + \mu\psi_1$ and, for the trigonometric functions:

$$\sin 2\psi \simeq \sin (2\psi_0 + \mu 2\psi_1) = \sin 2\psi_0 + \mu 2\psi_1 \cos 2\psi_0$$

$$\cos 2\psi \simeq \cos (2\psi_0 + \mu 2\psi_1) = \cos 2\psi_0 - \mu 2\psi_1 \sin 2\psi_0$$

and similarly for $\sin 4\psi$ and $\cos 4\psi$.

If one substitutes these values into (2.15) and (2.16), one finds that the approximation of the zero order, as previously, is

$$\rho_0(t) = \rho_0; \qquad \psi_0(t) = \varphi_0 - t \tag{2.17}$$

ρ_0 and φ_0 being arbitrary constants. As to ρ_0 it is determined, as before, by the condition of the elimination of a secular term in the following approximation, whereas φ_0 is determined from the initial conditions.

One has then the following d.e.:

$$\begin{aligned} d\rho/dt = \mu[\rho_0(1 - \cos 2\psi_0) - \tfrac{1}{4}\rho_0^2(1 - \cos 4\psi_0)] \\ + \mu^2[\rho_1(1 - \cos 2\psi_0) + 2\psi_1\rho_0 \sin 2\psi_0 \\ - \tfrac{1}{2}\rho_0\rho_1(1 - \cos 4\psi_0) + \psi_1\rho_0^2 \sin 4\psi_0] \end{aligned} \tag{2.18}$$

$$\begin{aligned} d\psi/dt = -1 + \mu(\tfrac{1}{2} \sin 2\psi_0 - \tfrac{1}{4}\rho_0 \sin 2\psi_0 - \tfrac{1}{8}\rho_0 \sin 4\psi_0) \\ + \mu^2(\psi_1 \cos 2\psi_0 - \tfrac{1}{4}\rho_1 \sin 2\psi_0 - \tfrac{1}{2}\rho_0\psi_1 \cos 2\psi_0 \\ - \tfrac{1}{8}\rho_1 \sin 4\psi_0 - \tfrac{1}{2}\rho_0\psi_1 \cos 4\psi_0) \end{aligned} \tag{2.19}$$

The d.e. for the first-order corrections are

$$d\rho_1/dt = \rho_0[(1 - \tfrac{1}{4}\rho_0) - (\cos 2\psi_0 - \tfrac{1}{4}\rho_0 \cos 4\psi_0)]$$
$$d\psi_1/dt = \tfrac{1}{2}(1 - \tfrac{1}{2}\rho_0) \sin 2\psi_0 - \tfrac{1}{8}\rho_0 \sin 4\psi_0$$

It is seen that the first term in the first equation is a secular one; its elimination determines $\rho_0 = 4$; with this value of ρ_0, the d.e. becomes

$$d\rho_1/dt = -4 \cos 2\psi_0 + 4 \cos 4\psi_0; \qquad d\psi_1/dt = -\tfrac{1}{2} \sin 2\psi_0 - \tfrac{1}{2} \sin 4\psi_0 \tag{2.20}$$

and the integration yields

$$\rho_1(t) = -2 \sin 2\psi_0 + \sin 4\psi_0; \qquad \psi_1(t) = \tfrac{1}{4} \cos 2\psi_0 + \tfrac{1}{8} \cos 4\psi_0 \tag{2.21}$$

For the second approximation we have

$$\begin{aligned} d\rho_2/dt &= \rho_1(1 - \cos 2\psi_0) + 2\psi_1\rho_0 \sin 2\psi_0 \\ &\quad - \tfrac{1}{2}\rho_0\rho_1(1 - \cos 4\psi_0) + \psi_1\rho_0^2 \sin 4\psi_0 \\ &= \rho_1(1 - \cos 2\psi_0) + 8\psi_1 \sin 2\psi_0 \\ &\qquad - 2\rho_1(1 - \cos 4\psi_0) + 16\psi_1 \sin 4\psi_0 \end{aligned} \tag{2.22}$$

Taking the values of ρ_1 and ψ_1 from (2.21), one calculates the various terms; these are of two types: the periodic terms and the secular ones. The periodic terms do not present any special interest and merely add harmonics; we omit their calculation. The secular terms are more important and give an idea as to the behavior of the approximation in the long run. One ascertains that the term $\rho_1(1 - \cos 2\psi_0)$ does not yield any secular term; the second term $8\psi_1 \sin 2\psi_0$ gives the secular term -2; the third one $-2\rho_1(1 - \cos 4\psi_0)$ gives the secular term $+1$; and the last one $16\psi_1 \sin 4\psi_0$ gives $+4$. The three terms together give $+3$; therefore

$$\frac{d\rho_2}{dt} = 3 + \text{periodic terms} \quad \text{and} \quad \rho_2(t) = 3t + \text{periodic terms}$$

On the basis of the second approximation, the variable ρ has a slight tendency to drift away from its stationary value $\rho_0 = 4$ but, as the latter *is stable*, this merely shows that the equilibrium point is slightly displaced from this value. The effect is, however, slight, since this secular term enters with the factor μ^2.

For the phase one has the d.e.

$$\frac{d\psi_2}{dt} = -\psi_1 \cos 2\psi_0 - \tfrac{1}{4}\rho_1 \sin 2\psi_0 - \tfrac{1}{8}\rho_1 \sin 4\psi_0 - 2\psi_1 \cos 4\psi_0 \tag{2.23}$$

If one carries out the calculation of these four terms, one finds that the third term $-\frac{1}{8}\rho_1 \sin 4\psi_0$ has no secular term. As to the other three, their respective values are $+\frac{1}{8}$, $-\frac{1}{4}$, $+\frac{1}{8}$; the secular term thus cancels out in the second approximation.

Thus, as far as the first two approximations are concerned, one can make the following conclusons:

(1) The amplitude has a fixed value very near to the value $\rho_0 = 4$; the slight difference is due to the displacement of the point of equilibrium on account of the secular term in the second approximation.

(2) As to the variable $\psi(t) = \psi_0(t) + \mu\psi_1(t) + \mu^2\psi_2(t) + \ldots$, it varies continuously in the same direction with small oscillations around the value of $\psi_0(t) = \varphi_0 - t$ (where φ_0 is an arbitrary constant).

This simple, although somewhat lengthy, calculation reveals an extremely complicated structure of energy fluctuations reducing to an infinite spectrum of even harmonics, of which only those of the order 2ψ and 4ψ are more or less important, inasmuch as their amplitudes are proportional to μ; for the higher harmonics (6ψ, 8ψ,...) the amplitudes are small of higher orders in μ.

One readily sees that, if μ is small (as we suppose), all these harmonics are negligible in comparison with the term $\rho_0 = 4$ yielded by the first approximation, and this is the reason the solution of the van der Pol equation is so simple in the first approximation ($r_0 = 2$). A similar situation appears in the angular velocity of rotation of the radius vector. The first approximation yields $d\psi_0/dt = -1$, that is, uniform rotation just as in the case of the harmonic oscillator, but beginning with the second approximation there appear perturbing *phase* modulations with small amplitudes $0(\mu^2)$.

Thus, although a van der Pol oscillator with a small value of μ appears as an almost ideal image of the harmonic oscillator, in reality it contains a germ of a complicated structure consisting of an infinite spectrum of amplitude and phase modulations which escapes observation only because this structure is very small if μ is small.

This complexity becomes apparent, however, when μ increases and becomes very large. The phenomena then become very complicated, and the infinite spectrum of even harmonics begins to dominate over the above-mentioned simple result ($\rho_0 \simeq 4$), when μ is small. It is obvious that for large μ the power series solution has no meaning and one has to use entirely different methods as we shall see in Part IV.

3. Lindstedt's method[1]

We consider a nearly linear d.e. of the form

$$\ddot{x} + \omega^2 x + \mu f(x) = 0 \tag{3.1}$$

[1] A. Lindstedt, *Mem. de l'Ac. Impér. de St. Petersburg* **31**, 1883.

which characterizes a conservative system with an unknown period T (frequency $\Omega = 2\pi/T$). In order to avoid dealing with an unknown period, we select a variable $z(\tau)$, which has the period 2π; clearly an angle satisfies this requirement. If such a variable is selected, the system has the period 2π in that variable but, as we had to change the time scale as well, the frequency is not Ω. In the new variables (the dependent variable z and the independent τ), (3.1) becomes:

$$\Omega^2\ddot{z} + \omega^2 z + \mu f(z) = 0 \tag{3.2}$$

As we changed both the dependent variable (from x to z) and the independent variable (from t to τ), we can set

$$z(\tau) = \sum_{n=0}^{\infty} \mu^n z_n(\tau); \qquad \Omega^2 = \sum_{n=0}^{\infty} \mu^n \alpha_n \tag{3.3}$$

It is clear that, for $\mu = 0$, one must have $\Omega^2 = \omega^2 = \alpha_0$, since in this case one has the harmonic oscillator whose period is 2π with the new choice of the time scale τ.

We have also

$$\begin{aligned} f(z) &= f(z_0 + \mu z_1 + \mu^2 z_2 + \dots) \\ &= f(z_0) + \mu z_1 f'(z_0) + \mu^2\left[z_2 f'(z_0) + \frac{z_1^2 f''(z_0)}{2}\right] + \dots \end{aligned} \tag{3.4}$$

The substitution of (3.3) and (3.4) into (3.2) results in a sequence of linear d.e.:

$$\begin{aligned} &\omega^2\ddot{z}_0 + \omega^2 z_0 = 0 \\ &\omega^2\ddot{z}_1 + \omega^2 z_1 = -f(z_0) - \alpha_1\ddot{z}_0 \\ &\omega^2\ddot{z}_2 + \omega^2 z_2 = -f'(z_0, z_1) - \alpha_2\ddot{z}_0 - \alpha_1\ddot{z}_1 \\ &\quad\cdots\cdots\cdots\cdots \\ &\omega^2\ddot{z}_{n+1} + \omega^2 z_{n+1} = F(z_0, z_1, \dots, z_n) - \alpha_{n+1}\ddot{z}_0 - \alpha_n\ddot{z}_1 - \dots - \alpha_1\ddot{z}_n \\ &\quad\cdots\cdots\cdots\cdots \end{aligned} \tag{3.5}$$

where $F(z_0, z_1, \dots, z_n)$ is a polynomial in $z_0, z_1, \dots, z_n$.

As the system is autonomous, the solution $z(\tau)$ is determined up to a translation on τ. One can thus always select φ in the $\tau = \Omega t + \varphi$ so as to have $\dot{z}(0) = 0$.

If $z_0, z_1, \dots, z_N$ and $\alpha_1, \alpha_2, \dots, \alpha_N$ are the solutions of the first $N+1$ equations of (3.5), it is clear that $z(\tau) = \sum_{n=0}^{N} \mu^n z_n(\tau)$ and $\Omega^2 = \sum_{n=0}^{N} \mu^n \alpha_n$ will satisfy (3.1) and may be regarded as the $(N+1)$th approximation.

The first equation (3.5) yields $z_0 = a \cos \tau$ (in view of $\dot{z}(0) = 0$), where a is an arbitrary constant.

There appears, however, an arbitrariness in the following approximations since we have introduced two expansions (3.3) in the same d.e. The essential feature of the method is that *this arbitrariness enables us to dispose of the available constants so as to eliminate gradually the secular terms in the subsequent approximations.*

Consider now the second equation (3.5)

$$\omega^2(\ddot{z}_1 + z_1) = -f(z_0) - \alpha_1\ddot{z}_0 = -f(a \cos \tau) + \alpha_1 a \cos \tau \tag{3.6}$$

The function $f(a \cos \tau)$ being developed in a Fourier series contains only the cosine terms, that is,

$$f(a \cos \tau) = \sum_{n=0}^{\infty} f_n(a) \cos n\tau = f_0(a) + f_1(a) \cos \tau + \sum_{n=2}^{\infty} f_n(a) \cos n\tau$$

and (3.6) becomes

$$\omega^2(\ddot{z}_1 + z_1) = -\sum_{n=2}^{\infty} f_n(a) \cos n\tau + [\alpha_1 a - f_1(a)] \cos \tau - f_0(a) \tag{3.7}$$

It is seen that the secular term is bound to appear in this case, since the variable z_1 has period 2π and $\cos \tau$ has the same period. The d.e. has thus a resonance solution increasing with τ. Since the constant α_1 is at our disposal, we determine it by the condition

$$\alpha_1 = \frac{f_1(a)}{a} \tag{3.8}$$

which eliminates the secular term in the first approximation and, in the meantime, determines the latter: z_1, α_1, since (3.6) becomes

$$\ddot{z}_1 + z_1 = -\frac{1}{\omega^2} f_0(a) - \frac{1}{\omega^2} \sum_{n=2}^{\infty} f_n(a) \cos n\tau \tag{3.9}$$

yielding the solution

$$z_1 = A \cos \tau - \frac{1}{\omega^2} f_0(a) - \frac{1}{\omega^2} \sum_{n=2}^{\infty} \frac{f_n(a) \cos n\tau}{n^2 - 1} \tag{3.10}$$

where A is a constant of integration; one can set $A = 0$ in order to simplify the solution.

Replacing z_0 and z_1 into the third equation (3.5) and eliminating again the resonance solution, one obtains the value of α_2.

The procedure becomes now clear, viz.: the subsequent approximations develop recurrently from the system (3.5) which determines the sequence

$z_0, z_1, z_2, \ldots$ of successive approximations but, as in each of them the resonance terms are eliminated, this results in another sequence $\alpha_1, \alpha_2, \ldots$ which determines the frequency Ω^2.

It is clear that, if one has determined two sequences $z_0, z_1, z_2, \ldots, z_n$ and $\alpha_0, \alpha_1, \alpha_2, \ldots, \alpha_n$ in this manner, one reaches the d.e.

$$\ddot{z}_{n+1} + z_{n+1} = \frac{1}{\omega^2}\left(b_0(a) + \sum_{n=2}^{\infty} b_n(a)\cos n\tau\right) + \frac{1}{\omega^2}(\alpha_{n+1}a + b_1(a)) \cos\tau$$

and, again, the elimination of the resonance term requires

$$\alpha_{n+1} = -b_1(a)/a \tag{3.11}$$

which results in the d.e.

$$\ddot{z}_{n+1} + z_{n+1} = \frac{1}{\omega^2}\left(b_0 + \sum_{n=2}^{\infty} b_n \cos n\tau\right) \tag{3.12}$$

yielding the solution

$$z_{n+1} = \frac{1}{\omega^2}\left(b_0(a) - \sum_{n=2}^{\infty} b_n(a)\frac{\cos n\tau}{n^2 - 1}\right) \tag{3.13}$$

As an example, consider the d.e.

$$\ddot{x} + x + \mu x^3 = 0 \tag{3.14}$$

Taking $z_0(\tau) = a\cos\tau$; $\omega^2 = 1$ and $\alpha_0 = 1$, one has

$$\ddot{z}_1 + z_1 = -z_0^3 - \alpha_1\ddot{z}_0 = -a^3\cos^3\tau + \alpha_1 a\cos\tau$$

$$= (\alpha_1 a - \tfrac{3}{4}a^3)\cos\tau - \frac{a^3}{4}\cos 3\tau$$

Hence, $\alpha_1 = \frac{3}{4}a^2$ and the first-order term is

$$z_1 = \tfrac{1}{32}a^3\cos 3\tau$$

Substituting α_1 and z_1 by their values, the third equation (3.5) gives, after the reduction of the trigonometric functions to multiple arguments:

$$\ddot{z}_2 + z_2 = (\alpha_2 a + \tfrac{3}{128}a^5)\cos\tau + \tfrac{21}{128}a^5\cos 3\tau - \tfrac{3}{128}a^5\cos 5\tau$$

whence $\alpha_2 = \frac{3}{128}a^4$ and the integration yields

$$z_2 = -\frac{21}{1024}a^3\cos 3\tau + \frac{a^5}{1024}\cos 5\tau$$

Hence the approximate solution satisfying the d.e. to the order of μ^3 is:

$$x \simeq a\cos(\omega t + \varphi) + \mu\frac{1}{32}a^3\left(1 - \mu\frac{21}{32}a^2\right)\cos 3(\omega t + \varphi) + \mu^2\frac{a^5}{1024}\cos 5(\omega t + \varphi) \quad (3.15)$$

where $\omega = 1$, a and φ being arbitrary constants of integration. As to the frequency Ω, it is obtained from another series into which enter the quantities α_i. This gives

$$\Omega^2 \simeq 1 + \tfrac{3}{4}\mu a^2 + \tfrac{3}{128}\mu^2 a^4$$

4. Theorem of Poincaré

As was mentioned in the Introduction, the heuristic assumption of Poisson concerning solutions for nonlinear d.e. in form of a power series was proved by Poincaré by the method of Cauchy.[2] In view of the importance of this theorem for all developments outlined in Part II, we indicate briefly the proof of this theorem with a slight simplification due to Goursat.[2]

What is essential here is the *analytic dependence* of the solution on the parameter μ; otherwise the theorem follows closely the "Calcul des Limites" of Cauchy. There exists another proof of this theorem due to K. O. Friedrichs[3]; for our purpose the proof given here is likely to be more familiar from the general theory of d.e.'s.

We consider a differential system of the first order depending on a parameter μ

$$\begin{aligned} \dot{x} &= f(t, x, y; \mu) = \sum a_{\alpha\beta\gamma} x^\alpha y^\beta \mu^\gamma \\ \dot{y} &= \varphi(t, x, y; \mu) = \sum b_{\alpha\beta\gamma} x^\alpha y^\beta \mu^\gamma \end{aligned} \quad (4.1)$$

We assume that the right-hand sides of these equations are entire series in x, y, and μ, whose coefficients $a_{\alpha\beta\gamma}$ and $b_{\alpha\beta\gamma}$ are continuous functions of t in the interval: $t_0 \leq t \leq t_1$. We assume further that the series converge for any t in this interval as long as $|x|$, $|y|$, and $|\mu|$ are less than a certain number $\rho > 0$ and, moreover, the series do not contain any constant terms so that, for $\mu = 0$, the system (4.1) has a particular solution $x = y = 0$.

[2] H. Poincaré, *Les méthodes nouvelles de la mécanique céleste* **T.1**, Gauthier-Villars, Paris, 1892; also E. Goursat, *Cours d'Analyse* **T.2**, Gauthier-Villars, Paris, 1918.

[3] K. O. Friedrichs, *J. Inst. of Math. and Mechanics*, New York University, New York, 1946.

We shall try to satisfy formally (4.1) by the solutions of the form

$$
\begin{aligned}
x(t) &= \mu x_1(t) + \mu^2 x_2(t) + \ldots + \mu^n x_n(t) + \ldots \\
y(t) &= \mu y_1(t) + \mu^2 y_2(t) + \ldots + \mu^n y_n(t) + \ldots
\end{aligned}
\tag{4.2}
$$

whose coefficients $x_i(t)$, $y_i(t)$ vanish for $t = t_0$.

If one replaces x and y in (4.1) by their expressions (4.2) and equates the coefficients of like powers of μ, for the first terms x_1 and y_1 one has the d.e.

$$
\begin{aligned}
\dot{x}_1 &= a_{100} x_1(t) + a_{010} y_1(t) + a_{001} \\
\dot{y}_1 &= b_{100} x_1(t) + b_{010} y_1(t) + b_{001}
\end{aligned}
\tag{4.3}
$$

With the initial conditions $x_1(t_0) = y_1(t_0) = 0$, these d.e. determine $x_1(t)$ and $y_1(t)$. The system (4.3) is, in fact, a variational system relative to the particular solution: $x = y = 0$.

In a similar manner for the terms $x_n(t)$ and $y_n(t)$, one has a corresponding system:

$$
\begin{aligned}
\dot{x}_n &= a_{100} x_n(t) + a_{010} y_n(t) + u_n \\
\dot{y}_n &= b_{100} x_n(t) + b_{010} y_n(t) + v_n
\end{aligned}
\tag{4.4}
$$

where u_n and v_n are certain polynomials depending on $a_{\alpha\beta\gamma}$, $b_{\alpha\beta\gamma}$ and on $x_i(t)$ and $y_i(t)$ for which $i < n$.

One ascertains that for any n all functions x_n and y_n, together with their derivatives $\dot{x}_n$ and $\dot{y}_n$ are continuous in (t_0, t_1); moreover, they all are determined by quadratures, if one knows the solution of (4.3).

The problem is to show that the expansions (4.2) converge and, thus, can actually represent a solution of (4.1).

From now on one can follow the argument of "Calcul des Limites" of Cauchy. We consider an auxiliary system:

$$
\begin{aligned}
\dot{X} &= F(t, X, Y, \mu) = \sum A_{\alpha\beta\gamma} X^\alpha Y^\beta \mu^\gamma \\
\dot{Y} &= \Phi(t, X, Y, \mu) = \sum B_{\alpha\beta\gamma} X^\alpha Y^\beta \mu^\gamma
\end{aligned}
\tag{4.5}
$$

where the coefficients $A_{\alpha\beta\gamma}$ and $B_{\alpha\beta\gamma}$ are certain *majorating* functions relative to $a_{\alpha\beta\gamma}$ and $b_{\alpha\beta\gamma}$ in the same interval (t_0, t_1).

We look again for a possibility of satisfying (4.5) *formally* by the series solutions of the form

$$
\begin{aligned}
X(t) &= \mu X_1(t) + \mu^2 X_2(t) + \ldots + \mu^n X_n(t) + \ldots \\
Y(t) &= \mu Y_1(t) + \mu^2 Y_2(t) + \ldots + \mu^n Y_n(t) + \ldots
\end{aligned}
\tag{4.6}
$$

which satisfy the same condition: $X_i(t_0) = Y_i(t_0) = 0$.

One has again the differential system (compare to 4.3):

$$\dot{X}_1 = A_{100}X_1 + A_{010}Y_1 + A_{001}$$
$$\dot{Y}_1 = B_{100}X_1 + B_{010}Y_1 + B_{001} \tag{4.7}$$

which determines X_1 and Y_1 that vanish for $t = t_0$. If one compares (4.7) with (4.3), one ascertains that X_1 and Y_1 are majorating functions as compared to x_1 and y_1, respectively; the same holds for $\dot{X}$, $\dot{Y}$ as compared to $\dot{x}$, $\dot{y}$.

The same argument holds in general for X_n, Y_n, $\dot{X}_n$, and $\dot{Y}_n$ compared to x_n, y_n, $\dot{x}_n$, and $\dot{y}_n$.

It is sufficient to show now that, by a proper choice of the majorating functions $A_{\alpha\beta\gamma}$ and $B_{\alpha\beta\gamma}$, the series (4.6), as well as those obtained by the differentiation with respect to t, are uniformly convergent in (t_0,t_1) if $|\mu|$ is sufficiently small.

Let M be the upper limit of $|f|$ and $|\varphi|$ in (4.1) when t is in (t_0,t_1) and $|x|$, $|y|$, and $|\mu|$ are less than, or equal to a certain number $\rho > 0$.

The coefficient of a term $x^\alpha y^\beta \mu^\gamma$ is clearly less than the corresponding coefficient in the development of

$$\frac{M(x + y + \mu)}{1 - \dfrac{(x + y + \mu)}{\rho}} \tag{4.8}$$

in powers of x, y, and μ.

One can take, therefore, as a majorating system:

$$\frac{dX}{dt} = \frac{dY}{dt} = \frac{M(X + Y + \mu)\left(1 + \dfrac{X + Y + \mu}{\rho}\right)}{1 - \left(\dfrac{X + Y + \mu}{\rho}\right)} \tag{4.9}$$

so that the problem reduces now to showing that a solution of this system vanishing for $t = t_0$ can be developed in a power series in μ in (t_0,t_1) for a sufficiently small $|\mu|$.

With the above initial conditions and setting $X + Y + \mu = \rho\tau$, the system (4.9) reduces to one single d.e.

$$\frac{d\tau}{dt} = \frac{2M\tau(1 + \tau)}{1 - \tau} \tag{4.10}$$

with the condition: $\tau = \mu/\rho$ for $t = t_0$, τ being real. Separating the variables, the solution is then the root of the equation: $\tau = \alpha(\tau + 1)^2$, where

$$\alpha = \frac{\rho\mu e^{2M(t-t_0)}}{(\rho + \mu)^2} \tag{4.11}$$

For $t = t_0$ this solution reduces to $\tau_0 = \mu/\rho$. On the other hand, for $\mu = 0$, one has $\alpha = 0$.

It is clear that the root of the equation in τ, vanishing for $\alpha = 0$, is an analytic function of α as long as $\alpha < \frac{1}{4}$.

It is sufficient therefore to take $|\mu|$ sufficiently small:

$$\frac{\rho\mu e^{2M(t-t_0)}}{(\rho + \mu)^2} < \frac{1}{4} \tag{4.12}$$

in order to guarantee the convergence of power series solutions (4.6) in (t_0,t_1). This guarantees also the convergence of the differentiated series. As the series solutions satisfy formally the d.e. and since they are convergent, they are the actual solutions of (4.1).

Summing up:

The solutions of (4.1) *are analytic functions of the parameter* μ *provided* $|\mu|$ *is sufficiently small.*

This theorem is of a fundamental importance for all that follows excluding Part IV. It means that if the nonlinear terms enter with a certain parameter μ, there exists a power series solution in terms of this parameter, provided it is sufficiently small.

Chapter 10

PERIODIC SOLUTIONS (POINCARÉ)

1. Introductory remarks

In this chapter we outline the theory of Poincaré[1] following closely the exposition of Malkin.[2] The early adaptation of this theory to autonomous systems was made by Andronov,[3] and its application to nonautonomous systems was indicated by Mandelstam and Papalexi in their theory of subharmonic resonance (Chapter 19). At a later time Malkin unified these early developments and gave a general outline of the theory of Poincaré in its applications to the theory of oscillations.

Most of the nonlinear problems with one degree of freedom encountered in the theory of oscillations are amenable to the d.e. of the form:

$$\ddot{x} + x + \mu F(t, x, \dot{x}) = 0 \tag{1.1}$$

Sometimes this equation is also written in the form

$$\ddot{x} + x + f(t) + \mu F(t, x, \dot{x}, \mu) = 0 \tag{1.1a}$$

In these equations $F(t, x, \dot{x})$ is an analytic function of its arguments and periodic in t (as well as $f(t)$ in (1.1a)) with a period which can be always assumed to be 2π with a proper choice of the time scale; μ is the parameter assumed to be a small number.

In applications, the explicit dependence on t appears generally under the

[1] H. Poincaré, *Les méthodes nouvelles de la mécanique céleste* **T.3**, Gauthier-Villars, Paris, 1892; also E. Goursat, *Cours d'Analyse* **T.2**, Gauthier-Villars, Paris, 1918.

[2] I. G. Malkin, *Certain Problems in the Theory of Nonlinear Oscillations* (in Russian), Moscow, English translation, U.S. Atomic Energy Commission, 1959.

[3] A. Andronov and S. Chaikin, *Theory of Oscillations* (original text in Russian), Moscow, 1937.

sign of trigonometric functions, therefore (1.1) appears frequently in the form

$$\ddot{x} + x + \lambda \cos \omega t + \mu f(x,\dot{x}) = 0 \tag{1.2}$$

where λ and ω are constants.

There are two principal cases to be investigated according to whether ω *is not* an integer or is an integer (or close to an integer). The first case is relatively simple and leads to the so-called *nonresonance oscillation* (Section 2). The second case corresponds to the resonance oscillation (either exact resonance or its neighborhood).

If the independent variable t (time) appears explicitly in F, one has a *nonautonomous* system (NA); if t does not enter explicitly in F the system is called *autonomous* (A).

As is known from the theory of d.e. in the case of (A) systems, one can replace t by $t + t_0$ (t_0 being arbitrary) and still have a solution; this is sometimes called the *translation* property of autonomous systems. For (NA) systems this does not hold and the period of oscillation either is in most cases equal to that of the external periodic excitation (that is, the time dependent term in the d.e.) or is in a rational ratio relative to it. As in (A) systems the external periodic excitation is absent, the period of oscillation (if it exists) depends on the parameters of the system; in other words, it is determined by the d.e. itself. This constitutes the principal difference between (NA) and (A) systems.

One may consider the broader problem of a system:

$$\dot{x}_s = X_s(t, x_1, \ldots, x_r; \mu); \qquad s = 1, 2, \ldots, r \tag{1.3}$$

where the right-hand sides depend analytically on a parameter μ. We may then write (1.3) in the form:

$$\dot{x}_s = X_s^{(0)}(t, x_1, \ldots, x_r) + \mu X_s^{(1)}(t, x_1, \ldots, x_r) + \mu^2 X_s^{(2)}(t, x_1, \ldots, x_r) + \cdots \tag{1.4}$$

If $\mu = 0$, one has the *generating solution* of which we spoke in the Introduction:

$$\dot{x}_s = X_s^{(0)}(t, x_1^{(0)}, \ldots, x_r^{(0)}) \tag{1.5}$$

and one may expect intuitively that, if the periodic solution exists, it must branch off this generating solution when μ ceases to be zero while still being small. As mentioned previously, this is a rather delicate point, since it is not at all certain that the solution exists if $\mu \neq 0$. It is precisely here that the problem of Poincaré intervenes.

In this chapter we shall investigate the question of the existence of periodic solutions when $\mu \neq 0$ but is small. The question of stability of such solutions is treated in Chapter 13.

2. Nonresonance oscillation of an (NA) system

We consider a nonlinear (NA) system with one degree of freedom, viz.:

$$\ddot{x} + k^2 x + f(t) = \mu F(t, x, \dot{x}; \mu) \tag{2.1}$$

where $f(t)$ is a continuous periodic function with period 2π and $F(t, x, \dot{x}; \mu)$ is a nonlinear function analytic in x and $\dot{x}$ and also periodic in t with period 2π.

It is proposed to find a periodic solution of (2.1). Inasmuch as $f(t)$ and F are periodic with period 2π, clearly the period of the solution of (2.1) (if it exists) is also either 2π or is in a rational ratio to 2π.

For $\mu = 0$, one obtains the periodic solution of the generating d.e., viz.:

$$x_0(t) = -\frac{a_0}{2k^2} - \sum_{n=1}^{\infty} \frac{a_n \cos bt + b_n \sin nt}{k^2 - n^2} = \varphi(t) \tag{2.2}$$

since $f(t)$ is periodic and k is not an integer. It is necessary to investigate under which conditions (2.1) has a periodic solution which, for $\mu \to 0$, becomes (2.2) the generating solution.

Poincaré introduces two parameters β_1 and β_2 defined as initial deviations between the nonlinear solution $x(t,\mu)$ and the generating solution $x_0(t,0) = \varphi(t)$; that is, $x(0,\mu) - \varphi(0) = \beta_1$; $\dot{x}(0,\mu) - \dot{\varphi}(0) = \beta_2$. Thus, by definition, one has:

$$\begin{aligned} x(0, \beta_1, \beta_2, \mu) - \varphi(0) &= \beta_1 \\ \dot{x}(0, \beta_1, \beta_2, \mu) - \dot{\varphi}(0) &= \beta_2 \end{aligned} \tag{2.3}$$

It is clear that the solution is periodic (with period 2π), provided the following conditions are fulfilled:

$$\begin{aligned} x(2\pi, \beta_1, \beta_2, \mu) - x(0, \beta_1, \beta_2, \mu) &= \psi_1(\beta_1, \beta_2, \mu) = 0 \\ \dot{x}(2\pi, \beta_1, \beta_2, \mu) - \dot{x}(0, \beta_1, \beta_2, \mu) &= \psi_2(\beta_1, \beta_2, \mu) = 0 \end{aligned} \tag{2.4}$$

It is also clear that, if (2.4) is fulfilled, x and $\dot{x}$ at the time $t = 2\pi$ are the same as for $t = 0$. Hence, taking for the initial conditions of the second interval $(2\pi,4\pi)$ the terminal conditions of the first interval, one obtains exactly the same conclusions in the second interval and the solution x is obviously periodic in such a case.

The conditions (2.4) are thus the necessary and sufficient ones for the periodicity of the solution, and it is now necessary to determine the parameters β_1 and β_2 as functions of μ so as to secure the fulfillment of (2.4). It is noted that $\beta_1 = \beta_1(\mu)$ and $\beta_2 = \beta_2(\mu)$ such that β_1 and β_2 approach zero when $\mu \to 0$ which follows from their very definition.

We have seen that, under the assumed condition of analyticity of F

with respect to x and $\dot{x}$, the solution $x(t, \beta_1, \beta_2, \mu)$ is also analytic in β_1, β_2, and μ. As these quantities vanish for $\mu = 0$, one can write the solution in the form of a series

$$x(t, \beta_1, \beta_2, \mu) = \varphi(t) + A\beta_1 + B\beta_2 + C\mu + \ldots \tag{2.5}$$

where A, B, and C are unknown functions of t. The solution can be continued for higher orders by adding terms $D\beta_1\beta_2 + E\beta_1\mu + F\beta_2\mu + G\mu^2 + \ldots$, but we shall limit ourselves to the first order. If one substitutes (2.5) into (2.1) and equates the coefficients of the like powers of $\beta_1, \beta_2, \mu, \ldots$, one obtains first

$$\ddot{A} + k^2A = 0; \qquad \ddot{B} + k^2B = 0 \tag{2.6}$$

with the initial conditions

$$A(0) = 1; \qquad \dot{A}(0) = 0; \qquad B(0) = 0; \qquad \dot{B}(0) = 1 \tag{2.7}$$

which determines: $A = \cos kt$; $B = \dfrac{1}{k}\sin kt$, so that (2.4) to the first order become

$$\psi_1(\beta_1, \beta_2, \mu) = [x] = (\cos 2k\pi - 1)\beta_1 + \frac{1}{k}(\sin 2k\pi)\beta_2 + [C]\mu + \ldots \tag{2.8}$$

$$\psi_2(\beta_1, \beta_2, \mu) = [\dot{x}] = -k(\sin 2k\pi)\beta_1 + (\cos 2k\pi - 1)\beta_2 + [\dot{C}]\mu + \ldots$$

where we use the notations $[x] = x(2\pi) - x(0)$ etc.

Conditions (2.8) are identically fulfilled for $\mu = \beta_1 = \beta_2 = 0$. Besides this, the Jacobian of the left-hand terms with respect to β_1 and β_2 is different from zero for $\mu = \beta_1 = \beta_2 = 0$. In fact, we have

$$\left[\frac{\partial(\psi_1,\psi_2)}{\partial(\beta_1,\beta_2)}\right]_{\beta_1=\beta_2=\mu=0} = (\cos 2k\pi - 1)^2 + \sin^2 2k\pi \neq 0 \tag{2.9}$$

Hence, on the basis of the implicit functions theory, one can assert that for *a sufficiently small* μ, *there exists only one solution* $\beta_1 = \beta_1(\mu)$; $\beta_2 = \beta_2(\mu)$ *which vanishes with* μ *and, besides this, this solution is analytic in* μ.

In view of this, one can write this solution as a series

$$x = x(t) = \varphi(t) + \mu x_1(t) + \mu^2 x_2(t) + \ldots \tag{2.10}$$

where $x_i(t)$ are certain periodic functions with period 2π. For their determination one replaces (2.10) into (2.1) and equates coefficients of the like powers of μ. One verifies that, for the first approximation term $x_1(t)$, one has the d.e.

$$\ddot{x}_1 + k^2x_1 = F_1 = F(t, \varphi, \dot{\varphi}, 0) \tag{2.11}$$

and, more generally, for $x_i(t)$ $(i \geq 1)$, one has the d.e.

$$\ddot{x}_i + k^2 x_i = F_i \tag{2.11a}$$

where F_i are periodic functions of φ, $x_1(t), \ldots, x_{i-1}(t)$.

If all $x_1, x_2, \ldots, x_{i-1}$ have been calculated and are periodic, F_i is then a known periodic function and, as k is not an integer, there exists one and only one periodic solution of the form (2.2) for all approximations x_i.

Thus, there exists one and only one series solution (2.10), and it converges if μ is small enough. In this case the formal solution so obtained is, in fact, the actual solution.

The procedure outlined in this section ceases to hold when k is either an integer or near to an integer. In fact, in such cases the denominator of (2.2) would either vanish or become very small, which inevitably would destroy the convergence of the series solution (2.10), since in applied problems the parameter μ is generally a small fixed number. In view of this it is necessary for the resonance case (or its vicinity) to apply a somewhat different procedure.

3. Resonance oscillation of an (NA) system

If k is near to an integer n, we assume that $n^2 - k^2 \sim 0(\mu)$ so that $n^2 - k^2 = \mu a$, where a is finite and $\mu \ll 1$; we assume that the coefficients of the nth harmonic of $f(t)$ are also $0(\mu)$; that is: $a_n = \mu a_n'$; $b_n = \mu b_n'$, a_n' and b_n' being finite. If one includes the terms $\mu a x$ and $\mu(a_n' \cos nt + b_n' \sin nt)$ in the term $\mu F(t, x, \dot{x}, \mu)$, (2.1) becomes

$$\ddot{x} + n^2 x + f'(t) = \mu F(t, x, \dot{x}, \mu) \tag{3.1}$$

where

$$f'(t) = f(t) - a_n \cos nt - b_n \sin nt = \frac{a_0}{2} + \sum_{j \neq n} [a_j \cos jt + b_j \sin jt].$$

Consider the generating equation

$$\ddot{x}_0 + n^2 x_0 + f'(t) = 0 \tag{3.2}$$

The general solution of (3.2) is

$$\begin{aligned} x_0 &= -\frac{a_0}{2n^2} - \sum_{j \neq n} \frac{a_j \cos jt + b_j \sin jt}{n^2 - j^2} + M_0 \cos nt + N_0 \sin nt \\ &= \varphi(t) + M_0 \cos nt + N_0 \sin nt \end{aligned} \tag{3.3}$$

where M_0 and N_0 are arbitrary constants.

This solution is periodic with period 2π for any values of M_0 and N_0.

Hence, in this case, the generating solutions form a family depending on two arbitrary constants, whereas in the nonresonance case there was only one isolated generating periodic solution.

Since in the resonance case there is a family of periodic solutions, one can try to determine the constants M_0 and N_0 in such a manner that (3.1) should have a periodic solution which becomes the generating solution for $\mu = 0$.

One starts again with the series solution of the form

$$x(t, \beta_1, \beta_2, \mu) = x_0(t) + A\beta_1 + B\beta_2 + C\mu + \mu(D\beta_1 + E\beta_2 + F\mu) + \dots \tag{3.4}$$

It is noted that all terms of higher orders vanish for $\mu = 0$, since in this case (3.4) becomes the solution of the linear generating equation which contains the initial conditions only linearly. For the periodicity of (3.4) we have again condition (2.4). This yields as previously: $\ddot{A} + n^2A = 0$; $\ddot{B} + n^2B = 0$ with the initial conditions: $A(0) = 1$; $\dot{A}(0) = 0$; $B(0) = 0$; $\dot{B}(0) = 1$, whence $A = \cos nt$; $B = \frac{1}{n} \sin nt$ so that $[A] = [\dot{A}] = [B] = [\dot{B}] \equiv 0$ and the conditions of periodicity acquire the form

$$\begin{aligned} \psi_1(\beta_1, \beta_2, \mu) &= \mu\{[C] + [D]\beta_1 + [E]\beta_2 + [F]\mu + \dots\} = 0 \\ \psi_2(\beta_1, \beta_2, \mu) &= \mu\{[\dot{C}] + [\dot{D}]\beta_1 + [\dot{E}]\beta_2 + [\dot{F}]\mu + \dots\} = 0 \end{aligned} \tag{3.5}$$

The problem is reduced now to the determination of the functions $\beta_1(\mu)$ and $\beta_2(\mu)$ solutions of (3.5) which vanish with μ. Equations (3.5) have no linear terms with β_1 and β_2; therefore the Jacobian in this case vanishes for $\mu = \beta_1 = \beta_2 = 0$ and we cannot use the argument of Section 2.

It is possible to use, however, the following reasoning: from the periodicity $\psi_1 = 0$ and $\psi_2 = 0$; one can drop the factor μ since for $\mu = 0$ the conditions of periodicity are automatically fulfilled.

It is necessary, therefore, to equate to zero the brackets in (3.5) (which clearly corresponds to the case: $\mu \neq 0$), but one has to determine $\beta_1(\mu)$ and $\beta_2(\mu)$ in such a manner that the vanishing of the brackets should exist for *any* μ provided it is small.

Clearly, the problem is now purely algebraic and one sees at once that the necessary condition for the fulfillment of the above condition is:

$$[C] = 0; \qquad [\dot{C}] = 0 \tag{3.6}$$

It is necessary, therefore, to determine the conditions under which (3.6) can be fulfilled. If one replaces (3.4) into (3.1) and equates the coefficients of μ, one has the d.e.

$$\ddot{C} + n^2C = F(t, x_0, \dot{x}_0, 0) \tag{3.7}$$

with the initial conditions $C(0) = \dot{C}(0) = 0$. The solution has then the form

$$C(t) = \frac{1}{n}\int_0^t F[\tau, x_0(\tau), \dot{x}(\tau), 0] \sin n(t - \tau)d\tau$$
$$\dot{C}(t) = \int_0^t F[\tau, x_0(\tau), \dot{x}_0(\tau), 0] \cos n(t - \tau)d\tau \tag{3.8}$$

Replacing x_0 and $\dot{x}_0$ by their values (3.3), and then letting $t = 2\pi$, one obtains two relations between the constants of integration M_0 and N_0 which fulfill the above conditions, viz.:

$$P(M_0,N_0) = \int_0^{2\pi} F[\tau, M_0 \cos n\tau + N_0 \sin n\tau + \varphi(\tau);$$
$$-M_0 n \sin n\tau + N_0 n \cos n\tau + \dot{\varphi}(\tau), 0] \sin n\tau d\tau = 0$$
$$Q(M_0,N_0) = \int_0^{2\pi} F(\ldots) \cos n\tau d\tau = 0 \tag{3.9}$$

Equations (3.9) are thus the conditions determining particular values of the constants M_0 and N_0 for which a correspondence between the generating solution (3.3) and the solution of the nonlinear d.e. (3.1) is guaranteed.

One can visualize this situation also as follows: For the generating d.e. (3.2) there exists an infinity of periodic solutions depending on two arbitrary constants of integration M_0 and N_0, but this multiplicity of solutions shrinks to only one periodic solution which is in the neighborhood of a definite generating solution corresponding to definite values of M_0 and N_0. This isolated periodic solution is precisely the solution of the nonlinear d.e. (3.1). We shall see in Chapter 19 that this approach is the foundation of the method which Mandelstam and Papalexi developed in connection with the theory of the subharmonic resonance.

Once the constants M_0 and N_0 have been determined so as to fulfill the conditions (3.6), the rest of the problem (equation (3.5)) does not present any difficulty. In fact, we have

$$\psi_1' = [D]\beta_1 + [E]\beta_2 + [F]\mu + \ldots = 0$$
$$\psi_2' = [\dot{D}]\beta_1 + [\dot{E}]\beta_2 + [\dot{F}]\mu + \ldots = 0 \tag{3.10}$$

and these equations can be solved for $\beta_1(\mu)$ and $\beta_2(\mu)$ for any arbitrary (but small) μ, if the Jacobian

$$J = \left[\frac{\partial(\psi_1',\psi_2')}{\partial(\beta_1,\beta_2)}\right]_{\beta_1=\beta_2=\mu=0} = \begin{vmatrix} [D] & [E] \\ [\dot{D}] & [\dot{E}] \end{vmatrix} \neq 0 \tag{3.11}$$

In fact, in such a case there is one and only one solution $\beta_i = \beta_i(\mu)$ for which $\beta_i(0) = 0$, and this solution is analytic in μ. If one substitutes this solution into $x(t, \beta_1, \beta_2, \mu)$, one obtains also a periodic solution of (3.1) which is analytic in μ.

For the actual calculation of the Jacobian (3.11) it is necessary to calculate $[D]$, $[E]$, etc. For this purpose the series of Poincaré (3.4) has to be placed into (3.1) and the terms with $\beta_1\mu$ and $\beta_2\mu$ are to be equated after this substitution.

It is to be noted that in this substitution the nonlinear function F in (3.1) has to be developed into the Taylor series around the generating solution, viz.:

$$\begin{aligned} F^*(t, \beta_1, \beta_2, \mu) &= F^*(t, 0, 0, 0) + F_{\beta_1}{}^*\beta_1 + F_{\beta_2}{}^*\beta_2 + F_\mu{}^*\mu + \dots \\ &= F(t, x_0, \dot{x}_0, 0) + (F_x A + F_{\dot{x}}\dot{A})\beta_1 + (F_x B + F_{\dot{x}}\dot{B})\beta_2 \\ &\quad + (F_x C + F_{\dot{x}}\dot{C} + F_\mu)\mu + \dots \end{aligned} \tag{3.12}$$

where the notations F_x, $F_{\dot{x}}$ mean partial derivatives of F with respect to x and $\dot{x}$ into which the generating solutions have been replaced after the differentiation (that is, $\beta_1 = \beta_2 = \mu = 0$ and $x = x_0$; $\dot{x} = \dot{x}_0$).

For the determination of D and E one has to use the d.e. which are obtained by equating the coefficients of $\beta_1\mu$ and $\beta_2\mu$ on each side after the substitution of the series (3.4) into the d.e. (3.1).

We omit these intermediate calculations which are similar to those which were carried out in connection with function C and give directly the result

$$[D] = -\frac{1}{n}\frac{\partial P}{\partial M_0}; \qquad [\dot{D}] = \frac{\partial Q}{\partial M_0}; \qquad [E] = -\frac{1}{n^2}\frac{\partial P}{\partial N_0}; \qquad [\dot{E}] = \frac{1}{n}\frac{\partial Q}{\partial N_0} \tag{3.13}$$

so that the Jacobian (3.11) has now the expression

$$\begin{vmatrix} [D] & [E] \\ [\dot{D}] & [\dot{E}] \end{vmatrix} = -\frac{1}{n^2}\frac{\partial(P,Q)}{\partial(M_0,N_0)} \tag{3.14}$$

Summing up these results one can say:

The periodic solution of the nonlinear d.e. (3.1) corresponds to a generating solution (3.3) if the constants M_0 and N_0 satisfy the equations (3.9).

In other words, out of the infinity of periodic solutions of the generating system depending on two arbitrary constants M_0 and N_0, the solution of the nonlinear d.e. appears only in the neighborhood of one single generating solution, namely, the one for which M_0 and N_0 satisfy the equations (3.9).

This constitutes an essential difference with the nonresonance case in which one solution (the generating one, $\mu = 0$) goes into the other ($\mu \neq 0$)

without any change in the order of multiplicity of solutions as this happens in the resonance case for $\mu = 0$.

4. Calculation of periodic solutions; example

We assume that the constants M_0 and N_0 in the generating solution (3.3) satisfy the conditions (3.9), which shows that there is a periodic solution of (3.1) which is analytic in μ.

In such a case one can look for a series solution of the form

$$x = x_0(t) + \mu x_1(t) + \mu^2 x_2(t) + \dots \tag{4.1}$$

where $x_0 = x_0(t)$ is the generating solution and $x_1(t)$, $x_2(t)$ are certain unknown periodic functions of t with period 2π.

As usual, one replaces (4.1) into (3.1) and, equating coefficients of like powers of μ, one obtains a sequence of d.e. In the first place,

$$\ddot{x}_1 + n^2 x_1 = F(t, x_0, \dot{x}_0, 0) \tag{4.2}$$

As n is now an integer, (4.2) either has no periodic solutions or all its solutions are periodic. In fact, we have

$$F(t, x_0, \dot{x}_0, 0) = \frac{a_{01}}{2} + \sum_{m=1}^{\infty} [a_{m1} \cos mt + b_{m1} \sin mt] \tag{4.3}$$

To any summation term of this series corresponds a periodic term in the solution of (4.2) of the form

$$\frac{a_{m1} \cos mt + b_{m1} \sin mt}{n^2 - m^2}$$

with the exception of the term with the nth harmonic which accounts for a secular term:

$$\frac{t}{2n} (a_{n1} \sin nt - b_{n1} \cos nt)$$

Hence, in order to avoid such terms it is necessary that $a_{n1} = b_{n1} = 0$, or more explicitly

$$\int_0^{2\pi} F(t, x_0, \dot{x}_0, 0) \cos nt dt = \int_0^{2\pi} F(t, x_0, \dot{x}_0, 0) \sin nt dt = 0 \tag{4.4}$$

These are precisely same form as the equations (3.9) which specify the values of M_0 and N_0. Hence, (4.2) has also a periodic solution and, thus

$$x_1(t) = \frac{a_{01}}{2n^2} + \sum_{m=1}^{\infty} \frac{a_{m1} \cos mt + b_{m1} \sin mt}{n^2 - m^2} + M_1 \cos nt + N_1 \sin nt$$

M_1 and N_1 being arbitrary constants. This solution will be also periodic with period 2π. It is clear that the procedure can be continued for $x_2(t), \ldots$. We leave out some details in the formal argument here but refer to Malkin's text.[2] This shows that, for any combination of values of M_0 and N_0 forming a simple solution (3.9), there exists a series solution (4.1) with periodic coefficients which formally satisfies the d.e. (3.1); if μ is small enough the series converges and represents the actual periodic solution.

As an example we consider the operation of an electron tube circuit of the standard (induction coupled) type on the grid of which is impressed external periodic excitation with frequency ω_1. The equation of the circuit is

$$L\frac{di}{dt} + Ri + \frac{1}{C}\int_0^t idt - M\frac{dI_a}{dt} = P\sin\omega_1 t \tag{4.5}$$

where i is the current in the oscillating circuit, and I_a is the plate current. One can introduce the variable $v = \dfrac{1}{C}\displaystyle\int_0^t idt$ and consider two forms of nonlinear characteristics, viz.:

$$I_a(v) = S_0 v - \tfrac{1}{3}S_2 v^3; \qquad I_a(v) = S_0 v + \tfrac{1}{3}S_2 v^3 - \tfrac{1}{5}S_4 v^5 \tag{4.6}$$

The first of these two characteristics is the "soft" one and the second, the "hard" one. Consider the first case; the d.e. is

$$\ddot{v} + \left(\frac{R}{L} - \frac{MS_0}{LC} + \frac{MS_2}{LC}v^2\right)\dot{v} + \omega_0^2 v = \frac{P}{LC}\sin\omega_1 t \tag{4.7}$$

Introducing as independent variable $\tau = \omega_1 t$, one has

$$\ddot{v}_\tau + \omega^2 v = (\alpha' + \gamma'^2 v^2)v_\tau + \lambda'\sin\tau$$

where subscript τ designates differentiation with respect to τ and ω, α', γ', and λ are certain constants. Finally, setting $\alpha' = \alpha\mu$, $\gamma'^2 = \gamma^2\mu$, we have

$$\ddot{v} + \omega^2 v = \mu(\alpha - \gamma^2 v^2)\dot{v} + \lambda'\sin\tau \tag{4.8}$$

where we have dropped the subscript τ, since no confusion is to be feared from now on. In this form the d.e. is that of van der Pol with a "forcing term,"

$$\ddot{x} + \omega^2 x = \mu(1 - x^2)\dot{x} + \lambda\sin\tau \tag{4.9}$$

In order to obtain (4.9) from (4.7) and (4.8), one has to set

$$x = \sqrt{\frac{MS_2}{MS_0 - RC}}\cdot v; \qquad \mu = \frac{\omega_0^2}{\omega_1}(MS_0 - RC);$$

$$\lambda = \frac{P\omega_0^2}{\omega_1^2}\sqrt{\frac{MS_2}{MS_0 - RC}}; \qquad \omega^2 = \frac{\omega_0^2}{\omega_1^2}$$

[2] See footnote [2], page 232.

We can investigate now the solution of (4.9) for nonresonance and for resonance.

(*a*) *Nonresonant oscillations.* We assume that ω^2 is not an integer. In such a case the generating d.e. is

$$\ddot{x}_0 + \omega^2 x_0 = \lambda \sin \tau$$

and its periodic solution is: $u_0 = \dfrac{\lambda}{\omega^2 - 1} \sin \tau$. Using the series solution we obtain for u_1 the d.e.

$$\ddot{u}_1 + \omega^2 u_1 = (1 - u_0{}^2)\dot{u}_0 = \left[\frac{\lambda}{\omega^2 - 1} - \frac{\lambda^3}{4(\omega^2 - 1)^3}\right] \cos \tau + \frac{\lambda^3}{4(\omega^2 - 1)^3} \cos 3\tau.$$

which has the periodic solution

$$u_1 = \left[\frac{\lambda}{(\omega^2 - 1)^2} - \frac{\lambda^3}{4(\omega^2 - 1)^4}\right] \cos \tau + \frac{\lambda^3}{4(\omega^2 - 1)^3(\omega^2 - 9)} \cos 3\tau \tag{4.10}$$

One can calculate similarly the other terms: $u_2, u_3, \ldots$.

(*b*) *Resonance oscillations.* We assume $\omega^2 = 1 + \mu a$; $\lambda = \mu\lambda_0$. The d.e. (3.5) becomes then:

$$\ddot{x} + x = \mu[\lambda_0 \sin \tau - ax + (1 - x^2)\dot{x}] \tag{4.11}$$

The generating solution $u_0 = M_0 \cos \tau + N_0 \sin \tau$ contains two arbitrary constants. The series solution yields

$$\ddot{u}_1 + u_1 = \lambda_0 \sin \tau - au_0 + (1 - u_0{}^2)\dot{u}_0 \tag{4.12}$$

$$\ddot{u}_2 + u_2 = -au_1 + (1 - u_0{}^2)\dot{u}_1 - 2u_0 u_1 \dot{u}_0 \tag{4.13}$$

or

$$\begin{aligned}\ddot{u}_1 + u_1 = {} & (N_0 - aM_0 - \tfrac{1}{4}N_0 M_0{}^2 - \tfrac{1}{4}N_0{}^3) \cos \tau \\ & + (\lambda_0 - M_0 - aN_0 + \tfrac{1}{4}M_0{}^3 + \tfrac{1}{4}M_0 N_0{}^2) \sin \tau \\ & + \tfrac{1}{4}N_0(N_0{}^2 - 3M_0{}^2) \cos 3\tau + \tfrac{1}{4}M_0(M_0{}^2 - 3N_0{}^2) \sin 3\tau\end{aligned}$$

This d.e. has a periodic solution only if

$$\begin{aligned} P(M_0, N_0)/\pi &= -aN_0 - M_0[1 - \tfrac{1}{4}(M_0{}^2 + N_0{}^2)] + \lambda_0 = 0 \\ Q(M_0, N_0)/\pi &= -aM_0 + N_0[1 - \tfrac{1}{4}(M_0{}^2 + N_0{}^2)] = 0 \end{aligned} \tag{4.14}$$

These equations determine M_0 and N_0 in the generating solution. If M_0 and N_0 have been so determined, then we have

$$u_1 = M_1 \cos \tau + N_1 \sin \tau + \tfrac{1}{32}N_0(3M_0^2 - N_0^2) \cos 3\tau + \tfrac{1}{32}M_0(3N_0^2 - M_0^2) \sin 3\tau \quad (4.15)$$

where M_1 and N_1 are arbitrary constants which are determined from the condition of periodicity of $u_2(\tau)$.

In a similar manner for the next approximation, one has

$$\ddot{u}_2 + u_2 = f(M_0, N_0, M_1, N_1) \cos \tau + g(M_0, N_0, M_1, N_1) \sin \tau + \dots \quad (4.16)$$

where f and g are known functions. Setting $f = 0$, $g = 0$, one expresses the condition of periodicity for $u_2(\tau)$. One finds that the equations for the determination of M_1 and N_1 are linear, as they should be from the general theory. Moreover, the determinant of these equations coincides with the Jacobian

$$\frac{\partial(P,Q)}{\partial(M_0,N_0)} = -1 - a^2 + (M_0^2 + N_0^2) - \frac{3}{16}(M_0^2 + N_0^2)^2 \quad (4.17)$$

which is also in agreement with the general theory. If one considers now the values of M_0 and N_0 determined by (3.9) and (4.14), one finds

$$\tan \varphi = \left(1 - \frac{1}{4}A^2\right)/a; \qquad A^2\left[a^2 + \left(1 - \frac{A^2}{4}\right)^2\right] = \lambda_0^2 \quad (4.18)$$

which determines A^2 from a cubic equation. Hence, depending on the amplitude λ_0 of the external periodic excitations, one may have either one or three values of A^2.

We do not continue this problem further inasmuch as it coincides at this point with a different approach to this question suggested by van der Pol and further developed by Andronov and Witt, as we shall see in Chapter 18.

5. Autonomous systems

As previously mentioned, one calls *autonomous* systems (or d.e.) those in which the independent variable t (time) *does not appear explicitly*. In the theory of d.e. the autonomous systems appear merely as a special case of more general nonautonomous ones.

In the theory of oscillations the studies started, on the other hand, from autonomous cases, inasmuch as the van der Pol equation, which appeared first in these studies, belongs to the autonomous type.

As we mentioned in the introductory remarks to this chapter, the difference between the nonautonomous and the autonomous systems is that the solutions of the former have period of the external periodic excitation (or, more generally, are in a rational ratio to this period), whereas those of the latter are determined by the parameters of the d.e. itself.

It is also known from the general theory that the period of a solution of an autonomous system in some cases does not remain constant but is a function of the amplitude; a variation of period with amplitude is illustrated, for instance, in the case of a pendulum for larger angles of deviation. Also, in an autonomous d.e. one can always replace t by $t + t_0$, where t_0 is an arbitrary constant (the phase), and still have a solution of the same equation. This means that one can select arbitrarily the time origin in view of this property of *translation* in time which remains arbitrary. This permits selecting the origin of time at the instant when dx/dt is zero. This introduces a difference as compared to the nonautonomous case as regards the functions $\beta_1(\mu)$ and $\beta_2(\mu)$ of Poincaré.

In fact, since $\dot{x}_0(0) = 0$, the generating solution of $\ddot{x}_0 + k^2 x_0 = 0$ is of the form

$$x_0 = M_0 \cos kt \tag{5.1}$$

If one assumes a certain generating solution for a fixed value of M_0, our problem is to find the solution of the original autonomous equation

$$\ddot{x} + k^2 x = \mu f(x, \dot{x}, \mu) \tag{5.2}$$

where, as previously, we assume that f is an analytic function in x, $\dot{x}$, and μ.

This solution $x(t, \beta, \mu)$ is specified by the initial conditions

$$x(0, \beta, \mu) = x_0(0) + \beta = M_0 + \beta; \qquad \dot{x}_0(0, \beta, \mu) = 0 \tag{5.3}$$

where we assume $\beta(\mu) \underset{\mu \to 0}{\longrightarrow} 0$ as before.

Thus, in the autonomous systems (we consider here only systems with one degree of freedom), instead of two parameters β_1 and β_2, there is only one parameter β, and there will appear another function $\tau = \tau(\mu)$—the nonlinear correction for period—which has a similar property of vanishing together with μ.

The determination of β and τ is obtained again from the conditions of periodicity, viz.:

$$\begin{gathered} x(T + \tau, \beta, \mu) - x(0, \beta, \mu) = x(T + \tau, \beta, \mu) - M_0 - \beta = 0 \\ \dot{x}(T + \tau, \beta, \mu) = 0 \end{gathered} \tag{5.4}$$

Expanding these expressions in terms of τ, one gets

$$x(T,\beta,\mu) + \dot{x}(T,\beta,\mu)\tau + \tfrac{1}{2}[-k^2x(T,\beta,\mu) + \ldots]\tau^2 + \ldots - M_0 - \beta = 0$$
$$\dot{x}(T,\beta,\mu) + [-k^2x(T,\beta,\mu) + \ldots]\tau = 0 \tag{5.5}$$

The series of Poincaré in this case is:

$$x(t,\beta,\mu) = x_0(t) + A\beta + \mu[C + D\beta + E\mu + \ldots] \tag{5.6}$$

which yields: $\ddot{A} + k^2A = 0$; $A(0) = 1$; $\dot{A}(0) = 0$; whence $A = \cos kt$. If one substitutes (5.6) into (5.5), one gets

$$\mu\left[C\left(\frac{2\pi}{k}\right) + D\left(\frac{2\pi}{k}\right)\beta + E\left(\frac{2\pi}{k}\right)\mu + \dot{C}\left(\frac{2\pi}{k}\right)\tau + \ldots\right] + \tfrac{1}{2}\tau^2(-M_0k^2 + \ldots) = 0 \tag{5.7}$$
$$\mu\left[\dot{C}\left(\frac{2\pi}{k}\right) + \ldots\right] + \tau(-M_0k^2 + \ldots) = 0$$

From the second equation (5.7), one has

$$\tau = \mu\left[\frac{1}{M_0k^2}\dot{C}\left(\frac{2\pi}{k}\right) + \ldots\right] \tag{5.8}$$

with this value of τ, the first equation (5.7) gives

$$\mu\left[C\left(\frac{2\pi}{k}\right) + D\left(\frac{2\pi}{k}\right)\beta + Q\mu + \ldots\right] = 0 \tag{5.9}$$

Q being some coefficient.

We encounter here a problem similar to that discussed in Section 3, viz.: we wish to determine $\beta = \beta(\mu)$ for any value of μ as long as it is small.

The necessary condition is, therefore,

$$C\left(\frac{2\pi}{k}\right) = 0 \tag{5.10}$$

If $D(2\pi/k) \neq 0$, the solution for β will be unique and analytic. For the calculation of the functions C and D, we have the d.e.

$$\ddot{C} + k^2C = f(x_0, \dot{x}_0\, 0); \qquad \ddot{D} + k^2D = f_{x0}\cos kt - f_{\dot{x}_0}\sin kt$$
$$C(0) = \dot{C}(0) = D(0) = \dot{D}(0) = 0$$

where f_{x_0} and $f_{\dot{x}_0}$ are the partial derivatives of the function f with respect to

x and $\dot{x}$ into which the generating solutions have been substituted after the differentiation. Whence

$$C = \frac{1}{k}\int_0^t f(x_0, \dot{x}_0, 0)\sin k(t-\tau')d\tau'$$
$$D = \frac{1}{k}\int_0^t [(f_{x_0}\cos k\tau' - kf_{\dot{x}_0}\sin k\tau')\sin k(t-\tau')]d\tau' \equiv \frac{\partial C}{\partial M_0} \tag{5.11}$$

Therefore, the condition appears ultimately (after a change of the variable) as

$$P(M_0) = \int_0^{2\pi} f(M_0\cos u, -kM_0\sin u, 0)\sin u\,du = 0 \tag{5.12}$$

As the condition $D(2\pi/k) \neq 0$, it can be written as $[\partial P(M_0)]/\partial M_0 \neq 0$. Thus to any single nonzero root of (5.12) corresponds a unique analytic solution of (5.4) and, therefore, of (5.2).

6. Calculation of periodic solutions of autonomous systems

Given the d.e. (5.2), one could try to satisfy it by a series solution of the form:

$$x(t) = x_0(t) + \mu x_1(t) + \mu^2 x_2(t) + \ldots \tag{6.1}$$

In the case of autonomous systems there is, however, a complication owing to the fact that their period is now: $T + \tau(\mu)$ and, thus, depends on μ. For that reason from the equation of periodicity, viz.:

$$x_0\left(t + \frac{2\pi}{k} + \tau\right) + \mu x_1\left(t + \frac{2\pi}{k} + \tau\right) + \ldots = x_0(t) + \mu x_1(t) + \ldots$$

one cannot conclude that $x_s[t + (2\pi/k) + \tau] = x_s(t)$, inasmuch as the period of the left-hand side depends on μ (though $\tau = \tau(\mu)$), while that of the right-hand side is independent of μ.

Thus, for instance, the development of $\sin(1+\mu)t$ yields

$$\sin(1+\mu)t = \sin t + \mu t\cos t - \frac{\mu^2 t^2}{2}\sin t + \ldots$$

which shows that the periodic function $\sin(1+\mu)t$ has nonperiodic coefficients. This is due to the fact that the period $2\pi/(1+\mu)$ in this case depends on μ. It is impossible, therefore, to form any idea regarding the periodicity of the solution from the form (6.1).

According to Krylov-Bogoliubov, this difficulty can be obviated by setting

$$T' = \frac{2\pi}{k} + \tau = \frac{2\pi}{k}(1 + h_1\mu + h_2\mu^2 + \ldots) \tag{6.2}$$

where T' is the period of the solution and h_i certain unknown constants. This amounts to replacing the variable t in (5.2) by means of a substitution

$$t = \frac{\tau}{k}(1 + h_1\mu + h_2\mu^2 + \ldots) \tag{6.3}$$

To the periodic solution of (5.2) corresponds that of the new d.e.:

$$\frac{d^2x}{d\tau^2} + x(1 + h_1\mu + h_2\mu^2 + \ldots)^2$$
$$= \frac{\mu}{k^2} f\left[x, k(1 + h_1\mu + \ldots)^{-1}\frac{dx}{d\tau}, \mu\right](1 + h_1\mu + \ldots)^2 \tag{6.4}$$

whose period is 2π.

We can now look for the periodic solution of (6.4) in the form of a series

$$x(t) = x^{(0)}\tau + \mu x_1(\tau) + \mu^2 x_2(\tau) + \ldots \tag{6.5}$$

As the period is now 2π and is independent of μ, the coefficients $x_s(\tau)$ are periodic functions with period 2π.

As $dx/dt = 0$ for $t = 0$, one must also have

$$\left(\frac{dx_{(0)}}{d\tau}\right)_{\tau=0} = \left(\frac{dx_i}{d\tau}\right)_{\tau=0}, \qquad i = 1, 2, \ldots \tag{6.6}$$

Since the periodic solution exists, we shall show that it is possible to determine a series (6.5) satisfying formally the d.e. and that this is the only series solution. The procedure, to some extent, reminds us of the one already encountered in Section 4.

We have first, the d.e. $\ddot{x}^{(0)} + x^{(0)} = 0$ (differentiations are with respect to τ) whose solution (in view of (6.6)) is: $x^{(0)} = M_0 \cos\tau$, M_0 arbitrary. Then we have the second d.e.:

$$\ddot{x}_1 + x_1 = -2h_1 M_0 \cos\tau + \frac{1}{k^2} f(M_0\cos\tau, -kM_0\sin\tau, 0)$$

This d.e. has a periodic solution only in the case when

$$P(M_0) = \int_0^{2\pi} f(M_0\cos\tau, -kM_0\sin\tau; 0)\sin\tau d\tau = 0 \tag{6.7}$$

$$-2h_1 M_0 + \frac{1}{\pi k^2}\int_0^{2} f(M_0\cos\tau, -kM_0\sin\tau, 0)\cos\tau d\tau = 0$$

The first of these d.e. determines the amplitude M_0 of the generating solution, and the second determines h_1. We have thus

$$x_1 = \varphi_1(\tau) + M_1 \cos \tau + N_1 \sin \tau$$

where $\varphi_1(\tau)$ is a certain periodic function of τ with period 2π, and M_1 and N_1 are arbitrary constants. These constants are determined from the initial condition (6.6) and also from conditions of periodicity of $x_2(t)$. This, in turn, determines the constants h_2.

We omit here the formal argument by which it is shown that if x_m has a periodic solution, it is of the form

$$x_m = \varphi_m(\tau) + M_m \cos \tau + N_m \sin \tau$$

where M_m and N_m are arbitrary constants, and $\varphi_m(\tau)$ are certain periodic functions. From the integration of the d.e. for x_m, together with the initial conditions, one determines M_{m-1}, N_{m-1}, and h_m. We refer for these details to Malkin.[2]

It is seen that the procedure reminds one of that of Lindstedt (Chapter 9). In both cases the existence of two constants of integration in each step of the approximation procedure serves to determine the periodicity of the solution, on one hand, and add a term to the series development of the unknown frequency, on the other hand.

As an example we consider a standard electron-tube oscillator with an inductive coupling. Its d.e. is the same as in Section 4 (equation (4.8)) with $\lambda' = 0$, viz.:

$$\ddot{v} + \omega^2 v = \mu(\alpha - \gamma^2 v^2)\dot{v} \tag{6.8}$$

We change first the independent variable by the substitution

$$t = \frac{\tau}{\omega}(1 + h_1\mu + h_2\mu^2 + \ldots)$$

which transforms (6.8) into the d.e.

$$\ddot{v}_\tau + v(1 + h_1\mu + \ldots)^2 = \frac{\mu}{\omega}(\alpha - \gamma^2 v^2)(1 + h_1\mu + \ldots)\dot{v}_\tau \tag{6.9}$$

where the subscript τ means: differentiation with respect to τ.

We shall try to satisfy this d.e. formally by a series

$$v = M_0 \cos \tau + \mu v_1(\tau) + \mu^2 v_2(\tau) + \ldots \tag{6.10}$$

where the $v_i(\tau)$ are periodic functions with period 2π for which hold the initial conditions: $\dot{v}_1(0) = \dot{v}_2(0) = \ldots = 0$.

We have

$$\ddot{v}_1 + v_1 + 2h_1M_0 \cos \tau = -\frac{1}{\omega}(\alpha - \gamma^2 M_0^2 \cos^2 \tau)M_0 \sin \tau$$

$$= \left(-\frac{\alpha M_0}{\omega} + \frac{\gamma^2 M_0^3}{4\omega}\right) \sin \tau + \frac{\gamma^2 M_0^3}{4\omega} \sin 3\tau + \dots \tag{6.11}$$

Equating to zero the coefficients of $\cos \tau$ and $\sin \tau$, we have

$$h_1 = 0; \qquad \frac{P(M_0)}{\pi} = -\frac{\alpha M_0}{\omega} + \frac{\gamma^2 M_0^3}{4\omega} = 0; \qquad M_0^2 = \frac{4\alpha}{\gamma^2}$$

This yields for v_1 the solution

$$v_1 = -\frac{\gamma^2 M_0^3}{32\omega} \sin 3\tau + M_1 \cos \tau + N_1 \sin \tau$$

where M_1 and N_1 are arbitrary constants.

For v_2 we have the d.e.

$$\ddot{v}_2 + v_2 = -2h_2 M_0 \cos \tau + \frac{\alpha}{\omega}\dot{v}_1 - \gamma^2 \frac{M_0^2}{\omega}\dot{v}_1 \cos^2 \tau + \frac{\gamma^2 M_0^2}{\omega} v_1 \sin 2\tau$$

$$= M_0\left(-2h_2 + \frac{\alpha^2}{8\omega^2}\right) \cos \tau + \frac{\gamma^2 M_0^2 M_1}{2\omega} \sin \tau + \dots \tag{6.12}$$

where the nonwritten terms do not contain $\cos \tau$ and $\sin \tau$. Equating to zero the coefficients of $\cos \tau$ and $\sin \tau$, we get

$$h_2 = \frac{\alpha^2}{16\omega^2}; \qquad M_1 = 0; \qquad N_1 = \frac{3\gamma^2}{32\omega} M_0^3$$

Thus, the periodic solution is of the form:

$$v = M_0 \cos \tau + \left[\frac{3\gamma^2}{32\omega} M_0^3 \sin \tau - \frac{\gamma^2}{32\omega} M_0^3 \sin 3\tau\right]\mu + \dots$$

$$\tau = \omega t\left[1 - \frac{1}{16}\frac{\alpha^2}{\omega^2}\mu^2 + \dots\right]; \qquad M_0 = \sqrt{4\alpha}/\gamma \tag{6.13}$$

The solution is real if $\alpha > 0$, in which case the state of rest is unstable. The period of oscillation is given by the formula

$$T' = \frac{2\pi}{\omega}\left(1 + \frac{\alpha^2}{16\omega^2}\mu^2 + \dots\right) \tag{6.14}$$

In this manner the two series solutions for $v(\tau)$ and $\tau(\mu)$ progress in parallel in each step of the approximation procedure as was explained in connection with the method of Lindstedt (Chapter 9).

7. Nonanalytic case

In all previous discussions we considered the d.e. whose right-hand sides are analytic in x, $\dot{x}$, and μ. It often happens that the above conditions are not fulfilled in applications and, instead of analyticity, the problem is limited to the existence of continuous partial derivatives of the first order.

In such cases it is impossible to look for the solution in the form (6.5), but it is necessary to try to determine it in the form: $x(t) = x_0(t), x_1(t), x_2(t), \ldots$, where $x_1(t)$, $x_2(t)$ are successive approximations.

Thus, for instance, in the case when the generating d.e.

$$\ddot{x}_0 + k^2 x_0 + f(t) = 0 \tag{7.1}$$

has a unique periodic solution $x_0(t) = \varphi(t)$, it is still possible to find the periodic solution of the nonlinear d.e. for μ sufficiently small by the method of successive approximations.

We assume $x_0(t)$ as the zero-order approximation; then for the first approximation we have

$$\ddot{x}_1 + k^2 x_1 + f(t) = \mu F(t, x_0, \dot{x}_0, \mu) \tag{7.2}$$

and so on. As we deal here with a nonresonance case, there is only one periodic solution which is gradually made more accurate by the successive approximations.

For the resonance oscillations we have the d.e.

$$\ddot{x} + n^2 x + f(t) = \mu F(t, x, \dot{x}, \mu) \tag{7.3}$$

where the Fourier development of $f(t)$ does not contain the nth harmonic. The generating solution, as previously, will have the form

$$x_0(t) = \varphi_0(t) + M_0 \cos nt + N_0 \sin nt \tag{7.4}$$

where $\varphi_0(t)$ is a particular solution of the generating equation. It can be shown that the generating solution ($\mu = 0$) goes into the solution of (7.3) if M_0 and N_0 satisfy the conditions

$$\begin{aligned} P(M_0,N_0) &= \int_0^{2\pi} F(t, x_0, \dot{x}_0, 0) \sin nt\,dt = 0 \\ Q(M_0,N_0) &= \int_0^{2\pi} F(t, x_0, \dot{x}_0, 0) \cos nt\,dt = 0 \end{aligned} \tag{7.5}$$

which shows that the Fourier expansion of F does not contain the nth harmonic.

Hence, just as in the analytic case, there exists one periodic solution of (7.3) if the Jacobian

$$J = \frac{\partial(P,Q)}{\partial(M_0,N_0)} \neq 0 \tag{7.6}$$

If M_0 and N_0 satisfy these conditions, we can determine the periodic solution of (7.3) by successive approximations.

As the first approximation we take

$$x_1(t) = \varphi_1 1(t) + M_1 \cos nt + N_1 \sin nt \tag{7.7}$$

where M_1 and N_1 are for the time being unknown. We proceed in the same manner with other approximations.

Consider for instance $x_2(t)$; as n is an integer, it is necessary that x_2 should be periodic and for this we must have

$$\begin{aligned} P_1(M_1, N_1, \mu) &= \int_0^{2\pi} F(t, x_1, \dot{x}_1, \mu) \sin nt dt = 0 \\ Q_1(M_1, N_1, \mu) &= \int_0^{2\pi} F(t, x_1, \dot{x}_1, \mu) \cos nt dt = 0 \end{aligned} \tag{7.8}$$

which determine M_1 and N_1. In view of (7.5), (7.8) are satisfied for $\mu = 0$, and $M_1 = M_0$, $N_1 = N_0$. Since (7.6) is fulfilled, there exists only one solution $M_1(\mu)$, $N_1(\mu)$ for $\mu \neq 0$, which becomes $M_1(0) = M_0$; $N_1(0) = N_0$.

Using M_1 and N_1 determined from the next approximation we have

$$x_2(t) = \varphi_2^* + M_2 \cos nt + N_2 \sin nt$$

where M_2 and N_2 are yet unknown. The procedure becomes thus apparent from now on. We refer to Malkin[2] where the reader will find further examples of these calculations.

[2] See footnote [2], page 232.

Chapter 11

OSCILLATIONS IN SYSTEMS WITH SEVERAL DEGREES OF FREEDOM

1. Introductory remarks

In the preceding chapter we were concerned with systems with one degree of freedom which are amenable to a d.e. of the second order or, which is the same, to an equivalent system of two d.e. of the first order.

One encounters occasionally systems with two degrees of freedom amenable to a differential system of the fourth order; this occurs generally when two oscillatory systems are coupled together in some manner.

In applications one rarely encounters systems beyond the fourth order. The mathematical treatment of systems with several degrees of freedom is usually conducted on a general basis of systems with n degrees of freedom inasmuch as, from a theoretical point of view, this generalization does not introduce any special difficulties. The difficulties appear, however, in applying the general theory to concrete examples, inasmuch as the calculations are inevitably more complicated. We omit here a number of proofs that can be found in Malkin's text. The subject matter can be readily followed from the contents of the preceding chapters inasmuch as the procedure is essentially the same, being merely generalized to systems of $2n$ d.e. of the first order.

The last section, concerning the method of averaging, represents an abstract of a relatively long chapter in a recent treatise of Bogoliubov and Mitropolsky.[1]

[1] N. N. Bogoliubov and J. A. Mitropolsky, *Asymptotic Methods in the Theory of Nonlinear Oscillations* (in Russian), Moscow, 1948.

We briefly recall certain facts from the theory of linear d.e., some of which were obtained in Chapter 5. Given a differential system

$$\dot{x}_s = \sum_{j=1}^{n} a_{sj}x_j; \qquad s = 1, 2, \ldots, n \tag{1.1}$$

where the a_{sj} are constants, its solution is related to the nature of the roots λ of its characteristic equation

$$D(\lambda) = \begin{vmatrix} a_{11} - \lambda & a_{12} & \cdots & a_{1n} \\ a_{21} & a_{22} - \lambda & \cdots & a_{2n} \\ \cdot & \cdot & \cdot & \cdot \\ a_{n1} & a_{n2} & \cdots & a_{nn} - \lambda \end{vmatrix} = 0 \tag{1.2}$$

If λ is a root of (1.2), the functions

$$x_s = A_s e^{\lambda t} \tag{1.3}$$

are solutions of (1.1); A_s is a system of constants determined by a homogeneous system of equations

$$a_{s1}A_1 + \ldots + (a_{ss} - \lambda)A_s + \ldots + a_{sn}A_n = 0; \qquad \text{for } s = 1, 2, \ldots, n \tag{1.4}$$

If all roots are distinct, one obtains n particular solutions forming a fundamental system. The general solution is then a linear combination of these particular solutions multiplied by arbitrary constants.

If one of the roots λ is multiple, then the system (1.1), in addition to the solution (1.3) has also solutions of the form

$$x_s = f_s(t)e^{\lambda t} \tag{1.5}$$

where the f_s are polynomials whose degree does not exceed $p - 1$, and p is the multiplicity of the root. It is also known that, if the system has a solution of the form (1.5), it has also a solution x_s of the form

$$x_s = e^{\lambda t}\frac{df_s(t)}{dt} \tag{1.6}$$

as one verifies by differentiating (1.1) and noting that the derivatives of any solution are also solutions. In such a case the solution has the form:

$$x_s = e^{\lambda t}f_s(t) + e^{\lambda \tau}\frac{df_s}{dt} \tag{1.7}$$

If p is the multiplicity of the root λ which does not eliminate at least one minor of the $(n - 1)$st order of (1.2), the system (1.1) admits solutions of the type (1.5) in which the degree of one of the polynomials reaches $(p - 1)$. If one replaces these polynomials by their subsequent derivatives, one obtains p particular solutions of the form

$$x_{si} = e^{\lambda t} \frac{d^{i-1}}{dt^{i-1}} f_s(t) \tag{1.8}$$

The last solution in this sequence will be, obviously, (1.3). We shall not elaborate further on this question but will refer the reader to standard texts for further details.

If λ is a complex root, say $\lambda = \mu + i\nu$ the quantities A_s and $f_s(t)$ are also complex, say $A_s = B_s + iC_s; f_s = P_s + iQ_s$, where B_s, C_s, P_s, and Q_s are real. In such a case there are two real solutions, either of the form

$$x_s = (B_s \cos \nu t - C_s \sin \nu t)e^{\mu t}; \qquad x_s = (B_s \sin \nu t + C_s \cos \nu t)e^{\mu t}$$

or of the form

$$x_s = (P_s \cos \nu t - Q_s \sin \nu t)e^{\mu t}; \qquad x_s = (P_s \sin \nu t + Q_s \cos \nu t)e^{\mu t}$$

2. Periodic solutions of homogeneous linear systems with constant coefficients

From the preceding it follows that (1.1) has periodic nontrivial solutions either when (1.2) has a zero root or, more generally, purely imaginary roots. In the first case, if (1.2) has a zero root of multiplicity p to which corresponds k groups of solutions, (1.1) will have also k groups of partial solutions of the form

$$x_{sj} = A_{sj} \qquad (s = 1, 2, \ldots, n; \quad j = 1, \ldots, k) \tag{2.1}$$

where the A_{sj} are constants. Other solutions will be polynomials in t. Solutions (2.1) may be regarded as periodic with an arbitrary period. If (1.2) has a pair of imaginary roots $\pm \nu i$ of multiplicity greater than one, the system (1.1) has $2k$ solutions

$$\begin{aligned} x_{sj} &= B_{sj} \cos \nu t - C_{sj} \sin \nu t \\ x_{sj}^* &= B_{sj} \sin \nu t + C_{sj} \cos \nu t \end{aligned} \tag{2.2}$$

$(s = 1, \ldots, n; j = 1, \ldots, k)$, B_{sj} and C_{sj} being constants; these solutions are periodic with period $\omega = 2\pi/\nu$. Other solutions corresponding to the roots $\pm \nu i$ of any (that is, if the multiplicity of $\pm \nu i$ exceeds k) are of the form $t^m \sin \nu t$, $t^m \cos \nu t$, that is, are not periodic.

System (1.1), however, may have periodic solutions of period ω which are not (2.2) if (1.2) has roots $\pm p\nu i$, where p is an integer. In fact, in such a case the system will have solutions of period ω/p which may be regarded as being still of period ω.

Assume that (1.2) has a zero root to which correspond k groups of solutions and r pairs of imaginary roots $\pm p_j \nu i$, where p_j are integers; we assume that to these roots correspond k_j groups of solutions. In such a case the system will have $m = k + 2k_1 + \ldots + 2k_r$ periodic solutions with period $\omega = 2\pi/\nu$. If we designate each of these solutions as $\varphi_{s\alpha}$, the system (1.1) will have periodic solution of period ω:

$$x_s = C_1\varphi_{s1} + C_2\varphi_{s2} + \ldots + C_m\varphi_{sm} \tag{2.3}$$

containing m arbitrary constants C_j.

We shall make a distinction between two possible cases: (1) nonresonance and (2) resonance. In the first case the trivial solution is the only periodic solution with period ω of (1.1). In the second case there are m linearly independent periodic solutions of (1.1) with period ω other than the zero solution.

We consider the system

$$\dot{y} + a_{1s}y_1 + \ldots + a_{ns}y_n = 0 \tag{2.4}$$

which is called the *adjoint system* to 1.1 (the latter may be considered as the adjoint of (2.4).

It can be shown that to the resonance of (1.1) corresponds the resonance of (2.4) and vice versa.

In the case of nonresonance if m' denotes the analogous of m, for the adjoint system we have $m' = m$.

We conclude this section with certain theorems concerning periodic solutions of linear nonhomogeneous d.e. with constant coefficients.

Given

$$\dot{x}_s = a_{s1}x_1 + \ldots + a_{sn}x_n + f_s(t) \tag{2.5}$$

where the a_{sj} are constants and the $f_s(t)$ are continuous periodic functions with period ω. The following theorems show under which conditions the system (2.5) has periodic solutions. For proofs we refer to Malkin.[2]

(1) *In a nonresonance case, the system* (2.5) *admits one and only one periodic solution for any choice of functions* $f_s(t)$.

(2) *In a resonance case* (*that is, when the characteristic equation has either a zero root or roots of the form* $\pm 2\pi pi/\omega$, p *integer*), *the system* (2.5) *has periodic solutions only in the case when the functions* $f_s(t)$ *satisfy the conditions*

$$\int_0^\omega \sum_{\alpha=1}^n f_\alpha(\tau)\psi_{\alpha i}(\tau)d\tau = 0; \qquad i = 1, \ldots, m \tag{2.6}$$

[2] I. G. Malkin, *Certain Problems in the Theory of Nonlinear Oscillations* (in Russian), Moscow, 1956, English translation.

where $\psi_{s1}(t), \ldots, \psi_{sm}(\tau)$ $(s = 1, \ldots, n)$ are periodic solutions (with period ω) of the adjoint system

$$\dot{y}_s + a_{1s}y_1 + \ldots + a_{ns}y_n = 0; \qquad s = 1, 2, \ldots, n \tag{2.7}$$

In the first case the determinant

$$\begin{vmatrix} x_{11}(\omega) - 1 & x_{12}(\omega) & \ldots & x_{1n}(\omega) \\ x_{21}(\omega) & x_{22}(\omega) - 1 & \ldots & x_{2n}(\omega) \\ \cdot & \cdot & \cdot & \cdot \\ x_{n1}(\omega) & x_{n2}(\omega) & \ldots & x_{nn}(\omega) - 1 \end{vmatrix}$$

is: $\neq 0$; in the second case it is: $=0$. Here x_{is} represents a fundamental system of solutions of (1.1) with

$$x_{ss}(0) = 1_s; \qquad x_{sj}(0) = 0 \qquad \text{for} \quad s \neq j$$

If the conditions (2.6) are fulfilled, (2.5) admits an infinity of periodic solutions of the form

$$x_s = M_1\varphi_{s1} + M_2\varphi_{s2} + \ldots + M_m\varphi_{sm} + \omega_s(t) \tag{2.8}$$

where the M_i are arbitrary constants and $\omega_s(t)$ is a particular periodic solution of (2.5).

Consider the example

$$\ddot{x} + k^2x = f(t) \tag{2.9}$$

where $f(t)$ is a continuous periodic function with period $\omega = 2\pi/k$. The usual conditions of periodicity are

$$\int_0^{2n/k} f(t) \cos ktdt = \int_0^{2\pi/k} f(t) \sin ktdt = 0 \tag{2.10}$$

In order to make use of the general condition (2.6) we write first (2.9) as a system

$$\dot{x}_1 = x_2; \qquad \dot{x}_2 = -k^2x_1 + f(t) \tag{2.11}$$

and form the system

$$\dot{y}_1 = k^2y_2; \qquad \dot{y}_2 = -y_1 \tag{2.12}$$

which is adjoint with the homogeneous part of (2.11). This system has periodic solutions

$$\psi_{11} = \sin kt; \qquad \psi_{21} = \frac{1}{k}\cos kt; \qquad \psi_{12} = \cos kt; \qquad \psi_{22} = -\frac{1}{k}\sin kt$$

If one substitutes these solutions into (2.6), one finds that the condition is fulfilled.

As a second example, consider the system

$$\dot{x}_1 = -kx_2 + f(t); \qquad \dot{x}_2 = kx_1 + F(t) \tag{2.13}$$

where $f(t)$ and $F(t)$ are again continuous periodic functions of period $2\pi/k = \omega$. The adjoint system in this case is

$$\dot{y}_1 = -ky_2; \qquad \dot{y}_2 = ky_1 \tag{2.14}$$

and one can set as solutions

$$\psi_{11} = \cos kt; \qquad \psi_{12} = \sin kt; \qquad \psi_{21} = \sin kt; \qquad \psi_{22} = -\cos kt$$

If one substitutes these functions into (2.6), the conditions for the existence of periodic solutions of (2.12) are

$$\begin{aligned} &\int_0^{2n/k} [f(t) \cos kt + F(t) \sin kt]dt = 0 \\ &\int_0^{2n/k} [f(t) \sin kt - F(t) \cos kt]dt = 0 \end{aligned} \tag{2.15}$$

3. Nonresonance oscillations of nonautonomous systems

Consider a system

$$\dot{x}_s = \sum_{j=1}^{n} a_{sj}x_j + f_s(t) + \mu F_s(t, x_1, \ldots, x_n, \mu); \qquad s = 1, 2, \ldots, n \tag{3.1}$$

where the $f_s(t)$ are continuous periodic functions with period 2π and the F_s are given for $\mu < \mu_0$, $\mu_0 > 0$ for $x_i \in G$, where G is a certain region of the variables x_i; we assume here that the F_s are analytic in μ and are continuously differentiable in the other variables up to an order k.

Consider a corresponding generating system

$$\dot{x}_s^{(0)} = \sum_{j=1}^{n} a_{sj}x_j^{(0)} + f_s(t) \tag{3.2}$$

and assume that the characteristic equation (of the homogeneous part) has neither zero nor purely imaginary roots $\pm pi$, p an integer. In such a case (3.2) admits one and only one periodic solution

$$x_s^{(0)} = \varphi_s(t) = \varphi_1(0)x_{s1}(t) + \ldots + \varphi_n(0)x_{sn}(t)$$
$$+ \int_0^t \sum_{\alpha=1}^{n} x_{s\alpha}(t-\tau)f_\alpha(\tau)d\tau, \qquad (s = 1, \ldots, n) \tag{3.3}$$

which we consider as the generating solution. Here x_{sj}, as before, is the fundamental system of solutions of (1.1) with $x_{ss}(0) = 1$; $x_{sj}(0) = 0$ for $s \neq j$.

We wish to look for periodic solutions of (3.1) with period 2π which approach $\varphi(t)$ when $\mu \to 0$. We shall show first the existence of just one such solution and we shall indicate later how to calculate it. From the theory of d.e. it is known that to every $(n + 1)$-tuple of numbers $(\beta_1, \ldots, \beta_n, \mu)$ corresponds a unique solution of (3.1)

$$x_s = x_s(t, \beta_1, \ldots, \beta_n, \mu), \qquad (s = 1, \ldots, n)$$

such that

$$x_s(0, \beta_1, \ldots, \beta_n, \mu) = \varphi_s(0) + \beta_s \tag{3.4}$$

holds. This solution is continuously differentiable with respect to $\beta_1, \ldots, \beta_n, \mu$, as can be verified, and it satisfies equations

$$x_s = C_1 x_{s1}(t) + \ldots + C_s x_{sn}(t) + \int_0^t \sum_{\alpha=1}^n x_{s\alpha}(t - \tau) f_\alpha(\tau) d\tau$$
$$+ \mu \int_0^t x_{s\alpha}(t - \tau) F_\alpha(\tau, x, \mu) d\tau \tag{3.5}$$

where $F_s(t, x, \mu)$ stands for $F_s(t, x_1, \ldots, x_n, \mu)$, and $x_s(t, \beta, \mu)$ stands for $x_s(t, \beta, \ldots, \beta_n, \mu)$ with

$$c_s = \varphi_s(0) + \beta_s$$

Hence if one takes into account (3.3), (3.5) may be written as

$$x_s(t, \beta_1, \ldots, \beta_n, \mu) = \beta_1 x_{s1}(t) + \ldots + \beta_n x_{sn}(t) + \varphi_s(t)$$
$$+ \mu \int_0^t \sum_{\alpha=1}^n x_{s\alpha}(t - \tau) F_\alpha(\tau, x(\tau, \beta, \mu), \mu) d\tau \tag{3.6}$$

for every $(n + 1)$-tuple $(\beta_1, \ldots, \beta_n, \mu)$. If one imposes now the condition of periodicity

$$x_s(2\pi, \beta_1, \ldots, \beta_n, \mu) - x_s(0, \beta_1, \ldots, \beta_n, \mu) = 0 \tag{3.7}$$

one obtains

$$\beta_1 x_{s1}(2\pi) + \ldots + \beta_n x_{sn}(2\pi) - \beta_s$$
$$+ \int_0^{2n} \sum_{\alpha=1}^n x_{s\alpha}(2\pi - \tau) F_\alpha(\tau, x, \mu) d\tau = 0 \tag{3.8}$$

that is, a system of n equations with $n + 1$ unknowns: $\beta_1, \ldots, \beta_n, \mu$.

This system is satisfied with $\beta_1 = \beta_2 = \ldots = \beta_n = \mu = 0$ and its Jacobian reduces to

$$\begin{vmatrix} x_{11}(2\pi) - 1 & x_{12}(2\pi) & \ldots & x_{1n}(2\pi) \\ x_{21}(2\pi) & x_{22}(2\pi) - 4 & \ldots & x_{2n}(2\pi) \\ \cdot & \cdot & \cdot & \cdot \\ x_{n1}(2\pi) & x_{n2}(2\pi) & \ldots & x_{nn}(2\pi) - 1 \end{vmatrix} \tag{3.9}$$

As we are in the nonresonant case, the determinant does not vanish and we may apply to (3.8) the implicit function theorem; one can then assert that in the neighborhood of $\mu = 0$ there exists a unique n-tuple continuous functions: $\beta_1 = \beta_1(\mu), \ldots, \beta_n = \beta_n(\mu)$. If for every μ sufficiently near to $\mu = 0$ we choose this n-tuple set of $\beta_1, \ldots, \beta_n$, the corresponding $x_s(t, \beta_1, \ldots, \beta_n, \mu)$ or, rather

$$x_s = x_s(t, \beta_1(\mu), \ldots, \beta_n(\mu), \mu) = \tilde{x}_s(t,\mu)$$

will have the following properties:

(1) It is a solution of (3.1).
(2) It satisfies (3.7), that is, it has period 2π.
(3) It satisfies (3.4) or, rather

$$\tilde{x}_s(0,\mu) = \varphi_s(0) + \beta_s(\mu)$$

so that $\tilde{x}_s(0,\mu) \to \varphi_s(0)$ for $\mu \to 0$ and, therefore

$$\tilde{x}_s(t,\mu) \to \varphi_s(t)$$

over the period and, hence, everywhere, as $\mu \to 0$.

We conclude that *for a sufficiently small μ there is one and only one periodic solution which becomes the generating solution for $\mu = 0$.*

For the effective calculation of solutions, one has to distinguish between the two cases, viz.: (a) the function $F(x_1, \ldots, x_n, \mu)$ is analytic with respect to its variables; (b) F is not analytic. For the first case the periodic solution is also analytic in μ (since $\beta_1(\mu), \ldots, \beta_n(\mu)$ are analytic in μ) and one has to look for a series solution of the form

$$x_s = \varphi_s(t) + \mu x_s^{(1)}(t) + \ldots \tag{3.10}$$

with the resulting Poincaré's procedure of the substitution of x_s, $\dot{x}_s$ etc., into the d.e. and identification of the coefficients with like powers of μ, etc.

In case (b), one starts with the generating solution and determines further approximations $x_s^{(j)}$ by the method of successive approximations

$$\dot{x}_s^{(j)} = a_{s1}x_1^{(j)} + \ldots + a_{sn}x_n^{(j)} + f_s(t) \\ + \mu F_s(t, x_1^{(j-1)}, \ldots, x_n^{(j-1)}, \mu) \quad (3.11)$$

It is to be assumed that the functions F_s have continuous first partial derivatives with respect to the x_i.

4. Resonance oscillations of nonautonomous systems

We shall assume now that the characteristic equation (1.2) has some zero roots or purely imaginary roots $\pm pi$, p being an integer. This gives rise to the *resonance oscillations*; it is to be noted that the *neighborhood* of these roots is also considered as belonging to resonance zones. We consider the generating system (3.2); it has a periodic solution if the condition (2.6) is fulfilled. One can show that this is actually so if the Fourier development of f_s does not contain either resonating terms with cos pt and sin pt or constant terms if there are zero roots. We assume thus that (2.6) is fulfilled.

In such a case the generating system has a family of periodic solutions

$$x_s^0 = x_s^{0*}(t) + M_1\varphi_{s1} + \ldots + M_m\varphi_{sm}; \qquad s = 1, \ldots, n \qquad (4.1)$$

where $x_s^{0*}(t)$ is a particular periodic solution and M_i are constants. We select one solution of this family as generating solution; for this it is necessary to show that for $\mu \to 0$ there is one solution of (3.1) tending to this particular generating solution.

Let $x_s(t, \beta_1, \ldots, \beta_n, \mu)$ be the solution of (3.5) with the initial conditions (3.4) or rather $x_s(0, \beta_1, \ldots, \beta_n, \mu) = \varphi_s(0) + \beta_s$ where

$$\varphi_s(t) = x_s^{0*}(t) + M_1^*\varphi_{s1} + \ldots + M_m^*\varphi_{sm} \qquad (4.2)$$

and the M_i^* indicate the choice of *one* particular solution (4.1).

The procedure of the preceding section is still applicable but the Jacobian of (3.8) now vanishes for $\beta_1 = \ldots = \beta_n = \mu = 0$.

It can be shown (we refer to Malkin's text) that the system (3.1) may have for a sufficiently small μ a periodic resonance solution which becomes the generating solution for $\mu = 0$ if, setting

$$P_i(M_1^*, \ldots, M_m^*) = \int_0^{2\pi} \sum_{\alpha=1}^{n} F_\alpha(\tau, \varphi, 0)\psi_{\alpha i}(\tau)d\tau \qquad (4.3)$$

the parameters M_i^* satisfy the equations

$$P_i(M_1^*, \ldots, M_m^*) = 0 \qquad (4.3a)$$

For any simple solution of these equations for which $\frac{\partial(P_1, \ldots, P_m)}{\partial(M_1^*, \ldots, M_m^*)} \neq 0$, the periodic solution exists and is unique. This condition requires that the functions $P_i(M_1, \ldots, M_m)$ possess first derivatives with respect to $M_1, \ldots, M_m$.

Here again, the subsequent calculations are different according to whether the functions F_s in (3.1) are analytic in $x_1, \ldots, x_n$ and μ or not. If they are analytic, one looks for a series solution of the form

$$x_s = \varphi_s + \mu x_s^{(1)}(t) + \mu^2 x_s^{(2)}(t) + \ldots$$

which, being substituted into the d.e., gives rise to a recurrent system of d.e. obtained, as usual, by equating the coefficients of like powers of the μ. One has thus:

$$\dot{x}_s^{(1)} \sum_{j=1}^{n} a_{sj} x_j^{(1)} + F_s(t, \varphi_1, \ldots, \varphi_n, 0) \tag{4.4}$$

$$\dot{x}_s^{(k)} = \sum_{j=1}^{n} a_{sj} x_j^{(k)} + \sum_{\alpha=1}^{n} \frac{\partial F_s(t, \varphi_1, \ldots, \varphi_n, 0)}{\partial \varphi_\alpha} x^{(k-1)} + F_s^{(k-2)}; \qquad k = 2, \ldots \tag{4.5}$$

where $F_s^{(k-2)}$ are entire rational functions with periodic coefficients of $x_s^{(j)}$ for which $j < (k - 2)$.

Equation (4.4) admits a periodic solution if

$$\int_0^{2\pi} \sum_{\alpha=1}^{n} F_\alpha(t, \varphi_1, \ldots, \varphi_n, 0) \psi_{\alpha i} dt = 0; \qquad i = 1, \ldots, m \tag{4.6}$$

which is again the condition (4.3a) determining the parameters M_i^* of the generating solution. As this is possible in our case, one obtains a periodic solution for $x_s^{(1)}$ of the form

$$x_s^{(1)} = x_s^{(1)*} + M_1^{(1)} \varphi_{s1} + \ldots + M_m^{(1)} \varphi_{sm}$$

where $x_s^{(1)*}$ is some periodic solution of (4.4) and the $M_s^{(1)}$ are arbitrary constants which can be selected so as to obtain conditions of periodicity for $x_s^{(2)}$, and so on.

If the functions $x_s^{(1)}, \ldots, x_s^{(k-2)}$ have been so determined and are periodic, one substitutes them into (4.5) and gets

$$\dot{x}_s^{(k)} = \sum_{j=1}^{n} a_{sj} x_j^{(k)} + \sum_{i=1}^{m} M_i^{(k-1)} \sum_{\alpha=1}^{n} \frac{\partial F_s(t, \varphi_1, \ldots, \varphi_n, 0)}{\partial \varphi_\alpha} \varphi_{\alpha i} + f_s^{(k)}(t) \tag{4.7}$$

For the periodicity one has again the condition

$$\sum_{i=1}^{m} M_i^{(k-1)} \int_0^{2\pi} \sum_{s,\alpha=1}^{n} \frac{\partial F_s(t, \varphi_1, \ldots, \varphi_n, 0)}{\partial \varphi_\alpha} \varphi_{\alpha i} \psi_{sj} dt + \int_0^{2\pi} \sum_{s=1}^{n} f_s^{(k)} \psi_{sj} dt = 0 \quad (4.8)$$

which in view of (4.3) and (4.1) reduces to

$$\sum_{i=1}^{m} \frac{\partial P_j(M_1^*, \ldots, M_m^*)}{\partial M_i^*} M_i^{(k-1)} + \int_0^{2\pi} \sum_{s=1}^{n} f_s^{(k)} \psi_{sj} dt = 0 \quad (4.9)$$

One has thus for the determination of constants a linear system with the determinant which coincides with $\dfrac{\partial(P_1, \ldots, P_m)}{\partial(M_1^*, \ldots, M_m^*)} \neq 0$ and which permits their determination.

There is, therefore, one and only one system of the series solution. As an example consider a system

$$\ddot{x}_s = \mu F_s(t, x_1, \ldots, x_n, \dot{x}_1, \ldots, \dot{x}_n, \mu) \equiv \mu F_s(t, x, \dot{x}, \mu) \quad (4.10)$$

Here $s = 1, \ldots, n$ and F_s are analytic in $x_1, \ldots, x_n, \dot{x}_1, \ldots, \dot{x}_n$ and μ, and periodic in t with period 2π, which we shall write as:

$$\ddot{x}_s = \mu F_s(t, x, \mu); \qquad s = 1, 2, \ldots, n \quad (4.11)$$

The generating system

$$\ddot{x}_s^0 = 0$$

has a zero root of multiplicity $2n$; the system has therefore a family of periodic solutions containing n arbitrary constants and has the form $x_s^0 = M_s$. We look for a periodic solution

$$x_s = M_s^* + \mu x_s^{(1)} + \mu^2 x_s^{(2)} + \ldots \quad (4.12)$$

where $x_s = x_s(t)$. The recurrent system is

$$\ddot{x}_s^{(1)} = F_s(t, M^*, 0, 0) \quad (4.13)$$

$$\ddot{x}_s^{(2)} = \sum_{\alpha=1}^{n} \left[\frac{\partial F_s(t, M^*, 0, 0)}{\partial M_\alpha^*} x_\alpha^{(1)} + \left(\frac{\partial F_s(t, M^*, \dot{x}, 0)}{\partial \dot{x}_\alpha} \right)_{\dot{x}=0} \dot{x}^{\alpha(1)} \right] + \left(\frac{\partial F_s(t, M^*, 0, \mu)}{\partial \mu} \right)_{\mu=0} \quad (4.14)$$

The periodicity of $x_s^{(1)}$ in view of (4.13) requires

$$\int_0^{2\pi} F_s(t, M^*, 0, 0)dt = 0; \qquad s = 1, \ldots, n \quad (4.15)$$

and we shall assume that the Jacobian of the left-hand side with respect to M_α^* is different from zero. In fact, with this condition the functions $\dot{x}_s^{(1)}$ will be periodic of the form

$$\dot{x}_s^{(1)} = \int_0^t F_s(t, M^*, 0, 0)dt + A_s^{(1)} \tag{4.16}$$

where the $A_s^{(1)}$ are constants. If these constants are determined by the formula

$$A_s^{(1)} = -\frac{1}{2\pi}\int_0^{2\pi} dt \int_0^t F_s(t, M^*, 0, 0)dt$$

the functions x_s will be also periodic of the form

$$x_s^{(1)} = x_s^{(1)*} + M_s^{(1)}$$

where the $M_s^{(1)}$ are constants. Equations (4.15) determine M^* in the generating solution, after which one calculates $x_s^{(1)}$.

For the constants $M_s^{(1)}$ one forms the conditions of periodicity of $x_s^{(2)}$ which are

$$\sum_{n=1}^{n} M_\alpha^{(1)} \int_0^{2\pi} \frac{\partial F_s(t, M^*, 0, 0)}{\partial M_\alpha^*} dt + \int_0^{2\pi} \left\{ \left(\frac{\partial F_s(t, M^*, 0, \mu)}{\partial \mu} \right)_{\mu=0} \right.$$
$$\left. + \sum_{\alpha=1}^{n} \left[\frac{\partial F_s(t, M^*, 0, 0)}{\partial M_\alpha^*} \cdot x_\alpha^{(1)*} + \left(\frac{\partial F_s(t, M^*, \dot{x}, 0)}{\partial \dot{x}_\alpha} \right)_{\dot{x}=0} \dot{x}_\alpha^{(1)*} \right] \right\} dt = 0 \tag{4.17}$$

Thus, for the determination of the constants $M_s^{(1)}$ one has a system of linear equations whose determinant is different from zero. As we have assumed, calculation of other approximations proceeds in a similar manner.

5. Periodic resonance solutions of nonautonomous systems with nonanalytic d.e.

If in the d.e. (3.1) one maintains all conditions except the analyticity of F_s, one can apply the method of successive approximations, as was mentioned in Section 4, in which case it is necessary to assume only that the F_s have partial derivatives of the first order in all variables.

The problem is to find a periodic solution of (3.1) which reduces for $\mu = 0$ to the generating solution

$$x_s^0 = \varphi_s(t) = x_s^{(0)*} + M_1^*\varphi_{s1} + \ldots + M_m^*\varphi_{sm} \tag{5.1}$$

The constants M_i^*, as previously, are solutions of equations

$$P_i(M_1^*,\ldots, M_m^*) = \int_0^{2\pi} \sum_{\alpha=1}^{n} F_\alpha(t, \varphi_1,\ldots, \varphi_n, 0)\psi_{\alpha i}(t)dt = 0; \quad i = 1,\ldots, m \tag{5.2}$$

assuming that the condition

$$\frac{\partial(P_1,\ldots, P_m)}{\partial(M_1^*,\ldots, M_m^*)} \neq 0 \tag{5.3}$$

is fulfilled. In such a case the periodic solution exists and is unique. The method of successive approximations consists in taking as the first approximation the function

$$x_s^{(1)} = x_s^{(0)*} + M_1^{(1)}\varphi_{s1} + \ldots + M_m^{(1)}\varphi_{sm}$$

where the $M_i^{(1)}(\mu)$ are, for the time being, unknown functions of μ for which $M_i^{(1)}(0) = M_i^*$.

The procedure can be continued for other approximations $x_2^{(2)},\ldots$. Thus if the functions $x_s^{(k-1)}$ are periodic, one can write

$$x_s^{(k-1)} = M_1^{(k-1)}\varphi_{s1} + \ldots + M_m^{(k-1)}\varphi_{sm} + x_s^{(0)*} + \mu x_s^{(k-1)*} \tag{5.4}$$

where the $M_i^{(k-1)}$ are arbitrary constants and $x_s^{(k-1)*}$ is some periodic solution of the equations

$$\dot{x}_s^{(k-1)*} = a_{s1}x^{(k-1)*} + \ldots + a_{sn}x_n^{(k-1)*} + F_s(t, x_1^{(k-2)},\ldots, x_n^{(k-2)}, \mu) \quad k = 2, 3,\ldots$$

The constants $M_i^{(k-1)}$ are determined from the conditions of periodicity for the $x_s^{(k-1)*}$

$$P_i^{(k)}(M_1^{(k-1)},\ldots, M_m^{(k-1)}, \mu) = \int_0^{2\pi} \sum_{\alpha=1}^{n} F_\alpha(t, x_1^{(k-1)},\ldots, x_n^{(k-1)}, \mu)\psi_{\alpha i}dt = 0 \tag{5.5}$$

These equations are satisfied identically for $\mu = 0$, $M_i^{(k-1)} = M_i^*$ and, as the Jacobian is different from zero, they yield for μ small one and only one solution $M_i^{(k-1)}(\mu)$ such that $M_i^{(k-1)}(0) = M_i^*$.

Malkin[2] illustrates the above procedure by an example due to Butenin relative to the behavior of the follow-up system shown in Fig. 11.1. The contact segments C, C', C'', and C''' are connected with the follow-up axis. An arm A is actuated by some control mechanism which releases a follow-up action through the motor M rotating the follow-up contact segments; G is a generator with a separate excitation.

[2] See footnote [2], page 255.

If y is the angle of the arm A and x is the angle of the follow-up system (motor M and associated with it rotation of contact segments), the difference $\psi = y - x$ releases the follow-up action, which has a tendency to reduce ψ to zero.

The system is specified by the d.e.

$$\begin{aligned} \theta\ddot{x} + n\dot{x} &= \alpha I \text{ (mechanical d.e. for the armature of } M) \\ RI &= E(\psi) - c\dot{x} - L\frac{dI}{dr} \text{ (the balance of electromotive forces)} \end{aligned} \tag{5.6}$$

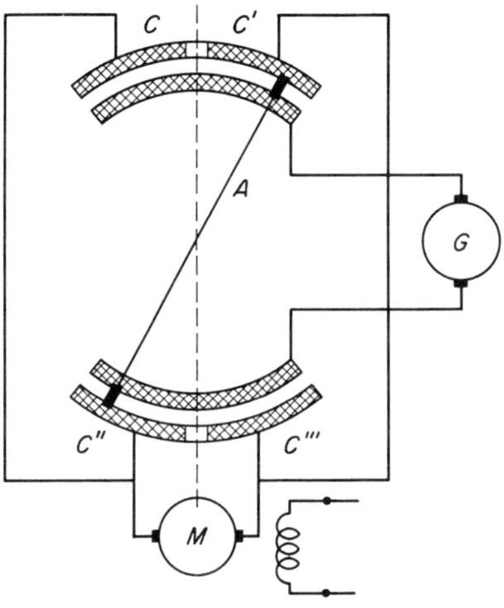

FIGURE 11.1

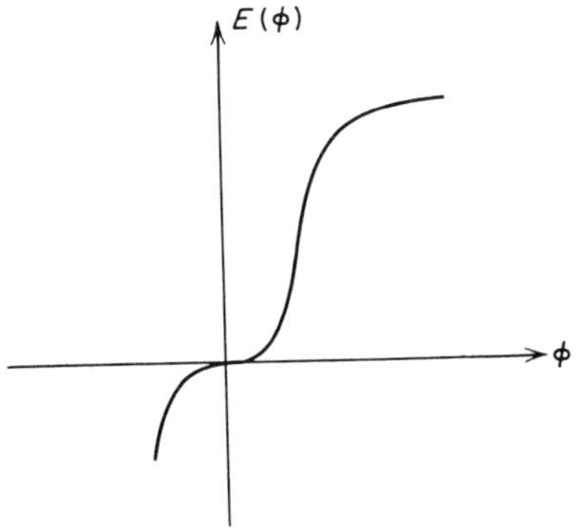

FIGURE 11.2

where θ is the reduced moment of inertia of the follow-up element; n is the coefficient of friction, α is a constant coefficient, R, L are the resistance and inductance in the armature of the motor, and I is the current; $E(\psi)$ is an electromotive force applied to the armature of the motor, and c is the coefficient entering into the expression of the counter e.m.f. The quantity $E(\psi)$ is a nonlinear function shown in Fig. 11.2. We assume, for instance, that the arm A executes a sinusoidal motion, say $y = A \sin \omega t$, and propose to establish the follow-up motion. Eliminating I between the two equations (5.6) and introducing the variable ψ, we have the d.e.

$$\dddot{\psi} + \frac{R\theta + Ln}{L\theta}\ddot{\psi} + \frac{Rn + c\alpha}{L\theta}\dot{\psi}$$

$$= -\frac{\alpha}{L\theta}E(\psi) - A\left(\omega^3 - \frac{Rn + c\alpha}{L\theta}\cdot\omega\right)\cos\omega t - \frac{R\theta + Ln}{L\theta}A\omega^2\sin\omega t \tag{5.7}$$

Changing the time scale, this equation can be written as

$$\dddot{\psi} + \frac{k^2}{\omega^2}\dot{\psi} = -E^*(\psi) - \frac{R\theta + Ln}{\omega L\theta}\ddot{\psi} - A\left(1 - \frac{k^2}{\omega^2}\right)\cos\tau - \frac{R\theta + Ln}{\omega L\theta} A \sin\tau \quad (5.8)$$

where $\tau = \omega t$; $k^2 = (Rn + c\alpha)/L\theta$; $E^*(\psi) = [\alpha/(\omega^3 L\theta)]E(\psi)$.

We express the nearly linear character of the problem by writing

$$\mu = \frac{R\theta + Ln}{\omega L\theta}; \qquad E^*(\psi) = \mu F(\psi) \quad (5.9)$$

assuming $\mu \ll 1$. Moreover, we assume oscillation near resonance so that k differs from ω by the order of μ. Expressing this: $k^2/\omega^2 = 1 + \mu b$ and taking into account (5.9), the d.e. (5.8) reduces finally to the form

$$\dddot{\psi} + \dot{\psi} = \mu[-F(\psi) - b\dot{\psi} - \ddot{\psi} + bA\cos\tau - A\sin\tau] \quad (5.10)$$

where the differentiations are with respect to τ. The characteristic equation of the generating system has one zero root and one pair of purely imaginary roots. The general solution of this system is:

$$\psi_0 = L + M\cos\tau + N\sin\tau$$

where L, M, and N are arbitrary constants. As generating solution we take:

$$\varphi = L^* + M^*\cos\tau + N^*\sin\tau$$

As the first approximation we take $\psi_1(t)$, which is the periodic solution of the d.e.

$$\dddot{\psi}_1 + \dot{\psi}_1 = \mu[-F(\varphi) + (bM^* + N^* - A)\sin\tau + (-bN^* + M^* + bA)\cos\tau]$$

The solution of this d.e. will be periodic if the Fourier development of its right-hand term does not contain either constant terms or terms with $\cos\tau$ and $\sin\tau$; whence

$$\int_0^{2\pi} F(\varphi)dt = 0; \qquad -\frac{1}{\pi}\int_0^{2\pi} F(\varphi)\cos\tau d\tau - bN^* + M^* + bA = 0$$
$$-\frac{1}{\pi}\int_0^{2n} F(\varphi)\sin\tau d\tau + bM^* + N^* - A = 0 \quad (5.11)$$

These three equations permit determining three constants L^*, M^*, and N^* in the generating solution. As $F(\psi)$ is odd, the first equation yields

$$L^* = 0$$

In such a case, designating $f(\psi) = \int F(\psi)d\psi$, we have

$$\int_0^{2\pi} F(\varphi)(N^* \cos\tau - M^* \sin\tau)d\tau = f(\varphi)\Big|_0^{2\pi} = 0$$

and, instead of the second and the third equations (5.11) we get two combinations. Multiplying the second by N^*, the third by $-M^*$ and adding, we have:

$$b(M^{*2} + N^{*2}) - AM^* - bAN^* = 0 \tag{5.12a}$$

and likewise multiplying the second by M^*, the third by N^* and adding:

$$-\frac{1}{\pi}\int_0^{2\pi} \varphi F(\varphi)d\tau + M^{*2} + N^{*2} + bAM^* - AN^* = 0 \tag{5.12b}$$

From these equations one determines M^* and N^* for the generating solution.

6. Oscillations of autonomous systems

Consider an autonomous system of d.e.,

$$\dot{x}_s = \sum_{j=1}^{n} a_{sj}x_j + \mu f_s(x_1,\ldots, x_n, \mu); \qquad s = 1,\ldots, n \tag{6.1}$$

We assume that the functions f_s possess continuous partial derivatives of the first-order relative to the variables $x_1,\ldots, x_n$ for μ sufficiently small.

We also assume that the generating system

$$\dot{x}_s = \sum_{j=1}^{n} a_{sj}x_j \tag{6.2}$$

has m independent particular periodic solutions of the same period T, corresponding to the roots $\pm p_j i(2\pi/T) = \pm i\omega$ of the characteristic equation; p_j is an integer.

Let the generating solution be

$$x_s^{(0)} = \varphi_s(t) = M_1^*\varphi_{s1} + \ldots + M_m^*\varphi_{sm} \tag{6.3}$$

where the M_1^* are constants. The problem consists in establishing conditions under which a periodic solution of (6.1) becomes (6.3) for $\mu \to 0$. In fact, if $x_s^*(t,\mu)$ is a function vanishing for $\mu \to 0$, the solution of (6.1) will be generally of the form

$$x_s(t) = M_1^*\varphi_{s1} + \ldots + M_m^*\varphi_{sm} + x_s^*(t,\mu) \tag{6.4}$$

The term with M_m^* can always be set to zero by a proper choice of time origin inasmuch as the system is autonomous. In fact, if $\varphi_{s,m-1}$ and φ_{sm}, correspond to the pair of roots $\pm i\omega$, the solution, as we saw, has the form

$$\varphi_{s,m-1}(t) = A_s \sin \omega t + B_s \cos \omega t; \qquad \varphi_{s,m} = A_s \cos \omega t - B_s \sin \omega t$$

and by replacing t by $t + h$, h constant, one can always determine this constant† so as to eliminate the coefficient of φ_{sm} and, thus, obtain the expression

$$x_s(t + h) = N_1\varphi_{s1}(t) + \ldots + N_{m-1}\varphi_{s,m-1}(t) + x_s^*(t + h, \mu) \quad (6.5)$$

On the other hand, although the number of independent parameters N_i is reduced by one in view of the "translation" property of autonomous systems, there appears instead another parameter—the *nonlinear frequency (or period) correction*, which we encountered already in the theory of autonomous systems with one degree of freedom (Chapter 10).

If T is the period of the generating solution, that of the autonomous system will be $T(1 + \mu\alpha)$, where generally $\alpha \neq 0$ and appears as an unknown in this problem.

Introducing a new independent variable τ related to the old one t by the relation

$$t = \tau(1 + \mu\alpha) \quad (6.6)$$

the problem will be to find a periodic solution with period T of the new differential system

$$\dot{x}_s = \sum_{j=1}^{n} a_{sj}x_j + \mu(1 + \mu\alpha)f_s(x_1, \ldots, x_n, \mu) + \mu\alpha \sum_{j=1}^{n} a_{sj}x_j \quad (6.7)$$

Let $\alpha = \alpha^*$ for $\mu = 0$. It can be shown that $M_1^*, \ldots, M_{m-1}^*$ and α^* satisfy the following system of equations

$$\int_0^T \sum_{\beta=1}^{n} f_\beta(\varphi_1, \ldots, \varphi_n, 0)\psi_{\beta i}dt + \alpha^*(A_{i1}M_1^* + \ldots + A_{i,m-1} - M_{m-1})$$
$$\equiv P_i(\alpha^*, M_1^*, \ldots, M_{m-1}^*) = 0; \qquad i = 1, \ldots, m \quad (6.8)$$

where $\psi_{sj}(t)$ are, as previously, periodic solutions of the adjoint system corresponding to the given system; moreover,

$$A_{ij} = T \sum_{\beta=1}^{n} \frac{d\varphi_{\beta j}}{dt} \cdot \psi_{\beta i} \quad (6.9)$$

are constants by property of adjoint systems as $d\varphi_{\beta j}/dt$ are clearly solutions of (6.2).

We omit the proof of this theorem and merely mention that the search

† It is sufficient to take h determined by the formula $\tan \omega h = -M_m^*/M_{m-1}^*$

for the periodic solution is conducted by the method of successive approximations (Malkin [2]) taking as approximation of the zero order, the generating solution (6.3), and as approximation of the kth order, the periodic solution of the system

$$\dot{x}_s^{(k)} = \sum_{j=1}^{n} a_{sj}x_j^{(k)} + \mu(1 + \alpha^{(k)}\mu)f_s(x_1^{(k-1)},\ldots, x_n^{(k-1)}, \mu) + \mu\alpha^{(k)} \sum_{j=1}^{n} a_{sj}x_j^{(k-1)} \quad (6.10)$$

If the functions f_s are analytic in x_i, it is possible to look for a periodic solution in the form of a series, viz.:

$$\begin{aligned} x_s(\tau) &= \varphi_s(\tau) + \mu x_s^{(1)}(\tau) + \mu^2 x_s^{(2)}(\tau) + \ldots \\ &= M_1^*\varphi_{s1}(\tau) + \ldots + M_{m-1}^*\varphi_{s,\, m-1}(\tau) + \mu x_s^{(1)}(\tau) \\ &\qquad + \mu^2 x_s^{(2)}(\tau) + \ldots \end{aligned} \quad (6.11)$$

where, without any loss of generality, the term with M_m^* in the generating solution has been set to zero again in view of the "translation" property of autonomous systems.

The functions $x^{(k)}(\tau)$ will be periodic of period T as this period does not depend now on μ.

For the function $x_s^{(1)}$ one has the equation

$$\begin{aligned} \dot{x}_s &= \sum_{j=1}^{n} a_{sj}x_j^{(1)} + f_s(\varphi_1,\ldots, \varphi_n; 0) + \alpha^*(a_{s1}\varphi_1 + \ldots + a_{sn}\varphi_n) \\ &= P_1(\alpha^*, M_1^*,\ldots, M_{m-1}^*) = 0; \qquad i = 1,\ldots, m \end{aligned} \quad (6.12)$$

For the periodicity one must have

$$\begin{aligned} &\int_0^T \sum_{\beta=1}^{n} f_\beta(\varphi_1,\ldots, \varphi_\beta, 0)\psi_{\beta i}d\tau + \alpha^*(A_{i1}M_1^* + \ldots + A_{i,\, m-1}M_{m-1}^*) \\ &\qquad = P_i(\alpha^*, M_1^*,\ldots, M_{m-1}^*) = 0; \qquad i = 1,\ldots, m \end{aligned} \quad (6.13)$$

where the A_{ij} are given by (6.9); assume that M_i^* and α^* are determined by these equations. If, moreover, one has

$$\frac{\partial(P_1,\ldots, P_m)}{\partial(\alpha^*, M_1^*,\ldots, M_{m-1}^*)} \neq 0 \quad (6.14)$$

then the periodic solution is guaranteed and one can set

$$x_s^{(1)}(\tau) = x_s^{(1)*}(\tau) + M_1^{(1)}\varphi_{s1}(\tau) + \ldots + M_{m-1}^{(1)}\varphi_{s,\, m-1}(\tau)$$

[2] See footnote [2], page 255.

where the first term on the right is some particular periodic solution and the $M_i^{(1)}$ are constants.

The constants $M_i^{(1)}$, as well as α_2, are determined from the conditions of periodicity of the $x_s^{(2)}$. For this, one follows the usual procedure, that is, one replaces the x_s by their expressions (6.11) and equates the coefficients with equal powers of μ, after which it is necessary to express the condition of periodicity which, in turn, requires a determination of constants from the next approximation.

The problem ultimately depends on the solution of m linear equations determining the quantities α_k and $M_i^{(k-1)}$, the determinant of which (in view of (6.14)) is different from zero.

7. Self-excited oscillations in coupled circuits

As an example consider the circuit shown in Fig. 11.3, consisting of two coupled oscillating circuits of which one is self-excited by the presence of

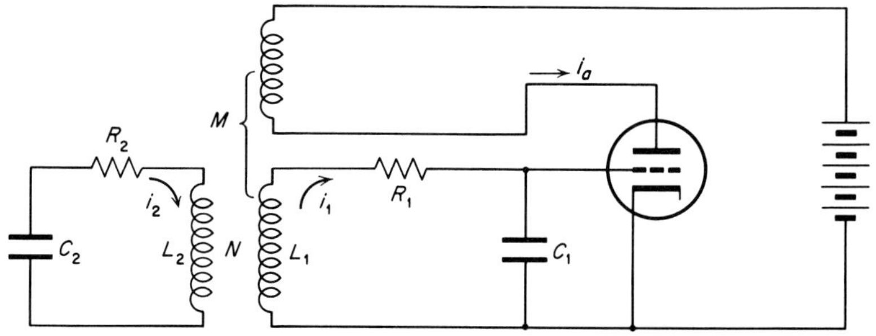

FIGURE 11.3

an electron tube. Neglecting grid current and anode reaction, the d.e. of the currents i_1, i_2, and i_a are

$$\begin{aligned} L_1 \frac{di_1}{dt} + R_1 i_1 + \frac{1}{C_1}\int_0^t i_1 dt &= M\frac{di_a}{dt} + \frac{di_2}{dt} \\ L_2 \frac{di_2}{dt} + R_2 i_2 + \frac{1}{C_2}\int_0^t i_2 dt &= N\frac{di_1}{dt} \end{aligned} \tag{7.1}$$

where M and N are the coefficients of mutual inductance shown in Fig. 11.3.

Introducing new variables $v_1 = \frac{1}{C_1}\int_0^t i_1 dt$; $v_2 = \frac{1}{C_2}\int_0^t i_2 dt$ with $n_1^2 = 1/L_1C_1$; $n_2^2 = 1/L_2C_2$; $\lambda_1 = NC_2n_1^2$; $\lambda_2 = NC_1n_2^2$, one gets

$$\begin{aligned} \ddot{v}_1 - \lambda_1\ddot{v}_2 + n_1^2 v_1 &= n_1^2\left(-R_1C_1\dot{v}_1 + M\frac{di_a}{dt}\right) \\ \ddot{v}_2 - \lambda_2\ddot{v}_1 + n_2^2 v_2 &= -n_2^2R_2C_2\dot{v}_2 \end{aligned} \tag{7.2}$$

If one approximates the characteristic by the polynomial

$$i_a = S_0v_1 - \tfrac{1}{3}S_2v_1^3$$

and if one introduces dimensionless voltages $x = v_1\sqrt{MS_2/(MS_0 - C_1R_1)}$; $y = v_2\sqrt{MS_2/(MS_0 - C_1R_1)}$ and takes as μ the quantity

$$\mu = n_1(MS_0 - C_1R_1)$$

equations (7.2) become

$$\ddot{x} - \lambda_1\ddot{y} + n_1^2x = \mu n_1(1 - x^2)\dot{x}; \qquad \ddot{y} - \lambda_2\ddot{x} + n_2^2y = -\mu\frac{n_2^2}{n_1}\cdot\delta\dot{y} \tag{7.3}$$

where $\delta = (C_2R_2)/(MS_0 - C_1R_1)$.

The general solution of the generating system

$$\ddot{x}_0 - \lambda_1\ddot{y}_0 + n_1^2x_0 = 0; \qquad \ddot{y}_0 - \lambda_2\ddot{x}_0 + n_2^2y_0 = 0 \tag{7.4}$$

is of the form

$$\begin{aligned} x_0 &= M_1\cos\omega_1t + M_2\sin\omega_1t + M_3\cos\omega_2t + M_4\sin\omega_2t \\ y_0 &= M_1k_1\cos\omega_1t + M_2k_1\sin\omega_1t + M_3k_2\cos\omega_2t + M_4k_2\sin\omega_2t \end{aligned} \tag{7.5}$$

where M_1, M_2, M_3, and M_4 are arbitrary constants and k_1 and k_2 are the values which take the quantity

$$k = \frac{\omega^2 - n_1^2}{\lambda_1\omega^2} = \frac{\lambda_2\omega^2}{\omega^2 - n_2^2} \tag{7.6}$$

for $\omega = \omega_1$ and $\omega = \omega_2$, and ω_1 and ω_2 are the roots of the characteristic equation

$$(1 - \lambda_1\lambda_2)\omega^4 + (n_1^2 + n_2^2)\omega^2 + n_1^2n_2^2 = 0 \tag{7.7}$$

From (7.5) it follows that the generating system has two families of periodic solutions, depending on two arbitrary constants and having the periods $2\pi/\omega_1$ and $2\pi/\omega_2$.

We take one of the generating solutions, say

$$x_0 = M^* \cos \omega t; \qquad y_0 = M^*k \cos \omega t \tag{7.8}$$

where ω and k are either ω_1, k_1 or ω_2, k_2.

We introduce the change of time scale in (7.3):

$$t = \tau(1 + \alpha^*\mu + \alpha_2\mu^2 + \ldots) \tag{7.9}$$

where α^*, α_2 are certain constants. The d.e. (7.3) become

$$\begin{aligned} \ddot{x} - \lambda_1\ddot{y} + n_1^2x &= \mu[-2\alpha^* n_1^2 x + n_1(1 - x^2)\dot{x}] + \ldots \\ \ddot{y} - \lambda_2\ddot{x} + n_2^2y &= \mu\left[-2\alpha^* n_2^2 y + \frac{n_2^2}{n_1}\delta\dot{y}\right] + \ldots \end{aligned} \tag{7.10}$$

where the differentiations are with respect to τ.

Looking for a periodic solution in the form of a series

$$x = M^* \cos \omega t + \mu x_1(\tau) + \ldots; \qquad y = kM^* \cos \omega\tau + \mu y_2(\tau) + \ldots$$

we have the d.e. for x_1 and y_1

$$\begin{aligned} &\ddot{x}_1 - \lambda_1\ddot{y}_1 + n_1^2x_1 \\ &\quad = -2\alpha^* n_1^2 M^* \cos \omega\tau - n_1(1 - M^{*2} \cos^2 \omega\tau)M^*\omega \sin \omega\tau \\ &\ddot{y}_1 - \lambda_2\ddot{x}_1 + n_2^2y_1 \\ &\quad = -2\alpha^* n_2^2 M^* k \cos \omega\tau + \frac{n_2^2}{n_1}\delta k M^* \omega \sin \omega\tau \end{aligned} \tag{7.11}$$

We now have to formulate the conditions of periodicity of x_1 and y_1. It is necessary to determine the coefficients P, Q, R, and S in the differential system

$$\begin{aligned} \ddot{x} - \lambda_1\ddot{y} + n_1^2x &= P \cos \omega\tau + Q \sin \omega\tau \\ \ddot{y} - \lambda_2\ddot{x} + n_2^2y &= R \cos \omega\tau + S \sin \omega\tau \end{aligned} \tag{7.12}$$

so that this system has a periodic solution. Letting $x = A \cos \omega\tau + B \sin \omega\tau$; $y = C \cos \omega\tau + D \sin \omega\tau$, we have

$$\begin{aligned} (n_1^2 - \omega^2)A + \lambda_1\omega^2C &= P; \qquad \lambda_2\omega^2A + (n_2^2 - \omega^2)C = R \\ (n_1^2 - \omega^2)B + \lambda_1\omega^2D &= Q; \qquad \lambda_2\omega^2B + (n_2^2 - \omega^2)D = S \end{aligned} \tag{7.13}$$

As ω^2 is a root of (7.7), the last equations have solutions only if

$$\lambda_2\omega^2P - (n_1^2 - \omega^2)R = \lambda_2\omega^2Q - (n_1 - \omega^2)S = 0 \tag{7.14}$$

which is the condition of periodicity.

Applied to (7.11), this yields

$$\alpha^* M^* = 0; \qquad n_1 \lambda_2 \omega^2 \left(1 - \frac{1}{4} M^{*2}\right) + \frac{n^2}{n_1}(n_1^2 - \omega^2)\delta k = 0$$

Hence, in view of (7.6), one has

$$\alpha^* = 0; \qquad M^{*2} = 4 - \frac{4\delta n_2^2(n_1^2 - \omega^2)}{n_1^2(n_2^2 - \omega^2)} \tag{7.15}$$

This gives the amplitude of the generating solution and the frequency correction in the first approximation.

8. Method of averaging

It is useful to supplement the preceding material, principally based on the theory of Poincaré, by a short outline of the *method of averaging* established by Krylov and Bogoliubov in connection with the asymptotic methods to which Chapters 14 and 15 are devoted. The method of averaging, as far as is known, was used for the first time by van der Pol in his early researches, but the full justification of the method appeared in the classical work of Krylov and Bogoliubov[3]; an improved version of this important development can be found in a recent text by N. N. Bogoliubov and J. A. Mitropolsky.[1] We shall give a brief outline of this work here in view of its importance in connection with the theory concerning many degrees of freedom.

It can be shown that in many cases the d.e. can be reduced to a form in which the right-hand terms are proportional to the small parameter μ. Such a form of the d.e. is designated by Krylov-Bogoliubov (K.B. for short) as the *standard form*, and many important generalizations of these authors arise from the discussion of d.e. in this particular form. We shall return to the same question in the following chapter in connection with almost periodic oscillations.

K.B. show that in many cases the d.e. are reducible to the standard form

$$\ddot{x}_k + \omega_k^2 x_k = \mu X_k(t, x_k, \dot{x}_k), \qquad k = 1, \ldots, n \tag{8.1}$$

Moreover, by a special change of dependent variables, one can reduce the system to the form:

$$\dot{x}_k = \mu X_k(t, x_1, \ldots, x_n), \qquad k = 1, 2, \ldots, n \tag{8.2}$$

[3] N. Krylov and N. N. Bogoliubov, *Introduction to Nonlinear Mechanics* (in Russian), Kiev, 1937.

[1] See footnote [1], page 252.

where

$$X_k(t, x_1,\ldots, x_n) = \sum_\nu e^{i\nu t} X_\nu(x_1,\ldots, z_n) \tag{8.3}$$

ν being the constant frequencies. One can consider also the cases when the d.e. in the standard form (8.2) have also terms with μ^2, μ^3,. . ., etc.

For systems with many degrees of freedom it is useful to simplify the notation by designating $x_1, x_2,\ldots, x_n$ by one letter x and dropping the indices and, instead of (8.2), writing in the vectorial form:

$$\dot{x} = \mu X(t,x) \tag{8.4}$$

where $X(t,x) = \sum_\nu e^{i\nu t} X_\nu(x)$. The formulas of differentiation of functions in this notation will be, for instance

$$\frac{dF}{dt} = \frac{\partial F}{dt} + \frac{\partial F}{dx}\frac{dx}{dt} = \frac{\partial F}{\partial t} + \left(\frac{dx}{dt}\frac{\partial}{\partial x}\right) F$$

where $\frac{\partial F}{\partial x}$ is the matrix $\left\|\frac{\partial F_k}{\partial x_q}\right\|$ attached to the vector $\frac{dx}{dt}$ and $\frac{dx}{dt}\frac{\partial}{\partial x}$ is an operational scalar product $\sum_{q=1}^{n} \frac{dx_q}{dt}\frac{\partial}{\partial x_q}$.

We assume that $F(t,x)$ is of the form

$$F(t,x) = \sum_\nu e^{i\nu t} F_\nu(x) \tag{8.5}$$

In such a case it is convenient to introduce the following operators

$$\begin{aligned} \underset{t}{M}[F(t,x)] &= F_0(x) \\ \tilde{F}(t,x) &= \sum_{\nu\neq 0} \frac{e^{i\nu\tau}}{i\nu} F_\nu(x) \\ \tilde{\tilde{F}}(t,x) &= \sum_{\nu\neq 0} \frac{e^{i\nu\tau}}{(i\nu)^2} F_\nu(x) \end{aligned} \tag{8.6}$$

from which it follows that

$$\frac{\partial\tilde{\tilde{F}}}{\partial t} = \tilde{F}, \qquad \frac{\partial\tilde{F}}{\partial t} = F - \underset{t}{M}[F] = F - F_0 \tag{8.7}$$

In these operational notations, the operator $\sim$ will be called the *integrating operator* and $\underset{t}{M}$ the *averaging operator* (considering x as constant and averaging with respect to t).

We consider the system (8.4) with $X(t,x)$ of the form (8.4). One can attempt to find an approximate solution of (8.4) by means of somewhat intuitive considerations. Thus, as $\dot{x}$ is small, in view of (8.2) where μ is a small quantity, x varies slowly but it is quite logical to expect that there may be rapidly varying terms inasmuch as they do not affect the slow variation of x in the long run. If the slowly varying term is ξ one can assume $x = \xi$ (that is, disregard the rapidly varying terms), in which case one has as the first approximation

$$\dot{x} = \mu X(t,x) = \mu X(y,\xi) = \mu \sum_{\nu} X_{\nu}(\xi) e^{i\nu t} \tag{8.8}$$

that is,

$$\dot{x} = \mu X_0(\xi) + \text{small oscillatory terms} \tag{8.9}$$

Assuming that these oscillatory terms do not influence the (relatively) slow variation of x, one obtains the d.e. of the first approximation

$$\dot{\xi} = \mu X_0(\xi) = \mu \underset{t}{M}[X(t,\xi)] \tag{8.10}$$

For the second approximation it is necessary to take into account the small oscillatory terms, that is, the terms: $\mu e^{i\nu t} X_{\nu}(\xi) = \mu P_{\nu}$ which accounts for the oscillations: $\mu P/i\nu$.

For the following approximations we have

$$x = \xi + \mu \sum_{\nu} \frac{P_{\nu}}{i\nu} = \xi + \mu \tilde{X}(t,\xi) \tag{8.11}$$

Substituting this expression for x into (8.4), we have

$$\dot{x} = \mu X(t, \xi + \mu \tilde{X}) \tag{8.12}$$

yielding

$$\dot{x} = \mu \underset{t}{M}[X(t, \xi + \mu \tilde{X})] + \text{small oscillatory terms}$$

If one neglects the effect of these small oscillatory terms on the steady variation of ξ, one can write

$$\dot{\xi} = \mu \underset{t}{M}[(Xt, \xi + \mu \tilde{X})] = \mu M[X(t,\xi) + \mu\left(\tilde{X} \frac{\partial}{\partial \xi}\right) X(t,\xi)] \tag{8.13}$$

This intuitive reasoning can be made more rigorous by the following argument. If, in the substitution of (8.11) into (8.4), one considers ξ as variable, one has

$$\dot{x} = \dot{\xi} + \mu \frac{\partial \tilde{X}}{\partial \xi} \dot{\xi} + \mu \frac{\partial \tilde{X}}{\partial t}; \qquad \text{where } \tilde{X} = \tilde{X}(t,\xi)$$

In this expression, in view of the property of the integrating operator, one has

$$\frac{\partial \tilde{X}}{\partial t} = X(t,\xi) - X_0(\xi)$$

Substituting x and $\dot{x}$ into (8.4), one obtains after certain rearrangements

$$\left(1 + \mu \frac{\partial \tilde{X}}{\partial \xi}\right)\dot{\xi} = \mu X_0(\xi) + \mu[X(t,\xi + \mu\tilde{X}) - \tilde{X}(t,\xi)] \qquad (8.14)$$

where 1 is the unit matrix. Multiplying on the left by $[1 + \mu(\partial X/\partial\xi)]^{-1}$, it is observed that the new variables ξ must satisfy equations of the form

$$\dot{\xi} = \mu\left[1 + \mu \frac{\partial \tilde{X}}{\partial \xi}\right]^{-1} X_0(\xi) + \mu\left[1 + \mu \frac{\partial \tilde{X}}{\partial \xi}\right]^{-1}[X(t,\xi + \tilde{X}\mu) - X(t,\xi)] \qquad (8.15)$$

On the other hand, the development in series yields

$$\left[1 + \mu \frac{\partial \tilde{X}}{\partial \xi}\right]^{-1} = 1 - \mu \frac{\partial \tilde{X}(t,\xi)}{\partial \xi} + \mu^2 \ldots$$

so that (8.15) gives

$$\dot{\xi} = \frac{d\xi}{dt} = \mu X_0(\xi) + \mu^2 \ldots \qquad (8.16)$$

Hence, if ξ satisfies (8.16), the right-hand side of which differs from the right-hand side of the equation

$$\dot{\xi} = \mu X_0(\xi) \qquad (8.17)$$

by quantities $O(\mu^2)$, the expression

$$x = \xi + \mu\tilde{X}(t,\xi) \qquad (8.18)$$

represents the exact solution of (8.4) up to terms of the first order. Then $x = \xi$ is the solution of equations of the first approximation and $x = \xi + \mu\tilde{X}(t,\xi)$ may be called the *improved first approximation.* If one substitutes the latter into the d.e., one finds that they are satisfied up to the second order of smallness.

It is seen that in order to go to higher approximations by this method it is necessary to be able to solve first the equations of the first approximation. In many applied problems, although this is possible, the accuracy with which the parameters are known does not justify additional work in building up higher approximations. If, however, it is desired to obtain higher approximations, the above procedure permits carrying this out, assuming that it is possible to obtain the first approximation in a simple manner, which is generally the case.

It is to be noted that by the very definition of the averaging operator $X_0(\xi) = \mu \underset{t}{M}[X(t,\xi)]$, equations of the first approximation can be written as

$$\dot{\xi} = \frac{d\xi}{dt} = \mu \underset{t}{M}[X(t,\xi)] \tag{8.19}$$

In this way the equations of the first approximation are obtained from the exact d.e.,

$$\dot{x} = \frac{dx}{dt} = \mu X(t,x) \tag{8.4}$$

by averaging the latter with respect to time appearing explicitly; in this process of averaging, ξ is treated as constant.

The formal process consisting in the replacement of the exact d.e. (8.4) by the "averaged out" d.e. (8.19) is called sometimes "the averaging principle." For the justification of this principle it is not necessary that $X(t,\xi)$ be of the form $X(t,x) = \sum_{\nu} e^{i\nu t} X_\nu(x)$, but the essential requirement is the existence of an average value

$$X_0(\xi) = \lim_{T\to\infty} \frac{1}{T} \int_0^T X(t,\xi)dt \tag{8.20}$$

The principle of averaging has been known for a very long time in connection with theory of gases and astronomical calculations, but its application to the theory of oscillations is relatively recent. Many of these developments are due to Krylov and Bogoliubov.[3]

In a similar way for the second approximation one tries to find a change of variables from x to ξ satisfying a d.e., of the form

$$\dot{\xi} = \frac{d\xi}{dt} = \mu X_0(\xi) + \mu^2 P(\xi) + \mu^3 \ldots \tag{8.21}$$

For this purpose one looks for an expression

$$x = \Phi(t, \xi, \mu) \tag{8.22}$$

which, for ξ satisfying a d.e. of the form

$$\dot{\xi} = \mu X_0(\xi) + \mu^2 P(\xi) \tag{8.23}$$

satisfies also

$$\frac{dx}{dt} = \mu X(t,x) \tag{8.24}$$

[3] See footnote [3], page 273.

with the accuracy $O(\mu^3)$. For ξ determined from the equation of the first approximation

$$\dot{\xi} = \mu X_0(\xi) \tag{8.25}$$

the expression

$$x = \xi + \mu \sum_{\nu \neq 0} \frac{e^{i\nu t}}{i\nu} X_\nu(\xi) = \xi + \mu \tilde{X}(t,\xi)$$

satisfies (8.25) with accuracy $O(\mu^2)$, one tries to determine (8.22) in the form

$$x = \xi + \mu \tilde{X}(t,\xi) + \mu^2 F(t,\xi) \tag{8.26}$$

where we take again $F(t,\xi) = \sum_k e^{ikt} F_k(\xi)$. Omitting a somewhat long argument at this point,[1] it can be shown that for ξ determined by the d.e.,

$$\dot{\xi} = \frac{d\xi}{dt} = \mu \underset{t}{M}[X(t,\xi)] + \mu^2 \underset{t}{M}\left[\left(\tilde{X}\frac{\partial}{\partial \xi}\right)\tilde{X}(t,\xi)\right] \tag{8.27}$$

the expression

$$x = \xi + \mu \tilde{X}(t,\xi) + \mu^2 \widetilde{\left(\tilde{X}\frac{\partial}{\partial \xi}\right) X(t,\xi)} - \mu^2 \frac{\partial \tilde{\tilde{X}}(t,\xi)}{\partial \xi} X_0(\xi) \tag{8.28}$$

satisfies the d.e. (8.4) with accuracy $O(\mu^3)$, where the symbol $\sim$ is the same as in (8.6).

We refer to Dragilev[4] in Chapter 4, where the complete theory of averaging up to the approximations of the nth order can be found. We shall apply this to an example.

We consider a physical pendulum whose fixed point is caused to oscillate in the vertical direction with small amplitude a and a relatively high frequency ω, such that the conditions

$$\omega > \omega_0 l/a; \qquad a/l \ll 1 \tag{8.29}$$

are fulfilled, where l is the length and $\omega_0 = \sqrt{g/l}$ is the free frequency of the pendulum.

It is known that under such conditions, the upper (generally unstable) position of equilibrium becomes stable.

The d.e. of the pendulum in this case is

$$\ddot{\theta} + \lambda\dot{\theta} + [(g - a\omega^2 \sin \omega t)/l] \sin \theta = 0 \tag{8.30}$$

[1] See footnote [1], page 252.

[4] A. D. Dragilev, *Prikl. Math. i Mechanika* (in Russian), **16**, 1949.

where θ is the angle of deviation from the lower position of equilibrium; $y = a \sin \omega t$ is the vertical displacement of the suspension. We assume that λ corresponds to a damped oscillatory motion if the suspension point is fixed, in which case, as is well known, $\lambda^2/4 < \omega^2$.

It is useful to change the time scale by introducing the "dimensionless time" defined by the relations

$$\tau = \omega t; \qquad \frac{d}{d\tau} = \frac{1}{\omega}\frac{d}{dt}; \qquad \frac{d^2}{d\tau^2} = \frac{1}{\omega^2}\frac{d^2}{dt^2}$$

With this new independent variable τ and setting $k = \omega_0/\omega$: a/l; $\alpha = (\lambda/2\omega_0)k$, the d.e. can be written as

$$\ddot{\theta} + 2\alpha \frac{a}{l} \dot{\theta} + \left[k^2\left(\frac{a}{l}\right)^2 - \left(\frac{a}{l}\right) \sin \tau\right] \sin \theta = 0$$

where the differentiations are now with respect to τ. If one sets $\mu = a/l$, one has finally

$$\ddot{\theta} + 2\mu a\dot{\theta} + (k^2\mu^2 - \mu \sin \tau) \sin \theta = 0 \qquad (8.31)$$

where α and k are less than 1.

In this form the d.e. is not yet in the standard form but, by a change of variables, it can be reduced to an equivalent system of two d.e. of the first order which will be in that form.

We introduce two new variables φ and Ω defined by relations

$$\theta = \varphi - \mu \sin \tau \sin \varphi; \qquad \dot{\theta} = \mu\Omega - \mu \cos \tau \sin \varphi \qquad (8.32)$$

where $\dot{\theta} = d\theta/d\tau$.

Differentiating the first of these two equations and comparing with the second, one has

$$(1 - \mu \sin \tau \cos \varphi)\dot{\varphi} = \mu\Omega \qquad (8.33)$$

Differentiating the second d.e. (8.32) and substituting into (8.31), one gets

$$\ddot{\theta} = (\mu \sin \tau - k^2\mu^2) \sin \theta - 2\alpha\mu\dot{\theta} \qquad (8.34)$$

Whence through some intermediate transformations, one obtains

$$\begin{aligned}\frac{d\Omega}{dt} &= (\sin (\varphi - \mu \sin \tau \sin \varphi) - \sin \varphi) \sin \tau \\ &\quad - k^2\mu \sin (\varphi - \mu \sin \tau \sin \varphi) \\ &\quad + \frac{\mu\Omega \cos \tau \cos \varphi}{1 - \mu \sin \tau \cos \varphi} - 2\alpha\mu(\Omega - \cos \tau \sin \varphi) \qquad (8.35)\end{aligned}$$

From (8.33) and (8.35) it is seen that φ and Ω satisfy two d.e. of the first order in the standard form

$$\frac{d\varphi}{dt} = \mu\Omega + \mu^2 \ldots;$$

$$\frac{d\Omega}{d\tau} = \mu(-\sin^2 t \sin\varphi \cos\varphi - k^2 \sin\varphi + \Omega \cos\tau \cos\varphi - 2\alpha\Omega + 2\alpha \cos\tau \sin\varphi) + \mu^2 \tag{8.35a}$$

We can now apply the averaging procedure noting that

$$\underset{\tau}{M}[\cos\tau] = 0; \qquad \underset{\tau}{M}[\sin^2\tau] = \tfrac{1}{2}$$

which gives equations of the first approximation

$$\frac{d\varphi}{d\tau} = \mu\Omega; \qquad \frac{d\Omega}{d\tau} = -\mu\left[\frac{1}{2}\sin\varphi\cos\varphi + k^2\sin\varphi + 2\alpha\Omega\right] \tag{8.36}$$

If we eliminate Ω between these equations, we obtain

$$\ddot{\varphi} + 2\mu\alpha\dot{\varphi} + \mu^2(k^2 + \tfrac{1}{2}\cos\varphi)\sin\varphi = 0 \tag{8.37}$$

It is noted that this d.e. does not contain the variables τ explicitly and represents a pendulous system with a fixed suspension point. It is seen directly from (8.37) that a stationary solution is possible for $\varphi = \pi$. The variational equation in this case is

$$\frac{d^2\delta\varphi}{d\tau^2} + 2\mu\alpha\frac{d\delta\varphi}{d\tau} + \mu^2\left(\frac{1}{2} - k^2\right)\delta\varphi = 0 \tag{8.38}$$

and as $\mu\alpha > 0$, the condition of stability is: $\frac{1}{2} - k^2 > 0$; that is,

$$\omega > \sqrt{2}\omega_0\frac{l}{a} \tag{8.39}$$

Thus, if the frequency of the vertical oscillations of the pendulum's suspension is large enough, the pendulum settles on the upper (otherwise unstable) position of equilibrium.

Thus, for instance, for $l = 40$ cm; $a = 2$ cm, the above formula gives: $\omega > 140\,\dfrac{\text{radians}}{\text{sec}}$; that is, if the frequency of oscillation is greater than 22.3 cycles per second, the abovementioned phenomenon takes place. It is noted that a considerable simplification in the final result (of the d.e. (8.37)) is possible here owing to a special choice of new variables φ and Ω introduced by equations (8.32).

For a complete mathematical presentation of the principle of averaging as well as for the justification of the asymptotic methods (chapters 14 and 15) the reader is referred to chapters 4 and 5 of Bogoliubov and Mitropolsky.[1]

[1] See footnote [1], page 252.

Chapter 12

ALMOST PERIODIC OSCILLATIONS IN NEARLY LINEAR SYSTEMS

1. Introductory remarks

In the preceding two chapters we investigated the principal properties of periodic oscillations, as well as the conditions for their existence in nearly linear systems. The terms "periodic solution" and "periodic oscillation" will not be considered here in any different sense, inasmuch as "solution" clearly relates to a d.e. and "oscillation" to the phenomenon governed by this d.e.

The question of periodicity is now sufficiently clear and, as one recalls, is defined by the existence of a number T, the period, such that $f(t + T) = f(t)$. If such a number T exists, we call the function $f(t)$ the periodic function.

One has difficulty if one tries to find something analogous in the case of a function $x(t) = a_1 \sin \omega_1 t + a_2 \sin \omega_2 t$, where the ratio ω_1/ω_2 is an irrational number. Intuitively one feels that one is in the presence of something which repeats itself, more or less, particularly if one waits long enough but one needs yet a more precise definition. Experimentally such facts are of a common occurrence in connection, for instance, with the so-called "Lissajou curves" in mechanics; in Chapter 21 we shall see other examples of such phenomena in connection with the so-called "retarded actions."

The difficulty in this case is in that it is impossible to define the period and proceed by the classical methods of Chapters 10 and 11.

The oscillatory phenomena of this kind are called *almost periodic* and we propose first to establish some preliminary starting points before formulating the definition of almost periodicity.[1]

[1] H. Bohr, *Acta Math.* **45**, 1925; **46**, 1925; **47**, 1926; A. S. Besicovich, *Almost Periodic Functions*, Cambridge, 1938.

We consider the set of all finite sums

$$s(x) = \sum_{n=1}^{N} a_n e^{i\lambda_n x} \tag{1.1}$$

where a_n are arbitrary complex and λ_n arbitrary real numbers. The set of functions $s(x)$ is "closed" by the addition of those functions $f(x) = u(x) + iv(x)$ which can be approximated by $s(x)$ *uniformly* for all x.

Thus for any $\varepsilon > 0$ there exists some $s(x)$ satisfying the inequality

$$|f(x) - s(x)| \leq \varepsilon \qquad \text{for } -\infty < x < +\infty \tag{1.2}$$

We call the function class obtained by this extension the *closure* of the set $[s(x)]$ and denote it by $H[s(x)]$ or, simply by H.

The main problem is to characterize the functions $f(x)$ of the class H by certain *structural properties* bearing no relationship to the concepts used in the definition of H. It is useful to give an example. Let us assume that

$$s(x) = \exp(i\lambda_1 x) + \exp(i\lambda_2 x) \tag{1.3}$$

We shall endeavor to obtain the closure H with this form of $s(x)$, which clearly means the sum of two periodic functions with periods $T_1 = 2\pi/|\lambda_1|$ and $T_2 = 2\pi/|\lambda_2|$ and, of course, with all integral multiples of these periods.

Two cases are of importance:

(1) λ_1/λ_2 is rational, in which case T_1/T_2 is also rational. In this case there exists a common period $T = n_1T_1 = n_2T_2$, and the sum $s(x)$ is purely periodic with period T.

(2) λ_1/λ_2 is irrational and, also, T_1/T_2 is irrational. In this case there is no common period and $s(x)$ *is not* periodic.

On the other hand, from the number theory (the Diophantine approximations) it is known that, given $\delta > 0$, there exists a pair of arbitrarily great integers n_1 and n_2 such that

$$|n_1T_1 - n_2T_2| < \delta \tag{1.4}$$

If τ is a number near enough to n_1T_1 and n_2T_2 (for example, between these two numbers), then τ is "almost" a period of $\exp(i\lambda_1 x)$ as well as of $\exp(i\lambda_2 x)$ and, accordingly, "almost" a period for their sum $s(x)$, the difference $s(x + \tau) - s(x)$ being small for all x.

One can now define the *translation number* τ as a real number such that, given an arbitrary function $f(x) = u(x) + iv(x)$ continuous in $(-\infty, +\infty)$, one has

$$|f(x + \tau) - f(x)| < \varepsilon \qquad \text{for } -\infty < x < +x \tag{1.5}$$

One could study functions $f(x)$ with the following property: for every

$\varepsilon > 0$, there exists a translation number $\tau(\varepsilon)$ of $f(x)$ and these numbers are arbitrarily great. This is not yet a desired definition, as shown by Bohr, because certain properties of invariance are not fulfilled.

By narrowing the assumption by the requirement that the numbers are *relatively dense* (that is, to any interval $\alpha < x < \alpha + L$, there exists at least one number τ of the set, where L is a certain length), one obtains the definition of an *almost periodic function* (A.P.F. for short), namely

A function $f(x)$ continuous for $-\infty < x + \infty$ will be called almost periodic when, given an $\varepsilon > 0$, there exists a relatively dense set of $\tau(\varepsilon)$.

In other words, to every $\varepsilon > 0$ a length $L = L(\varepsilon)$ exists such that each interval of length $L(\varepsilon)$ contains at least one translation number $\tau = \tau(\varepsilon)$.

A continuous periodic function of period T is a special case of an almost periodic function inasmuch as we can take the periods nT where $n = 0$, ± 1, $\pm 2, \ldots$, as translation numbers $\tau(\varepsilon)$ for each ε.

For an almost periodic $f(x)$ the same conclusion holds for $f(x + c)$, where c is an arbitrary real constant, and for $cf(x)$, where c is an arbitrary complex constant. If can also be shown that A.P.F. characterizes the class H previously defined.

From an intuitive point of view, the above definition means that a certain almost periodic phenomenon $x(t)$ is characterized by the following property: If one starts from some initial value $x(t_0)$, the phenomenon will repeat itself as nearly as desired if we wait long enough.

The concept of almost periodicity should not be confused with that of *near periodicity*, also sometimes encountered in applications but not possessing the above features of almost periodicity. Thus, for instance

$$x(t) = \sin t + \frac{\varepsilon}{1 + t^2} \tag{1.6}$$

would be an example of the near periodicity, which merely means that in the course of time the motion approaches a purely periodic one.

Finite Fourier polynomials with incommensurate frequencies are examples of almost periodic functions.

There exists a number of other properties of A.P.F. which we indicate:

(1) An A.P.F. is always bounded in the interval $(-\infty < x < +\infty)$.
(2) An A.P.F. is uniformly continuous in the interval $(-\infty < x < +\infty)$.
(3) A sum or product of A.P.F. is A.P.F.
(4) If an A.P.F. satisfies the condition $|f(x)| > \alpha > 0$, α being a positive constant, then $1/f(x)$ is also an A.P.F.
(5) The limit of an uniformly convergent sequence $f_1(x), f_2(z), \ldots$ of A.P.F. is also an A.P.F. Hence the almost periodicity is not only the

property of finite trigonometric sums but holds also for infinite trigonometric series as long as they converge uniformly.

(6) Any A.P.F. is characterized by its limit

$$g = \lim_{T\to\infty} \frac{1}{T}\int_0^T f(t)dt$$

which is called the average value of the A.P.F.

(7) For A.P.F. there exists a denumerable set $[\lambda_i]$ of frequencies such that for any λ not belonging to this set, one has

$$\lim \frac{1}{T}\int_0^T f(t)e^{-\lambda t}dt = 0$$

(8) For any A.P.F. there exists a denumerable set $[\omega_\alpha]$ of real numbers having the following property:

(a) Between a finite number of ω_α there exists no linear relations of the form $\sum n_\alpha\omega_\alpha = 0$ with integer coefficients not all zeros.
(b) Any number λ_i can be represented as a linear combination of a finite number of ω_α with integer coefficients. The set $[\omega_\alpha]$ is called the *base* of a given A.P.F.; for a periodic function the base consists of only one element; for some A.P.F. it consists of a finite number of elements.
(c) The base has the following important property: if $[\tau_m]$ is a sequence such that for any ω_α one has: $\exp\,(i\omega_\alpha T_m) \to 1$ for $m \to \infty$, then one has uniformly on real axis the relation

$$\left|f(t+\tau_m) - f(t)\right| \xrightarrow[m\to\infty]{} 0$$

2. Almost periodic solutions

Although the theory of A.P.F. has developed as a further generalization of properties of periodic functions, its subsequent development proceeded on its own basis without any connection with the theory of d.e. which appeared later in the work of Favard[2] who investigated systems of linear d.e. with almost periodic coefficients and established conditions under which such d.e. may have almost periodic solutions (A.P.S. for short).

With later developments in the theory of nonlinear oscillations it was natural to inquire whether such solutions are possible under certain conditions? A number of experimental facts seemed to point in that direction, as we shall see later. Thus, for instance, in the case of the so-called *synchronization* phenomenon (Chapter 18), there exists a range of the "forcing frequency" (that is, the frequency of the external periodic

[2] J. Favard, *Acta Math.* **51**, 1928.

excitation) in which the free (or, better, the autoperiodic) frequency of the self-excited oscillation is "entrained" by the forcing (or heteroperiodic) frequency, so that there exists only one frequency, which means that the oscillation is purely periodic. If this range is exceeded and the synchronization is lost, the two frequencies separate, thus giving rise to "beats" which means an almost periodic phenomenon.

In Chapter 21 we shall come across another example of almost periodic oscillations, caused by the so-called "retarded actions" in systems amenable to certain functional equations of the difference-differential type. Here again the oscillation is of an almost periodic kind, inasmuch as the frequencies of component oscillations are generally in an irrational ratio.

Finally the numerous phenomena of *modulated oscillations* very often lead to the same type of oscillations.

All this seems to indicate that certain well known nonlinear d.e. may exhibit occasionally A.P.S., and one is thus confronted with a general problem of establishing conditions under which such solutions may exist.

Some initial work in this direction was done by Bohr and Neugebauer[3] but, later on, Krylov and Bogoliubov[4] developed a special transformation which facilitated this task. Finally, Malkin[5] coordinated these various developments and introduced further generalizations of the theory which ultimately led to the establishment of conditions under which A.P.S. may be expected in a given differential system. The investigation of Malkin occupies nearly ninety pages in his book, and for that reason, we give here only a brief outline of his results, referring to the original text for the proofs of numerous theorems and other details.

The starting point for this investigation is the usual system of d.e.

$$\dot{x}_s = \sum_{=1}^{n} a_{sj}x_j + f_s(t); \qquad s = 1, 2, \ldots, n \tag{2.1}$$

which we encountered previously in connection with the study of periodic solutions. In this system, the first terms on the right-hand side of (2.1) characterize the autonomous part of the system (if $f_s(t) = 0$) and the term $f_s(t)$ is the forcing term so that the form (2.1) is a differential system of non-autonomous type.

It is obvious, on physical grounds, that A.P.S. in the case of non-autonomous systems can exist only when "beats" between the autoperiodic and heteroperiodic oscillations may be expected, as was just mentioned.

The subsequent analysis is purely formal, but it must be borne in mind

[3] H. Bohr and O. Neugebauer, *Nachr. Ges. Wissen.*, Göttingen.

[4] N. Krylov and N. Bogoliubov, *Ac. Sc.* (USSR), 1945.

[5] I. G. Malkin, *Certain Problems in the Theory of Nonlinear Oscillations* (in Russian), Moscow; English translation.

that, taking the d.e. in the form (2.1), one may expect almost periodic solutions, so that the aim of this sequence of theorems is merely a justification of this expected result.

Consider the characteristic equation for the homogeneous system (2.1) (that is, $f_s(t) = 0$)

$$D(\lambda)^1 = \begin{vmatrix} a_{11} - \lambda & a_{12} & \dots & a_{1n} \\ a_{21} & a_{22} - \lambda & \dots & a_{2n} \\ \cdot & \cdot & \cdot & \cdot \\ a_{n_1} & a_{n_2} & \dots & a_{nn} - \lambda \end{vmatrix} = 0 \tag{2.2}$$

There are two theorems of which (1) is due to Neugebauer and Bohr, and (2) to Malkin.

(1) *If $D(\lambda)$ has no roots with vanishing real parts,* (2.1) *has one, and only one, A.P.S. satisfying the inequality*

$$|x_s(t)| < AM \tag{2.3}$$

where M is the upper limit of $f_s(t)$ in the interval $(-\infty, +\infty)$ and A is a certain constant depending on a_{sj} but not on $f_s(t)$.

(2) *If $f_s(t)$ is a finite trigonometric sum, the system* (2.1) *may have an A.P.S. in the case when the real parts of the roots of* (2.2) *are zero, provided the following condition be fulfilled:*

$$\lim_{t\to\infty} \int_0^t \sum_{\alpha=1}^{n} f_\alpha(t)\psi_{\alpha r} dt = 0; \qquad r = 1, 2, \dots, m \tag{2.4}$$

where $\psi_{\alpha r}$ are A.P.F. of the adjoint system corresponding to the differential system:

$$\dot{y}_s + \sum_{j=1}^{n} a_{sj} y_j = 0 \tag{2.5}$$

It is useful to mention certain difficulties which appear if one tries to proceed in the same manner as with periodic solutions. In the first place, the usual procedure (of Poincaré) does not hold here, because it is impossible to determine the period.

One could think, however, that by proceeding formally (that is, by identifying the terms with like powers of μ) one can still obtain the series solution if one succeeds in showing that such a formal series converges.

It is not difficult to show, however, that in an almost periodic case such a formal series always diverges and thus cannot represent the solution. In order to show this we consider a d.e. of the form:

$$\ddot{x} + x = \mu(\gamma x^3 + A \sin t + B \sin \omega t) \tag{2.6}$$

where μ is a small parameter and ω is an irrational number.

If one tries to satisfy this d.e. by a series solution of the form: $x(t) = x_0(t) + \mu x_1(t) + \ldots$ and proceeds in the usual manner, one finds a sequence of d.e.

$$\ddot{x}_0 + x_0 = 0; \qquad \ddot{x}_1 + x_1 = \gamma x_0^3 + A \sin t + B \sin \omega t;$$
$$\ddot{x}_2 + x_2 = 3\gamma x_0^2 x_1, \ldots$$

As $\dot{x}$ does not enter in these d.e., the solution will have no cosine terms. We have thus $x_0 = M_0 \sin t$, where M_0 is an arbitrary constant determined by the condition of almost periodicity of x_1; this requires that in the d.e. for x_1, there should be no term with $\sin t$. This yields: $M_0^3 = -4A/3$; therefore for $x_1(t)$ one has the expression

$$x_1(t) = \frac{1}{32}\gamma M_0^3 \sin 3t + \frac{B}{1 - \omega^2} \sin \omega t + M_1 \sin t$$

where M_1 is another arbitrary constant; equating to zero the coefficient of $\sin t$ in the equation for $x_2(t)$, one has: $M_1 = \frac{1}{48}\gamma M_0^3$, after which one has

$$x_2(t) = \frac{3}{1024}\gamma M_0^2 A \sin 3t - \frac{3}{2048}\gamma M_0^2 A \sin 5t$$
$$+ \frac{3}{2(1-\omega^2)^2}\gamma M_0^2 B \sin \omega t - \frac{3\gamma M_0^2 B}{2(1-\omega^2)[1-(\omega-2)^2]} \sin(\omega - 2)t$$
$$+ \frac{3\gamma M_0^2 B}{2(1-\omega^2)[1-(\omega+2)^2]} \sin(\omega + 2)t + M_2 \sin t$$

where M_2 is still another constant, etc.

In this manner one builds gradually a series solution which satisfies *formally* the d.e. but, as in this case, the formal series does not converge, it cannot represent the actual solution. In fact, on the right-hand side of the equation for x_k there will be terms of the form $A_{mn} \sin(m\omega + n)t$, where m and n are integers of which the second may be positive or negative. For a sufficiently large k the equation for x_k will therefore have terms with large m and $|n|$ and to these terms in the d.e. will correspond in the solution $x(t)$ the terms of the form $\{A_{mn}/[1 - (m\omega + n)^2]\} \sin(m\omega + n)t$.

As was mentioned in Section 1, the expression $(m\omega + n)$, where ω is irrational and m and n are integers, can be made to differ from unity as little as desired by a proper choice of integers m and n (of which the latter may be negative).

The presence of these *small divisors*: $[1 - (m\omega + n)^2]$ may produce the divergence of the series formed in this manner. In spite of these difficulties, certain progress has been accomplished owing to the Krylov-Bogoliubov transformation, as we shall presently show.

3. Existence of A.P.S. in noncritical cases

We indicate first a relatively simple case mentioned by Malkin in connection with (NA) nearly linear systems

$$\dot{x}_s = \sum_{j=1}^{n} a_{sj}x_j + f_s(t) + \mu F_s(t, x_1, \ldots, x_n, \mu), \qquad s = 1, 2, \ldots, n \quad (3.1)$$

under the same assumptions as previously. We also assume that the characteristic equation of the generating system ($\mu = 0$), viz.:

$$D(\lambda) = \begin{vmatrix} a_{11} - \lambda & a_{12} & \cdots & a_{1n} \\ a_{21} & a_{22} - \lambda & \cdots & a_{2n} \\ \cdot & \cdot & \cdot & \cdot \\ a_{n1} & a_{n2} & \cdots & a_{nn} - \lambda \end{vmatrix} = 0 \quad (3.2)$$

has no purely imaginary roots; such roots will be called "critical roots" in what follows. Under these conditions the generating system admits one and only one A.P.S. $x_s = \varphi_s(t)$ which may be taken as the generating solution of (3.1).

One can apply the method of successive approximations by taking the generating solution as the first approximation. For the kth approximation one has then:

$$\dot{x}_s^{(k)} = \sum_{j=1}^{n} a_{sj}x_j^{(k)} + f_s(t) + \mu F_s(t, x_1^{(k-1)}, \ldots, x_n^{(k-1)}, \mu) \quad (3.3)$$

It is shown that for a sufficiently small μ, the sequence $\{x_s^{(k)}\}$ converges uniformly toward the A.P.F.: $x_s(t)$ satisfying (3.1). Malkin[5] formulates the theorem.

If (3.2) *has no critical roots, the system* (3.1) *has one and only one A.P.S. which approaches the generating solution when* $\mu \to 0$.

[5] See footnote [5], page 286.

4. Transformation of Krylov-Bogoliubov; standard systems

The case when critical roots are present is more complicated, but it may be reduced to the previously considered case owing to a transformation of Krylov and Bogoliubov.[4]

We assume first that the constants a_{sj} and the functions $f_s(t)$ in (3.1) are zero, in which case (3.1) becomes:

$$\dot{x}_s = \mu F_s(t, x_1, \ldots, x_n, \mu) \tag{4.1}$$

We call the d.e. in this form the *standard form*.

A number of systems can be reduced to the standard form by an appropriate change of variables. An example of such transformation was indicated in the last section of the preceding chapter.

Inasmuch as F_s depends on μ analytically one can write the preceding equation as

$$\dot{x}_s = \mu F_s^{(1)}(t, x_1, \ldots, x_n) + \mu^2 F_s^{(2)}(t, x_1, \ldots, x_n) + \ldots \tag{4.2}$$

In these notations the functions $F_s^{(j)}$ are of the form

$$F_s^{(j)} = A_{so}^{(j)} + \sum_p [A_{sp}^{(j)} \cos \nu_p t + B_{sp}^{(j)} \sin \nu_p t] \tag{4.3}$$

where $A_{sp}^{(j)}$ and $B_{sp}^{(j)}$ are functions of x_i only.

Krylov and Bogoliubov show that for any integer k one can find a system of functions $U_s^{(j)}(t, y_1, \ldots, y_n)$ such that, after the substitution:

$$x_s = y_s + \mu U_s^{(1)} + \ldots + \mu^k U_s^{(k)} \tag{4.4}$$

equations (4.2) are reduced to the form

$$\dot{y}_s = \mu Y_s^{(1)} + \ldots + \mu^k Y_s^{(k)} + \mu^{k+1} Y_s^* \tag{4.5}$$

where the functions $Y_s^{(1)}, \ldots, Y_s^{(k)}$ depend only on $y_1, \ldots, y_n$ and not on t, whereas $Y_s^* = Y_s^*(t, y_1, \ldots, y_n, \mu)$ depends also on t.

If one substitutes (4.4) into (4.2) and equates the coefficients of the like powers of μ, one has in general

$$\dot{U}_s^{(j)} + Y_s^{(j)} = R_s^{(j)}(t, y_1, \ldots, y_n) \tag{4.6}$$

where $R_s^{(j)}$ are entire functions of those of $U_r^{(i)}$, $\partial u_r^{(i)}/\partial y_\alpha$, $Y_r^{(i)}$, $F_r^{(i)}(t, y_1, \ldots, y_n)$ and $\partial F_r^{(i)}/\partial y_\alpha$, for which $i < j$. The essence of this transformation is that the substitution (4.4) is carried out so that the functions containing t appear in the terms of the order k in the transformed equations and do not change much the accuracy of the approximation.

[4] See footnote [4], page 286.

Let $f(t) = A_0 + \sum [A_p \cos \nu_p t + B_p \sin \nu_p t]$ be an A.P.F. and introduce the following notations:

$$M[f(t)] = A_0 = \lim_{T\to\infty} \frac{1}{T}\int_0^T f(t)dt; \quad J[f(t)] = \sum \frac{A_p \sin \nu_p t - B_p \cos \nu_p t}{\nu_p} \tag{4.7}$$

Then, $\int f(t)dt = M[f(t)]t + [f(t)] + \alpha$, where α is an arbitrary constant.

It can be shown that (4.6) is satisfied if one sets

$$Y_s^{(1)}(y_1,\ldots, y_n) = \lim_{T\to\infty} \frac{1}{T}\int_0^T F_s^{(1)}(t, y_1,\ldots, y)dt = -M[F_s^{(1)}(t, y_1,\ldots,y_n)]$$

$$U_s^{(1)}(t, y_1,\ldots, y_n) = J[F_s^{(1)}(t, y_1,\ldots, y_n)]$$

and so on for higher approximations.

5. Almost periodic solutions of standard systems

Consider a standard system:

$$\dot{x}_s = \mu F_s^{(1)}(t, x_1,\ldots, x_n) + \mu^2 F_s^{(2)}(t, x_1,\ldots, x_n) + \ldots; \qquad s = 1,\ldots, n \tag{5.1}$$

where the functions $F_s^{(j)} = A_{so}^{(j)}(x_1,\ldots, x_n) + \sum_p [A_{sp}^{(j)}(x_1,\ldots, x_n) \cos\nu_p t + B_{sp}^{(j)}(x_1,\ldots, x_n) \sin \nu_p t]$ and $A_{so}^{(j)}$, $A_{sp}^{(j)}$, and $B_{sp}^{(j)}$ are polynomials in x_i.

Using the Krylov-Bogoliubov transformation ($k = 1$)

$$x_s = y_s + \mu U_s(t, y_1,\ldots, y_n) \tag{5.2}$$

where $U_s = J[F_s^{(1)}(t, y_1,\ldots, y_n)]$, equations (5.1) become

$$\dot{y}_s = \mu Y_s(y_1,\ldots, y_n) + \mu^2 Y_s^*(t, y_1,\ldots, y_n, \mu) \tag{5.3}$$

where

$$Y_s(y_1,\ldots, y_n) = \lim_{T\to\infty} \frac{1}{T}\int_0^T F_s^{(1)}(t, y_1,\ldots, y_n, 0)dt \tag{5.4}$$

and Y_s^* is an A.P.F. of the same form as F_s.

Assume that the system of the first approximation

$$\dot{y}_s = \mu Y_s(y_1,\ldots, y_n) \tag{5.5}$$

has a particular solution $y_s = y_s^0$ corresponding to the equilibrium point so that

$$Y_s(y_1^0,\ldots, y_n^0) = 0 \tag{5.6}$$

The variational equations in this case are of the form

$$ {}_{s} = \sum_{j=1}^{n} p_{sj}\xi_j \tag{5.7}$$

where $p_{sj} = \partial Y_s/\partial y_j{}^0$ are constants, once $y_1,\ldots, y_n$ have been replaced by $y_1{}^0,\ldots, y_n{}^0$ after differentiations. If one assumes that the characteristic equation of (5.7) has no purely imaginary roots, one can show that (5.3) has, for a sufficiently small μ, an A.P.S. tending to $y_s = y_s{}^0$ for $\mu \to 0$.

By a somewhat lengthy argument which we omit here, Malkin proves the following theorem:

If all roots of the characteristic equation of the variational system (5.7) *have negative real parts, then the A.P.S. of* (5.1) *is asymptotically stable. If this equation has at least one root with a positive real part, it is unstable.*

6. Almost periodic oscillations when all roots are critical

Consider a nearly linear (N.A.) system:

$$\dot{x}_s = \sum_{j=1}^{n} a_{sj}x_j + f_s(t) + \mu F_s(t, x_1,\ldots, x_n, \mu); \qquad s = 1,\ldots, n \tag{6.1}$$

where the functions f_s and F_s satisfy the previously formulated conditions.

Assume the case when the characteristic equation of the system

$$\dot{x}_s = \sum_{j=1}^{n} a_{sj}x_j \tag{6.2}$$

has only critical roots. It can be shown that then all solutions of (6.2) will be A.P.S. If φ_{sj} is a fundamental system of solutions of (6.2) and ψ_{sj} that correspond to the adjoint system, all these functions will be finite trigonometric sums, as was previously mentioned.

The condition for an A.P.S. of the generating system ($\mu = 0$) is:

$$\lim_{T\to\infty} \frac{1}{T}\int_0^T \sum_{\alpha=1}^{n} f_\alpha(t)\psi_{\alpha i}dt = 0; \qquad i = 1,\ldots, n \tag{6.3}$$

If this condition is fulfilled, the generating solution will be an A.P.F. of the form

$$x_s = \sum_{j=1}^{n} M_j\varphi_{sj} + x_s{}^{0*}(t) \tag{6.4}$$

where M_j are arbitrary constants and $x_s{}^{0*}$ is some particular A.P.S. of the system. In view of the properties of $f_s(t)$, x_s^{0*} will be some finite trigonometric sums.

This result suggests solving (6.1) by using M_j as variables. This gives

$$\sum_{j=1}^{n} \dot{M}_j \varphi_{sj} = \mu F_s(t, x_1, \ldots, x_n, \mu) \tag{6.5}$$

where x_i on the right-hand side must be replaced by (6.4).

On the other hand, by the theorem of Liouville, we have the determinant

$$\varphi_{si}(t) = \varphi_{si}(0)e^{(a_{11}+ \ldots +a_{nn})t} = \varphi_{si}(0)e^{idt}$$

where, in view of the conditions imposed on the roots of (6.2), the quantity α is real. Hence, if (6.5) is solved with respect to $\dot{M}_i$, in the standard equations so obtained, viz.:

$$\dot{M}_s = \Phi_s(t, M_1, \ldots, M_n, \mu) \tag{6.6}$$

the functions Φ_s will have with respect to $t, M_1, \ldots, M_n, \mu$ the same form which the functions F_s had with respect to $t, x_1, \ldots, x_n, \mu$.

One can now assert that these equations have an A.P.S. which, for $\mu = 0$, becomes a certain number of constants M_s^* that correspond to (5.6).

In view of the definition of $Y_s(y_1, \ldots, y_n) = \lim\limits_{T\to\infty} \dfrac{1}{T}\displaystyle\int_0^T F_s(t, y_1, \ldots, y_n, 0)dt$, equations (5.6) have the form:

$$P_i(M_1, \ldots, M_n) = \lim_{T\to\infty} \frac{1}{T}\int_0^T \varphi_i(t, M_1, \ldots, M_n, 0)dt = 0 \tag{6.7}$$

In this case we have to assume that the characteristic equation has no purely imaginary roots, and in view of the form of the coefficients p_{sj} in (5.7), this characteristic equation has the form

$$\begin{vmatrix} \dfrac{\partial P_1}{\partial M_1} - \rho & \dfrac{\partial P_1}{\partial M_2} & \cdots & \dfrac{\partial P_1}{\partial M_n} \\ \dfrac{\partial P_2}{\partial M_1} & \dfrac{\partial P_2}{\partial M_2} - \rho & \cdots & \dfrac{\partial P_2}{\partial M_n} \\ \dfrac{\partial P_n}{\partial M_1} & \dfrac{\partial P_n}{\partial M_2} & \cdots & \dfrac{\partial P_n}{\partial M_n} - \rho \end{vmatrix} = 0 \tag{6.8}$$

Substituting the values M_i into (6.4), one obtains A.P.S. of the system (6.1) which, for $\mu = 0$, becomes the generating solution

$$x_s = M_1^*\varphi_{s1} + \ldots + M_n^*\varphi_{sn} + x_s^{0*}(t)$$

The A.P.S. so obtained is stable if all roots in (6.8) have negative real parts. If one of these roots has a positive real part, the solution is unstable.

7. General case; combination of critical and noncritical roots

Consider again a nonautonomous nearly linear system

$$\dot{x}_s = \sum_{j=1}^{n} a_{sj}x_j + f_s(t) + \mu F_s(t, x_1, \ldots, x_n, \mu); \qquad s = 1, \ldots, n \tag{7.1}$$

where f_s and F_s are functions satisfying previously stated conditions.

We assume now that the characteristic equation

$$\dot{x}_s = \sum_{j=1}^{n} a_{sj}x_j \tag{7.2}$$

has $m < n$ critical roots. The system (7.2) will have then m A.P.S. which will be designated as φ_{sj} and, likewise an adjoint system of (7.2) will have equally m A.P.S., called ψ_{sj}.

It is always possible to select the functions ψ_{sj} so as to have relations

$$\sum_{s=1}^{n} \varphi_{si}\psi_{sj} = \delta_{ij}; \qquad \delta_{ii} = 1; \qquad \delta_{ij} = 0 \qquad (i \neq j) \tag{7.3}$$

We shall assume that these conditions of normalization have been fulfilled and that the $f_s(t)$ satisfy also the condition

$$\lim_{T\to\infty} \frac{1}{T}\int_0^T \sum_{\alpha=1}^{n} f_\alpha(t)\psi_{\alpha 1}(t)dt = 0 \tag{7.4}$$

Under these conditions the generating system

$$\dot{x}_s^0 = \sum_{j=1}^{n} a_{sj}x_j^0 + f_s(t); \qquad s = 1, \ldots, n \tag{7.5}$$

will have the following general A.P.S.

$$x_s^0 = \sum_{j=1}^{m} M_j\varphi_{sj} + x_s^{0*}(t) \tag{7.6}$$

depending on m arbitrary constants M_j, where $x_s^{0*}(t)$ is some particular A.P.S. of (7.5) (see Section 6).

Malkin proves now the important theorem regarding A.P.S. of (7.1). In order to formulate that theorem (the proof of which requires ten pages in Malkin's book), we introduce the expressions

$$P_i(M_1, \ldots, M_n) = \lim_{T\to\infty} \frac{1}{T}\int_0^T \sum F_\alpha(t, x_1^0, \ldots, x_n^0, 0)\psi_{\alpha i}dt = 0 \tag{7.7}$$

where x_s^0, M_j, and $\psi_{\alpha i}$ have the previous meaning. We form also the equation:

$$\begin{vmatrix} \frac{\partial P_1}{\partial M_1} - \rho & \frac{\partial P_1}{\partial M_2} & \cdots & \frac{\partial P_1}{\partial M_m} \\ \frac{\partial P_2}{\partial M_1} & \frac{\partial P_2}{\partial M_2} - \rho & \cdots & \frac{\partial P_2}{\partial M_m} \\ \cdot & \cdot & \cdot & \cdot \\ \frac{\partial P_m}{\partial M_1} & \frac{\partial P_m}{\partial M_2} & \cdots & \frac{\partial P_m}{\partial M_m} - \rho \end{vmatrix} = 0 \tag{7.8}$$

and assume that this equation has no zero roots.

Now Malkin's result is as follows:

We assume that (7.5) *has a general A.P.S. of the form* (7.6) *and that the characteristic equation of* (7.2) *has exactly* $n - m$ *noncritical roots. Moreover, the quantities* M_i *satisfy the conditions* $P_i = 0$ *and, finally,* (7.8) *has neither zero nor purely imaginary roots; then, for a sufficiently small* μ, *the system* (7.1) *has A.P.S. which reduce to the generating A.P.S. as* $\mu \to 0$.

This theorem applies to the first approximation x_s^0; if higher approximations are desired, one has to apply the methods of successive approximations, and it is shown that these approximations converge if μ is sufficiently small.

It is mentioned that there are certain essential differences between these conditions for A.P.S. and the corresponding conditions for the periodic case, namely:

(1) It is essential to assume in this case that the characteristic equation of the generating system has $n - m$ noncritical roots, which was not necessary in the periodic case.

(2) In the periodic case it was sufficient to assume that equation (7.8) has no zero root; in the almost periodic case, a stronger requirement is necessary, namely, that (7.8) not only does not have zero roots, but does not have purely imaginary roots either.

It is stated that the theory of periodic solutions cannot be considered merely as a particular case of a more general theory of A.P.S., but is rather a special case requiring a separate treatment.

We shall not continue the review of other parts of Malkin's text devoted to the theory of A.P.S. of nonlinear nonautonomous systems, but shall indicate an example illustrating this subject.

8. Van der Pol equation with two forcing terms

As an example, we consider a system described by the d.e.

$$\ddot{x} + k^2 x = \mu(1 - x^2)\dot{x} + A \sin \omega_1 t + B \sin \omega_2 t \tag{8.1}$$

assuming that the ratio ω_1/ω_2 is irrational. In the case of one forcing term, it is known that the system approaches the synchronized state if the autoperiodic frequency is not far from one of the heteroperiodic frequencies. The phenomenon then consists in the "entrainment" of the autoperiodic frequency by the neighboring heteroperiodic one.

For the present case, as in the periodic one, two separate cases are to be considered: (a) the nonresonance oscillation, and (b) the resonance oscillation. The case (a) appears when none of the quantities $mk + m_1\omega_1 + m_2\omega_2$ approaches the order of smallness of μ where m, m_1, and m_2 are integers for which $|m| + |m_1| + |m_2| \leq 4$; $m \neq 0$.

For $\mu = 0$, the general solution of (8.1) has the form

$$\begin{aligned} x &= M_1 \cos kt + M_2 \sin kt + A^1 \sin \omega_1 t + B^1 \sin \omega_2 t \\ \dot{x} &= -kM_1 \sin kt + kM_2 \cos kt + A^1\omega \cos \omega_1 t + B^1\omega_2 \cos \omega_2 t \end{aligned} \tag{8.2}$$

where $A^1 = A/(k^2 - \omega_1^2)$; $B^1 = B/(k^2 - \omega_2^2)$ and M_1 and M_2 are arbitrary constants: $k \neq \omega_1$, $k \neq \omega_2$.

Our first concern is the calculation of the quantities P_i as defined by (7.7). In order to make this calculation simpler, we consider the quantities M_1 and M_2 as new variables (instead of x and $\dot{x}$). The procedure is essentially the method of the variation of constants which consists in the differentiation of M_1 and M_2 once the generating solution (where M_1 and M_2 are constants) are replaced in the nonlinear equation (8.1). This yields the system

$$\begin{aligned} \cos kt\dot{M}_1 + \sin kt\dot{M}_2 &= 0 \\ -k \sin kt\dot{M}_1 + k \cos kt\dot{M}_2 &= \mu(1 - x^2)\dot{x} \end{aligned}$$

so that the new system of d.e. is

$$\begin{aligned} \dot{M}_1 &= -\frac{\mu}{k}(1 - x^2)\dot{x} \sin kt \\ \dot{M}_2 &= \frac{\mu}{k}(1 - x^2)\dot{x} \cos kt \end{aligned} \tag{8.3}$$

Averaging out with respect to t, one has

$$\dot{M}_1 = \mu P_1(M_1,M_2); \qquad \dot{M}_2 = \mu P_2(M_1,M_2) \tag{8.4}$$

where

$$P_1 = -\frac{1}{k} \lim_{T\to\infty} \frac{1}{T} \int_0^T (1 - x^2)\dot{x} \sin ktdt$$
$$P_2 = \frac{1}{k} \lim_{T\to\infty} \frac{1}{T} \int_0^T (1 - x^2)\dot{x} \cos ktdt \tag{8.5}$$

It is noted that the system (8.4) is now autonomous. The functions P_1 and P_2 are the free terms in the expansion of the right-hand sides of (8.3) in the trigonometric series. Taking into account (8.2) one finds the required result:

$$P_1 = \tfrac{1}{8}M_1[4 - (M_1{}^2 + M_2{}^2) - 2A'^2 - 2B'^2]$$
$$P_2 = \tfrac{1}{8}M_2[4 - (M_1{}^2 + M_2{}^2) - 2A'^2 - 2B'^2] \tag{8.6}$$

The A.P.S. of (8.1) in the first approximation is determined by (8.2) in which M_1 and M_2 are constants satisfying the condition

$$P_1(M_1,M_2) = P_2(M_1,M_2) = 0 \tag{8.7}$$

assuming that the equation

$$D(\gamma) = \begin{vmatrix} \dfrac{\partial P_1}{\partial M_1} - \gamma & \dfrac{\partial P_1}{\partial M_2} \\ \dfrac{\partial P_2}{\partial M_1} & \dfrac{\partial P_2}{\partial M_2} - \gamma \end{vmatrix} = 0 \tag{8.8}$$

does not have any critical roots.

In this case equations (8.7) have solutions $M_1 = M_2 = 0$ and also an infinity of solutions corresponding to equating the square bracket to zero in (8.6). For the latter, however, the Jacobian $\partial(P_1,P_2)/\partial(M_1,M_2)$ is zero, as is easily seen, which shows that (8.8) has a zero root; for our purpose here these solutions are of no interest.

As to the solution $M_1 = M_2 = 0$, it corresponds to the A.P.S. of the d.e. (8.1) with the roots

$$\gamma_1 = \gamma_2 = \frac{1}{4}\left[2 - \frac{A^2}{(k^2 - \omega_1{}^2)^2} - \frac{B^2}{(k^2 - \omega_2{}^2)^2}\right] \tag{8.9}$$

For $\mu = 0$, this A.P.S. goes into the generating solution

$$x^0(t) = \frac{A^2}{k^2 - \omega_1{}^2} \sin \omega_1 t + \frac{B^2}{k^2 - \omega_2{}^2} \sin \omega_2 t \tag{8.10}$$

in which the frequency k is absent.

It is visible that A.P.S. is stable if the roots (8.9) are negative, which leads to the condition

$$\frac{A^2}{(k^2 - \omega_1{}^2)^2} + \frac{B^2}{(k^2 - \omega_2)^2} > 2 \tag{8.11}$$

This condition is obviously not fulfilled if k differs much from ω_1 and ω_2.

We investigate now the resonances case when k differs from ω_1 by a small quantity which we assume to be $O(\mu)$; we also assume that the amplitude corresponding to ω_1 is small.

Setting $k^2 = \omega_1{}^2 - \mu a$; $A = \mu\lambda$, we start with the d.e.

$$\ddot{x} + \omega_1{}^2 x = \mu(1 - x^2)\dot{x} + \mu a x + \mu\lambda \sin \omega_1 t + B \sin \omega_2 t \tag{8.12}$$

We compute again P_i with variables M_1 and M_2, and we obtain the system

$$\begin{aligned} \dot{M}_1 &= -\frac{\mu}{\omega_1} [(1 - x^2)\dot{x} + ax + \lambda \sin \omega_1 t] \sin \omega_1 t \\ \dot{M}_2 &= \frac{\mu}{\omega_1} [(1 - x^2)\dot{x} + ax + \lambda \sin \omega_1 t] \cos \omega_1 t \end{aligned} \tag{8.13}$$

where

$$\begin{aligned} x &\equiv M_1 \cos \omega_1 t + M_2 \sin \omega_1 t + \frac{B}{\omega_1{}^2 - \omega_2{}^2} \sin \omega_2 t \\ \dot{x} &= -M_1\omega_1 \sin \omega_1 t + M_2\omega_1 \cos \omega_1 t + \frac{B\omega_2}{\omega_1{}^2 - \omega_2{}^2} \cos \omega_2 t \end{aligned} \tag{8.14}$$

We arrive thus at the equations:

$$\begin{aligned} &2\omega_1 P_1(M_1, M_2) \\ &\quad = -\lambda + aM_2 + \omega_1 M_1\left[1 - \frac{B^2}{2(\omega_1{}^2 - \omega_2{}^2)} - \frac{1}{4}(M_1{}^2 + M_2{}^2)\right] \\ &2\omega_1 P_2(M_1, M_2) \\ &\quad = -aM_1 + \omega_1 M_2\left[1 - \frac{B^2}{2(\omega_1{}^2 - \omega_2{}^2)} - \frac{1}{4}(M_1{}^2 + M_2{}^2)\right] \end{aligned} \tag{8.15}$$

For the exact resonance ($a = 0$) there is a solution: $M_2 = 0$, $M_1 = M$, where M is given by equation

$$Q(M) = -\frac{1}{8} M^3 + \left[\frac{1}{2} - \frac{B^2}{4(\omega_1{}^2 - \omega_2{}^2)^2} M - \frac{\lambda}{2\omega_1}\right] = 0 \tag{8.16}$$

The roots of (8.8) are:

$$\gamma_1 = \frac{dQ(M)}{dM}; \qquad \gamma_2 = \frac{1}{2} - \frac{B^2}{4(\omega_1{}^2 - \omega_2{}^2)} - \frac{M^2}{8} = \frac{\lambda}{2\omega_1 M}$$

Hence, if (8.16) has one or three distinct real roots, (8.8) will have nonzero real roots. In such a case (8.12) has one or three A.P.S. reducing for $\mu = 0$ to the solution

$$x^0(t) = M \cos \omega_1 t + \frac{B}{\omega_1^2 - \omega_2^2} \sin \omega_2 t \tag{8.17}$$

of the generating system.

As (8.16) cannot have a triple root, one A.P.S. always exists. Stability conditions for these A.P.S. have the form

$$M < 0; \qquad \frac{dP(M)}{dM} < 0 \tag{8.18}$$

Hence stable A.P.S. are those which correspond to the negative roots of (8.16). This equation has, however, always one negative root; if this is the only real root, this root is stable as $P(-\infty) > 0$ and, for this root, $P^1(M) < 0$.

If (8.16) has three real negative roots, the A.P.S. corresponding to the largest and to the smallest roots are stable while the A.P.S. corresponding to the middle root is unstable.

9. Parameters of the generating system; nonresonance and resonance frequencies

We consider again (7.1) and the A.P.S. (7.6) of its generating system depending on m arbitrary constants $M_1, \ldots, N_m$.

We assume that the characteristic equation of (7.2) has m critical and $(n - m)$ noncritical roots. We denote the former roots: $\pm \lambda_1 i, \ldots, \lambda_q i$, where λ_i are free frequencies of the generating system; $m = 2q$ and $\nu_1, \ldots, \nu_N$ the frequencies which appear in the developments in trigonometric sums of f_s and F_s in terms of the variable t.

We adopt the following definition of the resonance frequency: We call λ_p the *resonance frequency* if, at least, one of the linear combinations:

$$m_1\lambda_1 + \ldots + m_p\lambda_p + \ldots + m_q\lambda_q + n_1\nu_1 + \ldots + n_N\nu_N \tag{9.1}$$

can become $0(\mu)$. The numbers $m_1, \ldots, m_q, n_1, \ldots, n_N$ are integers, $(m_p \neq 0)$ for which $|m_1| + |m_2| + \ldots + |n_1| + \ldots + |n_N| \leq r + 2$, where r is the maximum degree of F_s with respect to $x_1, \ldots, x_N$. Hence, if λ_p is not a resonance frequency, the roots are always simple. In such a case it can be shown that in the system

$$P_i(M_1, \ldots, M_m) = \lim_{T\to\infty} \frac{1}{T} \int_0^T \sum_{\alpha=1}^{n} F_\alpha(t, x_1^0, \ldots, x_n^0; 0)\psi_{\alpha i} dt = 0 \tag{9.2}$$

two of the constants $M_1 = M_2 = 0$.

This simplifies the calculations inasmuch as the parameters M_i corresponding to nonresonance frequencies are zero.

As an example, consider the d.e.

$$\ddot{x} + k^2 x = \mu(1 - x^2)\dot{x} + A \sin \omega_1 t + B \sin \omega_2 t \tag{9.3}$$

with ω_1/ω_2 being an irrational number. In this case k is not a resonance frequency and one can write the generating solution as

$$x^0 = \frac{A}{k^2 - \omega_1{}^2} \sin \omega_1 t + \frac{B}{k^2 - \omega_2{}^2} \sin \omega_2 t \tag{9.4}$$

For the characteristic equation (8.8) we form first the variational system, which yields

$$\ddot{\xi} + k^2 \xi = -2\mu x_0 \dot{x}_0 \xi + \mu(1 - x_0{}^2)\dot{\xi}$$

Changing the variable, $\xi = e^{\mu\gamma t}\sigma$, and limiting the calculation to the first order only, one has

$$\ddot{\sigma} + k^2 \sigma = -2\mu x_0 \dot{x}_0 \sigma + \mu(1 - x_0{}^2)\dot{\sigma} - 2\mu\gamma\dot{\sigma}$$

Conditions for the existence of an A.P.S. of the d.e. are:

$$\ddot{\sigma} + k^2 \sigma = -2x_0 \dot{x}_0 \sigma_0 + (1 - x_0{}^2)\dot{\sigma}_0 - 2\gamma\dot{\sigma}_0$$

where $\sigma_0 = M \cos kt + N \sin kt$. Equating to zero the coefficients of $\cos kt$ and $\sin kt$ on the right-hand side of the d.e. and taking into account (9.4), one obtains

$$\left[1 - \frac{A^2}{2(k^2 - \omega_1{}^2)^2} - \frac{B^2}{2(k^2 - \omega_2{}^2)^2} - 2\gamma\right] M = 0$$

$$\left[1 - \frac{A^2}{2(k^2 - \omega_1{}^2)^2} - \frac{B^2}{2(k^2 - \omega_2{}^2)^2} - 2\gamma\right] N = 0$$

which gives for the roots γ in (8.8) the values

$$\gamma_1 = \gamma_2 = \frac{1}{4}\left[2 - \frac{A^2}{(k^2 - \omega_1{}^2)^2} - \frac{B^2}{(k^2 - \omega_2{}^2)^2}\right]$$

This almost periodic solution becomes (for $\mu = 0$) the generating solution

$$x^0 = \frac{A^2}{k^2 - \omega_1{}^2} \sin \omega_1 t + \frac{B^2}{k^2 - \omega_2{}^2} \sin \omega_2 t$$

in which the frequency k is absent. The theory permits finding those A.P.S. of (9.3) which are "entrained" by frequencies of the external excitation.

We shall not reproduce here further properties of the nonresonance solutions which are investigated by Malkin in great detail.

10. Forced oscillations of a mono-rail car

As an example of the preceding theory, Malkin [5] refers to a recent paper of Boutenin regarding the behavior of a mono-rail car stabilized by means of a forced precession of a gyroscope suitably controlled and shown in Fig. 12.1. The mono-rail is shown as $ABCD$ containing inside a gyro frame supported in bearings MN; the gyro G is mounted within the frame on the MN axis. Designating by x the angle of rotation of the frame, by y that of the car, and by $M_1 \sin \nu t$ the external disturbing moment, the differential system is

$$
\begin{aligned}
A_0\ddot{x} - C\Omega\dot{y} - pbx &= -\gamma''\dot{x} + M \\
J_0\ddot{y} + C\Omega\dot{x} - Phy &= -\gamma^1\dot{y} + M_1 \sin \nu t
\end{aligned}
\tag{10.1}
$$

where A_0 and J_0 are certain effective moments of inertia (A_0 relative to the gyro and J_0 relative to the car), $C\Omega = K$ angular momentum of the gyro P and p, the weights of the car (without the weight E) and the weight of E, respectively, γ and γ'' coefficients of linear friction of the car and the frame.

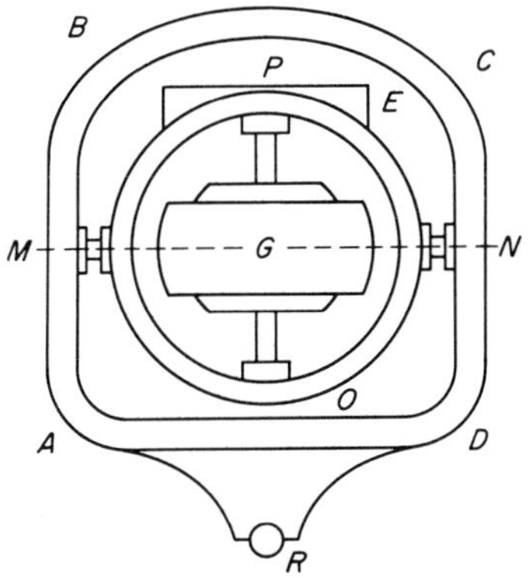

FIGURE 12.1

Finally M is the control moment which may be determined in any manner one wishes; it is assumed here that it is of the form

$$
\begin{aligned}
&M = (\alpha^1 - \beta^1\dot{x}^2)\dot{x} \\
&\alpha^1 > 0, \qquad \beta^1 > 0
\end{aligned}
$$

One introduces notations: $\gamma_1 = K/A_0$; $\gamma_2 = K/J_0$; $n_1{}^2 = pb/A_0$; $n_2{}^2 = Ph/J_0$ and assumes that the following dimensionless quantites are $\ll 1$,

$$\mu = (\alpha^1 - \gamma'')/A_0 n_1; \qquad \gamma^1/J_0 n_1; \qquad n_1\beta^1/A_0$$

The differential system can be written then as

$$
\begin{aligned}
\ddot{x} - \gamma_1\dot{y} - n_1{}^2 x &= \mu n_1(\dot{x} - \beta\dot{x}^3) = \mu f \\
\ddot{y} + \gamma_2\dot{x} - n_2{}^2 y &= -\mu n_1 \lambda \dot{y} + Q \sin \nu t = \mu F + Q \sin \nu t
\end{aligned}
\tag{10.2}
$$

where $\beta = \beta^1/(\alpha^1 - \gamma'')$; $\lambda = A_0\gamma^1/J_0(\alpha^1 - \gamma'')$; $Q = M_1/J_0$.

[5] See footnote [5], page 286.

The system (10.2) is investigated assuming first that one does not have the resonance condition expressible by a relation of the form

$$m_1\omega_1 + m_2\omega_2 + n\nu \simeq 0$$

where ω_1 and ω_2 are free oscillations of the linearized system and the external frequency, and m_1, m_2, and n are integers.

The general solution of the linear part of equation (10.2) is

$$\begin{aligned} x &= A_1 \sin \theta_1 + A_2 \sin \theta_2 + d \cos \nu t \\ \dot{x} &= A_1\omega_1 \cos \theta_1 + A_2\omega_2 \cos \theta_2 - d\nu \sin \nu t \\ y &= A_1k_1 \cos \theta_1 + A_2k_2 \cos \theta_2 + c \sin \nu t \\ \dot{y} &= -A_1\omega_1k_1 \sin \theta_1 - A_2\omega_2k_2 \sin \theta_2 + c\nu \cos \nu t \\ \theta_1 &= \omega_1 t + \alpha_1; \qquad \theta_2 = \omega_2 t + \alpha_2 \end{aligned} \tag{10.3}$$

where A_1, A_2, α_1, and α_2 are arbitrary constants, and the frequencies ω_1 and ω_2 satisfy the equation

$$\omega^4 + (n_1^2 + n_2^2 - \gamma_1\gamma_2)\omega^1 + n_1^2n_2^2 = 0 \tag{10.4}$$

Moreover, d and c have the following values:

$$d = \gamma_1\nu Q/[\nu^4 + (n_1^2 + n_2^2 - \gamma_1\gamma_2)\nu^2 + n_1^2n_2^2]$$

$$c = -(\nu^2 + n_1^2)Q/[\nu^4 + (n_1^2 + n_2^2 - \gamma_1\gamma_2)\nu^2 + n_1^2n_2^2]$$

For k_1 and k_2 hold relations:

$$k_1 = (\omega_1^2 + n_1^2)/\gamma_1\omega_1 = \gamma_2\omega_1/(\omega_1^2 + n_2^2)$$

$$k_2 = (\omega^2 + n_1^2)/\gamma_1\omega_2 = \gamma_2\omega_2/(\omega_2^2 + n_2^2)$$

If one introduces the new variables A_1, A_2, θ_1, θ_2 instead of x and y making use of (10.3), one obtains, after certain transformations and taking into account the above expressions for k_1 and k_2, the following system:

$$\begin{aligned} \dot{A}_1 &= -\mu(F^* \sin \theta_1 + k_2 f^* \cos \theta_1); \\ \dot{A}_2 &= \mu[F^* \sin \theta_1 + k_1 f^* \cos \theta_2] \\ \dot{\theta}_1 &= \omega_1 + \frac{\mu}{A_1}(-F^* \cos \theta_1 + k_2 f^* \sin \theta_1); \\ \dot{\theta}_2 &= \omega_2 + \frac{\mu}{A_2}(F^* \cos \theta_2 - k_1 f^* \sin \theta_2) \end{aligned} \tag{10.5}$$

where $F^* = -(\gamma_1 F)/(\omega_2^2 - \omega_1^2)$; $f^* = (\gamma_1 n_2 f)/n_1(\omega_2^2 - \omega_1^2)$.

Expanding the right-hand terms of these expressions into trigonometric sums and dropping all terms except the free terms, one obtains the following equations of the first approximation:

$$\dot{A}_1 = \mu b_1(c_1 - A_1{}^2\omega_1{}^2 - 2A_2{}^2\omega_2{}^2 - 2d^2\nu^2)A_1 = \mu R_1(A_1A_2)$$
$$\dot{A}_2 = \mu b_2(-c_2 + 2A_1{}^2\omega_1{}^2 + A_2{}^2\omega_2{}^2 + 2d^2\nu^2)A_2 = \mu R_2(A_1,A_2) \quad (10.6)$$
$$\dot{\theta}_1 = \omega_1; \qquad \dot{\theta}_2 = \omega_2$$

where $b_1 = (3\beta k_2 n_2 \gamma_1 \omega_1)/8(\omega_1{}^2 - \omega_2{}^2)$; $b_2(3\beta k_1 n_2 \gamma_1 \omega_2)/8(\omega_1{}^2 - \omega_2{}^2)$

$$c_1 = \frac{4}{3\beta}\left(1 - \frac{k_1 n_1 \lambda}{k_2 n_2}\right); \qquad c_2 = \frac{4}{3\beta}\left(1 - \frac{k_2 n_1 \lambda}{k_1 n_2}\right) \quad (10.7)$$

In the first approximation the oscillation is given by the following expressions

$$\begin{aligned} x &= A_1{}^* \sin(\omega_1 t + \varepsilon_1) + A_2{}^* \sin(\omega_2 t + \varepsilon_2) + d \cos \nu t \\ y &= A_1{}^* k_1 \cos(\omega_1 t + \varepsilon_1) + A_2{}^* k_2 \cos(\omega_2 t + \varepsilon_2) + c \sin \nu t \end{aligned} \quad (10.8)$$

where ε_1 and ε_2 are arbitrary constants, and $A_1{}^*$ and $A_2{}^*$ are the roots of two equations appearing as coefficients of μb_1 and μb_2 in (10.6). For free oscillations $Q = 0$ and, thus, $d = 0$, the system for $A_1{}^*$ and $A_2{}^*$ yields then

$$A_2{}^* = 0; \qquad A_1{}^{*2} = c_1/\omega_1{}^2 \quad (10.9)$$

To this solution correspond periodic oscillations with frequency ω_1 and the roots of the characteristic equation

$$\begin{vmatrix} \dfrac{\partial R_1}{\partial A_1} - \gamma & \dfrac{\partial R_1}{\partial A_2} \\ \dfrac{\partial R_2}{\partial A_1} & \dfrac{\partial R_2}{\partial A_2} - \gamma \end{vmatrix} = 0 \quad (10.10)$$

are here $\gamma_1 = -2b_1 A_1{}^* \omega_1{}^2$; $\gamma_2 = (2c - c_2)b_2$. The stability conditions for this solution are: $\omega_1{}^2 > \omega_2{}^2$; $2c_1 - c_2 < 0$.

The stationary equations have also a solution (for $d = 0$)

$$A_1{}^* = 0; \qquad A_2{}^{*2} = c_2/\omega_2{}^2$$

to which correspond periodic oscillations with frequency ω_2.

There is still another solution for $d = 0$ corresponding to

$$A_1{}^{*2} = (2c_2 - c_1)/3\omega_1{}^2; \qquad A_2{}^{*2} = (2c_1 - c_2)/3\omega_2{}^2 \quad (10.11)$$

To this solution corresponds almost periodic oscillations with frequencies ω_1 and ω_2. The characteristic equation corresponding to this solution is

$$\gamma^2 + 2\gamma(b_1 A_1{}^{*2}\omega_1{}^2 - b_2 A_2{}^{*2}\omega_2{}^2) + 12 b_1 b_2 A_1{}^{*2} A_2{}^{*2} \omega_1{}^2 \omega_2{}^2 = 0 \quad (10.12)$$

If this equation is real (that is, if the right-hand sides in (10.11) are positive), the corresponding oscillations are stable in the case of inequality:

$$3(b_1 A_1^{*2}\omega_1^2 - b_2 A_2^{*2}\omega_2^2) = b_1(2c_2 - c_1) - b_2(2c_1 - c_2) > 0$$

If one assumes $Q \neq 0$, in such a case equations

$$\begin{aligned}(c_1 - A_1^2\omega_1^2 - 2A_2^2\omega_2^2 - 2d^2\nu^2)A_1 &= 0\\ (-c_2 + 2A_1^2\omega_1^2 + A^2\omega_2^2 + 2d^2\nu^2)A_2 &= 0\end{aligned} \tag{10.13}$$

have a solution

$$A_1^* = 0; \qquad A_2^{*2}\omega_2^2 = c_2 - 2d^2\nu^2 \tag{10.14}$$

to which correspond almost periodic oscillations with frequencies ω_2 and ν.

Besides this there is also a solution

$$A_2^* = 0; \qquad A_1^{*2}\omega_1^2 = c_1 - 2d^2\nu^2 \tag{10.15}$$

to which correspond almost periodic oscillations with frequencies ω_1 and ν and, finally the solution

$$A_1^{*2}\omega_1^2 = \tfrac{1}{3}(2c_2 - c_1 - 2d^2\nu^2); \qquad A_2^{*2}\omega_2^2 = \tfrac{1}{3}(2c_1 - c_2 - 2d^2\nu^2) \tag{10.16}$$

to which correspond almost periodic oscillations with three frequencies ω_1, ω_2, and ν.

The conditions of stability are:

$$\begin{aligned}&\text{For (10.14):} \quad \omega_1^2 - \omega_2^2 < 0; \quad c_1 - 2c_2 + 2d^2\nu^2 > 0\\ &\text{For (10.15):} \quad \omega_1^2 - \omega_2^2 > 0; \quad 2c_1 - c_2 - 2d^2\nu^2 < 0\\ &\text{For (10.16):} \quad b_1(2c_2 - c_1 - 2d^2\nu^2) > b_2(2c_1 - c_2 - 2d^2\nu^2)\end{aligned} \tag{10.17}$$

Consider now the resonance case; for instance, $\nu = \omega_2$ and assume $Q = \mu Q^*$ as customary in the resonance cases (otherwise the generating system has no A.P.S.).

In this case, for $\mu = 0$, the solution of (10.2) can be written as:

$$\begin{aligned}x &= A\sin\theta + M_1\cos\omega_2 t + M_2\sin\omega_2 t\\ \dot{x} &= A\omega_1\cos\theta - M_1\omega_2\sin\omega_2 t + M_2\omega_2\cos\omega_2 t\\ y &= Ak_1\cos\theta - M_1k_2\sin\omega_2 t + M_2k_2\cos\omega_2 t\\ \dot{y} &= -A\omega_1k_1\sin\theta - M_1k_2\omega_2\cos\omega_2 t - M_2k_2\omega_2\sin\omega_2 t\\ \theta &= \omega_1 t + \alpha\end{aligned} \tag{10.18}$$

where A, M_1, M_2, and α are arbitrary constants.

Taking A, M_1, M_2, and θ as new variables, one has

$$
\begin{aligned}
&\dot{A}\sin\theta + \dot{M}_1\cos\omega_2 t \\
&\qquad + \dot{M}_2\sin\omega_2 r + A\cos\theta(\dot{\theta} - \omega_1) = 0 \\
&\dot{A}k_1\cos\theta - \dot{M}_1 k_2\sin\omega_2 t \\
&\qquad + \dot{M}_2 k_2\cos\omega_2 t - Ak_1\sin\theta(\dot{\theta} - \omega_1) = 0 \\
&\dot{A}\omega_1\cos\theta - \dot{M}_1\omega_2\sin\omega_2 t \\
&\qquad + \dot{M}_2\omega_2\cos\omega_2 t - A\omega_1\sin\theta(\dot{\theta} - \omega_1) = \mu f \\
&-\dot{A}\omega_1 k_1\sin\theta - \dot{M}_1\omega_2 k_2\cos\omega_2 t \\
&\qquad - \dot{M}_2\omega_2 k_2\sin\omega_2 t - A\omega_1 k_1\cos\theta(\dot{\theta} - \omega_1) = \mu F + \mu Q^*
\end{aligned}
\tag{10.19}
$$

Whence

$$
\begin{aligned}
\dot{M}_1 &= \mu(F_* \cos\omega_2 t - k_1 f^* \sin\omega_2 t) \\
\dot{M}_2 &= \mu(F_* \sin\omega_2 t + k_1 f^* \cos\omega_2 t) \\
\dot{A} &= \mu(-F_* \sin\theta - k_2 f^* \cos\theta) \\
\dot{\theta} &= \omega_1 + \mu(-F_* \cos\theta + k_2 f^* \sin\theta)
\end{aligned}
\tag{10.20}
$$

where

$$
F_* = \frac{\gamma_1 Q^*}{\omega_1^2 - \omega_2^2} + F^* \text{ and } F^* = -\frac{\gamma_1 F}{\omega_2^2 - \omega_1^2}; \qquad f^* = \frac{\gamma_1 n_2 f}{n_1(\omega_2^2 - \omega_1^2)}
$$

If one replaces in (10.12) on the right-hand sides x, $\dot{x}$, y, and $\dot{y}$ by their values (10.18) and averages out the variables, one obtains equations of the first approximation

$$
\begin{aligned}
\dot{M}_1 &= \mu b_2(-c_2 + M_1^2\omega_2^2 + M_2^2\omega_2^2 + 2A^2\omega_1^2)M_1 = \mu R_1(M_1, M_2, A) \\
\dot{M}_2 &= \mu b_2(-c_2 + M_1^2\omega_2^2 + M_2^2\omega_2^2 + 2A^2\omega_1^2)M_2 + \frac{\gamma_1 Q^*}{\omega_1^2 - \omega_2^2} \\
&= \mu R_2(M_1, M_2, A) \qquad (10.21) \\
\dot{A} &= \mu b_1(c_1 - 2M_1^2\omega_2^2 - 2M_2^2\omega_2^2 - A^2\omega_1^2)A = \mu R(M_1, M_2, A) \\
\dot{\theta} &= \omega_1
\end{aligned}
$$

The stationary solutions in the first approximation are

$$
\begin{aligned}
x &= A^*\sin(\omega_1 t + \varepsilon) + M_2^*\sin\omega_2 t \\
y &= A^* k_1\cos(\omega_1 t + \varepsilon) + M_2^* k_2\cos\omega_2 t
\end{aligned}
\tag{10.22}
$$

where ε is an arbitrary constant, and A^* and M_2^* are the roots of equations

$$b_2(-c_2 + M_2^2\omega_2^2 + 2A^2\omega_1^2)M_2 + \frac{\gamma_1 Q^*}{\omega_1^2 - \omega_2^2} = 0$$
$$(c_1 - 2M^2\omega_2^2 - A^2\omega_1^2)A = 0 \qquad (10.23)$$

These stationary solutions are stable if, for $M_1 = 0$, $M_2 = M_2^*$, and $A = A^*$.

The roots of the characteristic equation

$$D(\gamma) = \begin{vmatrix} \frac{\partial R_1}{\partial M_1} - \gamma & \frac{\partial R_1}{\partial M_2} & \frac{\partial R_1}{\partial A} \\ \frac{\partial R_2}{\partial M_1} & \frac{\partial R_2}{\partial M_2} - \gamma & \frac{\partial R_2}{\partial A} \\ \frac{\partial R}{\partial M_1} & \frac{\partial R}{\partial M_2} & \frac{\partial R}{\partial A} - \gamma \end{vmatrix} = 0 \qquad (10.24)$$

have negative real parts.

11. Physical aspects of A.P. oscillations

We have attempted to give in this chapter an outline of recent developments regarding A.P.S. in the theory of nearly linear d.e.; these developments, in turn, are based on the earlier work of Krylov and Bogoliubov.

There is a certain analogy between the treatment of almost periodic solutions and that of periodic ones (Chapters 10 and 11); thus, for instance the effort still centers on the determination of the generating solution which is approached by the solution of the nonlinear system where $\mu \to 0$.

In spite of this, the treatment of almost periodic cases is quite different from the periodic ones, as Malkin observes. Thus, in the almost periodic cases the characteristic equation of the generating system must have at least a certain number of roots other than purely imaginary roots, although this is not necessary in purely periodic cases, and so on.

Malkin states that at present it does not seem possible to consider the theory of periodic solutions as a particular case of more general theory of almost periodic solutions and that it is necessary to treat the former on its own merit as a *very special* case.

We shall not enter into this question of a purely formal nature but merely observe that this difficulty does not arise in the theory of oscillations.

In fact, in all oscillatory phenomena which we shall study later the interval in which almost periodic oscillations exist and that in which oscillation becomes periodic are formed together by a bifurcation value of a certain parameter. In this manner, when a periodic oscillation appears, the almost periodic disappears or vice versa so that, on formal grounds, it is impossible to consider that one phenomenon is a particular case of the other unless one wishes to consider this question from the standpoint of the bifurcation theory.

Thus, for instance, a regenerative amplifier amplifying a certain fixed frequency ω_0 may be regarded as producing a purely periodic oscillation. If, however, the parameter λ (the coefficient of coupling between the anode and the grid circuits) reaches a critical value $\lambda = \lambda_0$, an autoperiodic oscillation appears with a frequency ω, generally, not standing in any rational ratio with respect to ω_0; therefore an almost periodic oscillation begins in this manner.

In Chapter 18 we shall study the phenomenon of synchronization which consists in the existence of a certain zone in which the autoperiodic frequency (of the self-excited oscillation) is "entrained" by the heteroperiodic frequency (of the external periodic excitation). As long as the parameter determining this phenomenon remains within the interval corresponding to the zone of synchronization, only one frequency ω exists and the oscillation is periodic. If the parameter reaches its bifurcation value and goes beyond it, the two frequencies ω and ω_0 separate and an almost periodic oscillation results. These bifurcation effects are always reversible, as was mentioned in Chapter 7.

There are still more complicated situations as we shall learn later, but in all cases the periodicity and the almost periodicity are distinctly separated by the bifurcation point so that it is meaningless to consider that one aspect of the phenomenon (the periodicity) is a kind of "continuation" of the other aspect (the almost periodicity).

In the above example of a regenerative amplifier, the almost periodic "aspect" is characterized by the bifurcation of the first kind so that, topologically, the situation is also quite different from that of the amplifier as such.

In the second example (synchronization) the situation is also different between the zone of synchronization (periodicity) and the zone when the synchronization is lost (almost periodicity); here again the two cases are not comparable. In the first case the system has a singular point, whereas in the second case this singular point vanishes (Chapter 18).

In some other cases these transitions from periodicity to almost periodicity can be traced to complicated conditions of stability, but in all cases the two zones, *periodicity* and *almost periodicity*, *relate to entirely different situations*. On this basis the difference in the mathematical approach to these two cases is to be expected.

Chapter 13

DETERMINATION OF CHARACTERISTIC EXPONENTS

1. Determination of characteristic exponents on the basis of Poincaré's theory

The question of existence of periodic solutions (Chapters 10 and 11) is supplemented in this chapter by an investigation of stability on the basis of the general theory; this introduces the rather difficult question of determination of the characteristic exponents. The difficulty, as will be presently shown, is that the characteristic equation (for the general case of d.e. with periodic coefficients) contains coefficients depending on the *solutions* of the d.e., that is, ultimately, on the characteristic exponents that are unknown. One is thus in a kind of vicious circle from which the only issue is the method of approximations outlined in the last four sections of this chapter. It is possible, however, to proceed without the somewhat long calculations of these exponents by merely finding conditions under which they are all negative; very often this is sufficient if one is merely interested in ascertaining stability.

We recall (Chapter 5) that the variational equations give information only about asymptotic stability and orbital stability (or instability). The first holds if all characteristic exponents are negative (if they are real) or, more generally, have negative real parts. The instability occurs when at least one exponent has positive real part.

The theorems of this chapter are of importance for nonautonomous systems. If the system is autonomous, the situation is simpler; in particular, for autonomous systems of the second order which will be our first interest, Poincaré shows[1] that, inasmuch as one exponent is always

[1] H. Poincaré, *Les méthodes nouvelles de la mécanique céleste* **T.1**, Gauthier-Villars, Paris, 1892; also E. Goursat, *Cours d'Analyse* **T.2**, Gauthier-Villars, Paris, 1918.

zero, stability (or instability) is determined by the sign of the real part of the second exponent and this, as will be shown, is a simple matter.

Finally, if the real part of the second exponent (always for the second-order systems) vanishes, one has a very special case of *indifferent stability* which characterizes the harmonic oscillator.

In the general case, the reversal of a strict inequality implies the existence of an exponent with positive real part, and one obtains an instability criterion of equal validity for autonomous systems.

We follow closely the presentation of Malkin [2] throughout this chapter.

In Chapter 5 it was shown that the stability of a periodic motion depends on the variational system with periodic coefficients

$$\dot{y}_i = \sum_{j=1}^{n} p_{ij} y_i; \qquad i = 1, 2, \ldots, n \tag{1.1}$$

where

$$p_{ij} = \left(\frac{\partial X_i}{\partial x_j}\right)_0 \varphi(t) \tag{1.2}$$

are partial derivatives into which one substitutes the solution corresponding to the periodic motion after differentiation.

The characteristic equation is

$$\Delta(S) = \begin{vmatrix} y_{11}(\omega) - S & y_{12}(\omega) & \ldots & y_{1n}(\omega) \\ y_{21}(\omega) & y_{22}(\omega) - S & \ldots & y_{2n}(\omega) \\ \cdot & \cdot & \cdot & \cdot \\ y_{n1}(\omega) & y_{n2}(\omega) & \ldots & y_{nn}(\omega) - S \end{vmatrix} = 0 \tag{1.3}$$

where y_{ij} is the fundamental system of solutions with the initial conditions

$$y_{ij}(0) = \begin{cases} 1 & \text{for } i = j \\ 0 & \text{for } i \neq j \end{cases} \tag{1.4}$$

It was shown that the characteristic equation can be written also in the form:

$$S^n + A_1 S^{n-1} + \ldots + A_{n-1} S + A_n = 0 \tag{1.5}$$

but the principal difficulty here, as was just mentioned, is that, aside of A_n, all other coefficients of (1.5) are unknown.

It is to be noted that the coefficients A_i are *invariants* in the sense that they remain the same under a linear transformation of variables. To

[2] I. G. Malkin, *Certain Problems in the Theory of Nonlinear Oscillations* (in Russian), Moscow; English translation.

determine these invariants it is necessary to integrate the variational system, which is generally impossible.

As far as the problem of stability is concerned, it is not necessary to know the *exact values* of the roots S_i but only whether their moduli are greater or less than *one*. Hence, it is sufficient to determine the invariants A_i only with a certain approximation.

One can follow the general method of Poincaré by assuming that the right-hand sides of the d.e. depend analytically on the parameter μ. Thus, if the coefficient p_{ij} depends analytically on μ, the same is true for y_{ij} as well as for the invariants A_i.

We assume that the periodic solution also depends analytically on μ and is of the form

$$\varphi_i(t) = \varphi_i^{(0)}(t) + \mu\varphi_i^{(1)}(t) + \ldots \tag{1.6}$$

so that for $\mu = 0$, $\varphi_i(t) = \varphi_i^{(0)}(t)$ *is the generating solution.*

One can also write

$$p_{ij} = p_{ij}^{(0)} + \mu p_{ij}^{(1)} + \mu^2 p_{ij}^{(2)} + \ldots \tag{1.7}$$

where

$$p_{ij}^{(0)} = \left(\frac{\partial X_i}{\partial x_j}\right) x_i = \varphi(t); \qquad \mu = 0 \tag{1.8}$$

It is clear that the system

$$y_i^{(0)} = \sum_{j=1}^{n} p_{ij}^{(0)} y_j^{(0)}, \qquad i = 1, 2, \ldots, n \tag{1.9}$$

is the variational system for the generating solution. Moreover, if the functions p_{ij} depend on a parameter, y_{ij} and, therefore, the invariants A_i will also be functions of that parameter. We may assume that this dependence is analytic, as expressed by (1.6) and (1.7). One can connect now these considerations with the theory of Poincaré concerning the existence of periodic solutions (Chapter 10).

Let $\varphi_i^{(0)}(0) + \beta_i(\mu)$ be the initial conditions of a periodic solution and consider slightly modified initial conditions $\varphi_i^{(0)}(0) + \beta_i(\mu) + \gamma_i$, where γ_i are small numbers, the differential system in question being

$$\dot{x}_i = X_i^{(0)}(t, x_1, \ldots, x_n) + \mu X_i^{(1)}(t, x_1, \ldots, x_n) + \ldots \tag{1.10}$$

The solution clearly depends analytically on γ_i and one can write

$$x_i(t, \gamma_1, \ldots, \gamma_n) = x_i(t, 0, \ldots, 0) + \sum_{k=1}^{n} \Gamma_{ik}\gamma_k \tag{1.11}$$

where the quantities Γ_{ik} satisfy the system

$$\dot{\Gamma}_{ij} = \sum_{k=1}^{n} p_{ik} \Gamma_{jk} \tag{1.12}$$

with the initial conditions

$$\Gamma_{ij}(0) = \begin{cases} 1 & \text{for } i = j \\ 0 & \text{for } i \neq j \end{cases} \tag{1.13}$$

Hence, in our case

$$\Gamma_{ij} = y_{ij} \tag{1.14}$$

By forming the expressions $[x_i] = x_i(\omega, \gamma_1, \ldots, \gamma_n) - x_i(0, \gamma_1, \ldots, \gamma_n)$, we have

$$\left(\frac{\partial [x_i]}{\partial \gamma_j}\right)_{\gamma_1 = \ldots = \gamma_n = 0} \equiv y_{ij}(\omega) - y_{ij}(0) \tag{1.15}$$

On the other hand, the functions $x_i(t, \gamma_1, \ldots, \gamma_n)$ can be obtained from $x_i(t, \beta_1, \ldots, \beta_n, \mu)$ of the Poincaré theory by replacing β_i by $\beta_i(\mu) + \gamma_i$, giving

$$y_{ij}(\omega) - y_{ij}(0) = \frac{\partial \psi_i(\beta_1(\mu), \beta_2(\mu), \ldots, \beta_n(\mu))}{\partial \beta_j} \tag{1.16}$$

where ψ_i are the functions whose vanishing determines the periodicity of a solution (Chapter 10). This permits writing the characteristic equation (1.3) in the form originally indicated by Poincaré:

$$\begin{vmatrix} \dfrac{\partial \psi_1}{\partial \beta_1} + 1 - S & \dfrac{\partial \psi_1}{\partial \beta_2} & \cdots & \dfrac{\partial \psi_1}{\partial \beta_n} \\ \dfrac{\partial \psi_2}{\partial \beta_1} & \dfrac{\partial \psi_2}{\partial \beta_2} + 1 - S & \cdots & \dfrac{\partial \psi_n}{\partial \beta_2} \\ \cdot & \cdot & \cdot & \cdot \\ \dfrac{\partial \psi_n}{\partial \beta_1} & \dfrac{\partial \psi_n}{\partial \beta_2} & \cdots & \dfrac{\partial \psi_n}{\partial \beta_n} + 1 - S \end{vmatrix}_{\beta_i = \beta_i(\mu)} = 0 \tag{1.17}$$

It is to be noted that the determination of the functions ψ_i requires also the integration of the variational system of the generating solution.

It is possible now to express the conditions of stability in the form of inequalities. This does not require the actual calculation of the characteristic exponents but involves only the limits at which these exponents cease to be negative (that is, when the moduli of the roots cease to be less than *one*).

In a particular case of a system of the second order, the characteristic equation is:

$$S^2 - 2AS + B = 0 \tag{1.18}$$

For this equation to have only roots with moduli less than one, it is necessary that

$$|B| < 1 \tag{1.19}$$

This condition is sufficient if the roots are complex. If they are real, additional conditions are required. It is sufficient to set $S = (1 + \lambda)/(1 - \lambda)$ and require that in the quadratic equation so obtained the real parts of the roots should be negative. One obtains thus two additional conditions

$$2A + B + 1 > 0; \qquad -2A + B + 1 > 0 \tag{1.20}$$

With the Poincaré form (1.17) of the characteristic equation, one has

$$2A = 2 + \left(\frac{\partial\psi_1}{\partial\beta_1} + \frac{\partial\psi_2}{\partial\beta_2}\right)_{\beta_i=\beta_i(\mu)}; \qquad B = 1 + \left[\left(\frac{\partial\psi_1}{\partial\beta_1} + \frac{\partial\psi_2}{\partial\beta_2}\right) + J\right]_{\beta_i=\beta_i(\mu)} \tag{1.21}$$

where J is the Poincaré Jacobian:

$$J = \begin{vmatrix} \dfrac{\partial\psi_1}{\partial\beta_1} & \dfrac{\partial\psi_1}{\partial\beta_2} \\ \dfrac{\partial\psi_2}{\partial\beta_1} & \dfrac{\partial\psi_2}{\partial\beta_2} \end{vmatrix} \tag{1.22}$$

The conditions (1.20) are then:

$$J \geq 0; \qquad 4 + 2\left(\frac{\partial\psi_1}{\partial\beta_1} + \frac{\partial\psi_2}{\partial\beta_2}\right)_{\beta_i=\beta_i(\mu)} + J > 0 \tag{1.23}$$

The condition (1.19) can be written as

$$\int_0^\omega \left[\left(\frac{\partial X_1}{\partial x_1}\right) + \left(\frac{\partial X_2}{\partial x_2}\right)\right]_{x_i=\varphi_i(t)} dt < 0 \tag{1.24}$$

using the notations of Chapter 10. There are then *three* conditions of stability. If $\mu = 0$, the characteristic exponents of the solution become the characteristic exponents of the generating solution.

If μ is small (which we always assume), the criterion of stability requires that the characteristic exponents of the generating solution must be negative, in which case the stability is asymptotic. It is noted that here there is no necessity for *an actual determination* of the characteristic exponents inasmuch as this is implicitly contained in inequalities (1.19) and (1.20) which are equivalent to the assertion that the characteristic exponents are negative. The inequalities (1.23) and (1.24) merely specify the conditions (1.19) and (1.20) on the basis of the theory of Poincaré.

It is believed that this procedure was used for the first time in the theory of oscillations by L. Mandelstam and N. Papalexi in their theory of subharmonic resonance; in Chapter 19 we shall return to this question.

2. Stability of periodic solutions

We are now in a position to complete the investigation of *stability* of periodic solutions whose existence was studied in Chapter 10. Consider, for instance a nonautonomous system with one degree of freedom in a nonresonant case.

In this case the functions ψ_1 and ψ_2 (Chapter 10) are given by equations

$$\psi_1 = [x] = (\cos 2k\pi - 1)\beta_1 + \frac{1}{k}\sin 2k\pi\beta_2 + [C]\mu + \dots$$

$$\psi_2 = [\dot{x}] = -k\sin 2k\pi\beta_1 + (\cos 2k\pi - 1)\beta_2 + [\dot{C}]\mu + \dots$$

and the left-hand side of inequalities (1.23) are, respectively, $2(1 - \cos 2k\pi)$ and 4 if one limits oneself only to the first terms. Hence, for small μ these inequalities hold.

As to (1.24), it is here

$$\mu\int_0^{2\pi}\frac{\partial F(t, x, \dot{x}, \mu)}{\partial \dot{x}}\,dt < 0 \tag{2.1}$$

where x and $\dot{x}$ in $F(t, x, \dot{x}, \mu)$ must be replaced by their expressions in the power series

$$x(t) = x_0 + \mu x_1(t) + \mu^2 x_2(t) + \dots \tag{2.2}$$

For the first approximation one replaces x and $\dot{x}$ by the generating solution x_0 and $\dot{x}_0$ so that (2.1) becomes

$$\mu\int_0^{2\pi}\frac{\partial F(t, x_0, \dot{x}_0, 0)}{\partial \dot{x}_0}\,dt \leq 0 \tag{2.3}$$

where x_0 and $\dot{x}_0$ constitute the generating solution. Condition (2.3) is the only one required for stability since the other two are automatically fulfilled for small μ.

In the case of resonance (2.3) still holds, but the function x_0 is now given by the expression (3.3), Chapter 10. As the functions ψ_1 and ψ_2, (Section

2, Chapter 10) vanish for $\mu = 0$, the second condition (1.23) is fulfilled as inequality; as to $J \geq 0$, it can be written (leaving out terms of higher orders in μ) as

$$\mu \frac{\partial(P,Q)}{\partial(M_0,N_0)} > 0 \tag{2.4}$$

Equations (2.3) and (2.4) are the conditions of stability.

In an autonomous system with one degree of freedom (Section 6, Chapter 5) one of the roots of the characteristic equation is *one* (that is, the corresponding characteristic exponent vanishes); hence, the second root is B (in 1.19) which requires the fulfillment of inequality (1.24). If this condition is fulfilled (as inequality), there will still be stability but one cannot assert that this is an asymptotic stability. Here condition (1.24) has the form

$$\mu \int_0^\omega \frac{\partial f(x, \dot{x}, \mu)}{\partial \dot{x}} dt < 0 \tag{2.5}$$

In the condition (2.5), ω is the period of the solution and x and $\dot{x}$ correspond to this solution. Again, if one considers the first approximation, the condition (2.5) acquires the form

$$\mu \int_0^{2\pi/k} \frac{\partial f(M_0 \cos kt, \ -M_0 k \sin kt; 0)}{\partial \dot{x}} dt$$

$$= \frac{\mu}{k} \int_0^{2\pi} \frac{\partial f(M_0 \cos \alpha, \ -kM_0 \sin \alpha; 0)}{\partial \dot{x}} d\alpha \leq 0 \tag{2.6}$$

where M_0 is determined by

$$P(M_0) = \int_0^{2\pi} f(M_0 \cos u, \ -kM_0 \sin u; \theta) \sin u du \tag{2.7}$$

By integration by parts and after some transformations (2.6) can be written as

$$\mu \frac{dP(M_0)}{dM_0} \geq 0 \tag{2.8}$$

which is then the required condition of stability.

3. Determination of characteristic exponents by approximations

Although the method of the preceding two sections gives generally satisfactory results in applications, there are cases when it is necessary to go to higher approximations in the determination of characteristic exponents and also to establish the limits between the zones of stability and instability. For that purpose it is useful to introduce a parameter in the d.e. and to form a recursive system which permits one to carry out the calculations with any degree of accuracy.

It may be useful to mention in passing that the applications in which these calculations are encountered are very numerous. Thus, for instance, vibrations of driving rods in locomotives, critical zones of self-excitation of parametrically modulated circuits, etc., are within the scope of this problem.

We outline here the procedure due to Malkin, omitting some details. We consider a differential system with periodic coefficients

$$\dot{x}_i = \sum_{j=1}^{n} p_{ij}x_j; \qquad i = 1, 2, \ldots, n \tag{3.1}$$

where the periodic coefficients $p_{ij}(t)$ may depend on some parameters μ_i, $i = 1, 2., \ldots, k$ with respect to which they are analytic in a certain region

$$|\mu_i| \leq E_i \tag{3.2}$$

where $E_i > 0$ are certain fixed numbers; we assume also that the period ω of $p_{ij}(t)$ does not depend on μ_i. In such a case a solution $x_i = x_i(t, \mu_1, \ldots, \mu_k)$ will be also analytic in μ_i and, in view of the form of the characteristic equation (Section 5, Chapter 5), one reaches the following theorem of Liapounov.[3]

The coefficients of the characteristic equation of (3.1) *are analytic functions of the parameter* μ_i.

This results from the fact that the region of analyticity of p_{ij} coincides in this case with that of the coefficients of the characteristic equation.

This theorem can be used for an approximate calculation of coefficients of the characteristic equation and, hence, also for that of the characteristic exponents.

Without any loss of generality, we assume that there is only one parameter μ (not to be confused with μ of the preceding chapter) so that the coefficients (3.1) are of the form

$$p_{ij}(t) = q_{ij}(t) + \mu p_{ij}^{(1)}(t) + \mu^2 p_{ij}^{(2)} + \ldots; \qquad i = 1, 2, \ldots, n \tag{3.3}$$

where q_{ij} and $p_{ij}^{(1)}$, $p_{ij}^{(2)}, \ldots$ are continuous periodic functions with period ω and we assume that the series (3.3) converge for $|\mu| \leq E$.

[3] A. Liapounov, *C.R.* **123**, Paris, 1896.

We consider a fundamental system of solutions $x_{ij}(t,\mu)$ of (3.1) represented by power series expansions in terms of μ, viz.:

$$x_{ij} = x_{ij}^{(0)}(t) + \mu x_{ij}^{(1)}(t) + \mu^2 x_{ij}^{(2)}(t) + \ldots \tag{3.4}$$

converging for all values of t in the region $|\mu| < E$ with the initial conditions

$$\begin{aligned} x_{ij}^{(0)}(0) &= \begin{cases} 1 & \text{for } i = j \\ 0 & \text{for } i \neq j \end{cases} \\ x_{ij}^{(1)}(0) &= x_{ij}^{(2)}(0) = \ldots = 0 \end{aligned} \tag{3.5}$$

If one substitutes (3.4) into (3.1) and equates the coefficients of like powers of μ, one obtains a recursive system for the determination of $x_{ij}^{(0)}, x_{ij}^{(1)}, \ldots$, viz.:

$$\begin{aligned} \frac{dx_{ij}^{(0)}}{dt} &= \sum_{k=1}^{n} q_{ik} x_{kj}^{(0)} \\ \frac{dx_{ij}^{(1)}}{dt} &= \sum_{k=1}^{n} q_{ik} x_{kj}^{(1)} + \sum_{s=1}^{n} p_{is}^{(1)} x_{sj}^{(0)} \\ &\cdots\cdots\cdots \\ \frac{dx_{ij}^{(k)}}{dt} &= \sum_{k=1}^{n} q_{ik} x_{kj}^{(k)} + \sum_{s=1}^{n} \sum_{\beta=0}^{k-1} p_{is}^{(k-\beta)} x_{sj}^{(\beta)}; \quad i = 1, 2, \ldots, n \\ & \qquad\qquad\qquad\qquad\qquad\qquad\qquad\qquad k = 1, 2, \ldots, n \end{aligned} \tag{3.6}$$

It is observed that this system has the same terms with q. Assume that the integration of the system

$$\frac{dy_i}{dt} = \sum q_{ik} y_k \tag{3.7}$$

is known. For $\mu = 0$, this is the system (3.1). In such a case (3.6) permits a successive determination of all $x_{ij}^{(k)}$ beginning with $k = 0$ under the assumed initial conditions.

The series solution (3.4) is thus determined. If one sets $t = \omega$, one obtains the coefficients of the characteristic equation and, hence, the values of the characteristic exponents.

We have obtained a possibility of carrying out this approximation procedure due to the fact that for $\mu = 0$ we assume the solution of the system to be known. It is possible, however, to reduce more general cases to this particular case.

Assume, for instance, that we wish to determine coefficients of the characteristic equation of a differential system with periodic coefficients

$$\dot{x}_i = \sum_{k=1}^{n} \Gamma_{ik}(t)x_k \tag{3.8}$$

in which there are no parameters. We replace this system by

$$\dot{x}_i = \sum_{k=1}^{n} p_{ik}(t,\mu)x_k \tag{3.9}$$

where the function $p_{ik}(t,\mu)$ are so chosen that, for $\mu = 0$, the system (3.9) admits integration in a closed form (for example, becomes a system with constant coefficients), while for $\mu = \mu^*$ (where μ^* is a certain fixed number in the region of convergence of coefficients $p_{ik}(t,\mu)$), it becomes (3.8).

Let M be a maximum of $|\Gamma_{ik}(t)|$ for all i, k, and t and consider

$$\dot{x}_i = M \sum_{k=1}^{n} \frac{\Gamma_{ik}(t)}{M} x_k = M \sum \rho_{ik}(t)x_k \tag{3.10}$$

where $\rho_{ik}(t) = \Gamma_{ik}(t)/M$ so that $|\rho_{ik}(t)| < 1$.

We have thus:

$$\dot{x}_i = \mu \sum_{k=1}^{n} \rho_{ik}(t)x_k \tag{3.11}$$

For $\mu = 0$, we have by assumption a known solution, but for $\mu = M$ we have the original system. If M is small, one has a good convergence, assuming that $\mu = M$ is still within the limit of convergence of the coefficients $p_{ik}(t,\mu)$.

The method is particularly convenient when μ^* is small, that is, when the given system does not differ much from a system which can be integrated in a closed form inasmuch as in such a case only a few approximations will be sufficient.

4. Second-order systems; invariants A_i

As an example, we consider a system of the second order

$$\ddot{x} + p(t)x = 0 \tag{4.1}$$

and study first the d.e.

$$\ddot{y} = \mu p(t)y \tag{4.2}$$

where μ is a parameter. If $\mu = -1$, one has (4.1).

Let $f(t,\mu)$, $\varphi(t,\mu)$ be two particular solutions with the initial conditions

$$f(0,\mu) = 1; \qquad f'(0,\mu) = 0; \qquad \varphi(0,\mu) = 0; \qquad \varphi'(0,\mu) = 1 \tag{4.3}$$

On the other hand, we have

$$\begin{aligned} f(t,\mu) &= f_0(t) + \mu f_1(t) + \mu^2 f_2(t) + \dots \\ \varphi(t,\mu) &= \varphi_0(t) + \mu\varphi_1(t) + \mu^2\varphi_2(t) + \dots \end{aligned} \tag{4.4}$$

Substituting these expansions into (4.1), one gets

$$\ddot{f}_0 = 0; \qquad \ddot{\varphi}_0 = 0;\dots; \qquad \ddot{f}_n = pf_{n-1}; \qquad \ddot{\varphi}_n = p\varphi_{n-1}; \qquad n = 1, 2,\dots \tag{4.5}$$

Moreover,

$$\begin{aligned} &f_0(0) = 1; \quad f_0'(0) = 0; \quad \varphi_0(0) = 0; \quad \varphi_0'(0) = 1 \\ &f_n(0) = f_n'(0) = \varphi_n(0) = \varphi_n'(0) = 0; \quad n = 1, 2,\dots \end{aligned} \tag{4.6}$$

whence,

$$f_0(t) = 1; \qquad \varphi_0(t) = t;$$

$$f_n(t) = \int_0^t dt \int_0^t pf_{n-1}dt; \qquad \varphi_n(t) = \int_0^t dt \int_0^t p\varphi_{n-1}dt$$

As the coefficient A in (1.18) is

$$A = \tfrac{1}{2}[f(\omega) + \varphi'(\omega)]$$

this coefficient for (4.2) is

$$A^*(\mu) = 1 + \frac{1}{2}\sum_{n=1}^{\infty}[f_n(\omega) + \varphi_n'(\omega)]\mu^n \tag{4.7}$$

As this series converges for any values of μ, we can set, for instance, $\mu = -1$, which gives the coefficient A for (4.1)

$$A = A^*(-1) = 1 + \frac{1}{2}\sum_{n=1}^{\infty}[f_n(\omega) + \varphi_n'(\omega)](-1)^n \tag{4.8}$$

Suppose that $p(t)$ can have either negative or zero values, without being equal to zero identically. In such a case all functions f_n, f_n', φ_n, and φ_n' are negative for 1 odd and positive for n even; therefore all terms of (4.8) are positive. This conclusion results in the following theorem of Liapounov.

(I) *Under the just specified conditions, the characteristic equation has two real roots, of which one is greater than one and the other less than one.*

We give without proof[2] another theorem of Liaponuov:

(II) *If $p(t)$ can have either positive or zero values (without being equal to zero identically), then the characteristic equation of* (4.1) *has complex roots with moduli equal to one provided*

$$\omega \int_0^\omega p dt \leq 4 \tag{4.9}$$

5. Zones of stability

We consider now a second order d.e. of the form

$$\ddot{x} + \lambda^2(1 + \mu f)x = 0 \tag{5.1}$$

where $f = f(t)$ is a periodic function with period π (which can be obtained by a change of the independent variable). The quantity λ^2 is a parameter. We also assume that $f(t)$ can be expanded in a power series

$$f = f_1(t) + \mu f_2(t) + \mu^2 f_2(t) + \ldots \tag{5.2}$$

where the functions $f_i(t)$ do not depend on μ and are periodic with period π; the series (5.2) converges for $|\mu| < a$, a being a constant.

The parameter λ^2 will be assumed to vary, and we propose to investigate the regions or zones of stability as functions of λ. As we use λ^2, it is immaterial whether $\lambda > 0$ or $\lambda < 0$.

From the preceding it follows that those values of λ for which A (in 1.18) satisfies the inequality $A^2 < 1$, correspond to stability and those for which $A^2 > 1$, to instability. Hence, the separation of zones of stability and instability occurs at the thresholds at which the functions A have the values

$$A = +1 \tag{5.3}$$

$$A = -1 \tag{5.4}$$

The functions A are our unknown functions and we shall try to determine them in terms of other quantities by successive approximations.

From the general theory we have

$$A = \tfrac{1}{2}(x_1(\pi) + \dot{x}_2(\pi))$$

where x_1 and x_2 are two particular solutions satisfying the usual initial conditions.

[2] See footnote [2], page 309.

These solutions are entire functions of λ^2 and analytic functions of μ for $|\mu| < a$. We have thus

$$\begin{aligned} x_1 &= x_1^{(0)}(t) + \mu x_1^{(1)}(t) + \dots \\ x_2 &= x_2^{(0)}(t) + \mu x_2^{(1)}(t) + \dots \end{aligned} \tag{5.5}$$

Substituting (5.5) into (5.1) and taking into account (5.2), one obtains the system of d.e.:

$$\ddot{x}_i^{(0)} = -\lambda^2 x_i^{(0)}; \qquad \ddot{x}_i^{(1)} = -\lambda^2 x_i^{(1)} - \lambda^2 f_1(t) x_i^{(0)}; \dots \qquad i = 1, 2, \dots \tag{5.6}$$

with the initial conditions

$$x_1^{(0)}(0) = 1; \qquad \dot{x}_1^{(0)}(0) = 1; \qquad x_2^{(0)}(0) = 0; \qquad \dot{x}_2^{(0)}(0) = 1;$$
$$x_1^{(k)}(0) = \dot{x}_1^{(k)}(0) = x_2^{(k)}(0) = \dot{x}_2^{(k)}(0) = 0; \qquad k = 1, 2, \dots \tag{5.7}$$

This recursive system determines the functions $x_1^{(k)}$, $x_2^{(k)}$. In particular we have

$$x_1^{(0)} = \cos \lambda t; \qquad x_2^{(0)} = \frac{1}{\lambda} \sin \lambda t$$

so that

$$A = \cos \lambda\pi + \tfrac{1}{2}[x_1^{(1)}(\pi) + \dot{x}_2^{(1)}(\pi)]\mu + \dots \tag{5.8}$$

We determine now the thresholds (5.3) and (5.4). It is seen that (5.3) and (5.4) are satisfied for $\mu = 0$, $\lambda = n$, n being an integer. In particular for n even, equation (5.3) is satisfied and, for n odd, (5.4). One can expect that for $\mu \neq 0$ but small, these equations are still satisfied in the neighborhood of integer values of $\lambda = n$. We set $\lambda = n + \alpha$ and equate expressions (5.8) to $+1$ for n even, and to -1 for n odd. One obtains thus the following expressions

$$(-1)^{n+1}\left[\frac{\alpha^2\pi^2}{2!} - \frac{\alpha^4\pi^4}{4!} + \dots\right] + \frac{1}{2}\,[x_1(\pi) + \dot{x}_2(\pi)]_{\lambda=n+\alpha}\cdot\mu + \dots = 0 \tag{5.9}$$

The left-hand side of this equation is an analytic function of α and μ vanishing for $\alpha = \mu = 0$. As the first derivative of this expression with respect to α vanishes for $\alpha = \mu = 0$ and the second derivative is $\neq 0$, from the theory of implicit functions it follows that (5.9) admits two solutions for μ small. One of these solutions is an analytic function of μ and the other of $\sqrt{\mu}$.

If one substitutes these solutions $\alpha(\mu)$ into $\lambda = n + \alpha(\mu)$ and gives a sequence of integer values to the parameter λ, one can obtain a sequence of relations (5.3) and (5.4). As previously, for n even one has conditions (5.3) and for n odd, those given by (5.4).

It can be shown (we omit the proof, referring to Malkin,[2] that: (1) all such solutions are real, and (2) only the solutions in terms of developments according to the powers of μ are to be retained, as those corresponding to the powers of $\sqrt{\mu}$ do not give rise to real solutions and thus are to be ruled out.

In this manner one can construct a curve $A(\lambda)$ (Fig. 13.1) for a fixed μ with roots (5.3) and (5.4) marked on the λ axis. These roots λ_i' and λ_i'',

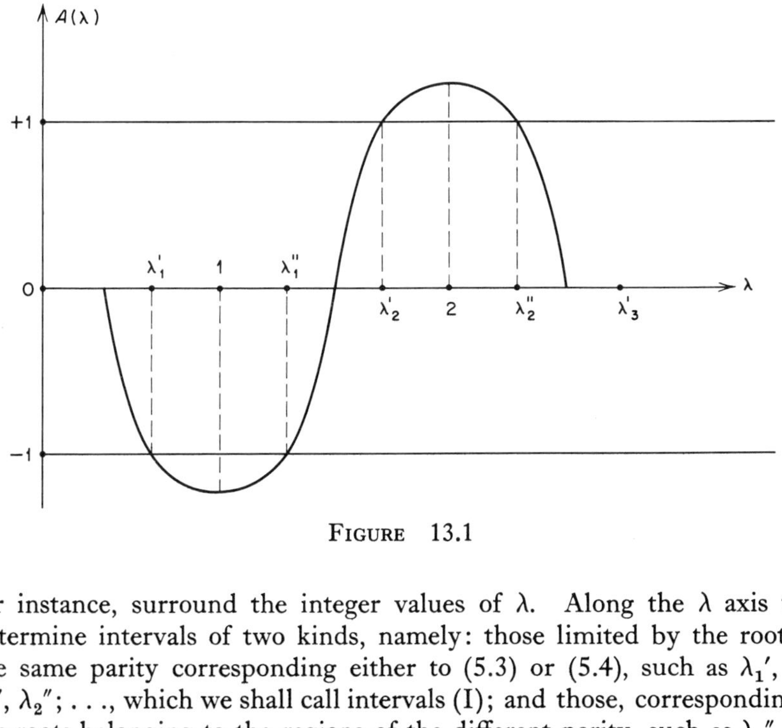

FIGURE 13.1

for instance, surround the integer values of λ. Along the λ axis they determine intervals of two kinds, namely: those limited by the roots of the same parity corresponding either to (5.3) or (5.4), such as λ_1', λ_1''; λ_2', λ_2''; . . ., which we shall call intervals (I); and those, corresponding to the roots belonging to the regions of the different parity, such as λ_1'', λ_2'; λ_2'', λ_3';. . ., which we shall call intervals (II).

The intervals (I) and (II) alternate, as is seen from Fig. 13.1. There may be special cases when two roots of the same parity coalesce, giving rise to a double root (with the disappearance of the corresponding interval I) but we shall be concerned only with the case of distinct roots.

It is clear that in each interval (I) one has $A^2 > 1$ and in (II), $A^2 < 1$.

[2] See footnote [2], page 309.

Thus the zones of stability are in (II) and those of instability in (I). In fact, in an interval II, the quantity A varies from -1 at the beginning of the interval to $+1$ at its end. As A cannot become $+1$ or -1 in the interval (II), $A^2 < 1$ in such intervals; in an interval (I), the quantity A is the same at the ends of the interval; thus, for instance, in $(\lambda_1', \lambda_1'')$ the terminal values of A are $+1$; hence, inside the intervals (I) one has either $A > +1$ or $A < -1$.

Summing up, if λ varies continuously, one has a sequence of alternate zones of stability and instability; the thresholds separating these zones are the roots of equations (5.3) and (5.4). The zones of stability are located in the intervals (II) and those of instability in (I) and all roots are in the neighborhood of integer values of λ. For $\mu = 0$, the roots coalesce on the integer values $\lambda = n$; for small values of μ the roots are analytic functions of μ.

6. Calculation of zones of stability

We consider the d.e. (5.1) as

$$\ddot{x} + \lambda^2[1 + \mu f(t,\mu)]x = 0 \tag{6.1}$$

where

$$f(t,\mu) = f_1(t) + \mu f_2(t) + \dots \tag{6.2}$$

and assume that for $\mu = 0$, a root becomes some integer n. As this root is an analytic function of μ, one can write

$$\lambda^2 = n^2 + \alpha_1\mu + \alpha_2\mu^2 + \dots \tag{6.3}$$

As was shown previously, (6.1) has a period of 2π if n is even and period π if n is odd. We inquire as to whether this solution is also analytic in μ.

As the coefficients in (6.1) are analytic in μ, any solution of this d.e. whose initial conditions do not depend on μ are also analytic in μ. If $x = x(t)$ is a solution corresponding to the initial conditions $x(0) = 1$, $\dot{x}(0) = 0$, the characteristic equation has a double root equal to $+1$ (n even) or -1 (n odd). Hence, solutions of (6.1) are either periodic or of the form $x(t) = t\varphi(t) + \psi(t)$, where φ and ψ are periodic. In such a case, $\varphi(t)$ determines the solution of (6.1) which is the periodic solution in question but, as the initial conditions do not depend on μ, this solution is analytic in μ as well as $\varphi(t)$.

If this solution is multiplied by $C(\mu)$, where $C(\mu)$ is an arbitrary function of μ (not necessarily analytic), one obtains another periodic solution. However, not every periodic solution is analytic in μ.

We consider an analytic periodic solution

$$x = x_0(t) + \mu x_1(t) + \mu^2 x_2(t) + \dots \tag{6.4}$$

where $x_i(t)$ are periodic and the series converges for μ sufficiently small. In order to guarantee this property, it is necessary to impose additional conditions determining a constant factor in this solution. If $x(0) \neq 0$, one can multiply it by a factor so as to have the initial condition of a value given in advance:

$$M = M_0 + \mu M_1 + \mu^2 M_2 + \dots \tag{6.5}$$

In this manner one obtains the condition $x_i(0) = M_i$, where M_i are constants satisfying the only condition that (6.5) converges. Thus one can select the first n constants M_i in an arbitrary manner.

If $x(0) = 0$, the derivative $\dot{x}_i(0) \neq 0$ and, again, one can impose the condition $\dot{x}_i(0) = N_i$, where N_i are arbitrary constants for which the series $N = N_0 + \mu N_1 + \mu^2 N_2 + \dots$ converges.

Hence, in one way or the other one can obtain as initial conditions either $x_i(0) = M_i$ or $\dot{x}_i(0) = N$ (we exclude the case when $x_0(0) = \dot{x}_0(0) = 0$). In each special case it is necessary to determine which of the two initial conditions, $x_i(0) = M_i$ or $\dot{x}_i(0) = N_i$, can be satisfied. If $x_0(0) \neq 0$, then the solution (6.4) does not vanish for $t = 0$ and one has $x_i(0) = M_i$. If $x_0(0) = 0$ but $\dot{x}_0(0) \neq 0$, then one can satisfy the condition: $\dot{x}_i(0) = N_i$. If, however, $x_0(0) = \dot{x}_0(0) = 0$, one has identically $x_0(t) = 0$. This case can be excluded, as it is always possible to divide (6.4) by a suitable power of μ so as to have a term independent of μ.

If one replaces the series (6.2) and (6.3) into (6.1), one has

$$\sum_{i=0}^{\infty} \ddot{x}_i \mu^i + \left(n^2 + \sum_{j=1}^{\infty} \alpha_j \mu^j\right)\left(1 + \sum_{j=1}^{\infty} f_j \mu^j\right) \sum_{i=0}^{\infty} x_i \mu^i = 0 \tag{6.6}$$

which yields a recursive system

$$\begin{aligned}
\ddot{x}_0 + n^2 x_0 &= 0 \\
\ddot{x}_1 + n^2 x_1 &= -n^2 f_1 x_0 - \alpha_1 x_0 \\
\ddot{x}_2 + n^2 x_2 &= -(n^2 f_1 + \alpha_1) x_1 - (n^2 f_2 + \alpha_1 f_1) x_0 - \alpha_2 x_0 \\
&\cdots\cdots\cdots\cdots\cdots\cdots \\
\ddot{x}_k + n^2 x_k &= -(n^2 f_1 + \alpha_1) x_{k-1} - \alpha_k x_0 + F_k(t, x_0, \dots, x_{k-2})
\end{aligned} \tag{6.7}$$

where F_k are linear functions of $x_0, \dots, x_{k-2}$ with periodic coefficients depending on $\alpha_1, \dots, \alpha_{k-1}$. This permits a recursive determination of x. For the first function x_0 one has

$$x_0 = A_0 \cos nt + B_0 \sin nt \tag{6.8}$$

In the second equation for x_1 the condition of periodicity is fulfilled if the Fourier development on the right-hand side has no terms with cos nt and sin nt; otherwise the secular terms due to the resonance solution would occur.

As f_1 is periodic with period π, its Fourier development proceeds with integer arguments of t (even, if n is even; odd, if n is odd). One has thus

$$\begin{aligned} n^2 f_1 \cos nt &= \sum_{m=1}^{\infty} (a_m \cos mt + b_m \sin mt) \\ n^2 f_1 \sin nt &= \sum_{m=1}^{\infty} (c_m \cos mt + d_m \sin mt) \end{aligned} \tag{6.9}$$

where

$$a_n = \beta/2; \qquad b_n = \gamma/2; \qquad c_n = \gamma/2; \qquad d_n = -\beta/2 \tag{6.10}$$

β and γ being the coefficients of cos $2nt$ and sin nt in the Fourier development of $n^2 f_1$. This development does not contain any constant term as the average value of $f(t)$ is zero.

From these expressions one obtains the coefficients of cos nt and sin nt on the right-hand side of equations for x_1, viz.:

$$(\beta + 2\alpha_1)A_0 + \gamma B_0 = 0; \qquad \gamma A_0 + (2\alpha_1 - \beta)B_0 = 0 \tag{6.11}$$

which must be satisfied by A_0 and B_0.

The condition for a nontrivial solution of (6.11) is that α_1 satisfy the quadratic equation

$$(2\alpha_1 + \beta)(2\alpha_1 - \beta) - \gamma^2 = 0 \tag{6.12}$$

whence

$$\alpha_1 = \neq \tfrac{1}{2}\sqrt{\beta^1 + \gamma^2} \tag{6.13}$$

Here there are obviously two cases according to $\beta^2 + \gamma^2 \neq 0$ or $= 0$. If $\beta^2 + \gamma^2 \neq 0$ one has two real roots for α_1. If one takes one of these roots, one of the quantities A_0 and B_0 can be taken arbitrarily.

Assume that A_0 is arbitrary; for this it is necessary that it be different from zero; this can be obtained if $\gamma \neq 0$; in this manner $x_0(0) \neq 0$ and one can, therefore, use the initial condition $x_i(0) = M_i$ for the calculation of $x_i(t)$. In particular, one can set $A_0 = 1$, then (6.11) permits determining B_0.

With such a choice of α_1, A_0, and B_0, the function $x_1(t)$ will be periodic. But in such a case the general solution

$$x_1 = x_1^*(t) + A_1 \cos nt + B_1 \sin nt$$

(where $x_1^*(t)$ is a particular solution) will be also periodic with two arbitrary constants A_1 and B_1. If one sets $A_1 = 0$, then $M_1 = x_1^*(0)$.

One proceeds in a similar manner for other approximations. Thus, for instance, for x_2 one equates to zero the coefficients of cos nt and sin nt; this results in equations

$$2A_0\alpha_2 + \gamma B_1 + 2p_2 = 0; \qquad 2B_0\alpha_2 + (2\alpha_1 - \beta)B_1 + 2q_2 = 0 \quad (6.14)$$

where p_2 and q_2 are coefficients of cos nt and sin nt of the development of $n^2f_1x_1^* + (\alpha_1 f_1 + n^2 f_2)x_0$.

We shall not pursue this calculation further but merely mention that, if one starts from a root of a quadratic equation in α_1 (equation (6.12)), one can obtain a formal expansion (6.3) for λ^2 for which (6.1) admits a periodic solution satisfying the prescribed initial conditions. To each root corresponds a formal expansion, and two periodic solutions result. On the other hand, to any integer n correspond two values of λ^2 which limit the corresponding region of stability. It follows, therefore, that the above-mentioned formal expansions for λ^2 represent precisely the limits of the zone of instability and are thus convergent.

There is still another case to be considered, when $\beta^2 + \gamma^2 = 0$; the only difference in this case is that the constants A_k and B_k are determined by the condition of periodicity of the $(k + 2)$ approximation and not by that of the $(k + 1)$ approximation. We shall not enter into further details of this case because the argument remains the same.

Malkin gives several examples illustrating the application of this method; we indicate here one relative to the parametric excitation of an electric circuit which we shall investigate in Chapter 20 by a different method. The d.e. in this case is

$$\ddot{x} + \lambda^2(1 + \mu \cos 2\tau)x = 0 \qquad (6.15)$$

where

$$\lambda^2 = \frac{4L - R^2C_0}{L^2C_0\omega^2}; \qquad \mu = \frac{4Lm}{4L^2 - R^2C_0}$$

This d.e. represents an oscillatory process in an electric circuit (with constants R, L, and C_0) with capacity C *modulated around its average value* C_0, the quantity μ being the *index of modulation*; it is assumed that $\mu \ll 1$.

We consider the zone of instability in the neighborhood of $\lambda = 2$ value and set

$$\lambda^2 = 4 + \alpha_1\mu + \alpha_2\mu^2 + \dots \qquad (6.16)$$

We shall try to satisfy (6.15) formally by a series solution

$$x = x_0 + \mu x_1 + \mu^2 x_2 + \dots \qquad (6.17)$$

with periodic coefficients (of period π); we have: $x_0 = A_0 \cos 2\tau + B_0 \sin 2\tau$ and a recursive sequence of d.e.

$$\ddot{x}_1 + 4x_1 = -4x_0 \cos 2\tau - \alpha_1 x_0; \qquad \ddot{x}_2 + 4x_2 = -4x_1 \cos 2\tau - \alpha_2 x_0$$

$$\ddot{x}_3 + 4x_3 = -4x_2 \cos 2\tau - \alpha_2 x_1 - \alpha_2 \cos 2\tau \cdot x_0 - \alpha_3 x_0 - \alpha_1 x_2 - \alpha_1 \cos 2\tau \cdot x_1$$

Equating to zero the coefficients of $\cos 2\tau$ and $\sin 2\tau$ in equation for x_1, one has: $\alpha_1 = 0$, A_0 and B_0 being arbitrary, which corresponds to the special case: $\beta^2 + \gamma^2 = 0$. With $\alpha_1 = 0$ we have the following approximation:

$$x_1 = -\frac{A_0}{2} + \frac{A_0}{6} \cos 4\tau + \frac{B_0}{6} \sin 4\tau + A_1 \cos 2\tau + B_1 \sin 2\tau \qquad (6.18)$$

where A_1 and B_1 are arbitrary. As A_0 and B_0 are still unknown, we go to the next approximation

$$\ddot{x}_2 + 4x_2 = (\tfrac{5}{3} - \alpha_2)A_0 \cos 2\tau - (\tfrac{1}{3} + \alpha_2)B_0 \sin 2\tau - 2A_1 \cos 4\tau - 2B_1 \sin 4\tau - \frac{A_0}{3} \cos 6\tau - \frac{B_0}{3} \sin 6\tau - 2A_1$$

The condition of periodicity results in $(\frac{5}{3} - \alpha_2)A_0 = 0$; $(\frac{1}{3} + \alpha_2)B_0 = 0$ which yields for α_2 two different values. In one case we have:

$$\alpha_2 = \tfrac{5}{3}; \qquad A_0 = 1; \qquad B_0 = 0$$

In the other case we have:

$$\alpha_2 = -\tfrac{1}{3}; \qquad A_0 = 0; \qquad B_0 = 1$$

In the first case: as $A_0 \neq 0$, one can set $A_1 = 0$ which gives

$$x_2 = \tfrac{1}{6}B_1 \sin 4\tau + \tfrac{1}{96} \cos 6\tau + B_2 \sin 2\tau \qquad (6.19)$$

B_2 being arbitrary. If one substitutes these approximations into x_3 and imposes again the conditions of periodicity, one has a system of equations for α_3 and B_1 which turn out to be: $B_1 = 0$; $\alpha_3 = 0$.

The following approximations are calculated in a similar way. In this particular case ($\gamma^2 + \beta^2 = 0$) the constants B_k appearing in x_k are determined not from the condition of periodicity for x_{k+1} but from those for x_{k+2}.

In the second case we have:

$$x_1 = \tfrac{1}{6} \sin 4\tau + A_1 \cos 2\tau;$$
$$x_2 = -\tfrac{1}{2}A_1 + \tfrac{1}{6}A_1 \cos 4\tau + \tfrac{1}{96} \sin 6\tau + A_2 \cos 2\tau \qquad (6.20)$$

The condition of periodicity for x_3 yields: $\alpha_3 = 0$, $A_1 = 0$. Thus, up to $0(\mu^3)$ the second region for ω (frequency of the capacity variation) for which (6.15) has unstable solutions is determined by inequalities

$$4 - \tfrac{1}{3}\mu^2 + \ldots \leq \lambda^2 \leq 4 + \tfrac{5}{3}\mu^2 + \ldots \tag{6.21}$$

From these inequalities it is seen that the necessary condition for the existence of this zone is

$$4L - R^2C_0 > 0 \tag{6.22}$$

In a similar way one can investigate the zones of stability around the value $\lambda = 3$. Setting $\lambda^2 = 9 + \alpha_1\mu + \alpha^2\mu^2 + \ldots$; $x_0 = A_0 \cos 3\tau - B_0 \sin 3\tau + \mu x_1 + \mu^2 x_2 + \ldots$, one obtains a recursive system

$$\begin{aligned}
\ddot{x}_1 + 9x_1 &= -9 \cos 2\tau \cdot x_0 - \alpha_1 x_0 \\
\ddot{x}_2 + 9x_2 &= -9 \cos 2\tau \cdot x_1 - \alpha_1(x_1 + \cos 2\tau \cdot x_0) - \alpha_2 x_0 \\
\ddot{x}_3 + 9x_3 &= -9 \cos 2\tau \cdot x_2 - \alpha_1(x_2 + \cos 2\tau \cdot x_1) \\
&\qquad - \alpha_2(x_1 + \cos 2\tau \cdot z_0) - \alpha_3 x_0 \\
&\ldots\ldots\ldots\ldots\ldots\ldots
\end{aligned} \tag{6.23}$$

The condition of periodicity for x_1 requires $\alpha_1 = 0$, A_0 and B_0 being arbitrary. For x_1 one has the expression

$$\begin{aligned}
x_1 = &-\tfrac{9}{16}A_0 \cos \tau - \tfrac{9}{16}B_0 \sin \tau \\
&+ \tfrac{9}{32}A_0 \cos 5\tau + \tfrac{9}{32}B_0 \sin 5\tau + A_1 \cos 3\tau + B_1 \sin 3\tau
\end{aligned}$$

where A_1 and B_1 are arbitrary.

Substituting α_1 and x_1 into the second equation (6.23), one has

$$\begin{aligned}
\ddot{x}_2 + 9x_2 = &\tfrac{9}{2}(\tfrac{9}{16}A_0 - A_1) \cos \tau - \tfrac{9}{2}(\tfrac{9}{16}B_0 + B_1) \sin \tau \\
&- \tfrac{9}{2}A_1 \cos 5\tau - \tfrac{9}{2}B_1 \sin 5\tau - \tfrac{81}{64}A_0 \cos 7\tau \\
&- \tfrac{81}{64} \sin 7\tau + (\tfrac{81}{64} - \alpha_2)A_0 \cos 3\tau + (\tfrac{81}{64} - \alpha_2)B_0 \sin 3\tau
\end{aligned} \tag{6.24}$$

which yields the conditions of periodicity for x_2:

$$(\tfrac{81}{64} - \alpha_2)A_0 = 0; \qquad (\tfrac{81}{64} - \alpha_2)B_0 = 0$$

thus

$$\alpha_2 = \tfrac{81}{64} \tag{6.25}$$

The expression for x_2 is

$$x_2 = \tfrac{9}{16}(\tfrac{9}{16}A_0 - A_1)\cos\tau - \tfrac{9}{16}(\tfrac{9}{16}B_0 + B_1)\sin\tau + \tfrac{9}{32}A_1\cos 5\tau + \tfrac{9}{32}B_1\sin 5\tau + \tfrac{81}{2560}A_0\cos 7\tau + \tfrac{81}{2560}B_0\sin 7\tau + A_2\cos 3\tau + B_2\sin 3\tau$$

where A_2 and B_2 are arbitrary.

Equation for x_3 after the substitution of values for x_1x_2, α_1, and α_2 is

$$\ddot{x}_3 + 9x_3 = -\tfrac{9}{2}(\tfrac{153}{512}A_0 - \tfrac{9}{16}A_1 + A_2)\cos\tau - \tfrac{9}{2}(\tfrac{153}{512}B_0 + \tfrac{9}{16}B_1 + B_2)\sin\tau - (\alpha_3 + \tfrac{729}{512})A_0\cos 3\tau + (\tfrac{729}{512} - \alpha_3)B_0\sin 3\tau - \tfrac{5103}{10240}A_0\cos 5\tau - \tfrac{5103}{10240}B_0\sin 5\tau - \tfrac{9}{2}A_2\cos 5\tau - \tfrac{9}{2}B_2\sin 5\tau + a_7\cos 7\tau + b_7\sin 7\tau + a_9\cos 9\tau + b_9\sin 9$$

where a_7, a_9, b_7, b_9 are definite coefficients which will not be needed. The conditions of periodicity are

$$(\alpha_3 + \tfrac{729}{512})A_0 = 0; \qquad (\alpha_3 - \tfrac{729}{512})B_0 = 0 \tag{6.26}$$

One has thus two branches of the periodic solution. For the first branch:

$$\alpha_3 = -\tfrac{729}{513}; \qquad B_0 = 0; \qquad A_0 = 1$$

One can now set $A_1 = A_2 = 0$ and omit the terms with $\cos 3\tau$. This yields the terms x_3

$$x_3 = -\tfrac{1377}{8192}\cos\tau\,\tfrac{9}{16}(\tfrac{9}{16}B_1 + B_2)\sin\tau + \tfrac{5103}{163840}\cos 5\tau + \tfrac{9}{32}B_2\sin 5\tau - (a_7/40)\cos 7\tau - (b_7/40)\sin 7\tau - (a_9/72)\cos 9\tau - (b_9/72)\sin 9\tau + B\sin 3\tau$$

B_3 being arbitrary. If one substitutes these approximations into x_4 and expresses the conditions of periodicity, one determines B_1 and α_4, viz.:

$$B_1 = 0; \qquad \alpha_4 = -\tfrac{235467}{327680}$$

For the second branch one has

$$\alpha_3 = \tfrac{729}{519}; \qquad B_0 = 1; \qquad A_0 = 0; \qquad A_1 = 0; \qquad \alpha_4 = \tfrac{260953}{327680}$$

The region of instability (that is, of the parametric excitation) is then:

$$9 + \tfrac{81}{64}\mu^2 - \tfrac{729}{512}\mu^3 - \tfrac{235467}{327680}\mu^4 + \cdots \leq \lambda^2 \leq 9 + \tfrac{81}{64}\mu^2 + \tfrac{729}{512}\mu^3 + \tfrac{260953}{327680}\mu^4 + \cdots$$

It is to be noted that the above calculation *relates only to a linear d.e. of Mathieu.*

Chapter 14

ASYMPTOTIC METHODS OF KRYLOV-BOGOLIUBOV-MITROPOLSKY

Autonomous systems

1. Introductory remarks

Independently of the methods of approximations based directly on the theory of Poincaré (chapters 10 and 11) and introduced in the theory of oscillations by Andronov and his school, there appeared another method due originally to N. Krylov and N. Bogoliubov (K.B. for short) which treats the same subject, that is, the existence and stability of periodic solutions of nearly linear d.e.

The first publication of these authors, which appeared in 1937, was used considerably in applied problems. Quite recently (1955) a treatise by N. Bogoliubov and J. Mitropolsky[1] extended this method, particularly its mathematical foundations. The authors call this method *asymptotic* in the sense of $\mu \to 0$. It should be noted that the term "asymptotic" in the theory of oscillations is frequently used also in the sense of $\mu \to \infty$, in which case the mathematical approach is entirely different, as will be seen in Part IV.

2. Successive approximations for the autonomous systems

We consider a nearly linear d.e. of the form

$$\ddot{x} + \omega^2 x = \mu f(x, \dot{x}) \tag{2.1}$$

If $\mu = 0$, one has the d.e. of the harmonic oscillator whose solution is

$$x = a \cos \psi; \qquad \psi = \omega t + \theta \tag{2.2}$$

[1] N. N. Bogoliubov and J. A. Mitropolsky, *Asymptotic Methods in the Theory of Nonlinear Oscillations* (in Russian), Moscow, 1958.

The amplitude a is constant and *the total phase* $\psi = \omega t + \theta$ increases monotonically with t. The term $\mu f(x,\dot{x})$ perturbs this simple situation as we saw previously. Physically that means that the fact that $\mu \neq 0$ accounts for certain absorptions and generations of energy which did not exist when $\mu = 0$.

On a purely formal basis, which will be justified later, one can assume that for $\mu \neq 0$ (but small), we have a relation

$$x = a \cos \psi + \mu u^{(1)}(a,\psi) + \mu^2 u^{(2)}(a,\psi) + \dots \tag{2.3}$$

where $u^{(1)}, u^{(2)},\dots,$ are certain periodic functions of ψ with period 2π. As to a and ψ themselves, we shall try to determine them from the equations

$$\begin{aligned} \dot{a} &= \mu A^{(1)}(a) + \mu^2 A^{(2)}(a) + \dots \\ \dot{\psi} &= \omega + \mu B^{(1)}(a) + \mu^2 B^{(2)}(a) + \dots \end{aligned} \tag{2.4}$$

The problem is to determine the functions $u^{(i)}$, $A^{(i)}$, and $B^{(i)}$ in such a manner that their substitution into (2.1) satisfies this d.e. with prescribed accuracy. As far as the procedure is concerned, we shall encounter again a recursive system as in the method of Poincaré, but the approach is different. In fact, the aim of the method is to determine certain periodic functions $u^{(i)}(a,\psi)$ which give the solution (2.3) under the conditions (2.4). These conditions have the form of d.e. in which there is only one dependent variable a (and not ψ). We shall omit for the moment the question of the physical significance of these assumptions; it will be clearer in the following chapter concerning the application of the same method to nonautonomous systems.

In the first place one must note that there is a certain arbitrariness in selecting the functions $u^{(i)}$. Suppose that one starts with some arbitrary functions $\alpha_1(a), \alpha_2(a),\dots, \beta_1(a), \beta_2(a),\dots,$ respectively, for $A^{(i)}$ and $B^{(i)}$ and replaces in (2.3) and (2.4) a and ψ by expressions

$$a = b + \mu\alpha_1(b) + \mu^2\alpha_2(b) + \dots; \qquad \psi = \varphi + \mu\beta_1(b) + \mu^2\beta_2(b) + \dots$$

One obtains then, instead of (2.3), similar equations but with different coefficients. It is clear that one has to impose some additional conditions in order to remove this arbitrariness at the start.

As one such condition one could require, for instance, that there be no first harmonic in $u^i(a,\psi)$; to some extent this is an intuitive approach by which one wishes to have the solution in the form of a Fourier series.

This assumption results in the conditions

$$\int_0^{2\pi} u^{(i)}(a,\psi) \cos \psi d\psi = 0; \quad \int_0^{2\pi} u^{(i)}(a,\psi) \sin \psi d\psi = 0, \quad i = 1, 2,\dots \tag{2.5}$$

In the following calculations we designate the partial differentiations by subscripts; thus $u_a^{(1)} = \partial u^{(1)}/\partial a$, etc.; we use dots for differentiations with respect to t; that is, $\dot{a} = da/dt$, etc. One has thus

$$\dot{x} = \dot{a}[\cos\psi + \mu u_a^{(1)} + \mu^2 u_a^{(2)} + \ldots] + \dot{\psi}[-a\sin\psi + \mu u_\psi^{(1)} + \mu^2 u_\psi^{(2)} + \ldots]$$

$$\begin{aligned}\ddot{x} = \ddot{a}[\cos\psi + \mu u_a^{(1)} + \mu^2 u_a^{(2)} + \ldots]\\ + \ddot{\psi}[-a\sin\psi + \mu u_\psi^{(1)} + \mu^2 u_\psi^{(2)} + \ldots]\\ + \dot{a}^2[\mu u_{aa}^{(1)} + \mu^2 u_{aa}^{(2)} + \ldots]\\ + 2\dot{a}\dot{\psi}[-\sin\psi + \mu u_{a\psi}^{(1)} + \mu^2 u_{a\psi}^{(2)} + \ldots]\\ + \dot{\psi}^2[-a\cos\psi + \mu u_{\psi\psi}^{(1)} + \mu^2 u_{\psi\psi}^{(2)} + \ldots]\end{aligned} \tag{2.6}$$

From (2.4) one obtains also expressions

$$\begin{aligned}\ddot{a} &= (\mu A_a^{(1)} + \mu^2 A_a^{(2)} + \ldots)(\mu A^{(1)} + \mu^2 A^{(2)} + \ldots)\\ &= \mu^2 A^{(1)} A_a^{(1)} + \mu^3 \ldots\\ \ddot{\psi} &= (\mu B_a^{(1)} + \mu^2 B_a^{(2)} + \ldots)(\mu A^{(1)} + \mu^2 A^{(2)} + \ldots)\\ &= \mu^2 A^{(1)} B_a^{(1)} + \mu^3 \ldots\end{aligned} \tag{2.7}$$

and likewise from (2.4) for $\dot{a}^2$, $\dot{a}\dot{\psi}$, and $\dot{\psi}^2$.

Substituting these various expressions into (2.6) and then $\dot{x}$, $\ddot{x}$ into the left-hand side of (2.1), it becomes

$$\begin{aligned}\ddot{x} + \omega^2 x = \mu[-2\omega A^{(1)}\sin\psi - 2\omega a B^{(1)}\cos\psi + \omega^2 u_{\psi\psi}^{(1)} + \omega^2 u^{(1)}]\\ + \mu^2[(A^{(1)}A_a^{(1)} - aB^{(1)2} - 2\omega a B^{(2)})\cos\psi\\ - (2\omega A^{(2)} + 2A^{(1)}B^{(1)} + A^{(1)}B_a^{(1)}a)\sin\psi + 2\omega A^{(1)}u_{a\psi}^{(1)}\\ + 2\omega B^{(1)}u_{\psi\psi}^{(1)} + \omega^2 u_{\psi\psi}^{(2)} + \omega^2 u^{(2)}] + \mu^3[\ \ ] + \ldots\end{aligned} \tag{2.8}$$

As to the right side of (2.1), it is:

$$\begin{aligned}\mu f(x,\dot{x}) = \mu f(a\cos\psi, -a\omega\sin\psi) + \mu^2[u^{(1)}f_x(a\cos\psi, -a\omega\sin\psi)\\ + (A^{(1)}\cos\psi - aB^{(1)}\sin\psi + \omega u_\psi^{(1)})\\ \times f_{\dot{x}}(a\cos\psi, -a\omega\sin\psi)] + \mu^3[\ \ ] + \ldots\end{aligned} \tag{2.9}$$

If one equates the like powers of μ on both sides, one obtains a recursive system

$$\begin{aligned}
\omega^2(u_{\psi\psi}^{(1)} + u^{(1)}) &= f^{(0)}(a,\psi) + 2\omega A^{(1)} \sin \psi + 2\omega a B^{(1)} \cos \psi \\
\omega^2(u_{\psi\psi}^{(2)} + u^{(2)}) &= f^{(1)}(a,\psi) + 2\omega A^{(2)} \sin \psi + 2\omega a B^{(2)} \cos \psi \\
&\cdots\cdots\cdots\cdots\cdots \\
\omega^2(u_{\psi\psi}^{(m)} + u^{(m)}) &= f^{(m-1)}(a,\psi) + 2\omega A^{(m)} \sin \psi + 2\omega a B^{(m)} \cos \psi
\end{aligned} \tag{2.10}$$

where

$$\begin{aligned}
f^{(0)}(a,\psi) &= f(a \cos \psi, -a\omega \sin \psi) \\
f^{(1)}(a,\psi) &= u^{(1)} f_x(a \cos \psi, -a\omega \sin \psi) + [A^{(1)} \cos \psi - aB^{(1)} \sin \psi + \omega u_{\psi}^{(1)}] \\
&\quad \times f_{\dot{x}}(a \cos \psi, -a\omega \sin \psi) + (aB^{(1)2} - A^{(1)} A_a^{(1)}) \cos \psi \\
&\quad + (2A^{(1)} B^{(1)} + A^{(1)} B_a^{(1)} \cdot a) \sin \psi - 2\omega A^{(1)} u_{a\psi}^{(1)} \\
&\quad - 2\omega B^{(1)} u_{\psi\psi}^{(2)}
\end{aligned} \tag{2.11}$$

It is clear that $f^{(k)}(a,\psi)$ is a periodic function of ψ with period 2π depending also on the amplitude a; its explicit expression is determined as soon as one determines $A^{(i)}(a)$, $B^{(i)}(a)$, $u^{(i)}(a,\psi)$ to the kth order.

Consider first the functions $f^{(0)}(a,\psi)$ and $u^{(1)}(a,\psi)$; their Fourier developments are

$$\begin{aligned}
f^{(0)}(a,\psi) &= g_0(a) + \sum_{n=1}^{\infty} [g_n(a) \cos n\psi + h_n(a) \sin n\psi] \\
u^{(1)}(a,\psi) &= v_0(a) + \sum_{n=1}^{\infty} [v_n(a) \cos n\psi + w_n(a) \sin n\psi]
\end{aligned} \tag{2.12}$$

Substituting these values into the first equation (2.10), one has

$$\begin{aligned}
\omega^2 v_0(a) &+ \sum_{n=1}^{\infty} \omega^2(1 - n^2)[v_n(a) \cos n\psi + w_n(a) \sin n\psi] \\
&= g_0(a) + [g_1(a) + 2\omega a B^{(1)}] \cos \psi + [h_1(a) + 2\omega A^{(1)}] \sin \psi \\
&\quad + \sum_{h=2}^{\infty} [g_n(a) \cos n\psi + h_n(a) \sin n\psi]
\end{aligned}$$

Equating coefficients of harmonics of the same order, one has

$$g_1(a) + 2\omega a B^{(1)} = 0; \qquad h_1(a) + 2\omega A^{(1)} = 0; \qquad v_0(a) = \frac{g_0(a)}{\omega^2};$$

$$v_n(a) = \frac{g_n(a)}{\omega^2(1 - n^2)}; \qquad w_n(a) = \frac{h_n(a)}{\omega^2(1 - n^2)}; \qquad n = 2, 3, \ldots \tag{2.13}$$

One has thus $A^{(1)}(a)$, $B^{(1)}(a)$ and all harmonics of the development of $u^{(1)}(a,\psi)$ except the first two: $v_1(a)$ and $w_1(a)$. However, in view of the requirement that none of $u^{(i)}(a,\psi)$ should have a fundamental harmonic, one has: $v_1(a) = 0$; $w_1(a) = 0$, so that

$$u^{(1)}(a,\psi) = \frac{g_0(a)}{\omega^2} + \frac{1}{\omega^2} \sum_{n=2}^{\infty} \frac{g_n(a) \cos n\psi + h_n(a) \sin n\psi}{1 - n^2} \tag{2.14}$$

Since $u^{(1)}(a,\psi)$, $A^{(1)}(a)$, and $B^{(1)}(a)$ are determined, one knows also $f^{(1)}(a,\psi)$; its Fourier development is

$$f^{(1)}(a,\psi) = g_0^{(1)}(a) + \sum_{n=1}^{\infty} [g_n^{(1)}(a) \cos n\psi + h_n^{(1)}(a) \sin n\psi]$$

Using the second equation (2.10) and conditions (2.5), one has also

$$\begin{gathered} g_1^{(1)}(a) + 2\omega a B^{(2)} = 0; \qquad h_1^{(1)}(a) + 2\omega A^{(2)} = 0; \\ u_2(a,\psi) = \frac{g_0^{(1)}(a)}{\omega^2} + \frac{1}{\omega^2} \sum_{n=2}^{\infty} \frac{g_n^{(1)}(a) \cos n\psi + h_n^{(1)}(a) \sin n\psi}{1 - n^2} \end{gathered} \tag{2.15}$$

The determination of the approximations of higher orders is thus sufficiently clear.

Summing up, the conditions (2.5) eliminate the fundamental harmonic in the functions $u^{(i)}(a,\psi)$ and this, in turn, guarantees the absence of secular terms in all successive approximations.

This new derivation due to N. Bogoliubov represents a considerable improvement as compared to the early K.B. derivation in which the first approximation was established by a direct argument and the higher-order approximations were introduced owing to an additional procedure resembling the Lindstedt method. Here all subsequent approximations develop recursively from a sequence of equations (2.10) obtained directly.

The apparent complexity of these calculations results from somewhat complicated substitution of the series development and subsequent differentiations but, once all this is completed, the rest amounts merely to the discussion of the first, and possibly, the second approximation and these are the only cases which are encountered in applied problems.

3. Differential equations of the first approximation

Consider the first approximation

$$x = a \cos \psi + \mu u^{(1)}(a,\psi) \tag{3.1}$$

$$\dot{a} = \mu A^{(1)}(a); \qquad \dot{\psi} = \omega + \mu B^{(1)}(a) \tag{3.2}$$

Let $\bar{A}^{(1)}$ and $\bar{B}^{(1)}$ be average values of $A^{(1)}(a)$ and $B^{(1)}(a)$ in an interval $(0,t)$, that is:

$$\Delta a = a(t) - a(0) \simeq \mu t\bar{A}^{(1)}; \quad \Delta(\psi - \omega t) = [\psi(t) - \omega t] - \psi(0) \simeq \mu t\bar{B}^{(1)}$$

It is seen that the time t during which a and $\psi = \omega t$ may acquire finite changes is of the order $1/\mu$. On the other hand, the d.e. (3.2) of the first approximation results from neglecting terms with μ^2 (in equation 2.4). An error of this order in $\dot{a}$ and $\dot{\psi}$ during t results in the error of order $\mu^2 t$ in the functions a and ψ. Hence, in the interval during which a and ψ undergo finite changes, errors in these quantities are of the order μ. Thus there is no necessity for keeping the term $\mu u^{(1)}(a,\psi)$ in (3,1) inasmuch as the error of using $x = a \cos \psi$ is of the same order. It is simpler to take $x = a \cos \psi$.

For the stationary state $\dot{a} = 0$, that is, $\mu A^{(1)}(a) + \mu^2 A^{(2)}(a) + \ldots = 0$; hence, if one neglects terms beginning with μ^2, the condition of the stationary state is

$$A^{(1)}(a) = 0 \tag{3.3}$$

Clearly the error in determining the stationary amplitude $a = a_0$ is of the *first* order of smallness.

Summing up, for the first approximation one takes $x = a \cos \psi$, where a and ψ are given by $\dot{a} = \mu A^{(1)}(a)$; $\dot{\psi} = \omega + \mu B^{(1)}(a)$. In a similar manner, for the second approximation one has $x = a \cos \psi + \mu u^{(1)}(a,\psi)$, where a and ψ are given by the d.e.

$$\dot{a} = \mu A^{(1)}(a) + \mu^2 A^{(2)}(a); \qquad \dot{\psi} = \omega + \mu B^{(1)}(a) + \mu^2 B^{(2)}(a) \tag{3.4}$$

From the general formulas of the preceding section, one has

$$\begin{aligned} A^{(1)}(a) &= -\frac{1}{2\pi\omega}\int_0^{2\pi} f(a \cos \psi, -a\omega \sin \psi) \sin \psi d\psi \\ B^{(1)}(a) &= -\frac{1}{2\pi a\omega}\int_0^{2\pi} f(a \cos \psi, -a\omega \sin \psi) \cos \psi d\psi \end{aligned} \tag{3.5}$$

$$u^{(1)}(a,\psi) = \frac{g_0(a)}{\omega^2} - \frac{1}{\omega^2}\sum_{n=2}^{\infty} \frac{g_n(a) \cos n\psi + h_n(a) \sin n\psi}{n^2 - 1} \tag{3.6}$$

where

$$\begin{aligned} g_n(a) &= \frac{1}{2\pi}\int_0^{2\pi} f(a \cos \psi, -a\omega \sin \psi) \cos n\psi d\psi \\ h_n(a) &= \frac{1}{2\pi}\int_0^{2\pi} f(a \cos \psi, -a\omega \sin \psi) \sin n\psi d\psi \end{aligned} \tag{3.7}$$

and analogous formulas for $A^{(2)}(a)$ and $B^{(2)}(a)$ which follow from the preceding section.

We return to a more detailed study of equations of the first approximation:

$$x = a \cos \psi; \qquad \dot{a} = \mu A^{(1)}(a); \qquad \dot{\psi} = \omega + \mu B^{(1)}(a) = \omega_1(a) \tag{3.8}$$

with expressions (3.5) for $A^{(1)}$ and $B^{(1)}$ this yields

$$\begin{aligned} \mu A^{(1)}(a) &= -\frac{\mu}{2\pi\omega} \int_0^{2\pi} f(a \cos \psi, \, -a\omega \sin \psi) \sin \psi d\psi \\ \omega_1(a) &= \omega - \frac{\mu}{2\pi a\omega} \int_0^{2\pi} f(a \cos \psi, \, -a\omega \sin \psi) \cos \psi d\psi \end{aligned} \tag{3.9}$$

It is to be noted that this form of equation of the first approximation was obtained earlier by van der Pol by using the solution of the form $x = M \cos t + N \cos t$ and by substituting this solution into the nearly linear d.e. Here it appears as the first approximation in the general method of the preceding section.

If one squares the second equation (3.9) and neglects the term with μ^2, one has

$$\omega_1^2(a) = \frac{1}{\pi a} \int_0^{2\pi} [\omega^2 a \cos \psi - \mu f(a \cos \psi, \, -a\omega \sin \psi)] \cos \psi d\psi \tag{3.10}$$

Setting: $F(x,\dot{x}) = \omega^2 x - \mu f(x,\dot{x})$, (3.10) can be written as

$$\omega_1^2(a) = \frac{1}{\pi a} \int_0^{2\pi} F(a \cos \psi, \, -a\omega \sin \psi) \cos \psi d\psi \tag{3.11}$$

Since $\int_0^{2\pi} \omega^2 a \cos \psi \sin \psi d\psi = 0$, one can also write

$$A^{(1)}(a) = -\frac{1}{2\pi\omega} \int_0^{2\pi} F(a \cos \psi, \, -a\omega \sin \psi) \sin \psi d\psi \tag{3.12}$$

Equations (3.11) and (3.12) give $\omega_1(a)$ and $A^{(1)}(a)$ in terms of $F(x,\dot{x})$ avoiding somewhat unsymmetrical formulas (3.9).

The preceding theory of the first approximation follows from the general theory. It is possible, however, to derive the d.e. of the first approximation in a different way as was done originally (1937) by Krylov and Bogoliubov. In fact, suppose we start with the solution $x = a \cos \psi$, where $a = a(t)$ and $\psi = \omega t + \theta(t)$ and impose the condition that $\dot{x}$ should have the form: $\dot{x} = -a\omega \sin \psi$. Proceeding as previously we have

$$\dot{x} = \dot{a} \cos \psi - a\dot{\theta} \sin \psi - a\omega \sin \psi \tag{3.13}$$

In view of the form which we wish to have for $\dot{x} = -a\omega \sin \psi$, one has an additional condition

$$\dot{a} \cos \psi - a\dot{\theta} \sin \psi = 0 \tag{3.14}$$

If one differentiates now $\dot{x} = -a\omega \sin \psi$, considering a and θ as variables,

$$\ddot{x} = -\dot{a}\omega \sin \psi - a\omega\dot{\theta} \cos \psi - a\omega^2 \cos \psi \tag{3.15}$$

If one substitutes $\ddot{x}$, $\dot{x}$, and x into the nearly linear d.e., one gets

$$-\omega\dot{a} \sin \psi - a\omega\dot{\theta} \cos \psi = \mu f(a \cos \psi, -a\omega \sin \psi) \tag{3.16}$$

The system (3.14) and (3.16) can be solved for $\dot{a}$ and $\dot{\theta}$ and one has

$$\begin{aligned} \dot{a} &= -\frac{\mu}{\omega} f(a \cos \psi, -a\omega \sin \psi) \sin \psi \\ \dot{\theta} &= -\frac{\mu}{a\omega} f(a \cos \psi, -a\omega \sin \psi) \cos \psi \end{aligned} \tag{3.17}$$

As $\dot{a}$ and $\dot{\theta}$ are small in view of the small factor μ, a and θ are slowly varying functions of time.

The right-hand side of (3.17) can be represented by trigonometric series:

$$\begin{aligned} -\frac{\mu}{\omega} f(a \cos \psi, -a\omega \sin \psi) \sin \psi &= \mu \sum_k [f_{k1}^{(1)}(a) \cos k\psi + f_{k2}^{(1)}(a) \sin k\psi] \\ -\frac{\mu}{a\omega} f(a \cos \psi, -a\omega \sin \psi) \cos \psi &= \mu \sum_k [f_{k1}^{(2)}(a) \cos k\psi + f_{k2}^{(2)}(a) \sin k\psi] \end{aligned} \tag{3.18}$$

We can consider a and θ as consisting of slowly varying components and of small rapidly oscillating terms; call the first $\bar{a}$ and $\bar{\theta}$ and assume in the first approximation $a = \bar{a}$, $\theta = \bar{\theta}$; hence $\bar{\psi} = \omega t + \bar{\theta}$. Then:

$$\begin{aligned} \dot{\bar{a}} &= -\frac{\mu}{\omega} f(\bar{a} \cos \bar{\psi}, -\bar{a}\omega \sin \bar{\psi}) \sin \bar{\psi} \\ &= \mu \sum_{k=0}^{\infty} [f_{k1}^{(1)}(\bar{a}) \cos k\bar{\psi} + f_{k2}^{(1)}(\bar{a}) \sin k\bar{\psi}] \\ \dot{\bar{\theta}} &= -\frac{\mu}{\bar{a}\omega} f(\bar{a} \cos \psi, -\bar{a}\omega \sin \bar{\psi}) \cos \bar{\psi} \\ &= \mu \sum_{k=0}^{\infty} [f_{k1}^{(2)}(\bar{a}) \cos k\bar{\psi} + f_{k2}^{(2)}(\bar{a}) \sin k\bar{\psi}] \end{aligned} \tag{3.19}$$

otherwise

$$\begin{aligned} \dot{\bar{a}} &= \mu f_{01}^{(1)}(\bar{a}) + \text{small rapidly oscillating terms} \\ \dot{\bar{\theta}} &= \mu f_{01}^{(2)}(\bar{a}) + \text{small rapidly oscillating terms} \end{aligned} \tag{3.20}$$

On physical grounds one can assume that the slow process is not influenced by small rapid oscillations, and we can consider the d.e. without these rapid oscillations.

In fact, carrying out the averaging and introducing ψ, one obtains

$$\begin{aligned} \dot{a} &= -\frac{\mu}{2\pi\omega}\int_0^{2\pi} f(a\cos\psi,\ -a\omega\sin\psi)\sin\psi d\psi \\ \dot{\psi} &= \omega - \frac{\mu}{2\pi a\omega}\int_0^{2\pi} f(a\cos\psi,\ -a\omega\sin\psi)\cos\psi d\psi \end{aligned} \tag{3.21}$$

These are the d.e. of the first approximation in the form in which they were originally obtained by Krylov and Bogoliubov and in which they are generally used in applications.

One can improve the accuracy of the first approximation by taking into account the small oscillating terms just mentioned. These terms are $\mu f_{k1}^{(i)}(\bar{a})\cos k\bar{\psi}$ and $\mu f_{k2}^{(i)}(\bar{a})\sin k\bar{\psi}$. This results in more accurate values for a and θ, viz.:

$$\begin{aligned} a &= \bar{a} + \mu\sum_{k\neq 0}\frac{1}{k}[f_{k1}^{(1)}(\bar{a})\sin k\bar{\psi} + f_{k2}^{(1)}(\bar{a})\cos k\bar{\psi}] \\ \theta &= \bar{\theta} + \mu\sum_{k\neq 0}\frac{1}{k}[f_{k1}^{(2)}(\bar{a})\sin k\bar{\psi} + f_{k2}^{(2)}(\bar{a})\cos k\bar{\psi}] \end{aligned} \tag{3.22}$$

This may be regarded as an *improved first approximation*. Of one substitutes these expressions into the formulas of the first approximation:

$$x = a\cos\psi; \qquad \dot{x} = -a\omega\sin\psi$$

one obtains the explicit form of the first approximation

$$x = a\cos\psi + \mu u^{(1)}(a,\psi)$$

For the second approximation determining a and θ with the accuracy of the order μ^2 (inclusive), one has to replace (3.22) into the right-hand side of the d.e. and average out with respect to time appearing explicitly. One can apply this method to conservative systems (for example, frictionless pendulum) of the form: $m\ddot{x} + p(x) = 0$, where $p(x)$ is a nearly linear function of the form

$$p(x) = kx + \mu\Phi(x)$$

Using the preceding notations, we have $\omega^2 = k/m$; $f(x,\dot{x}) = -\Phi(x)/m$ and, for the first approximation, one has to expand $\Phi(a \cos \psi)$ into a Fourier series which has here only cosine terms, that is

$$\Phi(a \cos \psi) = \sum_{n=0}^{\infty} C_n(a) \cos n\psi$$

From the general formula one has

$$g_n(a) = -C_n(a)/m; \qquad h_n(a) = 0$$

and

$$A^{(1)}(a) = 0; \qquad B^{(1)}(a) = \frac{1}{2\omega m a} C_1(a)$$

One has for the first approximation: $x_1 = a \cos \psi$

$$\dot{a} = 0; \qquad \dot{\psi} = \omega + \frac{\mu C_1(a)}{2\omega m a} = \omega_1(a)$$

This shows that $a = a_0 = $ const and $\psi = \omega_1(a_0)t + \theta$.

In the first approximation, the oscillation is still harmonic but the nonlinearity is felt in that now that frequency depends on the amplitude. For approximations of higher order, harmonics appear and the dependence of frequency on the amplitude becomes more complicated. We shall not go into a detailed investigation of this case but refer to the treatise of Bogoliubov and Mitropolsky where the reader can find numerous examples of these calculations.

4. Nonlinear damping

We consider a d.e. of the form

$$m\ddot{x} + kx = \mu F(\dot{x}) \tag{4.1}$$

where $F(\dot{x})$ may be regarded as characterizing a nonlinear friction term. We have to use the general formulas of Section 2.

In the first place we have

$$\frac{1}{m} F(-a\omega \sin \psi) = \sum_{n=0}^{\infty} F_n(a\omega) \cos n\left(\psi + \frac{\pi}{2}\right)$$

Comparing this with the general form of expansions

$$\begin{aligned} f^{(0)}(a,\psi) &= g_0(a) + \sum_{n=1}^{\infty} [g_n(a) \cos n\psi + h_n(a) \sin n\psi] \\ u^{(1)}(a,\psi) &= v_0(a) + \sum_{n=1}^{\infty} [v_n(a) \cos n\psi + w_n(a) \sin n\psi] \end{aligned} \tag{4.2}$$

one has:

$$g_n(a) = F_n(a\omega)\cos(n\pi/2); \qquad h_n(a) = -F_n(a\omega)\sin(n\pi/2) \tag{4.3}$$

whence

$$A^{(1)}(a) = \frac{1}{2\omega}F_1(a\omega); \qquad B^{(1)}(a) = 0 \tag{4.4}$$

Taking into account relations, $x = a\cos\psi$; $\dot{a} = \mu A^{(1)}(a)$; $\dot{\psi} = \omega + \mu B^{(1)}(a)$, we have the first approximation:

$$x = a\cos\psi; \qquad \dot{a} = \frac{\mu}{2\omega}F_1(a\omega); \qquad \dot{\psi} = \omega; \qquad \omega = \sqrt{k/m} \tag{4.5}$$

It is seen that the frequency remains the same as in the linear case but the amplitude decreases.

For the second approximation one has

$$u^{(1)}(a,\psi) = -\frac{1}{\omega^2}\sum_{\substack{n\neq 1\\ n=0}}^{\infty}\frac{F_n(a\omega)\cos n(\psi + \pi/2)}{n^2 - 1} \tag{4.6}$$

Moreover, for $A^{(2)}(a)$ and $B^{(2)}(a)$ one has the following expressions (omitting some intermediate calculations)

$$A^{(2)}(a) = 0;$$

$$B^{(2)}(a) = \frac{F_1(a\omega)}{8\omega^3 a}\frac{dF_1(a\omega)}{da} - \frac{F_1{}^2(a\omega)}{4\omega^3 a^2} - \frac{1}{2\omega^3 a^2}\sum_{n=2}^{\infty}\frac{n^2 F_n{}^2(a\omega)}{n^2 - 1}$$

The second approximation has then the form

$$x = a\cos\psi - \frac{\mu}{\omega^2}\sum_{\substack{n\neq 1\\ n=0}}^{\infty}\frac{F_n(a\omega)\cos n(\psi + \pi/2)}{n^2 - 1} \tag{4.7}$$

$$\dot{a} = \mu F_1(a\omega)/2\omega; \qquad \dot{\psi} = \omega + \mu^2 B^{(2)}(a) \tag{4.8}$$

Thus, for instance, for $\ddot{x} + \lambda\dot{x} + \omega^2 x = 0$, $\lambda < 2\omega$ one has $F(\dot{x}) = -k\dot{x}$; $\lambda = \mu k$; $F(\dot{x}) = -k\dot{x}$; $F_1(a\omega) = -k\omega a$; $F_n(a\omega) = 0$ so that the second approximation is

$$x = a\cos\psi; \qquad \dot{a} = -\lambda a/2; \qquad \dot{\psi} = \omega\left[1 - \frac{1}{8}\left(\frac{\lambda}{\omega}\right)^2\right]$$

For the amplitude this approximation gives the exact value $a = a_0 e^{-\lambda t/2}$ and for the frequency it gives a correction corresponding to two first terms of the expansion of the exact formula (which is known here)

$$\omega\sqrt{1 - \frac{1}{4}\left(\frac{\lambda}{\omega}\right)^2} = \omega\left[1 - \frac{1}{8}\left(\frac{\lambda}{\omega}\right)^2 - \frac{1}{128}\left(\frac{\lambda}{\omega}\right)^4 + \ldots\right]$$

which is to be expected since we do not take here any terms beyond the second order.

Consider a d.e. of the form:

$$\ddot{x} \pm \alpha\dot{x}^2 + \omega^2 x = 0; \qquad \alpha = \mu k \tag{4.9}$$

where the sign plus must be taken for $\dot{x} > 0$, and minus for $\dot{x} < 0$. In this case $F(\dot{x}) = -k\dot{x}|\dot{x}|$. As previously one has to find nth term in the expansion $F(a \cos \psi)$ which yields relations

$$F_0(a) = F_2(a) = \ldots = F_{2q}(a) = 0 \qquad q = 0, 1, 2, \ldots$$

$$F_1(a) = -\frac{8ka^2}{3\pi}; \ldots; \quad F_{2q+1}(a) = \frac{8ka^2(-1)^{q+1}}{\pi(2q+1)[(2q+1)^2 - 4]}$$

The second approximation is

$$x = a \cos \psi - \frac{8\alpha a^2}{\pi} \sum_{q=1}^{\infty} \frac{\sin (2q+1)\psi}{(2q+1)[(2q+1)^2 - 1][(2q+1)^2 - 4]}$$

$$= a \cos \psi - \frac{\alpha a^2}{15\pi}\left(\sin 3\psi + \frac{1}{21} \sin 5\psi + \ldots\right)$$

with a and ψ given by

$$\dot{a} = -\frac{4\alpha\omega}{3\pi} a^2; \qquad \dot{\psi} = \omega\left[1 - \frac{4\alpha^2 a^2}{\pi^2} C\right] \tag{4.10}$$

and

$$C = 8 \sum_{q=1}^{\infty} \frac{1}{[(2q+1)^2 - 1][(2q+1)^2 - 4]^2} = 0.0407$$

Integrating the first equation (4.10), one gets

$$\frac{1}{a} - \frac{1}{a_0} = \frac{4\alpha\omega}{3\pi} t$$

whence

$$a = \frac{a_0}{1 + \left(\dfrac{4\alpha\omega a_0}{3\pi}\right)t} \tag{4.11}$$

is the law for the decay of the amplitude on the basis of the first approximation theory.

If one substitutes (4.11) into the second equation (4.10), one obtains

$$\psi = \omega t - \frac{3C\alpha a_0}{\pi}\left[1 - \frac{1}{1 + \dfrac{4\alpha\omega a_0}{3\pi} \cdot t}\right] + \psi_0 \tag{4.12}$$

The d.e. can be reduced to a quadrature which cannot be integrated in a closed form, although it is still possible to derive the law for the decay of amplitudes. In this manner one can verify the accuracy of the asymptotic method by comparison with the data derived from the exact theory.

This applies also to all cases in which the exact integration is known, as, for instance, for the d.e. of the pendulum in the general form

$$m\ddot{x} + \frac{g}{l}\sin x = 0 \tag{4.13}$$

Readers interested in the quantitative results will find a number of tables in the Bogoliubov-Mitropolsky treatise giving a comparison between the results of exact methods and those obtained on the basis of the asymptotic theory for the first and the second approximations. In the case of the d.e. (4.9) and for $4\alpha a_0 = 1$, the difference between the exact values of decreasing amplitudes and those obtained by the theory of the first approximation is only of the order of 1%; for $4\alpha a_0 = 0.1$, it is of the order of 0.4%. For the second approximation, the error is considerably less.

Although this result is to be expected on theoretical grounds, it is still more persuasive when the comparative results are presented in a form of tables which thus permits a better appreciation of the method in cases when no explicit integration is possible as in the example of the following section.

5. Self-excited systems

Consider the nearly linear d.e. (2.1) in the form

$$\ddot{x} + \omega^2 x = \mu f(x)\dot{x} \tag{5.1}$$

In order to bring this d.e. within the scope of the general method, we can start with the d.e.

$$m\ddot{y} + ky = \mu F(\dot{y}) \tag{5.2}$$

with $\dot{x} = y$ and a differentiation having set $\dot{y} = x$. One then obtains

$$m\ddot{x} + kx = \mu F'(x)\dot{x} \tag{5.3}$$

(where $F' = (d/dy)F$), that is, of the form (5.1).

It is necessary to consider the Fourier expansion of $f(a\cos\psi)a\omega\sin\psi$; $\psi = \omega t + \theta$. Define $F^*(x) = \int_0^x f(x)dx$ and consider its development

$$F^*(a\cos\psi) = \sum_{n=0}^{\infty} F_n^*(a)\cos n\psi \tag{5.4}$$

Comparing with (4.4), one gets

$$A^{(1)}(a) = \tfrac{1}{2}F_1^*(a), \qquad B^{(1)}(a) = 0 \tag{5.5}$$

Hence, in the first approximation,

$$x = a\cos\psi; \qquad \dot a = \frac{\mu}{2} F_1^*(a); \qquad \dot\psi = \omega \tag{5.6}$$

For the second approximation, one has

$$x = a\cos\psi + \frac{\mu}{\omega}\sum_{n=2}^{\infty}\frac{nF_n^*(a)\sin\psi}{n^2 - 1} \tag{5.7}$$

where a and ψ are determined by the d.e.

$$\dot a = \frac{\mu}{2} F_1^*(a); \qquad \dot\psi = \omega + \mu^2 B^{(2)}(a) \tag{5.8}$$

and

$$B^{(2)}(a) = -\frac{1}{8a\omega} F_1^*(a)\frac{dF_1^*(a)}{da} - \frac{1}{2\omega a^2}\sum_{n=2}^{\infty}\frac{n^2F_n^*(a)}{n^2 - 1} \tag{5.9}$$

In the case of the van der Pol equation

$$\ddot x - \mu(1 - x^2)\dot x + x = 0$$

$f(x) = 1 - x^2$; $F^*(x) = x - x^3/3$ which gives

$$F^*(a\cos\psi) = a\left(1 - \frac{a^2}{2}\right)\cos\psi - \frac{a^3}{12}\cos 3\psi$$

We have thus:

$$F_1^*(a) = a\left(1 - \frac{a^2}{4}\right); \quad F_3^* = -\frac{a^3}{12}; \quad F_n^*(a) = 0 \quad \text{for } n \neq 1, \neq 3$$

For the first approximation one has:

$$x = a\cos\psi; \qquad \dot a = \frac{\mu a\omega}{2}\left(1 - \frac{a^2}{4}\right); \qquad \dot\psi = \omega \tag{5.10}$$

The d.e. for $\dot a$ is integrated by a quadrature if one multiplies both sides by a. Omitting this calculation, one has

$$a = a_0 \exp\left(\tfrac{1}{2}\mu\omega t\right)/\sqrt{1 + \tfrac{1}{4}a_0^2(\exp(\mu\omega t) - 1)} \tag{5.11}$$

For $t \to \infty$, $a \to a_0 = 2$, so that in the first approximation

$$x = 2\cos(t + \theta) \tag{5.12}$$

For the second approximation, one has

$$x = a\cos\psi - \frac{\mu a^3}{32}\sin 3\psi \tag{5.13}$$

where a and ψ are given by the d.e.

$$\dot{a} = \frac{\mu\omega a}{2}\left(1 - \frac{a^2}{4}\right); \qquad \dot{\psi} = \omega - \mu^2\left(\frac{1}{8} - \frac{a^2}{8} + \frac{7a^4}{256}\right) \tag{5.14}$$

Hence, if one uses the result of the first approximation $a = a_0 = 2$, the second approximation yields

$$x = 2\cos(\omega t + \theta) - \frac{\mu}{4}\sin 3(\omega t + \theta); \qquad \omega = 1 - \frac{\mu^2}{16} \tag{5.15}$$

6. Stationary amplitudes and their stability

In the preceding section we always encountered the amplitude equation in the form

$$\dot{a} = \Phi(a) \tag{6.1}$$

where $\Phi(a) = \mu A^{(1)}(a) - \mu^2 A^{(2)}(a) + \ldots$

This d.e., at least theoretically, is reduced to quadratures, but integrations may be still difficult in certain cases and it is useful to investigate the properties of functions $\Phi(a)$ with a view to determining their real roots. If $\Phi(a) > 0$, this means that no stationary state exists. From a physical point of view the amplitude a cannot grow indefinitely, but this generally means that the form of the d.e. changes for large a, as we mentioned in Section 10, Chapter 3.

The interesting cases are associated with positive roots $a = a_0$. It is clear that the stability of the stationary amplitude is given by

$$\Phi_a(a^*) < 0 \tag{6.2}$$

which means that the derivative of $\Phi(a)$ with respect to a must be negative for $a = a^*$, a^* being the stationary value of a, in general, and $a_0, a_1, \ldots$ used below are stationary values of particular amplitudes.

This result is obtained from the variational equation; it is sufficient to replace a by $a_0 + \delta a$ in (6.1) which gives the condition (6.2).

In order to investigate the condition of the state of rest, we set $a_0 \simeq 0$ and use:

$$\Phi_a(0) > 0 \tag{6.3}$$

which is obviously *the condition of self-excitation.* The conditions (6.2) and (6.3) considered together give rise to a usual topological configuration (Chapter 1) which consists in instability of the state of rest and in stability of the periodic motion.

If μ is sufficiently small, it is clear that the first approximation dominates the rest to such an extent that the addition of higher approximation terms does not modify the qualitative character of solutions but merely modifies

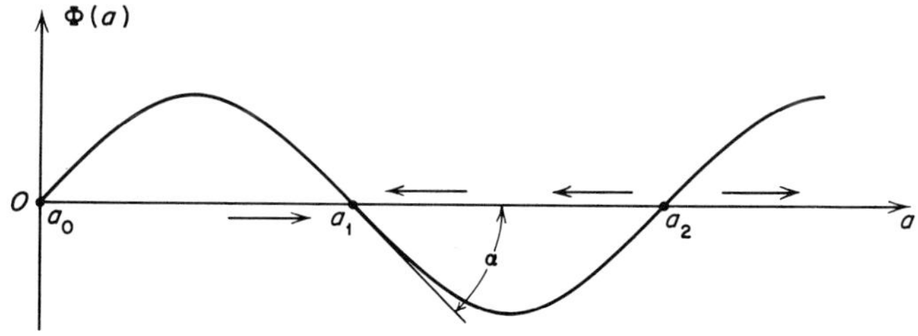

FIGURE 14.1

its quantitative nature slightly, assuming that there are no critical thresholds in the neighborhood of these solutions.†

One can represent these relations graphically in the (a,Φ) plane, Fig. 14.1. Thus for instance if $\Phi(a)$ has a stable root, $a = a_1$, $\Phi_a(a) < 0$, the amplitude a returns to this root a_1 if disturbed; if the root is unstable, the opposite effect takes place. Figure 14.1 shows the curve $\Phi(a)$ plotted against a. The roots of $\Phi(a)$ are the points where $\Phi(a)$ cuts the abscissa axis. The condition of stability $\Phi_a(a) < 0$ is clearly: $\tan \alpha < 0$, where α is the angle between the tangent to $\Phi(a)$ and the a axis. For the curve $\Phi(a)$ shown, the point $a = a_1$ is obviously stable, whereas a_2 and O are unstable. This means that the oscillation with amplitude $a = a_1$ is able to maintain itself while a_0 and a_2 are unstable.

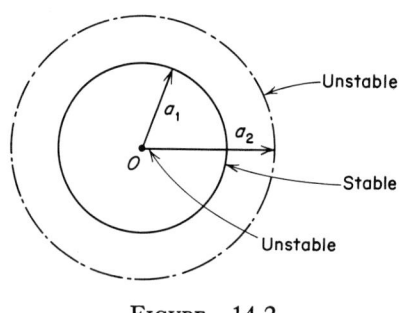

FIGURE 14.2

The theory of approximations gives thus the same result which has previously been analyzed topologically (Chapter 3), viz.: an unstable singular point O is surrounded by an stable limit cycle $a = a_1$, which, in turn, is surrounded by a unstable cycle $a = a_2$ (Fig. 14.2); in other words

† This corresponds to what we have called "noncritical structure" (Section 9, Chapter 7).

the configuration is *USU*, Chapter 7. One can, likewise, ascertain a similar situation in the case of other polycyclic configurations. One can also take, on the axis of the ordinates, the ratio $\Phi(a)/a$; Fig. 14.3 shows a polycyclic configuration consisting of stable cycles $a_1, a_3, \ldots$, with the intermediate unstable cycles $a_2, a_4, \ldots$, the origin O being an unstable singularity.

The theory of approximations permits a quantitative analysis of such situations. In fact, the exact form of the function $\Phi(a)$ is

$$\Phi(a) = \mu A^{(1)}(a) + \mu^2 A^{(2)}(a) + \ldots + \mu^n A^{(n)}(a) \tag{6.4}$$

hence, the stationary amplitude $a = a^*$ is given by the expression

$$A^{(1)}(a^*) + \mu A^{(2)}(a^*) + \ldots + \mu^{n-1} A^{(n)}(a^*) = 0 \tag{6.5}$$

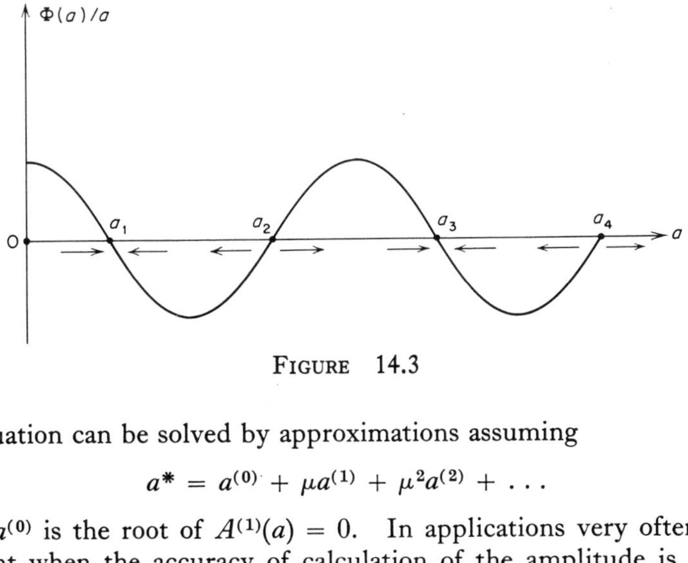

FIGURE 14.3

this equation can be solved by approximations assuming

$$a^* = a^{(0)} + \mu a^{(1)} + \mu^2 a^{(2)} + \ldots \tag{6.6}$$

where $a^{(0)}$ is the root of $A^{(1)}(a) = 0$. In applications very often this is sufficient when the accuracy of calculation of the amplitude is less important than the knowledge that it *exists* and when one is satisfied with the value of its fundamental harmonic. The solution is then merely $a = a^{(0)}$ and the limit cycles are circles with the radii corresponding to these values of a. If it is desired to have the amplitude with a greater approximation, one calculates the next term $a^{(1)}$ in (6.6), which is

$$a^{(1)} = -A^{(2)}(a_0^*)/A^{(1)\prime}(a_0^*) \tag{6.7}$$

where $A^{(1)\prime}(a_0^*)$ is the derivative of $A^{(1)}(a)$ with respect to a at the point $a = a_0^*$, and so on.

Likewise the stability or instability of a stationary state ($a = a^*$) is determined by the condition

$$A^{(1)\prime}(a^*) + \mu A^{(2)\prime}(a^*) + \ldots + \mu^{n-1}A_n{}'(a^*) \lessgtr 0 \tag{6.8}$$

Here we come across a situation mentioned in connection with the fixed point theorem (Section 6, Chapter 3), namely: if μ is sufficiently small and the system is far from critical thresholds (that is, is structurally stable), the stability of the first approximation decides the question, provided $A^{(1)\prime}(a^*)$ is sufficiently great. If this quantity is very small or zero, the determination of stability depends on the next term $A^{(2)\prime}(a^*)$ and, again, it is to be assumed that $|A^{(2)\prime}(a^*)|$ is sufficiently great to be able to neglect the rest of the series.

One can also apply the conclusions of the bifurcation theory (Chapter 7) to these subsequent approximations. We can assume, for instance that the d.e. contains a parameter λ; this parameter is different from the parameter μ associated with the nonlinear function $f(x,\dot{x})$. If λ is varied, the solution varies as we have investigated previously (in Chapter 2 in the discussion of conservative systems and in Chapter 7 in connection with more general forms of d.e.).

The preceding conclusions are to be modified in that, instead of $\Phi(a)$, we have now $\Phi(a; \lambda)$.

We consider, for example, the state of rest ($a = 0$). The stability of the state of rest, as we saw, is given by $\Phi_a(0,\lambda) < 0$. Assume that for small values of λ we have this condition but for larger values (say $\lambda > \lambda_0$) we have $\Phi_a(0,\lambda) > 0$. Clearly $\lambda = \lambda_0$ is the *bifurcation* value of the parameter for which the stability of the state of rest is changed and becomes instability for $\lambda > \lambda_0$. In Chapter 7 we encountered similar situations. If this passage through $\lambda = \lambda_0$ value of the parameter corresponds to a continuous variation of a beginning with $a = 0$, this is obviously the case of a *soft self-excitation*; if this passage of λ through $\lambda = \lambda_0$ is accompanied by a jump in the amplitude a, this is a *hard self-excitation*, using the terms defined in Chapter 3.

A graphical construction of the type just mentioned permits exploring further these bifurcation phenomena. It is interesting to note that such constructions have been used by engineers on a more or less intuitive basis, long before the advent of the modern theory of oscillations.

Let us assume that the function $\Phi(a,\lambda)$ appearing in the amplitude d.e.

$$\dot{a} = \Phi(a,\lambda) \tag{6.9}$$

has the form

$$\Phi(a,\lambda) = \left[\Phi(a) - \frac{a}{\lambda}\right]\psi(a,\lambda) \tag{6.10}$$

where $\psi(a,\lambda)$ is some positive function and $\Phi(a)$ does not depend on λ. The amplitude a is stationary if $\Phi(a,\lambda) = 0$; this requires that the curve $y_1 = \Phi(a)$ and the straight line $y_2 = a/\lambda$ have a point of intersection. Referring to Fig. 14.4 the curve y_1 has the form shown; its slope at the origin is maximum and decreases with a. As to the straight line y_2 its slope for small λ is greater than ξ_0 so that y_1 and y_2 have no point of intersection and, thus, no stationary state: $\Phi(a,\lambda) = 0$ is possible. When the parameter λ reaches the bifurcation value $\lambda = \lambda_0$ for which the slope of y_2 becomes equal to ξ_0, the system is at the limit of self-excitation; for $\lambda > \lambda_0$ there will be a point of intersection M of y_1 and y_2 corresponding to that particular value of λ. For this point the stationary amplitude is Om. This case corresponds to the *soft* self-excitation since the stationary amplitude a starts from zero and increases smoothly with λ.

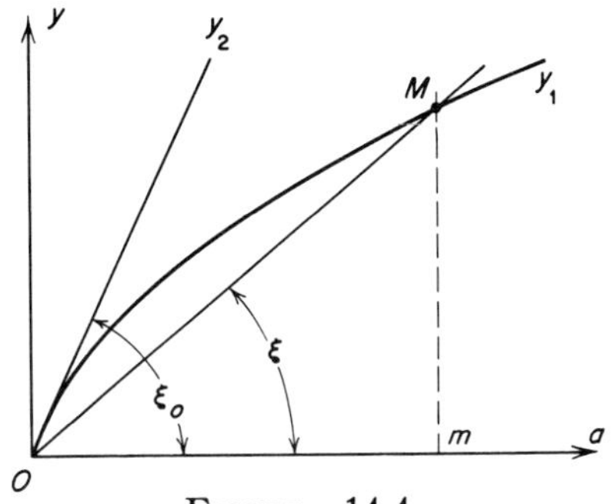

FIGURE 14.4

Figure 14.5 shows the case of a *hard* self-excitation. The difference with

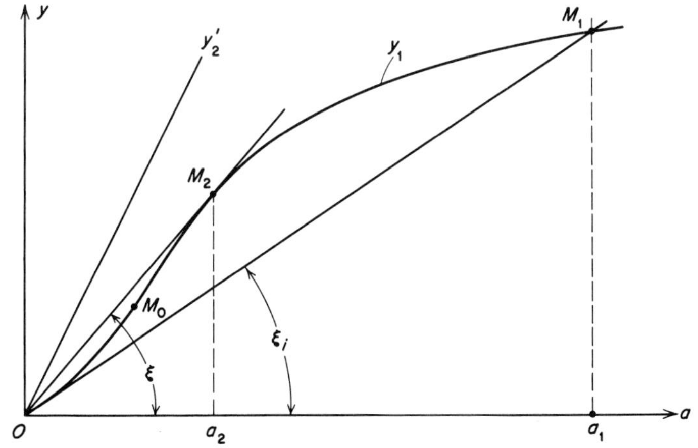

FIGURE 14.5

the preceding case is that the curve y_1 has now an inflexion point at M_0. If one repeats the preceding argument concerning the clockwise rotation of the straight line y_2 with the increasing λ, one finds that the contact between

y_1 and y_2 occurs only at M_2 to which corresponds a *finite* amplitude Oa_2; the self-excitation thus starts *abruptly* with an amplitude Oa_2 and increases thereafter monotonically with λ as in the previously considered case of the soft self-excitation. The portion OM_0M_2 of the curve y_1 is unstable and in this range the oscillation cannot exist because, for any point of this range, the slope of y_1 is greater than that of y_2 (see Chapter 7).

If one plots a as a function of λ, the situation (characterizing a hard self-excitation) appears as shown in Fig. 14.6: for increasing λ the amplitude follows the path $OA_1A_2A_3$; if λ begins to decrease, the path will be

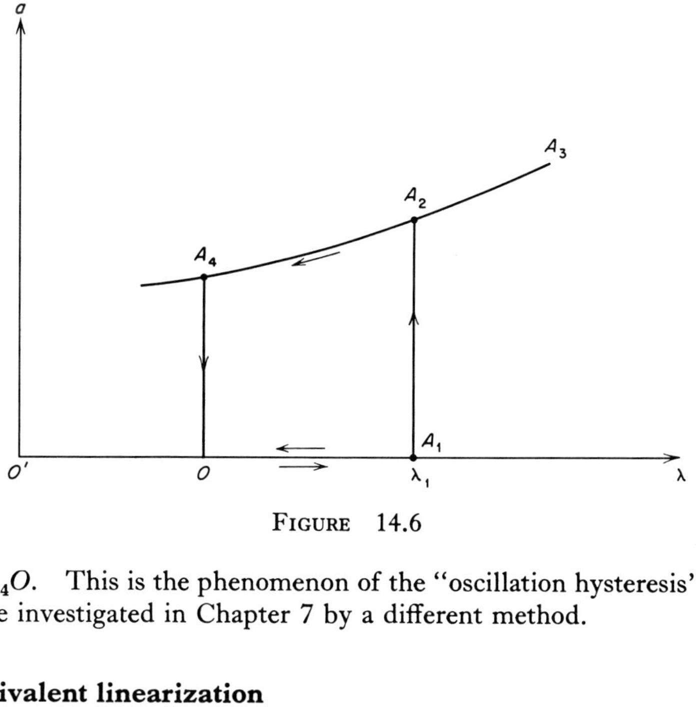

FIGURE 14.6

$A_3A_2A_4O$. This is the phenomenon of the "oscillation hysteresis" which we have investigated in Chapter 7 by a different method.

7. Equivalent linearization

In applied problems equations of the first approximation play generally an important role inasmuch as the *qualitative* aspect of a given problem appears already in the first approximation, as was mentioned in the Introduction to Part II.

Krylov and Bogoliubov have indicated a method, *the method of equivalent linearization*, in which a given nearly linear d.e. can be replaced by *an equivalent linear* d.e. with the property that the solutions of the two equations can be made to differ from each other by an error of the order μ^2.

Although it may seem that nothing of special interest can be gained by this method, inasmuch as a nearly linear d.e. can be solved directly by approximations with any accuracy, there are cases when the method is valuable. This happens, for instance, in the theory of modern control systems where certain linear loops are interlinked with nonlinear (generally nearly linear) ones. If one applies to the nonlinear loops of the system the method of equivalent linearization, one can treat the whole system as linear, which simplifies the problem.

Consider, for instance, one of these nearly linear loops of the form

$$m\ddot{x} + kx = \mu f(x,\dot{x}) \tag{7.1}$$

where m and k are positive constants.

If one follows the theory of the first approximation, one starts with a solution of the form

$$x = a \cos \psi \tag{7.2}$$

where a and ψ are given by the d.e. of the first approximation:

$$\dot{a} = \frac{\mu}{2\pi\omega m} \int_0^{2\pi} f(a \cos \psi, -a\omega \sin \psi) \sin \psi d\psi \tag{7.3}$$

$$\dot{\psi} = \omega_e(a)$$

where

$$\omega_e^2(a) = \omega^2 - \frac{\mu}{\pi m a} \int_0^{2\pi} f(a \cos \psi, -a\omega \sin \psi) \cos \psi d\psi \tag{7.4}$$

$\omega = \sqrt{k/m}$ being the linear frequency (for $\mu = 0$).

It is recalled that the solution $x = a \cos \psi$ is the fundamental harmonic of the true nonlinear series solution

$$x = a \cos \psi + \mu u^{(1)}(a,\psi) + \mu^2 u^{(2)}(a,\psi) + \dots$$

which can be calculated by approximations as was shown previously.

We define the following functions

$$\begin{aligned} \lambda_e(a) &= \frac{\mu}{\pi a \omega} \int_0^{2\pi} f(a \cos \psi, -a\omega \sin \psi) \sin \psi d\psi \\ k_e(a) &= k - \frac{\mu}{\pi a} \int_0^{2\pi} f(a \cos \psi, -a\omega \sin \psi) \cos \psi d\psi \end{aligned} \tag{7.5}$$

The equations of the first approximation become then

$$\dot{a} = -\frac{\lambda_e(a)}{2m} a; \qquad \dot{\psi} = \omega_e(a) \qquad \text{with } \omega_e^2(a) = k_e(a)/m \tag{7.6}$$

Equations (7.5) define the so-called *linearized parameters* $\lambda_e(a)$ and $k_e(a)$.

In order to see that the linearized d.e. (7.6) have a solution differing from the solution of the nonlinear d.e. by a rest $0(\mu^2)$, we differentiate (7.2) taking into account (7.3); this gives

$$\dot{x} = -a\omega_e(a) \sin \psi - \frac{\lambda_e(a)}{2m} a \cos \psi \tag{7.7}$$

and, differentiating once more:

$$\ddot{x} = -a\omega_e{}^2(a) \cos \psi + \frac{\lambda_e(a)}{m} a\omega_e(a) \sin \psi + \frac{\lambda_e{}^2(a)}{4m^2} a \cos \psi$$
$$+ \frac{\lambda_e(a)}{2m} a^2 \frac{d\omega_e(a)}{da} \sin \psi + \frac{d\lambda_e(a)}{da} \frac{a}{2m} \frac{\lambda_e(a)}{2m} a \cos \psi \tag{7.8}$$

This can be written also (taking into account $x = a \cos \psi$ and also (7.7) as

$$\ddot{x} = -\frac{k_e(a)}{m} x - \frac{\lambda_e(a)}{m} \dot{x} - \frac{\lambda_e{}^2(a)}{4m^2} x$$
$$+ \frac{\lambda_e(a)}{2m} a^2 \frac{d\omega_e(a)}{da} \sin \psi + \frac{1}{2m} \frac{d\lambda_e(a)}{da} a \frac{\lambda_e(a)}{2m} x \tag{7.9}$$

Substituting this value into (7.1) and in view of (7.5), one has

$$m\ddot{x} + \lambda_e(a)\dot{x} + k_e(a)x = 0(\mu^2) \tag{7.10}$$

Thus, the first approximation of the nearly linear d.e. (7.1) satisfies also the linear d.e. (7.10) with accuracy of the order μ^2; this is precisely the accuracy with which the first approximation determines the solution of the nearly linear equation.

From this point of view the first approximation of the nearly linear d.e. (7.1) and the linearized d.e. (7.10) *are equivalent.*

It is seen that the process of *linearization* has introduced two parameters: (1) $\lambda_e(a)$ which may be called the *equivalent coefficient of damping*, and (2) $k_e(a)$, the *equivalent coefficient of restoring* force. It is noteworthy that inasmuch as both $\lambda_e(a)$ and $k_e(a)$ are functions of the amplitude a, the direct solution of the linearized equation is likely to be more complicated than the original nearly linear d.e.

It is to be recalled, however, that the theory of oscillations is concerned primarily with a *stationary state.* If so, the quantities $\lambda_e(a)$ and $k_e(a)$ *are constants* defined by the relations

$$\begin{aligned} \lambda_e(a) &= \frac{\mu}{\pi a\omega} \int_0^{2\pi} f(a \cos \psi, -a\omega \sin \psi) \sin \psi d\psi \\ k_e(a) &= -\frac{\mu}{\pi a} \int_0^{2\pi} f(a \cos \psi, -a\omega \sin \psi) \cos \psi d\psi \end{aligned} \tag{7.11}$$

of which the first is the same as the first equation (7.5) and the second merely defines a corrective term by which the coefficient of the nonlinear restoring force $k_e(a)$ differs from the corresponding coefficient k of the linear case.

One can proceed now by a linear argument by defining the decrement $\delta_e(a)$ and the frequency $\omega_e(a)$ by the usual formulas

$$\delta_e(a) = \frac{\lambda_e(a)}{2m}; \qquad \omega_e^2(a) = \frac{k_e(a)}{m} \tag{7.12}$$

The d.e. are now linear

$$\dot{a} = -\delta_e(a); \qquad \dot{\psi} = \omega \tag{7.13}$$

and it is sufficient to replace $\delta_e(a)$ and ω by the above formulas and thus obtain a solution with accuracy of μ^2.

The procedure so far is purely formal and one may ask the question: How could one guess that the new parameters (7.11) would produce such an interesting result, namely, that the two d.e. happen to have the same solution up to $O(\mu^2)$?

The authors of the method offer a justification to what may appear as a "guess" in the solution of these equivalent parameters (7.11). There are two different ways of justifying this initial "guess" on physical grounds: (1) consideration of energy, and (2) consideration of the harmonic balance.

The two justifications of the "plausible guess" (7.11) are called: (1) the principle of energy balance and (2) the principle of harmonic balance.

Strictly speaking if one is satisfied with a purely formal result (7.10) there is no necessity for these physical "principles" but, if one wishes to *understand* why $\lambda_e(a)$ and $k_e(a)$ have been *so defined*, these principles throw an interesting light on the procedure and permit having a deeper insight into the general *method of averaging* which is at the very foundation of the whole theory.

We consider first the *principle of the energy balance*, using the well known fact in the theory of alternating currents where the power $W = EI$ is considered in terms of its components: the energy component, $W_e = EI \cos \varphi$, and the wattless component, $W_w = EI \sin \varphi$, φ being the phase difference between the vectors E and I (one uses also the term "power factor" referring to $\cos \varphi$).

Consider first the energy balance. In the nearly linear d.e. the force capable of producing a work is $F = \mu f(x,\dot{x})$; in the linearized equation such a force is $F_e = [-k_e(a)x + \lambda_e(a)\dot{x}]$. If we postulate that in both cases the average power per cycle is the same, we can write

$$\mu \int_0^T f(x,\dot{x})\dot{x}dt = -\lambda_e(a) \int_0^T \dot{x}^2 dt \tag{7.14}$$

since the average work per cycle produced by the restoring force $-k_e(a)x$ is obviously zero.

On the basis of the first approximation: $x = a \cos t(\omega t + \theta)$; $\dot{x} = -a\omega \sin(\omega t + \theta)$, where a and θ may be regarded as constant during the time interval $(0, T)$. Since ω is the frequency, $(T = 2\pi/\omega)$, the expression (7.14) becomes

$$-\mu\omega \int_0^{2\pi/\omega} f[a \cos(\omega t + \theta), -a\omega \sin(\omega t + \theta)]a \sin(\omega t + \theta)dt$$
$$= -\lambda_e(a) \int_0^{2\pi/\omega} a^2\omega^2 \sin^2(\omega t + \theta)dt \quad (7.15)$$

The right-hand term in this equation is: $-\lambda_e(a)a^2\pi\omega$. Thus

$$\lambda_e(a)\pi\omega a^2 = \mu a \int_0^{2\pi} f(a \cos \psi, -a\omega \sin \psi) \sin \psi d\psi$$

and this is precisely the expression that defines λ_e (equation (7.11)). In a similar manner one can give a physical interpretation to the equivalent term $k_e(a)$ in (7.11) if one makes use of the concept of the *reactive power* which, as is recalled, is·

$$P_r = \frac{1}{T} \int_0^T E(t)i^*(t)dt$$

where $E(t)$ is the electromotive force and i^* is the reactive component of current: $i^*(t) = i(t - T/4)$. We can use this concept also in a mechanical case.

Since the average (active) power per cycle is

$$\frac{1}{T} \int_0^T F(t)\dot{x}(t)dt$$

the corresponding reactive power is

$$\frac{1}{T} \int_0^T F(t)\dot{x}\left(t - \frac{T}{4}\right)dt \quad (7.16)$$

We have to equate the average reactive powers corresponding to the nonlinear force $\mu f[x(t), \dot{x}(t)]$ on one hand, and to the linearized force $k_e(a)x(t) + \lambda_e(a)\dot{x}(t)$ on the other. This gives

$$\mu \frac{1}{T} \int_0^T f[x(t), \dot{x}(t)]\dot{x}\left(t - \frac{T}{4}\right)dt$$
$$= -\frac{1}{T} \int_0^T [k_e(a)x(t) + \lambda_e(a)\dot{x}(t)]\dot{x}\left(t - \frac{T}{4}\right)dt \quad (7.17)$$

It is clear that $\lambda_e(a)$ and $k_e(a)$ are of the same order. We can now use again the data of the first approximation for x and $\dot{x}$ and obtain

$$\mu \frac{1}{T} \int_0^T f[x(t), \dot{x}(t)]\dot{x}\left(t - \frac{T}{4}\right)dt = \frac{a\omega\mu}{2\pi} \int_0^{2\pi} f(a \cos \psi, -a\omega \sin \psi) \cos \psi d\psi$$

It is sufficient to set: $f[x(t), \dot{x}(t)] = k_e(a)x(t) + \lambda_e(a)\dot{x}(t)$ to obtain

$$k_e(a) = -\frac{\mu}{\pi a} \int_0^{2\pi} f(a \cos \psi, -a\omega \sin \psi) \cos \psi d\psi \tag{7.18}$$

which is the second formula in (7.11).

The principle of the harmonic balance consists in equating the expression $F = J \cos(\omega t + \varphi)$ (representing the fundamental harmonic) to the equivalent linear force $F_e = J_e \cos(\omega t + \varphi_e)$ calculated by the foregoing procedure. The identification of terms results in the relations of definition:

$$J_e = J; \qquad \varphi_e = \varphi$$

In fact, for a harmonic oscillation the equivalent linear force is

$$-k_e(a)a \cos(\omega t + \theta) + \omega\lambda_e(a)a \sin(\omega t + \theta) \tag{7.19}$$

and the fundamental harmonic of the nonlinear force is

$$\left[\frac{1}{\pi} \int_0^{2\pi} f(\quad) \cos \psi d\psi\right] \cos(\omega t + \theta) + \left[\frac{1}{\pi} \int_0^{2\pi} f(\quad) \sin \psi d\psi\right] \sin(\omega t + \theta) \tag{7.20}$$

where $f(\quad) = f(a \cos \psi, -a\omega \sin \psi)$.

If one equates (7.19) and (7.20), one gets

$$ak_e(a) = -\frac{1}{\pi} \int_0^{2\pi} f(\quad) \cos \psi d\psi; \qquad \omega\lambda_e(a)a = \frac{1}{\pi} \int_0^{2\pi} f(\quad) \sin \psi d\psi \tag{7.21}$$

which gives again the values (7.11) for $\lambda_e(a)$ and $k_e(a)$.

The method of equivalent linearization requires a calculation of equivalent parameters in the first place. Once this is done the d.e. can be treated as a linear one.

Assume, for instance, that we wish to investigate the oscillation of a weight suspended on a nonlinear spring specified by a relation

$$F = f(x)$$

Then, for a harmonic oscillation $x = a\cos(\omega t + \varphi)$, the fundamental harmonic of the "spring constant" is

$$\frac{\cos(\omega t + \theta)}{\pi}\int_0^{2\pi} f(a\cos\varphi)\cos\varphi d\varphi$$

Hence, by the above principle of harmonic balance

$$k_e(a) = \frac{1}{\pi a}\int_0^{2\pi} f(a\cos\varphi)\cos\varphi d\varphi \tag{7.22}$$

which gives the frequency of the linearized system

$$\omega_e(a) = \sqrt{k_e(a)/m}$$

Since $k_e(a) = c + k(a)$, c being the linear spring constant $[c \gg k(a)]$, this gives the formula

$$\omega_e(a) = \sqrt{\frac{c}{m}}\left[1 + \frac{1}{2}\frac{k(a)}{c}\right] \tag{7.23}$$

valid in the first approximation (that is, error $O(\mu^2)$).

If the system is acted on by a small nonlinear damping of the form $\Phi = \Phi(\dot{x})$ and the oscillation is not far from being harmonic, the fundamental harmonic of the damping action is

$$\frac{\sin(\omega t + \theta)}{\pi}\int_0^{2\pi} \Phi(-a\omega\sin\varphi)\sin\varphi d\varphi \tag{7.24}$$

The principle of the harmonic balance introduces the equivalent linear damping

$$\Phi_e = \lambda_e(a)\dot{x} \tag{7.25}$$

where

$$\lambda_e(a) = -\frac{1}{a\omega\pi}\int_0^{2\pi} \Phi(-a\omega\sin\varphi)\sin\varphi d\varphi \tag{7.26}$$

With this value $\lambda_e(a)$ the original nonlinear system can be treated as a linear one with decrement

$$\delta_e(a) = \lambda_e(a)/2m \tag{7.27}$$

As another example, consider an electron-tube circuit of a standard (inductively coupled) type; if i is the current in the oscillating circuit and I_a the plate current, one has the well known d.e.

$$L\frac{di}{dt} + Ri + \frac{1}{c}\int idt = M\frac{dI_a}{dt} \tag{7.28}$$

where L, R, and C are the constants of the oscillating circuit, and M is the coefficient of mutual induction between the plate and the oscillating circuit. The nonlinear element of the scheme is the electron-tube characteristic

$$I_a = f(E) = f(E_0 + e)$$

where E_0 is a constant voltage and e is the oscillating grid voltage. If $e = a \cos(\omega t + \theta)$, the fundamental harmonic of I_a is

$$\frac{\cos(\omega t + \theta)}{\pi} \int_0^{2\pi} f(E_0 + a \cos \varphi) \cos \varphi d\varphi$$

One can thus replace the nonlinear relation by the equivalent linear

$$I_a = Se \tag{7.29}$$

which defines the *average transconductance* S of the tube

$$S(a) = \frac{1}{\pi a} \int_0^{2\pi} f(E_0 + a \cos \varphi) \cos \varphi d\varphi \tag{7.30}$$

One can write (7.28) as

$$L \frac{di}{dt} + Ri + \frac{1}{c} \int idt = MS \frac{de}{dt} \tag{7.31}$$

If one neglects the anode reaction, the voltage on the grid is

$$e = \frac{1}{c} \int idt$$

This gives finally the d.e.

$$LC\ddot{e} + (RC - MS)\dot{e} + e = 0 \tag{7.32}$$

which can be treated as a linear d.e., that is:

$$\omega = \frac{1}{\sqrt{LC}}; \qquad \delta_e(a) = \frac{RC - MS(a)}{2LC}$$

It is seen that for a stationary state the oscillation establishes itself for such a value of $a = a_0$ for which the transconductance $S(a_0) = RC/M$. In this case the decrement vanishes and the system operates as a harmonic oscillator, at least in the first approximation. In Chapter 9 we have seen that in reality the mechanism of oscillation is more complicated due to the presence of an infinite spectrum of even harmonics which are, however, so small that the first approximation (in which they are neglected) is generally sufficient if μ is sufficiently small.

The authors give a number of other applications of the method of equivalent linearization. As previously mentioned, this method has been used recently in connection with the theory of nonlinear control systems.

Chapter 15

ASYMPTOTIC METHODS OF KRYLOV-BOGOLIUBOV-MITROPOLSKY

Nonautonomous systems

1. Introductory remarks

Sections 2, 3, 4, 5 of this chapter relate to the application of the Krylov-Bogoliubov methods to nonautonomous systems in the new version of Bogoliubov and Mitropolsky,[1] the last three sections present a brief review of a recent monograph of Mitropolsky [2] concerning nonstationary processes. The material contained in these two references is so voluminous that it was possible to abstract only a very small part concerning the salient points of these methods. It was necessary to omit the greater part of the examples, which constitute, perhaps, the most important part of this work for those who are interested primarily in applications.

The essence of these methods is the same as that of the preceding chapter, that is, the successive approximations are developed *formally* by a recursive procedure for a fixed number m of terms, but for $\mu \to 0$; this permits determining the accuracy of approximations to any desired degree $O(\mu^m)$. On the other hand, the question of convergence of expansions for $m \to \infty$ is not considered.

The difference between the subject treated in the preceding chapter and in this one is that for the nonautonomous systems there appears an additional variable νt (dependence on t explicitly) which relates this investigation to the rather difficult subject of nonlinear resonance. As this subject is inherently welded, so to speak, with the asymptotic method,

[1] N. Bogoliubov and J. Mitropolsky, *Asymptotic Methods in the Theory of Nonlinear Oscillations* (in Russian), Moscow, 1958.

[2] J. Mitropolsky, *Nonstationary Processes* (in Russian), *Ac. Nauk* (USSR), 1955.

we follow the exposition of the authors although the problem of resonance per se is presented only in Chapter 19. This situation remains, however, the same as in Chapter 10 where the consideration of resonance was also inevitable if one follows the theory of Poincaré in its application to the nonautonomous systems.

The subject of the last three sections in this chapter follows the general method of the preceding chapter but here appears another variable τ, *the slow time*, which complicates calculations still further; for that reason, it was necessary to condense the presentation still more. Although these exceedingly long calculations may appear somewhat disappointing to a beginner, one must have in mind that, once they are completed, the matter reduces merely to writing down the conditions for each particular case in which one is interested.

In view of the presence of additional variables, the series expansions appear here in the form of double Fourier series which are taken in the exponential form. Aside of this, the general method of approach is similar to that of the preceding chapter.

For the same reason as in Chapter 10 dealing with Poincaré's method, it is necessary to separate studies of nonresonance oscillations from those taking place at resonance. In the latter case the phase angle (between autoperiodic and heteroperiodic oscillations) appears as a new variable in the series expansions which are more complicated for that reason.

As regards the last three sections of this chapter devoted to a brief outline of Mitropolsky's extension of the asymptotic method, it was impossible to go into a survey of their numerous applications which constitute perhaps the most interesting part of this monograph. In fact the existing methods are limited to the investigation of series solutions of d.e. under the assumption of *fixed* parameters so that the behavior of the system when the parameters *vary* may be quite different. In this way a *passage* through a resonance zone may be quite different from the *stationary state* of resonance and depends largely on *the rate* at which this passage is effected. Thus, for instance, torsional oscillations in mechanical problems appear differently in stationary and nonstationary (transient) states, and so on. All this opens a new approach to certain applied problems and requires supplementing the existing theory of nonlinear oscillations, built on the concept of stationary periodic solutions, by extensions arising from the variable parameters. At the same time these new extensions take care also of different *modulations* on the same basis on which the earlier theory deals with the periodic solutions. Owing to this a variety of new problems, such as those dealing with phase or frequency modulations, are merged into the general theory of asymptotic methods.

2. Formulation of the problem

We consider now a nonautonomous nearly linear d.e. containing t explicitly

$$\ddot{x} + \omega^2 x = \mu f(\nu t, x, \dot{x}) \tag{2.1}$$

We suppose that the explicit dependence on time (term with νt) appears in a periodic term (period 2π) whose right-hand side is a trigonometric polynomial:

$$f(\nu t, x, \dot{x}) = \sum_{n=-N}^{N} e^{in\nu t} f_n(x,\dot{x}) \tag{2.2}$$

where each $f_n(x,\dot{x})$ is a polynomial in x and $\dot{x}$.

Considered as representing a certain oscillator, (2.1) characterizes an oscillator with unit mass, free frequency ω acted on by a small nonlinear perturbation term μf with the explicit (periodic) dependence on t through a trigonometric term with argument νt.

If one proceeds with the general method outlined in the preceding chapter, starting with the solution $x = a \cos(\omega t + \varphi)$ and $\dot{x} = -a\omega \sin \times (\omega t + \varphi)$, one obtains ultimately terms containing $\sin(n\nu + m\omega)t$ and $\cos(n\nu + m\omega)t$, where n and m are integers. Such terms will appear also in the expressions for $u^{(1)}, u^{(2)}, \ldots$.

If one of these *combination frequencies* $n\nu + m\omega$ happens to be close enough to the free frequency ω of the system, one can expect that the amplitude will grow just as in an ordinary (linear) resonance.

We reach thus a more or less obvious conclusion that the cumulative effects (in which the amplitude grows) in nonlinear systems appear not only when $\nu = \omega$ (as in the linear systems) but also when $n\nu + m\omega = \omega$, that is, when

$$\nu \simeq \frac{p}{q}\,\omega \tag{2.3}$$

where p and q are sufficiently small relatively prime integers.

We introduce the following classification:

(1) $p = q = 1$, that is, $\nu \simeq \omega$; this is the *fundamental* (or "ordinary") resonance.
(2) $q = 1$; $\nu \simeq p\omega$; this is the so-called *subharmonic* (or "the fractional order," or a "demultiplication") resonance.
(3) $p = 1$; $\nu = \omega/q$; this is a *superharmonic* resonance which does not present any particular interest in what follows.

The most interesting and important feature of nonlinear systems is the subharmonic resonance (2) to which we shall return in Chapter 19, Part III.

As the ratio p/q is any rational fraction and the resonances occur in the neighborhood of p/q, it may seem that they form an everywhere dense set. In reality this is not so inasmuch as resonances arise only if p and q are *sufficiently small* relatively prime integers and, besides this, there are additional conditions to be fulfilled, which eliminate some of these cases, as we shall see later. In Part III we shall go into a more detailed investigation of these conditions as the result of which it will appear that only in a few special cases it was possible to establish completely the necessary and sufficient conditions for a *physical existence* of a subharmonic resonance. By the term "physical existence" we shall mean both the mathematical existence of stationary amplitude and phase and their stability. For the time being we shall be concerned mostly with the formal aspect of the argument.

What is important at this stage is the fact that a relatively small perturbing force μf may, under certain conditions, account for a relatively large effect as far as the increase of amplitude is concerned.

This plausible assumption permits approaching the problem in a more definite manner. In fact, if a small force produces a finite effect, this means that the work of this force per cycle is conserved (in the form of energy) and is added to the work during the preceding cycles.

The expression for the virtual work in a harmonic condition: $x_0(t) = a \cos (\omega t + \varphi)$ is obviously

$$\delta x_0 = \cos (\omega t + \varphi)\delta a - a \sin (\omega t + \varphi) \cdot \delta\varphi \tag{2.4}$$

where we assume that this work changes the amplitude (δa) as well as the phase ($\delta\varphi$).

On the other hand, the "external force": $\mu f(\nu t, x_0, \dot{x}_0)$, in which x and $\dot{x}$ are replaced by their values x_0 and $\dot{x}_0$, can be represented as a Fourier series in terms of the "combination frequencies" $\lambda_{mn} = n\nu + m\omega$ with a view to calculating the virtual work:

$$\mu f(\nu t, x_0, \dot{x}_0)\delta x_0 = \delta W \tag{2.5}$$

In the process of averaging over a sufficiently long time it can be shown that there will remain only relatively small combination frequencies λ_{nm} whose "secular effect" on the cumulative process will be apparent under some special conditions. This means that under the assumed pattern of the formation of the cumulative effects, only such resonances are possible for which $\delta W \simeq 0$.

There appear thus two separate cases: (1) nonresonance oscillation, and (2) resonance oscillation. For the case (1), one has the condition

$$n\nu + m\omega \neq \omega \tag{2.6}$$

From the number theory, it follows that, given any irrational ratio ν/ω, one can always find such integers n and m that the expression

$$n\nu + (m - 1)\omega \tag{2.7}$$

can approach zero as closely as we please. As we do not wish that expression (2.6) should approach zero (otherwise we would be in the case of resonance which we wish to avoid at present), it is necessary to have a rational ratio p/q and, moreover, to impose later an additional condition.

3. Successive approximations in nonresonance cases

The derivation of a recursive system of approximations follows a method similar to that used for the autonomous systems in the preceding chapter. However, certain differences will appear later.

If $\mu = 0$, (2.1) has a harmonic solution, $x = a \cos \psi$, and the amplitude and phase d.e. are, respectively: $\dot{a} = 0$; $\dot{\psi} = \text{const.}$

If $\mu \neq 0$, we look again for a solution of the form

$$x = a \cos \psi + \mu u^{(1)}(a, \psi, \nu t) + \mu^2 u^{(2)}(a, \psi, \nu t) + \dots \tag{3.1}$$

The only difference with (2.3) of the preceding chapter is that the $u^{(i)}$ depend now on νt and are periodic with period 2π with respect to both angular variables.

The amplitude a and the phase ψ are still given by (2.4) of the preceding chapter, namely:

$$\begin{aligned} \dot{a} &= \mu A^{(1)}(a) + \mu^2 A^{(2)}(a) + \dots \\ \dot{\psi} &= \omega + \mu B^{(1)}(a) + \mu^2 B^{(2)}(a) + \dots \end{aligned} \tag{3.2}$$

The right-hand sides of these d.e. depend only on a as, in the absence of resonance, there is no stationary relation between the phase of the external periodic excitation and that of the oscillation. The situation will be different, however, in the case of resonance when this phase relation becomes stationary. Thus, for the nonresonance case the matter is still similar to that discussed in Section 2 of the preceding chapter, with the exception that the functions $u^{(i)}$ depend now on νt in addition to: a and ψ.

The fundamental requirement (2.5) of the preceding chapter remains the same as before; this guarantees the absence of "small divisors" in the functions $u^{(i)}$. We omit the calculations of $\dot{x}$ and $\ddot{x}$ which are the same as previously with the exception that now there will appear additional terms with $\dot{u}^{(1)}$ and $\dot{u}^{(2)}$, $\ddot{u}^{(2)}$,. . . in view of the explicit dependence on t of these functions.

The left-hand side of (2.1) after substitutions becomes:

$$\ddot{x} + \omega^2 x = \mu[u_{\psi\psi}{}^{(1)}\omega^2 + \ddot{u}^{(1)} + 2u_{\psi t}{}^{(1)}\cdot\omega + \omega^2 u^{(1)} - 2a\omega B^{(1)}\cos\psi - 2\omega A^{(1)}\sin\psi] + \mu^2[u_{\psi\psi}{}^{(2)}\omega^2 + \ddot{u}^{(2)} + 2u_{\psi t}{}^{(2)}\omega + \omega^2 u^{(2)} - 2a\omega B^{(2)}\cos\psi - 2\omega A^{(2)}\sin\psi + (A^{(1)}A_n{}^{(1)} - aB^{(1)2})\cos\psi - (aA^{(1)}B_a{}^{(1)} + 2A^{(1)}B^{(1)})\sin\psi + 2\omega A^{(1)}u_{a\psi}{}^{(1)} + 2\omega B^{(1)}u_{\psi\psi}{}^{(1)} + 2u_{at}{}^{(1)}A^{(1)} + 2u_{\psi t}{}^{(1)}B^{(1)}] + \mu^3[\quad] + \ldots \quad (3.3)$$

The right-hand side of (2.1) can be written as

$$\mu f(\nu t, x, \dot{x}) = \mu f(\nu t, a\cos\psi, -a\omega\sin\psi) + \mu^2[f_x(\nu t, a\cos\psi, -a\omega\sin\psi)u^{(1)} + f_{\dot{x}}(\nu t, a\cos\psi, -a\omega\sin\psi)(A^{(1)}\cos\psi - aB^{(1)}\sin\psi + u_\psi{}^{(1)}\cdot\omega + \dot{u}^{(1)})] + \mu^3\ldots \quad (3.4)$$

In order that the series solution (3.1) should satisfy the d.e. (2.1) with accuracy $O(\mu^{m+1})$ it is necessary to equate the coefficients of like powers of μ up to the order μ^m.

This results in a recursive system of d.e., viz.:

$$\begin{aligned} \omega^2 u_{\psi\psi}{}^{(1)} + 2\omega u_{\psi t}{}^{(1)} + \ddot{u}^{(1)} + \omega^2 u^{(1)} &= f_0(a, \psi, \nu t) + 2a\omega B^{(1)}\cos\psi + 2\omega A^{(1)}\sin\psi \\ \omega^2 u_{\psi\psi}{}^{(2)} + 2\omega u_{\psi t}{}^{(2)} + \ddot{u}^{(2)} + \omega^2 u^{(2)} &= f_1(a, \psi, \nu t) + 2a\omega B^{(2)}\cos\psi + 2\omega A^{(2)}\sin\psi \end{aligned} \quad (3.5)$$

. . . .

where

$$f_0(a, \psi, \nu t) = f(\nu t, a\cos\psi, -a\omega\sin\psi)$$

$$f_1(a, \psi, \nu t) = f_x(\nu t, a\cos\psi, -a\omega\sin\psi)u^{(1)} + f_{\dot{x}}(\nu t, a\cos\psi, -a\omega\sin\psi) \times [A^{(1)}\cos\psi - aB^{(1)}\sin\psi + u_\psi{}^{(1)}\omega + \dot{u}^{(1)}] + (aB^{(1)2} - A^{(1)}A_a{}^{(1)})\cos\psi + (A^{(1)}B_a{}^{(1)}a + 2A^{(1)}B^{(1)})\sin\psi - 2\omega B^{(1)}u_{\psi\psi}{}^{(1)} - 2A^{(1)}u_{at}{}^{(1)} - 2B^{(1)}u_{\psi t}{}^{(1)} - 2\omega A^{(1)}u_{a\psi}{}^{(1)} \quad (3.6)$$

. . . .

The functions $f_k(a, \psi, \nu t)$ are periodic with period 2π with respect to both arguments ψ and νt and depend also on a.

Since we have here periodic functions of two arguments, with period 2π, one can use a double Fourier series in a complex form.

For a periodic function of one argument x one can write the Fourier expansion

$$f(x) = \sum_{n=-\infty}^{\infty} c_n e^{inx} \tag{3.7}$$

with

$$c_n = \frac{1}{2\pi} \int_0^{2\pi} f(\sigma) e^{-in\sigma} d\sigma \tag{3.8}$$

which yields the following relations between c_i, a_i, and b_i of the usual Fourier series (in real notations)

$$f(x) = \frac{a_0}{2} + \sum_{n=1}^{\infty} [a_n \cos nx + b_n \sin nx]$$

namely

$$c_n = \frac{a_n - ib_n}{2}; \qquad c_{-n} = \frac{a_n + ib_n}{2} \tag{3.9}$$

For a function $f(x,y)$ periodic (with period 2π) both with respect to x and y, one can consider formally $f(x,y)$ as a function of x which gives:

$$f(x,y) = \sum_{n=-\infty}^{\infty} c_n(y) e^{inx} \tag{3.10}$$

with

$$c_n(y) = \frac{1}{2\pi} \int_0^{2\pi} f(\sigma, y) e^{-in\sigma} d\sigma \tag{3.11}$$

The coefficient $c_n(y)$, in turn, can be expanded as a complex Fourier series

$$c_n(y) = \sum_{m=-\infty}^{\infty} c_{nm} e^{imy} \tag{3.12}$$

with

$$c_{nm} = \frac{1}{2\pi} \int_0^{2\pi} c_n(\eta) e^{-im\eta} d\eta = \frac{1}{4\pi^2} \int_0^{2\pi} \int_0^{2\pi} f(\xi, \eta) e^{-i(n\xi + m\eta)} d\xi d\eta \tag{3.13}$$

If one substitutes (3.12) into (3.10), one has

$$f(x,y) = \sum_{n=-\infty}^{\infty} \sum_{m=-\infty}^{\infty} c_{nm} e^{i(nx+my)} \tag{3.14}$$

One can now undertake the determination of $A^{(1)}(a)$, $B^{(1)}(a)$, and $u^{(1)}(a, \psi, \nu t)$ from the first d.e. (3.5) and, in the first place, develop $f_0(a, \psi, \nu t)$ into a double Fourier series with respect to both arguments. We have

$$f_0(a, \psi, \nu t) = \sum_n \sum_m f_{nm}{}^{(0)}(a) e^{i(n\nu t + m\psi)} \tag{3.15}$$

In view of (3.13) we have:

$$f_{nm}{}^{(0)} = \frac{\nu}{4\pi^2} \int_0^{2\pi} \int_0^{2\pi} f(\nu t, a \cos \psi, -a\omega \sin \psi) e^{-i(n\nu t + m\psi)} dt d\psi \tag{3.16}$$

One can also do this with respect to $u^{(1)}(a, \psi, \nu t)$, viz.:

$$u^{(1)}(a, \psi, \nu t) = \sum_n \sum_m \bar{f}_{nm}(a)^{i(n\nu t + m\psi)} \tag{3.17}$$

Substituting (3.15) and (3.17) for f_0 and $u^{(1)}$ into the first equation (3.5), one gets

$$\sum_n \sum_m [\omega^2 - (n\nu + m\omega)^2] \bar{f}_{nm}(a) e^{i(n\nu t + m\psi)}$$
$$= 2a\omega B^{(1)} \cos \psi + 2\omega A^{(1)} \sin \psi + \sum \sum f_{nm}{}^{0}(a) e^{(in\nu t + m\psi)} \tag{3.18}$$

It is necessary to determine $\bar{f}_{nm}$, $A^{(1)}$ and $B^{(1)}$ so that $u^{(1)}$ has no resonance terms. This condition is fulfilled if one has:

$$2a\omega B^{(1)} \cos \psi + 2\omega A^{(1)} \sin \psi$$
$$= \sum_{[\omega^2 - (n\nu + m\omega)^2] = 0} \sum f_{nm}{}^{(0)}(a) \exp [i(n\nu t + m\psi)] \tag{3.19}$$

the double summation extending for all m and n for which $nq + (m \pm 1p) = 0$. Equating the coefficients of the like harmonics in (3.18) one has

$$\bar{f}_{nm}(a) = \frac{f_{mn}{}^{(0)}(a)}{\omega^2 - (n\nu + m\omega)^2} \tag{3.20}$$

for all n and m satisfying the condition: $\omega^2 - (n\nu + m\omega)^2 \neq 0$. This condition is equivalent to: $n^2 + (m^2 - 1)^2 \neq 0$ (that is, $n \neq 0, m \neq \pm 1$). If one substitutes (3.20) into (3.17) and sets $\nu t = \theta$ (the angular variable), one has:

$$u^{(1)}(a, \psi, \theta) = \frac{1}{4\pi^2} \sum_{\substack{n \quad m \\ n^2 + (m^2 - 1)^2 \neq 0}} \sum \frac{e^{(in\theta + m\psi)}}{[\omega^2 - (n\nu + m\omega)^2]}$$
$$\times \int_0^{2\pi} \int_0^{2\pi} f_0(a, \psi, \theta) e^{-i(n\theta + m\psi)} d\theta \cdot d\psi \tag{3.21}$$

Returning to the trigonometric functions and equating the coefficients of the like harmonics in (3.19), one has

$$
\begin{aligned}
A^{(1)}(a) &= -\frac{1}{4\pi^2\omega}\int_0^{2\pi}\int_0^{2\pi} f_0(a, \psi, \theta)\sin\psi d\theta d\psi \\
B^{(1)}(a) &= -\frac{1}{4\pi^2 a\omega}\int_0^{2\pi}\int_0^{2\pi} f_0(a, \psi, \theta)\cos\psi d\theta d\psi
\end{aligned}
\tag{3.22}
$$

Once $u^{(1)}(a, \psi, \theta)$, $A^{(1)}(a)$ and $B^{(1)}(a)$ are determined, one can also determine $f_1(a, \psi, \theta)$ by a further generalization of the Fourier expansion of a periodic function of N arguments $x_1, x_2, \ldots, x_N$, viz.:

$$
f(x_1, x_2, \ldots, x_N) = \sum_{n_1, n_2, \ldots, n_N = -\infty}^{\infty} c_{n_1, n_2, \ldots, n_N} e^{i(n_1x_1 + n_2x_2 + \cdots + n_Nx_N)}
$$

with the corresponding Fourier determination of

$$
\begin{aligned}
c_{n_1, n_2, \ldots, n_N} = \frac{1}{(2\pi)^N}\int_{0(N)}^{2\pi}\cdots\int_0^{2\pi} f(\xi_1, \ldots, \xi_N) \\
\times \exp[-i(n_1\xi_1 + \ldots + n_N\xi_N)]d\xi_1, \ldots, d\xi_N
\end{aligned}
$$

In this manner one determines $u^{(2)}(a, \psi, \theta)$; $A^{(2)}(a)$ and $B^{(2)}(a)$ and so on: the formulas become rather complicated even for the second order but the procedure is quite definite.

From (3.22) it follows that in the equations of the first approximation appears only the free term $f_0(x,\dot{x})$ of the expansion of the perturbing force $f(\theta, x, \dot{x})$, so that (in view of (3.6)) one has

$$
f_0(x,\dot{x}) = \lim_{T\to\infty}\frac{1}{T}\int_0^T f(\tau, x, \dot{x})d\tau \tag{3.23}
$$

For the d.e. of the first approximation it is sufficient, therefore, to average the perturbing term μf with respect to the time appearing explicitly. One thus falls back on the determination of $A^{(1)}(a)$ and $B^{(1)}(a)$ in the preceding chapter.

It follows that the nonresonance treatment of the nonautonomous systems leads to the same procedure as for the autonomous systems except for the use of the multiple Fourier series instead of the ordinary Fourier series; this merely results from the periodicity of solution in terms of more than one argument.

The second approximation is

$$
x = a\cos\psi + \mu u^{(1)}(a, \psi, \theta) \tag{3.24}
$$

where $u^{(1)}(a, \psi, \theta)$ is given by (3.21). As was just mentioned, the effect of the external periodic excitation in this case is felt only in the second approximation; this means that the "combination frequencies" (like 2.6) appear only in the second approximation; the amplitudes of these *combination harmonics* are $0(\mu)$.

Considering now the case of a stationary oscillation (in 3.24), one has:

$$a = \text{const}; \qquad \psi = \omega(a)t + \theta; \qquad \theta = \text{const} \tag{3.25}$$

The oscillation of x consists of an "autoperiodic oscillation" with frequency $\omega(a)$, "heteroperiodic oscillation" with frequencies $n\nu$ ($n = 1, 2, 3,\ldots$) and "combination oscillations" with frequencies $n\nu \pm m\omega$ ($n, m = 1, 2, 3,\ldots$). The amplitudes of the latter increase with the approach to a corresponding resonance when the divisor $\omega^2 - (n\nu \pm m\omega)^2$ becomes small.

If the free (or "autoperiodic") oscillations are absent ($a = 0$), (3.24) becomes

$$x = \mu \sum_{n=1}^{\infty} \frac{A_n \cos n\theta + B_n \sin n\theta}{\omega^2 - n^2\nu^2} \tag{3.26}$$

where

$$A_n = \frac{1}{\pi}\int_0^{2\pi} f_0(\theta, 0, 0) \cos n\theta d\theta; \qquad B_n = \frac{1}{\pi}\int_0^{2\pi} f_0(\theta, 0, 0) \sin n\theta d\theta$$

Here one has a purely heteroperiodic (or forced) oscillation. It was mentioned previously that the term with $f_0(x,\dot{x})$ which appears in the first approximation (and corresponds to the first term of the expansion $f(\theta, x, \dot{x})$) determines the stability of the autoperiodic oscillation. In fact, if the equivalent decrement (Section 7, Chapter 14)

$$\lambda_e{}^*(a) > 0 \tag{3.27}$$

then $a(t)_{t\to\infty} \to 0$.

The equivalent decrement is given by the expression

$$\lambda_e{}^*(a) = \frac{1}{4\pi^2\omega}\int_0^{2\pi}\int_0^{2\pi} f(\theta, a\cos\psi, -a\omega\sin\psi)\sin\psi d\theta d\psi$$

$$= \frac{1}{2\pi\omega}\int_0^{2\pi} f_0(a\cos\psi, -a\omega\sin\psi)\sin\psi d\psi \tag{3.28}$$

If the external periodic excitation is absent (that is, the function f in (2.1) does not depend on t explicitly), one has the usual condition of self-excitation

$$\lambda_e(a) < 0 \tag{3.29}$$

where

$$\lambda_e(a) = \frac{1}{2\pi\omega}\int_0^{2\pi} f(0, a\cos\psi, -a\omega\sin\psi)\sin\psi d\psi \tag{3.30}$$

It may happen (depending exclusively on the form of the function f) that a system which is self-excited in the absence of the external periodic excitation loses its self-excitation as soon as external periodic excitation is applied. This corresponds to the so-called *asynchronous quenching*. There may be also an opposite effect of *asynchronous excitaton.*

As an example, consider the van der Pol equation

$$\ddot{x} + \mu(x^2 - 1)\dot{x} + x = E \sin \nu t \tag{3.31}$$

If one sets $x = y + U \sin \nu t$ where $U = E/(1 - \nu^2)$ (we consider a nonresonance case), (3.31) becomes

$$\ddot{y} + y = \mu[1 - (y + U \sin \nu t)^2](\dot{y} + U\nu \cos \nu t) \tag{3.32}$$

In the first approximation, the solution of this d.e. is

$$y = a \cos (t + \theta); \qquad \theta = \text{const} \tag{3.33}$$

where a is given by the d.e.

$$\dot{a} = \frac{\mu a}{2}\left(1 - \frac{a^2}{4} - \frac{U^2}{2}\right) \tag{3.34}$$

If $U^2 < 2$, the system is self-excited and there exists a stationary state with amplitude:

$$a^2 = 4 - 2U^2 \tag{3.35}$$

For $U^2 > 2$, $a(t)_{t\to\infty} \to 0$, which is the phenomenon of *asynchronous quenching*.

In the second approximation, the solution of (3.31) yields

$$\begin{aligned} x = a \cos \psi &+ \mu \frac{U\nu(4 - U^2 - 2a^2)}{4(1 - \nu^2)} \cos \theta + \mu \frac{U^3\nu}{4(1 - 9\nu^2)} \cos 3\theta \\ &+ \mu \frac{Ua^2(2 + \nu)}{4(1 + \nu)(3 + \nu)} \cos (\theta + 2\psi) \\ &+ \mu \frac{Ua^2(2 + \nu)}{4(1 - \nu)(3 - \nu)} \cdot \cos (\theta - 2\psi) + \mu \frac{U^2a(2 + \nu)}{16\nu(1 + \nu)} \sin (2\theta + \psi) \\ &+ \mu \frac{U^2a(1 - 2\nu)}{16\nu(1 - \nu)} \sin (2\theta - \psi) - \frac{a^3}{32} \sin 3\psi \end{aligned} \tag{3.36}$$

where a and ψ satisfy the d.e.

$$\begin{aligned} \dot{a} &= \frac{\mu a}{2}\left(1 - \frac{a^2}{4} - \frac{U^2}{2}\right); \\ \dot{\psi} &= 1 - \mu^2\left(\frac{1}{8} - \frac{a^2}{8} + \frac{7a^4}{256}\right) + \mu^2 \frac{U^2(5\nu - 1)}{8(1 - \nu^2)} \\ &+ \mu^2 \frac{U^2a^2(7\nu^4 - 40\nu^2 + 32\nu - 9)}{32(9 - \nu^2)(1 - \nu^2)} + \mu^2 \frac{U^4(1 + 4\nu - 8\nu^2)}{64(1 - \nu^2)} \end{aligned} \tag{3.37}$$

In the second approximation, in addition to the forced oscillations with frequencies ν and 3ν, there are also oscillations with frequency 3ω as well as the combination frequencies $\nu \pm 2\omega$; $2\nu \pm \omega$. Besides, if $U^2 < 2$, the heteroperiodic oscillation becomes unstable; if $U^2 > 2$ the heteroperiodic oscillation is the only one possible in the stationary state (since the autoperiodic oscillation vanishes), so that, ultimately, the oscillation is

$$x = \mu \frac{U\nu(4 - U^2)}{4(1 - \nu^2)} \cos \theta + \mu \frac{U^3\nu}{4(1 - 9\nu^2)} \cos 3\theta \tag{3.38}$$

4. Successive approximations for resonance oscillations

We assume now that

$$\omega \simeq \frac{p}{q} \nu \tag{4.1}$$

where p and q are relatively prime. Two cases are possible: (I) when (4.1) corresponds to the equality, that is, the *exact* resonance; and (II) when (4.1) corresponds to an *approximate equality*, that is, the *neighborhood* of the resonance, at the limit of which the resonance zone transforms itself into a nonresonance zone (the matter treated in the preceding section).

One can define by $\mu\Delta$ the *deviation* between ω^2 (square of the free frequency when $\mu = 0$) and $\left(\frac{p}{q}\nu\right)^2$, i.e.

$$\omega^2 = \left(\frac{p}{q}\nu\right)^2 + \mu\Delta \tag{4.2}$$

The basic d.e. can be written then as

$$\ddot{x} + \left(\frac{p}{q}\nu\right)^2 x = \mu[f(\nu t, x, \dot{x}) - \Delta x] \tag{4.3}$$

which amounts to the transfer of the deviation $\mu\Delta$ to the right-hand side. We look again for the solution of the form (compare with (3.1))

$$x = a \cos \psi + \mu u^{(1)}(a, \psi, \nu t) + \mu^2 u^{(2)}(a, \psi, \nu t) \tag{4.4}$$

where a and ψ are functions of t and $\psi = (p/q)\,\nu t + \theta$, θ being the phase; it is clear that for the *exact* resonance $\theta = \text{const}$.

In the nonresonance case, as was mentioned (beginning of Section 13), there is no stationary relation between the phase of the external periodic excitation and that of the oscillation; here, on the contrary this circumstance plays an important role and we can set

$$\theta = \psi - \frac{p}{q} \nu t \tag{4.5}$$

This means that in the d.e. for a and ψ the terms $A^{(1)}, \ldots, B^{(1)}, \ldots$ will depend not only on a, as previously, but also on θ, so that we have

$$\begin{aligned} \dot{a} &= \mu A^{(1)}(a,\theta) + \mu^2 A^{(2)}(a,\theta) + \ldots \\ \dot{\psi} &= \frac{p}{q}\nu + \mu B^{(1)}(a,\theta) + \mu^2 B^{(2)}(a,\theta) + \ldots \end{aligned} \tag{4.6}$$

Inasmuch as on the right-hand side appears the phase angle θ and not the *total phase* ψ, it is useful to keep the variable θ everywhere so that instead of (4.4), (4.6) we can write

$$x = a \cos\left(\frac{p}{q}\nu t + \theta\right) + \mu u^{(1)}\left(a, \theta, \frac{\nu}{q} t\right) + \mu^2 u^{(2)}\left(a, \theta, \frac{\nu}{q} t\right) \tag{4.7}$$

$$\begin{aligned} \dot{a} &= \mu A^{(1)}(a,\theta) + \mu^2 A^{(1)}(a,\theta) + \ldots \\ \dot{\theta} &= \mu B^{(1)}(a,\theta) + \mu^2 B^{(2)}(a,\theta) + \ldots \end{aligned} \tag{4.8}$$

where, for the sake of convenience, we shall use θ instead of ψ, since

$$\frac{p}{q}\nu t + \theta = \psi \tag{4.9}$$

but it must be recalled that now we have two partial derivatives $\psi_t = \dot{\psi}$ and ψ_θ in carrying out calculations.

As previously we have first

$$\begin{aligned} \dot{x} = {} & [\cos\psi + \mu u_a^{(1)} + \mu^2 u_a^{(2)} + \ldots]\dot{a} + [-a\sin\psi + \mu u_\theta^{(1)} + \ldots]\dot{\theta} \\ & + \left[-a\frac{p}{q}\nu\sin\psi + \mu\dot{u}^{(1)} + \mu^2\dot{u}^{(2)} + \ldots\right] \end{aligned} \tag{4.10}$$

$$\begin{aligned} \ddot{x} = {} & [\cos\psi + \mu u_a^{(1)} + \mu^2 u_a^{(2)} + \ldots]\ddot{a} + [\mu u_{aa}^{(1)} + \mu^2 u_{aa}^{(2)} + \ldots]\dot{a}^2 \\ & + 2[-\sin\psi + \mu u_{a\theta}^{(1)} + \mu^2 u_{a\theta}^{(2)} + \ldots]\dot{a}\dot{\theta} \\ & + 2\left[-\frac{p}{q}\nu\sin\psi + \mu u_{at}^{(1)} + \mu^2 u_{at}^{(2)} + \ldots\right]\dot{a}\dot{\theta} \\ & + [-a\sin\psi + \mu u_\theta^{(1)} + \mu^2 u_\theta^{(2)}]\ddot{\theta} \\ & + [-a\cos\psi + \mu u_{\theta\theta}^{(1)} + \mu^2 u_{\theta\theta}^{(2)} + \ldots]\dot{\theta}^2 \\ & + 2\left[-a\frac{p}{q}\nu\cos\psi + \mu u_{\theta t}^{(1)} + \mu^2 u_{\theta t}^{(2)} + \ldots\right]\dot{\theta} \\ & + \left[-a\left(\frac{p}{q}\nu\right)^2\cos\psi + \mu\ddot{u}^{(1)} + \mu^2\ddot{u}^{(2)} + \ldots\right] \end{aligned} \tag{4.11}$$

Likewise from (4.8) we have

$$\ddot{a} = \mu^2[A^{(1)}A_a^{(1)} + B^{(1)}A_\theta^{(1)}] + \mu^3 \ldots; \qquad \dot{a}^2 = \mu^2 A^{(1)2} + \mu^3 \ldots;$$

$$\dot{\theta}^2 = \mu^2 B^{(1)2} + \mu^3 \ldots; \qquad \ddot{\theta} = \mu^2[A^{(1)}B_a^{(1)} + B^{(1)}B_\theta^{(1)}] + \mu^3 \ldots;$$

$$\dot{a}\dot{\theta} = \mu^2 A^{(1)}B^{(1)} + \mu^3 \ldots$$

The rest of the formal procedure is the same as before, viz.: one forms the left-hand side of the d.e. (4.3), then carries out the development of its right-hand side according to the powers of μ and, finally, equates the coefficients of like powers of μ which results in a recursive system of the d.e. of successive approximations. One has thus:

$$\begin{aligned} \ddot{u}^{(1)} + \left(\frac{p}{q}\nu\right)^2 u^{(1)} &= f_0(a, \nu t, \psi) + 2\frac{p}{q}\nu A^{(1)} \sin\psi \\ &\quad + (2ap\nu/q)B^{(1)} \cos\psi - \Delta a \cos\psi \\ \ddot{u}^{(2)} + \left(\frac{p}{q}\nu\right)^2 u^{(2)} &= f_1(a, \nu t, \psi) \\ &\quad + [(2p\nu/q)A^{(2)} + aA^{(1)}B_a^{(1)} + aB^{(1)}B_\theta^{(1)} + 2A^{(1)}B^{(1)}] \sin\psi \\ &\quad + [(2ap\nu/q)B^{(2)} - A^{(1)}A_a^{(1)} - B^{(1)}A_\theta^{(1)} + aB^{(1)2}] \cos\psi \\ &\ldots\ldots\ldots\ldots \end{aligned} \tag{4.12}$$

where

$$\begin{aligned} f_0(a, \nu t, \psi) &= f[\nu t, a\cos\psi, -(ap\nu/q)\sin\psi] \\ f_1(a, \nu t, \psi) &= f_x[\nu t, a\cos\psi, -(ap\nu/q)\sin\psi]u^{(1)} \\ &\quad + f_{\dot{x}}[\nu t, a\cos\psi, -(ap\nu/q)\sin\psi] \\ &\quad \times (A^{(1)}\cos\psi - aB^{(1)}\sin\psi + \dot{u}^{(1)}) \\ &\quad - \Delta u^{(1)} - 2A^{(1)}u_{at}^{(1)} - 2B^{(1)}u_{\theta t}^{(1)} \end{aligned} \tag{4.13}$$

As previously, $f_k(a, \nu t, \psi)$ are periodic with period 2π with respect to both angular variables νt and ψ; as to $A^{(i)}(a,\theta)$ and $B^{(i)}(a,\theta)$, as will appear later, they are periodic in θ with period 2π.

The next step is the integration of the first d.e. (4.12). One begins by representing $u^{(i)}$ and f_0 in the form of a Fourier series

$$\begin{aligned} u^{(i)}(a, \nu t, \psi) &= \sum_n \sum_m u_{nm}^{(i)} \exp[i(n\nu t + m\psi)] \\ f_0(a, \nu t, \psi) &= \sum_n \sum_m f_{nm}^{(0)}(a) e^{i(n\nu t + m\psi)} \\ f_{nm}^{(0)}(a) &= \frac{1}{4\pi^2} \int_0^{2\pi} \int_0^{2\pi} f(a, \theta, \psi) e^{-i(n\theta + m\psi)} d\theta d\psi \end{aligned} \tag{4.14}$$

Substituting these expressions into the first d.e. (4.12) and identifying the terms corresponding to the same harmonics, one has

$$u_{nm}{}^{(1)}(a) = f_{nm}{}^{(0)}(a)/[(p\nu/q)^2 - (n\nu + mp\nu/q)^2] \tag{4.15}$$

for all n and m for which the denominator is different from zero, which is equivalent to the condition

$$nq + (m \pm 1)p \neq 0 \tag{4.16}$$

There is also a relation between $A^{(1)}(a,\theta)$ and $B^{(1)}(a,\theta)$ given by

$$(2p\nu/q)A^{(1)} \sin \psi + ((2ap\nu/q)B^{(1)} - \Delta a) \cos \psi + \sum_{\substack{n \quad m \\ [nq+(m\pm 1)p=0]}}\sum e^{i[n\nu t+m\psi]} f_{nm}{}^{(0)}(a) = 0 \tag{4.17}$$

the double summation extending for all m and n for which $nq + (m \pm 1)p = 0$.

If one substitutes (4.15) into the first equation (4.14), one obtains the explicit expression for $u^{(1)}(a, \nu t, \psi)$, viz.:

$$u^{(1)}(a, \nu t, \psi) = \frac{1}{4\pi^2}\sum_n \sum_m \frac{e^{i(n\nu t+m\psi)}}{(p\nu/q)^2 - (n\nu + mp\nu/q)^2} \times \int_0^{2\pi}\int_0^{2\pi} f(a, \nu t, \psi) e^{-i(n\nu t+m\psi)} \cdot d(\nu t) d\psi \tag{4.18}$$

In the double summation the indices n and m must be taken for all values for which $nq + (m \pm 1)p \neq 0$. In (4.18) appear complex exponential functions

$$\exp i\left[(nq + mp)\frac{\nu t}{q} + m\theta\right] = \exp i\left[\mp\left(\frac{p\nu}{q}t + \theta\right) + (m \pm 1)\theta\right]$$
$$= \left\{\cos\left[\left(\frac{p\nu}{q}\right)t + \theta\right] \mp i \sin\left[\left(\frac{p\nu}{q}\right)t + \theta\right]\right\} e^{i(m\pm 1)\theta}$$

When, instead of (4.16), we have $nq = (m \pm 1)p = 0$, $m \pm 1$ is divisible by q and we designate it as σ; then equating the coefficients of $\cos \psi$ and $\sin \psi$ in (4.17) one gets

$$\begin{aligned} A^{(1)}(a,\theta) &= \frac{q}{4\pi^2 \nu p}\sum_\sigma \exp(iq\sigma\theta) \\ &\quad \times \int_0^{2\pi}\int_0^{2\pi} f_0(a, \nu t, \psi) \exp(-iq\sigma\theta) \sin \psi d(\nu t) d\psi \\ B^{(1)}(a,\theta) &= \frac{\Delta}{2}\frac{q}{p\nu} - \frac{q}{4\pi^2 a\nu p}\sum_\sigma \exp(iq\sigma\theta) \\ &\quad \times \int_0^{2\pi}\int_0^{2\pi} f_0(a, \nu t, \psi) \exp(-iq\sigma\theta) \cos \psi d(\nu t) d\psi \end{aligned} \tag{4.19}$$

Thus for the first approximation of the resonance case, the solution of the d.e. (4.3) is

$$x = a \cos\left[\left(\frac{p\nu}{q}\right)t + \theta\right] \tag{4.20}$$

where a and θ are given by the d.e.

$$\begin{aligned} \dot{a} &= -\frac{\mu q}{4\pi^2 \nu p} \sum_{\sigma} e^{ips\theta} \int_0^{2\pi} \int_0^{2\pi} f_0(a, \nu t, \mu) e^{-iq\sigma\theta} \sin \psi d(\nu t) d\psi \\ \dot{\theta} &= \frac{\mu \Delta q}{2p\nu} - \frac{\mu q}{4\pi^2 a\nu p} \sum_{\sigma} e^{iq\sigma\theta} \int_0^{2\pi} \int_0^{2\pi} f_0(a, \nu t, \psi) e^{iq\sigma\theta} \cos \psi d(\nu t) d\psi \end{aligned} \tag{4.21}$$

taking into account that $\mu\Delta$ is of the first order of smallness. Once $u^{(1)}$, $A^{(1)}$, and $B^{(1)}$ are determined, one can calculate $f_1(a, \nu t, \psi)$ and, therefore $A^{(2)}(a,\theta)$ and $B^{(2)}(a,\theta)$.

One can also extend this analysis to cases when the deviation Δ is not too small; this permits investigating the function region between the resonance zone and the nonresonance zone. This is the most general case from which both the nonresonance and the resonance cases appear as special cases.

The solution is taken in the form

$$x = a \cos \psi + \mu u^{(1)}(a, \nu t, \psi) + \mu^2 u^{(2)}(a, \nu t, \psi) + \dots \tag{4.22}$$

and two other associated d.e. are

$$\begin{aligned} \dot{a} &= \mu A^{(1)}(a,\theta) + \mu^2 A^{(2)}(a,\theta) + \dots \\ \dot{\theta} &= \omega - \frac{p\nu}{q} + \mu B^{(1)}(a,\theta) + \mu^2 B^{(2)}(a,\theta) + \dots \end{aligned} \tag{4.23}$$

where $\omega = p\nu/q$ is not small.

In these formulas $u^{(1)}(a, \nu t, \psi)$, $u^{(2)}(a, \nu t, \psi)$ have period 2π with respect to both angular variables ψ and νt and $A^{(i)}(a,\theta)$ and $B^i(a,\theta)$; $i = 1, 2, \dots$ are periodic with period 2π with respect to the angular variable θ.

For the determination of these functions one could use the procedure of direct differentiations and substitutions into the fundamental d.e. with the subsequent equating terms with equal powers of μ as was done previously. It is more convenient, however, to use the method of the harmonic balance (Chapter 14). One replaces the expression for the fundamental harmonic: $x = a \cos \psi$; $\psi = (p/q)\nu t + \theta$ into the basic d.e. (2.1), taking into account (4.23), and one equates the fundamental harmonics on both sides of (2.1) after these substitutions.

For the second approximation for the determination of the fundamental harmonic on the left side of (2.1) one must take into account the terms with μ^2 and for $f(\nu t, x, \dot{x})$ one introduces $\mu u^{(1)}(a, \nu t, \psi)$.

For the details of these calculations we refer to Bogoliubov and Mitropolsky[1], pages 176–177; we indicate here only the result of this procedure in connection with the first approximation:

$$\left(\omega - \frac{p}{q}\nu\right)\frac{\partial A^{(1)}}{\partial\theta} - 2a\omega B^{(1)}$$

$$= \frac{1}{2\pi^2}\sum_{\sigma}\exp(i\sigma q\theta)\int_0^{2\pi}\int_0^{2\pi} f_0(a, \theta, \psi)\exp(-iq\sigma\theta')\cos\psi d\theta d\psi \tag{4.24}$$

$$\left(\omega - \frac{p}{q}\nu\right)a\frac{\partial B^{(1)}}{\partial\theta} + 2\omega A^{(1)}$$

$$= -\frac{1}{2\pi^2}\sum_{\sigma}\exp(i\sigma q\theta)\int_0^{2\pi}\int_0^{2\pi} f_0(a, \theta, \psi)\exp(-iq\sigma\theta')\sin\psi d\theta d\psi$$

It is noted that the right-hand sides of these equations are periodic in θ and can be represented as $\sum k_n(a)\exp(in\theta)$. In view of this for the determination of $A'(a,\theta)$ and $B'(a,\theta)$ one can use expressions of a similar form which amounts to simple trigonometric operations.

For the first approximation one has

$$\dot{a} = \mu A^{(1)}(a,\theta); \qquad \dot{\theta} = \omega - \frac{p\nu}{q} + \mu B^{(1)}(a,\theta) \tag{4.25}$$

where $A^{(1)}(a,\theta)$ and $B^{(1)}(a,\theta)$ are periodic solutions of the system (4.24).

Equations (4.25) generally cannot be integrated in a closed form because their right-hand sides depend on a and θ. However, the qualitative aspect of solution can be established from the theory of Poincaré. This amounts to the determination of singular points of the system:

$$A^{(1)}(a,\theta) = 0; \qquad \omega - \frac{p}{q}\nu + \mu B^{(1)}(a,\theta) = 0 \tag{4.26}$$

There may be also periodic solutions. This leads to two principal forms of stationary oscillations: (a) those which correspond to the constant solution (singular points) and (b) oscillations corresponding to the periodic solution.

In the first case and in the first approximation, oscillations occur with frequency equal *exactly* to $(p/q)\nu$; this may be called *the exact* resonance. For higher-order approximations in addition to the fundamental frequency $(p/q)\nu$ are present also other harmonics of the "divided"† frequency ν/q.

[1] See footnote [1], page 356.

† "Subharmonics" but we shall introduce this term only in Chapter 19.

If the system has a constant solution $a = 0$, one obtains clearly the non-resonance case; this may be regarded as a *heteroperiodic solution*, since the autoperiodic component vanishes.

For the stationary synchronous state the question of stability reduces as usual to the investigation of the variational equations

$$\begin{aligned}\frac{d\delta a}{dt} &= \mu A_a^{(1)}(a_0,\theta_0)\delta a + \mu A_\theta^{(1)}(a_0,\theta_0)\delta\theta \\ \frac{d\delta\theta}{dt} &= \mu B_a^{(1)}(a_0,\theta_0)\delta a + \mu B_\theta^{(1)}(a_0,\theta_0)\delta\theta\end{aligned} \tag{4.27}$$

which leads to the characteristic equation

$$S^1 - \mu(A_a^{(1)} + B_\theta^{(1)})S + \mu^2(A_a^{(1)}B_\theta^{(1)} - A_\theta^{(1)}B_a^{(1)}) = 0 \tag{4.28}$$

with the conditions of stability:

$$A_a^{(1)} + B_\theta^{(1)} < 0; \qquad A_a^{(1)}B_\theta^{(1)} - A_\theta^{(1)}B_a^{(1)} > 0 \tag{4.29}$$

If no singular point exists, there are beats of two oscillations with frequencies ω and $(p\nu/q) + \Delta\omega$. As an example consider the operation of an ordinary (induction coupled) electron-tube circuit on the grid of which (in addition to the feedback from the oscillating circuit) is impressed an electromotive force $F \cos \nu t$.

The d.e. of the circuit is

$$\ddot{e} + \omega^2 e = -\omega^2\left[\frac{L}{R}\dot{e} - (M - DL)\dot{I}_a\right] \tag{4.30}$$

where e is the grid voltage; L, R are, respectively, inductance and resistance of the oscillating circuit; M is the coefficient of mutual inductance (between the anode and the grid circuit); D is the transconductance; $\omega^2 = 1/LC$ is the free frequency of the oscillating circuit and

$$I_a = f(E_0 + F\cos\nu t + e) \tag{4.31}$$

is the anode current considered as a nonlinear function of the grid voltage.

Thus, for instance, if $f(E_0 + u)$, $u = e + F\cos\nu t$ is a cubic polynomial, viz.:

$$f(E_0 + u) = f(E_0) + S_0 u + S_1 u^2 - S_2 u^3, \qquad S_2 > 0 \tag{4.32}$$

and all terms on the right-hand side of (4.30) are small, one can apply the preceding formulas.

Thus, for $p = 1$, $q = 2$, that is, $\omega = \nu/2$ and in the first approximation one has

$$e = a\cos\left(\frac{\nu}{2}t + \theta\right)$$

where a and θ are given by the d.e.

$$\dot{a} = \delta_0\left[-\frac{3S_2}{4S_{\text{cr.}}}a^3 + \left(\frac{S_0 - S_{\text{cr.}} - \frac{3}{2}S_2F^2}{S_{\text{cr.}}}\right)\cdot a\right] + \frac{aS_1F\delta_0}{2S_{\text{cr.}}}\cos 2\theta$$

$$\dot{\theta} = \left(\omega - \frac{\nu}{2}\right) - \frac{S_1F\delta_0}{2S_{\text{cr.}}}\cdot\sin 2\theta \tag{4.32a}$$

where

$$\delta_0 = \frac{1}{2RC}; \qquad S_{\text{cr.}} = \frac{L}{R(M - DL)}$$

Setting $m = 3S_2/4S_{\text{cr.}}$; $r = (S_0 - S_{\text{cr.}} - \frac{3}{2}S_2F^2)/S_{\text{cr.}}$; $s = (S_1F\delta_0)/2S_{\text{cr.}}$; $k = \omega - (\nu/2)$, (4.32a) is

$$\dot{a} = a[-\delta_0 ma^2 + (r + s\cos 2\theta)]; \qquad \dot{\theta} = k - s\sin 2\theta \tag{4.33}$$

where m and s are positive constants.

It is noted that there may be a point of equilibrium $\dot{\theta} = 0$ if $|k| < |s|$, in which case $\theta = \theta_0 = \frac{1}{2}\arcsin(k/s)$. The stability of equilibrium is determined from the variational equation which is here $d\delta\theta/dt = -2s\cos\theta_0\delta\theta$ which requires that $\cos\theta_0 > 0$. All depends on the existence of a positive root of the bracket in the first equation (4.33). As $\delta_0 > 0$ and (we assume) $m > 0$, then

$$a_0 = \sqrt{\frac{r + s\cos 2\theta_0}{\delta_0 m}} \tag{4.34}$$

is a stationary amplitude assuming that $r + s\cos 2\theta_0 > 0$. In such a case the system operates in a "synchronous" condition; that is, there exists an oscillation

$$e = a_0\cos\left(\frac{\nu}{2}t + \theta_0\right) \tag{4.35}$$

with a_0 and θ_0 determined as previously explained.

It may happen, however, that a_0 does not exist; this occurs if $r + s\cos\theta_0 < 0$, that is, for a sufficiently large value of F. The only possible stationary state is then $a_0 = 0$, in which case the only oscillation possible is the *heteroperiodic* one, that is, one produced by the external periodic excitation $e = F\cos\nu t$.

If $k = \omega - (\nu/2)$ is large, (so that $|k| > |s|$), $\dot{\theta}$ keeps the same sign. One can integrate the second equation (4.33) and represent θ as

$$0 = \Delta\omega t + \Phi(\Delta\omega t + \theta) \tag{4.36}$$

where $\Phi(\theta)$ is periodic with period 2π. In this case θ does not vary uniformly but undergoes a phase modulation.

Here again, the equilibrium of the amplitude may take place either for $a = 0$ or for $a = a_0$, depending on the existence of a real positive root $a = a_0$ of the first equation (4.33). If such a root $a = a_0$ exists and is stable, the phase modulation produces a corresponding amplitude modulation owing to the term $s \cos 2\theta$ in the bracket of (4.33).

Summing up:

If $k = \omega - (\nu/2)$ (the deviation between the "free" frequency and the "forced" frequency) is sufficiently small, there may be a synchronous operation with frequency $\nu/2$ provided a_0 exists. It is clear that, if F is sufficiently large (and $r < 0$), the amplitude $a = a_0$ will cease to exist, in which case the only stationary condition is $a = 0$, that is, a heteroperiodic state.

If k is large enough, both amplitude and phase are "modulated" quantities, which means that both the autoperiodic (self-excited) and the "heteroperiodic" (forced) oscillations exist at the same time ("the beats") but, again, if the condition for the existence of the autoperiodic amplitude is not fulfilled, the heteroperiodic oscillation exists alone.

5. External periodic excitation of a nonlinear oscillator; jumps of amplitude

We consider an oscillatory system specified by the d.e.†

$$m\ddot{x} + kx = \mu f(x,\dot{x}) + \mu E \sin \nu t \tag{5.1}$$

and investigate its fundamental resonance ($p = q = 1$, Section 2).

In the first approximation we look for a solution of the form

$$x = a \cos (\nu t + \theta) = a \cos \gamma \tag{5.2}$$

where a and θ are given by the d.e.

$$\begin{aligned} \dot{a} &= -\frac{\mu}{2\pi\omega m}\int_0^{2\pi} f_0(a,\gamma) \sin \gamma d\gamma - \frac{\mu E}{m(\omega + \nu)} \cos \theta \\ \dot{\theta} &= \omega - \nu - \frac{\mu}{2\pi\omega a m}\int_0^{2\pi} f_0(a,\gamma) \cos \gamma d\gamma + \frac{\mu E}{ma(\omega + \nu)} \sin \theta \end{aligned} \tag{5.3}$$

where $f_0(a,\gamma) = f(a \cos \gamma, \; -a\omega \sin \gamma)$.

† It is assumed that the amplitude of the external periodic excitations is small; if such assumptions cannot be made and we have the d.e.: $m\ddot{x} + k\dot{x} = \mu f(x,\dot{x}) + E \sin \nu t$, the change of variable $x = y + [E/(k - \nu^2 m)] \sin \nu t$ removes the term with $E \sin \nu t$.

We introduce the notations

$$\lambda_e(a) = \frac{\mu}{\pi a \omega} \int_0^{2\pi} f_0(a,\gamma) \sin \gamma d\gamma;$$
$$k_e(a) = k - \frac{\mu}{\pi a} \int_0^{2\pi} f_0(a,\gamma) \cos \gamma d\gamma \tag{5.4}$$

which may be regarded as equivalent parameters (Section 7, Chapter 14) if $E = 0$. Equations (5.3) can be written as

$$\dot{a} = -\delta_e(a)a - \frac{\mu E}{m(\omega + \nu)} \cos \theta;$$
$$\dot{\theta} = \omega_e(a) - \nu + \frac{\mu E}{ma(\omega + \nu)} \sin \theta \tag{5.5}$$

where $\delta_e(a) = \lambda_e(a)/(2m)$ and $\omega_e(a) = \sqrt{k_e(a)/m}$ may be considered as equivalent decrement and frequency, respectively.

The stationary state exists if (5.5) has a singular point, that is, if

$$\delta_e(a)a + \frac{\mu E}{m(\omega + \nu)} \cos \theta = 0; \quad \omega_e(a) - \nu + \frac{\mu E}{ma(\omega + \nu)} \sin \theta = 0 \tag{5.6}$$

Up to $0(\mu^2)$ these expressions can be written as

$$2m\nu a\delta_e(a) = -\mu E \cos \theta; \qquad ma[\omega_e^2(a) - \nu^2] = -\mu E \sin \theta \tag{5.7}$$

whence, upon the elimination of θ, one has

$$m^2a^2[(\omega_e^2(a) - \nu^2)^2 + 4\nu^2\delta_e^2(a)] = \mu^2E^2 \tag{5.8}$$

which determines the amplitude and the phase in (5.2).

In the resonance case ($\omega = \nu$) the system (5.3) can be written as

$$2\nu\dot{a} = -2\nu a\delta_e(a) - \frac{\mu E}{m} \cos \theta;$$
$$2\nu a\dot{\theta} = [\omega_e^2(a) - \nu^2]a + \frac{\mu E}{m} \sin \theta \tag{5.9}$$

If one sets the right-hand side of these equations as $R(a,\theta)$ and $\Phi(a,\theta)$, respectively, the stationary state is given again by two equations

$$aR(a,\theta) = \Phi(a,\theta) = 0 \tag{5.10}$$

The conditions of stability are (we use subscripts a and θ to designate the partial derivatives with respect to these variables)

$$aR_a(a_0,\theta_0) + \Phi_\theta(a_0,\theta_0) < 0;$$
$$R_a(a_0,\theta_0)\Phi_\theta(a_0,\theta_0) - R_\theta(a_0,\theta_0)\Phi_a(a_0,\theta_0) > 0 \tag{5.11}$$

where a_0 and θ_0 are the roots common to two equations (5.10).

If one takes into account the first equation (5.9), the first condition (5.11) of stability becomes

$$aR_a + \Phi_\theta = -2\nu a \frac{d[a\delta_e(a)]}{da} - 2\nu a\delta_e(a) = -2\nu \frac{d[a^2\delta_e(a)]}{da} < 0 \tag{5.12}$$

Moreover $2\nu a^2\delta_e(a) = \dfrac{a^2\lambda_e(a)}{m}\nu = \dfrac{2\nu}{m\omega^2}W(a)$ where we define

$$W(a) = \frac{1}{2\pi}\int_0^{2\pi} \mu f(a\cos\gamma, -a\omega\sin\gamma)a\omega\sin\gamma d\gamma \tag{5.13}$$

which may be regarded as *average power* dissipated for period by the oscillation (5.2) and, in normal cases, it is obvious that $W(a)$ increases monotonically with a so that $W_a(a) > 0$. Thus, the first condition (5.11) is always fulfilled.

As to the second condition (5.11), we differentiate (5.10) with respect to ν, taking into account the fact that R and Φ depend also explicitly on ν which gives

$$R_a a_\nu + R_\theta\theta_\nu = -R_\nu; \qquad \Phi_a a_\nu + \Phi_\theta\theta_\nu = -\Phi_\nu \tag{5.14}$$

where $a_\nu = da/d\nu$ and $\theta_\nu = d\theta/d\nu$. The second condition of stability is then

$$a_\nu(R_a\Phi_\theta - R_\theta\Phi_a) = \Phi_\nu R_\theta - R_\nu\Phi_\theta \tag{5.15}$$

where

$$\begin{aligned} R_\theta &= \frac{\mu E}{m}\sin\theta; \qquad \Phi_\theta = \frac{\mu E}{m}\cos\theta; \\ R_\nu &= -2\delta_e(a)a; \qquad \Phi_\nu = -2\nu a \end{aligned} \tag{5.16}$$

so that the right-hand side in (5.15) becomes

$$\Phi_\nu R_\theta - R_\nu\Phi_\theta = 2a\left(-\nu\frac{E}{m}\sin\theta + \delta_e(a)\frac{E}{m}\cos\theta\right)\mu \tag{5.17}$$

Taking into account (5.14)

$$(\Phi_\nu R_\theta - R_\nu\Phi_\theta) = 2\nu a^2[(\omega_e^2(a) - \nu^2) - 2\delta_e^2(a)] \tag{5.18}$$

we have

$$(R_a\Phi_\theta - \Phi_a R_\theta)a_\nu = 2\nu a^2[(\omega_e^2(a) - \nu^2) - 2\delta_e^2(a)] \tag{5.19}$$

The problem of stability depends on the fulfillment of the second condition (5.11), that is, whether the system (5.9) *does not* or *does* have a saddle point. In the first case the system is stable and, in the second, unstable.

This condition for stability requires the fulfillment of two partial conditions.

$$\begin{aligned} \omega_e^2(a) &> \nu^2 + 2\delta_e^2(a) \qquad \text{for } a_\nu > 0 \\ \omega_e^2(a) &< \nu^2 + 2\delta_e^2(a) \qquad \text{for } a_\nu < 0 \end{aligned} \tag{5.20}$$

If $\delta_e(a)$ is a small quantity $0(\mu^2)$, the above criteria are simplified, viz.:

$$(1)\ a_\nu > 0; \qquad \omega_e(a) > \nu; \qquad (2)\ a_\nu < 0; \qquad \omega_e(a) < \nu \tag{5.21}$$

These conditions are convenient for the graphical analysis of $a(\nu)$ as is often done in connection, for instance, with Duffing's equation, which we shall examine subsequently.

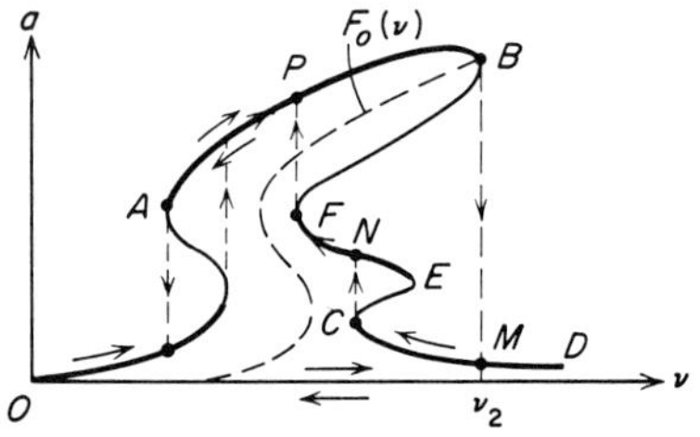

FIGURE 15.1

If one constructs the function $a = F(\nu)$ solution of (5.8) and the curve $a = F_0(\nu)$ for the *exact resonance* $[\omega_e(a) = \nu]$, called sometimes, *the backbone curve* (broken line in Fig. 15.1)), then the branches to the left of $a = F_0(\nu)$, in which a increases with ν, are stable; on the branches to the right of $a = F_0(\nu)$, the contrary is true. The stable branches are shown in heavy line in Fig. 15.1. This representation is convenient for the analysis of stability of amplitudes in nonlinear systems. Thus, for instance, if one starts with $\nu = \nu_1$ (Fig. 15.2) and increases the frequency, the amplitude a follows the stable branch AB. For $\nu = \nu_2$ (point B), the amplitude drops suddenly onto the lower branch CD (point M) and follows it (from M to D) if ν continues to increase. If, however, ν decreases (say,

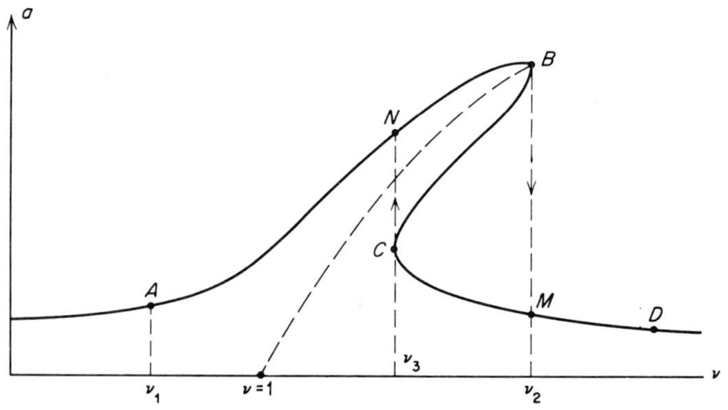

FIGURE 15.2

from $\nu = \nu_2$), the lower branch will be followed up to C; at this point there will be an upward jump CN on the stable branch AB.

It is observed that these phenomena relate closely to the theory of bifurcations (Chapter 7), the forced frequency ν appearing here as the parameter λ of the general theory having a number of bifurcation values at which there appear "exchanges of stability," using a term of Poincaré.

This leads to the so-called "hysteresis phenomena" often observed in nonlinear systems, as was mentioned in Chapter 7. As an example, consider the Duffing equation with a forcing term:

$$m\frac{d^2x_1}{dt_1^{\,2}} + \mu b\frac{dx_1}{dt_1} + cx_1 + \mu d x_1^{\,3} = \mu\bar{E}\sin\nu_1 t_1 \tag{5.22}$$

We introduce the change of the variables

$$x_1 = \sqrt{d/c}\,x; \qquad t_1 = \sqrt{m/c}\,t$$

and set: $\delta = b/\sqrt{cm}$; $E = \bar{E}/\sqrt{cd}$; $\nu = \nu_1\sqrt{m/c}$; $E_1 = \mu E$. The transformed equation will be

$$\ddot{x} + \mu\delta\dot{x} + x + \mu(d/c)^2x^3 = E_1\sin\nu t$$

Taking $d = c$, one obtains the d.e.

$$\ddot{x} + x = \mu f(x,\dot{x}) + \mu E\sin\nu t \tag{5.23}$$

where $f(x,\dot{x}) = -\delta\dot{x} - x^3$.

Equations (5.5) are here

$$\dot{a} = -\frac{1}{2}\delta a - [E_1/(1+\nu)]\cos\theta; \qquad \dot{\theta} = 1 - \nu + \frac{3}{8}a^2 + \frac{E_1}{a(1+\nu)}\sin\theta \tag{5.24}$$

For the fundamental (synchronous) resonance one has the relations

$$\delta a + E_1\cos\theta = 0; \qquad a\left[\left(1 + \frac{3a^2}{8}\right)^2 - \nu^2\right] + E_1\sin\theta = 0 \tag{5.25}$$

From (5.4) we have

$$\delta_e(a) = \frac{\delta}{2}; \qquad \omega_e(a) = 1 + \frac{3a^2}{8} \tag{5.26}$$

If one eliminates θ between the two equations (5.25), relation (5.8) becomes

$$a^2\left\{\left[\left(1 + \frac{3a^2}{8}\right)^2 - \nu^2\right]^2 + \nu^2\delta^2\right\} = E_1^{\,2} \tag{5.27}$$

that is

$$\nu = \sqrt{\omega_e^{\,2}(a) \pm \sqrt{(E_1/a)^2 - \delta^2}} \tag{5.28}$$

The curve $\nu = \omega_e(a)$ is given by $\nu = 1 + (3a^2/8)$ and the resonance curves are given by (5.28). It is noted that for this construction one considers ν as a function of a (Fig. 15.2).

As was explained in connection with Fig. 15.1, the branches ANB and DMC are stable and CB unstable. The hysteresis cycle is clearly $MCNBM$.

One can also carry out the second approximation, which merely introduces harmonics into the stationary oscillations and slightly changes the curves of resonance.

It is to be noted that the first approximation reveals only the fundamental resonance ($p = 1$, $q = 1$). The subharmonic resonances can be detected only in approximations of higher orders; we shall enter into their more detailed study in Chapter 19, and shall limit the discussion here to the question of stability of amplitudes.

We have followed closely the Krylov-Bogoliubov method (in a later exposition by N. Bogoliubov and J. Mitropolsky) but the same subject of "jumps" of resonance has been also treated independently by J. J. Stoker.[3] Considerable experimental data on these phenomena has been obtained by Ludeke.[4]

6. Nonstationary processes; slow time

The asymptotic method has been recently (1955) applied by J. A. Mitropolsky[2] to the investigation of nonstationary phenomena, that is, to a field which has been entirely neglected from the beginning of modern developments in the theory of oscillations.

In general, by the term "nonstationary" phenomena (or oscillations) is meant all cases in which the coefficients of d.e. vary slowly in time. The phenomena of this nature are not necessarily periodic.

Perhaps the best known type of these phenomena is that resulting when the parameters of an oscillating system with a relatively high (for example, radio) frequency are *modulated* by a much slower (for example, audio) frequency. The same situation occurs in the frequency modulated phenomena.

In mechanics one encounters a number of problems of this nature; a typical problem is that of a pendulum with a variable length. Similar problems of modulation occur also in mechanics; we refer to Mitropolsky[2] for details.

[3] J. J. Stoker, *Nonlinear Vibrations*, Interscience Publishers, New York, 1950; R. Reissig, *Wiss. Zeitsch. der Humboldt Un.*, No. 2, Jg. V (1955/56).

[4] C. A. Ludeke, *J. Appl. Phys.*, July, 1946.

[2] See footnote [2], page 356.

A number of these problems which are usually studied from the point of view of a stationary state acquire a different aspect when approached from the standpoint of a nonstationary condition. Thus, for instance, a resonance effect is considerably attenuated if the zone of resonance is just *passed* instead of allowing the oscillation to build up for a fixed value of the frequency at some point in this zone.

All this leads to a number of questions which so far have been disregarded in the classical approach.

In order to simplify such problems, it is necessary to assume that the rate of variation of parameters is much slower than the frequency of the motion considered at a fixed value of these parameters. If one assumes, for instance, that both processes (that is, oscillation and the corresponding parameter variation) are periodic, this amounts to assuming that, although for the oscillation the period is T, for the parameter variation it is T/μ, μ being small. This, clearly, requires the introduction of a different time scale for the parameter variation.

In some cases the parameter variation does not need to be periodic as this happens, for instance, in the case of a pendulum of a variable length; in this particular case the variation of the length is a nonperiodic function of time. One can consider a general problem specified by a d.e. of the form

$$\frac{d}{dt}[m(\tau)\dot{x}] + c(\tau)x = \mu F(\tau, \theta, x, \dot{x}); \qquad \dot{x} = \frac{dx}{dt} \tag{6.1}$$

where m is a mass (or moment of inertia, or coefficient of inductance) and c is the coefficient of restoring force (that is, "spring constant" in mechanical problems; "elastance," that is, $1/c$, in electrical problems, etc.), but here we have to make an assumption that m and c may also depend on the *slow time* τ and are, therefore, $m(\tau)$ and $c(\tau)$. The same duality of independent variables (the ordinary, or "rapid" time t and the "slow" time τ) appears on the right-hand side of (6.1); we merely have written the angular coordinate θ corresponding to the normal time t for the sake of having period 2π.

Summing up, the problem is now specified in terms of two independent variables t and τ with an additional relation $d\tau/dt = \mu$ when differentiations are to be carried out with respect to t, as usual. Consider, for instance, a d.e.

$$\mu^2 \frac{d^2x}{d\tau^2} + p(\tau)x = 0 \tag{6.2}$$

in terms of the "slow time" τ. Introducing $t = \tau/\mu$, it becomes

$$\ddot{x} + p(\mu t)x = 0 \tag{6.3}$$

where $\ddot{x} = d^2x/dt^2$, as usual. Thus, with τ as independent variable and μ appearing before the second derivative, the d.e. is transformed into (6.3) where p is now slowly varying function of the *ordinary* time t.

It is obvious that the functions $m(\tau)$ and $c(\tau)$ in (6.1) are always positive; otherwise the problem would have no physical meaning. Moreover, in differentiations, one has:

$$\frac{dm(\tau)}{dt} = \frac{dm(\tau)}{d\tau}\frac{d\tau}{dt} = \mu\frac{dm(\tau)}{d\tau} \tag{6.4}$$

likewise for $c(\tau)$. A differentiation of a function of τ with respect to t always introduces a small factor μ and this is to be taken into account in the approximation procedure where the terms are arranged as coefficients of like powers of μ.

We can assume that, for a certain interval $(0 \leq t \leq T)$, the slow time τ is in the interval $(0 \leq \tau \leq (T/\mu))$ and in this interval the functions $m(\tau)$, $c(\tau)$, $(d\theta/d\tau) = \nu(\tau)$ and $F(\tau, \theta, x, \dot{x})$ are continuously differentiable an indefinite number of times for all finite values of their arguments.

One immediate conclusion results from these assumptions, viz.: even if the parameters $m(\tau)$ and $c(\tau)$ vary periodically, nothing indicates that the solution $x(t)$ will be periodic. We are now sufficiently acquainted with the theory of d.e. with periodic coefficients to be able to notice this circumstance. In fact, the frequency $\omega(\tau)$ defined in the conventional manner as $\omega(\tau) = \sqrt{c(\tau)/m(\tau)}$ does not now remain constant, and all familiar concepts of the theory of oscillations governed by d.e. with *fixed* parameters cease to hold here. One can retain the concept of frequency (either $\nu(\tau) = d\theta/d\tau$ or $\omega(\tau) = \sqrt{c(\tau)/m(\tau)}$) as being a function of the slow time τ.

A few examples of d.e. with slowly varying parameters illustrate these new concepts inherent in the use of "slow time."

The d.e. of a pendulum with a variable length is

$$\frac{d}{dt}\left[ml^2(\tau)\frac{d\theta}{dt}\right] + gl(\tau)\sin\theta = 0 \tag{6.5}$$

where $l(\tau)$ is a slowly varying length l of the pendulum.

Likewise, it can be shown that a nonstationary process of an approach to a stationary condition in electron-tube circuits can be brought to a d.e. of the form

$$\ddot{x} + x = \mu(\tau)f(\tau, x, \dot{x}) + \lambda_0 \sin n(\tau)t$$

which is also of the type (6.1). A number of other processes in the theory of clystrons, frequency-modulated circuits, etc., reduce also to the d.e. of the same type; to the same d.e. belong certain mechanical phenomena, for instance, longitudinal oscillations of cables of a variable length observed in connection with the operation of mine hoists, that is, problems involving varying constraints, for which, as is known from theoretical mechanics, the kinetic energy may have linear terms, in addition to the terms of zero or of the second order in velocities.

Consideration of these (generally parasitic) modulations opens a number of problems which were discussed under the assumption of fixed parameters in the d.e. Such problems are to be supplemented by the investigation of the effects of these slow modulations arising from changes in the parameter values.

7. Successive approximations for nonstationary processes; slow time

If $\mu = 0$ in (6.1) and m and c are constant, we have the d.e.

$$m\ddot{x} + cx = 0 \tag{7.1}$$

whose solution is

$$x = a \cos (\omega t + \varphi) \tag{7.2}$$

$\omega = \sqrt{c/m}$ and a and φ are arbitrary constants determined by initial conditions.

In the general case there appear a great variety of possible solutions, most of which are related to the so-called nonlinear (or subharmonic) resonance, which we shall study in Part III. Here the situation is more complicated inasmuch as the solutions are not periodic.

As we are interested here only in *formal* solutions of (6.1) by approximations, we may disregard the physical significance of the problem for the moment and concentrate on the derivation of successive approximations, This follows the usual argument (Chapter 14) with the difference that the calculation is more complicated in view of the presence of the additional variable τ.

We can look for the solution of (6.1) of the form

$$x = a \cos (s\varphi + \psi) + \mu u^{(1)}(\tau, a, \theta, s\varphi + \psi) + \mu^2 u^{(2)}(\tau, a, \theta, s\varphi + \psi) + \dots \tag{7.3}$$

where $u^{(1)}$ and $u^{(2)}$ are periodic functions of θ and $s\varphi + \psi$ with period 2π, $\varphi = (1/r)\theta$; s and r are small relatively prime integers. The variable $s\varphi + \psi$ is introduced for the purpose of taking into account the passage of the solution through the various kinds of subharmonic resonances. As this matter will be treated only in Part III, we may simplify this exposition by assuming the fundamental resonance: $s = 1, r = 1$; under this assumption $s\varphi + \psi = \theta + \psi$.

As previously, we assume that the amplitude a and the phase ψ are given by the d.e.

$$\begin{aligned} \dot{a} &= \mu A^{(1)}(\tau, a, \psi) + \mu^2 A^{(2)}(\tau, a, \psi) + \ldots \\ \dot{\psi} &= \omega(\tau) - \nu(\tau) + \mu B^{(1)}(\tau, a, \psi) + \mu^2 B^{(2)}(\tau, a, \psi) \end{aligned} \tag{7.4}$$

where $\omega(\tau) = \sqrt{c(\tau)/m(\tau)}$; $\nu(\tau) = d\theta/dt$; $\tau = \mu t$. We set $\Delta\omega(\tau) = \omega(\tau) - \nu(\tau)$, that is, the difference between the frequency of the external periodic excitation $\nu(\tau)$ and the instantaneous frequency of the system $\omega(\tau)$.

The problem is to find such functions $u^{(1)}, u^{(2)}, \ldots, A^{(1)}, A^{(2)}, \ldots, B^{(1)}, B^{(2)}, \ldots$ that the series (7.3), after the substitution of a and ψ, as given by the series (7.4) should be the actual solution of (6.1).

The applicability of the method does not depend on the *actual convergence* of the series (7.3) and (7.4) (which may be even divergent), but *depends on their asymptotic properties for* $\mu \to 0$. We shall consider the series (7.3) as a formal solution necessary for the construction of the asymptotic approximation.

$$x^{(m)} = a \cos(\theta + \psi) + \mu u^{(1)}(\tau, a, \theta, \theta + \psi) + \ldots + \mu^m u^{(m)}(\tau, a, \theta, \theta + \psi) \tag{7.5}$$

where a and ψ are determined from d.e.

$$\begin{aligned} \dot{a} &= \mu A^{(1)}(\tau, a, \psi) + \ldots + \mu^m A^{(m)}(\tau, a, \psi) \\ \dot{\psi} &= \Delta\omega(\tau) + \mu B^{(1)}(\tau, a, \psi) + \ldots + \mu^m B^{(m)}(\tau, a, \psi) \end{aligned} \tag{7.6}$$

The problem of determination of $u^{(1)}, u^{(2)}, \ldots, A^{(1)}, A^{(2)}, \ldots, B^{(1)}, B^{(2)}, \ldots$ has a certain degree of arbitrariness, inasmuch as any replacement of a and ψ by new variables b and ψ_1 related by equations

$$a = b + \mu\alpha_1(b) + \mu^2\alpha_2(b) + \ldots; \qquad \psi = \psi_1 + \mu\beta_1(\psi_1) + \mu^2\beta_2(\psi_1) + \ldots$$

results in similar expressions but with different coefficients.

One can impose, however, an additional condition, as we did in Chapter 14, which consists in the requirement that the fundamental harmonic should be absent in all functions $u^{(1)}, u^{(2)}, \ldots$.

This requirement is expressed by the conditions

$$\int_0^{2\pi} u^{(i)}(\tau, a, \theta, \gamma) \cos \gamma d\gamma = 0; \qquad \int_0^{2\pi} u^{(i)}(\tau, a, \theta, \gamma) \sin \gamma d\gamma = 0; \qquad i = 1, 2, \ldots, m \tag{7.7}$$

where $\gamma = \theta + \psi$.

Physically this means that we take as amplitude a *the full amplitude* of the fundamental harmonic so that no resonance terms can appear in $u^{(1)}$, $u^{(2)}$ for any τ in the interval $0 \leq \tau \leq \mathfrak{T}$; $\mathfrak{T} = T/\mu$, where T is some finite value.

In order to determine $u^{(1)}, u^{(2)}, \ldots, A^{(1)}, A^{(2)}, \ldots, B^{(1)}, B^{(2)}, \ldots$ under the condition (7.7), one has to determine $\ddot{a}$, $\ddot{\psi}$ and also the function F in (6.1). For the sake of simplification we introduce the same notations for partial differentiations as previously, viz.: $A_a, A_\psi, A_\tau, \ldots$ means differentiations with respect to a, ψ, τ. Moreover, in differentiations with respect to t in the derivatives depending on τ (the slow time) appears the factor $d\tau/dt = \mu$, which transfers the corresponding term to the next higher order in μ, etc.; we recall that $\Delta\omega = \omega(\tau) - \nu(\tau)$ in this notation; finally by dots (for example, $\dot{x}, \ddot{x}, \dot{a}, \ddot{a}, \dot{\psi}, \ddot{\psi}, \ldots$) we shall mean always differentiations with respect to t (that is, dx/dt, d^2x/dt^2, da/dt, $d^2a/dt^2, \ldots$). One has then

$$\begin{aligned}\ddot{a} &= \mu\Delta\omega A_\psi^{(1)} + \mu^2[A_a^{(1)}A^{(1)} + A_\psi^{(1)}B^{(1)} + A_\tau^{(1)} + \Delta\omega A_\psi^{(2)}] + \mu^3\ldots\\ \ddot{\psi} &= \mu[(\Delta\omega)_\tau + (\Delta\omega)B_\psi^{(1)}] + \mu^2[B_a^{(1)}A^{(1)} + B_\psi^{(1)}B^{(1)}\\ &\qquad + B_\tau^{(1)} + \Delta\omega B_\psi^{(2)}] + \mu^3\ldots\end{aligned} \tag{7.8}$$

With these expressions, one obtains for $\dot{x}$ and $\ddot{x}$ the following expressions:

$$\begin{aligned}\dot{x} = {} & -a\omega\sin\gamma + \mu[A^{(1)}\cos\gamma - B^{(1)}a\sin\gamma + u_\theta^{(1)}\nu + u_\gamma^{(1)}\omega]\\ & + \mu^2[A^{(2)}\cos\gamma - B^{(2)}a\sin\gamma + u_\tau^{(1)} + u_a^{(1)}A^{(1)}\\ & \qquad + u_\psi^{(1)}B^{(1)} + u_\theta^{(2)}\nu + u_\gamma^{(2)}\omega] + \mu^3\ldots\end{aligned}$$

$$\begin{aligned}\ddot{x} = {} & -a\omega^2\cos\gamma + \mu\{[\Delta\omega A_\psi^{(1)} - 2a\omega B^{(1)}]\cos\gamma\\ & - [\Delta\omega aB_\psi^{(1)} + 2a\omega A^{(1)}]\sin\gamma - [\omega_\tau a\sin\gamma + u_{\theta\theta}^{(1)}\nu^2\\ & + 2u_{\theta\gamma}^{(1)}\nu\omega + u_{\gamma\gamma}^{(1)}\omega^2]\} + \mu^2\{[\Delta\omega A_\psi^{(2)} - 2a\omega B^{(2)}]\cos\gamma\\ & - [\Delta\omega aB_\psi^{(2)} + 2\omega A^{(2)}]\sin\gamma + [A_a^{(1)}A^{(1)} + A_\psi^{(1)}B^{(1)}\\ & + A_\tau^{(1)} - aB^{(1)2}]\cos\gamma - [2A^{(1)}B^{(1)} + aB_a^{(1)}A^{(1)}\\ & + aB_\psi^{(1)}B^{(1)} + aB_\tau^{(1)}]\sin\gamma + 2u_{\tau\gamma}^{(1)}\omega + 2u_{\tau\theta}^{(1)}\\ & + 2u_{a\theta}^{(1)}\nu A^{(1)} + 2u_{a\gamma}^{(1)}\omega A^{(1)} + 2u_{\theta\gamma}^{(1)}\nu B^{(1)} + 2u_{\gamma\gamma}^{(1)}\omega B^{(1)}\\ & + u_\gamma^{(1)}B_\psi^{(1)}\Delta\omega + \Delta\omega u_a^{(1)}A_\psi^{(1)} + u_\theta^{(1)}\nu_\tau + u_\gamma^{(1)}\omega_\tau\\ & \qquad + u_{\theta\theta}^{(2)}\nu^2 + 2u_{\theta\gamma}^{(2)}\nu\omega + u_{\gamma\gamma}^{(2)}\omega^2\} + \mu^3\ldots\end{aligned} \tag{7.9}$$

If one substitutes x and $\dot{x}$ into the right-hand side of (6.1) and develops it into a Taylor's series, one obtains

$$\begin{aligned}\mu F(\tau, \theta, x, \dot{x}) = {} & \mu F(\tau, \theta, a\cos\gamma, -a\omega\sin\gamma) + \mu^2[F_x u^{(1)} + Fu_{\dot{x}}(A^{(1)}\cos\gamma\\ & - B^{(1)}a\sin\gamma + u_\theta^{(1)}\nu + u_\gamma^{(1)}\omega)] + \mu^3[\quad] + \ldots\end{aligned} \tag{7.10}$$

The next step, as usual, is the identification of coefficients with like powers of μ. We omit here the very long calculations that follow and merely mention that this permits determining the functions $u^{(i)}(\tau, a, \theta, \gamma)$, $A^{(i)}(\tau, a, \psi)$, $B^{(i)}(\tau, a, \psi)$.

Again a double Fourier series is used. Thus, for instance,

$$F^{(0)}(\tau, a, \theta, \gamma) = \sum_{n,m=-\infty}^{\infty} F_{nm}{}^{(0)}(\tau,a)e^{i(n\theta+m\gamma)} \tag{7.11}$$

where

$$F_{nm}{}^{(0)}(\tau,a) = \frac{1}{4\pi^2}\int_0^{2\pi}\int_0^{2\pi} F^{(0)}(\tau, a, \theta, \gamma)e^{-i(n\theta+m\gamma)}d\theta d\gamma \tag{7.12}$$

The next step is the determination of the coefficients. In the general case (when r and s are necessarily 1) we find that in order to introduce the requirement that $u^{(1)}$ should not have the first harmonic in the angular variable γ, the summation determining $u^{(1)}$

$$u^{(1)}(\tau, a, \theta, \gamma) = \sum_{n,m=-\infty}^{\infty} g_{nm}(\tau,a)e^{i(n\theta+m\gamma)} \tag{7.13}$$

should be carried out only for such values of n and m for which

$$n\theta + m\gamma \neq \pm\gamma + s_1\psi \tag{7.14}$$

where in $\gamma = s\varphi + \psi$ the quantity $\varphi = \theta/r$. This amounts to the requirement $g_{nm}(\tau,a) = 0$ for all n and m satisfying the condition

$$nr + s(m \pm 1) = 0 \tag{7.15}$$

Substituting $F^{(0)}$ and $u^{(1)}$ into the d.e. and equating the coefficients of the like harmonics, one determines $g_{nm}(\tau,a)$, which, finally, determines the function $u^{(1)}$

$$u^{(1)}(\tau, a, \theta, \gamma) = \frac{1}{4\pi^2 m}\sum_{n,m=-\infty}^{\infty} e^{i(n\theta-m\gamma)}\frac{1}{\omega^2-(m\omega+n\nu)^2}\int_0^{2\pi}\int_0^{2\pi} F_0 e^{-i(n\theta+m\gamma)}\cdot d\theta\cdot d\gamma \tag{7.16}$$

the summation runs along all indices for which $nr + s(m \pm 1) \neq 0$.

If one equates the coefficients of the fundamental harmonics of γ, one obtains the coefficients $A^{(1)}(\tau, a, \psi)$ and $B^{(1)}(\tau, a, \psi)$. We do not reproduce all these calculations but merely indicate the results.

As the first approximation one takes

$$x = a\cos\gamma = a\cos(s\varphi + \psi) \tag{7.17}$$

where a and ψ are determined from equations of the first approximation

$$\dot{a} = \mu A^{(1)}(\tau, a, \psi); \qquad \dot{\psi} = \omega(\tau) - \frac{s}{r}\nu(\tau) + \mu B^{(1)}(\tau, a, \psi) \quad (7.18)$$

where $A^{(1)}$ and $B^{(1)}$ are given by formulas

$$A^{(1)}(\tau, a, \psi) = \frac{1}{2\pi^2 m}\sum e^{i\sigma r\psi}\left[\frac{(r\omega - s\nu)\sigma i \int_0^{2\pi}\int_0^{2\pi} F_0 e^{i\sigma r\psi}\cos\gamma d\gamma d\theta - 2\omega \int_0^{2\pi}\int_0^{2\pi} F_0 e^{-i\sigma r\psi}\sin\gamma d\gamma d\theta}{4\omega^2 - (r\omega - s\nu)^2\sigma^2}\right] - \frac{a}{2\pi m\omega}\frac{d(m\omega)}{d\tau} \quad (7.19)$$

$$B^{(1)}(\tau, a, \psi) = -\frac{1}{2\pi^2 ma}\sum e^{i\sigma r\psi}\left[\frac{(r\omega - s\nu)\sigma i \int_0^{2\pi}\int_0^{2\pi} F_0 e^{-i\sigma r\psi}\sin\gamma d\gamma d\theta + 2\omega \int_0^{2\pi}\int_0^{2\pi} F_0 e^{-i\sigma r\psi}\cos\gamma d\gamma d\theta}{4\omega^2 - (r\omega - s\nu)^2\sigma^2}\right]$$

where $$\gamma = s\varphi + \psi$$

One proceeds thus to the second approximation but we shall not reproduce these long calculations.

8. Oscillations of a pendulum with a variable length

In his monograph Mitropolsky gives examples of application of the general method to a number of problems. We give here one concerning oscillations of a pendulum with a variable length. The d.e. in this case is

$$\frac{d}{dt}[ml^2(\tau)\cdot\dot{\theta}] + 2\delta\frac{d}{dt}[l(\tau)\theta] + mgl(\tau)\sin\theta = 0 \quad (8.1)$$

where m is the mass, θ the angle of deviation of the pendulum from the vertical, 2δ the coefficient of damping, g the acceleration of gravity, $l(\tau)$ the length of the pendulum varying slowly as a function of the "slow time" τ so that $\tau = \mu t$.

For small oscillation one can use the first two terms of the development of $\sin\theta$; then (8.1) can be written

$$\frac{d}{dt}[ml^2(\tau)\dot{\theta}] + mgl(\tau)\theta = \mu f(\tau, \theta, \dot{\theta}; \mu) \quad (8.2)$$

where

$$\mu f(\tau, \theta, \dot{\theta}; \mu) = -\left[2\delta l(\tau)\cdot\dot{\theta} - \frac{mgl(\tau)\theta^3}{6} + 2\mu\delta\frac{dl}{d\tau}\theta\right] \tag{8.3}$$

In the first approximation,

$$\theta = a\cos\psi$$

where a and ψ are given by the system of d.e.

$$\dot{a} = -\frac{\delta a}{ml(\tau)} - \frac{3\mu(dl/d\tau)a}{4(\tau)}, \qquad \dot{\psi} = \omega(\tau) - \frac{\omega(\tau)a^2}{16} \tag{8.4}$$

with $\omega(\tau) = \sqrt{g/l(\tau)}$.

Integrating the first d.e. with the initial condition $a(0) = a_0$, one has

$$a = a_0[l(0)/l(\tau)]^{3/4}\exp\left[-\frac{\delta}{m}\int_0^t\frac{dt}{l(\tau)}\right] \tag{8.5}$$

Replacing this value into the second equation, one has

$$\psi = \int_0^t \omega(\tau)\left\{1 - \frac{a_0^2}{16}\left(\frac{l(0)}{l(\tau)}\right)^{3/2}\exp\left[-\frac{2\delta}{m}\int_0^t\frac{dt}{l(\tau)}\right]\right\}dt \tag{8.6}$$

If one sets $l = \text{const}$ in these expressions, one obtains the usual formulas

$$a = a_0e^{-(\delta/ml)t}; \qquad \psi = \omega\left[t + \frac{a_0^2}{16\lambda}(e^{-(2\delta/ml)t} - 1)\right] + \varphi$$

and, therefore

$$\theta = a_0e^{-(\delta/ml)t}\cos\left\{\sqrt{\frac{g}{l}}\left[t + \frac{a_0^2}{16\lambda}(e^{-(2\delta/ml)t} - 1)\right] + \varphi\right\} \tag{8.7}$$

where $\lambda = 2\delta/ml$ and φ is the initial phase. From these expressions one can form an idea as to the effect of the parameter variation in the d.e.

For instance, if the length l of the pendulum varies linearly with the time, that is,

$$l(\tau) = l_0 + l_1\tau = l_0 + \frac{dl}{d\tau}\cdot t \tag{8.8}$$

and amplitude a and the phase ψ are given by the expressions

$$a = a_0\left(\frac{l_0}{l_0 + \mu l_1 t}\right)^{\left(\frac{\delta}{\mu m l_1} + \frac{3}{4}\right)}; \qquad l_1 = dl/d\tau \tag{8.9}$$

$$\psi = \int_0^t\sqrt{\frac{g}{l_0 + \mu l_1 t}}\left\{1 - \frac{a_0^2}{16}\left[\frac{l_0}{l_0 + \mu l_1 t}\right]^{\left(\frac{2\delta}{\mu m l_1} + \frac{3}{2}\right)}\right\}dt;$$

The amplitude of oscillation will not vary exponentially but inversely proportional to a power function of time. If $\delta > 0$ and $l_1 > 0$ for $\delta < 0$, $l_1 > 0$, with $-\delta/(\mu l_1 m) < \frac{3}{4}$, the oscillation will be damped.

We refer to Mitropolsky[2] for other applications of this theory which could not be presented in this chapter.

[2] See footnote [2], page 356.

Chapter 16

STROBOSCOPIC METHOD

1. Introductory remarks

In Chapters 10 and 11 we reviewed the theory of Poincaré concerning the *existence* of periodic solutions and in Chapter 13, the stability of periodic solutions. From the latter chapter it was seen that the determination of the characteristic exponents, in general, is a rather difficult problem for the reasons explained there. This difficulty is inherent in the d.e. with periodic coefficients (Section 6, Chapter 5).

There is, however, a possibility (and this forms the subject of this chapter) of obviating this difficulty by replacing the original differential system, generally containing time explicitly, by an auxiliary system of an autonomous type having the property that the existence as well as stability of a singular point of this auxiliary system is the criterion for the existence and stability of a periodic solution of the original system.[1] The method is based on the transformation theory of d.e. (see, for instance, Levinson [2]) and its development is due largely to discussions with M. Schiffer during the early period of this work. The method turned out to be very convenient in applied problems studied in Part III.

We shall give the exposition of this method to the first approximation. Approximations of higher orders are obtained by the perturbation method.

2. Transformation of points and regions; planes (ψ) and (φ); stroboscopic image

The derivation of the stroboscopic d.e. is outlined in the following section but it is useful to indicate first an intuitive approach to this prob-

[1] N. Minorsky, *C.R. Ac. Sc.* (Paris), t. 232, 1951; Rendiconti, *Acc. Sc.*, Bologna, 1952; *Cahiers de physique*, No. 119, Paris, 1960.

[2] N. Levinson, *Am. of Math.* (2), **45**, 1944.

lem. We shall be concerned with a transformation $\mathfrak{T}$ of a point in a plane as a result of a certain operation; in our case this operation is specified by the d.e. in question.

In order to illustrate this idea we can consider one of the trajectories of the harmonic oscillator specified by some initial conditions $x(0)$, $y(0)$ which determines the initial coordinates of the representative point $R = R_0$ in the phase plane. We may define the transformation by letting the d.e. $\ddot{x} + x = 0$ operate for some time, say 2π. In view of the fact that 2π is a period, R having been at $t = 0$ at the point R_0 will return to the same point. This we can indicate in the form of an *identical transformation*

$$\mathfrak{T}(R_0) = R_0 \tag{2.1}$$

One can consider this situation also in an intuitive manner. The trajectory of the harmonic oscillator is a circle (whose radius is determined by the initial conditions) described by R with a uniform angular velocity 1. This may be visualized as a wheel rotating uniformly with this velocity, and we may mark some point R on the wheel that gives an image of a continuous angular motion with period 2π.

Suppose we illuminate the wheel by stroboscopic flashes occurring once for each period 2π. We shall not see the continuous motion any more but shall see only a fixed point R_0. Thus we shall actually "see" the result of the transformation ($\mathfrak{T}$) but not the continuous motion.

We may define two planes (which in reality are the same plane): (1) the plane (ψ) in which we observe a continuous motion (a continuous illumination), and (2) the plane (φ) in which we observe only a fixed point R_0 (a stroboscopic illumination). This physical interpretation is particularly simple but one can also use a formal argument. One could, for instance, consider, instead of a phase-plane diagram (a circle traversed with a constant angular velocity), a circular cylinder with a helix of a constant pitch traced on its surface, which gives a third dimension t. Along the t axis we trace planes perpendicular to the t axis at equal distance 2π from each other. The point R following the helix will cross those planes at $t = 0, 2\pi, 4\pi, \ldots$, and the intersections of the helix with these planes at these instants when projected on a plane parallel to these planes will obviously give again a fixed point R_0.

Instead of a point R_0 one could consider a certain area (representing a certain continuous region of initial conditions) and, in this particular *identical* transformation, this area will be transformed into exactly the same area. Here again in the (ψ) plane we shall "see" a bundle of trajectories and in the (φ) plane we shall "see" only the invariant area.

For an identical transformation arising from the d.e. $\ddot{x} + x = 0$, this is so trivial that it is hardly worth mentioning but, for operations performed by other d.e., a variety of different situations may occur as we shall see later.

As the next case in the order of difficulty we may consider a van der Pol oscillator. In the first approximation, as we know, this oscillator is characterized by a stable limit cycle of radius 2 to which approach spiral trajectories both from inside and from outside, the rotation on trajectories being uniform at least in the first approximation. If we represent again what happens by a rotating wheel, the motion along the radius is non-uniform, approaching the value $r = r_0 = 2$ with an exponentially decaying velocity, but the wheel itself rotates with a constant velocity.

In the (ψ) plane we shall see a spiral C, Fig. 16.1, approaching the circle $r = r_0$ in an asymptotic manner; in the (φ) plane we shall see a succession of discrete points A_1, A_2,. . . gradually approaching the limit point A_0 for which $r = r_0$, both from inside and from the outside. The "slow motion" of these stroboscopic points occurs along the radius since, in the first approximation, the angular velocity of rotation is constant. If, however, one takes into account the second approximation, this slow motion of the stroboscopic points will occur not exactly along a radius but along a curve C' slightly different from the radius, so that the ultimate fixed point will be at A_0' slightly different from A. If we wish to think in terms of the transformation theory, with the van der Pol equation (with small μ) as "operator" of the transformation, we can write

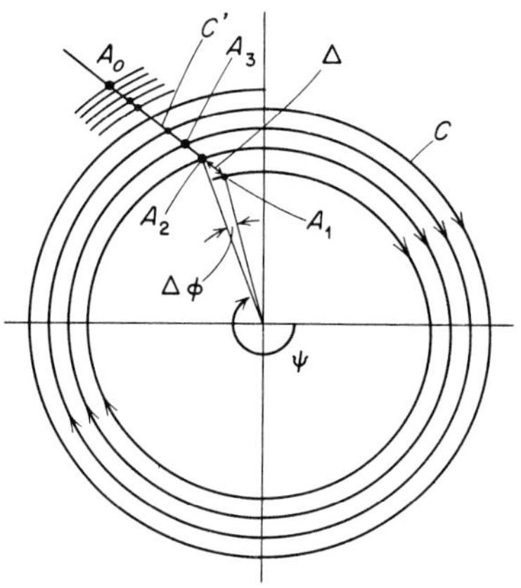

FIGURE 16.1

$$\mathfrak{T}^n(R_0)_{n\to\infty} = R^* = 2 \tag{2.2}$$

For the phase (φ), if we wish to limit the investigation to the first approximation only, we have an identical transformation

$$\mathfrak{T}^n(\varphi_0) = \varphi_0 \tag{2.3}$$

but if higher-order terms are taken into account, we have

$$\mathfrak{T}^n(\varphi_0)_{n\to\infty} = \varphi^* \tag{2.4}$$

These two examples give sufficient insight into the meaning of the point transformation effected by a d.e. acting as an operator defining the transformation. In the first case (the d.e. $\ddot{x} + x = 0$), we have an identical transformation (2.1), and in the second case (the d.e. $\ddot{x} + \mu(x^2 - 1)\dot{x} + x = 0$), we have an asymptotic transformation (2.2), (2.3), or (2.4) which establishes a limit point (R^*,φ^*) in the (φ) plane.

We note in passing that both planes (ψ) and (φ) are in reality the same phase plane in which the things are "seen" under a different light; the plane (ψ) is that in which a "constant illumination" is used and (φ) is merely what appears under a "stroboscopic illumination." We shall confine our remarks exclusively to the (φ) plane, but this correspondence between the two planes must be kept in mind in order to follow the argument.

If a fixed point exists in the (φ) plane, a trajectory of the (ψ) plane passes always through this point. This, in turn, means that *this trajectory is periodic.* Thus, if by some argument we show that a fixed point exists in the (φ) plane, this amounts to the assertion that the corresponding motion in the (ψ) plane is periodic. Likewise, if one can show that the fixed point in the (φ) plane is *stable*, the periodic motion in the (ψ) plane is also stable. This is a purely intuitive argument at this stage but, as we shall see, its implications are justified also by a formal reasoning.

The stroboscopic points $A_1, A_2, \ldots$ form a discrete sequence, but in our optical analogy, this sequence appears as a pseudocontinuous sequence of a "slow motion" owing to a persistence of vision. This slow motion may either approach a limit or not.

We have thus a definite physical pattern which will enable us to carry out the stroboscopic transformation in two steps:

1. We replace the original d.e. (whose solutions are trajectories in the (ψ) plane) by two sequences of difference equations which, from the knowledge of r_n and φ_n (in the nth step of the transformation) determine r_{n+1} and φ_{n+1} for the $(n + 1)$st step. This amounts to establishing a discrete sequence of points $A_1, \ldots, A_n, A_{n+1} \ldots$ in the (φ) plane.
2. One can also be guided by the "persistence of vision" which accounts for a (pseudo) continuous slow motion. Thus, one can pass from the differential equations (in the first step of the argument) to *the stroboscopic d.e.*

The preceding conclusion can be then formulated as follows:

The existence of a stable singular point in the (φ) *plane corresponds to the existence of a stable periodic solution (motion) in the* (ψ) *plane.*

The stroboscopic method, as will be shown, is particularly useful for the nonautonomous systems inasmuch as their stroboscopic counterparts are always autonomous; therefore the difficult problem of stability of a periodic solution (Chapter 13) is reduced to a much simpler problem of stability of a singular point (Chapter 1). Before entering into this question more fully in the next section, it is useful to consider the various forms of point transformations.

1. The most important and also frequently encountered case is when

$$\mathfrak{T}^n(R_0)_{n\to\infty} = R^* \tag{2.5}$$

which amounts, as we saw previously, to

$$r^n(r_0)_{n\to\infty} = r_0^* \quad \text{and} \quad \varphi^n(\varphi_0)_{n\to\infty} = \varphi^* \tag{2.6}$$

We shall give an example of this situation in Section 4; a number of similar examples will be found in Part III dealing with nonlinear oscillations.

2. Occasionally $\mathfrak{T}^n(r_0)_{n\to\infty} = r^*$, but there is no similar limit transformation for φ. This means that in the plane (φ) "we see" a vector of a constant length r_0^* rotating with a certain (generally small) angular velocity $\dot{\varphi}$. The only case that one encounters in applied problems of a nearly linear type is when $\dot{\varphi} = \eta$, $\eta \ll 1$ being a small constant. Such a situation occasionally arises in the first approximation; in some other cases (for example, the van der Pol equation), it manifests itself only in the second approximation. All depends on the *form* of the d.e. which appears here as an operator of the point transformation. Whenever such a slow rotation $\dot{\varphi}$ occurs, the periodic solution does not have the period 2π but has a neighboring period $2\pi \pm \varepsilon$, ε being a small constant.

This case does not present any difficulty inasmuch as, instead of considering the plane (φ), we may introduce a plane (φ') rotating with respect to (φ) with angular velocity η and thus obtain a fixed point in the (φ') plane. In a physical language, we can say: we can "stop" the rotating stroboscopic image if we space our stroboscopic flashes (in time) not by the intervals 2π but by intervals $2\pi \pm \epsilon$.

3. It may also occur that $\mathfrak{T}^n(\varphi_0)_{n\to\infty} = \varphi_0^*$, but r does not approach any limit going, for instance, to infinity. We "see" in the (φ) plane the point R moving along a radius $\varphi = \varphi_0^*$ beyond any bound. We shall encounter this case when a linear d.e. of Mathieu acts as operator of the transformation. A very interesting experiment of Mandelstam and Papalexi, of which we shall speak in Part III, illustrates this somewhat special case.

4. It may happen that neither r nor φ approach any limit but *oscillate* about some limits; this means that R describes a small curve around a fixed point R_0. We shall return to this question later, but it is sufficient to say here that this indicates the presence of *almost periodic oscillations* inasmuch as both amplitude and phase undergo *modulations*.

5. Neither r nor φ approach any limit nor oscillate around some limits; this case is clearly of no interest and merely means that no stationary state exists.

It is useful to add that these intuitive considerations are possible if there is a well-defined sequence of the stroboscopic points $A_1, A_2, \ldots, A_n, A_{n+1}, \ldots$ which may form a definite pattern resulting in a possibility of "seeing a slow motion" in the (φ) plane. This is possible if both r and φ change but little during each period 2π; this circumstance, in turn, is related to the *near linearity* of the problem. It is clear that, if a system is *strongly nonlinear*, the stroboscopic points would be scattered so that it would be impossible to talk about the polygon $A_1, A_2, \ldots$.

3. Stroboscopic differential equations

We consider a nearly linear nonautonomous system of the form

$$\dot{x} = X(x, y, t); \quad \dot{y} = Y(x, y, t) \tag{3.1}$$

It is more convenient to introduce the variables defined by equations

$$\rho = r^2 = x^2 + \dot{x}^2 = x^2 + y^2; \qquad \psi = \arctan(y/x) \tag{3.2}$$

This is accomplished in the usual manner by forming two combinations $x\dot{x} + y\dot{y} = \frac{1}{2}(d\rho/dt)$ and $x\dot{y} - y\dot{x} = \rho(d\psi/dt)$ and by replacing $x = r\cos\psi$ and $y = r\sin\psi$ in the right sides of the d.e. (3.1).

We note in passing that the variable ρ generally is more convenient than r inasmuch as in physical problems it is a measure of *energy*, as we have noted previously. The introduction of the variable ρ is not always possible and depends on the form of the right-hand terms in (3.1). Very often it is more convenient to use the variables r and φ in the stroboscopic d.e. Finally, in some cases the polar coordinates are not convenient and one has to deal directly with the cartesian variables x and y. We shall postpone the detailed study of these various cases until a later section and assume here that the introduction of the variables ρ and ψ is possible.

In these variables the system (3.1) takes the form

$$\frac{d\rho}{dt} = F(\rho, \psi, t); \qquad \frac{d\psi}{dt} = G(\rho, \psi, t) \tag{3.3}$$

where F and G are periodic functions with period 2π in t.

As we wish to consider the case of a nearly linear system, the d.e. (3.3) *are in the neighborhood* of the d.e. of the harmonic oscillator whose d.e. in these variables are

$$\frac{d\rho}{dt} = 0; \qquad \frac{d\psi}{dt} = -1 \tag{3.4}$$

The significance of these d.e. is obvious. In fact, the first shows that the total energy ρ of the oscillator is constant: $\rho = \rho_0$, which is clearly the energy integral; and the second merely shows that the angular velocity of the radius vector on a trajectory (which is a circle $\rho = \rho_0$ here) is constant and is equal to -1, that is, takes place clockwise.

If the d.e. (3.3) are nearly linear, they can be put in the form

$$\frac{d\rho}{dt} = \mu f(\rho, \psi, t) + \ldots; \qquad \frac{d\psi}{dt} = -1 + \mu g(\rho, \psi, t) + \ldots \tag{3.5}$$

as, for $\mu = 0$, they reduce to (3.4). The functions f and ρ are again periodic with period 2π in t.

We apply the usual procedure of integration by the series of the form

$$\rho(t) = \rho_0(t) + \mu\rho_1(t) + \ldots; \qquad \psi(t) = \psi_0(t) + \mu\psi_1(t) + \ldots \tag{3.6}$$

and limit ourselves to the first approximation only, since μ is small. As regards the approximation of the zero order, it is clearly the solution of the harmonic oscillator. This yields

$$\rho_0(t) = \rho_0 = \text{const}; \qquad \psi_0(t) = \varphi_0 - t \tag{3.7}$$

where ρ_0 and φ_0 are the initial values $\rho(0) = \rho_0$; $\psi(0) = \varphi_0$. The first-order corrective terms $\rho_1(t)$ and $\psi_1(t)$ are obtained from the perturbation series and are

$$\rho_1(t) = \int_0^t f(\rho_0, \varphi_0 - \sigma, \sigma)d\sigma; \qquad \psi_1(t) = \int_0^t g(\rho_0, \varphi_0 - \sigma, \sigma)d\sigma \tag{3.8}$$

so that the solution in the first approximation is

$$\rho(t) = \rho_0 + \mu\rho_1(t); \qquad \psi(t) = \varphi_0 - t + \mu\psi_1(t) \tag{3.9}$$

We may use these expressions for evaluating $\rho(t)$ and $\psi(t)$ for $t = 2\pi$, $4\pi, \ldots$, but we cannot do that for $t \to \infty$ inasmuch as the higher-order terms in the expressions (3.6), which we have neglected, may impair the accuracy of the approximation in the long run.

To avoid this, we adopt the following procedure: instead of letting t vary indefinitely in the expressions for $\rho_1(t)$ and $\psi_1(t)$, we vary it only during an interval 2π and calculate the variations $\rho_1(2\pi)$ and $\psi_1(2\pi)$ during one interval 2π. With the new values of $\rho(2\pi)$ and $\psi(2\pi)$, which we consider for the second interval as $\rho(0)$ and $\psi(0)$, we calculate the new values of ρ and ψ at the end of the second interval, and so on. In this manner we vary t only by finite intervals 2π in each interval and adjust the initial conditions so that the terminal conditions of the $(n-1)$st interval become the initial conditions for the nth interval. This avoids the cumulative error that might otherwise occur due to the presence of higher-order terms omitted in the first approximation, if we let $t \to \infty$. We obtain a transformation

$$\rho_1(2\pi) = K'(\rho_0,\varphi_0); \qquad \psi_1(2\pi) = L'(\rho_0,\varphi_0) \tag{3.10}$$

where $K'(\rho_0,\varphi_0)$ and $L'(\rho_0,\varphi_0)$ are the integrals (3.8) between 0 and 2π. It is noted that the independent variable t has already disappeared inasmuch as K' and L' are certain numbers, functions of the initial conditions ρ_0 and φ_0 (that is, of a point T_0 in the phase plane). As $\rho(t) = \rho_0 + \mu\rho_1(t)$ and $\psi(t) = \psi_0 + \mu\psi_1(t)$ and, in view of the remark just made, the result obtained can be specified by two transformations

$$\rho(2\pi) = K(\rho_0,\varphi_0); \qquad \varphi(2\pi) = L(\rho_0,\varphi_0) \tag{3.11}$$

Transformation (3.11) may be regarded as a manifold of initial values subject to the transformation. Starting from some initial values ρ_0, φ_0, we obtain the new initial values (for the second period T). This is a purely *spatial* transformation which gives a sequence of points (ρ_0',φ_0'), (ρ_0'',φ_0''), . . . starting with some point (ρ_0,φ_0).

We shall try to relate the preceding argument to the existence of a fixed point in the (φ) plane and, for that purpose, it is useful to give an example. Suppose that, instead of ρ and φ, we have cartesian coordinates x and y and a certain transformation, say, $x(T) = x_0y_0$; $y(T) = x_0/y_0$, T being the period, say, 2π.

If we start, for instance, with $x_0 = 1$, $y_0 = 2$, we have $x(T) = 2$; $y(T) = \frac{1}{2}$. This defines completely the transformation. In fact, for $x(2T)$ and $y(2T)$ we start with the initial conditions $x_0 = 2$, $y_0 = \frac{1}{2}$ so that

$$x(2T) = 1; \qquad y(2T) = 4$$

and so on for $x(3T)$, $y(3T)$,. . ..

If a transformation is given, say, $x(T) = f_1(x_0,y_0)$; $y(T) = f_2(x_0,y_0)$, one can explore different situations in the phase plane when T goes through a sequence of integer values $T = 1, 2, 3,\ldots$.

One can, for instance, take an arbitrary area, say, a small square with coordinates of its four corners (x_0,y_0), (x_0',y_0'), (x_0'',y_0'') and (x_0''',y_0''') and, under the prescribed transformation $x(T) = f_1(x_0,y_0)$; $y(T) = f_2(x_0,y_0)$ investigate how, starting from each four corners, the transformation progresses for T, $2T$, $3T$,. . .. The initial square (as defined by the coordinates of its four corners) will be thus transformed into some other area limited by four points $(\bar{x}_0,\bar{y}_0)$, $(\bar{x}_0',\bar{y}_0)$, $(\bar{x}_0'',\bar{y}_0'')$, and $(\bar{x}_0''',\bar{y}_0''')$, if one proceeds as shown, the bars indicating the result of a repeated number of transformations. The new area may either expand or contract, depending on the transformation used. The situation is a purely formal one at this stage.

It may be worth mentioning at this point that we touched on a similar question in Chapter 2 when we spoke about the invariance of areas in the Hamiltonian variables in conservative systems. This is a purely formal result of the transformation theory.

We shall try to introduce a more definite requirement, namely: *the original area* (*the square*) *must contract indefinitely* under the effect of repeated transformations. In fact, if we succeed in imposing such a condition, the transformation applied to different points of a certain area and repeated indefinitely will ultimately end at a *fixed point*. Since we have lost our variable t, in integrating between 0 and 2π, we have to reestablish something analogous to it in order *to follow* the process of repeated iterations. From a purely *spatial* manifold of the initial conditions, we have to introduce a kind of *temporal* element which could guide the path of the transformation to a fixed point which is our aim.

We return to (3.11) which can be written in the form (3.9) as

$$\rho' = \rho + \mu\rho_1; \qquad \varphi' = \varphi + \mu\psi_1 \tag{3.12}$$

where ρ' and φ' are the values of ρ and φ at the end of the transformation (2π), and ρ and φ mean these values at the beginning of the interval; we drop 2π as the angles are determined only modulo 2π. Thus $\rho' - \rho = \Delta\rho$ and $\varphi' - \varphi = \Delta\varphi$ is the effect of the transformation during one period 2π. As in integrations (3.8) appear generally the factor 2π, we can write

$$\Delta\rho = 2\pi\mu K(\rho_0,\varphi_0); \qquad \Delta\varphi = 2\pi\mu L(\rho_0,\varphi_0)\ldots \tag{3.13}$$

It is obvious that, if the transformation leads ultimately to a fixed point, $\Delta\rho = \Delta\varphi = 0$.

It is convenient to introduce here the temporal element (which we have lost in integration (3.8) between 0 and 2π) by defining as element of the new ("stroboscopic") time τ the expression

$$\Delta\tau = 2\pi\mu \tag{3.14}$$

The transformation equations (3.13) acquire now a more familiar form

$$\frac{\Delta\rho}{\Delta\tau} = K(\rho_0,\varphi_0); \qquad \frac{\Delta\varphi}{\Delta\tau} = L(\rho_0,\varphi_0) \tag{3.15}$$

These may be regarded as difference equations by which, starting from (ρ_0,φ_0) we determine the increments $\Delta\rho/\Delta\tau$ and $\Delta\varphi/\Delta\tau$ which we add to ρ_0 and φ_0, so as to obtain the initial conditions ρ_0', φ_0' for the next interval 2π, 4π, and so on. In our physical analogy this process amounts to determining the successive stroboscopic points $A^{(2)}$, $A^{(3)}$,... starting from the first one $A^{(1)}$ (Fig. 16.1).

We may still be guided in our analogy by introducing a passage to the limit; in fact, in this analogy, the persistence of vision produces the impression of a slow (quasi-continuous) motion along the trajectory C' (Fig. 16.1) instead of a set of discrete points $A^{(1)}$, $A^{(2)}$,... appearing successively after short time intervals 2π. This merely amounts to replacing the polygon $A^{(1)}$, $A^{(2)}$,... formed by stroboscopic points by a continuous curve C' and by considering (approximately) $\Delta\rho$, $\Delta\varphi$, and $\Delta\tau$ in (3.15) as $d\rho$, $d\varphi$, and $d\tau$. It is clear that we can do that if the total duration of the process is long enough compared to one period 2π and, moreover, $\Delta\tau$ is sufficiently small, as is seen from (3.14); the approximation is the better, the smaller is μ.

We obtain the *stroboscopic d.e.*

$$\frac{d\rho}{d\tau} = K(\rho,\varphi); \qquad \frac{d\varphi}{d\tau} = L(\rho,\varphi) \tag{3.16}$$

which will be important in the sequel.

It is noted that these d.e. are of the *autonomous type* and this enables us to find the condition for the existence of the fixed point inasmuch as the coordinates of this point are the same as those of the singular point (3.16). The problem of establishing the existence of a periodic solution of the original nonautonomous system (3.1) is thus reduced to ascertaining the existence of a singular point of (3.16). Likewise the question of stability of the periodic solution of (3.1) is reduced to the investigation of stability of the singular point of (3.16) which is a relatively simple problem.

There is an important link in the chain of these arguments worth emphasizing here, namely the passage from (3.13) to (3.15). In fact, (3.13) is still a purely topological transformation of the manifold of the initial conditions which merely gives a point (ρ_0',φ_0') if a point (ρ_0,φ_0) is given; in other words, to any point (ρ_0,φ_0) of the plane it gives a *corresponding point* (ρ_0',φ_0'); this is a purely *planar* transformation of one point into another point or of one area into another area.

As regards (3.15), there now appears a *temporal element* owing to the introduction of the concept of the *stroboscopic time* τ defined by (3.14); this permits "following" one particular transformation starting from a point (ρ_0,φ_0) by discrete steps $2\pi\mu$. It is clear, however, that the element $\Delta\tau$ so introduced has the significance of *time*, inasmuch as from the very beginning we spoke about "a period 2π" meaning also that 2π is an angle, which merely amounts to saying that the frequency is *one*.

The necessity for assuming μ small in this argument is also clear. In fact, we assumed that the time 2π is small as compared to the total duration of the process. If μ is small, $\Delta\tau$ satisfies the condition of smallness (with respect to the total duration) still better, and the replacement of $\Delta\tau$ by $d\tau$ is justified on this basis as a reasonable approximation (of order μ^2).

Thus, in this case the smallness of μ is required on the topological basis, viz.: the subsequent points $A^{(1)}$, $A^{(2)}$,. . . should be near enough to each other to be able to make the passage to the limit (3.15) → (3.16) without too much error.

There is still another point to be mentioned. We have assumed the transformation of the form (3.12) which can be written also as

$$(\mathfrak{T}): \rho' = \rho + \mu K'(\rho,\varphi); \qquad \varphi' = \varphi + \mu L'(\rho,\varphi) \tag{3.17}$$

This enabled us to obtain the stroboscopic system (3.16) and determine the fixed point of the transformation as the singular point of (3.16), viz.:

$$K(\rho^*,\varphi^*) = L(\rho^*,\varphi^*) = 0 \tag{3.18}$$

The question arises as to how the result obtained from the transformation (3.17) is related to a more accurate result that results from a more precise transformation

$$(\mathfrak{T}'): \rho' = \rho + \mu K'(\rho,\varphi) + \mu^2 K''(\rho,\varphi); \quad \varphi' = \varphi + \mu L'(\rho,\varphi) + \mu^2 L''(\rho,\varphi) \tag{3.19}$$

in which the terms with μ^2 are taken into account.

It is clear that, instead of the roots ρ^* and φ^* of (3.12) we have to find the roots of equations

$$K'(\rho,\varphi) + \mu K''(\rho,\varphi) = 0; \qquad L'(\rho,\varphi) + L''(\rho,\varphi) = 0 \tag{3.20}$$

The difficulty now is in that K' and L' have been computed by linearization, but K'' and L'' are yet unknown; their determination would require a complete integration, which is generally impossible. It should be noted, however, that one can still proceed by the general theory of approximations even if one does not know the functions K'' and L''; as long as these functions are bounded and μ is sufficiently small, the original result: $K' = L' = 0$ may be still a good approximation for the full equations (3.20). We may consider (3.20) as a system of two equations in which ρ and φ are functions of μ, that is, $\rho(\mu)$ and $\varphi(\mu)$ such that for $\mu \to 0$, $\rho(\mu) \to \rho^*$ and $\varphi(\mu) \to \varphi^*$.

We can impose this condition on $\rho(\mu)$ and $\varphi(\mu)$ provided the Jacobian

$$J = \left|\frac{\partial(K' + \mu K'', L' + \mu L'')}{\partial(\rho,\varphi)}\right|_{\mu=0,\ \rho=\rho^*,\ \phi=\phi^*} \neq 0 \tag{3.21}$$

This is the argument of Poincaré which we encountered in Chapter 10. Developing (3.21) we have

$$\begin{vmatrix} \dfrac{\partial K'}{\partial \rho} + \mu\dfrac{\partial K''}{\partial \rho} & \dfrac{\partial L'}{\partial \rho} + \mu\dfrac{\partial L''}{\partial \rho} \\ \dfrac{\partial K'}{\partial \varphi} + \mu\dfrac{\partial K''}{\partial \varphi} & \dfrac{\partial L'}{\partial \varphi} + \mu\dfrac{\partial L''}{\partial \varphi} \end{vmatrix}_{\mu=0} = \begin{vmatrix} \dfrac{\partial K'}{\partial \rho} & \dfrac{\partial L'}{\partial \rho} \\ \dfrac{\partial K'}{\partial \varphi} & \dfrac{\partial L'}{\partial \varphi} \end{vmatrix} \neq 0 \tag{3.22}$$

If this condition is fulfilled, the transformation $(\mathfrak{T})$ in (3.17) gives a good approximation and the use of $(\mathfrak{T}')$ (equation (3.19)) merely adds a very small correction so that one can be satisfied with a similar result yielded by $(\mathfrak{T})$. If, however (3.21) is not fulfilled the result yielded by $(\mathfrak{T})$ may be quite different from a correct result of application of $(\mathfrak{T}')$. In such a case the problem becomes more difficult.

We merely mention this point; in applications which follow we shall be concerned mostly with *normal* problems for which $(\mathfrak{T})$ is sufficient.

4. Application of the stroboscopic method to the Mathieu oscillators

We shall illustrate the application of the stroboscopic method in connection with two oscillators governed by the d.e.

$$\ddot{x} + (1 + a\cos 2t)x = 0 \tag{4.1}$$

and

$$\ddot{x} + b\dot{x} + (1 + a\cos 2t)x + cx^3 = 0 \tag{4.2}$$

These are the Mathieu d.e., the first one (4.1) being a linear one and the second (4.2)—nonlinear on account of the term cx^3. We assume that a, b, and c are small quantities of the same order.

This problem has a definite physical meaning, as we shall see in Part III in Chapter 20 dealing with the phenomenon of the so-called *parametric excitation.*

The smallness of a, b, and c is required by the condition of the near linearity as is generally the case of all analytic methods of approximations. Consider first the linear case (4.1). Written as an equivalent system this d.e. gives

$$\dot{x} = y; \qquad \dot{y} = -x - ax\cos 2t$$

Forming the two combinations $x\dot{x} + y\dot{y} = \frac{1}{2}(d\rho/dt)$; $x\dot{y} - y\dot{x} = \rho(d\psi/dt)$ (with $\rho = r^2 = x^2 + \dot{x}^2 = x^2 + y^2$; $\psi = \arctan(y/x)$ $x = r\cos\psi$ $y = r \sin\psi$) we have

$$\frac{d\rho}{dt} = -a\rho\sin 2\psi\cos 2t; \qquad \frac{d\psi}{dt} = -1 - a\cos^2\psi\cos 2t \tag{4.3}$$

With the series solutions (3.6) and, in view of the assumed smallness of a, the approximation of the zero order is again (3.7), that is

$$\rho_0(t) = \rho_0; \qquad \psi_0(t) = \varphi_0 - t \tag{4.4}$$

ρ_0 and φ_0 being two arbitrary constants. Substituting (4.4) into (4.3), one obtains the d.e. for the first-order corrective terms $\rho_1(t)$, $\psi_1(t)$, viz.:

$$\frac{d\rho_1}{dt} = -\rho_0\sin 2(\varphi_0 - t)\cos 2t; \qquad \frac{d\psi_1}{dt} = -\cos^2(\varphi_0 - t)\cos 2t \tag{4.5}$$

whence

$$\rho_1(2\pi) = -\rho_0 \sin 2\varphi_0 \int_0^{2\pi} \cos^2 2t dt = -\tfrac{1}{2}2\pi\rho_0 \sin 2\varphi_0$$

and

$$\rho(2\pi) = \Delta\rho = -\tfrac{1}{2}2\pi a \cdot \rho_0 \sin 2\varphi_0 \tag{4.6}$$

Setting $2\pi a = \Delta\tau$, the preceding equation becomes

$$\frac{\Delta\rho}{\Delta\tau} = -\frac{1}{2}\rho_0 \sin 2\varphi_0 \tag{4.7}$$

This gives the first sequence of difference equations (3.13) for ρ_0, ρ_0', ... The passage to the continuous variable yields the first stroboscopic d.e.

$$\frac{d\rho}{d\tau} = -\frac{1}{2}\rho \sin 2\varphi \tag{4.8}$$

In a similar way the second d.e. (4.5) yields $\psi_1(2\pi) = -\tfrac{1}{4}2\pi \cos 2\varphi_0$, and

$$\Delta\varphi = \psi(2\pi) = -\tfrac{1}{4}2\pi a \cos 2\varphi_0 \tag{4.9}$$

This gives the second sequence (3.13) in the form

$$\frac{\Delta\varphi}{\Delta\tau} = -\frac{1}{4}\cos 2\varphi_0 \tag{4.10}$$

which, at the limit, yields the second stroboscopic d.e.

$$\frac{d\varphi}{d\tau} = -\frac{1}{4}\cos 2\varphi \tag{4.11}$$

The d.e. (4.8) and (4.11) give the stroboscopic system in the first approximation. As no confusion is to be feared from now on, we change the notations and assign the subscript 0 to the *stationary values.*

It is observed that the system (4.8), (4.11) has no singular point as $\sin 2\varphi$ and $\cos 2\varphi$ cannot be zero at the same time.† On the other hand, from (4.11) it is clear that the (φ) phase has a point of equilibrium φ_0 when $\cos 2\varphi_0 = 0$. This occurs when $\varphi_0 = \pi/4$ or $\varphi_0 = 3\pi/4$ which requires the investigation of stability for these two values of φ_0, since φ_0 is defined only modulo 2π.

The variational d.e. for (4.11) is:

$$\frac{d\delta\varphi}{d\tau} = -\frac{1}{4}\cos 2(\varphi_0 + \delta\varphi) \simeq \frac{1}{2}\sin 2\varphi_0\delta\varphi \tag{4.12}$$

† We exclude $\rho = 0$ which is the position of equilibrium.

Since $\cos 2\varphi_0 = 0$, $\sin 2\varphi_0 = \pm 1$ and it is seen that the condition of stability requires that $\sin 2\varphi_0$ should be -1, which means that $\varphi_0 = 3\pi/4$ is a stable phase, whereas $\varphi_0 = \pi/4$ is unstable. Thus, from any value the phase approaches the stationary value $\varphi_0 = 3\pi/4$ and settles on it.

The first stroboscopic d.e. (4.8) then becomes

$$\frac{d\rho}{d\tau} = \frac{1}{2}\rho \tag{4.13}$$

which shows that ρ increases exponentially, that is, without any bound. It is seen that a linear d.e. of Mathieu sets up a transformation of the type (3) in the classification of Section 2 as we have already mentioned.

If one wishes to go beyond the first approximation, one has to use more terms in the series solutions

$$\rho(t) = \sum_{n=0}^{\infty} a^n \rho_n(t); \qquad \psi(t) = \sum_{n=0}^{\infty} a^n \psi_n(t) \tag{4.14}$$

One can do that by replacing ρ and ψ in (4.3) by the series (4.4) and collecting the terms with equal powers of the small parameter a. This applies also to the trigonometric function. For instance, for the next approximation the term $\sin 2\psi$ is replaced by $\sin 2(\psi_0 + a\psi_1) = \sin 2\psi_0 \times \cos 2a\psi_1 + \cos 2\psi_0 \sin 2a\psi_1 \simeq \sin 2\psi_0 + a2\psi_1 \cos 2\psi_0$, and the first d.e. (4.3) becomes

$$\frac{d(\rho_0 + a\rho_1 + a^2\rho_2 + \ldots)}{dt} = -a(\rho_0 + a\rho_1 + \ldots)(\sin 2\psi_0 + a2\psi_1 \cos 2\psi_0 + \ldots) \cos 2t$$

likewise for the second equation (4.3). If one collects the terms with equal powers of a, one obtains two sequences of the d.e. in $d\rho_2/dt$, $d\rho_3/dt, \ldots$, $d\psi_2/dt$, $d\psi_3/dt, \ldots$ by which additional terms in the stroboscopic d.e. can be computed.

If the parameter a is small, these higher-order approximations are generally of no special interest in applications inasmuch as they modify but little the quantitative character of the solution without changing anything in its qualitative aspect, provided, of course, the condition (3.22) holds, which is generally true in applied problems.

Only in some special cases difficulties may arise and the first approximation may fail to give an answer. We have already mentioned that such difficulty arises when the Jacobian of (3.21) vanishes, in which case the singular point $\rho = \rho^*$; $\varphi = \varphi^*$ is not an elementary one, which we studied in Chapter 1. Likewise, when frequency correction is zero in the first approximation, one is obliged to go to the second approximation for $d\varphi/dt$ (Section 5, below).

In what follows we shall be mostly concerned with *normal* cases and, unless stated to the contrary, shall deal with the first approximation only under the assumption of the smallness of parameters.

As another application of the stroboscopic method, we consider the d.e. (4.2). It is to be noted that in the standard form (4.1) of the Mathieu d.e., the term $b\dot{x}$ does not appear inasmuch as it can always be eliminated by the well known transformation of the dependent variable. This is impossible, however, for a nonlinear d.e. of Mathieu and, for that reason, one is obliged to keep this term. As there are now three small parameters, it is more convenient to have the series (4.14), where we use μ instead of a.

Following the preceding calculation, we have

$$\begin{aligned} \frac{d\rho}{dt} &= -2b\rho \sin^2 \psi - a\rho \sin 2\psi \cos 2t - 2c\rho^2 \cos^3 \psi \sin \psi \\ \frac{d\psi}{dt} &= -1 - \frac{1}{2} b \sin 2\psi - a \cos^2 \psi \cos 2t - c\rho \cos^4 \psi \end{aligned} \tag{4.15}$$

Since a, b, and c are small of the first order, the zero-order approximation still remains (4.4). As to the first-order corrections $\rho_1(t)$ and $\psi_1(t)$, they are given by the d.e.

$$\begin{aligned} \frac{d\rho_1}{dt} &= -B\rho_0 - A\rho_0 \sin 2\varphi_0 \cos^2 2t \\ \frac{d\psi_1}{dt} &= -\frac{1}{2} A \cos 2\psi_0 \cos^2 2t - \frac{3}{8} C\rho_0 \end{aligned} \tag{4.16}$$

where $A = a/\mu$; $B = b/\mu$ and $C = c/\mu$ and the terms, whose integration between 0 and 2π gives zero, are omitted; we assume a, b, c, and μ positive.

We have thus:

$$\rho_1(2\pi) = -\rho_0\left(B \cdot 2\pi + A \sin 2\varphi_0 \int_0^{2\pi} \cos^2 2t dt\right)$$
$$= -\tfrac{1}{2}\rho_0 2\pi(2B + A \sin 2\varphi_0)$$

and therefore $\rho(2\pi) = \mu\rho_1(2\pi)$. Setting $\rho(2\pi) = \Delta\rho$; $2\pi\mu = \Delta\tau$, one has

$$\frac{\Delta\rho}{\Delta\tau} = -\frac{1}{2} \rho_0(2B + A \sin 2\varphi_0) \tag{4.17}$$

and the first stroboscopic d.e. is

$$\frac{d\rho}{d\tau} = -\frac{1}{2} \rho[2B + A \sin 2\varphi] = K(\rho,\varphi) \tag{4.18}$$

As to the second, we obtain from the second d.e. (4.16)

$$\psi(2\pi) = \Delta\varphi = -2\pi\mu\cdot\tfrac{1}{4}[A\cos 2\varphi_0 + \tfrac{3}{2}C\rho_0]$$

that is,

$$\frac{\Delta\varphi}{\Delta\tau} = -\frac{1}{4}\left(A\cos 2\varphi_0 + \frac{3}{2}C\rho_0\right)$$

The second stroboscopic d.e. is

$$\frac{d\varphi}{d\tau} = -\frac{1}{4}\left[A\cos 2\varphi + \frac{3}{2}C\rho\right] = L(\rho,\varphi) \tag{4.19}$$

The d.e. (4.18) and (4.19) form the stroboscopic system corresponding to (4.2). We change the notation, attaching the subscript 0 to the coordinates of the singular point which exists if the terms in brackets of (4.18) and (4.19) can be equated to zero simultaneously.

From (4.18) we have

$$\sin 2\varphi_0 = -\frac{2B}{A} \tag{4.20}$$

and, as $|\sin 2\varphi_0| \leq 1$, one must have $2B \leq A$.

From (4.19) follows

$$\cos 2\varphi_0 = -\frac{3C}{2A}\rho_0 \tag{4.21}$$

From $\sin^2 2\varphi_0 + \cos^2 2\varphi_0 = 1$, we obtain

$$\rho_0 = \frac{2}{3C}\sqrt{A^2 - 4B^2} \tag{4.22}$$

The reality of ρ_0 imposes again the condition $2B \leq A$.

It is necessary to ascertain also the stability of the stationary values ρ_0 and φ_0. The simplest is to write the characteristic equation in the form (Chapter 1)

$$S^2 - (K_\rho + L_\varphi)S + (K_\rho L_\varphi - K_\varphi L_\rho) = 0 \tag{4.23}$$

where K_ρ, K_φ, L_ρ, and L_φ are partial derivatives of the functions $K(\rho,\varphi)$ and $L(\rho,\varphi)$ in (4.18) and (4.19), into which the stationary values of ρ_0 and φ_0 are replaced after the differentiation. Carrying out this calculation, we have

$$K_\rho = 0; \qquad K_\varphi = \frac{2}{3C}(A^2 - 4B^2); \qquad L_\rho = -\frac{3}{8}C; \qquad L_\varphi = -B$$

and (4.23) becomes

$$S^2 + BS + \tfrac{1}{4}(A^2 - 4B^2) = 0 \tag{4.24}$$

The singular point *is not* a saddle point if $2B \leq A$ but is a stable singularity (either a node or a focus), since $B > 0$.

We conclude that the condition

$$2B \leq A \tag{4.25}$$

is the necessary and sufficient condition for the existence of a stable singular point in the (φ) plane† and, therefore, for the existence of a stable periodic solution of the d.e. (4.2) with amplitude ρ_0 given by (4.22), and the phase, by (4.20). The physical significance of the d.e. (4.1) and (4.2) will be explained in Chapter 20 in connection with the phenomenon of the so-called parametric excitation, as was just mentioned.

5. Application of the method to autonomous systems; second approximation

The stroboscopic method is particularly advantageous in connection with nonautonomous systems inasmuch as it permits replacing a difficult problem of stability of a periodic solution (Chapter 13) by a much simpler problem of stability of equilibrium (that is, of a singular point).

In cases when an autonomous system has no frequency (or period) correction in the first approximation, the second stroboscopic d.e. reduces to $d\varphi/d\tau = 0$ and in such a case it is necessary to go to higher approximations in order to have a more precise representation of what happens in the stroboscopic plane.

As an example we consider the van der Pol equation

$$\ddot{x} + \mu(x^2 - 1)\dot{x} + x = 0 \tag{5.1}$$

With the variables ρ and ψ, the equivalent system is

$$\begin{aligned} \dot{\rho} &= \mu\rho(1 - \cos 2\psi) - \tfrac{1}{4}\mu\rho^2(1 - \cos 4\psi) \\ \dot{\psi} &= -1 + \tfrac{1}{2}\mu \sin 2\psi - \mu\rho(\tfrac{1}{4}\sin 2\psi + \tfrac{1}{8}\sin 4\psi) \end{aligned} \tag{5.2}$$

Since μ is small, the approximation of the zero order remains the same as before

$$\rho_0(t) = \rho_0; \qquad \psi_0(t) = \varphi_0 - t \tag{5.3}$$

where ρ_0 and φ_0 are integration constants.

The d.e. for the first-order corrections $\rho_1(t)$ and $\psi_1(t)$ are

$$\dot{\rho}_1 = \rho_0(1 - \tfrac{1}{4}\rho_0) - \rho_0(\cos 2\psi_0 - \tfrac{1}{4}\rho_0 \cos 4\psi_0) \tag{5.4}$$

† The above conditions result from the fact that $J = \partial(K,L)/\partial(\rho,\varphi) \neq 0$.

we have thus by integration

$$\rho_1(2\pi) = \rho_0(1 - \tfrac{1}{4}\rho_0)2\pi \tag{5.5}$$

since the second term on the right-hand side of (5.4) cancels out in integration between 0 and 2π. Since $\rho(2\pi) = \mu\rho_1(2\pi)$, setting $\rho(2\pi) = \Delta\rho$ and $2\pi\mu = \Delta\tau$, we have a difference equation

$$\frac{\Delta\rho}{\Delta\tau} = \rho_0\left(1 - \frac{1}{4}\rho_0\right) \tag{5.6}$$

which permits calculating ρ_0', ρ_0'',. . . after each interval 2π. If one passes to a continuous variable by assuming the small quantity $\Delta\tau$ as $d\tau$ at the limit, one obtains the stroboscopic d.e.

$$\frac{d\rho}{d\tau} = \rho\left(1 - \frac{1}{4}\rho\right) \tag{5.7}$$

which shows that there is a stationary value $\rho^* = 4$ (that is, $r^* = 2$) which is the well known result.

We apply the same procedure to the second equation (5.2) which yields

$$\psi_1 = \tfrac{1}{2}\sin 2\psi_0 - \rho_0(\tfrac{1}{4}\sin 2\psi_0 + \tfrac{1}{8}\sin 4\psi_0) \tag{5.8}$$

where ψ_0 is given by (5.3). It is clear that here $\psi_1(2\pi) = 0$ since the trigonometric functions cancel out in the integration between 0 and 2π. In the first approximation we have thus $\psi(2\pi) = \psi_0(2\pi) + \mu\psi_1(2\pi) = \psi_0(2\pi)$; whence $\Delta\varphi(2\pi) = 0$, since ψ is determined only modulo 2π. This means that the variation of ψ (that is, in our previous notations: φ) is zero over one period which one can write formally as $\Delta\varphi/\Delta\tau = 0$ or, at the limit, as

$$\frac{d\varphi}{d\tau} = 0 \tag{5.9}$$

Thus, if one limits the calculation to the first approximation only, the stroboscopic system is given by (5.7) and (5.8).

In the stroboscopic plane (φ) the amplitude is $\rho = \rho^* = 4$, but the phase remains arbitrary. In other words, any point on the circle $\rho^* = 4$ may be a fixed point. This result is, however, only approximate inasmuch as the first approximation does not give yet a complete information as regards the second equation (5.2).

For that reason, it is useful to go to the second approximation. We start, therefore, from a more general form of equations (5.2) by replacing ρ by $\rho_0 + \mu\rho_1$ and ψ by $\psi_0 + \mu\psi_1$ and keeping only the terms of the order μ^2 (that is, neglecting the terms with μ^3, μ^4,. . .).

A simple, but somewhat long, calculation results in the d.e.

$$\begin{aligned}
\frac{d\rho}{dt} &= \mu[\rho_0(1 - \tfrac{1}{4}\rho_0) - \rho_0 \cos 2\psi_0 + \tfrac{1}{4}\rho_0{}^2 \cos 4\psi_0] \\
&\quad + \mu^2[\rho_1(1 - \tfrac{1}{2}\rho_0) - \rho_1 \cos 2\psi_0 + 2\rho_0\psi_1 \sin 2\psi_0 \\
&\qquad\qquad + \tfrac{1}{2}\rho_0\rho_1 \cos 4\psi_0 - \rho_0{}^2\psi_1 \sin 4\psi_0] \\
\frac{d\psi}{dt} &= -1 + \mu[\tfrac{1}{2} \sin 2\psi_0 - \tfrac{1}{4}\rho_0 \sin 2\psi_0 - \tfrac{1}{8}\rho_0 \sin 4\psi_0] \\
&\quad + \mu^2[\psi_1(1 - \tfrac{1}{2}\rho_0) \cos 2\psi_0 - \tfrac{1}{4}\rho_1 \sin 2\psi_0 - \tfrac{1}{2}\rho_0\psi_1 \cos 4\psi_0 \\
&\qquad\qquad - \tfrac{1}{8}\rho_1 \sin 4\psi_0]
\end{aligned} \tag{5.10}$$

The zero-order approximation remains the same:

$$\rho_0(t) = \rho_0; \qquad \psi_0(t) = \varphi_0 - t \tag{5.11}$$

The first-order correction $\rho_1(t)$ is given by the d.e.

$$\frac{d\rho_1}{dt} = \rho_0\left(1 - \frac{1}{4}\rho_0\right) - \rho_0 \cos 2\psi_0 + \frac{1}{4}\rho_0{}^2 \cos 4\psi_0 \tag{5.12}$$

where the first term must vanish in order to avoid a secular term; this gives $\rho_0 = 4$ as previously; we have thus

$$\frac{d\rho_1}{dt} = -4 \cos 2(\varphi_0 - t) + 4 \cos 4(\varphi_0 - t)$$

and the integration yields

$$\rho_1(t) = -2 \sin 2\psi_0 - \sin 4\psi_0 + C_1 \tag{5.13}$$

In a similar manner one obtains

$$\psi_1(t) = -\tfrac{1}{4} \cos 2\psi_0 - \tfrac{1}{8} \cos 4\psi_0 + C_2 \tag{5.14}$$

The integration constants C_1 and C_2 are determined by the conditions: $\rho_1(0) = \psi_1(0) = 0$. For the second-order corrective terms $\rho_2(t)$ and $\psi_2(t)$ we have the d.e.

$$\begin{aligned}
\frac{d\rho_2}{dt} &= -\rho_1 + \rho_1 \cos 2\psi_0 + 8\psi_1 \sin 2\psi_0 + 2\rho_1 \cos 4\psi_0 - 16\psi_1 \sin 4\psi_0 \\
\frac{d\psi_2}{dt} &= -\psi_1 \cos 2\psi_0 \quad \frac{1}{4}\rho_1 \sin 2\psi_0 - 2\psi_1 \cos 4\psi_0 - \frac{1}{8}\rho_1 \sin 4\psi_0
\end{aligned} \tag{5.15}$$

It is seen that the first equation (for ρ_2) has no secular terms; therefore in the integration between 0 and 2π, one has $\rho_2(2\pi) = 0$ and, for similar reasons from (5.13), $\rho_1(2\pi) = 0$; thus $\rho_0 = 4$ holds, at least, up to the order μ^3. The situation is different, however, for the second equation (5.15) where, in the first term, appears $\cos^2 2\psi_0$, in the second, $\sin^2 2\psi_0$, etc., which results in

$$\Delta\varphi = \varphi(2\pi) = \mu^2 a + 0(\mu^3) \tag{5.16}$$

There appears a small positive term $\mu^2 a$ which accounts for a slow rotation of the stroboscopic point on a circle of radius $\rho_0 = 4$ with angular velocity $\mu^2 a(t)$.

In the first approximation the stroboscopic image of the van der Pol equation is a circle of radius 4 with an arbitrary phase (in view of $d\varphi/dt = 0$) but, in the second approximation, it is the same circle but traversed with a very small angular velocity ψ_2. We are thus in case 2 (Section 2); there is no fixed point in the (φ) plane but such a point exists in the (φ') plane rotating with respect to (φ) with average angular velocity $\mu^2 a(t)$. This is again an approximation inasmuch as $a(t)$ fluctuates between some limits.

6. Further properties of the stroboscopic transformation

Case 5 (Section 2) may also occur in some problems. As an example, assume that the stroboscopic system is of the form

$$\frac{d\rho}{d\tau} = A(\rho) + \beta \sin\varphi; \qquad \frac{d\varphi}{d\tau} = C + D\cos\varphi \tag{6.1}$$

where B, C, and D are positive constants: $C > D$ and B sufficiently small; $A(\rho)$ is a polynomial in ρ with a constant term, say A^*.

Since $C > D$, $d\varphi/d\tau$ keeps the same sign and φ varies continuously in the same direction, although with a variable velocity $d\varphi/d\tau$. The trigonometric functions $\sin\varphi$ and $\cos\varphi$ vary between the limits $+1$ and -1 and the last term of the polynomial $A^* + B\sin\varphi$ oscillates between the values $A^* + B$ and $A^* - B$ (we assume for simplicity: $A^* \geq B$). The oscillation of the last term (free of ρ) in the polynomial $A(\rho)$ produces an oscillation of its roots; if the equilibrium point is at some root, say ρ_0, this root will oscillate between some limits around its value $\rho = \rho_0$. If one had $B = D = 0$, this would be case 2 (Section 2) and a fixed point would exist in a plane (φ') rotating with respect to (φ) with $d\varphi/d\tau = C$. With $B \neq 0$ and $D \neq 0$, both amplitude and phase are now *modulated* so that in the stroboscopic plane there will be a small trajectory around a fixed point; this trajectory is not necessarily a closed curve but may be of an *ergodic* nature, that is, passing as near as desired to some point in this neighborhood if the time is sufficiently long; only when the amplitude and phase have modulations of *commensurate periods* will the curve have a re-entrant path. Clearly, the ergodic case characterizes an almost periodic solution (Chapter 12).

Two additional remarks are noteworthy. For the sake of simplification we shall call the original d.e. (either (3.1) or (3.3)), the system (A) and the corresponding stroboscopic system (3.16), the system (B).

1. To a given (A) corresponds always one and only one (B) but the inverse is not true; to a given (B), in general, corresponds an infinity of systems (A). This is due to the fact that the stroboscopic transformation is *not a topological one* (that is, 1:1, continuous). In fact, to an arc of a trajectory in the (ψ) plane corresponds just one point in the (φ) plane.

2. It may happen that (B) exhibits the presence of more than one stable singular point, which shows that (A) has also more than one stable periodic solution or, topologically, has more than one stable limit cycles. This, as we saw, is quite possible, for instance, in the case of the polycyclic structures (for example, of a "concentric" type); only one of them is "actual," the others being only "virtual" (Chapter 3). A situation of this kind will be encountered in (Chapter 23).

It is to be mentioned that the property of the stroboscopic plane (φ) permits one also to have an insight into the order of multiplicity of periodic solutions for three major classes of oscillatory systems, viz.: (1) harmonic oscillator, (2) autonomous systems (for example, a van der Pol oscillator) (3) nonautonomous systems (for example, nonlinear non-autonomous oscillator of Mathieu, Section 5).

As any fixed point in the plane (φ) (or in the plane (φ')) characterizes a periodic solution of the original system, there is a *double infinity* of solutions of (1) which is merely a different way of saying that periodic solutions of the harmonic oscillator depend on two arbitrary constants of integration. First, there is a continuous family of circles of radius ρ and, on each of these circles, there is another continuous family of solutions differing from each other by the phase. The double infinity of solutions is represented by the infinity of fixed points in a finite region of the plane (φ).

As regards (2), the periodic solutions reduce here to a *simple infinity* (infinity of points on a closed curve) since ρ (or r) ceases to be arbitrary and *is determined by the d.e.*; as to the phase, it is still arbitrary so that, on the circle $\rho_0 = 4$ (or $r_0 = 2$) (in the case of the van der Pol oscillator) there is an infinity of possible periodic solutions differing from each other only by the phase.

Finally for (3), the number of periodic solutions shrinks to one, rarely to two and, generally, only to a finite number. In the case of a nonlinear Mathieu oscillator we established the existence of one single periodic solution with amplitude ρ_0 and phase φ_0. This is due to the fact that, in case (3), *both amplitude and phase are determined by the d.e.* It is also possible that there is no periodic solution at all, as was found in the case of a linear Mathieu oscillator.

Summing up, the periodic solutions which form a double infinity for (1), and a simple infinity for (2), shrink to isolated solutions (or no solutions) for (3). It must be noted that this classification relates only to *mathematical* solutions. In physical applications these considerations become more or less evanescent inasmuch as "multiplicity of solutions differing from each other only by the phase" does not have any special physical meaning unless *it is compared* to some fixed reference phase. Thus, periodic solutions of nonautonomous systems are exceptions, whereas those of autonomous and conservative systems are the rule.

One can also say that in case (1) neither the amplitude nor the phase are determined by the d.e.; in case (2) only the amplitude is determined by the d.e. but not the phase; in case (3) bóth amplitude and phase are determined by the d.e.

The expression "amplitude is determined by the d.e." is clearly equivalent to the previously defined concept of "orbital stability," whereas the expression "phase is determined by the d.e." is equivalent to the nonautonomous character of the system in question; in fact, for an autonomous system the translation property applies and the phase is indeterminate.

7. Existence and stability of the fixed point

In Section 3 we used the argument of the transformation theory and indicated the form (3.12) of the transformation in which the quantities φ_1 and ψ_1 appear as increments (of the first order since we are concerned here with the first approximation) of ρ and ψ during the time 2π. After this we introduced the "stroboscopic time" defined by its element $\Delta\tau = 2\pi\mu$ which enabled us to obtain this result in the form of two difference equations (3.15); this, in turn, resulted in the stroboscopic d.e. (3.16) through the obvious passage to the limit: $\Delta\rho \to d\rho$; $\Delta\varphi \to d\varphi$; $\Delta\tau \to d\tau$. In our optical analogy this is equivalent to the *persistence of vision* which assimilates a discrete sequence of points $A_1, A_2, \ldots$ to a continuous curve C'. Once the stroboscopic system (3.16) is established, one can investigate the existence and stability of its *singular point* instead of speaking about the fixed point of the transformation in the (ψ) plane. In this manner there appears a convenient "short cut" which is useful in applications, as we shall see in a number of cases in Part III.

This argument contains, however, a certain heuristic element by which a purely *spatial* transformation (3.12) is ultimately adapted for a *temporal* representation appearing in the stroboscopic system (3.16).

It is possible, however, to use directly the transformation theory as was done by Gomory[3] whose proof follows.† Given a differential system:

$$\dot{x} = y + \mu f(x, y, t, \mu); \qquad \dot{y} = -x + \mu g(x, y, t, \mu) \tag{7.1}$$

where f and g are analytic functions of the indicated variables containing a small parameter μ; the functions f and g are periodic in t with period 2π. For $\mu = 0$ these d.e. become those of the harmonic oscillator.

Following Poincaré's theory (Chapter 10), we can assume a solution of the form:

$$\begin{aligned} x(t, x_0, y_0, \mu) &= \varphi_0(t, x_0, y_0) + \mu\varphi_1(t, x_0, y_0) + \mu^2\varphi_2(t, x_0, y_0) + \dots \\ y(t, x_0, y_0, \mu) &= \psi_0(t, x_0, y_0) + \mu\psi_1(t, x_0, y_0) + \mu^2\psi_2(t, x_0, y_0) + \dots \end{aligned} \tag{7.2}$$

where x_0 and y_0 are the initial conditions. Since (7.1) for $\mu = 0$ represents the harmonic oscillator, we have

$$\varphi_0(2\pi, x_0, y_0) = x_0; \qquad \psi_0(2\pi, x_0, y_0) = y_0 \tag{7.3}$$

Setting $t = 2\pi$ in (7.2), we have:

$$\begin{aligned} x(2\pi, x_0, y_0, \mu) &= x_0 + \mu\varphi_1(2\pi, x_0, y_0) + \mu^2\varphi_2(2\pi, x_0, y_0) + \dots \\ y(2\pi, x_0, y_0, \mu) &= y_0 + \mu\psi_1(2\pi, x_0, y_0) + \mu^2\psi_2(2\pi, x_0, y_0) + \dots \end{aligned} \tag{7.4}$$

These expressions give the position of the representative point R after the time 2π. Consider now the stroboscopic system (3.16) which can be written as:

$$\frac{dx_0}{d\tau} = \frac{\varphi_1}{2\pi}(2\pi, x_0, y_0); \qquad \frac{dy_0}{d\tau} = \frac{\psi_1}{2\pi}(2\pi, x_0, y_0) \tag{7.5}$$

In fact, by their very nature, the stroboscopic d.e. assimilate the discrete variation of the initial conditions after each period 2π to a continuous change of these initial conditions; this approximation is possible, as was mentioned previously, when one period 2π is small as compared to the total duration of the process which we assume.

One has to prove that, if the point $(\hat{x}, \hat{y})$ of (7.5) is a stable singular point (a node or a focus), then (7.1) has a stable periodic solution: $x(t, \bar{x}_0(\mu), \bar{y}_0(\mu), \mu)$; $y(t, \bar{x}_0(\mu), \bar{y}_0(\mu), \mu)$ such that

$$(\bar{x}_0(\mu), \bar{y}_0(\mu)) \to (\hat{x}, \hat{y}) \qquad \text{for } \mu \to 0 \tag{7.6}$$

[3] R. Gomory, unpublished communication to the author, 1956; see also M. Urabe, *J. Science*, University of Hiroshima, 1956; Masataka, Yorinaga, *ibid.*, 1960.

† Notations below are different from those in the preceding sections but no confusion is to be feared.

One can specify this by writing:

$$\begin{aligned} \varphi_1(2\pi, x_0, y_0) &= a(x_0 - \hat{x}) + b(y_0 - \hat{y}) + \dots \\ \psi_1(2\pi, x_0, y_0) &= c(x_0 - \hat{x}) + d(y_0 - \hat{y}) + \dots \end{aligned} \tag{7.7}$$

where the nonwritten terms are, at least, of the second degree in x_0, y_0.

We wish to consider the case of a positive damping for the system containing only linear terms and $\begin{vmatrix} a & b \\ c & d \end{vmatrix} \neq 0$; this means that the root λ of the characteristic equation is of the form $\lambda = \alpha + i\beta$ with $\alpha < 0$.

For periodicity one must have

$$x(2\pi, x_0, y_0, \mu) - x_0 = 0; \qquad y(2\pi, x_0, y_0, \mu) - y_0 = 0 \tag{7.8}$$

This may occur for such a special point $(\bar{x}_0(\mu), \bar{y}_0(\mu))$ which satisfies (7.8). The condition for this is

$$\begin{aligned} a(x_0 - \hat{x}) + b(y_0 - \hat{y}) + \mu\varphi_2 &= 0 \\ c(x_0 - \hat{x}) + d(y_0 - \hat{y}) + \mu\psi_2 &= 0 \end{aligned} \tag{7.9}$$

Under this condition one can assert that there exists a point

$$\begin{aligned} \bar{x}_0(\mu) - \hat{x} &= \alpha_1\mu + \alpha_2\mu^2 + \dots \\ \bar{y}_0(\psi) - \hat{y} &= \beta_1\mu + \beta_2\mu^2 + \dots \end{aligned} \tag{7.10}$$

for which (7.8) holds and for $\mu = 0$; this point is:

$$(\bar{x}_0, \bar{y}_0) = (\hat{x}, \hat{y}) \tag{7.11}$$

This establishes the existence of a periodic solution.

We investigate now the question of stability. In the neighborhood of the fixed point $(\bar{x}_0, \bar{y}_0)$ the transformation is:

$$\begin{aligned} (x - \bar{x}_0) &= A(x_0 - \bar{x}_0) + B(y - \bar{y}_0) + \text{higher-order terms} \\ (y - \bar{y}_0) &= C(x_0 - \bar{x}_0) + D(y - \bar{y}_0) + \text{higher-order terms} \end{aligned} \tag{7.12}$$

It is clear that if the modulus $|\lambda|$ of the root λ of the characteristic equation

$$\begin{vmatrix} A - \lambda & B \\ C & D - \lambda \end{vmatrix} = 0$$

is less than one, a point near to $(\bar{x}_0, \bar{y}_0)$ will approach it under the effect of repeated transformations; if $|\lambda| > 1$, this point is unstable. In fact, we have

$$x(2\pi, x_0, y_0, \mu) = x_0 + \mu\varphi_1(2\pi, x_0, y_0) + \mu^2\varphi_2(2\pi, x_0, y_0) + \cdots$$
$$\bar{x} = \bar{x} + \mu\varphi_1(2\pi, \bar{x}, \bar{y}) + \mu^2\varphi_2(2\pi, \bar{x}, \bar{y}) + \cdots$$

thus

$$x(2\pi, x_0, y_0, \mu) - \bar{x} = (x_0 - \bar{x})$$
$$+ \mu\left[\frac{\partial\varphi_1}{\partial x}\bigg|_{(\bar{x},\bar{y})}(x_0 - \bar{x}) + \frac{\partial\varphi_1}{\partial y}\bigg|_{(\bar{x},\bar{y})}(y_0 - \bar{y}) + \cdots\right] + \mu^2[\quad] + \cdots \tag{7.13}$$

But

$$\frac{\partial\varphi_1}{\partial x}\bigg|_{(\bar{x},\bar{y})} \doteq \frac{\partial\varphi_1}{\partial x}\bigg|_{(x,\hat{y})} + O(\mu) = a + O(\mu)$$

$$\frac{\partial\varphi_1}{\partial y}\bigg|_{(\bar{x},\bar{y})} = \frac{\partial\varphi_1}{\partial y}\bigg|_{(\hat{x},\hat{y})} + O(\mu) = b + O(\mu)$$

since $\bar{x} - \hat{x}, \bar{y} - \hat{y} = O(\mu)$. Hence

$$x(2\pi, x_0, y_0, \mu) - \bar{x} = (1 + \mu a + O(\mu^2))(x_0 - \bar{x})$$
$$+ [\mu b + O(\mu^2)](y_0 - \bar{y}) + (x_0 - \bar{x}, y_0 - \bar{y})_2 \tag{7.14}$$

$$y(2\pi, x_0, y_0, \mu) - \bar{y} = (\mu c + O(\mu^2))(x_0 - \bar{x})$$
$$+ [1 + \mu d + O(\mu^2)](y_0 - \bar{y}) + (x_0 - \bar{x}, y_0 - \bar{y})_2 \tag{7.15}$$

The characteristic equation therefore is

$$\begin{vmatrix} 1 + \mu a + O(\mu^2) - \lambda & \mu b + O(\mu^2) \\ \mu c + O(\mu^2) & 1 + \mu d + O(\mu^2) - \lambda \end{vmatrix} = 0 \tag{7.16}$$

If $\mu \neq 0$, one can write it as

$$\begin{vmatrix} a + O(\mu) - \left[\dfrac{\lambda - 1}{\mu}\right] & b + O(\mu) \\ c + O(\mu) & d + O(\mu) - \left[\dfrac{\lambda - 1}{\mu}\right] \end{vmatrix} = 0 \tag{7.17}$$

If $\mu \to 0$, the coefficients approach the values: a, b, c, and d. On the other hand, by our hypotheses $(\lambda - 1)/\mu = \delta(\mu) + i\gamma(\mu)$, where $\delta(\mu) \to \alpha < 0$; $\gamma(\mu) \to \beta$ and

$$\lambda = 1 + \mu[\delta(\mu) + i\gamma(\mu)]; \qquad \lambda = 1 + \mu[\delta(\mu) - i\gamma(\mu)]$$

$$\lambda\bar{\lambda} = |\lambda|^2 = 1 + 2\mu\delta(\mu) + \mu^2[\delta^2(\mu) + \gamma^2(\mu)]$$
$$= 1 + \mu\{2\delta(\mu) + \mu[\delta^2(\mu) + \gamma^2(\mu)]\}$$

Since we can choose $\mu < \mu_0$, $\varphi(\mu) < \alpha/2$ and then choose μ so small that

$$\mu[\delta^2(\mu) + \gamma^2(\mu)] < |\alpha|$$

the term in brackets will be negative and

$$|\lambda| < 1 \tag{7.18}$$

This means that for μ sufficiently small and positive, $(\bar{x}_0(\mu), \bar{y}_0(\mu))$ is a stable point.

Chapter 17

GENERALIZATION OF NYQUIST'S DIAGRAM FOR NONLINEAR SYSTEMS

1. Introductory remarks

Although the Nyquist diagram is probably sufficiently well known to need no further elaboration here, it is useful to outline its principle as an aid to extending its use to nonlinear systems.

This diagram describes the behavior of a linear system with a feedback when the frequency of the system varies from 0 to ∞; its principal purpose is to establish the condition of stability.

If one assumes first that the system in question is *open* (that is, the feedback connection is removed), the application of a harmonic input signal: $x_{in} = \exp(j\omega t)$ results in the corresponding output signal

$$x_{out} = A(\omega) \exp [j\omega t + \varphi(\omega)] = A(\omega) \exp [\varphi(\omega)] x_{in} \qquad (1.1)$$

This means that x_{out} is a certain vector (in the complex plane) determined by the known laws of the alternating current theory.

If the feedback connection is restored, that is, the system is *closed*, the effect of the retroaction may either increase or decrease the input signal and this, in turn, will produce either increasing or decreasing oscillatory phenomenon at the output.

It is useful to introduce the concept of stability by certain a posteriori considerations. Suppose that we open the system when it is in a steady oscillatory state. In the closed condition one has:

$$x_{out} = -x_{in} = \bar{A}(\bar{\omega}) \exp(j\bar{\omega} t) \qquad (1.2)$$

One can consider this condition as a threshold *of stability* for which the energy introduced through the feedback is just sufficient to maintain oscillatory level, provided one does not exceed the limit

$$x_{out}/x_{in} = -1 \qquad (1.3)$$

which is the expression for "closing." This simple physical consideration permits formulating the following criterion due to Nyquist.

If a linear oscillatory open system is stable and if its amplitude-phase characteristic (1) *does not include the point* $(-1, j0)$ *of the complex plane, the corresponding closed system is also stable when frequency* ω *varies in the interval* $0, \infty$.

It has been shown that for $0 < \omega < \infty$ the path of the vector (1.1) is always a closed curve and that the above theorem has a definite physical meaning. In fact, if the retroaction (feedback) from x_{out} is so great that the phenomenon becomes cumulative, there is no equilibrium in the above sense. In terms of the engineering practice the Nyquist criterion is the formulation of the concept of the *regenerative amplification* which can exist only up to the limit beyond which the regenerative amplification becomes *self-excitation* and there is no possibility of using the linear theory.

We shall not give here the proof of Nyquist's theorem because it can be found easily in standard textbooks; it is sufficient to say that it is based on the application of Cauchy's theorem (in the complex domain).

It is seen that as far as fundamentals are concerned the situation seems to be somewhat obscure; the Nyquist diagram is based on the theory of the complex variable whereas it is desired to adapt it to the real domain where everything, beginning with singularities, is different from the complex domain. Moreover the concept of the limit cycle which is fundamental for nonlinear oscillations (the real domain) simply *does not exist* in the theory of d.e. in the complex domain and other definitions and concepts.

However, occasionally in applications physicists or engineers use certain algorithms without bothering about their formal justifications. If the thing "works," this is considered as correct procedure and it is up to mathematicians to justify it formally.

Apparently something of this kind occurs here. Under certain assumptions of a somewhat postulational character it can be shown that this purely linear tool, the Nyquist diagram, can be "twisted" so to speak into the nonlinear domain at the cost of certain additional complications. In this manner practically all that is known from the standard nonlinear theory can be obtained also from the generalized Nyquist diagram. The whole question is then: Is this worth while? In fact, we shall see that the practical difficulties, such as constructing a *family* of Nyquist diagrams (instead of one single diagram needed in the linear domain), are so great that it seems to be simpler to follow the general theory because the principal advantage of the diagram—its simplicity—is lost on account of these complications.

Although the development of analytical methods renders the use of the generalized Nyquist diagram somewhat obsolete, brief mention of these attempts is still of interest. The approach to this generalization can be made if one tries to establish the difference between what is linear and what is nonlinear from a *physical point of view*. The fundamental idea is that in all nonlinear oscillations the amplitude and the frequency are interrelated, whereas in the linear case they are independent of each other. If one admits this as a (somewhat hidden) postulate, one can formulate (heuristically) the following theorem:

In the nonlinear case, instead of one single Nyquist diagram (as in normal linear cases), there exists a continuous family of such diagrams depending on amplitude as parameter.

This may be regarded as a plausible assumption from which one starts the argument without attempting to analyze the theoretical foundations.

This physical approach appeared independently in the work of Theodorchik[1] (USSR) and Blaquière[2] (France), the former preceding the latter by about five years; the two developments follow different procedures but the result is ultimately the same. The work of Theodorchik is illustrated by numerous examples and, in its theoretical part, it uses the data of the first approximation with which we are now familiar. In the development of Blaquière more attention is attached to the theoretical part which is presented in a form of an operational method but, unfortunately, the applications of this method are less worked out and it is difficult to judge the merits of this method.

For these reasons we shall outline in some detail the procedure of Theodorchik and shall mention the second approach briefly in the last section.

It must be noted that these attempts to generalize the Nyquist diagram for nonlinear applications should not be considered as an elaboration of a new method but rather as a *representation* in a special phase plane of the variables: a (amplitude) and ω (frequency), instead of the conventional phase plane. This accounts for a somewhat special form of variational equations; thus, for instance, one comes across the question of *stability of the frequency* (instead of the phase), etc. In other words, nothing essentially new is gained but some known results appear differently due to this special representation.

It is to be noted also that the principal advantage of the usual (linear)

[1] K. F. Theodorchik, *Auto-oscillatory Systems* (in Russian), Moscow, 1948.

[2] A. Blaquière, *J. de Phys. et Radium* (8), **13**, 1952, *Mecanique non linéaire, Memorial des Sciences Mathematiques* **CXLI**, 1960.

Nyquist diagram, its *simplicity*, is somewhat lost because of this additional parameter a, the amplitude which introduces certain complications.

2. Theory of generalized Nyquist diagrams (Thoedorchik)

Theodorchik[1] approaches the question of Nyquist's diagram from a purely physical point of view, as was mentioned previously.

The usual treatment of a linear system with a feedback is reduced to a four-pole scheme, the stability of which can be predicted by open-circuit measurements, as was mentioned in Section 1.

If one applies to the input terminals of a four-pole scheme an electromotive force $u = \text{Re}\,(u_0 e^{ipt})$† from a source whose internal impedance is equal to the output impedance of the four-pole scheme and measures the amplitude v_0 and the phase ψ at the output terminals (always in open circuit with the feedback removed) for $0 < p < \infty$, where p is the frequency, these open circuit data are sufficient to predict the performance of the four-pole scheme when the feedback connection is restored.

The measurements permit thus determining two factors: (1) amplification factor (or "gain"): $\mu(p) = (v_0/u_0)$; and (2) the phase angle $\psi(p)$. These two quantities can be reduced to one if one defines the *complex* amplification factor

$$\mu^*(p) = \mu(p)e^{i\psi(p)} \tag{2.1}$$

If p varies, $\mu^*(p)$ describes a locus in the complex plane and, from the known relation in the circuits $\mu^*(0) = \mu^*(\infty)$, it is noted that the curve $\mu^*(p)$ is closed when p varies from 0 to ∞.

This results in the Nyquist theorem:

If the closed curve $\mu^*(p)$ contains the point $\mu = 1$,‡ $\psi = 0$ in its interior, the four-pole system (with feedback) is statically unstable. If this point (1,0) *is outside the loop $\mu^*(p)$, it is stable.*

Theodorchik is not interested in the manner in which the Nyquist diagram was originally obtained but considers rather its physical significance which enables him to formulate immediately a corresponding *interpretation*, namely: "statically unstable" means "self-excited" and "statically stable," on the contrary, means: "not self-excited."

Thus all underlying mathematical considerations, such as the complex

[1] See footnote [1], page 418.

† We conserve here Theodorchik's notations.

‡ We use here Theodorchik's notations which are somewhat different from those in use in the U.S.A.; for unstance, instead of writing $\mu = -1$, one writes here $\mu = +1$, which does not change the meaning of the following conclusions.

domain versus the real domain, etc., are completely by-passed and the theory starts from the physical concept of the Nyquist diagram under the above interpretation.

The next step in the argument is to introduce the concept of nonlinearity by postulating that in this case, instead of one single diagram, there exists a *family* of such diagrams depending on the amplitude a (or u_0) as a parameter. With this heuristic assumption the nonlinear coefficient μ^* will have the form

$$\mu^*(u_0,p) = \mu(u_0,p)e^{i\psi(u_0,p)} \tag{2.2}$$

Still another step in this argument is to introduce a distinction between the "soft" and the "hard" systems which, in the analytical theories, is characterized by the degree of the polynomial appearing as the coefficient of x. In this interpretation a "soft system" is characterized by the property: $\mu(u_0,p) \rightarrow 0$ as u_0 increases. In such a case the family of closed loops μ^* is contained inside a limit loop $\mu^*(0,p)$ and the criterion of the static stability (no self-excitation) is that the point $\mu = 1$, $\psi = 0$ is *outside* the loop $\mu^*(0,p)$. The condition of stability can be specified as follows: the system is stable for all frequencies for which $\psi(0,p) = 0$, $p = \omega_i$ corresponds to $\mu(0,\omega_i) < 1$. If this condition is not fulfilled (that is, if the point (1,0) is *inside* the loop $\mu^*(0,p)$ the system is unstable.

The next step in this "fitting" of the two-parameter Nyquist diagram into the known facts of the theory of nonlinear oscillations is to consider more carefully the mechanism of self-excitation. It is noted that the limit diagram for $u_0 \simeq 0$ (that is, the point of its intersection with the real axis) may be regarded as the unstable singular point of the analytic theory.

In this interpretation the increasing amplitudes during the transient period of self-excitation are equivalent to the motion of the representative point R *toward* the origin traversing gradually the family of Nyquist's loops. This transient period ends when R reaches the point (1,0) on the μ axis (in reality this approach is asymptotic so that the above statement means: close enough to the point 1, 0).

Once this point is reached, which means the *stationary state*, the amplitude a_0 and the frequency ω_0 are determined by two equations

$$\psi(a_0,\omega_0) = 0; \qquad \mu(a_0,\omega_0) = 1 \tag{2.3}$$

and the self-excitation from rest is then given by the condition

$$\mu(u_0,p) = \mu(0,\omega_0) > 1 \tag{2.4}$$

It is seen that the result of this "fitting" the known phenomena of self-excitation into the Nyquist diagram representation is reminiscent of familiar features of the analytic theory. In fact, equations (2.3) determine the conditions of the stationary state for which the phase ψ and the coefficient of amplification μ acquire, respectively, the values 0 and 1; and the condition (2.4) merely states that, for $a = 0$, the coefficient of amplification is in the region where the amplitudes grow. The difference here is the fact that, instead of the coordinates: amplitude *and phase* of the analytic theory, these "coordinates" are: amplitude *and frequency*. This involves, obviously, a different graphical representation.

It is apparent that all this does not constitute any *theory* properly speaking, but merely amounts to an *interpretation* of the two-parameter family of Nyquist's diagrams on the basis of known conclusions of the analytic theory.

The fact that the frequency appears now as a parameter raises a new question, namely: the *stability of the frequency*. If $\psi > 0$ during the transient process of self-excitation and the phase angle increases, the frequency also increases; conversely if $\psi < 0$ and ψ increases, the frequency decreases. The question of stability of the frequency plays an analogous role as stability of the phase in the analytic theory (for example, in the examples of the stroboscopic method, Chapter 16).

As regards the stability of the amplitude, it reduces to the criterion

$$\left(\frac{\partial \mu}{\partial u_0}\right)_{u_0=\omega_0} < 0 \tag{2.5}$$

which again reminds us, at least formally, of the criterion of stability

$$\Phi_\rho(\rho_0) < 0$$

which we encountered often in the theory of the first approximation (for example, Section 6, Chapter 14).

It is to be noted that the topology here is peculiar to the Nyquist diagram but, otherwise, the procedure is essentially the same as that in the classical theory.

For instance, if one takes the stroboscopic method as a kind of a comparison system, we have the stroboscopic d.e.

$$\frac{d\rho}{d\tau} = R(\rho,\varphi); \qquad \frac{d\varphi}{d\tau} = \Phi(\rho,\varphi) \tag{2.6}$$

and the condition of the stationary state: ρ_0, φ_0 is obtained from two equations:

$$R(\rho_0,\varphi_0) = 0; \qquad \Phi(\rho_0,\varphi_0) = 0 \tag{2.7}$$

In the Theodorchik's adaptation of Nyquist's diagram for the nonlinear cases, the stationary state is given by equations

$$\psi(a_0,\omega_0) = 0; \qquad \mu(a_0,\omega_0) = 1 \tag{2.8}$$

from which the stationary state a_0, ω_0 is determined in a similar manner. One sees that the known results guide this identification but the results remain in terms of the Nyquist diagram. If μ does not vary monotonically with u_0 but follows, for instance, curve A, Fig. 17.1, the results are different and lead to the interpretation of a "hard" self-excitation (Fig. 17.2). In such a case there are two limiting Nyquist loops L and L' representing μ^*. If the point (1,0) is inside the smaller loop L shown in shading, the self-excitation occurs as previously explained; the point R

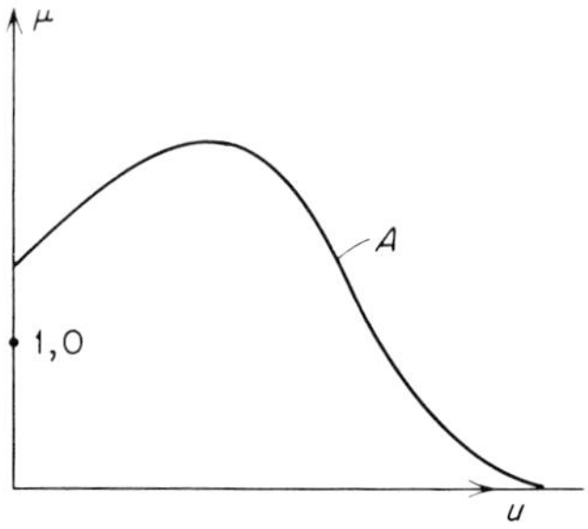

FIGURE 17.1

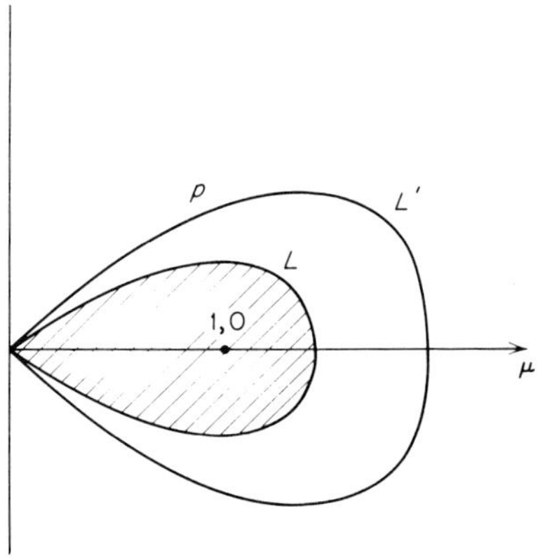

FIGURE 17.2

approaches first the boundary and from there settles at the point (1,0). If however, the point (1,0) is in the region between the two loops, the state of rest is stable but, if an impulse transfers R and L', the condition of self-excitation occurs in a previously described manner and this gives the interpretation of a "hard" self-excitation.

3. Stationary state of self-excitation

The identification of conditions of stationary state with equations (2.8) of the generalized Nyquist diagram permits applying this procedure to the investigation of the various nonlinear phenomena which are treated in Part III by the general theory.

Assume, for instance, that the complex coefficient μ^* is of the form

$$\mu^* = \mu(\cos\psi + i\sin\psi) = \frac{g_1 + ig_2}{r_1 + ir_2} = \frac{g_1r_1 + g_2r_2 + j(r_1g_2 - r_2g_1)}{r_1^2 + r_2^2} \quad (3.1)$$

where g and r are known functions of amplitude, frequency and, possibly, some other parameters. We look for the points of the intersection of the curves μ^* with the real and imaginary axes.

The real values of μ^* are obtained as roots of equation

$$\mu \sin \psi = \frac{r_1 g_2 - r_2 g_1}{r_1^2 + r_2^2} = 0 \tag{3.2}$$

and the imaginary ones are determined by the equation

$$\mu \cos \psi = \frac{g_1 r_1 + g_2 r_2}{r_1^2 + r_2^2} \tag{3.3}$$

The modulus $|\mu^*|$ is given by the expression

$$|\mu^*| = \frac{\sqrt{(r_1 g_2 - r_2 g_1)^2 + (r_1 g_1 + r_2 g_2)^2}}{r_1^2 + r_2^2} \tag{3.4}$$

Theodorchik applied this method to a number of problems and found, practically, all results that have been established previously by analytical methods, at least as far as the first approximation is concerned.

As an example of this procedure consider an ordinary electron-tube oscillator with an inductive coupling between the anode and the oscillating circuits. In the notations of Theodorchik† the d.e. is:

$$\ddot{x} + 2\delta\dot{x} + \omega_0^2 x = -M\omega_0^2 \frac{dI_a}{dt} \tag{3.5}$$

where I_a is the anode current, M the coefficient of the mutual inductance of the coupling, ω_0 the frequency of the oscillating current, and δ the decrement.

Using the previous notations, the grid voltage is

$$u = \operatorname{Re} u_0 e^{ipt} \tag{3.6}$$

If one takes I_a in the form $I_a = S_0 u - \frac{1}{3}S_2 u^3$, (3.5) becomes

$$\ddot{x} + 2\delta\dot{x} + \omega_0^2 x = -ipu_0 M\omega_0^2(S_0 - \tfrac{1}{4}S_2 u_0^2)e^{ipt} \tag{3.7}$$

keeping on the right-hand side only the fundamental frequency.

† Theodorchik prefers to write the van der Pol equation with a soft characteristic in the form $\ddot{x} + [2\delta + M\omega_0^2 S_0 - M\omega_0^2 S_2 x^2]\dot{x} + \omega_0^2 x = 0$. In terms of electrical circuits $\delta = R/2L$ is the decrement of a linear circuit. Then, if one defines $\delta_0 = \delta + \frac{1}{2}M\omega_0^2 S_0$, the van der Pol equation with these notations becomes

$$\ddot{x} + 2(\delta_0 + \delta_2 x^2)\dot{x} + \omega_0^2 x = 0$$

where $\delta_2 = M\omega_0^2 S_2$. We shall not use these notations later and it is sufficient to consider δ and δ_1 as some constants because this is rather irrelevant to the argument in this case.

One looks for a solution of the form: $x = x_0 e^{ipt}$. After the substitution of this expression in (3.7), one gets

$$\begin{aligned} \mu^*(u_0 p) &= x_0/u_0 = [-ipM^2\omega_0(S_0 - \tfrac{1}{4}S_2u_0^{\,2})]/[(\omega_0^{\,2} - p^2) + 2ip\delta] \\ \mu &= [-Mp\omega_0^{\,2}(S_0 - \tfrac{1}{4}S_2u_0^{\,2})]/\sqrt{(\omega_0^{\,2} - p^2)^2 + (2p\delta)^2} \end{aligned} \tag{3.8}$$

$$\begin{aligned} \sin\psi &= (\omega_0^{\,2} - p^2)/\sqrt{}\,; \qquad \cos\psi = -2p\delta/\sqrt{}\,; \\ \sqrt{} &= \sqrt{(\omega_0^{\,2} - p^2)^2 + (2p\delta)^2} \end{aligned} \tag{3.9}$$

One obtains thus for frequencies the following equations

$$p(\omega_0^{\,2} - p^2) = 0; \qquad \text{hence, either } p = 0 \quad \text{or} \quad p = \omega_0$$

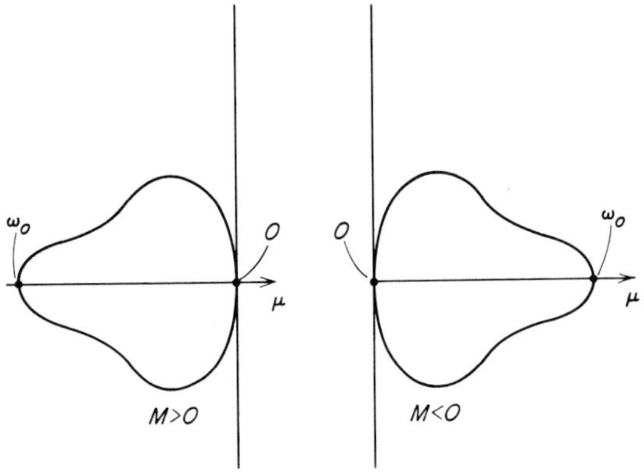

FIGURE 17.3

to which correspond the real values of μ^*

$$\mu(u_0,0) = 0; \qquad \mu(u_0,\omega_0) = -M\omega_0^{\,2}(S - \tfrac{1}{4}S_2u_0^{\,2})/2\delta \tag{3.10}$$

From the expressions $g_1r_1 + g_2r_2 = 0$ and $(g_2/r_1)_\Omega = -(g_1/r_2)_\Omega = \Delta_\Omega$, where Ω_k is the frequency corresponding to the intersection of the loop with the imaginary axis, one finds $p = \Omega = 0$ and, therefore, $\Delta_\Omega = 0$. This gives the diagram shown in Fig. 17.3. It is seen that, for $M > 0$, the feedback is negative and oscillations are impossible; for $M < 0$ the self-excitation is possible.

The condition (2.4) in this case is

$$|M|\omega_0^{\,2}S/2\delta > 1$$

and from (2.3) one obtains the equation for the stationary amplitude

$$[|M|\omega_0^2(S_0 - \tfrac{1}{4}S_2a_0^2)]/2\delta = 1$$

$$a_0 = 2\sqrt{[|M|\omega_0^2S_0 - 2\delta]/|M|\omega_0^2S_2} = \sqrt{\delta_0/\delta} \qquad (3.11)$$

For a hard self-excitation one has to take the nonlinear characteristic

$$I_a = S_0u + \tfrac{1}{3}S_2u^3 - \tfrac{1}{5}S_4u^5$$

One then obtains a diagram in Fig. 17.2 and, as we mentioned previously, according to the location of the point (1,0) one has either soft or hard self-excitation.

4. Interaction of nonlinear oscillations

Another interesting application of the generalized Nyquist diagram relates to the impossibility of a biharmonic state in nonlinear systems with soft characteristic. This phenomenon has been discovered by van der Pol and we give a more detailed investigation of this question in Chapter 23 by the classical theory. We mention it merely as an example of the use of the generalized Nyquist diagram.

Let us assume first that the condition (2.4) of self-excitation be fulfilled for two frequencies: ω_1 and ω_2. Under this assumption one has

$$\mu(0,\omega_1) > 1 \qquad \text{and} \qquad \mu(0,\omega_2) > 1 \qquad (4.1)$$

In order to see whether these two frequencies can coexist, we assume that

$$u = a_1 \cos \omega_1 t + a_2 \cos \omega_2 t \qquad (4.2)$$

and that there is no rational ratio between ω_1 and ω_2. It is sufficient to assume a soft characteristic: $I_a = Su - \tfrac{1}{3}S_2u^3$ and calculate the real coefficient of amplification for both cases.

Omitting the intermediate calculations one obtains the following expressions

$$\begin{aligned} \mu(\omega_1, a_1, a_2) &= |M|[S - \tfrac{1}{4}S_2(a_1^2 + 2a_2^2)]/2LC\left[\delta + \delta_1 \frac{\nu^2 - \omega_1^2}{\nu_1 - \omega_2^2}\right] \\ \mu(\omega_2, a_1, a_2) &= |M|[S - \tfrac{1}{4}S_2(2a_1^2 + a_2^2)]/2LC\left[\delta + \delta_1 \frac{\nu^2 - \omega_2^2}{\nu_1 - \omega_2^2}\right] \end{aligned} \qquad (4.3)$$

where ν^2 and ν_1^2 are certain constants (characterizing free frequencies of a system with two degrees of freedom), and δ and δ_1 are corresponding decrements. From the mere inspection of formulas (4.3) it is seen that the existence of the *asynchronous frequencies* (that is, those for which the ratio ω_1/ω_2 is an irrational number), spoils the conditions of self-excitation for both oscillations since, as the amplitudes a_1 and a_2 begin to grow, both coefficients $\mu(\omega_1, a_1, a_2)$ and $\mu(\omega_2, a_1, a_2)$, which were at the start greater than one, begin to decrease.

A moment will be reached when one will have, for example, relations

$$\mu(\omega_1, a_1, a_2) > 1 \qquad \text{and} \qquad \mu(\omega_2, a_1, a_2) = 1 \tag{4.4}$$

This means that the oscillations with frequency ω_1 is still growing, whereas that with frequency ω_2 is just about to disappear which happens when $\mu(\omega_2, a_1, a_2) < 1$. The final stationary state will be then characterized by the presence of only one frequency ω_1. In terms of the generalized Nyquist diagram this will result in conditions

$$\mu(\omega_1, a_{10}, 0) = 1; \qquad \mu(\omega_2, a_{10}, 0) < 1 \tag{4.5}$$

If the parameters of the system are changed, the oscillation with frequency ω_1 will disappear, whereas that with frequency ω_2 will reach the stationary state

$$\mu(\omega_2, 0, a_{20}) = 1; \qquad \mu(\omega_1, 0, a_{20}) < 1 \tag{4.6}$$

One can develop this argument further and find the phenomenon of quenching of one oscillation by the other (Section 2, Chapter 23).

5. Stability

From the preceding it follows that the d.e. appear in the form

$$\frac{da}{dt} = N_1 a\delta(\mu - 1); \qquad \frac{d\varphi}{dt} = N_2\delta\psi \tag{5.1}$$

which is a generalization of the results indicated in Section 3. For the stationary state we have

$$\mu(a, \omega, \ldots, \varepsilon) = 1; \qquad \psi(a, \omega, \ldots, \varepsilon) = 0 \tag{5.2}$$

where a is the amplitude, ω is the frequency, and ε is a parameter. If the value of this parameter is changed from its stationary value $\varepsilon = \varepsilon_0$, one obtains conditions of stability in the usual form resulting from the variational equations. Replacing instead of the stationary values $a = a_0$, $\omega = \omega_0$, etc., the perturbed values $a = a_0 + da$, $\omega = \omega_0 + d\omega_0$, etc., and limiting results only to the linear terms in da, $d\omega$,. . ., $d\varepsilon$, one has the linear system in terms of $da/d\varepsilon$ and $d\omega/d\varepsilon$, namely,

$$\frac{da}{d\varepsilon} = \frac{\begin{vmatrix} \left(\frac{\partial\mu}{\partial\omega}\right)_0 & \left(\frac{\partial\mu}{\partial\varepsilon}\right)_0 \\ \left(\frac{\partial\psi}{\partial\omega}\right)_0 & \left(\frac{\partial\psi}{\partial\varepsilon}\right)_0 \end{vmatrix}}{\begin{vmatrix} \left(\frac{\partial\mu}{\partial a}\right)_0 & \left(\frac{\partial\mu}{\partial\omega}\right)_0 \\ \left(\frac{\partial\psi}{\partial a}\right)_0 & \left(\frac{\partial\psi}{\partial\omega}\right)_0 \end{vmatrix}} \qquad \frac{d\omega}{d\varepsilon} = \frac{\begin{vmatrix} \left(\frac{\partial\mu}{\partial\varepsilon}\right)_0 & \left(\frac{\partial\mu}{\partial a}\right)_0 \\ \left(\frac{\partial\psi}{\partial\varepsilon}\right)_0 & \left(\frac{\partial\psi}{\partial a}\right)_0 \end{vmatrix}}{\begin{vmatrix} \left(\frac{\partial\mu}{\partial a}\right)_0 & \left(\frac{\partial\mu}{\partial\omega}\right)_0 \\ \left(\frac{\partial\psi}{\partial a}\right)_0 & \left(\frac{\partial\psi}{\partial\omega}\right)_0 \end{vmatrix}} \tag{5.3}$$

Following the argument of Chapter 1 (equation (6.8)), one finds that the conditions of stability are

$$\left(\frac{\partial\mu}{\partial a}\right)_0 + \left(\frac{\partial\psi}{\partial\omega}\right)_0 < 0 \tag{5.4}$$

$$\left(\frac{\partial\mu}{\partial a}\right)_0\left(\frac{\partial\psi}{\partial\omega}\right)_0 - \left(\frac{\partial\mu}{\partial\omega}\right)_0\left(\frac{\partial\psi}{\partial a}\right)_0 > 0 \tag{5.5}$$

The first condition (5.4) shows that the singular point is stable and (5.5) indicates that the singular point in question *is not* a saddle point. In formulas (5.3), (5.4), and (5.5), the notations $(\partial\mu/\partial\omega)_0$,. . ., $(\partial\psi/\partial\omega)_0$ mean the partial derivatives of μ and ψ with respect to a, ω, and ε which appear in (5.2) into which the stationary values are substituted after the differentiations.

It is seen that the question of stability reduces again to the classical procedure of the variational equations. In this theory the variables μ and ψ replace x and y of the general theory.

One can simplify these criteria on the basis of certain considerations of the order of magnitude. For instance, the quantity $(\partial\psi/\partial a)$ in applications is always very small and is zero in the first approximation for a number of problems. Likewise $(\partial\mu/\partial\omega)_0$ is also small in systems operating not far from resonance, which is generally the case. Under these conditions the criterion (5.5) reduces to

$$\left(\frac{\partial\mu}{\partial a}\right)_0\left(\frac{\partial\psi}{\partial\omega}\right)_0 > 0 \tag{5.6}$$

which splits into two conditions

$$\left(\frac{\partial\mu}{\partial a}\right)_0 < 0; \qquad \left(\frac{\partial\psi}{\partial\omega}\right)_0 < 0 \tag{5.7}$$

Equations (5.3) become then

$$\frac{da}{d\varepsilon} \simeq -\left[\left(\frac{\partial\mu}{\partial\varepsilon}\right)_0 \Big/ \left(\frac{\partial\mu}{\partial a}\right)_0\right]; \qquad \frac{d\omega}{d\varepsilon} = -\left[\left(\frac{\partial\psi}{\partial\epsilon}\right)_0 \Big/ \left(\frac{\partial\psi}{\partial\omega}\right)_0\right] \tag{5.8}$$

These relations show the dependence of amplitude and frequency on parameter ε.

6. Retarded actions

An interesting application of the generalized Nyquist diagrams was made by Theodorchik to cases in which the so-called *retarded actions* appear. Although we shall enter into this matter more fully in Chapter 21, we outline it here in order to give a further illustration of the use of this method.

Consider a d.e.

$$\ddot{x} + 2(\delta + \delta_2 x^2)\dot{x} + \omega_0{}^2 x + \nu^2 x_\tau = 0 \tag{6.1}$$

The notation x_τ is equivalent to $x(t - \tau)$; that is, although $\ddot{x}$, $\dot{x}$, and x could be written $\ddot{x}(t)$, $\dot{x}(t)$, and $x(t)$ emphasizing the fact that they relate to the time t, the term with $x_\tau = x(t - \tau)$ is a *retarded one* and relates to the past time $t - \tau$, τ being a *time-lag*. Such equations are called *difference-differential* equations, but we postpone their study to Chapter 21. We consider here only the relatively simple case in which we look only for one simple harmonic solution which permits proceeding in an elementary manner.

One can write (6.1) in the form

$$\ddot{x} + \omega^2 x = (\omega^2 - \omega_0{}^2)x - 2(\delta + \delta_2 x^2)\dot{x} - \nu^2 x_\tau = \sum \mathfrak{F} \tag{6.2}$$

We have merely added and subtracted: $+\omega^2 x - \omega^2 x$ to (6.1) and regrouped the terms so as to have the form (6.2) of the d.e. If one substitutes into (6.2) the harmonic solution $x = a \sin \omega t$ where both a and ω are unknown quantities, it can be shown, that, according to the general theory, this equation can always be brought to the form

$$\ddot{x} + \omega^2 x = F[a(t), \omega] \cos \omega t + f[a(t), \omega] \sin \omega t + \text{const} + \text{harmonics} \tag{6.3}$$

If one limits the calculation to the terms of the fundamental frequency ω, one obtains for the first approximation (we omit here some details)

$$a(t) = \frac{1}{2\omega}\int_0^t F[a(\xi), \omega]d\xi; \qquad 0 = \frac{1}{2\omega}\int_0^t f[a(\xi), \omega]d\xi \tag{6.4}$$

For the transient condition the first equation gives

$$\frac{da}{dt} = \frac{1}{2\omega} F(a,\omega) \tag{6.5}$$

and for the stationary state one has

$$F(a,\omega) = 0; \qquad f(a,\omega) = 0 \tag{6.6}$$

If one calculates the coefficients of $\cos \omega t$ and $\sin \omega t$ in the expression $\sum \mathfrak{F}$ in (6.2), one has for (6.5) the d.e.

$$\frac{da}{dt} = \frac{a}{2\omega}\left(\nu^2 \sin \omega\tau - 2\delta\omega - \frac{1}{2}\,\delta_2 \omega a^2\right) \tag{6.7}$$

and for (6.6) the following equations

$$\begin{aligned} (\omega^2 - \omega_0{}^2)a - \nu^2 a \cos \omega\tau &= 0 \\ a[\nu^2 \sin \omega\tau - 2\delta\omega - \tfrac{1}{2}\delta_2 \omega a^2] &= 0 \end{aligned} \tag{6.8}$$

The equilibrium $a = 0$ is stable if

$$\nu^2 \sin \omega\tau - 2\delta\omega < 0 \tag{6.9}$$

and unstable if this expression is positive. If $a \neq 0$, the first equation (6.8) gives

$$\omega^2 = \omega_0{}^2 + \nu^2 \cos \omega_0 \tau \tag{6.10}$$

and the second equation gives the stationary amplitude

$$a = \sqrt{(2\nu^2 \sin \omega\tau - 4\delta\omega)/\delta_2 \omega} \tag{6.11}$$

If the condition of self-excitation is fulfilled, the stationary amplitude always exists. If, however, one changes the parameter ν^2, the oscillations can maintain themselves as long as

$$\nu^2 > (2\delta\omega)/\sin \omega\tau \tag{6.12}$$

An electron-tube oscillator may be regarded from this point of view as a retarded system in which the feedback (at least as far as the fundamental harmonic is concerned) produces a time-lag of $\pi/2$, the period being 2π. In fact, under this approximation one has

$$\nu^2 x_{\omega\tau = T/4} = \nu^2 a \cos(\omega t - \pi/2) = -\nu^2 a \sin \omega t = +(\nu^2 \dot{x})/\omega \tag{6.13}$$

In view of this (6.1) acquires the form

$$\ddot{x} + 2[(\delta - (\nu^2/2\omega)) + \delta_2 x^2]\dot{x} + \omega_0{}^2 x = 0 \tag{6.14}$$

If one compares this d.e. with the general form, one sees that

$$\frac{\nu^2}{2\omega} = |M|\omega_0{}^2S_0 \tag{6.15}$$

We shall limit ourselves to a short review of the theory which *postulates* the existence of a simple harmonic solution: $x = a \sin \omega t$ and neglects harmonics. It is clear that, by this, we limit our investigation to the first approximation only.

In Chapter 21 we shall return to this question from the standpoint of the exact theory and it will be shown that in general the problem is more complicated and leads to an infinite spectrum of harmonics, whose frequencies are not in commensurate ratios.

If, however, one limits oneself to the fundamental harmonic, the above method leading to the d.e. (6.2) with the resulting reduction of the problem to two algebraic equations (6.6) is sufficient if the problem is nearly linear, inasmuch as the fundamental harmonic dominates the others.

Theodorchik gives a number of examples in which this method can be used, such as operation of an electric bell, oscillations of thermostats, of Helmholtz resonators, short-wave generators, etc.

Returning now to the corresponding Nyquist diagram, one finds that the effect of a retardation manifests itself in the appearances of a term of the form

$$\gamma(u_0,p) = -p\tau(u_0,p) \tag{6.16}$$

so that, instead of (2.2), the complex coefficient of amplification is now

$$\mu^*(u_0, p, \tau) = \mu(u_0,p)e^{i[\psi(u_0,p)-\gamma(u_0,p)]} \tag{6.17}$$

This shows that each radius vector is merely turned over an angle γ and, by a more detailed analysis, one reaches the following conclusions:

(a) All frequencies tend to decrease.

(b) For some frequencies conditions of self-excitation become better; for some others, they become worse.

(c) A new frequency may appear and the corresponding amplitudes may become sufficiently large that they cannot be neglected even in the first approximation.

If one assumes that the time-lag τ is constant, one can conclude that there appear infinitely many frequencies which disappear if $\tau = 0$. It is simpler, however, to deal directly with the exact theory based on properties of solutions of difference–differential equations, as will be shown in Chapter 21.

7. Concluding remarks

It is useful to say also a few words about the theory of Blaquière[2] mentioned previously. In this approach use is made of the operational calculus, which is too long to be entered into here.

The existence of a family of Nyquist's loops depending on the parameter a (the amplitude) is also assumed in this theory but the subsequent development is different. More specifically, Blaquière considers especially the transient state of a nonlinear oscillation and shows that the original solution of the d.e., which is practically linear when the amplitude is small, becomes gradually "distorted" by the nonlinear terms. In terms of this theory this amounts to a "distortion" of the originally existing Hilbert's space by the effect of the nonlinear operator. By Hilbert's space is meant the functional representation of the solution in the form of a Fourier series satisfying the d.e., and by its "distortion" is meant the modification of the coefficients of this series as the transient trajectory of the oscillator traverses the family of the subsequent linear Nyquist's loops. The actual nonlinear trajectory is considered as a kind of envelope of linear trajectories so traversed that there exists at any moment a state of tangency with some linear trajectory. It is seen that the transient state is more important in this approach than in other theories with which we have been concerned so far. The fundamental problem of the stationary state does not differ much, however, from that in other approaches, namely: the stationary amplitude and frequency are still determined from two algebraic equations (for example, (2.3)) as in all other theories of the first approximation.

In one respect the work of Blaquière specifies more clearly the true nature of this generalization, namely: it introduces essentially curvilinear coordinates in the Nyquist phase plane by defining "equi-amplitude" and "equi-pulsation" (or equi-frequency) curves whose intersection gives the operating point for which the stationary state exists if conditions of stability are fulfilled. However, in contrast with this simplification, the stability conditions are more complicated and we do not propose to go into this question here.

We merely mention these few salient points because the entire development is yet in a somewhat preliminary state, particularly in view of a lack of any, more or less uniform, relationship with known nonlinear phenomena. In part III these phenomena will be treated by the classical theory.

There remains still the fundamental question: What will be the place of the generalized Nyquist's diagram in the theory of oscillations assuming that the various difficulties existing at present are ultimately eliminated?

[2] See footnote [2], page 418.

In the first place, one must bear in mind that, by its very nature, this diagram does not constitute any special *method* in the theory of oscillations but is merely a very interesting graphical interpretation of conditions existing in linear systems with a feedback connection. By this diagram one can visualize the *behavior of the solution of the d.e.* when the parameter (the frequency) varies.

In the nonlinear domain this interpretation is much more complicated since, instead of one parameter, there are two such parameters; what is particularly difficult is the fact that these two parameters are in reality *interdependent*. As the result of this, the nonlinear extension of this diagram complicates the problem considerably, instead of simplifying it. In fact, the main advantage of the diagram—its simplicity—(in the linear case) is lost as soon as one tries to use it in nonlinear problems. Since the two parameters ω and a are connected together through the d.e., this d.e. is not permitted to be more or less in the background, as in the linear case when the amplitude does not play any role; therefore one can concentrate on the behavior of the solution directly in terms of the frequency.

Since in the nonlinear case the principal advantage of the diagram is lost anyhow, the question arises as to why it is necessary to build up a graphical representation which on one hand is not simple and on the other hand requires the analytical solution (at least the first approximation).

Such questions are inevitable if one considers the nonlinear problems in their *normal* form, that is, when the existence of a stationary state is the only point of interest. In fact, in the previous examples it was necessary to solve first the nonlinear problem by the standard procedure and only then to identify the conditions of the stationary state, with the point (1,0) of the Nyquist diagram.

It must be noted, however, that this somewhat negative attitude toward the extension of the diagram is probably not quite justified since its primary use, very likely, will be not in cases when a stationary state (in the usual sense of this term) exists but rather in cases when *it does not exist*.

In fact, it does not seem to be of any particular interest to look for a stationary state of a self-excited system by means of the Nyquist diagram inasmuch as it is much easier to establish this directly from the equations of the first approximation. On the other hand, the diagram can still be useful in the investigation of a nonlinear *passive* system under a certain periodic (or almost periodic) excitation. Thus, for instance, the problem of nonlinear filters belongs to this class and, as control problems are closely related to such questions, it is likely that such extensions of Nyquist's diagram may be of interest particularly in view of a gradually increasing importance of nonlinear control systems.

PART III

OSCILLATIONS OF NEARLY LINEAR SYSTEMS

INTRODUCTION

We shall investigate in this Part III the principal nonlinear oscillatory phenomena under the fundamental assumption of their near-linearity, in which case the analytical methods of approximations outlined in Part II are applicable.

The assumption of a near linearity as a rule is sufficiently well justified in a great majority of nonlinear oscillations. In fact, the phenomena with "a strong nonlinearity" (large μ) constitute a somewhat restricted class of the so-called *relaxation oscillations* encountered most frequently in electrical circuits. As the methods of analytic approximations do not hold here and an entirely different method of attack is necessary, we postpone this subject to Part IV.

As regards nearly linear oscillations, it would be, perhaps, an exaggeration to say that their treatment is merely a straightforward application of analytical methods of approximations we encountered in Part II. The reason is due to the fact that what is generally called "phenomenon" (in a physical sense) is, as a rule, only a somewhat idealized condition or "aspect" of a more complicated situation involving some other aspects as well.

Thus, for instance, it is customary to consider the phenomenon of synchronization by itself, and likewise that of nonlinear or subharmonic resonance; but these two phenomena generally take place together and are, therefore, merely two different *aspects* of a more complicated real phenomenon. In fact, the studies of resonance are directed primarily to the establishment of conditions for the existence and stability of the *amplitude*, whereas those of synchronization are related to the investigation of a possibility of existence of a fixed phase relation between the autoperiodic and the heteroperiodic oscillations.

Although from a mathematical point of view it is somewhat meaningless to speak about two different "phenomena" in this case, from a physical standpoint it is *convenient* to introduce this distinction which simplifies the problem by inviting attention to features in which one is primarily interested.

The theory of nonlinear oscillations (as contrasted with the theory of nonlinear d.e.) developed historically as the result of attempts to explain various physical phenomena considered in this, somewhat simplified, manner.

One reaches the somewhat paradoxical conclusion that the same d.e. may explain entirely different "physical phenomena" according to the different *regions* of its parameter space. This circumstance is generally not considered in the theory of nonlinear d.e. which are at the foundation of these studies but, on the contrary, become very important when one tries to obtain a physical insight into these questions.

For example, if one considers a frequently encountered d.e. of the form

$$\ddot{x} + f(x)\dot{x} + x = \lambda \sin \omega t \qquad \text{(III-1)}$$

and investigates the effect of the parameter ω (the frequency of the external periodic excitation) one readily sees that a number of "phenomena" appear according to the value of this parameter.

It is noted first, that, for $\lambda = 0$, the frequency of the left-hand term (considered as representing an oscillator) is *one*. If, however, $\lambda \neq 0$ and ω has integer values, say $\omega = 2, 3, \ldots$, the d.e. (III-1) gives rise to the theory of the subharmonic resonance. The problem here is to establish the existence and stability of the periodic solution with period 2π while the forcing term has the period $2\pi/\omega$. No synchronization "aspect" is involved in this formulation of the problem and, for that reason, the theory so formed relates to the conditions of occurrence of an *exact* subharmonic resonance.

But one can approach the same problem from a different angle involving the question of synchronization. Here one can avoid the necessity of considering the subharmonic resonance and investigate only the fundamental resonance: $\omega = 1$, but as one is now interested in *synchronization*, it is necessary to "arrange" the d.e. somewhat differently, viz.:

$$\ddot{x} + f(x)\dot{x} + (1 + \varepsilon)x = \lambda \sin t; \qquad \varepsilon \ll 1 \qquad \text{(III.2)}$$

In this formulation, for $\lambda = 0$, the frequency of the left-hand term is $\sqrt{1 + \varepsilon}$, whereas that of the forcing term is *one*. The problem now consists in establishing the conditions under which the solution of (III.2) has frequency equal to *one*, in spite of the fact that for $\lambda = 0$, the autoperiodic oscillation has a somewhat different frequency; the quantity ε is sometimes called "detuning."

One could begin the same problem obviously by considering the d.e.

$$\ddot{x} + f(x)\dot{x} + x = \lambda \sin (1 + \eta)t \tag{III.3}$$

and show that the oscillation can, under certain conditions, have frequency $(1 + \eta)$ instead of 1, which amounts to the same thing.

One could also attack a more general problem of the d.e. (III.1) by considering ω as a variable parameter. When ω has integer values $\omega = 2, 3$, there *may be* occasionally subharmonic solutions with frequency 1. In the *neighborhood* of these values (that is, $\omega = 2 + \eta, 3 + \eta$) there may be still subharmonic solutions but any such neighborhood involves inevitably the question of synchronization (in this case this would be "synchronization on subharmonics" using a technical term).

Outside these regions of synchronization, there appear "beats" or interference between the heteroperiodic and autoperiodic oscillations and, if the latter disappears for some reason, there remains only, generally, a very small heteroperiodic oscillation.

Finally, it may happen also that the resonance oscillation may disappear *within the range* of synchronization. In such a case the amplitude suddenly drops to a very small value without any intermediate beats on two frequencies.

It is seen thus that the phenomena present themselves in a very complicated manner in spite of the fact that the d.e. remains the same. The variety of different cases here depends on the different ranges of one single parameter ω.

Even if one considers a relatively simple case when the resonance oscillation is absent (that is, when the ratio between the heteroperiodic and autoperiodic frequencies is sufficiently far away from a rational number), there appear the so-called *asynchronous actions*, the term "asynchronous" emphasizing the lack of any subharmonic effect. In some cases the asynchronous action manifests itself in that the heteroperiodic oscillation extinguishes the autoperiodic one; in some other cases it, on the contrary, releases it.

Here again the d.e. is the same (III.1), but these asynchronous actions appear again in special regions of the parameter space. The first of these two effects, the *asynchronous quenching*, appears generally when ω is large but otherwise quite arbitrary. The second effect, the *asynchronous excitation*, does not involve ω but depends on the parameters of the function $f(x)$, that is, on the coefficients of its polynomial representation.

We have touched these questions in order to emphasize the fact that in the theoretical study of nonlinear oscillations very little progress could be accomplished without a continuous guidance by experimental evidence. Lord Rayleigh, van der Pol, Appleton, and other early investigators used analytical methods in conjunction with the knowledge of the order of magnitude of different parameters, which ultimately permitted a localization of these various phenomena in their respective regions of the parameter space, specifying their principal features, neglecting the others which are less important.

Summing up, the study of the various kinds of nonlinear oscillations which we are about to undertake should not be considered as a purely formal application of the methods of approximations but these approximations appear rather as a tool in hands of a physicist once the knowledge of a physical phenomenon has been established.

Referring specifically to the contents of the various chapters of Part III, Chapter 18 discusses the phenomenon of synchronization. The beginning of the chapter deals with the theory of van der Pol supplemented by the topological analysis of Andronow-Witt. As this matter is discussed in existing texts, it is somewhat abridged here. The application of the stroboscopic method is given in Section 5 of Chapter 18; it presents certain advantages in obtaining quantitative results at the cost of additional calculations.

Chapter 19 deals with the subharmonic resonance. The beginning of the chapter is devoted to a somewhat abridged outline of the classical theory of Mandelstam and Papalexi, omitting details of electron-tube circuits by which these authors verified their theory. The important work of Krylov and Bogoliubov is touched but slightly in this chapter since its fundamentals have been presented in greater detail in Chapters 14 and 15. The application of the stroboscopic method, on the other hand, is presented more fully, because it is necessary to introduce some additional extensions of this method. The essential feature of this extension is in that it is often necessary to operate here with the cartesian components $x(t)$, $y(t)$ of the solution instead of the variables $\rho(t)$ and $\psi(t)$, as was indicated in Chapter 16.

Chapter 20 concerns the phenomenon of the so-called parametric excitation. This subject has become more important in view of recent experimental material which has permitted the introduction of certain generalizations.

In Chapter 21 are treated oscillations amenable to difference–differential equations. Recently this field has developed very rapidly in connection with parasitic oscillations frequently observed in modern automatic control systems. Although there has been considerable theoretical development in this field, most of this material relates to *linear* difference–differential equations which are of relatively little use here. For this reason Chapter 21 aims primarily at the formulation of applied problems rather than at its final solution which does not exist at present. Certain connections with the classical theory are indicated, however, under an additional restriction of smallness of time-lags.

Chapter 22 deals with some additional studies of the van der Pol–Liénard equation considered in the parameter space of coefficients of the polynomial representation of functions $f(x)$ and $g(x)$; Chapter 23 relates to the question of interaction of nonlinear oscillations.

Chapter 24 discusses the so-called asynchronous actions and Chapter 25 deals with a somewhat new class of phenomena discovered by Theodorchik, concerning the d.e. with coefficients exhibiting "inertial" nonlinearities. This is as yet a little explored field of "hereditary" phenomena which is outlined briefly with the aid of some plausible simplifications. Owing to this, some conclusions can still be obtained in the first approximation.

Present knowledge of these phenomena is far from being uniform. Thus, for instance, the manifestations of the synchronization and the subharmonic resonance effects are apparently better understood than those of other phenomena, but the difficulties of their quantitative treatment are yet too great for applied problems.

As regard some other phenomena, even the mathematical fundamentals are not yet available as was mentioned in Chapters 21 and 25, for instance. In view of this, it often becomes necessary to introduce certain physical idealizations with a view to bringing the problem within the scope of the differential equations of the first approximation.

All this merely emphasizes the fact that not all gaps between the mathematical treatment and the corresponding physical facts have been bridged as yet, but such situations are inevitable in a new science which has not reached its complete codification.

Chapter 18

SYNCHRONIZATION

1. Introductory remarks

The phenomenon of *synchronization* or "entrainment of frequency" was the first to be studied among other nonlinear phenomena. Apparently, it was observed for the first time by Huygens (1629–1695) who reported that two clocks, which were slightly "out of step" with each other when hung on a wall, became synchronized when fixed on a thin wooden board. These effects were rediscovered more than two centuries later in electrical circuits by a number of physicists among whom one should mention Lord Rayleigh,[1] Vincent,[2] Möller,[3] Appleton,[4] van der Pol,[5] and others. The last two authors developed the theory of this phenomenon; in the following section we indicate briefly these results in the van der Pol version which permits establishing a connection with a topological extension of this theory by Andronov and Witt[6]; this approach was extended by Stoker.[7]

The synchronization effect can readily be observed in electronic circuits. If one applies to the grid of an electron-tube oscillator oscillating, say, with frequency ω_0, an extraneous electromotive force of frequency ω, one observes the "beats" of the two frequencies. If the frequency ω approaches the frequency ω_0, the frequency of the beats decreases but this happens only up to a certain value of the difference $|\omega - \omega_0|$ after which the beats disappear suddenly and there remains only the frequency ω. Everything happens as if the "free" (autoperiodic) frequency ω_0 were "entrained" by the extraneous (heteroperiodic) frequency ω. Figure 18.1 shows the

[1] Lord Rayleigh, *Theory of Sound*, Vol. 1, London, 1894.
[2] J. H. Vincent, *Proc. Phys. Soc.* (London) **32**, 1919.
[3] H. G. Möller, *Zeitsch. drahtl. T. und T.* **17**, 1921.
[4] E. V. Appleton, *Proc. Cambridge Phil. Soc.* (London) **21**, 1922.
[5] B. van der Pol, *Phil. Mag.* **43**, 1922.
[6] A. Andronov and A. Witt, *Arch. für Electroth.* **24**, 1930.
[7] J. J. Stoker, *Nonlinear Vibrations*, Interscience Publishers, New York, 1950.

difference $|\omega - \omega_0|$ plotted against ω. On the basis of the linear theory the variation of $|\omega - \omega_0|$ should follow the path $ABCDE$, whereas in reality it follows the path $ABB'CD'DE$.

The phenomenon of synchronization is, perhaps, the best known among other nonlinear phenonema. It has been observed not only in electrical systems but also in mechanical, acoustical, electroacoustical systems and, finally, in control systems using relays. Moreover, this effect has been produced artificially in some cases in order to secure a very accurate synchronization of clocks or electric motors with frequency of quartz oscillators.

In the following section are given the fundamentals of the early Appleton–van der Pol theory; we use the version of the last-mentioned author because it is a little more convenient for the topological representation of Andronov–Witt outlined in Sections 3 and 4.

In Section 5 is outlined the theory of synchronization on the basis of the stroboscopic method; this permits connecting the theory of this phenomenon with the general theory of d.e. without using any special form of solution like (2.5). This method permits proceeding directly with the theory of approximation without involving the topological arguments, as the result of which the procedure is better adapted for a quantitative work. There is, however, a certain disadvantage, as in all quantitative methods, consisting of computational difficulties when the problem is ultimately brought to an algebraic formulation.

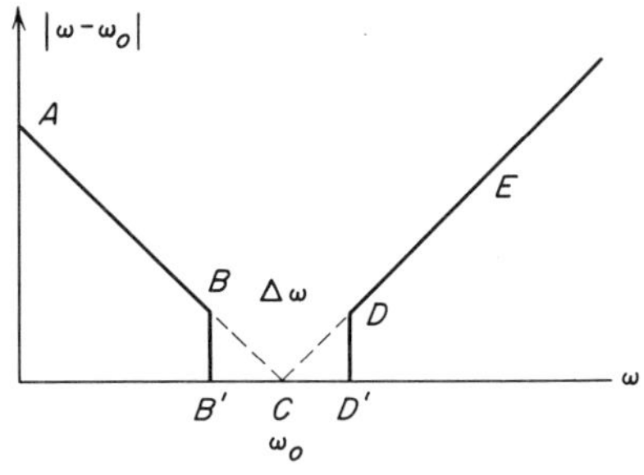

FIGURE 18.1

The last section deals with certain physical and engineering applications of the synchronization effect.

2. Theory of van der Pol

The d.e. of an electron-tube oscillator with an inductive coupling is

$$L\frac{di}{dt} + Ri + \frac{1}{C}\int idt = M\frac{dI}{dt} \tag{2.1}$$

where i is the current in the oscillating circuit (having constant parameters L, R, and C), I is the current in the anode circuit of the electron tube, and M is the coefficient of mutual inductance between the anode and the oscillating circuit whose frequency is ω_0. If one introduces an external periodic excitation $E_0 \sin \omega_1 t$ on the grid (in addition to the usual feedback connection), the d.e. takes the form

$$L\frac{di}{dt} + Ri + \frac{1}{C}\int i dt - M\frac{dI}{dt} = E_0 \sin \omega_1 t \tag{2.2}$$

The nonlinear element of the circuit is here $I = f(e_g)$: the anode current I, considered as a function of the grid voltage e_g, and the usual approximation of the nonlinear function by the polynomial is taken here in the form

$$I = f(e_g) \simeq Se_g(1 - e_g^2/3V_s) \tag{2.3}$$

where S is the mutual conductance, and V_s is the so called "saturation voltage." As usual, one neglects the secondary effects, such as grid voltage, anode reaction, etc. Using the notations

$$v = \frac{e_g}{V_s} = \int \frac{i dt}{CV_s}; \qquad \alpha = \frac{MS}{LC} - \frac{R}{L};$$

$$\gamma = \frac{MS}{3LC}; \qquad B = \frac{E_0}{V_s}; \qquad \omega_0^2 = \frac{1}{LC}$$

the d.e. (2.2) takes the form

$$\ddot{v} - \alpha\dot{v} + \gamma\dot{v}^3 + \omega_0^2 v = B\omega_1^2 \sin \omega_1 t \tag{2.4}$$

which will now be investigated.

Van der Pol assumes the solution in the form

$$v = b_1 \sin \omega_1 t + b_2 \cos \omega_1 t \tag{2.5}$$

where b_1 and b_2 are slowly varying functions of time. If one substitutes this solution into (2.4), one obtains the following d.e.

$$\begin{aligned} 2\dot{b}_1 + zb_2 - \alpha b_1(1 - b^2/a_0^2) &= 0 \\ 2\dot{b}_2 - zb_1 - \alpha b_2(1 - b^2/a_0^2) &= -B\omega_1^2 \end{aligned} \tag{2.6}$$

where $z = 2(\omega_0 - \omega_1)$; $b^2 = b_1^2 + b_2^2$; $a_0^2 = 4\alpha/3\gamma$.

It is obvious that if $b_1(t)$ and $b_2(t)$ are constant, the solution $v(t)$ is periodic with frequency ω_1. One is inclined to call $v(t)$ the "forced" oscillation, but in the nonlinear case this term is rather misleading and it is preferable to call it *heteroperiodic* oscillation (or solution). If b_1 and b_2 are slowly varying functions of time, the solution $v(t)$ is almost periodic (Chapter 12).

It is noted that the d.e. in the form (2.6) are autonomous, being of the form

$$\dot{b}_1 = P(b_1,b_2); \qquad \dot{b}_2 = Q(b_1,b_2) \tag{2.7}$$

In the derivation of these d.e., van der Pol assumes that $\ddot{b}_1$ and $\ddot{b}_2$ are negligible since b_1 and b_2 are slowly varying functions. This also presupposes that the parameters α and γ in the original d.e. are small.

Under these assumptions one reaches the following important conclusion: the periodic solution of (2.4) exists if the system (2.7) has a singular point. For the state of equilibrium $\dot{b}_1 = \dot{b}_2 = 0$, which results in the equation of the stationary state

$$z^2 + \alpha^2(1 - b^2/a_0^2)^2 = B^2\omega_1^2/b^2 \tag{2.8}$$

As both b_1 and b_2 are stationary, (2.8) represents the condition of the heteroperiodic oscillation. This equation, with certain transformation of variables, reduces to the form

$$x^2 + (1 - y)^2 = E/y \tag{2.9}$$

where $y = b^2/a_0^2$; $E = \omega_1^2 B/a_0$; $x = z/\alpha$.

We shall leave the argument of van der Pol at this point and consider the topological representation of the synchronization effect by Andronov and Witt.[6]

3. Topological analysis of Andronov and Witt

If one introduces the notations

$$x = b_1/a_0; \qquad y = b_2/a_0; \qquad a = z/\alpha; \qquad A = -B\omega_1/a_0\alpha;$$
$$r^2 = x^2 + y^2; \qquad \tau = \alpha t/2$$

the d.e. (2.6) becomes

$$dx/d\tau = x(1 - r^2) - ay; \qquad dy/d\tau = ax + y(1 - r^2) + A \tag{3.1}$$

These are still the van der Pol d.e. but written in a different form and the procedure remains the same as before, namely, one determines the singular point and discusses its stability. The coordinates of the singular point are here

$$x_0 = -a\rho/A; \qquad y_0 = +\rho(1 - \rho)/A \tag{3.2}$$

where $\rho = r_0^2$. As to ρ, it is determined from the cubic equation

$$\rho[a^2 + (1 - \rho)^2] = A^2 \tag{3.3}$$

This equation is discussed graphically considering a and ρ as variables and A as a parameter. The family of curves represented by (3.3) is shown in Fig. 18.2. For sufficiently small values of A, the curve consists of two branches M_1 and M_1'. For an increasing A the branch M_1 increases in

[6] See footnote [6], page 438.

size while the branch M_1' rises until the two branches join, as shown by the curve M. If A increases further, the curve consists of only one branch M_2. The curves of the family are symmetrical with respect to the ρ axis.

For the investigation of stability one replaces x and y in (3.1) by $x_0 + \xi$, $y_0 + \eta$, where ξ and η are small variations, and obtains the variational equations (the derivation of which we omit here) leading to the following characteristic equation:

$$S^2 - 2(1 - 2\rho)S + [(1 - \rho)(1 - 3\rho) + a^2] = 0 \tag{3.4}$$

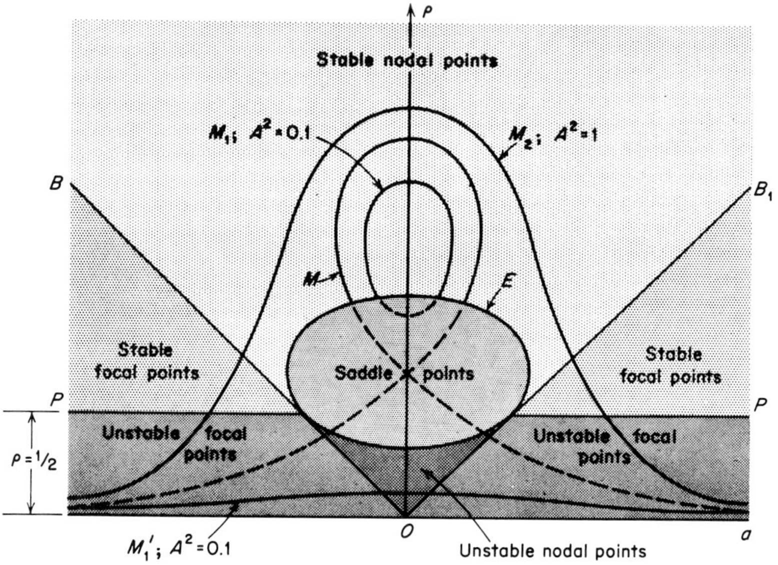

FIGURE 18.2

which is easily discussed. If the last term is negative, one has saddle point. The region of these singularities is thus confined to the ellipse

$$(1 - \rho)(1 - 3\rho) + a^2 = 0 \tag{3.5}$$

The real roots are separated from the complex ones by straight lines $\rho + a = 0$; $\rho - a = 0$ (lines OB and OB_1 in Fig. 18.2). One ascertains that these lines are tangent to the ellipse E at the value of the ordinate $\rho = 1/2$. The condition of stability is clearly $(1 - 2\rho) < 0$ so that the straight line $\rho = 1/2$ separates the regions of stability (above) from instability (below).

The division of the (a,ρ) plane into the various regions of stability gives a convenient way of ascertaining at a glance what may be expected in a given oscillatory system under the different values of detuning as well as the external periodic excitation (parameter A). As the variable a increases from small values, there appears first the region in which there are three real roots, the discriminant of the cubic equation being negative. One ascertains that only one root is stable. For larger values of a the discriminant changes sign and there is only one real root. As long as this root is stable, the synchronization still exists but, as soon as this root enters the region of instability, the synchronization is lost. At this point the functions b_1 and b_2 of the van der Pol theory cease to be constant and the almost periodic condition sets in.

It was shown in Chapter 3 that the d.e. (3.1) admit a limit cycle, as is easily ascertained by applying the Poincaré-Bendixson theorem. The question of the appearance of limit cycle in this case has been analyzed by Andronov and Witt[6] and elaborated still further by Stoker.[7]

Summing up, the theory of synchronization outlined in this and in the preceding sections is based on a very elegant procedure of van der Pol to reduce an essentially nonautonomous system to the autonomous form owing to the form (2.5) of this solution. Once this reduction has been accomplished, the rest of the problem does not present any difficulty. It is to be noted also that this procedure is still within the scope of the small parameter method of Poincaré as it appears somewhat implicitly from the fact that the second derivatives $\ddot{b}_1$ and $\ddot{b}_2$ are neglected by van der Pol.

4. Conditions of the stationary state of synchronization

During the synchronization the autoperiodic oscillation is suppressed and there remains only the heteroperiodic oscillation. The solution of the d.e. (2.4) is then

$$v = b_1 \sin \omega_1 t + b_2 \cos \omega_2 t = b \sin (\omega_1 t + \varphi) \tag{4.1}$$

where $\cos \varphi = b_1/b$, $\sin \varphi = b_2/b$, $b = \sqrt{b_1^2 + b_2^2}$, and φ is the phase of the oscillation relatively to the external voltage.

One has thus

$$\tan \varphi = b_2/b_1 = y_0/x_0 = (1 - \rho)/a \tag{4.2}$$

The amplitude of the oscillation is

$$b = \sqrt{b_1^2 + b_2^2} = a_0 \sqrt{x_0^2 + y_0^2} = k \sqrt{4\alpha/3} \tag{4.3}$$

where $k = (\rho/a)\sqrt{a^2 + (1 - \rho)^2}$. As a_0 is the amplitude of the generating

[6] See footnote [6], page 438.
[7] See footnote [7], page 438.

solution of (2.4) when $B = 0$, the amplitude of the heteroperiodic solution during the synchronization is equal to the autoperiodic amplitude affected by the factor k.

5. Theory of synchronization by the stroboscopic method

The synchronization is characterized by the entrainment of the autoperiodic frequency by the heteroperiodic one. One can start the investigation either from the d.e.

$$\ddot{x} + (cx^2 - a)\dot{x} + x = e \sin (1 + \gamma)t$$

or from the d.e.

$$\ddot{x} + (cx^2 - a)\dot{x} + (1 + \gamma)x = e \sin t \tag{5.1}$$

In the first case the frequency *one* (*ife* $= 0$) is "entrained" to the frequence $(1 + \gamma)$(*ife* $\neq 0$); in the second case this entrainment occurs from frequency $\sqrt{1 + \gamma}$ to the frequency *one*. As in the first case the integrations are to be carried out between 0 and $2\pi/(1 + \gamma)$, and in the second case, between 0 and 2π, the latter is simpler so that we shall start with the d.e. (5.1). It is to be noted that the coefficients c, a, and e in (5.1) are assumed to be small in what follows. As regards c and a, this is the usual assumption of near-linearity. As to e, this requires an explanation. In fact, in applications e is always a finite quantity. If e is finite the amplitude will increase considerably since small nonlinear terms (with c and a) are unable to limit it at a finite value. It can be shown, however, that by changing the variable x for x/μ (μ small), one obtains the *asymptotic* form of the d.e. in which the coefficient of $\sin t$ is again small as is assumed directly in (5.1), which is thus quite general for all conditions.

We recall the essential points in the stroboscopic reduction. The equivalent system is $\dot{x} = y$; $\dot{y} = ay - cx^2y - x - \gamma x + e \sin t$. We form two combinations: $x\dot{x} + y\dot{y} = r(dr/dt)$; $r = \sqrt{x^2 + y^2}$; $\psi = \arctan (y/x)$; and $x\dot{y} - y\dot{x} = r^2(d\psi/dt)$.

One obtains two d.e. in terms of r and ψ:

$$\begin{aligned} \frac{dr}{dt} &= ar \sin^2 \psi - \frac{1}{4} cr^3 \sin^2 2\psi - \frac{\gamma r}{2} \sin 2\psi + e \sin \psi \sin t \\ \frac{d\psi}{dt} &= -1 + \frac{a}{2} \sin 2\psi - cr^2 \cos^3 \psi \sin \psi + \frac{e}{r} \cos \psi \sin t - \gamma^2 \cos^2 \psi \end{aligned} \tag{5.2}$$

Assuming that a, c, γ, and e are small $0(\mu)$, in the series solution for the first d.e. (5.2): $r(t) = r_0(t) + \mu r_1(r)$, one has obviously $r_0(t) = r_0 =$ const, whereas the second equation yields: $\psi_0(t) = \varphi_0 - t$, where r_0 and φ_0 are the initial conditions. For the first-order corrective terms $r_1(t)$ and $\psi_1(t)$, one has the d.e.

$$\begin{aligned}\frac{dr_1}{dt} &= Ar_0 \sin^2 \psi_0 - \frac{1}{4} Cr_0^3 \sin^2 2\psi_0 - \frac{\Gamma r_0}{2} \sin 2\psi_0 + E \sin \psi_0 \sin t \\ \frac{d\psi_1}{dt} &= \frac{A}{2} \sin 2\psi_0 - Cr_0^2 \cos^2 \psi_0 \sin \psi_0 + \frac{E}{r_0} \cos \psi_0 \sin t - \Gamma \cos^2 \psi_0\end{aligned} \tag{5.3}$$

where $A = a/\mu$, $C = c/\mu$, $\Gamma = \gamma/\mu$, and $E = e/\mu$.

In order to obtain the variations $r_1(2\pi)$ and $\psi_1(2\pi)$ of r_1 and ψ_1 after one period 2π, one integrates (5.3) between 0 and 2π recalling that $dt = -d\psi_0$. Some terms vanish and the remaining ones are

$$\begin{aligned}r_1(2\pi) &= 2\pi[\tfrac{1}{2}Ar_0 - \tfrac{1}{8}Cr_0^3 - \tfrac{1}{2}E \cos \varphi_0] \\ \psi_1(2\pi) &= 2\pi[E \sin \varphi_0/2r_0 - \Gamma/2]\end{aligned} \tag{5.4}$$

Clearly $r(2\pi) = \mu r_1(2\pi)$ and $\psi(2\pi) = \mu\psi_1(2\pi)$; changing notations: $r(2\pi) = \Delta r$; $\psi(2\pi) = \Delta\varphi$; $2\pi\mu = \Delta\tau$ (the stroboscopic time) and, passing from difference equations to the d.e. (that is, $\Delta r \to dr$; $\Delta\varphi \to d\varphi$; and $\Delta\tau \to d\tau$), one obtains the stroboscopic system

$$\begin{aligned}\frac{dr}{d\tau} &= -\frac{1}{8} C(r^3 - pr + q \cos \varphi) = R(r,\varphi); \qquad p = \frac{4A}{C}; \qquad q = \frac{4E}{C} \\ \frac{d\varphi}{d\tau} &= \frac{1}{2}\left[\frac{E \sin \varphi}{r} - \Gamma\right] = \Phi(r, \varphi)\end{aligned} \tag{5.5}$$

The condition of synchronization is expressed by the singular point of the system (5.5). The difficult point of the problem is that the variables r and φ do not separate; moreover, the variable r is yielded by the real root of the cubic equation which involves an additional difficulty (the change of sign of the discriminant).

It is preferable to establish first the conditions for $\Gamma = 0$, that is, for the *fundamental resonance of the order one.* To simplify the discussion we assume that the parameters have been so chosen that the cubic equation has only one real root.

If $\Gamma = 0$, $d\varphi/d\tau = 0$ for $\sin \varphi_0 = 0$ (from now on we shall indicate by the subscript zero the stationary quantities because no confusion is to be feared with the previous notations), that is, for $\cos \varphi_0 = \pm 1$. The variational equation for the second equation (5.5) shows that the phase is stable for $\cos \varphi_0 = -1$.

The usual procedure (Section 4, Chapter 5) for the investigation of stability yields here

$$R_r = -\frac{1}{8} C(3r_0^2 - p); \qquad R_\varphi = \frac{1}{8} Cq \sin \varphi;$$
$$\Phi_r = -\frac{E}{2r^2} \sin \varphi; \qquad \Phi_\varphi = \frac{E}{2r} \cos \varphi \tag{5.6}$$

These partial derivatives are to be evaluated at the point: $\rho = \rho_0$; $\varphi = \varphi_0 = \pi$. The characteristic equation is then

$$S^2 + \frac{1}{2}\left[\frac{1}{4} C(3r_0^2 - p) + \frac{E}{2r_0}\right]S + \frac{1}{8r_0} CE(3r_0^2 - p) = 0 \tag{5.7}$$

From the explicit form of the roots of the cubic equation, it follows that $3r_0^2 = p > 0$, and the resonance point $(r_0, \varphi_0 = \pi)$ is stable, which is the well known fact.

We consider now the question of synchronization: $\Gamma \neq 0$. In particular we study the system

$$R(r, \varphi, \Gamma) = 0; \qquad \Phi(r, \varphi, \Gamma) = 0 \tag{5.8}$$

which has the solution $r = r_0$; $\varphi = \pi$; $\Gamma = 0$, as has been just shown.

The Jacobian of $R(r, \varphi, \Gamma)$ and $\Phi(r, \varphi, \Gamma)$ with respect to r and φ at the point $(r_0, \pi, 0)$ is

$$\begin{vmatrix} -\frac{1}{8}C(3r_0^2 - p) & 0 \\ 0 & -\frac{E}{2r_0} \end{vmatrix} \neq 0 \tag{5.9}$$

then, for a Γ sufficiently small, there are two functions: $r = r(\Gamma)$ and $\varphi = \varphi(\Gamma)$ which satisfy the system (5.8) and are such that

$$\lim_{\Gamma\to 0} r(\Gamma) = r_0; \qquad \lim_{\Gamma\to 0} \varphi(\Gamma) = \pi \tag{5.10}$$

Since for $\Gamma = 0$ the stability conditions are satisfied, they will be satisfied also for a sufficiently small Γ. This establishes the existence of the synchronized state but not its vanishing for a larger Γ.

If $\Gamma \neq 0$, the characteristic equation has the form

$$S^2 + \frac{1}{2}\left[\frac{1}{4}(3r_0^2 - p) + \frac{E}{2r_0}\cos\varphi_0\right]S + \frac{EC}{16}\left[\frac{q}{r^2}\sin^2\varphi - (3r_0^2 - p)\cos\varphi_0\right] = 0 \tag{5.11}$$

In this expression $\sin\varphi_0 = \Gamma r_0/E > 0$; $\cos\varphi_0 < 0$ (since for $\Gamma = 0$, $\cos\varphi_0 = -1$), and (5.11) can be written

$$S^2 + \frac{1}{2}\left[\frac{1}{4}(3r_0{}^2 - p) - \frac{E}{2r_0}|\cos\varphi_0|\right]S + \frac{EC}{16}\left[\frac{q}{r^2}\sin^2\psi + (3r_0{}^2 - p)|\cos\varphi_0|\right] \tag{5.12}$$

In this form the characteristic equation shows that near the point $(r_0, \pi, 0)$ there is still a stable synchronized state if the coefficient of S is positive; this is possible since $3r_0{}^2 - p > 0$ by our assumption provided that the condition $3r_0{}^2 - p > (2E/r_0)|\cos\varphi_0|$ is fulfilled.

On the other hand, the condition for the existence of one single real root is:

$$\Delta = q^2\cos^2\varphi/4 - p^3/27 > 0 \tag{5.13}$$

and as $|\cos\varphi|$ decreases with increasing Γ, it is clear that Δ decreases and may become negative for a sufficiently large Γ. When this happens, instead of one single real root there will be three such roots and the factor $3r_0{}^2 - p$ may become negative† with the resulting loss of stability and the disappearance of the synchronized state when the coefficient of S in (5.12) vanishes.

We have been able to carry out only a qualitative discussion; in order to ascertain this quantitatively, it is clear that one has to use a numerical approximation procedure.

For instance, one will start with the cubic equation $r^3 - pr - q = 0$ (for a chosen combination of parameters A, C, and E) and determine the real positive root $r = r_0'$. With this value $r = r_0'$ one determines $\varphi = \varphi_0'$ from the second equation (5.5) and then one determines again the real positive root $r = r_0''$ of the cubic equation $r^3 - pr - q|\cos\varphi_0'| = 0$. One continues the procedure for increasing Γ until, for a sufficiently large Γ, calculation will indicate that the coefficient of S in (5.12) is near zero, which shows that the threshold of stability is reached. The procedure is long and tedious because the variables r and φ do not separate and one has to use successive approximations.

Unfortunately, in all nonlinear problems one encounters similar difficulties as soon as one tries to obtain quantitative results. If one wishes to explore only the qualitative aspect of this phenomenon, the matter is considerably simpler, as can be seen from Sections 2 and 3 of this chapter.

† It is recalled that, for $\Delta > 0$, the real root is $r_0 = r_0' = \pm 2\sqrt{p/3}\cosh(u/3)$ and for $\Delta < 0$ the three roots (instead of $\cosh(u/3)$) contain factors $\cos(u/3)$, $\cos(\pi + u)/3$ and $\cos(\pi - u)/3$, so that the absolute values $|r_0''|$ of roots in this case are smaller than $|r_0'|$ for $\Delta > 0$.

A remark is noteworthy: this problem is a nonautonomous one and, strictly speaking, its solutions cannot be represented in a phase plane but should be considered in the three-dimensional space (x, y, t).

This circumstance manifests itself in the present case by the fact that the initial conditions influence the behavior of the solution.[8] This is apparent if in (5.1) one puts the right-hand term in the form $e \sin (t + \pi) = -e \sin t$ instead of $e \sin t$, as we did. If one carries out these calculations, one finds instability where one formerly had stability.

However, through the transient period of the phase adjustment the phenomenon will ultimately swing into the state which we have analyzed in this section for $\Gamma = 0$.

From a physical point of view this initial transition for $\Gamma = 0$ is accomplished, so to speak, automatically, the system seeking its stable resonance point $(r_0, \pi, 0)$. If, from this point, one introduces the "detuning" of frequencies $(\Gamma \neq 0)$, the phenomenon of synchronization properly speaking takes place; the variational equations are "modified" as compared to their pure resonance values ($\sin \varphi_0 = 0$, $\cos \varphi_0 = -1$) which, through somewhat complicated relationships just outlined, influences the characteristic equation (5.12). When the latter goes through the threshold of its stability, the singular point of the system (5.5) disappears and with it vanishes also the synchronized state.

6. Mutual synchronization

In the preceding section it was assumed that the synchronization is due to the *action* of an external source on the oscillator but *the reaction* of the oscillator on the source was disregarded. In physical terms this means that the external source has a sufficiently high energy (or power) level to be able to neglect this reaction.

The situation is different if one considers the problem of a *mutual synchronization* of two oscillators, which we may assume to be identical in all respects except for a certain small relative *detuning* of their frequencies.

It must be noted that the problem of this nature is at the same time that of a coupled system, that is, a system with two degrees of freedom of which it forms a particular case. However, since this is also the problem of synchronization of a somewhat particular type, we investigate it here.[9]

As an example, we consider the system of two d.e. of the form

$$\begin{aligned} \ddot{x}_1 + \omega_1^2 x_1 &= (a - b x_1^2)\dot{x}_1 + \varepsilon Q_1 \ddot{x}_2 \\ \ddot{x}_2 + \omega_2^2 x_2 &= (a - b x_2^2)\dot{x}_2 + \varepsilon Q_2 \ddot{x}_1 \end{aligned} \qquad (6.1)$$

[8] N. Minorsky, *Rend. del Semin. Mat. di Torino*, 1954.

[9] Ch. Hayashi, *Forced Oscillations in Nonlinear Systems*, Nippon Printing Co., Osaka, 1953.

which represents two identical van der Pol oscillators coupled together inductively with a certain detuning of their frequencies to the values ω_1 and ω_2.

As the derivation of these d.e. is well known, we omit it here and merely indicate the significance of the symbols:

$$Q_1 = M\omega_1^2 C_2 = (M/L)(C_2/C_1); \qquad Q_2 = M\omega_2^2 C_1 = (M/L)(C_1/C_2)$$

$$C_1 = C_0 - \Delta C; \qquad C_2 = C_0 + \Delta C; \qquad \Delta C/C_0 \ll 1$$

$$k_1 + k_2 = \omega_1^2 - \omega_2^2; \qquad \omega_1^2 = 1 + k_1; \qquad \omega_2^2 = 1 - k_2; \qquad \gamma = \frac{M}{L}$$

This means that we consider two identical oscillators loosely coupled to each other (which is indicated by the small factor ε) and slightly "detuned" in frequencies so that $\omega_1^2 = 1 + k_1$ and $\omega_2^2 = 1 - k_2$, k_1 and k_2 being small positive numbers; this detuning is effected with respect to the frequency ω_0 which can be assumed to be $\omega_0 = 1$. The positive constants a and b are assumed to be also small and within the scope of the nearly linear theory. In view of these assumptions it will be sufficient to discuss the problem only within the limits of the theory of the first approximation.

We apply the stroboscopic reduction omitting the details, since the only difference between this problem and the previous ones is in the existence of two degrees of freedom.

In what follows we shall use double subscripts such as r_{10}, r_{11}, ψ_{10}, $\psi_{11}, \ldots$, etc., the first subscript relates to the degree of freedom and the second one to the order of approximation; for instance, r_{10} will mean: the first degree of freedom and the zero order of approximation, and so on.

There is a slight difference in calculations due to the fact that the detuning in one degree (the first one) involves $\omega_1^2 = 1 + k_1$ and in the second degree, $\omega_2^2 = 1 - k_2$. In this manner it is sufficient to calculate the data for the first degree and to introduce the above correction only in one term of the second degree.

As this calculation has been carried out on many occasions we omit it here and indicate directly the stroboscopic equations for r_1 and r_2, where $r_1 = \sqrt{x_1^2 + y_1^2}$; $r_2 = \sqrt{x_2^2 + y_2^2}$; $y = \dot{x}$, namely:

$$\frac{dr_1}{d\tau} = -\sigma[r_1^3 - pr_1 + n_2 r_2 \sin\varphi]; \qquad \frac{dr_2}{d\tau} = -\sigma[r_2^3 - pr_2 - n_1 r_1 \sin\varphi] \tag{6.2}$$

where $\sigma = b/8$; $p = 4A/B$; $n_1 = 4\varepsilon Q_1/b$; $n_2 = 4\varepsilon Q_2/b$; $\varphi = \varphi_1 - \varphi_2$; $A = a/\mu$, $B = b/\mu$, μ being the parameter of the series solution.

The interesting feature of this reduction is that the phase angles φ_{10}

and φ_{20} (which appear in the zero-order approximation: $\psi_{10}(t) = \varphi_{10} - t$; $\psi_{20}(t) = \varphi_{20} - t$) do not appear in each degree of freedom but appear in both degrees by their difference $\varphi = \varphi_1 - \varphi_2$. In this manner the coupling manifests itself mainly in the phase.

The second stroboscopic equations (for the phase φ) are here

$$\frac{d\varphi_1}{d\tau} = -\frac{1}{2}\left[K_1 + EQ_1 \frac{1}{\lambda} \cos(\varphi_{10} - \varphi_{20})\right]$$

$$\frac{d\varphi_2}{d\tau} = \frac{1}{2}[K_2 - EQ_2\lambda \cos(\varphi_{10} - \varphi_{20})] \tag{6.3}$$

where $\lambda = r_1/r_2$.

These two equations can be combined into one equation which gives

$$\frac{d\varphi}{d\tau} = -\frac{1}{2}[K - E \cos\varphi(Q_2\lambda^2 - Q_1)/\lambda] \tag{6.4}$$

where $K = K_1 + K_2 = \dfrac{k_1 + k_2}{\mu} = \dfrac{\omega_1^2 - \omega_2^2}{\mu}$.

This reduction of two d.e. (6.3) into one equation (6.4) is not always possible, however. If it is possible, the stroboscopic system, instead of being of the fourth order (6.2 and 6.3) becomes of the third order (6.2 and 6.4).

If this reduction to the third order is possible (and this is the principal part of the problem), everything happens as if the two frequencies $\omega_1^2 = 1 + k$ and $\omega_2^2 = 1 - k$ (assuming $k_1 = k_2 = k \ll 1$) were "entrained" to the same common frequency $\omega_0^2 = 1$.

The singular point of the stroboscopic system in such a case is given by equations

$$\frac{dr_1}{d\tau} = 0; \qquad \frac{dr_2}{d\tau} = 0; \qquad \frac{d\varphi}{d\tau} = 0 \tag{6.5}$$

The determination of the first two conditions does not present any particular difficulty (except rather complicated calculations in the general case), but the last one can be fulfilled only under a special condition which we shall examine.

Before proceding further, it is useful to emphasize the significance of this condition $d\varphi/d\tau = 0$. As $\varphi = \varphi_1 - \varphi_2$, it is clear that $d\varphi/d\tau = 0$ exists if the difference $(d\varphi_1/d\tau) - (d\varphi_2/d\tau) =$ const. The other alternative $d\varphi_1/d\tau = 0$ and $d\varphi_2/d\tau = 0$ requires too special conditions to be of interest here. In other words, although nonlinear frequency correction exists in each component degree of freedom, it cancels out in the final result so that, if all three conditions (6.5) are fulfilled, the period remains 2π in the first approximation.

This justifies a posteriori the stroboscopic reduction in which integrations are performed between 0 and 2π.

The condition $d\varphi/d\tau = 0$ is obtained from (6.4) and yields

$$\cos \varphi = f(\lambda) = \frac{K\lambda}{E(Q_2\lambda^2 - Q_1)} \tag{6.6}$$

The function $f(\lambda)$ can take any values ($f(0) = 0; f(\infty) = 0; f(\sqrt{Q_1/Q_2}) = \infty$). In order to represent $\cos \varphi$ by the function $f(\lambda)$ it is necessary to

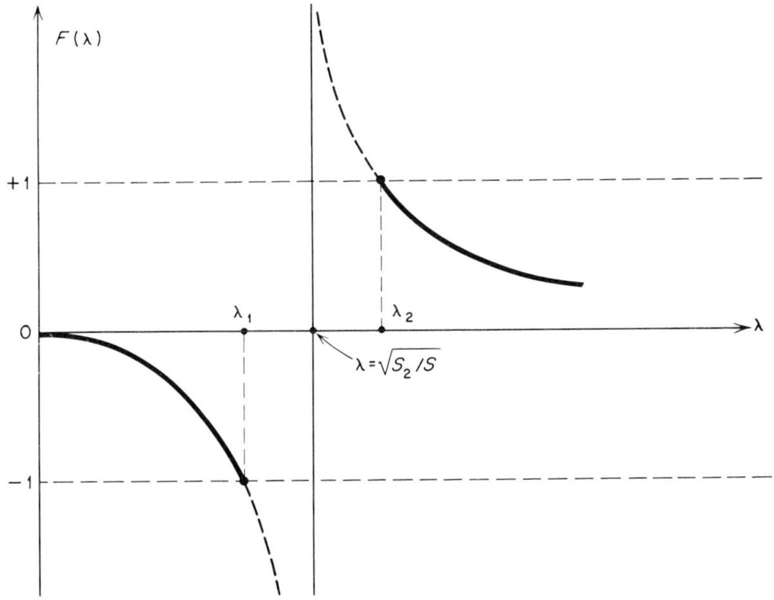

FIGURE 18.3

consider it in the intervals in which $-1 \leq f(\lambda) \leq 1$. If then equation (6.6) is satisfied, the requirement $d\varphi/d\tau = 0$ can be fulfilled and this combined with two other conditions $dr_1/d\tau = 0$ and $dr_2/d\tau = 0$ establishes the existence of the synchronized state of the two oscillators. The graph of the function $f(\lambda)$ is shown in Fig. 18.3. For small values of λ, $f(\lambda) < 0$ and for large λ it is positive. For $\lambda_0 = \sqrt{Q_1/Q_2}$ the function undergoes the discontinuity from $-\infty$ to $+\infty$ when λ traverses the value $\lambda = \lambda_0$ increasing.

The useful part of the curve $f(\lambda)$ is limited by the values $f(\lambda_1) = +1$ and $f(\lambda_2) = -1$.

The values $\lambda = \lambda_1$ and $\lambda = \lambda_2$ obtained from (6.6) are:

$$\lambda_1 = \frac{F}{2} + \sqrt{\frac{F^2}{4} + D}; \qquad \lambda_2 = -\frac{F}{2} + \sqrt{\frac{F^2}{4} + D} \tag{6.7}$$

where $F = K/EQ_2$; $D = Q_1/Q_2$.

In the interval (λ_2, λ_1) equation (6.6) cannot be fulfilled so that there is no root of the right-hand side of (6.4). In this interval synchronization cannot exist since $d\varphi/d\tau \neq 0$. In the intervals $(0, \lambda_2)$ and (λ_1, ∞), on the contrary, condition $d\varphi/d\tau = 0$ can exist.

The interval in which the synchronization is lost is

$$\Delta\lambda = \lambda_1 - \lambda_2 = F = \frac{K}{EQ_2} \tag{6.8}$$

It is seen that $\Delta\lambda$ decreases if K decreases or if EQ_2 increases, which is sufficiently clear on physical grounds, since a certain amount of work is required to bring two frequencies together and this work is the smaller, the smaller is the detuning K or the larger is the coupling EQ_2.

It is necessary also to take into account equations (6.2) since the stationary values r_{10} and r_{20} in (6.4) are the same as in (6.2). From the latter equations one obtains

$$(r_{10}/r_{20})^2 = \lambda^2 = (p\lambda - n_2 \sin \varphi_0)/(p\lambda + n_1\lambda^2 \sin \varphi_0) \tag{6.9}$$

which can be written as

$$(n_1 \sin \varphi_0)\lambda^4 + p\lambda^3 - p\lambda + n_2 \sin \varphi_0 = 0 \tag{6.10}$$

This equation results from two equations (6.2), but it is necessary to express also that $\sin \varphi_0$ appearing in (6.8) corresponds to $\cos \varphi_0$ calculated from (6.4); this can be expressed from the relation $\sin^2 \varphi_0 + \cos^2 \varphi_0 = 1$, but the calculations are long and we do not reproduce them here.

In the symmetrical case ($\lambda = 1$; $r_{10} = r_{20}$), conditions are much simpler and (6.8) reduces to:

$$(n_1 + n_2) \sin \varphi_0 = 0 \tag{6.11}$$

As $(n_1 + n_2) \pm 0$, one must have $\sin \varphi_0 = 0$, that is, $\cos \varphi_0 = \pm 1$; on the other hand, we have $f(1) = k/(s_2 - s_1)$ and as $(s_2 - s_1) < 0$, one has $\cos \varphi_0 = \cos (\varphi_{10} - \varphi_{20}) = -1$ which shows that for the symmetrical equilibrium state $\varphi_0 = \varphi_{10} - \varphi_{20} = \pi$. This means that the individual phases φ_{10} and φ_{20} of oscillators are in opposition to each other.

Under this condition equations (6.2) show that the stationary amplitudes r_{10} and r_{20} of each oscillator are equal and are:

$$r_{10}^{\,2} = r_{20}^{\,2} = p = 4a/b$$

For $a = b$, this yields

$$r_{10} = r_{20} = 2$$

which is the well known result.

Summing up, in the symmetrical case of two identical van der Pol oscillators, with a symmetrical detuning, each oscillator oscillates with the same amplitude $r_0 = r_{10} = r_{20} = 2$ as if it were alone. The condition of coupling imposes the opposition of the individual phases of φ_{10} and φ_{20} but the period continues to be 2π in spite of the fact that the individual frequencies are made to be different (being $\omega_1^2 = 1 + k$ for one oscillator and $\omega_2^2 = 1 + k$ for the other one).

If the condition of synchronization $d\varphi/d\tau = 0$ cannot be fulfilled, that is, $(d\varphi_1/d\tau) \neq (d\varphi_2/d\tau)$, $\varphi = \varphi_1 - \varphi_2$ ceases to be constant and varies continuously (although not uniformly); this results in similar variations of $\sin\varphi$ and $\cos\varphi$ between the limits $+1$ and -1. In other words, both amplitudes and phases undergo *modulations* and the oscillation becomes almost periodic. From the quantitative point of view the problem becomes complicated in view of these cross-modulations.

In the general case when the synchronization disappears the problem is amenable to a differential system of the fourth order since in addition to two amplitude equations (6.2) one has also two phase equations (6.3) without any possibility of combining the latter into one single equation (6.4).

The question of stability of the stationary state reduces to the variational equations which are obtained from (6.2) and (6.4) if one replaces the stationary values r_{10}, r_{20}, and φ_0 by $r_{10} + \delta r_1$, $r_{20} + \delta r_2$ and $\varphi_0 + \delta_\varphi$, where δr_1, δr_2, and $\delta\varphi$ are small perturbations. One has to perform the variation of $\lambda = r_{10}/r_{20}$ in the same manner.

For equations (6.2) the variational equations are

$$\frac{d\delta r_1}{d\tau} = [-3\sigma r_{10}^2]\delta r_1 + [\sigma(p + n_2 \sin\varphi_0)]\delta r_2 + [-\sigma n_2 r_{20} \cos\varphi_0]\delta\varphi$$
$$\frac{d\delta r_2}{d\tau} = [\sigma(p + n_1 \sin\varphi_0)]\delta r_1 + [-3\sigma r_{20}^2]\delta r_2 + [\sigma n_1 r_{10} \cos\varphi_0]\delta\varphi \tag{6.12}$$

For the phase in the synchronized condition the variational equation is

$$\frac{d\delta\varphi}{d\tau} = \frac{1}{2}[\gamma \cos\varphi_0(+)/r_{10}^2 r_{20}]\delta r_1 - \frac{1}{2}[\gamma\cos\varphi_0(+)/r_{10}r_{20}^2]\delta r_2 + \frac{1}{2}[\gamma \sin\varphi_0(-)/r_{10}r_{20}]\delta\varphi \tag{6.13}$$

where $(+) = \omega_1^2 r_{20}^2 + \omega_2^2 r_{10}^2$ and $(-) = \omega_1^2 r_{20}^2 - \omega_2^2 r_{10}^2$.

The characteristic equation of the synchronized system is

$$\begin{vmatrix} A_{11}-S & A_{12} & A_{13} \\ A_{21} & A_{22}-S & A_{23} \\ A_{31} & A_{32} & A_{33}-S \end{vmatrix} = D(S) = 0 \qquad (6.14)$$

where

$$\begin{aligned} &A_{11} = -3\sigma r_{10}; \quad A_{12} = \sigma(p - n_2 \sin\varphi_0); \quad A_{13} = -\sigma n_2 r_{20} \cos\varphi_0 \\ &A_{21} = \sigma(p + n_1 \sin\varphi_0); \quad A_{22} = -3\sigma r_{20}^2 \\ &A_{23} = \sigma n_1 r_{10} \cos\varphi_0; \quad A_{31} = \tfrac{1}{2}[\gamma \cos\varphi_0(+)/r_{10}^2 r_{20}]; \\ &A_{32} = -\tfrac{1}{2}[\gamma \cos\varphi_0(+)/r_{10} r_{20}^2]; \quad A_{33} = \tfrac{1}{2}[\gamma \sin\varphi_0(-)/r_{10} r_{20}] \end{aligned} \qquad (6.15)$$

In the developed form $D(s)$ is

$$S^3 + B_2 S^2 + B_1 S + B_0 = 0 \qquad (6.16)$$

where B_0, B_1, and B_2 are functions of A_{ij}; $i, j = 1, 2, 3$.

In the symmetrical case, $r_{10} = r_{20} = r_0$; $\lambda = 1$; $\sin\varphi_0 = 0$; $\cos\varphi_0 = \pm 1$. As in this case $\cos\varphi_0$ appears only as $\cos^2\varphi_0$, its sign is immaterial. The characteristic equation in this case becomes

$$S^3 + (6\sigma r_0^2)S^2 + [\sigma^2(9r_0^2 - p^2) + \tfrac{1}{2}\sigma\gamma(\omega_1^2 + \omega_2^2)(n_1 + n_2)]S + \tfrac{1}{2}\sigma^2\gamma(3r_0^2 - p)(\omega_1^2 + \omega_2^2)(n_1 + n_2) = 0 \qquad (6.17)$$

By Hurwitz's theorem (6.14) has no roots with real positive parts if

$$(1)\ B_2 > 0; \qquad (2)\ B_0 > 0; \qquad (3)\ B_2 B_1 - B_0 > 0 \qquad (6.18)$$

As condition (1) is always fulfilled, conditions (2) and (3) reduce to one single condition:

$$3r_0^2 - p > 0 \qquad (6.19)$$

On the other hand, for the synchronized state one has $r_0 = \sqrt{p}$ and it is seen that (6.19) is always fulfilled and the system is stable.

We do not attempt to carry out calculations of stability for an asymmetrical condition $\lambda \neq 1$ in view of their complexity but, on physical grounds it seems plausible to assume that any such asymmetrical state is unstable as long as two oscillators are identical and the detuning of their frequencies is symmetrical.

The situation would be different if the oscillators would have different parameters and their detuning were also different. In such an asymmetrical case calculations would be more complicated although the physical nature of the synchronization would be the same, namely: for a sufficiently large detuning the synchronized state disappears and the stroboscopic system becomes of the fourth order giving rise to an almost periodic oscillation.

For a coupled system involving more than two oscillators, the synchronized state is impossible in view of the impossibility of fulfilling the condition $d\varphi/d\tau = 0$ for the *whole* system, although for a partial synchronized coupling of two oscillators of the system this does not seem to be impossible as shown in this analysis.

7. Other forms of synchronization

In the analysis of the synchronization effect it is customary to develop the theory on the basis of the d.e. of the electron-tube circuits. The only reason for this, as in other investigations of nonlinear phenomena, is the simplicity with which the d.e. can be formed on the basis of the Kirchhoff laws. This does not mean that the synchronization is a special property of the electron-tube circuits. On the contrary, the numerous experimental data seem to indicate that this phenomenon is rather general and appears in any system possessing appropriate nonlinear features regardless of its physical nature.

We mentioned previously that the first discovery of the phenomenon of synchronization by Huygens (1629–1695) was made in connection with the synchronization of the periods of the two clocks hung on a thin wall. Similarly, Lord Rayleigh notes that two organ pipes of slightly different frequencies, when placed near to each other, vibrate at the same frequency but, if their frequencies are sufficiently far apart, one hears the frequency of the beats. Later on Lord Rayleigh[1] improved his experiment by moving the pipes sufficiently far from each other and obtaining the same result by means of an appropriate coupling through an acoustic resonator.

A detailed investigation of synchronization of clocks in the presence of the escapement mechanism was carried out by T. Haag[10]; this is precisely the domain in which the synchronization effect was observed for the first time by Huygens.

Phenomenon of acoustic synchronization was investigated by S. Chaikin and K. Theodorchik.[11] These authors used the arrangement shown in Fig. 18.4.

A telephone T is inserted in the anode circuit of an electron-tube oscillator and a microphone M in its grid circuit. The elements T and M are coupled acoustically by means of two armatures A_1 and A_2 fixed on the

[1] See footnote [1], page 438.
[10] J. Haag, *Ann. Ecole Norm. Supér.*, Paris, Sér. 3, 64, 1947.
[11] K. F. Theodorchik and S. Chaikin, *J. Tech. Phys.* (USSR) **2**, 1932.

same rod R, which is centralized by means of a spring not shown; a suitable damper is also attached on R.

The mechanical system A_1RA_2 is described by a linear d.e. of the 2nd order and has a frequency ω. As to the electronic oscillator, it has its own frequency ω_0 of the self-excited oscillations. If the difference $\omega - \omega_0$ is not too small, the beats of the two frequencies are observed; but if $|\omega - \omega_0|$ becomes sufficiently small, both frequencies coalesce into one, frequency ω. The nonlinear frequency ω_0 is thus "entrained" by the external frequency and it is found that the ratio $(\omega - \omega_0)/\omega$ is proportional to the ratio a/a_0, where a is the amplitude of oscillations of the mechanical system driven by the acoustic pressure waves emitted by telephone, and a_0

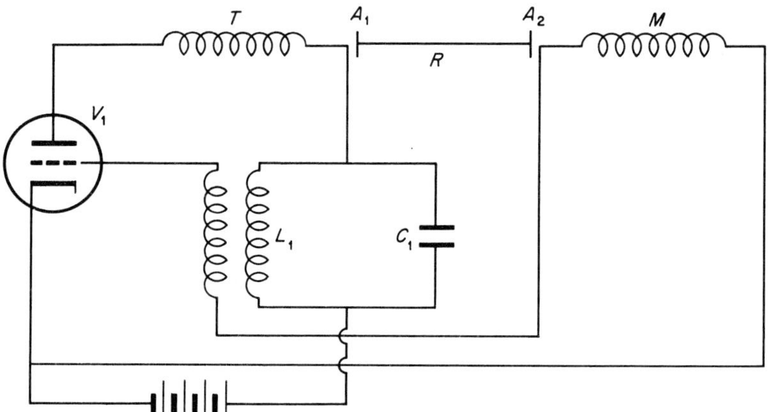

FIGURE 18.4

is the amplitude of the self-sustained oscillation of the electron-tube oscillator.

This method has been applied to measure acoustic intensity by observing the magnitude of the zone of synchronization, knowing a_0 and determining the proportionality factor by calibration.

In recent years the synchronization phenomenon has been used extensively for a very accurate speed control of electric motors, such as those used in high precision clocks and similar devices in which an extreme accuracy of speed control is necessary. The usual scheme consists of a quartz oscillator circuit with an additional circuit for *frequency demultiplication*. The latter, as will be shown in Chapter 19, depends on the so-called *subharmonic resonance* involving the subharmonics (harmonics of a fractional ratio). As the quartz oscillator maintains its frequency with a very high degree of accuracy, this accuracy is preserved in the frequency demultiplication process and thus appears on the low-frequency end where it is used to drive a synchronous motor. Ultimately, the scheme amounts to a *synchronization on a subharmonic* of a very low order, using a technical term.

In order to secure a reliable synchronization, the latter is produced, so to speak, *synthetically*. A number of schemes exist in this connection. We shall mention one of the earlier schemes due to Kaden[12] which is sufficiently simple and illustrates well the principle used (Fig. 18.5).

An electron tube V_1 operates as an oscillator with frequency ω_1, having C_1 and L_1 as constants of its oscillating circuit. The coefficient of

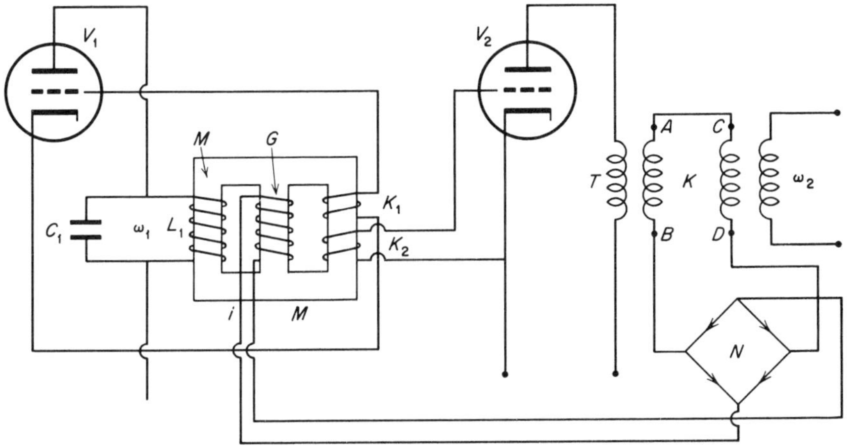

FIGURE 18.5

inductance L_1 can be varied within certain limits, owing to a saturated iron core on which the coil L_1 is wound; the degree of saturation of the core can be varied by means of changes in electric current i flowing through an auxiliary coil G wound on the middle leg of the magnetic circuit M shown.

There are two other coils: coil K_1, producing the feedback voltage to the grid of V_1 which merely maintains an oscillation with frequency ω_1; and coil K_2, transmitting oscillation with frequency ω_1 to the second tube V_2 working as amplifier. The output of V_2 is coupled (through a transformer T) to the branch AB of the synchronizing network K; the latter has also a second branch CD into which the second frequency ω_2 is transmitted inductively.

The synchronizing network $BACD$ is closed on a bridge N formed by

[12] H. Kaden, *ENT* **16**, 1939.

rectifying elements; the direction of rectification is indicated by arrows. The diagonal points of the bridge are connected to the saturation coil of the iron core M.

In the circuit $BACD$ there are thus two induced voltages: E_1, with frequency ω_1 induced in the AB branch; and E_2, with frequency ω_2 induced in the CD branch.

We consider the case when the difference $\omega_2 - \omega_1 = \omega$ is small. The vector diagram is shown in Fig. 18.6; we can assume that the vector E is fixed; then E_2 rotates with frequency $\Delta\omega$ in one direction or other, depending on the sign of the difference $\omega_2 - \omega_1$.

The resultant vector E_r is the voltage between B and D and represents, therefore, to a certain scale the rectified current i flowing through the coil G.

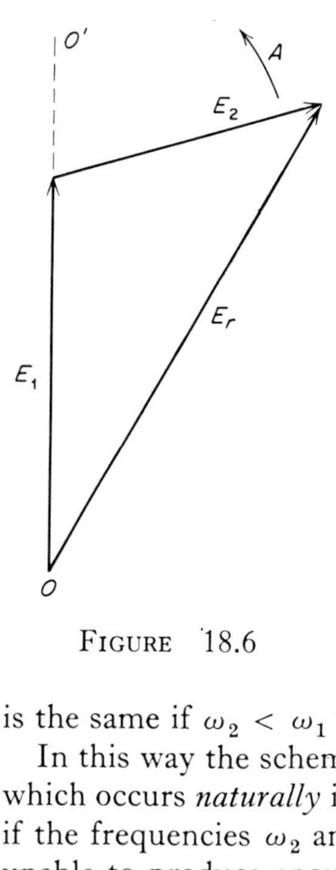

FIGURE 18.6

As $\omega_1 = 1/\sqrt{L_1(i)C_1}$, where $L_1(i)$ is a nonlinear function of i (decreasing with an increasing i or vice versa), it is apparent that

$$\frac{d\varphi}{dt} = \omega_2 - \omega_1 = \Delta\omega$$

where φ is the magnetic flux through the coil. If, initially, $\Delta\omega > 0$, this means that $\omega_2 > \omega_1$ and that E_2 rotates in the direction of the arrow A (that is, toward the advance) around the end of E_1 as center. The resultant voltage E_r is thus increased and so is the current i owing to the rectifier scheme. This reduces that value of $L_1(i)$ and increases the frequency ω_1 until the initial difference $\Delta\omega$ is reduced to zero. Owing to this arrangement, the equilibrium point $\omega_2 - \omega_1 = \Delta\omega = 0$ is stable. The argument is the same if $\omega_2 < \omega_1$ with an opposite rotation of the vector E_2.

In this way the scheme just mentioned produces exactly the same effect which occurs *naturally* in the phenomenon of synchronization. Likewise, if the frequencies ω_2 and ω_1 are far apart from each other, the circuit is unable to produce enough variation of $L_1(i)$ to keep the two frequencies "locked" together; in such a case the vector E_2 rotates continuously (although not uniformly) around the end of the vector E_1, always in the same direction. Here, again, one has the phenomenon of "beats" of the two frequencies ω_2 and ω_1 that occurs in the phenomenon of the natural synchronization, once the zone of synchronization has been exceeded.

In this scheme the origin of the synchronizing effect is in the nonlinearity of the inductance L_1; if this inductance were linear, it would be impossible to produce this result.

Summing up, in all such schemes, whatever their nature may be, the origin of synchronization phenomenon lies in the existence of a point of stable equilibrium when the two frequencies coalesce. This means that the phenomenon in such a case has a natural tendency to approach this point of coalescence of the two frequencies.

Chapter 19

NONLINEAR RESONANCE

1. Introductory remarks

Phenomena of nonlinear resonance exhibit a far greater variety than those of an ordinary (linear) resonance owing to the existence of the so-called "subharmonics," that is, harmonics of a *fractional order* which constitute the essential property of nonlinear systems.

Historically, the study of subharmonics or, more generally, of *combination tones* (or frequencies) preceded the establishment of modern theories of nonlinear resonance; in Section 2 we indicate briefly the nature of subharmonics from a physical point of view.

The theory of resonance appeared first as a component part of the general theory long before a study of physical manifestations of resonance was undertaken. Thus, Poincaré, in his early (1892) studies of nonlinear d.e.,[1] observes that "there exist periodic solutions with period $2k\pi$ distinct from solutions with period 2π and coalescing with the latter for $\mu = 0$, which clearly indicates the bifurcation of subharmonic solutions at this point.

In Chapters 10 and 14 we had occasion to go into this matter in some detail by two different methods. In particular in Chapter 10, we saw complications which appear when k approaches an integer value. Likewise in Chapter 15 dealing with the asymptotic methods of Krylov-Bogoliubov, it was seen that the condition of resonance leads to an essential modification of the theory in that there appears a fixed relation between the phases of the external periodic excitation and that of the oscillation itself, which results in a modification of *the form* of the asymptotic series.

A fundamental investigation of the theory of the subharmonic resonance with its physical implications was undertaken by L. Mandelstam and N.

[1] H. Poincaré, *Les méthodes nouvelles de la mécanique céleste* **T.1**, Gauthier-Villars, Paris, 1892; also E. Goursat, *Cours d'Analyse* **T.2**, Gauthier-Villars, Paris, 1918.

Papalexi[2]; this study is reviewed in Section 3 where its connection with the theory of Poincaré (Chapter 10) is apparent. In fact, these authors take as generating solution, the solution the nonhomogeneous linear ($\mu = 0$) d.e. and develop the theory on this basis. The existence of a subharmonic solution is obtained relatively simply in this case but for stability there are certain complications in view of the fact that the d.e. is a nonautonomous one. The authors avoid, however, the difficulty of calculation of the characteristic exponents by establishing sufficient conditions for stability in a form of certain inequalities as was explained in Chapter 13. Most of the treatment concerns "potentially self-excited systems" and electronics plays an important role in verifying the conclusions of the theory developed in detail for the subharmonic resonance of the order $\frac{1}{2}$ and, to a lesser degree, for that of the order $\frac{1}{3}$. This theory served as a basis for numerous experimental developments in USSR (between 1933 and 1940) and recent researches of Ch. Hayashi[3] follow also this method.

A different approach to this question was followed by Krylov-Bogoliubov in their early (1937) work[4] which was developed further in a recent treatise of Bogoliubov-Mitropolsky (1955)[5] reviewed in Chapters 14 and 15. This work follows the "asymptotic method" of these authors; the features of resonance are related in this theory to the existence of a fixed phase between the heteroperiodic and autoperiodic oscillations. This idea appeared in the earlier work of Krylov and Bogoliubov and, in fact, is related to the question of synchronization.

What complicates the theory to some extent is the fact that resonance properly speaking is generally accompanied by synchronization. The two phenomena, in reality, form merely two distinct features or "aspects" of the same effect: namely, the resonance feature relates to the amplitude and synchronization to the phases that are "locked" together during the resonance as was mentioned in the Introduction to Part III.

In addition to these two principal theories of the subharmonic resonance, there appeared recently a third one as the result of discussions with M. Schiffer. This approach is based on the application of the stroboscopic method and the starting point is the same as in Mandelstam–Papalexi theory, that is, one takes as a generating solution the solution of the linear ($\mu = 0$) nonhomogeneous equation but the subsequent treatment is

[2] L. Mandelstam and N. Papalexi, *Zeitschr. für Physik* **73**, 1932.

[3] Ch. Hayashi, *Forced Oscillations in Nonlinear Systems*, Nippon Printing Co., Osaka, 1953.

[4] N. Krylov and N. Bogoliubov, *Introduction to Nonlinear Mechanics* (in Russian), 1937.

[5] N. Bogoliubov and J. Mitropolsky, *Asymptotic Methods in the Theory of Nonlinear Oscillations* (in Russian), Moscow, 1958.

different inasmuch as the stroboscopic transformation is introduced immediately after the formulation of conditions of periodicity (Poincaré). This results, as usual, in an ultimate autonomous system. The problem of the determination of the stationary state reduces to that of determining real roots common to two algebraic equations in terms of the integration constants appearing in the generating solution. Although this method, at least theoretically, leads to a simple problem of determination of singular points and their stability, the practical difficulties of carrying out calculations lie in the last (algebraic) step of the problem as will be seen from examples in Section 6.

2. Subharmonics

The existence of harmonics (or "ultraharmonics" as some authors call them in order to distinguish them from "subharmonics") is sufficiently well known from the linear theory and needs no further mention here.

It is likely that Laplace was first to observe that in the case of celestial motions with frequencies ω_1 and ω_2, disturbances with frequencies $\omega = m\omega_1 + n\omega_2$ (m and n being integers) are occasionally observed.

It was, however, Helmholtz[6] who definitely discovered the existence of subharmonics in his theory of physiological acoustics. In particular, he showed that the ear often hears sounds of frequencies that are not contained in the incoming acoustic radiation; he showed also that the reason for that is due to nonlinearity of the d.e. governing the oscillations of the tympanic membrane.

The conclusions of Helmholtz can be demonstrated in a simpler manner by means of the well known properties of the electron tubes, whose nonlinear characteristic has the form

$$i_a = f(v) \tag{2.1}$$

where i_a is the anode current and v is the grid voltage. If one uses the customary approximation of $f(v)$ by a polynomial, for instance:

$$i_a = a_1 v + a_2 v^2 + a_3 v^3 \tag{2.2}$$

and assumes for the sake of simplicity that v consists of two sinusoidal oscillations of the same amplitude but of different frequencies:

$$v = k(\sin \omega_1 t + \sin \omega_2 t) \tag{2.3}$$

the substitution of this expression for v into (2.2), after a reduction of trigonometric functions to a finite Fourier polynomial, shows that in

[6] H. Helmholtz, *Sensation of Tone*, Longmans, Green, London, 1895.

addition to the frequencies ω_1 and ω_2 there appear also terms with frequencies

$$2\omega_1, 2\omega_2, 3\omega_1, 3\omega_2, \omega_1+\omega_2, \omega_1-\omega_2,$$
$$2\omega_1+\omega_2, 2\omega_1-\omega_2, 2\omega_2+\omega_1, 2\omega_2-\omega_1$$

The first four frequencies are the ordinary harmonics (or "ultraharmonics") but the last six are the "combination tones." Those of them whose frequency is *lower* than the lowest of the impressed frequencies (ω_1 or ω_2) *are subharmonics.* By a proper choice of ω_1 and ω_2, as well as of the *form* of the nonlinear characteristic, one can obtain subharmonics of a relatively low (fractional) order and, by cascading some of such arrangements, one can obtain subharmonics of an exceedingly low order. In this way one obtains that which is called *frequency demultiplication,* to which we referred in Section 7, Chapter 18, when we referred to the synchronization of an electric motor with the frequency of a quartz oscillator.

If one considers the existence of numerous subharmonics in a nonlinear system, one can form an intuitive idea as to how a subharmonic resonance occurs. In fact, if one of these subharmonics has a frequency near to that of the free oscillations of the system, an oscillation of this particular frequency will be singled out relatively to the other subharmonics; this may happen either as the result of interaction of a subharmonic with the external frequency or on account of a mutual interaction of harmonics.

Such an intuitive approach, however, does not lead anywhere, inasmuch as what is generally given is not a subharmonic but the d.e., and the ascertaining of whether a given d.e. has a definite subharmonic, which, in addition, is *stable,* constitutes precisely the principal part of the problem.

The problem, as we will see, generally splits into two parts: (1) proof of existence of a certain subharmonic (of a given order, say, $\frac{1}{2}$, $\frac{1}{3}$, etc.) and (2) that of its stability. Only if these two conditions are fulfilled can one be certain that a corresponding subharmonic resonance actually exists.

A remark is noteworthy: from the form $\omega = m\omega_1 + n\omega_2$, where m and n are *some* integers (positive or negative), one could conclude that "combination frequencies" are densely distributed in the continuous spectrum of frequencies. If so, the subharmonic resonances are also distributed more or less continuously. This conclusion is not true, however; in fact, Krylov and Bogoliubov have shown in their earlier study (1937) that the resonances become "washed out" for increasing m and n. Hence, only those of them are actually observable which occur for relatively small integer values of m and n.

In this manner the condition of subharmonic resonance appears each time when the following approximate relation is fulfilled

$$\omega_0 = m\omega_1 \pm n\omega_2 \tag{2.4}$$

where m and n are small integers and ω_0 is the "free" frequency of the system.

In spite of this limitation, the variety of subharmonic resonances is still considerable, and whether they actually *exist* (that is, exist *mathematically* and *are stable*) constitutes the principal problem of this chapter.

We mention in passing that, as far as known, no applications of these phenomena exist at present, very likely on account of the fact that the problem is still not quite thoroughly explored.

3. Theory of L. Mandelstam and N. Papalexi

Consider a nearly linear d.e. of the form

$$\ddot{x} + x = \mu f(x,\dot{x}) + \lambda_0 \sin nt \tag{3.1}$$

which represents a nonautonomous system in view of the external periodic excitation with period $2\pi/n$. The problem is to establish the conditions for the existence of a subharmonic oscillation with period 2π. It is noted that this problem has been already investigated in Chapter 10 on the basis of the general theory; we take it up here in a somewhat special form due to Mandelstam and Papalexi in order to establish a closer connection with applications. We omit details relative to electron-tube circuits by which this theory was confirmed later and consider only its essential points.

If $\mu = 0$, (3.1) is a linear nonhomogeneous d.e. whose solution is of the form

$$x(t) = a_0 \sin t - b_0 \cos t + \frac{\lambda_0}{1 - n^2} \sin nt \tag{3.2}$$

where the first two terms on the right side represent the free oscillation and the last one represents the forced oscillation, a_0 and b_0 being two constants of integration. One has thus an infinity of periodic solutions depending on two parameters a_0 and b_0.

The argument is clearly the same as in the theory of Poincaré ($\lambda_0 = 0$), namely: try to establish the conditions under which a periodic solution still exists for $\mu \neq 0$ but small: this merely amounts to selecting the generating solution in the form (3.2) instead of $x(t) = a \sin t - b \cos t$ corresponding to the harmonic oscillator $\ddot{x} + x = 0$.

A generating solution of this kind is possible if, in turn, it is possible to demonstrate that the solution of (3.1) can be represented in the same form as (3.2) with functions $a(\mu)$ and $b(\mu)$ instead of a_0 and b_0 such that

$$a(\mu) \to a_0; \qquad b(\mu) \to b_0 \quad \text{when} \quad \mu \to 0 \tag{3.3}$$

Once the possibility of such a generating solution is ascertained, it is necessary to show that it is *stable*. The second part of the proof (stability) is more difficult than the first part (existence) as will appear in what follows.

The authors use a double transformation of variables; one introduces first the variable: $z = x - [\lambda_0/(1 - n^2)] \sin nt$ which is substituted into (3.1) and then the second change of variables:

$$u = \dot{z} \cos t + z \sin t; \qquad v = \dot{z} \sin t - z \cos t \tag{3.4}$$

results finally in the equivalent system

$$\begin{aligned} \dot{u} &= (\ddot{z} + z) \cos t = \mu\psi(u, v, t) \cos t \\ \dot{v} &= (\ddot{z} + z) \sin t = \mu\psi(u, v, t) \sin t \end{aligned} \tag{3.5}$$

where $\psi(u, v, t) = f(u \sin t - v \cos t + [\lambda_0/(1 - n^2)] \sin nt;\ u \cos t + v \sin t + [n\lambda_0/(1 - n^2)] \cos nt)$. The function $\psi(u, v, t)$ is periodic with period 2π. It is seen from this expression for $\psi(u, v, t)$ that the x which appears in $f(x,\dot{x})$ of (3.1) is of the form:

$$x = u(\mu) \sin t - v(\mu) \cos t + \frac{\lambda_0}{1 - n^2} \sin nt \tag{3.6}$$

so that what we have just called $a(\mu)$ and $b(\mu)$ in (3.3) appears now as $u(\mu)$ and $v(\mu)$ in these notations. One has to show then that $u(\mu) \rightarrow a_0$ and $v(\mu) \rightarrow b_0$, when $\mu \rightarrow 0$.

From (3.5) one can write:

$$\begin{aligned} u &= u^0(\mu) + \mu \int_0^t \psi(u, v, \tau) \cos \tau d\tau \\ v &= v^0(\mu) + \mu \int_0^t \psi(u, v, \tau) \sin \tau d\tau \end{aligned} \tag{3.7}$$

The constants of integration u^0 and v^0 relate to the nearly linear problem; but for the linear one ($\mu = 0$), they are a_0 and b_0.

One can, therefore, following Poincaré, define two constants α and β by relations

$$u^0 = a_0 + \alpha(\mu); \qquad v^0 = b_0 + \beta(\mu) \tag{3.8}$$

$\alpha(\mu)$ and $\beta(\mu)$ in these notations playing the role of the parameters $\beta_1(\mu)$ and $\beta_2(\mu)$ (Chapter 10) of the theory of Poincaré.

On the other hand, the functions $u(\mu)$ and $v(\mu)$ can be expanded into series arranged according to the ascending powers of parameters which, in view of (3.8), gives:

$$\begin{aligned} u &= a_0 + \alpha + \mu C_1(t) + \mu\alpha D_1(t) + \mu\beta E_1(t) + \mu^2 G_1(t) + \dots \\ v &= b_0 + \beta + \mu C_2(t) + \mu\alpha D_2(t) + \mu\beta E_2(t) + \mu^2 G_2(t) + \dots \end{aligned} \tag{3.9}$$

If, on the other hand, one expands the function $\psi(u, v, t)$ in Taylor's series around the values a_0 and b_0 and compares (3.7) and (3.9), one finds the following expressions for the coefficients $C_1(t)$, $D_1(t)$,. . .

$$C_1(t) = \int_0^t \psi(a_0, b_0, \tau) \cos \tau d\tau; \quad C_2(t) = \int_0^t \psi(a_0, b_0, \tau) \sin \tau d\tau$$

$$D_1(t) = \int_0^t \left[\frac{\partial \psi}{\partial u}\right] \cos \tau d\tau; \qquad D_2(t) = \int_0^t \left[\frac{\partial \psi}{\partial u}\right] \sin \tau d\tau \tag{3.10}$$

$$E_1(t) = \int_0^t \left[\frac{\partial \psi}{\partial v}\right] \cos \tau d\tau; \qquad E_2(t) = \int_0^t \left[\frac{\partial \psi}{\partial v}\right] \sin \tau d\tau$$

where the symbols $\left[\frac{\partial \psi}{\partial u}\right]$ and $\left[\frac{\partial \psi}{\partial v}\right]$ designate the partial derivatives of ψ with respect to u and v for $\mu = \alpha = \beta = 0$.

If u and v are periodic, $u(2\pi) - u(0) = 0$; $v(2\pi) - v(0) = 0$, and one has from (3.9):

$$\begin{aligned} C_1(2\pi) + \alpha D_1(2\pi) + \beta E_1(2\pi) + \mu G_1(2\pi) + \dots &= 0 \\ C_2(2\pi) + \alpha D_2(2\pi) + \beta E_2(2\pi) + \mu G_2(2\pi) + \dots &= 0 \end{aligned} \tag{3.11}$$

Since μ, $\alpha(\mu)$, and $\beta(\mu)$ are small and, moreover, these equations are to be satisfied for any small μ, the first condition is

$$\begin{aligned} C_1(2\pi) &= \int_0^{2\pi} \psi(a_0, b_0, \tau) \cos \tau d\tau = 0 \\ C_2(2\pi) &= \int_0^{2\pi} \psi(a_0, b_0, \tau) \sin \tau d\tau = 0 \end{aligned} \tag{3.12}$$

This permits determining the constants a_0 and b_0. For the determination of α and β, one has the system

$$\begin{aligned} \alpha D_1(2\pi) + \beta E_1(2\pi) + \mu G_1(2\pi) &= 0 \\ \alpha D_2(2\pi) + \beta E_2(2\pi) + \mu G_2(2\pi) &= 0 \end{aligned} \tag{3.13}$$

if one limits the expansion to the first order only. These equations permit determining the unknowns α and β approaching zero together with μ provided

$$\Delta = \begin{vmatrix} D_1(2\pi) & E_1(2\pi) \\ D_2(2\pi) & E_2(2\pi) \end{vmatrix} \neq 0 \tag{3.14}$$

This proves the *existence* of the subharmonic solution under the condition (3.14) so that

$$u(\mu) \to a_0; \qquad v(\mu) \to b_0$$
$$\mu \to 0 \qquad\qquad \mu \to 0$$

as was previously set forth.

The second part of the proof is somewhat more involved as the d.e. in this case is a nonautonomous one, but the complication inherent in the calculation of the characteristic exponents is avoided here by merely formulating conditions in the form of inequalities which show that these exponents are negative (see beginning of Chapter 13).

If one introduces in equation (3.5): $u = u_0 + \eta$; $v = v_0 + \xi$, where u_0 and v_0 are periodic and η and ξ are perturbation functions, the variational equations are

$$\begin{aligned} \dot{\eta} &= (\mu\psi_u \cos t)\eta + (\mu\psi_v \cos t)\xi \\ \dot{\xi} &= (\mu\psi_u \sin t)\eta + (\mu\psi_v \sin t)\xi \end{aligned} \tag{3.15}$$

Since u^0 and v^0 satisfy the d.e. (3.5), the system (3.15) has periodic coefficients (Section 5, Chapter 5). If $\eta_1(t)$, $\xi_1(t)$ and $\eta_2(t)$, $\xi_2(t)$ are two sets of solutions forming a fundamental system, one can assume the initial conditions: $\eta_1(0) = 1$, $\xi_1(0) = 0$; $\eta_2(0) = 0$; $\xi_2(0) = 1$.

Since $\eta_1(t + 2\pi)$, $\xi_1(t + 2\pi)$,. . . are also solutions, one can write

$$\begin{aligned} \eta_1(t + 2\pi) &= a\eta_1(t) + b\eta_2(t); \qquad \xi_1(t + 2\pi) = a\xi_1(t) + b\xi_2(t) \\ \eta_2(t + 2\pi) &= c\eta_1(t) + d\eta_2(t); \qquad \xi_2(t + 2\pi) = c\xi_1(t) + d\xi_2(t) \end{aligned} \tag{3.16}$$

For $t = 0$ and, in view of the initial conditions, one has

$$\eta_1(2\pi) = a; \qquad \xi_1(2\pi) = b; \qquad \eta_2(2\pi) = c; \qquad \xi_2(2\pi) = d \tag{3.17}$$

Reducing (3.16) to the canonical form, one has $\eta_1(t + 2\pi) = S_1\eta_1(t)$, . . .; this is possible if S is a root of the characteristic equation

$$F(S) = \begin{vmatrix} a - S & b \\ c & d - S \end{vmatrix} = \begin{vmatrix} \eta_1(2\pi) - S & \xi_1(2\pi) \\ \eta_2(2\pi) & \xi_2(2\pi) - S \end{vmatrix} = 0 \tag{3.18}$$

that is:

$$F(S) = S^2 + pS + q = 0 \tag{3.19}$$

where

$$p = -[\eta_1(2\pi) + \xi_2(2\pi)] \quad \text{and} \quad q = [\eta_1(2\pi)\xi_2(2\pi) - \eta_2(2\pi)\xi_1(2\pi)] \tag{3.20}$$

If one sets $\mu = 0$, the variational system yields $d\eta/dt = 0$ and $d\xi/dt = 0$;

that is, η and ξ remain equal to their initial values. Hence, for $\mu = 0$, $p = -2$, $q = +1$, stability exists if the real parts of the characteristic exponents h_1 and h_2 are negative or, which is the same, $|e^{2\pi h_1}| < 1$ and $|e^{2\pi h_2}| < 1$. This requires that the moduli of the roots S_1 and S_2 should be less than *one*. On the other hand, (3.19) has roots with moduli less than unity only if

$$p > -2; \qquad 1 + p + q > 0 \tag{3.21}$$

This follows from equations: $1 + p + q = (S_1 + 1)(S_2 - 1)$; $p = -(S_1 + S_2)$.

The conditions of stability can be thus fulfilled only when the first nonvanishing derivatives of p and $p + q$ with respect to μ are positive for $\mu = 0$. If either $(dp/d\mu)$ or $[d(p + q)]/d\mu$ are zero, one has to investigate the sign of either $(d^2p/d\mu^2)$ or $[d^2(p + q)]/d\mu^2$ in order to ascertain the condition of stability. This is precisely the point we mentioned owing to which the actual calculation of the characteristic exponents is avoided.

In order to determine the sign of these derivatives, one replaces η and ξ in (3.15) by η_1 and ξ_1 (and, likewise by η_2, ξ_2) and integrates between 0 and 2π, which gives

$$\begin{aligned} \eta_1(2\pi) - \eta_1(0) &= \eta_1(2\pi) - 1 \\ &= \mu \int_0^{2\pi} \psi_u \eta_1 \cos t\,dt + \mu \int_0^{2\pi} \psi_v \xi_1 \cos t\,dt \\ \xi_1(2\pi) - \xi_1(0) &= \xi_1(2\pi) \\ &= \mu \int_0^{2\pi} \psi_u \eta_1 \sin t\,dt + \mu \int_0^{2\pi} \psi_v \xi_1 \sin t\,dt \end{aligned} \tag{3.22}$$

where ψ_u and ψ_v are partial derivatives of ψ in (3.5) or (3.15).

Differentiating these equations with respect to μ and, after that, passing to the limit $\mu = 0$, with the above initial conditions, one obtains

$$\begin{aligned} \left[\frac{d\eta_1(2\pi)}{d\mu}\right]_{\mu=0} &= \int_0^{2\pi} \psi_u \cos t\,dt = D_1(2\pi); \\ \left[\frac{d\xi_1(2\pi)}{d\mu}\right]_{\mu=0} &= \int_0^{2\pi} \psi_u \sin t\,dt = D_2(2\pi) \\ \left[\frac{d\eta_2(2\pi)}{d\mu}\right]_{\mu=0} &= \int_0^{2\pi} \psi_v \cos t\,dt = E_1(2\pi); \\ \left[\frac{d\xi_2(2\pi)}{d\mu}\right]_{\mu=0} &= \int_0^{2\pi} \psi_v \sin t\,dt = E_2(2\pi) \end{aligned} \tag{3.23}$$

One obtains thus

$$\left[\frac{dp}{d\mu}\right]_{\mu=0} = -[D_1(2\pi) + E_2(2\pi)]; \qquad \left[\frac{d(p+q)}{d\mu}\right]_{\mu=0} = 0;$$

$$\left[\frac{d^2(p+q)}{d\mu^2}\right]_{\mu=0} = 2\begin{vmatrix} D_1(2\pi) & E_1(2\pi) \\ D_2(2\pi) & E_2(2\pi) \end{vmatrix} \tag{3.24}$$

so that the conditions of stability (from (3.19)) are:

$$D_1(2\pi) + E_2(2\pi) < 0; \qquad \begin{vmatrix} D_1(2\pi) & E_1(2\pi) \\ D_2(2\pi) & E_2(2\pi) \end{vmatrix} > 0 \tag{3.25}$$

If one compares these conditions with the condition (3.14) of a subharmonic solution, it is seen that the second condition (3.25) is the same but more restrictive inasmuch as "different from zero" is replaced here by "greater than zero." But the first condition (3.25) of stability is an additional one not appearing in the conditions for the existence of subharmonic solution. Thus, in this case the stability conditions (3.25) are both necessary and sufficient not only for stability but also for the existence of a subharmonic solution.

As previously mentioned, the method does not require an actual calculation of the characteristic exponents but merely investigates the sign of these exponents in the neighborhood of $\mu = 0$ by means of derivatives with respect to μ, once the signs of p and q in the characteristic equation (3.19) have been ascertained directly from the variational equations.

4. Application to the resonance of the order $\frac{1}{2}$

Mandelstam and Papalexi investigated very thoroughly the fractional (or subharmonic) resonance of the order $\frac{1}{2}$, which is the most frequently encountered case. We abridge somewhat their presentation as it contains a considerable amount of data relative to the calculation of various constants of electronic circuits by which this theory was checked. The reader can easily find these details in the original paper; here we are interested mostly in conclusions. The nonlinear element of the system is here the anode current I_a of the electron tube which is represented by a polynomial.

$$I_a = f_1(x) = I_{a0} + \alpha x + \beta x^2 + \gamma x^3 + \delta x^4 + \varepsilon x^5 \tag{4.1}$$

The term δx^4 is disregarded in calculation. If the nonlinear characteristic is of a "hard" type, this polynomial is of the fifth degree; for a "soft" self-excitation it is sufficient to set $\varepsilon = 0$. We can consider first the latter case. The coefficient β characterizes the asymmetrical term (the one which does not change its sign when the grid voltage reverses); normally this term is very small and it is logical to consider it as a small parameter of the general theory. The reduction of the electronic data (which we omit) gives the following values

$$f(x,\dot{x}) = (k + 2x + \gamma_1 x^2 + \varepsilon_1 x^4)\dot{x} + (\xi/\beta)x \tag{4.2}$$

$$\gamma_1 = 3\gamma/\beta; \qquad \varepsilon_1 = 5\varepsilon/\beta; \qquad \xi = (\omega^2 - 4\omega_0^2)/4\omega_0^2;$$

$$\mu = \beta/(1 + \xi); \qquad k = [\alpha - 2\theta(1 + \xi)]/\beta \tag{4.3}$$

As to the coefficients in (4.1), they are:

$$\alpha = (\alpha_0 V_0)/I_0; \qquad \beta = (\beta_0 V_0^2)/I_0; \qquad \gamma = (\gamma_0 V_0^2)/I_0; \qquad \varepsilon = (\varepsilon_0 V_0^2)/I_0$$

the original (experimental) characteristic being: $i_a = i_{a0} + \alpha_0 V + \beta_0 V^2 + \gamma_0 V^3 + \varepsilon_0 V^5$, where V is the grid voltage, ω the frequency of the external excitation, ω_0 the frequency (free) of the system, and n the *order* of the resonance; thus, for the *exact* resonance $\xi = 0$.

For $n = 2$, the solution (3.6) can be written as

$$x(t) = a \sin t - b \cos t - \frac{\lambda_0}{3} \sin 2t \tag{4.4}$$

As was shown in the preceding section, for $a(\mu)$ and $b(\mu)$ one has to solve two equations

$$\int_0^{2\pi} \psi(a, b, \tau) \cos \tau d\tau = 0; \qquad \int_0^{2\pi} \psi(a, b, \tau) \sin \tau d\tau = 0 \tag{4.5}$$

In these integrals the function ψ is $f(x,\dot{x})$ where, instead of x and $\dot{x}$, one substitutes their expressions (4.4). We omit the intermediate calculations and give only their result in the form of two algebraic equations

$$a\left[k + \frac{\gamma_1}{4}\left(a^2 + b^2 + \frac{2}{9}\lambda_0^2\right)\right] = b\left(\frac{\lambda_0}{3} + \frac{\xi}{\beta}\right);$$

$$b\left[k + \frac{\gamma_1}{4}\left(a^2 + b^2 + \frac{2}{9}\lambda_0^2\right)\right] = a\left(\frac{\lambda_0}{3} - \frac{\xi}{\beta}\right) \tag{4.6}$$

From these equations one obtains

$$X^2 = a^2 + b^2 = -\frac{2}{9}\lambda_0^2 - \frac{4}{\gamma_1}[k \pm \sqrt{\lambda_0^2/9 - \xi^2/\beta^2}] \tag{4.7}$$

where $X = \sqrt{a^2 + b^2}$ is the amplitude of the subharmonic oscillation and

$$\tan \varphi = \frac{b}{a} = \left[k + \frac{\gamma_1}{4}\left(X^2 + \frac{2\lambda_0^2}{9}\right)\right] \Big/ \left(\frac{\lambda_0}{3} + \frac{\xi}{\beta}\right) \tag{4.8}$$

is its phase.

For the question of stability one has to form the criteria (3.25) by calculating the functions of Poincaré: $D_1(2\pi), \ldots$. Since, by (4.4) x and $\dot{x}$

are expressed in terms of a and b, these calculations are obvious and, omitting them, the criteria (3.25) reduce to the following inequalities:

$$k + \frac{\gamma_1}{2}\left(X^2 + \frac{\lambda_0^2}{9}\right) < 0; \qquad \gamma_1\left[k + \frac{\gamma_1}{4}\left(X^2 + \frac{2\lambda_0^2}{9}\right)\right] > 0 \quad (4.9)$$

If $X = 0$ (that is, the subharmonic oscillation is absent), (4.9) gives the stability condition of the small heteroperiodic oscillation which is of no special interest.

From (4.7) and (4.9) one can obtain another combination, namely:

$$\mp\sqrt{\lambda_0^2/9 - \xi^2/\beta^2} + \frac{\gamma_1}{4}X^2 < 0; \qquad \gamma_1(\mp\sqrt{\lambda_0^2/9 - \xi^2/\beta^2}) > 0 \quad (4.10)$$

For a "soft" characteristic γ_1 is generally negative, one has then:

$$X^2 = \frac{4}{|\gamma_1|}(k + \sqrt{\lambda_0^2/9 - \xi^2/\beta^2} + \gamma_1\lambda_0^2/18) \quad (4.11)$$

which gives X^2 as a function of λ_0 and ξ. The condition of reality gives:

$$\sqrt{\lambda_0^2/9 - \xi^2/\beta^2} > -(k + \gamma_1\lambda_0^2/18) \quad (4.12)$$

If one considers X^2 as a function of ξ, the values of ξ, for which X^2 vanishes, determine the interval in which the subharmonic resonance exists. In this interval $(\xi_1,\xi_2)X^2$ goes through a maximum; for these details we refer to the paper of Mandelstam and Papalexi and give here only the conclusions.

If the characteristic is soft (that is, $\delta = \varepsilon = 0$ in (4.1)), the curve of the resonance has the appearance shown in Fig. 19.1 for a given value of λ_0.

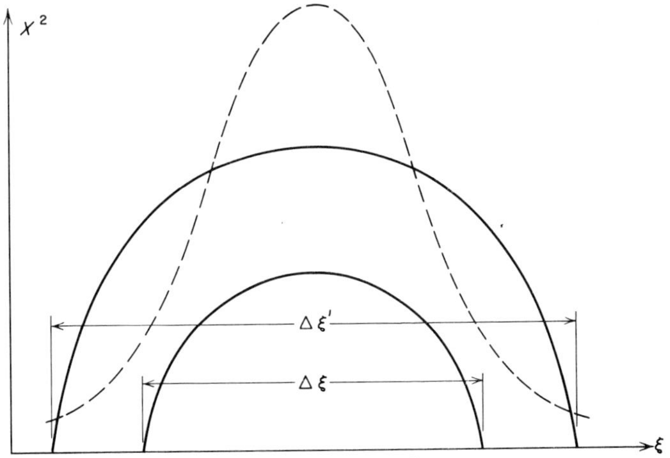

FIGURE 19.1

If λ_0 increases, the maximum of X^2 increases with the incidental increase of the interval of resonance; this increase, however, is not monotonic with λ_0, as in the linear case, inasmuch as, for a sufficiently large λ_0, the resonance oscillation suddenly disappears. The behavior of the resonance oscillation is thus entirely different from that of the linear resonance.

In the case of a "hard" characteristic calculations are similar but are much longer. In this case one has to use the full polynomial (4.2) which means that (4.4) has to be raised to the 4th power and then multiplied by $\dot{x}$ before the integration. Equations analogous to (4.6) in this case are:

$$\begin{aligned} a\left\{k + \frac{\gamma_1}{4}\left(X^2 + \frac{2}{9}\lambda_0^2\right) + \frac{\varepsilon_1}{8}\left[X^4 + \lambda_0^4/81 + \frac{\lambda_0^2}{9}(5X^2 + 4b^2)\right]\right\} &= b\left(\frac{\lambda_0}{3} + \frac{\xi}{\beta}\right) \\ b\left\{k + \frac{\gamma_1}{4}\left(X^2 + \frac{2}{9}\lambda_0^2\right) + \frac{\varepsilon_1}{8}\left[X^4 + \lambda_0^4/81 + \frac{\lambda_0^2}{9}(5X^2 + 4a^2)\right]\right\} &= a\left(\frac{\lambda_0}{3} - \frac{\xi}{\beta}\right) \end{aligned} \tag{4.13}$$

In order to simplify the discussion, it is assumed that $X^2 \gg (\lambda_0^2/9)$ on the basis of some experimental considerations. In such a case the preceding equations are simplified and are:

$$\begin{aligned} a\left(k + \frac{\gamma_1}{4}X^2 + \frac{\varepsilon_1}{8}X^4\right) &= \left(\frac{\lambda_0}{3} + \frac{\xi}{\beta}\right)b \\ b\left(k + \frac{\gamma_1}{4}X^2 + \frac{\varepsilon_1}{8}X^4\right) &= \left(\frac{\lambda_0}{3} - \frac{\xi}{\beta}\right)a \end{aligned} \tag{4.14}$$

which gives for X^2 the quadratic equation:

$$\frac{\varepsilon_1}{8}X^4 + \frac{\gamma_1}{4}X^2 + [k \pm \sqrt{\lambda_0^2/9 - \xi^2/\beta^2}] = 0 \tag{4.15}$$

Conditions of stability of the subharmonic solution are here:

$$k + \frac{\gamma_1}{4}X^2 + \frac{1}{8}\varepsilon_1 X^4 < 0; \quad (\varepsilon_1 X^2 + 2\gamma_1)(\varepsilon_1 X^4 + 2\gamma_1 X^2 + 8k) > 0 \tag{4.16}$$

Combining these conditions, one obtains the stable subharmonic solution

$$X^2 = -\gamma_1/\varepsilon_1 + \sqrt{\gamma_1^2/\varepsilon_1^2 - 8/\varepsilon_1(k + \sqrt{\lambda_0^2/9 - \xi^2/\beta^2})} \tag{4.17}$$

which is subject to the additional condition of the reality of X^2.

These long calculations result in the curve of resonance shown in Fig. 19.2. The only difference as compared to curves of Fig. 19.1 is that the resonance in this case appears and disappears abruptly for certain critical values of detuning; all other features remain the same as in the case previously studied.

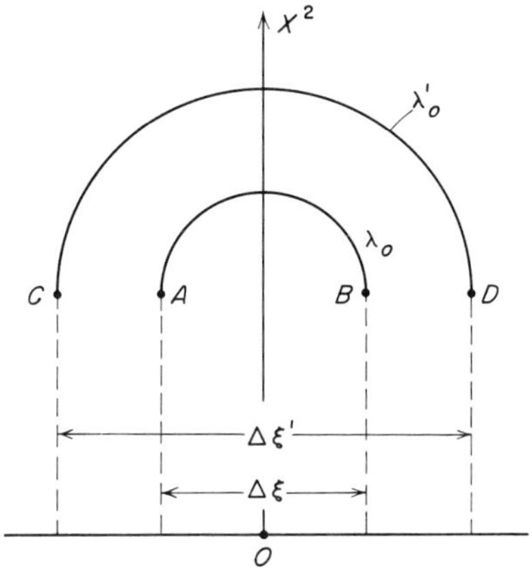

FIGURE 19.2

As far as known, most of this work was conducted in connection with the various resonances observed in electronic circuits inasmuch as the theory of these authors was particularly well adapted for this purpose. As we mentioned, the recent (1954) experimental work of C. Hayashi followed closely this theory for the investigation of subharmonic resonances of higher orders without going, however, into a detailed investigation of the various factors as did Mandelstam and Papalexi in their original research in connection with the resonance of the order $\frac{1}{2}$.

5. Subharmonic resonance by the stroboscopic method

We consider again a nonautonomous nearly linear d.e.

$$\ddot{x} + x + \mu f(x,\dot{x}) = \sin nt \tag{5.1}$$

and the corresponding solution for $\mu = 0$,

$$\begin{aligned} x_0(t) &= A \sin t + B \cos t + \frac{1}{1 - n^2} \sin nt \\ \dot{x}_0(t) = y_0(t) &= A \cos t - B \sin t + \frac{n}{1 - n^2} \cos nt \end{aligned} \tag{5.2}$$

Thus,

$$x_0(0) = B; \qquad y_0(0) = A + \frac{n}{1 - n^2} \tag{5.3}$$

Forming the conditions of periodicity (of Poincaré) and, taking as initial conditions, $x_1(0) = 0$; $y_1(0) = 0$, one finds easily

$$\begin{aligned} x_1(t) &= -\int_0^t \sin (t - \tau) f[x_0(\tau), y_0(\tau)] d\tau \\ y_1(t) &= -\int_0^t \cos (t - \tau) f[x_0(\tau), y_0(\tau)] d\tau \end{aligned} \tag{5.4}$$

as was explained in Chapter 18.

The problem, therefore, is to determine the constants A and B, so that the nonlinear solution be periodic with period 2π. In fact, since the external periodic excitation here has period $2\pi/n$ (n integer) and the system has a stable periodic solution with period 2π, clearly this is a *subharmonic solution of order n* (or of a *fractional order* $1/n$, which is merely a different term).

We shall investigate the first approximation only, assuming that μ is small. As was mentioned on several occasions, higher-order approximations do not add anything new from a qualitative point of view but merely add small corrections $0(\mu^2)$ at the cost of extremely long calculations which we shall avoid.

There are two distinct stages in this procedure: (1) *exact* subharmonic resonance; and (2) existence of a subharmonic solution in *the neighborhood* of the exact subharmonic resonance. The first problem is relatively simple inasmuch as all integrations in this case are to be carried out between 0 and 2π; whereas the second is more complicated since these integrations are between 0 and $T = 2\pi + \varepsilon$, ε being the detuning. We indicate here both possibilities but give examples of calculation only in connection with the exact resonance. The calculation of the *neighborhood* in which a given subharmonic resonance can maintain itself does not present any basic difficulty but merely requires very long and tedious calculations which, likely, can be better produced by computing machines if a detailed study of this question is deemed desirable. We note in passing that inasmuch as here we are not limited to any particular type of the problem (for example, self-excited, potentially self-excited, or passive systems), the discussion can be conducted in general terms and the procedure holds in all cases.

If a subharmonic solution (of order n) exists, it is to be periodic and we begin by formulating the condition of periodicity (Poincaré). In order to approach the problem in its full generality, we consider rather the case (2) of detuning. The problem is then: the external periodic excitation takes place with period $T = 2\pi/n$ but this period does not stand in an *exact* rational ratio with period 2π of the system, the relation being $T = 2\pi + \varepsilon$, ε being a small detuning. One must, therefore, show that the system may still exhibit a subharmonic oscillation with period 2π. It is clear that this requires two distinct "aspects" of the real phenomenon (see Introduction, Part III): (a) synchronization, (b) resonance. The stage (a) has been investigated in the preceding chapter where it was shown that, generally, if the detuning is small, the oscillation is synchronized with a nearest submultiple of the external frequency. If one assumes this, there remains still the question as to whether the amplitude can maintain itself under this condition of "detuning" (stage (b)); we concentrate, therefore, our attention only on this stage (b) of the phenomenon.

In the first approximation $x(t) = x_0(t) + \mu x_1(t)$; $y(t) = \dot{x}(t) = y_0(t) + \mu y_1(t)$ so that we have

$$x(T) = x_0(T) + \mu x_1(T) = A \sin T + B \cos T + \frac{1}{1 - n^2} \sin nT + \mu x_1(T)$$

$$y(T) = y_0(T) + \mu y_1(T) = A \cos T - B \sin T + \frac{n}{1 - n^2} \cos nT + \mu y_1(T) \tag{5.5}$$

and $x(0) = B$; $y(0) = A + n/(1 - n^2)$.

The first step in the stroboscopic method is to introduce the difference equations which are here

$$x(T) - x(0) = \Delta\xi = A \sin T + B (\cos T - 1) + \frac{1}{1 - n^2} \sin nT + \mu x_1(T)$$

$$y(T) - y(0) = \Delta\eta = A (\cos T - 1) - B \sin T + \frac{n}{1 - n^2} (\cos nT - 1) + \mu y_1(T) \tag{5.6}$$

where $x_1(T)$ and $y_1(T)$ are the integrals (5.4) taken between 0 and T; for simplicity, we set $x_1(T) = M(A,B)$; $y_1(T) = N(A,B)$, since, obviously these integrals are functions of A and B.

We can divide (5.6) by $\mu T = \Delta\tau$ resulting in

$$\frac{\Delta\xi}{\Delta\tau} = \frac{A \sin T}{\mu T} + \frac{B (\cos T - 1)}{\mu T} + \frac{n}{1 - n^2} \frac{\sin nT}{\mu T n} + \frac{1}{T} M(B,A)$$

$$\frac{\Delta\eta}{\Delta\tau} = -\frac{B \sin T}{\mu T} + \frac{A (\cos T - 1)}{\mu T} + \frac{n}{1 - n^2} \frac{\cos nT - 1}{\mu T} + \frac{1}{T} N(B,A) \tag{5.7}$$

When the system traverses the sequence of its periods, in each period the constants A and B undergo a change and, in order to emphasize this circumstance, we set

$$x(0) = B = \xi; \qquad y(0) = A = \eta - \frac{n}{1 - n^2} \tag{5.8}$$

The usual passage to the limit ($\Delta\tau \to d\tau$) from the difference equations to the stroboscopic system (Chapter 16) results in the d.e.

$$\frac{d\xi}{d\tau} = \xi\frac{\cos T - 1}{\mu T} + \eta\frac{\sin T}{\mu T} + \frac{n}{1-n^2}\left(\frac{\sin nT}{\mu nT} - \frac{\sin T}{\mu T}\right) + P(\xi,\eta)$$

$$\frac{d\eta}{d\tau} = -\xi\frac{\sin T}{\mu T} + \eta\frac{\cos T - 1}{\mu T} + \frac{n}{1-n^2} \times \left(\frac{\cos nT - 1}{\mu T} - \frac{\cos T - 1}{\mu T}\right) + Q(\xi,\eta) \tag{5.9}$$

where

$$P(\xi,\eta) = \frac{1}{T}M\left(\xi, \eta - \frac{n}{1-n^2}\right); \qquad Q(\xi,\eta) = \frac{1}{T}N\left(\xi, \eta - \frac{n}{1-n^2}\right).$$

If the detuning ε is small, $\sin T = 0(\mu)$; $(\cos T - 1) = 0(\mu^2)$; setting $(\sin T)/\mu T = p$, the system (5.9) can be written as

$$\frac{d\xi}{d\tau} = p\eta + P(\xi,\eta); \qquad \frac{d\eta}{d\tau} = -p\xi + Q(\xi,\eta) \tag{5.10}$$

If, however, one considers that the detuning is zero, $T = 2\pi$ and the preceding equations become

$$\frac{d\xi}{d\tau} = \frac{1}{2\pi}M\left(\xi, \eta - \frac{n}{1-n^2}\right); \quad \frac{d\eta}{d\tau} = \frac{1}{2\pi}N\left(\xi, \eta - \frac{n}{1-n^2}\right) \tag{5.11}$$

The condition of the stationary state requires

$$M\left(\xi_0, \eta_0 - \frac{n}{1-n^2}\right) = 0; \qquad N\left(\xi_0, \eta_0 - \frac{n}{1-n^2}\right) = 0 \tag{5.12}$$

The d.e. (5.11) are much simpler than (5.9), and in the following section we shall give examples of application of this method. In Section 7 the general procedure (equation (5.9)) is indicated.

6. Applications of the stroboscopic method

We propose to explore the possibility of a subharmonic resonance of the order $\frac{1}{2}$ in connection with the d.e.

$$\ddot{\varphi} - \mu(\alpha - \beta\varphi^2)\dot{\varphi} + \varphi = e\sin 2t; \qquad \alpha > 0, \qquad \beta > 0 \tag{6.1}$$

Setting $\varphi = ex$, this d.e. becomes

$$\ddot{x} - \mu(\alpha - \beta e^2x^2)\dot{x} + x = \sin 2t \tag{6.2}$$

where, as usual, μ is a small parameter.

As a zero-order solution $x_0(t)$ we take the solution of the d.e.: $\ddot{x}_0 + x_0 = \sin 2t$, which gives

$$\begin{aligned} x_0(t) &= A \sin t + B \cos t - \tfrac{1}{3} \sin 2t \\ y_0(t) &= A \cos t - B \sin t - \tfrac{2}{3} \cos 2t \end{aligned} \tag{6.3}$$

with the initial conditions: $x_0(0) = B$; $y_0(0) = A - \frac{2}{3}$; we assume also that $x_1(0) = y_1(0) = 0$, so that the first-order corrective terms x_1 and y_1 are

$$\begin{aligned} x_1(t) &= \int_0^t \sin (t - \tau)(\alpha - \beta e^2 x_0{}^2)\dot{x}_0 d\tau \\ y_1(t) &= \int_0^t \cos (t - \tau)(\alpha - \beta e^2 x_0{}^2)\dot{x}_0 d\tau \end{aligned} \tag{6.4}$$

We consider here the exact subharmonic resonance (detuning zero) so that integration in (6.4) is between 0 and 2π.

The conditions of periodicity of the subharmonic solution is

$$\begin{aligned} x(2\pi) - x(0) &= x_0(2\pi) - x_0(0) + \mu x_1(2\pi) = 0 \\ y(2\pi) - y(0) &= y_0(2\pi) - y_0(0) + \mu y_1(2\pi) = 0 \end{aligned} \tag{6.5}$$

and since x_0 and y_0 are periodic with period 2π, the conditions of the stationary state reduce merely to $x_1(2\pi) = 0$, $y_1(2\pi) = 0$. We consider, however, first *the approach* to such a state; this yields (after the passage from difference equations to the stroboscopic d.e.) the system

$$\begin{aligned} \frac{d\xi}{d\tau} &= -\frac{1}{2\pi} \int_0^{2\pi} \sin \tau(\alpha - \beta e^2 x_0{}^2)\dot{x}_0 d\tau \\ \frac{d\eta}{d\tau} &= +\frac{1}{2\pi} \int_0^{2\pi} \cos \tau(\alpha - \beta e^2 x_0{}^2)\dot{x}_0 d\tau \end{aligned} \tag{6.6}$$

It is convenient to integrate these integrals by parts observing that

$$(\alpha - \beta e^2 x_0{}^2)\dot{x}_0 = \frac{d}{d\tau}\left(\alpha x_0 - \frac{1}{3}\beta e^2 x_0{}^3\right)$$

As the integrated parts vanish on both limits, the stroboscopic system takes the form

$$\frac{d\xi}{d\tau} = \frac{1}{2\pi}\int_0^{2\pi} \cos \tau\, g(x_0) d\tau; \qquad \frac{d\eta}{d\tau} = \frac{1}{2\pi}\int_0^{2\pi} \sin \tau\, g(x_0) d\tau \tag{6.7}$$

where $g(x_0) = \alpha x_0 - \frac{1}{3}\beta e^2 x_0{}^3$, $x_0(\tau)$ being given by (6.3).

A simple, although somewhat long, calculation gives

$$\int_0^{2\pi} g(x_0)\cos\tau d\tau = \pi\{\alpha B - \gamma[\tfrac{1}{4}(B^3 + A^2B) + \tfrac{1}{18}B]\}$$
$$\gamma = \beta e^2 \quad (6.8)$$
$$\int_0^{2\pi} g(x_0)\sin\tau d\tau = \pi\{\alpha A - \gamma[\tfrac{1}{4}(A^3 + AB^2) + \tfrac{1}{18}A]\}$$

One can conduct calculation either in variables ξ and η or in B and A, since, in view of the initial conditions, $x_0(0) = B = \xi; y_0(0) = A - \frac{2}{3} = \eta$, so that $d\xi/d\tau = dB/d\tau$ and $d\eta/d\tau = dA/d\tau$. We take, for instance, the variables B and A, in which case the system (6.7) becomes

$$\begin{aligned}\frac{dB}{d\tau} &= \frac{1}{2}B\left\{\alpha - \gamma\left[\frac{1}{4}(A^2 + B^2) + \frac{1}{18}\right]\right\}\\ \frac{dA}{d\tau} &= \frac{1}{2}A\left\{\alpha - \gamma\left[\frac{1}{4}(A^2 + B^2) + \frac{1}{18}\right]\right\}\end{aligned} \quad (6.9)$$

and one sees at once that the polar coordinates simplify the matter. We set $\rho = A^2 + B^2$; $\psi = \arctan(A/B)$. The simple way of doing this is to introduce two combinations of equations (6.9), as we did previously, on a number of occasions, viz.:

$$B\frac{dB}{d\tau} + A\frac{dA}{d\tau} = \frac{1}{2}\frac{d\rho}{d\tau} \quad \text{and} \quad B\frac{dA}{d\tau} - A\frac{dB}{d\tau} = \rho\frac{d\psi}{d\tau}$$

The first combination yields

$$\frac{d\rho}{d\tau} = \rho\left[\left(\alpha - \frac{1}{18}\beta e^2\right) - \frac{1}{4}\beta e^2\rho\right] = \Phi(\rho) \quad (6.10)$$

As to the second, it gives merely $d\psi/d\tau = 0$.†

The interest thus centers on the d.e. (6.10) which shows that there exists a stationary solution

$$\rho_0 = \frac{4}{\beta e^2}\left(\alpha - \frac{1}{18}\beta e^2\right) \quad (6.11)$$

provided $\alpha \geqslant \frac{1}{18}\beta e^2$, since ρ is essentially positive. Thus the subharmonic

† This circumstance is due to the fact that we have taken the last term on the left side of (6.1) as φ. In a more general case when, instead of φ, one would have $g(\varphi)$, the second stroboscopic equation $d\psi/d\tau$ would not be zero identically in the first approximation (see Chapter 22).

resonance *exists* as long as the amplitude e of the external excitation is *below* a certain limit given by inequality

$$e \leq \sqrt{\frac{18\alpha}{\beta}} \tag{6.12}$$

It is necessary to show also that ρ_0 is *stable*. In this case the condition of stability is very simple and is expressed by the inequality (Section 6, Chapter 14)

$$\Phi'_\rho(\rho_0) < 0 \tag{6.13}$$

If one differentiates the right-hand side of (6.10) with respect to ρ and substitutes for ρ its value ρ_0 given by (6.11), one finds easily that the condition (6.13) is fulfilled.

Thus the subharmonic resonance of order $\frac{1}{2}$ exists in this case as long as the amplitude e is not too high (condition (6.12)). We find thus the same result which Mandelstam and Papalexi obtained by a different method.

In the form (6.1) or (6.2) with $\alpha > 0$ and $\beta > 0$, the d.e is the van der Pol equation, but it is interesting to explore the different cases when the condition $\alpha > 0$, $\beta > 0$ is not fulfilled; in such a case this will not be a van der Pol equation but some other equation whose physical nature can easily be ascertained.

In the first place, if both $\alpha < 0$ and $\beta < 0$, this merely amounts to the change of sign before μ, and we have

$$\ddot{x} + \mu(|\alpha| - |\beta|e^2x^2)\dot{x} + x = \sin 2t \tag{6.14}$$

This is a kind of an "inverted" van der Pol equation; considered as an oscillator, this is an oscillator with positive damping for small x and a negative damping for larger x. In order to explore whether the subharmonic resonance in such a system is possible we have to replace α and β in the preceding formulas by $-|\alpha|$ and $-|\beta|$. We have thus

$$\rho_0 = \frac{4}{|\beta|e^2}\left(|\alpha| - \frac{1}{18}|\beta|e^2\right) \tag{6.15}$$

the criterion (6.13) remains the same, but the condition of stability is not fulfilled.

In the two remaining cases when either $\alpha > 0$, $\beta < 0$ or $\alpha < 0$, $\beta > 0$, the same argument shows that the subharmonic amplitude does not exist.

For the subharmonic resonance of the order $\frac{1}{3}$ (for the same d.e. (5.1))

$$\begin{aligned} x_0 &= A\sin t + B\cos t - \tfrac{1}{8}\sin 3t \\ y_0 &= A\cos t - B\sin t - \tfrac{3}{8}\cos 3t \end{aligned} \tag{6.16}$$

and the calculation of $\int_0^{2\pi} f(x_0) \cos \tau d\tau$ and $\int_0^{2\pi} f(x_0) \sin \tau d\tau$ is to be carried out with the new value of x_0.

We shall not reproduce this calculation here but merely note that, unlike the symmetrical system (6.9), one obtains here a rather unsymmetrical system.

$$\begin{aligned} \frac{dB}{d\tau} &= \frac{1}{2} B\alpha - \frac{1}{3} \beta e^2 \left[\frac{3}{4}(A^2 + B^2) - \frac{3}{16} A + \frac{3}{256}\right] \\ \frac{dA}{d\tau} &= \frac{1}{2} \alpha A - \frac{1}{6} \beta e^2 \left[\frac{21}{32}(A^2 + B^2) + \frac{3}{256} A\right] \end{aligned} \tag{6.17}$$

which precludes the use of the variable ρ which was helpful in the case of the system (6.9). The determination of the roots $A = A_0$, $B = B_0$ common to both equations (6.17) is here more difficult inasmuch as it hinges on the algebraic solution of two equations of a rather complicated type which are obtained by equating to zero the right-hand sides of equations (6.17).

This is a typical kind of difficulty which one encounters in the determination of the subharmonic resonance by this method, but similar difficulties arise also in other methods. The difficulty is not in the theoretical fundamentals of these methods but rather in carrying out the actual calculations when it becomes necessary to determine roots $A = A_0$ and $B = B_0$ common to two algebraic equations. If this reduces to a simple case which we encountered in the case of the subharmonic resonance of the order $\frac{1}{2}$, the answer is given at once, but in the case of the resonance of the order $\frac{1}{3}$ the algebraic problem is more difficult and a considerable amount of work is to be done before one is in a position to answer with certainty as to whether this particular resonance exists or not.

The preceding examples relate to the subharmonic solutions of the van der Pol equation which represents by its very nature an *energy absorbing* physical oscillatory system.

The method applies, however, to purely *passive* (energy dissipating systems). As an example we investigate the existence of a subharmonic resonance $\frac{1}{3}$ in the case of the Duffing equation by considering the d.e. of the form:

$$\ddot{\xi} + a\dot{\xi} + c\xi^3 + \xi = e \sin 3t \tag{6.18}$$

which, by means of the change of variable $\xi = ex$ is transformed into the d.e.

$$\ddot{x} + x + \mu(\alpha\dot{x} + \gamma e^2 x^3) = \sin 3t \tag{6.19}$$

where $a = \mu\alpha$ and $c = \mu\gamma$.

Following the same procedure one has

$$x_0(t) = A \sin t + B \cos t - \tfrac{1}{8} \sin 3t$$
$$y_0(t) = A \cos t - B \sin t - \tfrac{3}{8} \cos 3t \qquad (6.20)$$

Whence $\ddot{x}_1 + x_1 = -(\alpha\dot{x}_0 + \gamma e^2 x_0{}^3) = -f(x_0,\dot{x}_0)$. The conditions of periodicity are here

$$x_1(2\pi) - x_1(0) = \Delta x_1(0) = \mu \int_0^{2\pi} \sin \tau f(x_0,\dot{x}_0)d\tau = 0$$
$$y_1(2\pi) - y_1(0) = \Delta y_1(0) = -\mu \int_0^{2\pi} \cos \tau f(x_0,\dot{x}_0)d\tau = 0 \qquad (6.21)$$

The calculation reduces to the evaluation of integrals $\int_0^{2\pi} \sin \tau f(x_0,\dot{x}_0)d\tau$ and $\int_0^{2\pi} \cos \tau f(x_0,\dot{x}_0)d\tau$ which are to be equated to zero. We omit this calculation and give the result. The ultimate algebraic system for the determination of A_0 and B_0 is:

$$\alpha A + \tfrac{3}{4}\gamma e^2[B(A^2 + B^2) - \tfrac{1}{4}AB + \tfrac{1}{32}B] = 0$$
$$-\alpha B + \tfrac{3}{4}\gamma e^2[(A - \tfrac{1}{8})(A^2 + B^2) + \tfrac{1}{32}A] = 0 \qquad (6.22)$$

The real roots $A = A_0$ and $B = B_0$ common to these two equations give the stationary state $\dfrac{dx(0)}{d\tau} = \dfrac{d\xi}{d\tau} = 0$; $\dfrac{dy(0)}{d\tau} = \dfrac{d\eta}{d\tau} = 0$ of the stroboscopic d.e. The amplitude of the subharmonic oscillation is then: $\sqrt{A_0{}^2 + B_0{}^2}$ and its phase: $\varphi_0 = \arctan A_0/B_0$.

In this case no simple reduction is possible and the difficulty is again of a purely algebraic character. In the first place one notes that neither $A = 0$, $B \neq 0$ nor $A \neq 0$, $B = 0$ satisfy the system (6.22); the case $A = B = 0$ is to be rejected anyhow.

There remains thus the case $A \neq 0$, $B \neq 0$. One can set $A = \lambda B$ where λ is any finite number; it appears here as an undetermined parameter.

If one substitutes $A = \lambda B$ in both equations and cancels B throughout, one obtains a system of two quadratic equations,

$$a_0 B^2 + a_1 B + a_2 = 0; \qquad b_0 B^2 + b_1 B + b_2 = 0 \qquad (6.23)$$

where the coefficients $a_0, \ldots, b$ are functions of λ. By expressing that these two equations have one common root by one of the existing procedures (for example, Sylvester, Cauchy, etc.), this determines the value of $\lambda = \lambda_0$. It is sufficient to solve one of the equations for $B = B_0$ in order to obtain also $A_0 = B_0\lambda_0$. The procedure is long and tedious but it is straightforward.

As regards the equations of stability, it does not present any special difficulty, once the stroboscopic system is written in the form

$$\begin{aligned}\frac{dB}{d\tau} &= \left\{-\alpha B + e^2\left[\frac{3}{4}A^3 + \frac{3}{4}AB^2 + \frac{3}{32}(A^2 + B^2) + \frac{3}{256}A\right]\right\}\\ \frac{dA}{d\tau} &= \left\{\alpha A + e^2\left[\frac{3}{4}B^3 + \frac{3}{4}A^2B - \frac{3}{16}AB + \frac{3}{256}B\right]\right\}\end{aligned} \tag{6.24}$$

The principal amount of work is, however, in the actual numerical calculation of roots common to two algebraic equations.

7. Zones of subharmonic resonance

The phenomena of subharmonic resonance are observed not only for the *exact* fractional ratios, say, $\frac{1}{2}$, $\frac{1}{3}$,. . . between the free ($e = 0$) frequency of the system and that of the external periodic excitation but also for a certain *detuning* of the former. This circumstance has been mentioned already in connection with the theory of Mandelstam and Papalexi where the detuning is used as a parameter against which the resonance curves are plotted.

In the general case relative *to the neighborhood* of an exact subharmonic resonance the two phenomena take place simultaneously, viz.: there is the entrainment of frequency to the exact subharmonic value and there are also modifications in the curve of resonance which exists only for a certain interval of detuning.

As the phenomenon of synchronization has been investigated in the preceding chapter, we concentrate our attention here on the resonance part of the phenomenon. This requires the investigation of the system (5.9) instead of (5.11).

In spite of the asymptotic character of the differential system (5.9) obtained under the assumption that $\sin T \simeq 0(\mu)$; $\cos T - 1 \simeq 0(\mu^2)$, the calculations are considerably more complicated than when $T = 2\pi$.

In the first place, in the integration by parts (now between the limits 0 and T, and not between 0 and 2π), the integrated part does not vanish and for $M(A,B)$ and $N(A,B)$ we have now the expressions:

$$\begin{aligned}M(A,B) &= -\sin T(\alpha B - \tfrac{1}{3}\beta B^3)\Big|_0^T + \int_0^T \cos(T-\tau)(\alpha x_0 - \tfrac{1}{3}\beta x_0^3)d\tau\\ N(A,B) &= -(\cos T - 1)(\alpha B - \tfrac{1}{3}\beta B^3)\Big|_0^T + \\ &\qquad \int_0^T \sin(T-\tau)(\alpha x_0 - \tfrac{1}{3}\beta x_0^3)d\tau\end{aligned} \tag{7.1}$$

In the case of the exact resonance, all integrations were between 0 and 2π, which accounted for vanishing of a considerable number of terms; now none of these terms vanishes.

Thus, for instance, the term $\int_0^T x_0^3 \cos \tau d\tau$ gives rise to the polynomial of the form

$$A^3a_1 + B^3a_2 + 3AB^2a_3 + 3A^2Ba_4 + 3mA^2a_5 + 3mB^2a_6 + 3mABa_7 + 3m^2Aa_8 + 3m^2Ba_9 + m^3a_{10} \quad (7.2)$$

where $m = (1/(1 - n^2))$ and the coefficients a_i, $i = 1,\ldots, 10$, are

$$a_1 = \tfrac{1}{4}\Big|\sin^4 t\Big|_0^T$$

$$a_2 = \Big|\tfrac{3}{8}t + \tfrac{3}{16}\sin 2t + \tfrac{1}{4}\sin t \cos^3 t\Big|_0^T$$

$$a_3 = -\tfrac{1}{4}\Big|\cos^4 t\Big|_0^T$$

$$a_4 = \Big|-\tfrac{1}{32}\sin 4t + \tfrac{1}{8}t\Big|_0^T$$

$$a_5 = \int_0^T [\tfrac{1}{8}\sin(n+1)t + \tfrac{1}{8}\sin(n-1)t - \tfrac{1}{8}\sin(n+3)t - \tfrac{1}{8}\sin(n-3)t]dt$$

$$a_6 = \int_0^T [\tfrac{1}{8}\sin(n+3)t + \tfrac{1}{8}\sin(n-3)t + \tfrac{3}{8}\sin(n+1)t + \tfrac{3}{8}\sin(n-1)t]dt \quad (7.3)$$

$$a_7 = \int_0^T [\tfrac{1}{4}\cos(n-3)t - \tfrac{1}{4}\cos(n+3)t + \tfrac{1}{4}\cos(n-1)t - \tfrac{1}{4}\cos(n+1)t]dt$$

$$a_8 = \int_0^T [\tfrac{1}{4}\sin 2t - \tfrac{1}{8}\sin 2(n+1)t + \tfrac{1}{8}\sin 2(n-1)t]dt$$

$$a_9 = \int_0^T [\tfrac{1}{4} - \tfrac{1}{4}\cos 2nt + \tfrac{1}{4}\cos 2t - \tfrac{1}{8}\cos 2(n-1)t - \tfrac{1}{8}\cos 2(n+1)t]dt$$

$$a_{10} = \int_0^T [\tfrac{3}{8}\sin(n+1)t + \tfrac{3}{8}\sin(n-1)t - \tfrac{1}{8}\sin(3n+1)t - \tfrac{1}{8}\sin(3n-1)t]dt$$

There is still another similar polynomial for $\int_0^T x_0^3 \sin \tau d\tau$ with analogous coefficients, say $b_1, b_2, \ldots, b_{10}$. We consider here the general case but for particular cases, one must set $n = 1, 2, 3, \ldots$ and take the corresponding m and n.

It is seen that the problem reduces to the determination of real roots $A = A_0$ and $B = B_0$ common to two algebraic equations with 10 terms whose coefficients $a_1, \ldots, a_{10}, b_1, \ldots, b_{10}$ are ultimately functions of T, that is, of the detuning.

In this manner a region of detuning ε (that is, of T in which the right-hand terms of (5.9) have real roots $A = A_0$, $B = B_0$ common to both equations) is that in which a given subharmonic resonance can exist; if these roots cease to be common to the two equations for some value of detuning, this indicates the limit at which the subharmonic resonance can maintain itself. To this is added, of course, the investigation of stability.

The problem is very complicated if attempted by ordinary procedure but, presumably, is within the reach of computing machines if such a thorough investigation of the subharmonic resonance is ever to be attempted.

8. Comparison of different methods

The preceding survey of the principal methods shows that the phenomena of nonlinear resonance exhibit a striking contrast between the simplicity of their physical nature considered in terms of subharmonics and very long calculations required in order to obtain quantitative results. This probably explains a scarcity of information regarding resonances of orders other than the order of $\frac{1}{2}$ which is the only one relatively well explored.

The experimental work, on the other hand, has progressed considerably in recent years particularly in the interesting researches of Ch. Hayashi[3] (electrical oscillations) and K. Ludeke[7] (mechanical oscillations). There is also a considerable experimental evidence on this subject published earlier by the Russian school of physicists under the guidance of L. Mandelstam and N. Papalexi (1932–1940) and more recently in a number of other publications.

As it is impossible to enter into a detailed survey of this experimental work and its relation to the various theories, we shall endeavor merely to correlate the various methods.

The Mandelstam–Papalexi (M.P.) theory, in view of its adaptation to

[3] See footnote [3], page 461.
[7] K. Ludeke, *J. Appl. Phys.*, July, 1946.

the electronic circuits, continues to play an important role in studies of the subharmonic resonance in electron-tube circuits. Very often investigators are more concerned with the question of *existence* of the exact resonance than with a more tedious calculation of the effect of detuning in the range in which the resonance persists.

As the question of nonlinear resonance was also investigated in Chapter 10 (in connection with Malkin's work) and in Chapters 14 and 15 (in connection with the asymptotic methods of Krylov–Bogoliubov–Mitropolsky) where it is organically welded with the general theory, it is useful to correlate these various developments.

In the M.P. method the procedure follows closely the theory of Poincaré (Chapter 10) as was mentioned in Section 1. The problem of the existence of a subharmonic solution is established from the identification of the results with the series solution which determines the functions $C_1(2\pi)$, $C_2(2\pi)$, $D_1(2\pi)$, $D_2(2\pi)$, $E_1(2\pi)$, and $E_2(2\pi)$ of the Poincaré theory.

The problem of stability reduces to the determination of the sign of the characteristic exponents since the authors operate directly with the d.e. in their original nonautonomous form. The calculation of these exponents is avoided, however; and, instead, it is shown that, for μ small, it is possible to formulate conditions of stability in the form of inequalities, which guarantees the existence of negative characteristic exponents. The rest of the theory is merely its adaptation to the electron-tube circuits by which it has been checked experimentally.

The method is thus closely connected to the subject of Chapters 10 and 13 and the complication is rather in the effective calculations of integrals (4.6) and in the discussion of results so obtained, particularly when the nonlinear function is approximated by a polynomial of the 5th degree. These difficulties of carrying through long and tedious calculations exist, however, in other methods.

In the extension of the stroboscopic method outlined in Section 5, the beginning is the same as in the M.P. method, viz.: the generating solution is sought in the form of the solution of the linear nonhomogeneous d.e. (3.1), but, after this, the method consists in formulating the conditions of periodicity (of Poincaré) per one period (2π in the case of the exact subharmonic resonance and $2\pi + \varepsilon$ in the case when a detuning ε exists). This leads finally to the stroboscopic procedure in which the constants A and B of the linear solution appear as variables of the transformation; at the limit $\Delta\tau \rightarrow d\tau$, these variables become continuous which leads to the stroboscopic d.e.

The problems of existence and stability reduce thus to analogous problems relative to the singular point of the stroboscopic system. In some cases when it is possible to introduce the variable $\rho = x^2 + \dot{x}^2$ in the final result, the determination of the stationary state is considerably simplified as in the case of the d.e. (6.1), but this is rather an exception than the rule and, in general, the principal difficulty is in the calculation of stationary values appearing here as real positive roots common to two algebraic equations. The problem of stability, on the contrary, is relatively simple, since the discussion of the variational equations does not present any difficulty once the stationary values $A = A_0$ and $B = B_0$ are determined.

Thus, both the M.P. method and the stroboscopic method start in a similar manner. The practical difficulties in obtaining quantitative results are, however, different in these two methods. In the M.P. method the existence of a subharmonic solution is relatively simple, but the question of stability is more difficult inasmuch as one operates with a nonautonomous system. In the stroboscopic method, the problem of stability is simple enough but the major part of calculations concerns the determination of stationary values $A = A_0$ and $B = B_0$, as was just mentioned.

In the K.B.M.† approach to the theory of resonance (Chapters 14 and 15), the solution is sought in the form (2.3), Chapter 14, where a and ψ in the nonresonance case are determined by the d.e. (2.4), Chapter 14, namely:

$$\begin{aligned} \dot{a} &= \mu A^{(1)}(a) + \mu^2 A^{(2)}(a) + \dots \\ \dot{\psi} &= \omega + \mu B^{(1)}(a) + \mu^2 B^{(2)}(a) + \dots \end{aligned} \tag{8.1}$$

In the resonance case for the same form of the solution, a and ψ are determined by the d.e. (4.6), Chapter 15, namely:

$$\begin{aligned} \dot{a} &= \mu A^{(1)}(a,\theta) + \mu^2 A^{(2)}(a,\theta) + \dots \\ \dot{\psi} &= \frac{p}{q}\nu + \mu B^{(1)}(a,\theta) + \mu^2 B^{(2)}(a,\theta) + \dots \end{aligned} \tag{8.2}$$

where $\theta = \psi - (p/q)\nu t$ may be regarded as phase difference between the autoperiodic and heteroperiodic oscillations. It is clear that, in the nonresonance case, θ has no physical significance since the two frequencies are different; but, in the resonance case, the variable θ acquires a physical meaning inasmuch as, for the stationary condition, it approaches also a stationary value which implicitly amounts to assuming the existence of a synchronized state in the presence of a certain detuning. This is the reason why the variable θ appears in the d.e. (8.2) and not in (8.1). This circumstance complicates somewhat the formal argument as was mentioned in Chapter 15. Moreover in the resonance case the variables do not

† Krylov-Bogoliubov-Mitropolsky.

separate in the d.e. of the K.B.M. theory, but it is possible to determine the stationary values since the ultimate system is still in Poincaré's form, viz.:

$$\dot{a} = f(a,\theta); \qquad \dot{\theta} = g(a,\theta) \tag{8.3}$$

and the stationary values are determined as real roots common to two algebraic equations

$$f(a_0,\theta_0) = g(a_0,\theta_0) = 0 \tag{8.4}$$

Thus, the K.B.M. procedure in its final stage becomes analogous to the stroboscopic method, although the intermediate procedure is entirely different in both cases.

As to the procedure, the stroboscopic method occupies, in some respects, an intermediate position between the M.P. and the K.B.M. methods. With the former it has common starting points; and with the latter it has a common algebraic problem at the end of calculations, leading to the determination of the stationary state.

In the K.B.M. method and in the stroboscopic method, the practical difficulty is in the determination of roots common to both algebraic equations ((8.4) or in similar equations of the stroboscopic system). This comparison concerns, however, only the calculating procedure in the two methods but, in reality, the M.P. method and the stroboscopic method are similar and are derived directly from the theory of Poincaré, however, the stroboscopic method transforms the initial nonautonomous system to the autonomous form, while the M.P. method operates directly with the nonautonomous system.

We have made this comparison in order to emphasize the evolution of the theory, as well as its practical implications, with a view to obtaining quantitative results concerning resonance.

Chapter 20

PARAMETRIC EXCITATION

1. Introductory remarks

It has been known for a long time that, if a parameter of an oscillatory system is made to vary at frequency $2f$, f being the free frequency of the system, the system begins to oscillate with the frequency f. Lord Rayleigh produced an experiment[1] in which a stretched string was attached at one end to a prong of a tuning fork capable of vibrating in the direction of the string. It was observed that when the fork vibrated with frequency $2f$, lateral vibrations of the string were built up with frequency f.

In later years Brillouin[2] and Poincaré[3] studied this effect in electric circuits. Similar experiments were produced in 1934 by Mandelstam and Papalexi with a specially designed oscillating circuit which they call "parametric generator".[4] These latter experiments were particularly valuable in throwing light on the physical nature of this phenomenon, as will be mentioned below.

The term *parametric excitation*, or action, is generally used in connection with these phenomena to emphasize the fact that the effect is due to the variation of the parameter with a double frequency.

The parametric effect is apparently a frequent occurrence in physics, but its systematic study is relatively recent. The old problem of pendulum with a variable length belongs to this class, although it is generally treated in theoretical mechanics on the basis of the theory of the variable constraint which is difficult to generalize for electrical phenomena. Finally, the operation of a swing, which consists in a periodic raising and lowering of

[1] Lord Rayleigh, *Phil. Mag.*, April, 1883.
[2] L. Brillouin, *Eclairage Electrique*, April, 1897.
[3] H. Poincaré, *ibid.*, March, 1907.
[4] L. Mandelstam and N. Papalexi, *J. Tech. Phys.* (USSR), 1934.

the center of gravity of the body of the person on the swing, belongs also to the same class of phenomena.

These phenomena are generally amenable to the d.e. with periodic coefficients, more specifically, to the d.e. of Mathieu. The theory of this linear d.e. is of relatively little help in studies of these phenomena which are essentially nonlinear in most cases.

In fact, the Mandelstam-Papalexi experiments have shown that, if the circuit of their "parametric generator" has a small amount of resistance (that is, a practically linear circuit), the amplitude of oscillation grows indefinitely until the insulation is destroyed by an excessive voltage. If, however, a nonlinear conductor is inserted in series with the circuit, a stable stationary condition is reached with amplitude determined by the d.e.

As the treatment of both cases—the linear and the nonlinear—has been given by means of the stroboscopic method in Chapter 16, we do not reproduce it here, but merely mention the conclusions.

In the linear case the phase stabilizes itself on the stable value $\varphi_0 = 3\pi/4$ for which the amplitude is unstable. This is also in agreement with the Mandelstam-Papalexi experiments which show that, in this case, the amplitude grows indefinitely until the circuit is destroyed by an excessive voltage.

In a nonlinear case, on the contrary, the amplitude reaches a fixed stationary value which is again in agreement with the experimental results.

The physical significance of the threshold (4.25) (Chapter 16) is also clear; it means that a cumulative phenomenon is possible if the energy input introduced into the system owing to a periodic variation of the parameter is greater than the energy dissipated by the system.

Summing up, the cumulative character of the phenomenon is sufficiently clear on this basis since the stability of the phase is always associated with the instability of the amplitude. The difference between the linear and nonlinear cases is that this instability is permanent in the former case, whereas in the second case it decreases with the increasing amplitude and vanishes for a certain fixed value of amplitude for which the oscillation becomes stationary.

2. General form of the differential equation of parametric action

The case given in Chapter 16 relates to the most frequently encountered form of parametric action, but in recent years a number of other phenomena were discovered which, in one way or the other, are related to the periodic variation of a parameter in a d.e.

We shall indicate one such case, known under the name of the phenomenon of Bethenod,[5] which gave rise to some discussions before its nature was understood.

It is useful to consider the d.e. of the parametric action in a somewhat more general form than that investigated in Chapter 16, namely:

$$\ddot{x} + b\dot{x} + x + (a - cx^2)x \cos 2t + ex^3 = 0 \tag{2.1}$$

The essential feature of this d.e. is that a certain polynomial containing even powers of x appears as a coefficient of the term $x \cos 2t$. Nothing especially interesting is gained by considering more complicated polynomials, and the form (2.1) is sufficiently general. We shall again suppose that all coefficients are small, which will permit establishing conclusions from the first approximations.

Since the derivation of the stroboscopic d.e. is sufficiently clear from Chapter 16, we omit it here and indicate the form of these equations:

$$\frac{d\rho}{d\tau} = \frac{1}{2}\rho[(C\rho - 2A)\sin 2\varphi - 4B]; \quad \frac{d\varphi}{d\tau} = \frac{1}{4}\left[(C\rho - A)\cos 2\varphi - \frac{3}{2}E\rho\right] \tag{2.2}$$

where $A = a/\mu$; $B = b/\mu$; and $E = e/\mu$.

It is noted that the case $C = 0$ gives the d.e. (4.2), Chapter 16 which has been discussed.

The system (2.2) has a singular point defined by the expressions

$$\sin 2\varphi_0 = 4B/(C\rho_0 - 2A); \qquad \cos 2\varphi_0 = 3E\rho_0/2(C\rho_0 - A) \tag{2.3}$$

and the stationary amplitude ρ_0 is obtained from the expression $\sin^2 2\varphi_0 + \cos^2 2\varphi_0 = 1$, in which the values (2.3) for $\sin 2\varphi_0$ and $\cos 2\varphi_0$ are replaced. This results in the quartic equation for ρ_0:

$$4(C\rho_0 - A)^2(C\rho_0 - 2A)^2 - 64B^2(C\rho_0 - A)^2 - 9E^2\rho_0^2(C\rho_0 - 2A)^2 = 0 \tag{2.4}$$

The variational equations result in the characteristic equation

$$S^2 + BS + Q = 0 \tag{2.5}$$

where

$$Q = -\tfrac{1}{4}[(C\rho_0 - A)^2 \sin^2 2\varphi_0 - 2B(C\rho_0 - A)\sin 2\varphi_0 + \tfrac{1}{2}C\rho_0(C\rho_0 - 2A)\cos^2 2\varphi_0 - \tfrac{3}{4}E\rho_0(C\rho_0 - 2A)\cos 2\varphi_0] \tag{2.6}$$

with $\rho_0 > 0$ obtained from (2.4).

[5] J. Bethenod, *C. R. Ac. Sc.* (Paris) **207**, 1937; N. Minorsky, *Colloque Intern. de Porquerolles,* France, 1951.

The singular point is stable if

$$B > 0; \qquad Q > 0 \tag{2.7}$$

If the system is dissipative ($B > 0$), the first condition (2.7) is fulfilled so that the determination of stability reduces to showing that $Q > 0$. In view of this, the determination of a stable periodic solution of (2.1) reduces to the determination of such positive root ρ_0 of (2.4) which, upon its substitution into (2.6), makes Q positive.

In the general case, when all four parameters A, B, C, and E are different from zero, the computational problem is rather complicated but, if one of these four parameters is zero, the discussion is facilitated by the possibility of using a representation in a three-dimensional space of the remaining three parameters. In the meantime one can ascertain better the physical nature of these complicated phenomena.

We shall consider, therefore, the following cases:

$C = 0$; the parameter space of the variables A, B, and E; this case corresponds to a linear parametric modulation and the well known experiments of Mandelstam and Papalexi were carried out under such condition.

$A = 0$; it will be shown that the parametric excitation does not exist in such a case for the reason which will appear later. These two cases are relatively simple and will be outlined in this section.

The remaining two cases, $B = 0$ and $E = 0$, are more complicated and are investigated in the following section.

Case: $C = 0$. The d.e. becomes

$$\ddot{x} + b\dot{x} + x(1 + a\cos 2t) + ex^3 = 0 \tag{2.8}$$

From (2.4) one finds:

$$\rho_0 = \tfrac{2}{3}E\sqrt{A^2 - 4B^2} \tag{2.9}$$

The stationary amplitude exists if

$$A > 2B; \qquad E \neq 0 \tag{2.10}$$

One has also

$$\sin 2\varphi_0 = -2B/A; \qquad \cos 2\delta_0 = -3E\rho/2A \tag{2.11}$$

If one replaces these values into (2.6) where one sets $C = 0$, one has

$$Q = \tfrac{9}{16}E^2\rho^2 > 0 \tag{2.12}$$

which shows that the stationary amplitude ρ_0 is always stable as long as $E \neq 0$.

As regards the first condition (2.10), its physical significance can be readily ascertained. In fact, assume that, in addition to $c = 0$, one has also $b = e = 0$. In such a case one has a linear Mathieu equation: $\ddot{x} + (1 + a \cos 2t)x = 0$, which we investigated in Chapter 16. There it was shown that the stable phase is $\varphi_0 = 3\pi/4$, to which corresponds an unstable amplitude as was also demonstrated by the Mandelstam–Papalexi experiments. It is obvious that the coefficient a (or A) is a measure of the energy absorbed by the system just as b (or B) is a measure of the energy dissipated.

We can consider now the parameter space (A, E, B). Inasmuch as the first condition (2.7) requires $B > 0$, it is sufficient to consider only the half-space for $B > 0$. Besides this, it is necessary to eliminate the volume limited by the planes $A = \pm 2B$ passing through the E axis which expresses geometrically the first condition (2.10). In the remainder of the upper half-space to each point (A, E, B) is attached a value of ρ_0 defined by (2.9), and for any such point there exists a stable stationary state.

One can investigate in a similar manner the corresponding subcases if one fixes the value of one of the three variables which leads to a planar representation of the remaining two variables. Thus, for instance, if one sets $A = A_0$, one obtains the corresponding relation in a (B,E) plane, $\rho_0 = \frac{2}{3}E\sqrt{A_0^2 - 4B^2}$ which may be regarded as a curve in that plane. As ρ_0 represents the total energy stored in the oscillation, this gives the information about the equal energy levels if, in the above equation ρ_0, is considered constant. The same procedure can be applied to other subcases, namely: $B = B_0$; $\rho_0 = \rho_0(A,E)$ and $E = E_0$; $\rho_0 = \rho_0(A,B)$.

Case: $A = 0$. The parametric space in this case is (B, C, E). In view of the previously explained significance of the coefficient A, one can expect that the oscillation cannot exist in this case because the system does not absorb energy necessary to cover its dissipation. If one carries out the calculation for Q, one finds that it is negative, which shows the instability and, therefore, the absence of the self-maintained oscillation.

3. Special cases—bifurcation surfaces; stability of the state of rest

The two remaining cases, $B = 0$ and $E = 0$, are more interesting and we shall investigate them in more detail.

Case: $B = 0$. In this case the system is nondissipative, although it is not conservative since the parametric action accounts for the energy input into the system. Under that condition the amplitude can be still limited by the nonlinearity; we shall see that, in fact, the presence of two nonlinearities (one associated with the coefficient C and the other with E) may account for a somewhat special situation.

It is to be noted also that the characteristic equation (2.5) shows that the only singular point here is the center if $Q > 0$. This means that in such a case the stability may exist, although it is not an asymptotic one. In view of this, any small perturbation deviating the system from its singular point (in the φ plane) accounts for a small closed trajectory which manifests itself in the ψ plane as a small modulation (Chapter 16). In what follows we shall investigate the stability disregarding the fact that it is not asymptotic here.

From the second equation (2.2) one has

$$\rho_0 = 2A \cos 2\varphi_0/(2C \cos 2\varphi_0 - 3E) \tag{3.1}$$

In this expression one has to consider two cases: $\cos 2\varphi_0 = 1$ and $\cos 2\varphi_0 = -1$. As regards the stability condition: $Q > 0$, it is determined here (omitting constant positive factors) by the expression:

$$(C\rho_0 - 2A)(3E \cos 2\varphi_0 - 2C) > 0 \tag{3.2}$$

It is convenient to consider again the parameter space of the variables A, C, E with four octants corresponding to $A > 0$; the octants corresponding to $A < 0$ are of no interest here, as was previously mentioned. For each of these four octants it is necessary to express ρ and Q as positive; one has thus two conditions and the one which dominates the other is thus the *sufficient condition* for the existence of a stationary stable state.

The results of these simple calculations are indicated in the following two tabulations, of which case I relates to $A > 0$, $\cos 2\varphi_0 = 1$ and case II to $A > 0$, $\cos 2\varphi_0 = -1$. Each tabulation gives data for each of the four octants, namely: (1) $C > 0, E > 0$; (2) $C > 0, E < 0$; (3) $C < 0, E < 0$; and (4) $C < 0, E > 0$. For each octant are indicated: the values of the stationary amplitude ρ_0 (the energy), if it exists, and the sufficient condition of stability. If the latter is not indicated, this means that stability exists for any point of the octant in question. When one of the variables (C or E) is negative, we indicate it as $-C$, $-\bar{E}$. With these conventions we have

	Case I	Case II
(1)	$\rho_0 = 2A/(2C - 3E)$; $C > 3E$	$\rho_0 = 2A/(2C + 3E)$
(2)	$\rho_0 = 2A/(2C + 3\bar{E})$	$\rho_0 = 2A/(2C - 3\bar{E})$; $C > 3\bar{E}$
(3)	$\rho_0 = 2A/(3\bar{E} - 2\bar{C})$; $3\bar{E} > 2\bar{C}$	ρ_0 does not exist
(4)	ρ_0 does not exist	$\rho_0 = 2A(3E - 2\bar{C})$; $3E > 2\bar{C}$

It is noted that in some octants ρ_0 does not exist; this happens each time that one obtains for $\rho_0 = r^2$ a negative value; from a physical point of view this means that there is no equilibrium point. In some octants in which the sufficient condition is not indicated, a stable stationary state exists for any point in this octant. In the octants in which the sufficient condition is indicated, the corresponding inequality *considered as equation* determines a certain surface Σ (a plane here) in the parameter space which separates the two regions; one in which the stable stationary state exists and the other one in which it does not exist. Thus, for instance in case I (1) the sufficient condition is $C > 3E$. The equation $C = 3E$ is thus a Σ surface; any point of this surface corresponds to the critical condition (appearance or disappearance of the oscillation).

It is to be noted also that there is a tendency toward an internal resonance in the system when two nonlinearities (one associated with the coefficient C and the other with E) cancel each other. Thus, for instance, if $\rho_0 = 2A/(2C - 3E)$, it is clear that $\rho_0 \to \infty$ if $C \to 3E/2$. In reality the stability condition limits this approach so that the oscillation disappears *before* the resonance has a chance to build itself up.

A similar study can be carried out for $E = 0$, but here the results are less interesting; one finds that, except for some special regions of the parameter space, there is no stability.

This investigation of the existence and stability of the stationary state must be supplemented by the investigation of the self-excitation. For this purpose it is sufficient to investigate the stability of the state of rest. If the state of rest is unstable and there exists a stable stationary state, it is logical to expect that the oscillation will develop from the state of rest and approach the stationary condition. The situation is, to some extent, similar to that designated as configuration: *US* in Chapter 7. In reality, in the present case this is merely a formal procedure based on the use of the stroboscopic (φ) plane since the system (2.1) is not autonomous and, for that reason, one cannot use the topological concepts of the theory of Poincaré.

With this restriction in mind, we investigate now the condition of self-excitation of the system represented by (2.1). Assuming in the stroboscopic system (2.2) ρ as a small quantity, one has the following d.e.:

$$d\rho/d\tau = -\tfrac{1}{2}\rho(2A \sin 2\varphi + 4B); \qquad d\varphi/d\tau = -\tfrac{1}{4}A \cos 2\varphi \tag{3.3}$$

These stroboscopic d.e. are valid only in the neighborhood of the state of rest; they contain only the terms with A and B (that is, only the linear terms) because in the neighborhood of the state of rest the nonlinear terms are negligible for the obvious reasons. The problem is thus considerably simplified, and we can use the argument outlined in Chapter 16 on the Mathieu oscillators. Without repeating it here, we merely mention that the stability of the phase requires $\sin 2\varphi_1 = -1$, where by the subscript 1 we specify the condition near the state of rest. With this value of the stable phase, the amplitude equation is:

$$d\rho/d\tau = \rho(A - 2B) \tag{3.4}$$

so that the condition of self-excitation is

$$A > 2B \tag{3.5}$$

4. Phenomenon of Bethenod

An interesting application of the d.e. (2.1) is the so-called *phenomenon of Bethenod*[5] which is essentially as follows.

If one provides a physical pendulum (Fig. 20.1) with a piece of soft iron P and places the pendulum above a coil C coaxial with the pendulum as shown and carrying alternating current, it is observed that under certain conditions, the pendulum starts oscillating and reaches a stationary state with a constant amplitude.

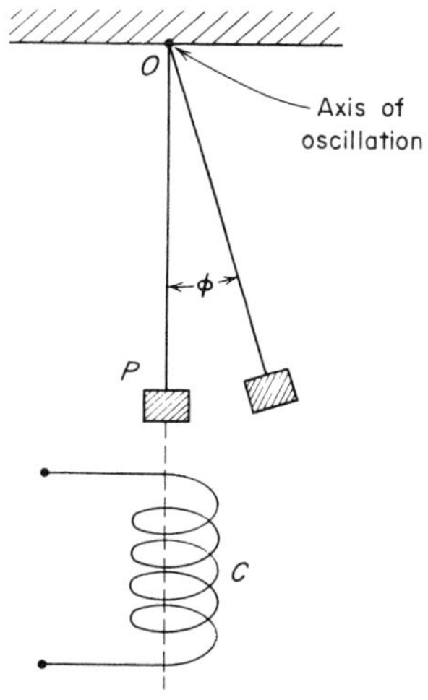

FIGURE 20.1

Several theories of this phenomenon were discarded until Y. Rocard[6] formulated the d.e. which seemed to be plausible on physical grounds, but it was impossible to integrate them in the form in which they were established originally.

These d.e. are

$$\frac{dL(\theta)i}{dt} + zi = E \sin \omega t \tag{4.1}$$

$$J\ddot{\theta} + D\dot{\theta} + C\theta = d/d\theta[\tfrac{1}{2}L(\theta)i^2] \tag{4.2}$$

Equation (4.1) relates to the electric circuit formed by the inductance $L(\theta)$ of the coil, C; z is the constant impedance of the rest of the circuit; and $E \sin \omega t$ is the external periodic excitation applied to the circuit. The inductance $L(\theta)$ is clearly a variable quantity depending on the angle θ of the pendulum swinging above the coil C and thus modifying the reluctance of its magnetic circuit. Equation (4.2) is the d.e. of the pendulum having the moment of inertia J, coefficient of linear damping D, and the restoring couple constant C. We take the d.e. of the pendulum as linear, inasmuch as this circumstance is not essential in the

[5] See footnote [5], page 490.

[6] Y. Rocard, *Dynamique générale des oscillations*, Masson, Paris, 1943.

theory of this phenomenon. What *is* essential is the form of the nonlinear function $L(\theta)$, as will be shown later.

The term on the right side of the d.e. (4.2) is the ponderomotive action of the electromagnetic field which, by Maxwell's theory, is the derivative with respect to the coordinate of the intrinsic energy $\frac{1}{2}Li^2$ stored in the coil.

The physical significance of the d.e. (4.1) and (4.2) is very simple: as the pendulum swings, the inductance $L(\theta)$ undergoes variations and the ponderomotive action of the process appears as a mechanical reaction applied to the pendulum. The whole question is: Can the process of this kind result in the maintenance of oscillations?

In their present form (4.1) and (4.2) cannot be integrated and, for that reason, we adopt the following argument.[7] It will be assumed that the stationary motion (that is, the oscillation of the pendulum) exists with some unknown frequency Ω. In such a case the d.e. (4.1) gives a periodic solution of the form $\theta = \theta_0 \cos \Omega t$, θ_0 and Ω being unknown at present. If one substitutes this solution into (4.2) under the assumption of a certain function $L(\theta)$, one obtains a certain nonlinear d.e. whose form will be indicated later. If it is possible to show that this d.e. possesses a periodic solution, this solution will be precisely the one in which we are interested here, and the quantities θ_0 and Ω will be determined by integrating this d.e. in the first approximation, assuming the smallness of the various parameters.

As regards the function $L(\theta)$, on physical grounds, it is clear that this function is maximum for $\theta = 0$ since the magnetic reluctance is maximum for this value of θ. As $|\theta|$ increases, the function $L(\theta)$ decreases, but the rate of this decrease diminishes for larger angles since the pendulum recedes into the region of the stray flux which has a tendency to become constant for larger values of $|\theta|$. Thus, on physical grounds, the form

$$L(\theta) = L_0 - a_2\theta^2 + a_4\theta^4 \tag{4.3}$$

answers these requirements. It will be assumed that $b_2 = a_2/L_0$; $b_4 = a_4/L_0$ are small quantities of the first order.

Under these assumptions a simple but somewhat lengthy calculation, which we omit here, reduces the electrical d.e. (4.1) to the form

$$\begin{aligned} di/dt - (\beta + \alpha_2 \sin 2\Omega t + \alpha_4 \sin 4\Omega t)i \\ = M \sin \omega t + \tfrac{1}{2}N \sin (\omega + \Omega)t + \tfrac{1}{2}N \sin (\omega - \Omega)t, \end{aligned}$$

where $\beta = z/L_0$; $\alpha_2 = (b_2\theta_0{}^2 + b_4\theta_0{}^4)\Omega$; $\alpha_4 = \frac{1}{2}b_2\theta_0{}^4\Omega$; $M = m(1 +$

[7] N. Minorsky, *C. R. Ac. Sc.* (Paris) **231**, 1950; *J. Appl. Phys.* **22**, 1951; *J. Franklin Inst.* **254**, 1952.

$b_2\theta_0^2/2)$; $N = mb_2\theta_0^4\Omega$; $m = E/L_0$ which is a linear d.e of the form $di/dt + f(t)i = g(t)$ whose solution is

$$i(t) = \exp[-F(t)] \int_0^t g(t) \exp[F(t)]dt; \qquad F(t) = \int_0^t f(t)dt$$

If one carries out this calculation taking into account the order of magnitude of the different terms, the solution is of the form

$$i(t) = i_0 \cos \omega t + i_2 \cos(\omega + 2\Omega)t + i_2' \cos(\omega - 2\Omega)t + i_4 \cos(\omega + 4\Omega)t + \ldots \quad (4.4)$$

In this expression the constants $i_0, i_2, i_2', i_4, \ldots$, are easily calculated in terms of the fixed parameters L_0, ω, z, b_2, and b_4, but for what follows we shall not need these expressions. It is sufficient to note that i_0 is of the zero order (that is, finite), i_2 and i_2' are of the first order of smallness, i_4 and i_4' are of the second order, so that for the first approximation it is sufficient to retain only the first three terms of the trigonometric series (4.4).

One can now calculate the coupling terms on the right side of (4.2). Since i enters by its square, one has to square the trigonometric polynomial (4.4) limited to the first three terms. This results in a constant term $q_0 = \frac{1}{2}(i_0^2 + i_2^2 + i_2'^2)$ and also in terms containing the frequency ω. Since this frequency ω (of alternating current) is very large as compared to the frequency of the pendulum Ω, the effect of the terms with ω does not produce any average action on the pendulum. However, the terms corresponding to the double cross-products of the terms with $\cos(\omega + \Omega)t$ and $\cos(\omega - \Omega)t$ result in the terms of the form $i_0 i_2 \cos 2\Omega t$ and $i_0 i_2' \cos 2\Omega t$, in which the high frequency ω does not appear. These terms of the double frequency 2Ω account precisely for the parametric action. Once this point has been ascertained, one can drop all terms except those with $\cos 2\Omega t$, and the calculation of the coupling terms results in the expression of the form

$$\frac{d}{d\theta}\left[\frac{1}{2} Li^2\right] = -L_0[q_0 b_2 \theta + 2q_0 b_4 \theta^3 + (q_2 b_2 + 2q_4 b_4 \theta^2)\theta \cos 2\Omega t]$$

where $q_2 = i_0(i_2 + i_2')$.

If one substitutes this expression into (4.2) and rearranges the terms with the change of the independent variable from t to $\tau = \Omega t$, one obtains the d.e.

$$\ddot{\theta} + b\dot{\theta} + \theta + e\theta^3 + (a + c\theta^2)\theta \cos 2\tau = 0 \quad (4.5)$$

where $b = D/J\Omega^2$; $e = 2\alpha_4 q_0/J\Omega^2$; $a = \alpha_2 q_2/J\Omega^2$; $c = 2\alpha_4 q_2/J\Omega^2$; and $\Omega^2 = (C + \alpha_2 q_0)/J$.

The last expression gives $\Omega = \sqrt{(C + \alpha_4 q_0)/J}$, which shows that the mechanical frequency $\Omega_0 = \sqrt{C/J}$ of the pendulum is modified by the ponderomotive action of the electromagnetic origin.

It is noted that the d.e. (4.5) is the d.e. (2.1). Inasmuch as (2.1) has a periodic solution, this is the solution in which we are interested here. Since we have already analyzed the d.e. (2.1) in the stroboscopic form, we do not repeat the argument here but merely mention that the existence of a stable periodic solution requires, in addition to the consistency of equations (2.3) and (2.4), the fulfillment of the condition that $Q > 0$. One ascertains easily that the state of rest ($\rho = \rho_1 \simeq 0$) can be unstable (equation (3.5)), in which case the pendulum starts spontaneously from rest so that the position of equilibrium $\theta = 0$, which is stable for $E = 0$, becomes unstable in the case when $E \neq 0$.

It is important to note that, in this case, there exists no rational ratio between the two frequencies ω and Ω. At the early stage of the investigation of this effect it was thought that its origin was in the phenomenon of the subharmonic resonance which would require a rational ratio of the two frequencies. Very accurate measurements made by J. Haag at the French Institute of Chronometry revealed definitely that no such rational ratio exists, which is in agreement with the aforementioned theory of the parametric action in which no such ratio is required.

5. Origin of the parametric action

In Section 4, Chapter 16, as well as in the preceding sections of this chapter, it was assumed that $\omega = 2$ in the d.e.

$$\begin{aligned} \ddot{x} + (1 + a\cos\omega t)x &= 0 \\ \ddot{x} + b\dot{x} + (1 + a\cos\omega t)x + cx^3 &= 0 \end{aligned} \tag{5.1}$$

which are the d.e. (4.1) and (4.2), but merely written with ω instead of $\omega = 2$. The assumption, $\omega = 2$, to some extent was imposed by the experimental evidence of Melde[8] and Lord Rayleigh[1] at the beginning of these studies.

If one introduces the variables $\rho = x^2 + \dot{x}^2 = x^2 + y^2$ and $\psi = \arctan(y/x)$, for instance in the first equation (4.1) as we did in Section 4, Chapter 16, this equation is replaced by the equivalent system of two d.e. of the first order

$$\frac{d\rho}{d\tau} = -a\rho \sin 2\psi \cos \omega t; \qquad \frac{d\psi}{d\tau} = -1 + \frac{1}{2}a(1 + \cos 2\psi)\cos\omega t \tag{5.2}$$

[8] F. Melde, *Poggend. Ann.* **109**, 1860.
[1] See footnote [1], page 488.

and, since we assume that a is small (of the first order), the zero-order term in the solution by series: $\rho(t) = \rho_0(t) + a\rho_1(t) + a^2\rho_2(t) + \ldots$; $\psi(t) = \psi_0(t) + a\psi_1(t) + \ldots$ is still

$$\rho_0(t) = \rho_0 = \text{const}; \qquad \psi_0(t) = \varphi_0 - t \tag{5.3}$$

as before, ρ_0 and φ_0 being the initial conditions. If one limits oneself to the first approximation only, the first-order corrective terms are given by the d.e.

$$\frac{d\rho_1}{d\tau} = -\rho_0 \sin 2\psi_0 \cos \omega t; \qquad \frac{d\psi_1}{d\tau} = -\frac{1}{2}\cos \omega t - \frac{1}{2}\cos 2\psi \cos \omega t$$

Taking into account (5.3) and, after certain trigonometric transformations, one obtains the changes in ρ_1 and ψ_1 after the time 2π

$$\Delta\rho_1 = \rho_1(2\pi) = -\frac{1}{2}\rho_0 \left[\sin 2\varphi_0 \int_0^{2\pi} \cos(\omega - 2)\tau d\tau + \cos 2\varphi_0 \times \int_0^{2\pi} \sin(\omega - 2)\tau d\tau + \sin 2\varphi_0 \int_0^{2\pi} \cos(\omega + 2)\tau d\tau - \cos 2\varphi_0 \int_0^{2\pi} \sin(\omega + 2)\tau d\tau\right]$$

$$\Delta\psi_1 = \psi_1(2\pi) = -\frac{1}{2}\int_0^{2\pi} \cos \omega\tau d\tau - \frac{1}{4}\left[\cos 2\varphi_0 \int_0^{2\pi} \cos(\omega - 2)\tau d\tau - \sin 2\varphi_0 \int_0^{2\pi} \sin(\omega - 2)\tau d\tau + \cos 2\varphi_0 \int_0^{2\pi} \cos(\omega + 2)\tau d\tau + \sin 2\varphi_0 \int_0^{2\pi} \sin(\omega + 2)\tau d\tau\right]$$

Carrying out these calculations and noting that $\Delta\rho = a\Delta\rho_1$; $\Delta\psi = a\Delta\psi_1 = \Delta\varphi$, one has

$$\begin{aligned}\Delta\rho &= -\frac{1}{4\pi}\rho_0(2\pi a)\left[\left(\frac{2\omega}{\omega^2 - 4}\right)\sin 2\varphi_0 \sin 2\pi\omega + \left(\frac{4}{\omega^2 - 4}\right)\cos 2\varphi_0(1 - \cos 2\pi\omega)\right]\\ \Delta\varphi &= -\frac{1}{4\pi\omega}\sin 2\pi\omega(2\pi a) - \frac{1}{8}(2\pi a)\left[-\left(\frac{4}{\omega^2 - 4}\right)\sin 2\varphi_0(1 - \cos 2\pi\omega) + \left(\frac{2\omega}{\omega^2 - 4}\right)\cos 2\varphi_0 \sin 2\pi\omega\right]\end{aligned} \tag{5.4}$$

If one divides these equations by $2\pi a = \Delta\tau$, one gets

$$\frac{\Delta\rho}{\Delta\tau} = -\frac{1}{2\pi}\rho_0\frac{1}{\omega^2 - 4}[\omega\sin 2\varphi_0\sin 2\pi\omega + 2\cos 2\varphi_0(1 - \cos 2\pi\omega)]$$

$$\frac{\Delta\rho}{\Delta\tau} = -\frac{1}{4\pi\omega}\sin 2\pi\omega - \frac{1}{4\pi(\omega^2 - 4)}[-2\sin 2\varphi_0(1 - \cos 2\pi\omega) + \omega\cos 2\varphi_0\sin 2\pi\omega] \tag{5.5}$$

whence the stroboscopic system

$$\frac{d\rho}{d\tau} = -\frac{1}{2\pi}\frac{\rho}{\omega^2 - 4}(\omega m\sin 2\varphi + 2n\cos 2\varphi) \tag{5.6}$$

$$\frac{d\varphi}{d\tau} = -\frac{1}{4\pi\omega}m - \frac{1}{4\pi(\omega^2 - 4)}(\omega m\cos 2\varphi - 2n\sin 2\varphi) \tag{5.7}$$

where $m = \sin 2\pi\omega$; $n = 1 - \cos 2\pi\omega$.

These are the stroboscopic d.e. when $\omega \neq 2$. But if $\omega = 2$, as is usually assumed in the elementary theory of parametric action, the matter is considerably simplified and one finds the d.e. that were obtained by assuming directly $\omega = 2$. It is useful, however, to obtain the principal parametric resonance ($\omega = 2$) from the general equations (5.6) and (5.7) in which ω is not necessarily $\omega = 2$. In these equations appear factors $m/(\omega^2 - 4) = \sin 2\pi\omega/(\omega^2 - 4)$ and $n/(\omega^2 - 4) = (1 - \cos 2\pi\omega)/(\omega^2 - 4)$ and the application of the Hopital rule shows that $\lim \left|\frac{\sin 2\pi\omega}{\omega^2 - 4}\right|_{\omega\to 2} = \frac{\pi}{2}$ and $\lim \left|\frac{1 - \cos 2\pi\omega}{\omega^2 - 4}\right|_{\omega\to 2} = 0$, which gives the d.e. (4.8) and (4.11), Chapter 16, where we assumed $\omega = 2$. If the parametric resonance exists *in the neighborhood* of $\omega = 2$ in a steady state, one must have simultaneously $d\rho/d\tau = 0$ and $d\varphi/d\tau = 0$. We investigate first the condition $d\varphi/d\tau = 0$, assuming $\omega = 2 + \varepsilon$, $\varepsilon \ll 1$ being a small *detuning*. Equating to zero the right-hand side of (5.7) one obtains to the first order

$$4\varepsilon m/\omega = 2(n\sin 2\varphi - m\cos 2\varphi) = 2\lambda\sin(2\varphi - \gamma) \tag{5.8}$$

where $\lambda = \sqrt{n^2 + m^2}$; $\sin\gamma = 2m/\lambda$; $\cos\gamma = 2n/\lambda$.

If ε is small of the first order, m is also $O(\varepsilon)$, so that the left-hand side is $O(\varepsilon^2)$ and, since λ is $O(\varepsilon)$, $\sin(2\varphi - \gamma)$ must be $O(\varepsilon)$, that is, small; hence $2\varphi_0 - \gamma \simeq 0$ or π. Since $\gamma = \arctan(m/n)$, for ε small, $\gamma \simeq \pi/2$ (since $m \simeq O(\varepsilon)$ and $n \simeq O(\varepsilon^2)$); this yields $\varphi_0 \simeq \pi/4$ at the limit $\varepsilon \to 0$ but this, as was seen in Section 4, Chapter 16, is an unstable phase; there is still another possibility: $\gamma = 3\pi/2$ in which case $\varphi_0 \simeq 3\pi/4$ and this, as we saw, corresponds to the region of stability of the phase for the exact parametric resonance: $\omega = 2$. Thus the presence of a small detuning merely displaces the equilibrium point of the phase by a small amount. The essential point is that the position of equilibrium for the phase can always be determined from (5.8) if the detuning is given, which then determines m, n, λ, and γ.

As regards the amplitude equation (5.6) for ε small, $n \simeq O(\varepsilon^2)$ and $\cos 2\varphi_0$ is small in the neighborhood of $\varphi_0 = 3\pi/4$ so that one can write approximately

$$\frac{d\rho}{d\tau} = +\frac{1}{\pi}\frac{\rho}{\omega^2 - 4}\sin 2\pi\omega \tag{5.9}$$

Since $\sin 2\varphi_0 \simeq -1$ in this neighborhood, $d\rho/d\tau$ is also in the neighborhood of its value given previously.

It is clear, however, that for a larger detuning $d\rho/d\tau$ may vanish when

$$\tan 2\varphi_0 = -2n/\omega m$$

as this follows from (5.6), and may become negative if $\tan \varphi_0 < -(2n/\omega m)$. This shows that the parametric excitation may disappear if the detuning is greater than a certain limit since its existence requires that $(d\rho/d\tau) > 0$ (in the linear case). Although in the above argument it was assumed that $\varepsilon > 0$, one sees easily that for $\varepsilon < 0$ it remains the same.

Thus the effect of the detuning is much the same as in the ordinary subharmonic resonance where the oscillation may exist for a certain zone of detuning, as was shown in Chapter 19, and is not confined to the *exact* resonance only.

The manifestations of the two phenomena—the subharmonic resonance and the parametric resonance—are different. In the former, the resonance oscillation is produced directly by the *external periodic excitation*; in the latter, it is due to the periodic variation of one of the parameters *in the absence* of any direct external excitation.

The physical nature of the two phenomena is, however, the same. In the subharmonic resonance, the energy is introduced *directly* into the system by the external excitation and appears there with a subharmonic frequency; in the parametric resonance the energy is introduced *not directly* but through the degree of freedom of the parameter variation but the subharmonic process remains the same; the degree of freedom of the variable parameter absorbs the energy with frequency $2f$ and transmits it to the principal degree of freedom where the oscillation develops in the form of an oscillation with frequency f. In both cases the amplitude of the resonance oscillation decreases with the increasing detuning and vanishes for a certain critical value of the latter.

It may seem, therefore, that subharmonic resonance exhibits a greater variety of oscillations (resonances of order $\frac{1}{2}$, $\frac{1}{3}$,. . .) whereas the parametric oscillation exhibits only the resonance $\frac{1}{2}$. This is due only to the fact that here one operates with the first-order approximation in which this particular resonance appears; the other resonances, appearing in approximations of higher orders, are very small. In the subharmonic resonance, on the other hand, the various resonance may appear in the first order, depending on the form of the d.e. This circumstance inherent in the parametric excitation is due to a particular feature of the d.e. of the Mathieu-Hill type; thus by its very nature, the parametric resonance is a somewhat special form of the more general subharmonic resonance. Parametric excitation (or resonance) is generally characterized by the existence of terms $x \cos 2t$, $x^3 \cos 2t$, $x \cos 4t$,. . . in the d.e., whereas in the nonlinear (or subharmonic) resonance the explicit dependence on time appears usually in the form of a term $\lambda \sin nt$, λ being a constant as we saw in Chapter 19.

Since both forms of resonance have a common feature, the two phenomena can coexist under certain conditions. As far as known, this question has not been investigated as yet, although no special difficulty would seem to prevail.

If one starts with the d.e.:

$$\ddot{x} + b\dot{x} + (a - cx^2)x \cos 2t + ex^3 = d \cos n(t - \delta) \qquad (5.10)$$

where one assumes, as usual, the smallness of a, b, c, and e. If $d = 0$, one has the case of the parametric excitation which was investigated in Section 2. If $d \neq 0$, one has to add to this the study of the subharmonic resonance, the only interesting cases here being when $n = 1$ and $n = 2$.

Calculations do not present any difficulty, but are rather long and are not reproduced here. It is obvious that the phase δ of the external periodic excitation is of importance in this case. In view of the presence of the right-hand term, it is preferable to use the variable r (instead of ρ) in the stroboscopic transformation. With the variables r and ψ the equivalent system is

$$\begin{aligned}
\dot{r} &= \tfrac{1}{2}br(1 - \cos 2\psi) - \tfrac{1}{2}ar \cos 2\psi \cos 2t + \tfrac{1}{8}cr^3(2 \sin 2\psi \\
&\quad + \sin 4\psi) \cos 2t - \tfrac{1}{8}er^3(2 \sin 2\psi + \sin 4\psi) \\
&\qquad + d \sin \psi \cos n(t - \delta) \\
\dot{\psi} &= -1 - \tfrac{1}{2}b \sin 2\psi - \tfrac{1}{2}a(1 + \cos 2\psi) \cos 2t + \tfrac{1}{8}cr^3(3 + 4 \cos 2\psi \\
&\quad + \cos 4\psi) \cos 2t - \tfrac{1}{8}er^2(3 + 4 \cos 2\psi + \cos 4\psi) \\
&\qquad + \frac{d}{r} \cos \psi \cos n(t - \delta)
\end{aligned} \qquad (5.11)$$

If one assumes also that d is small (of the same order as a, b, c, and e), the zero order solution is: $r_0(t) = r_0$; $\psi_0(t) = \varphi_0 - t$, r_0 and φ_0 being constants.

It is sufficient to substitute these values into (5.11) and to carry out the usual stroboscopic transformation separately for two cases $n = 1$ and $n = 2$. In the final result it is sufficient to investigate the influence of the parameter δ on the solution and its stability.

6. Parametric excitation in electrical circuits

It is useful to supplement the preceding theory by certain physical considerations regarding the parametric effect in electrical circuits where the relations are particularly simple.

Consider, for example, an oscillating circuit comprising a variable capacity capable of being changed between $C_{max} = C_0 + \Delta C$ and $C_{min} = C_0 - \Delta C$. The "parametric generator" of Mandelstam and Papalexi, to which we referred previously, consists essentially of such arrangement.

As was mentioned previously, a system of this kind is amenable to the d.e. of Mathieu but very often the qualitative nature of what happens in such a case can be better ascertained, as was shown by Meissner[9] by the d.e. of Hill with a rectangular ripple. We recall that these d.e. are of the form

$$\ddot{x} + M(t)x = 0 \tag{6.1}$$

In the Mathieu equation $M(t) = a + b \cos 2t$ and in the Hill equation $M(t)$ is a Fourier series. In the Hill-Meissner equation one has

$$M(t) = \omega^2 + \frac{4}{\pi} a^2\left(\cos t - \frac{1}{3}\cos 3t + \frac{1}{5}\cos 5t \ldots\right) \tag{6.2}$$

which represents a rectangular "ripple." This means that it is assumed that the capacity variation between maximum and minimum is produced discontinuously at certain instants. This, in turn, amounts to considering, instead of the Hill-Meissner d.e., an alternate sequence of two linear d.e.

$$\ddot{x} + (a^2 + b^2)x = 0 \qquad \text{and} \qquad \ddot{x} + (a^2 - b^2)x = 0 \tag{6.3}$$

with $a^2 > b^2$

We shall first discuss the physical aspect of the phenomenon and, once this point is clarified, we consider its representation in the phase plan following the argument of Mandelstam and Papalexi.

Assume that we begin changing capacity, as just mentioned, by decreasing it discontinuously at $t = 0$. Since there are always some residual

[9] E. Meissner, *Schweitz-Bauzeitung* **72**, 1918.

charges on the condenser and the initial stored energy is purely electrostatic, a sudden reduction of capacity requires an impulsive work $\Delta Cq^2/2C^2$, q being a small initial charge on condenser. In the next quarter cycle the energy will be purely electromagnetic $Li^2/2$ with the zero potential difference across the condenser; at this instant $t = T/4$ we re-establish the capacity (again discontinuously) from C_2 to C_1 ($C_2 = C_{min}$; $C_1 = C_{max}$) without doing any work. It is noted that the amount of energy just added still remains in the system as we assume it to be conservative (or rather, nondissipative); therefore the energy stored in the system is now greater than it was prior to $t = 0$. At the instant $t = T/2$ we reduce again the capacity to C_2 which adds another increment of energy, and at the time $t = 3T/4$ we re-establish the capacity to C_1 without doing any work, etc. Thus the capacity will be always reduced at the instants $t = 0, T/2, T, \ldots$ with the additions of energy to the system and re-established to its maximum value at the instants $t = T/4, 3T/4, \ldots$ without doing any work.

In terms of the Hill–Meissner equation a^2 corresponds to C_0 and b^2 to ΔC, and the process just described is the solution of this d.e. Since, however, this equation is nothing but an alternate sequence of two linear d.e., it is simpler to attack it on this basis which gives a simple phase-plane representation.

We consider the d.e. of an oscillating circuit

$$L_0\ddot{q} + (1/C)q = 0$$

and assume, as previously, C varying in steps between C_1 and C_2. Using the notations $\gamma = \Delta C/C$, $\gamma \ll 1$ and $1/L_0C_0 = \omega^2$, the preceding equation can be written as

$$\ddot{q} + (1 \mp \gamma)\omega^2 q = 0 \tag{6.4}$$

where the minus sign corresponds to C_1 and the plus to C_2. This equation may be regarded as an alternate sequence of two d.e.

$$\ddot{q} + \alpha_1 q = 0; \qquad \ddot{q} + \alpha_2 q = 0 \tag{6.5}$$

where the independent variable has been changed from t to $\tau = \omega t$ and where $\alpha_1 = 1 + \gamma$; $\alpha_2 = 1 - \gamma$.

Solutions of each of these two d.e. can be represented in the phase plane (Fig. 20.2) by families Γ_1 and Γ_2 of ellipses and, since $\alpha_1 > 1$ and $\alpha_2 < 1$, the family Γ_1 corresponds to C_2 (the "vertical" ellipses) and Γ_2 to C_1 (the "horizontal" ellipses). We use the terms "vertical" and "horizontal" to designate the orientation of the larger axis of the ellipse. The two families serve as a kind of curvilinear coordinate system to represent the motion of the representative point R, because the subsequent changes from C_1 to C_2, and vice versa, operate at the frequency of the "ripple." It is useful to trace the curves of the families in the initial stage but, in ascertaining the resultant motion of R, they can be omitted, as in Fig. 20.2. The abscissa axis serves to represent the quantity q (the charge) and the axis of the ordinates, the current $\dot{q}$.

Referring to Fig. 20.2, if one starts, for instance, with the point $A(q_0,0)$ after the capacity has been reduced to the minimum C_2, the arc AB of the vertical ellipse will be followed and, at the point $B(0,\dot{q}_0)$, the capacity is increased to C_1, which transfers R on the horizontal ellipse through B so that the arc BC of that ellipse will be followed until, at C, the capacity is reduced to C_2 which transfers R at the point C on the vertical ellipse passing through C up to the point D, and so on. It is seen that the trajectory of the point R will be composed of pieces of the elliptic arcs but, *on the average*, the radius vector will increase continuously in the course of time.

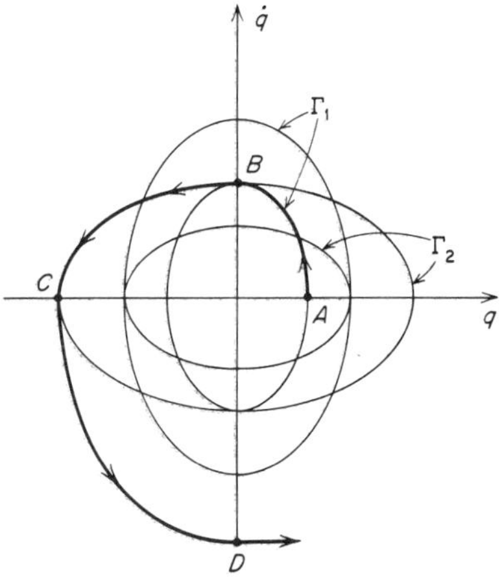

FIGURE 20.2

As r^2 represents the total energy stored in the oscillating circuit, it is seen that the energy of the system will steadily increase as the result of the operation of the ripple in C. *The system is thus nonconservative, although it is not a dissipative one.*

One notes also that, if the time of switching from C_2 to C_1, or vice versa, is changed by $T/2$, the effect would be just opposite; that is, instead of *adding* the energy to the system, the ripple, on the contrary, will *withdraw*, as can be observed from Fig. 20.2.

This graphical representation, while being simple, does not provide any criterion as to which of these two possibilities the physical system will "select." In the analytical treatment of this case the answer to this question is given by the criterion of stability, which shows that the phenomenon always occurs so as to increase its energy, inasmuch as the stable phase ($\varphi_0 = 3\pi/4$) corresponds to this condition.

As regards the *inverse parametric effect* when the ripple *withdraws* the energy, as far as known it has never been detected directly for the simple reason that the phase corresponding to this action is unstable. This does not mean yet that it cannot be produced artificially by *imposing* the desired phase by an external feedback arrangement. An arrangement of this kind may be useful in extinguishing forced oscillations of certain structures by the inverse parametric effect consisting in applying properly timed axial stresses.

7. Autoparametric excitation

In the early studies of parametric excitation it was customary to distinguish between the heteroparametric and autoparametric excitations. In the former case, parameter variations are produced by an *external* periodic excitation (of a double frequency), and in the latter, by the system itself.

In later years the only cases which turned out to be of interest are those that correspond to the heteroparametric excitation (which now is simply called "parametric excitation," dropping the word "hetero").

It is useful, however, to emphasize the physical significance of this classification although, mathematically, it amounts to the same thing. For instance, in Lord Rayleigh's experiment, the vibrating tuning fork may be regarded as an *external* source of a periodic excitation (with frequency $2f$) which modifies the tension in the stretched string at this frequency $2f$. However, with respect to the string itself, the fluctuating tension may be regarded as a variable parameter which, through the interplay of the Mathieu equation, produces lateral vibration of the string with frequency f. The phenomenon here is obviously of a heteroparametric type inasmuch as the tuning fork in this case appears as an *external* source of energy which produces the parameter variation in the parametrically excited system (the string).

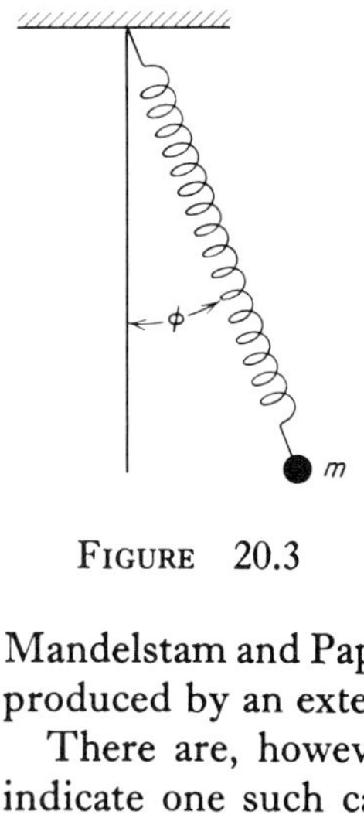

FIGURE 20.3

The same applies to the parametric generator of Mandelstam and Papalexi, where the variations of a parameter (C or L) are produced by an external source of energy.

There are, however, some cases in which this is less evident, and we indicate one such case reported by Gorelik and Witt[10] which illustrates well the significance of the autoparametric excitation. These authors investigated the motion of a physical pendulum suspended on a string and capable of oscillating in a plane (Fig. 20.3). Let m be mass of the bob; l_0 the length of the pendulum in the absence of the dynamical load (centrifugal force); r its length under load; k its spring constant; and g the acceleration of gravity. The system has two degrees of freedom: the angle φ of the pendulum and the elongation z of the spring.

The kinetic energy of the pendulum is:

$$T = \tfrac{1}{2}m(\dot{r}^2 + r^2\dot{\varphi}^2) \tag{7.1}$$

[10] G. Gorelik and A. Witt, *J. Tech. Phys.* (USSR) **3**, 1933.

and the potential energy

$$V = \frac{k}{2}(r - l_0)^2 - mgr\left(1 - \frac{\varphi^2}{2}\right) \tag{7.2}$$

where $1 - \frac{1}{2}\varphi^2 \simeq \cos \varphi$, the first term corresponding to the elasticity of the suspension and the second to gravity.

Using notations $L = r_0 + mg/k$; $z = (r - l)/l$, one can write

$$T = \frac{1}{2} mL^2(\dot{z}^2 + \dot{\varphi}^2 + 2z\dot{\varphi}^2); \qquad V = \frac{1}{2} mL^2\left(\frac{k}{m} z^2 + \frac{g}{l}\varphi^2 + \frac{g}{l} z\varphi^2\right) \tag{7.3}$$

where z and φ are small quantities which we assume to be of the first order.

The Lagrangian equations for both degrees of freedom are:

$$\begin{aligned} \ddot{z} + \frac{k}{m} z + \left(\frac{g}{2l}\varphi^2 - \dot{\varphi}^2\right) &= 0 \\ \ddot{\varphi} + \frac{g}{l}\varphi + \left(\frac{g}{l} z\varphi + 2\dot{z}\dot{\varphi} + 2z\ddot{\varphi}\right) &= 0 \end{aligned} \tag{7.4}$$

It is seen that, if the terms in parentheses are zero, the first equation represents the oscillation of the pendulum along its length with frequency $\omega_2 = \sqrt{k/m}$; and the second, the φ oscillation with frequency $\omega_\varphi = \sqrt{g/z}$ for a rigid pendulum. The terms in parentheses constitute thus a kind of a *nonlinear coupling* between the two degrees of freedom.

The interesting case arises when $\omega_2 = (p/q)\omega_\varphi$, the other cases being of no special interest. One can see, in fact, that when the pendulum oscillates with frequency ω_φ in the φ degree of freedom, the centrifugal force stretching the spring passes through *two periods* when the pendulum completes only *one period* (in the φ degree). Thus the oscillation in the φ degree of freedom induces a double frequency oscillation in the z degree.

Assume that initially $\varphi \simeq 0$ and the stretched spring has been released for $t = 0$. The z oscillation is, clearly, $z = z_0 \cos \omega_z t$. If one replaces this value of z into the second equation (7.4) and rearranges the terms, one has

$$(1 + 2z_0 \cos \omega_z t)\ddot{\varphi} - (2\omega_z \sin \omega_z t)\dot{\varphi} + \omega_\varphi^2(1 + z_0 \cos \omega_z t)\varphi = 0 \tag{7.5}$$

This is a linear d.e. with periodic coefficients and it can be reduced to the standard form of the Mathieu equation.

In view of what was said previously, the phase (in terms of the stroboscopic method) will establish itself on its stable value at which the amplitude

φ is unstable, and the φ oscillation will gradually build up from its initial $\varphi \simeq 0$ value.

There is, however, a difference here as compared to the Mandelstam–Papalexi case of the heteroperiodic excitation. In the latter case there is an *outside* supply of energy (in the form of a mechanism producing the parameter variation); here the system is of an autoperiodic type inasmuch as the parametric building up of the φ oscillation occurs at the expense of the parametric decay of the z oscillation, the system being conservative.

Ultimately the φ oscillation will be built up to some extent at the expense of the z oscillation that will be reduced, which amounts to a redistribution of energy contents stored in the two degrees of freedom through the mechanism of the autoparametric excitation.

The phenomenon, however, is not reversible. In fact, if one takes for $t = 0$, $\varphi = \varphi_0$, $z = 0$, in the φ degree, there will be established the oscillation: $\varphi = \varphi_0 \cos \omega_\varphi t$. If one replaces φ by this value in the first equation (7.4), one has

$$\ddot{z} + \omega_z^2 z = \frac{\omega_\varphi^2 \varphi_0}{4} (1 - 3 \cos 2\omega_\varphi t) \tag{7.6}$$

which is a d.e. of the harmonic oscillator acted on by an external periodic excitation with frequency $2\omega_\varphi = \omega_2$. This is an ordinary resonance for the z degree of freedom. The energy of the φ degree will be transferred into z degree which will ultimately oscillate alone. The asymmetry in this case is due to the fact that in the parametric excitation the energy is always *absorbed* in the degree of freedom of the parameter at a double frequency $2f$ and transferred into the degree of freedom of the principal oscillation (with frequency f), but the inverse process is impossible as was previously shown. This inverse process occurs, therefore, in the manner of an ordinary resonance, as was mentioned in connection with equation (7.6).

The situation would be different if, instead of an elastic pendulum just mentioned, this would be a rigid pendulum with a mass m moving up and down the length of the pendulum with frequency $\omega_z = 2\omega_\varphi$ by a source of an external power (for example, by an electric motor). In such a case an ordinary heteroparametric oscillation would develop just as in the case of a swing or Lord Rayleigh's arrangement, and a stationary amplitude would be reached in view of the nonlinearity of the problem for larger amplitudes.

Another interesting case of an autoparametric oscillation was produced by Sekerska[11] who passed an alternating current of 50 cycles frequency through a stretched wire capable of oscillating with frequency of 50 cycles.

[11] Sekerska, see paper by L. Mandelstam, N. Papalexi, *et al.*, *J. Tech. Phys.* (USSR), 1934.

As the heating effect of the current flowing through the wire in this case is of 100 cycles frequency, the tension in the wire associated with heating effect is of the same frequency. On the other hand, the stretch of the spring varies at the 100-cycle frequency so that one obtains precisely the same case of the heteroparametric oscillation as Lord Rayleigh observed by causing variations of tension in the wire by means of a tuning fork.

It is seen that the difference between what was called in the past heteroparametric and autoparametric excitations is more or less evanescent if one introduces a parameter variation by some sort of external power; it becomes real, as in the Gorelik-Witt experiments, when the system is conservative. In the latter case the autoparametric action always robs the high-frequency ($2f$) system of its energy content and transfers this energy into the low-frequency (f) system, but the inverse process is impossible on the basis of the parametric action and takes place simply in accordance with ordinary resonance by which the energy distributes itself between the two degrees of freedom. The real reason for this, as was shown previously, is due to the fact that the energy is *absorbed* in the high-frequency ($2f$) end (of the parameter variation) and is transferred into the low-frequency end (of the principal oscillation), but not vice versa. This, in turn, depends on the fact that the phase φ (in the stroboscopic analysis) is $3\pi/4$ and not $\pi/4$, the latter phase being unstable.

8. Parametric excitation by the asymptotic method

The asymptotic methods of Krylov–Bogoliubov–Mitropolsky reviewed in Chapters 14 and 15 permit obtaining successive approximations for the parametric action, as we indicate here, briefly referring for further details to the recent treatise of these authors.[12]

In the linear case of a d.e. of Mathieu

$$\ddot{x} + \omega^2(1 - h \cos \gamma t)x = 0 \tag{8.1}$$

where h is the index of modulation and, in the case of $\omega = \nu/2$, the first approximation is

$$x = a \cos \left(\frac{\nu}{2} t + \theta\right) \tag{8.2}$$

where a and θ are given by the d.e.

$$\dot{a} = -\frac{ah\omega^2}{2\nu} \sin 2\theta; \qquad \dot{\theta} = \omega - \frac{\nu}{2} - \frac{h\omega^2}{2\nu} \cos 2\theta \tag{8.3}$$

[12] N. Bogoliubov and J. Mitropolsky, *Asymptotic Methods in the Theory of Nonlinear Oscillations* (in Russian), Moscow, 1958.

with the variables $u = \cos\theta$, $v = a\sin\theta$, one can reduce the system (8.3) to the form

$$\dot{u} = \left[-\frac{h\omega^2}{2\nu} - k\right]v; \qquad \dot{v} = \left[-\frac{h\omega^2}{2\nu} + k\right]u \tag{8.4}$$

where $k = \omega - \nu/2$.

The system, being autonomous, the nature of its solution reduces to the investigation of roots of the characteristic equation (Chapter 1)

$$S^2 - \left(\frac{h^2\omega^4}{4\nu^2}\right) + k^2 = 0 \tag{8.5}$$

thus, the general solution of (8.4) is of the form

$$u = C_1 e^{St} + C_2 e^{-St};$$
$$v = C_1[(-h\omega^2/2\nu + \kappa)/S]\, e^{-St}) + C_2[(h\omega^2/2\nu - \kappa)/S]\, e^{-St}) \tag{8.6}$$

C_1 and C_2 being two constants of integration.

The amplitude a and the phase θ are given by relations

$$a^2 = u^2 + v^2; \qquad \theta = \arctan(v/u) \tag{8.7}$$

If $S = \sqrt{h^2\omega^4/4\nu^2 - k^2}$ is imaginary, a is a bounded function of t. If S is real, a increases exponentially.

The condition of reality of S is $(h\omega^2/2\nu) > k$ or, to the first order,

$$\frac{h\omega}{4} > k \tag{8.8}$$

since $\nu = 2\omega + 0(h)$. Thus, if ν is in the interval,

$$2\omega(1 - h/4) < \nu < 2\omega(1 + h/4) \tag{8.9}$$

the amplitude increases exponentially; the authors call this case: the parametric resonance.

One can build up higher approximations following the general procedure of Chapters 14 and 15.

Thus, for the second approximation (equations (4.7) and (4.8), Chapter 15) one has:

$$x = a\cos\left(\frac{\nu}{2}t + \theta\right) - \frac{ah\omega}{8\left(\omega + \frac{\nu}{2}\right)}\cos\left(\frac{3}{2}\nu t + \theta\right) \tag{8.10}$$

where a and θ are determined from the d.e.

$$\dot{a} = -\frac{ah\omega^2}{2\gamma}\sin 2\theta; \qquad \dot{\theta} = k + \frac{h^2(\omega + \gamma)}{32\left(\omega + \frac{\gamma}{2}\right)} - \frac{h\omega^2}{2\gamma}\cos 2\theta \tag{8.11}$$

The same change of variables: $u = \cos\theta$; $v = a\sin\theta$ reduces this system to

$$\dot{u} = \left[-\frac{h\omega^2}{2\nu} - k - \frac{h^2(\omega+\nu)\omega}{32\left(\omega+\frac{\nu}{2}\right)}\right]v; \quad \dot{v} = \left[-\frac{h\omega^2}{2\nu} + k + \frac{h^2(\omega+\nu)\omega}{32\left(\omega+\frac{\nu}{2}\right)}\right]u \tag{8.12}$$

The roots of the characteristic equation in this case are

$$S = \pm\sqrt{h^2\omega^4/4\nu^2 - \left[k + \frac{h^2(\omega+\nu)\omega}{32(\omega+\nu)2}\right]^2} \tag{8.13}$$

The zone of instability is now:

$$2\omega[1 - h/4 - h^2/64] < \nu < 2\omega[1 + h/4 - h^2/64] \tag{8.14}$$

If a = const in the (8.2) or in (8.11), the solution is periodic with period $4\pi/\nu$. It is necessary for this that S in the above expressions be zero, which thus determines the relation between ω and h.

For the first approximation, this gives

$$2\omega/\nu = 1 \pm h/4 \tag{8.15}$$

and, for the second,

$$2\omega/\gamma = 1 \pm h/4 + 5h^2/64 \tag{8.15a}$$

With the same degree of accuracy one can write, for the first approximation,

$$4\omega^2/\nu^2 = 1 \pm h/2 \tag{8.16}$$

and, for the second,

$$4\omega^2/\nu^2 = 1 \pm h/2 + 7h^2/32 \tag{8.16a}$$

These relations may be regarded as curves in the $(4\omega^2/\nu^2, h)$ plane which limit the regions of periodicity of solutions of (8.1). The periodic solutions with period $4\pi/\nu$ are thus for the first and for the second approximations, respectively,

$$x_1 = a_0\cos\left(\frac{\nu}{2}t + \theta_0\right); \quad x_2 = a_0\cos\left(\frac{\nu}{2}t + \theta_0\right) - \frac{a_0 h}{16}\cos\left(\frac{3\nu}{2}t + \theta_0\right) \tag{8.17}$$

The authors show that the asymptotic method gives the same results as the classical perturbation method by which Mathieu's functions C_n and S_n are developed in Fourier series.

The preceding results were derived in the case when $p = 1$ in the expression $\omega \simeq (\gamma/2)p$, but the authors develop calculations also when $p = 2$ and $p = 3$, the procedure being the same but the calculations more complicated. These higher-order parametric resonances are less marked and require a greater index of modulation h than in the usually encountered case: $p = 1$.

In the nonlinear case the method remains the same but the results are somewhat different. Consider, for instance, the d.e.

$$\ddot{x} + \omega^2(1 - h \cos \nu t)x + 2\delta\dot{x} + \gamma x^3 = 0 \tag{8.18}$$

which is the d.e. (4.2), Chapter 16 but merely written with different notations.

For the parametric resonance of order $\frac{1}{2}$, the solution in the first approximation is taken in the form

$$x = a \cos\left(\frac{\nu}{2}t + \theta\right) \tag{8.19}$$

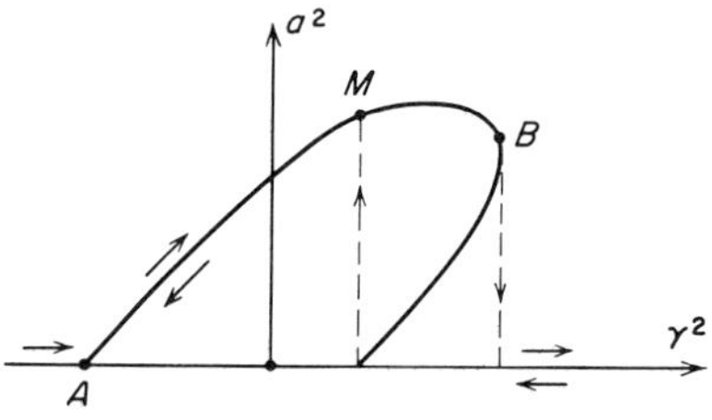

FIGURE 20.4

where, according to the general theory (Chapter 15) a and θ must satisfy the differential system

$$\dot{a} = -\delta a - \frac{ah\omega^2}{2\nu} \sin 2\theta;$$

$$\dot{\theta} = k + \frac{3\gamma a^2}{4\nu} - \frac{h\omega^2}{2\nu} \cos 2\theta \tag{8.20}$$

and, for the stationary state, one equates the right-hand sides of these equations to zero. Eliminating θ, one obtains

$$a^2 = \frac{4}{3\gamma}\left[(\nu/2)^2 - \omega^2 \mp \frac{1}{2}\sqrt{h^2\omega^4 - 4\nu^2\delta^2}\right] \tag{8.21}$$

In the (ν^2, a^2) plane (Fig. 20.4) this gives the resonance curves ABC; the branch AB is stable and BC unstable so that the phenomenon exhibits the usual hysteresis character, depending on the direction in which ν^2 varies.

For the determination of the interval of the resonance, one has to set to zero the right-hand side of (8.21). In the first approximation this gives the interval of the parametric resonance

$$\omega^2 - \frac{1}{2}\sqrt{h^2\omega^4 - 16\omega^2\delta^2} < \left(\frac{\nu}{2}\right)^2 < \omega^2 + \frac{1}{2}\sqrt{h^2\omega^4 - 16\omega^2\delta^2} \tag{8.22}$$

so that the width Δ of the resonance zone is

$$\Delta = \omega\sqrt{h^2\omega^2 - 16\delta^2} \tag{8.23}$$

Thus, the parametric excitation (or resonance) exists if

$$h\omega > 4\delta \tag{8.24}$$

One notes that this is the same conclusion that was obtained previously by the stroboscopic method. The conclusion in both methods is the same: the energy input into the system through the parameter variation must be greater than the energy dissipated by the system in order that the parametric resonance could develop.

Chapter 21

OSCILLATIONS CAUSED BY RETARDED ACTIONS

1. Introductory remarks

Recent developments in the theory and use of automatic control systems resulted in studies of an entirely different class of oscillations amenable to the so-called *difference-differential equations* (d.d.e. for short). The latter appear, however, as a special class of more general *functional* equations, such as integro-differential equations and a number of other equations of a still more complicated type.

A d.e. in its physical interpretation may be regarded as a mathematical tool which, from the knowledge of *the present*, gives a means for obtaining the future of a phenomenon by proceeding in small steps from a given initial condition; the past is not involved in this argument.

In the functional equations, on the other hand, the past exerts its influence on the present and, hence, on the future. Here this influence is organically welded in the argument and is not of a formal character (reversing the time). In the integro-differential equations the *whole past* exerts its influence in the determination of the present (and the future). In a more restricted case of d.d.e. which form the object of this chapter, only a certain *finite interval* of the immediate past is involved in the determination of the present.

The entire subject has grown considerably in recent years and it is beyond the scope of this chapter to discuss this matter in detail, the only aim here being to outline briefly some connections between certain types of these oscillations and some aspects of the theory of d.d.e.

These connections are yet poorly defined and not all that results from the existing theory of these functional equations is useful in the theory of oscillations; conversely, not all that is required in the latter can yet be offered by the former. Thus, for instance, most of the mathematical work done so far concerns the *linear* d.d.e., whereas the oscillatory phenomena amenable to these equations are essentially of a nonlinear type, as we shall see later. The difficulties of using these d.d.e. in a linear form are much the same as those we encountered in oscillations governed by ordinary d.e. As regards the nonlinear extention of d.d.e., it is still in an early stage, limited mostly to the theorems of existence, and not sufficient for applied problems aiming primarily at the determination of conditions of the stationary state.

These phenomena appear at certain critical thresholds just as all other nonlinear oscillations do when a d.e. depends on a parameter. This suggests a study of these equations with a variable parameter.

In spite of the two complications imposed by applied problems—namely, (1) nonlinearity and (2) variable parameter—the establishment of a first contact between the theory and the observed facts is still possible if one restricts the problem to a special case when *the time lag is also small.* In this case it is possible to bring the problem within the realm of the theory of the first approximation. The results so obtained seem to be sufficient to account for the observed facts, but the problem is still somewhat restricted for the above reasons.

The principal difficulty in studies of d.d.e. is in the linear problem itself which, as will be shown, is of a special *transcendental* character. In the preceding theory (of Poincaré) the linear problem was particularly simple resulting, generally, in a simple generating solution. Here, on the contrary, the linear problem leads always to an infinite spectrum of frequencies with which such a system can oscillate. The determination of this spectrum requires a corresponding determination of zeros of certain analytic functions.

If one is interested only in connections between the theory and problems of oscillations, most of these difficulties can be set aside, inasmuch as what is of interest in such a case is not the *entire* infinite spectrum of frequencies but only a few of them (most frequently only one) that are actually responsible for the appearance of these oscillations in practice. This simplifies the problem considerably.

Historically the discovery of d.d.e. was due apparently to Laplace[1] and Condorcet.[2] Practically nothing was done throughout the nineteenth century; and only after the First World War did the study of these equations gain momentum, principally on account of applied problems that appeared

[1] P. S. Laplace, *Mem. de Math. et Phys., Ac. Royale,* 1774, 1779, 1782.
[2] A. N. Condorcet, *ibid.,* 1771.

at that time.[3] This progress continues in an ever-increasing manner. The reader is referred to recent texts on this subject.[4, 5]

In this chapter we are obliged to restrict the subject considerably, leaving out most of the mathematical foundations of the theory and concentrating rather on certain applied questions necessary for the establishment of a contact with the physical nature of these "retarded oscillations."

2. Difference-differential equations arising in applications

A sufficiently general form of a homogeneous linear d.d.e. is

$$x^{(n)}(t) + a_{n-1}x^{(n-1)}(t - h_{n-1}) + \dots + a_1\dot{x}(t - h_1) + a_0 x(t - h_0) = 0 \tag{2.1}$$

where x is the unknown function, t is the independent variable (the time), a_i are constant coefficients (in applications they are always real numbers), and $h_{n-1}, h_{n-2}, \dots, h_1, h_0$ are real constants; if they are positive, they are usually called *time-lags*.

In these notations $x^{(i)}(t - h_i)$ is the derivative of the ith order considered at the *retarded time* $(t - h_i)$. One can consider, therefore, a d.d.e. as a linear combination of a certain number of *retarded derivatives*.

If $h_{n-1} = \dots = h_1 = h_0 = 0$, the d.d.e. becomes an ordinary d.e. The form (2.1), although not the most general one, is still too complicated from the standpoint of applications, which have not progressed far enough as yet.

The most commonly encountered d.d.e. are of the form

$$\ddot{x}(t) + a_1\dot{x}(t - h) + a_0 x(t) = 0 \tag{2.2}$$

$$\ddot{x}(t) + a_1\dot{x}(t) + a_0 x(t - h) = 0 \tag{2.3}$$

If one assigns to these d.d.e. the obvious physical meaning, (2.2) is a d.d.e. with a *retarded damping*, and (2.3) is a d.d.e. with a *retarded restoring force*. These d.d.e. are studied in more detail in the following section.

In order to simplify the notations we shall attach a subscript h to the

[3] E. Schmidt, *Math. Annalen* **70**, 1911.

[4] R. Bellman, Report 256, Project Rand, 1954; W. Hahn, *Math. Annalen* **131**, 1956; E. Pinney, *Ordinary Difference-Differential Equations*, University of California Press, 1958.

[5] A. D. Myshkis, *Usp. Math. Nauk* (in Russian) **4**, 1949; *Am. Math. Soc.*, translation No. 55, 1951.

retarded terms, writing x_h, $\dot{x}_h$ instead of $x(t - h)$ and $\dot{x}(t - h)$, and (2.2) and (2.3) will be written as

$$\ddot{x} + a_1\dot{x}_h + a_0 x = 0 \tag{2.4}$$

$$\ddot{x} + a_1\dot{x} + a_0 x_h = 0 \tag{2.5}$$

One can obviously consider d.d.e. of a *mixed type* in which both nonretarded and retarded terms are present so that, for instance, one can have

$$\ddot{x} + a_1'\dot{x} + a_1\dot{x}_h + a_0 x = 0 \tag{2.6}$$

$$\ddot{x} + a_1\dot{x} + a_0' x + a_0 x_h = 0 \tag{2.7}$$

The first of these equations is characterized by a nonretarded (natural) damping $a_1'\dot{x}$ as well as by the retarded one: $a_1\dot{x}_h$. The second equation (2.7) has a nonretarded restoring force (or moment) $a_0'x$ and a retarded one $a_0 x_h$.

The term "nonretarded" does not require further explanation, inasmuch as this is the usual significance of terms $\ddot{x}(t)$, $\dot{x}(t)$, $x(t)$ encountered in ordinary d.e. As to the "retarded" terms, they are often encountered in control problems where a certain control action is produced *artificially*. In such a case, in view of inevitable time-lags in a control system, this action does not relate to the instant t at which it is supposed to be exerted, but to a past instant $t - h$, (that is, appears with a time-lag).

Since in the modern control systems (involving electron-tube circuits) the intensity of control action can be adjusted, it is preferable to emphasize this circumstance by attaching the symbol λ, a variable parameter, as a coefficient of retarded terms. In these notations (2.6) and (2.7) can be written

$$\ddot{x} + a_1'\dot{x} + \lambda\dot{x}_h + a_0 x = 0 \tag{2.8}$$

$$\ddot{x} + a_1\dot{x} + a_0' x + \lambda x_h = 0 \tag{2.9}$$

Equation (2.8) represents, for instance, a certain dynamical (electrical or mechanical) system in which an artificially produced damping $\lambda\dot{x}_h$ is added in order to increase an insufficient natural damping $a_1'\dot{x}$ and, likewise, (2.9) represents a system with a natural ($a_0'x$) and artificially produced (λx_h) restoring forces.

For instance, in a problem of automatic steering (in azimuth) of a craft, the latter has no preference to follow one direction rather than the other, which means that $a_0' = 0$ in (2.9); as to the artificially produced restoring force, this is generally accomplished by connecting the rudder with a direction indicating instrument through a control system but, in view of a time lag such an action is again retarded and corresponds to the term λx_h in (2.9).

All these effects were of little importance in the past when the controlled craft or missiles were moving at relatively slow speeds. However, with increasing velocities, these effects become more important and, at times, even critical owing to a sudden release of oscillations leading to instability. Such oscillations are generally unpredictable on the basis of the d.e. of motion; their presence becomes, however, clear from the study of a "neighboring" d.d.e. but the zone separating the two equations is generally so small that this is often overlooked in applications.

3. Characteristic equation; neighborhood of a harmonic solution

A d.e. (for instance, with constant coefficients) has the property that the degree of its characteristic equation is always equal to the order of the d.e. The fundamental property of a d.d.e. is that its characteristic equation is always of an *infinite degree* whatever its order may be. This means that a d.d.e. (*considered as an ordinary d.e.*) may have an infinity of roots of its characteristic equation and it may happen that some of these roots have positive real parts, which leads to a *self-excitation* of oscillations with frequencies corresponding to the imaginary parts of these roots. It is precisely in this connection that the interest to these d.e. appeared in applied problems.

One readily ascertains this peculiarity of d.d.e. if one develops a retarded quantity as a Taylor series in terms of the corresponding nonretarded quantities. Thus, for instance

$$\dot{x}_h = \dot{x}(t-h) = \dot{x}(t) - \frac{h}{1!}\ddot{x}(t) + \frac{h^2}{2!}\dddot{x}(t) - \cdots$$

$$= \dot{x}\left(1 - \frac{h}{1!}(\ddot{x}/\dot{x}) + \frac{h^2}{2!}(\dddot{x}/\dot{x}) - \cdots\right) \tag{3.1}$$

If one tries to satisfy a d.d.e. by a solution of the form

$$x = x_0 e^{zt}, \qquad z = \alpha + i\omega \tag{3.2}$$

it is clear that $\ddot{x}/\dot{x} = z$, $\dddot{x}/\dot{x} = z^2, \ldots$, and (3.1) becomes

$$\dot{x}_h = \dot{x}\left(1 - \frac{hz}{1!} + \frac{h^2z^2}{2!} - \cdots\right) = \dot{x}e^{-hz} \tag{3.3}$$

If one substitutes this expression for $\dot{x}_h$ into (2.4), for instance, one obtains an *algebraico-transcendental* characteristic equation

$$f(z) = z^2 + a_1 z e^{-hz} + a_0 = 0 \tag{3.4}$$

and the problem consists in determining the zeros of the entire function

$f(z)$. It is seen that the above passage to the limit simplifies the problem involving a characteristic equation of an infinitely high degree (if one considers a d.d.e. as an ordinary d.e. of an infinitely high order) but, introducing instead, a problem of a transcendental nature.

The characteristic equation of a linear d.d.e. can be thus written at once. For example, for a d.d.e. of a mixed type

$$\ddot{x} + a_1'\dot{x} + a_1\dot{x}_h + a_0'x + a_0x_h = 0$$

the characteristic equation is

$$f(z) = z^2 + a_1'z + a_1ze^{-hz} + a_0' + a_0e^{-hz} = 0$$

and so on.

Considerable attention has been given to this problem in recent years; we refer in particular to a publication of Pontriagin[6] which will be mentioned later.

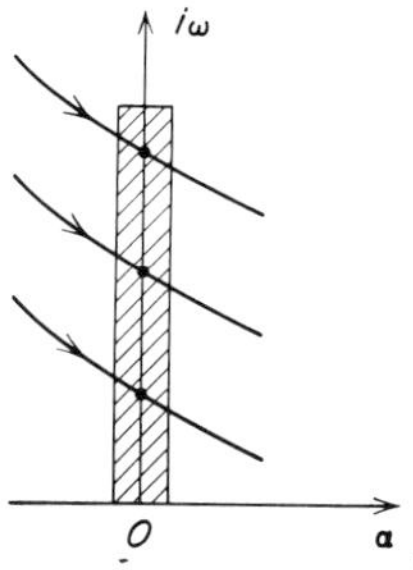

FIGURE 21.1

In applied problems the situation is somewhat simpler, but requires some additional extensions. The simplification in applied problems arises from the fact that, instead of determining the totality of zeros of an analytic function $f(z)$, it is often sufficient in applications to determine only one or two such zeros in a limited area of the complex plane (Fig. 21.1). For instance, if a system (for example, control system) normally operates with a frequency, say, 1 (in some normalized scale), it is obviously useless to try to determine zeros of $f(z)$ whose imaginary parts are of the order, say, 100 or higher, although the frequencies of the order, say, 5 may be still of interest. In a similar way, on physical grounds, the energy absorption ($\alpha < 0$) or dissipation ($\alpha > 0$) in such oscillations, as a rule, is not very large and, if the order is approximately known, it is always possible to determine a region (shaded area in Fig. 21.1) in which the location of zeros of $f(z)$ may be of interest.

On the other hand, these problems have an additional complication which generally is less investigated in the theory of d.d.e., namely, the effect of the variable parameter λ, as we mentioned previously. It is clear that, if λ varies, the zeros of $f(z)$ move in some manner in the complex plane (Fig. 21.1) and it may happen that, for some special *harmonic* value $\lambda = \lambda_1$ of the parameter λ, the path of a zero may cross the imaginary axis $i\omega$. When this happens, the analytic function $f(z)$ has a purely imaginary root $z_1 = i\omega_1$ which means that for this particular *harmonic value* $\lambda = \lambda_1$ of the parameter λ, there exists a purely harmonic (or sinusoidal) solution of the d.d.e.

[6] L. Pontriagin, *Bull. Ac. Sc.* (USSR), Math. Series **6**, 1942.

These harmonic values are of interest in that they permit exploring what happens not only *exactly* for these values but also *in their neighborhood.* The fact that, on physical grounds, the real part α of the roots is generally small, permits specifying the problem still closer and this, as will appear later, gives sufficient information regarding the connection between the theory of linear d.d.e. and corresponding physical facts.

As an example of a transcendental problem of this nature we consider first the d.d.e. (2.5) which may be regarded, for instance, as a control problem in presence of a retarded restoring force (or moment). We have thus

$$\ddot{x} + p\dot{x} + \lambda x_h = 0 \tag{3.5}$$

with the characteristic equation

$$f(z) = z^2 + pz + \lambda e^{-hz} \tag{3.6}$$

Substituting $z = \alpha + i\omega$ in this equation and separating the real and the imaginary parts, one has

$$\begin{aligned} \alpha^2 - \omega^2 + p\alpha + \lambda e^{-h\alpha} \cos \beta &= 0 \\ 2\alpha\omega + p\omega - \lambda e^{-h\alpha} \sin \beta &= 0 \end{aligned} \tag{3.7}$$

where $\beta = \omega h$; whence

$$\begin{aligned} \cos \beta &= (\omega^2 - p\alpha - \alpha^2)/\lambda e^{-h\alpha} \\ \sin \beta &= (p\omega + 2\alpha\omega)/\lambda e^{-h\alpha} \end{aligned} \tag{3.8}$$

From these two equations one obtains two derived equations

$$\text{cotan } \beta = (\omega^2 - p\alpha - \alpha^2)/\omega(p + 2\alpha);$$

$$(\omega^2 - p\alpha - \alpha^2)^2 + \omega^2(p + 2\alpha)^2 = \lambda^2 e^{-2h\alpha} \tag{3.9}$$

If $\alpha = 0$, one has

$$\cos \beta_1 = \omega_1{}^2/\lambda_1; \qquad \sin \beta_1 = p\omega_1/\lambda_1 \tag{3.10}$$

$$\text{cotan } \beta_1 = \omega_1/p = (1/ph)\,\beta_1 \tag{3.11}$$

$$\beta^4 + h^2p^2\beta^2 - \lambda^2h^4 = 0 \tag{3.12}$$

The subscripts 1 are attached in order to indicate that these quantities correspond to the harmonic values ($\alpha = 0$). Equation (3.11) can be solved graphically as intersection of two curves (Fig. 21.2) $y_1 = \text{cotan } \beta$ and $y_2 = (1/ph)\beta$. The curve y_1 is a multivalued function of β and we indicate it in strips 1, 2,. . . of intervals $\pi/2$, Fig. 21.2. The roots are indicated as β', β'', β''', but in view of (3.10), the root β'' is to be rejected and only the roots in the first, fifth, ninth,. . . strips are to be retained. As

(3.11) does not contain the variable parameter λ, the roots β', β'',. . . may be regarded as *fixed roots* (depending only on the fixed parameters p and h). These roots give, therefore, the harmonic frequencies ω_1', ω_1''',. . . of the transcendental spectrum of the linear problem.

There remains yet the second equation (3.12) whose positive root is

$$\beta_{11}(\lambda) = +\sqrt{-\frac{p^2h^2}{2} + \sqrt{\frac{p^4h^4}{4} + \lambda^2h^4}} \tag{3.13}$$

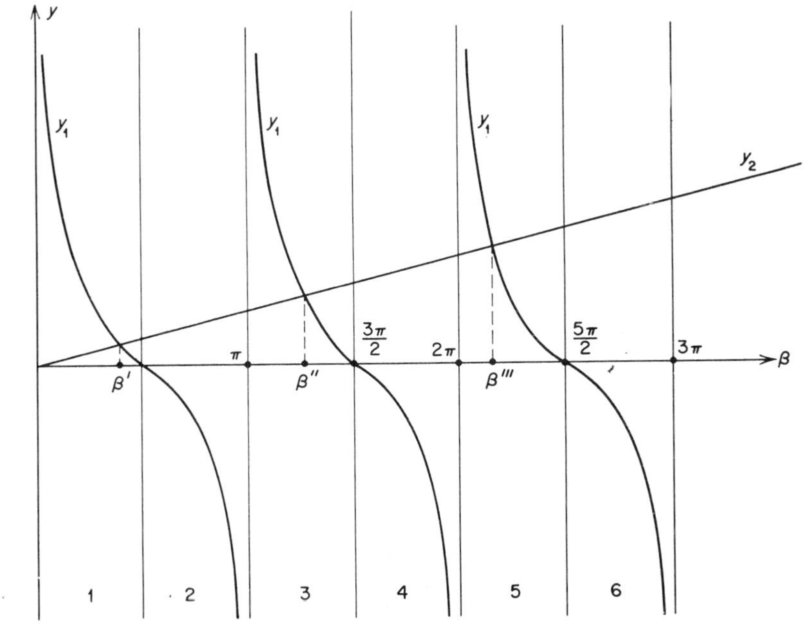

FIGURE 21.2

This root is a function of λ: $\beta_{11}(\lambda)$ which increases monotonically with λ starting with $\beta_{11}(0) = 0$.

Hence, as λ increases continuously from the value $\lambda = 0$, the root $\beta_{11}(\lambda)$ moves also continuously along the abscissa axis and, for a certain value, $\lambda = \lambda_1$ may coincide with one of the fixed roots, say β'. At this point

$$\beta' = \beta_{11}(\lambda_1) \tag{3.14}$$

and similarly for other fixed roots β''',. . . for other values λ_1''',. . ..

Whenever such a coincidence (3.14) occurs, the two equations: (3.11) and (3.12) have a common harmonic root $\beta_1 = h\omega_1$ and the analytic function $f(z)$ has a purely imaginary root $i\omega_1$, so that $f(i\omega_1) = 0$.

If λ continues to increase, there will appear other coincidences between the fixed roots and the moving one so that to a discrete sequence $\lambda_1, \lambda_1', \lambda_1'', \ldots$ of the parameter values will correspond also a discrete sequence of harmonic frequencies $\omega_1, \omega_1', \omega_1'', \ldots$, thus forming a *transcendental spectrum*, the ratios ω_1/ω_1', ω_1'/ω_1'' being generally incommensurate.

Consider now what happens *in the neighborhood* of one such harmonic root (3.14). For this purpose we have to add a small increment Δz to the harmonic value $z_1 = i\omega_1$ so that $z = i\omega_1 + \Delta z$. Substituting this into $f(z)$, we have:

$$(z_1 + \Delta z)^2 + p(z_1 + \Delta z) + (\lambda_1 + \Delta\lambda)e^{-h(z_1+\Delta z)} = 0 \qquad (3.15)$$

Assuming $\Delta\lambda$ and $\Delta z = \Delta\alpha + i\Delta\omega$ as small quantities of the first order and carrying out calculations only to this order, one has

$$\frac{\Delta\alpha}{\Delta\lambda} = \frac{1}{\lambda_1}\frac{p\omega_1^2 + h\lambda_1^2}{(p - h\omega_1^2)^2 + \omega_1^2(2 + hp)^2} = M$$

$$\frac{\Delta\omega}{\Delta\lambda} = \frac{1}{\lambda_1}\frac{\omega_1 p^2 + 2\omega_1^3}{(p - h\omega_1^2)^2 + \omega_1^2(2 + hp)^2} = N \qquad (3.16)$$

where M and N are positive constants. Hence, if $\Delta\lambda > 0$, both $\Delta\alpha$ and $\Delta\omega$ are also positive. But for $\Delta\lambda = 0$, one has a harmonic root, that is, $\alpha = 0$; hence for $\Delta\lambda > 0$, one has also $\alpha > 0$; and for $\Delta\lambda < 0$, $\alpha < 0$.

This means that, for $\lambda > \lambda_1$, $\alpha > 0$, the system *absorbs* the energy; for $\lambda < \lambda_1$, $\alpha < 0$, the system, on the contrary, dissipates it but, since the d.d.e. is linear, there is nothing that would limit this energy absorption for $\Delta\lambda > 0$ or draining of energy for $\Delta\lambda < 0$. Therefore, it is obvious that, on physical grounds, such a linear retarded system is unstable, since the only point at which it can remain in the stationary state is when the parameter λ has *exactly* a harmonic value which is meaningless on physical grounds.

Thus from the standpoint of stability of self-excited oscillations, a linear d.d.e. is unable to account for the observed facts, just as it was impossible to account for the existence of self-sustained oscillations on the basis of an ordinary linear d.e. On the other hand, as regards the transcendental frequency spectrum of these harmonics, the theory of linear d.d.e gives a correct account, as will be mentioned later.

Hence, if one tries to fit the oscillations appearing in retarded systems into the framework of the linear theory of d.d.e., one has exactly the same difficulty that was experienced in the theory of ordinary d.e. when one tried to fit self-sustained oscillations into a similar linear pattern.

Obviously, the only issue from this situation is to investigate the *nonlinear* d.d.e. In fact, all observed oscillations of this kind start spontaneously from rest as soon as a certain threshold value of a parameter is reached; moreover, they generally exist not only for *one isolated* value of the parameter (as indicated by the linear theory), *but for a certain interval* of these parameter values; finally, oscillation persists with a definite stationary amplitude for a given value of parameter.

All this presents a familiar picture of nonlinear, self-excited oscillations. Before considering this question for the nonlinear d.d.e. in the next section, we shall briefly outline the procedure for determining the roots of the characteristic equation corresponding to the d.d.e. (2.8), for instance. This problem of a *retarded damping* is also of considerable interest in applications.[7] This equation was encountered under special conditions of antirolling stabilization systems,[8] as will be mentioned later.

It is necessary, therefore, to investigate here a mixed d.d.e. containing both nonretarded and retarded damping for obvious physical reasons, viz.:

$$\ddot{x} + p\dot{x} + \lambda\dot{x}_h + \omega_0{}^2 = 0 \tag{3.17}$$

The characteristic equation in this case is

$$z^2 + pz + \lambda z e^{-hz} + \omega_0{}^2 = 0 \tag{3.18}$$

As previously, replacing $z = \alpha + i\omega$, separating the real and the imaginary parts, one has the two following equations (compare with (3.11) and (3.12)):

$$\alpha^3 + (p + \omega \operatorname{cotan} \beta)\alpha^2 + (\omega^2 + \omega_0{}^2) + \omega[(\omega^2 - \omega_0{}^2) \operatorname{cotan} \beta + p\omega] = 0 \tag{3.19}$$

$$(\alpha + \alpha p - \omega^2 + \omega_0{}^2)^2 + \omega^2(2\alpha + p)^2 = \lambda^2(\alpha^2 + \omega^2)e^{-2h\alpha}$$

Setting $\alpha = 0$ in these equations one has the harmonic relations

$$\sin \beta_1 = \frac{\omega_1{}^2 - \omega_0{}^2}{\lambda\omega_1}; \qquad \cos \beta_1 = -\frac{p}{\lambda}; \qquad \tan \beta_1 = \frac{\omega_0{}^2 - \omega_1{}^2}{p\omega_1} \tag{3.20}$$

$$\beta^4 - (2\beta_0{}^2 + \lambda^2 h^2 - p^2h^2)\beta^2 + \beta_0{}^4 = 0 \tag{3.21}$$

recalling that $\beta = \omega h$. The third equation (3.20) gives the "fixed roots" by construction similar to that which was explained in connection with Fig. 21.2. Here we have two curves (Fig. 21.3).

$$y_1 = \tan \beta \quad \text{and} \quad y_2 = n/\beta - m\beta; \qquad n = \omega_0{}^2h/p \quad \text{and} \quad m = 1/ph$$

[7] N. Minorsky, *C. R. Ac. Sc.* (Paris) **226**, 1948; *J. Appl. Phys.* **19**, 1948.
[8] N. Minorsky, *J. Appl. Mech.*, 1942.

The second equation (3.20) allows only the roots which are in the quadrants in which $\cos \beta_1 < 0$. Moreover, there is an additional restriction imposed by the first equation (3.20). In fact, $y_2 = 0$ when $\beta = \sqrt{n/m} = \omega_0 h = \beta_0$. The point β_0 of intersection of the curve y_2 with the β axis represents thus (up to a constant factor h) the synchronous undamped frequency (Fig. 21.3). To the left of this point $\beta_1^2 - \beta_0^2 < 0$, so that $\sin \beta_1 < 0$ and, to the right, $\beta_1^2 - \beta_0^2 > 0$; hence $\sin \beta_1 > 0$.

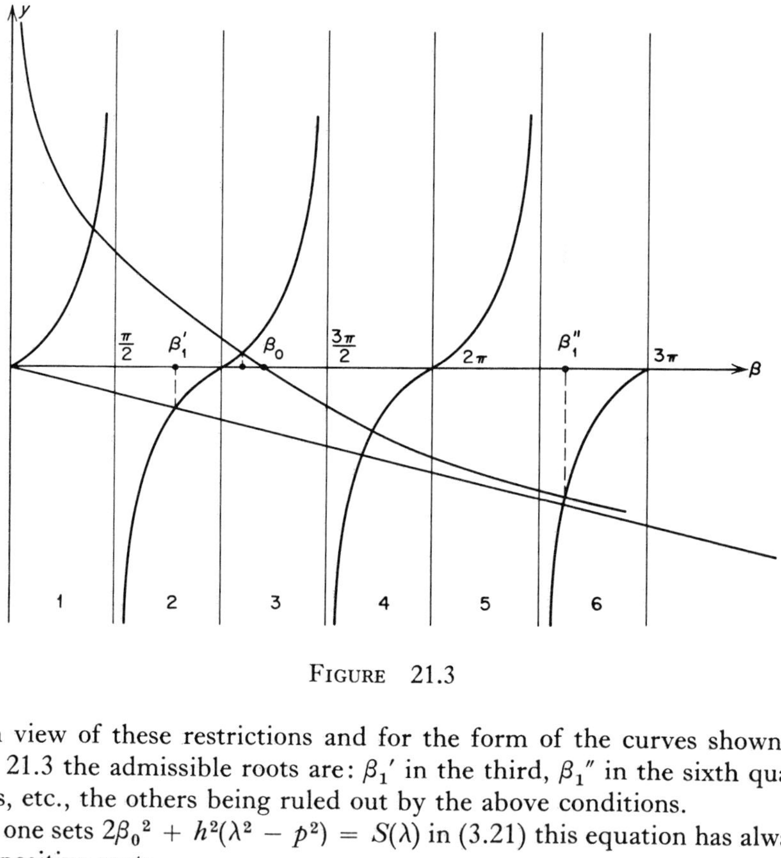

FIGURE 21.3

In view of these restrictions and for the form of the curves shown in Fig. 21.3 the admissible roots are: β_1' in the third, β_1'' in the sixth quadrants, etc., the others being ruled out by the above conditions.

If one sets $2\beta_0^2 + h^2(\lambda^2 - p^2) = S(\lambda)$ in (3.21) this equation has always two positive roots

$$\beta_{11}^2 = S/2 \pm \sqrt{S^2/4 - \beta_0^2} \tag{3.22}$$

Since the quantity under the square root reduces to the expression

$$h^2(\lambda^2 - p^2)[4\beta_0^2 + h^2(\lambda^2 - p^2)]$$

which is positive because $\lambda \geq p$ by the second equation (3.20).

If the parameter λ increases continuously beginning with the value $\lambda = p$, the roots $(\beta_{11})_1$ and $(\beta_{11})_2$ vary also monotonically; the root $(\beta_{11})_1$ (corresponding to the plus sign in (3.22) increases while $(\beta_{11})_2$ decreases, and, as was explained previously, one of these two "moving" roots may coincide with one of the fixed roots. When such a coincidence occurs, the characteristic equation (3.18) has a purely imaginary root $z_1 = i\omega_1$ whose frequency is determined by this graphical construction.

The peculiarity of this system is that there are here *two* roots moving in opposite directions from the point $\beta = \beta_0$, and which of these two roots will encounter first a fixed root is a matter of the configuration of the latter roots (that is, depends on fixed parameters of the system). Thus a relatively slow frequency ("floating") may give way suddenly to a rather high frequency or vice versa. These somewhat erratic phenomena have a simple explanation on the basis of such frequency diagrams.

In this case, as in the previously analyzed one, no further information can be obtained from these diagrams for the reasons previously set forth, viz.: The question of a self-sustained oscillation of this kind cannot be obtained from the *linear* d.d.e., although the transcendental frequency spectrum determined in this manner generally gives a fairly good approximation to the observed frequencies.

4. Advanced versus retarded actions

Although we referred always to the "retarded actions" ($h > 0$), the conclusions are similar in the case of *advanced* actions ($h < 0$). The latter are always possible in modern systems of automatic controls.

In fact from the very beginning of the theory of control action,[9] the effect of higher time derivatives components in a control system has been ascertained and used for the purpose of a modification of *actual* parameters of a system to be controlled so as to obtain more appropriate *effective* parameters of the controlled system.

For instance, if one has, say, three control actions: (1) x, the departure (for example, from a desired equilibrium point); (2) $\dot{x}$, the rate of departure; and (3) $\ddot{x}$, the rate of the rate (that is, acceleration of departure) and if one combines them linearly, viz.: $x + m\dot{x} + n\ddot{x}$, the combination so founded *may* represent a certain number of terms of a Taylor expansion or not, according to the magnitude of the coefficients m and n. In fact, if a retarded process (limited to the first three terms of its Taylor expansion) is

$$x(t - h) = x(t) - \frac{h}{1!}\dot{x}(t) + \frac{h^2}{2!}\ddot{x}(t)$$

[9] N. Minorsky, *J. Franklin Inst.* **232,** 1941.

and if it is intended to be compensated for by a control action $f(x) = x(t) + mx(t) + nx(t)$, it is necessary that $x(t - h) + f(x) = 0$, which requires that $m = -h$; $n = h^2/2!$, etc. If the coefficients m and n do not correspond exactly to these relations, the controlled system may still exhibit either a time-lag ($h > 0$) of an advance of the phase ($h < 0$) according to whether the control system is *undercompensated* or *overcompensated* by these derivatives' actions. In such systems with time-derivatives control components, an *advance* is just as possible as a *time-lag*. Nothing is changed in the preceding discussions except that one has to introduce $h < 0$ instead of $h > 0$, with a corresponding change in the argument.

An interesting case arises if a dynamical system of the second order has two retarded actions with the same time-lag h, which gives rise to a d.d.e. of the form

$$\ddot{x} + \lambda \dot{x}_h + \rho x_h = 0 \tag{4.1}$$

which is a particularly simple case. The coefficients λ and ρ may be assumed as variable parameters of the preceding theory if one wishes to carry out a complete discussion as we did previously.

In this case the characteristic equation is

$$z^2 + \lambda z e^{-hz} + \rho e^{-hz} = 0 \tag{4.2}$$

or, which is the same,

$$z^2 e^{hz} + \lambda z + \rho = 0 \tag{4.3}$$

but the latter equation may be regarded as a characteristic equation of the d.d.e.

$$\ddot{x}_{-h} + \lambda \dot{x} + \rho x = 0 \tag{4.4}$$

which is a d.d.e. of a system with an "advanced inertia." There is no difficulty in discussing the properties of such a system following the same procedure as previously, but we shall not enter into this matter here inasmuch as we shall encounter these properties of "advanced" systems in the nonlinear case which is of a greater interest than the linear retarded (or advanced) systems.

5. Nonlinear problem; stationary state; frequency correction and stability

The exact theory of nonlinear d.d.e. has not yet progressed sufficiently for use in applications. Most of the work accomplished so far[10] is limited to the theorems of existence but this, in spite of its considerable theoretical interest, does not give a tool for dealing with applied problems. It is

[10] F. H. Brownell, *Contribution to Theory of Nonlinear Oscillations*, No. 20, 1950.

necessary, therefore, to restrict the problem to a point at which one can bring it within the scope, at least, of the first approximation where some preliminary conclusions may be still obtained.

In all preceding chapters, we made use of the method of small parameters. affecting nonlinear terms which made it possible to bring problems within the scope of series approximations.

It seems possible however to extend this general method of attack also to the d.d.e. in the asymptotic case if *one assumes that the time-lag h is also small* as will be now shown.

Consider a nonlinear d.d.e. of the form

$$\ddot{x} + a\dot{x} + \lambda(x_h + \mu x_h{}^3) = 0 \tag{5.1}$$

This is the d.d.e. (3.5) to which we add a small nonlinear term $\mu x_h{}^3$. If $\mu = 0$ we have a liner d.d.e. which was studied previously. If $h = 0$ this is an ordinary nearly linear d.e.

Since the linear transcendental problem is sufficiently clear from Section 3, we can concentrate our attention *on one particular frequency* of the spectrum and investigate what happens in its neighborhood. For the linear problem, as was seen in Section 3, to a harmonic root $z_1 = i\omega_1$ corresponds a harmonic value $\lambda = \lambda_1$ of the parameter. In the neighborhood of this value for $\lambda = \lambda_1 - \Delta\lambda(\Delta\lambda \ll 1)$, $\alpha < 0$ and the oscillation disappears; for $\lambda = \lambda_1 + \Delta\lambda$, $\alpha > 0$ and amplitude grows indefinitely and this, as was mentioned, does not give any correct representation of such phenomena since, for $\lambda = \lambda_1 + \Delta\lambda$, one generally observes a stable stationary amplitude.

We propose now to investigate the problem *locally* in the neighborhood of a harmonic frequency ω_1 which, by a proper change of the independent variable, can be always reduced to: $\omega_1 = 1$ and, therefore $\beta_1 = h$. As to the parameter λ, its scale being arbitrary, one can also assume that the corresponding harmonic value: $\lambda_1 = 1$.

We consider thus the following d.d.e.

$$\ddot{x} + a\dot{x} + x_h + \varepsilon x_h{}^3 = 0 \tag{5.2}$$

where a, ε, and h will be assumed as small positive quantities of the first order.

We can now apply the stroboscopic method with a view to obtaining the first approximation. There appears, however, a difficulty in this case in that, owing to the existence of a nonlinear frequency correction in the first approximation, the period is not 2π. It is preferable to begin with the second equation (in ψ); with notations used previously the equivalent system corresponding to (5.2) is:

$$\dot{x} = y; \qquad \dot{y} = -ay - x_n - \varepsilon x_h{}^3 \tag{5.3}$$

Forming the combination: $x\dot{y} - y\dot{x} = \rho(d\psi/dt)$ and using the formula of the definition:

$$x_h = xe^{-hz} = xe^{-h\alpha}e^{-i\beta} \tag{5.4}$$

where both h and α are $0(\mu)$, $\exp(-h\alpha) \simeq 1$, one has to the first order (since $x = r\cos\psi$)

$$x_h \simeq r\cos(\psi - \beta) \tag{5.5}$$

This formula shows that the retardation is noticeable only in the phase. It is recalled that $\beta = \omega h$ and, since ω is finite and $h \sim 0(\mu)$, β is also $0(\mu)$ so that in the first approximation $\cos\beta \simeq 1$, $\sin\beta \simeq \beta$, which gives

$$\begin{aligned}
xx_n &= r\cos\psi\, r\cos(\psi - \beta) = \rho\cos\psi\,(\cos\psi + \beta\sin\psi)\\
&= \tfrac{1}{2}(1 + \cos 2\psi + \beta\sin 2\psi)\\
yx_h &= r\sin\psi r\cos(\psi - \beta) = \rho\sin\psi(\cos\psi + \beta\sin\psi)\\
&= \tfrac{1}{2}[\sin 2\psi + \beta(1 - \cos 2\psi)] \qquad (5.6)\\
xx_h{}^3 &= \rho^2\cos\psi(\cos\psi + \beta\sin\psi)^3 = \rho^2(\cos^4\psi + 3\beta\cos^3\psi\sin\psi)\\
yx_h{}^3 &= \rho^2\sin\psi(\cos\psi + \beta\sin\psi)^3 = \rho^2(\sin\psi\cos^3\psi + 3\beta\cos^2\psi\sin^2\psi)
\end{aligned}$$

The formula $\rho\dfrac{d\psi}{dt} = x\dot{y} - y\dot{x}$ upon the substitution of values (5.6) gives

$$\frac{d\psi}{dt} = \frac{d\psi_0}{dt} + \mu\frac{d\psi_1}{dt} = -1 - \left[\frac{1}{2}(a + \beta)\sin 2\psi + \varepsilon\rho\cos^4\psi\right] \tag{5.7}$$

For this approximation of the zero order we assume that the period is 2π, which gives $\psi_0(t) = \varphi_0 - t$, where φ_0 is the initial value of $\psi_0(t)$.

For the first-order corrective term, one has

$$\frac{d\psi_1}{dt} = -\left[\frac{1}{2}(A + B)\sin 2\psi_0 + E\rho_0\cos^4\psi_0\right] \tag{5.8}$$

where $A = a/\mu$, $B = \beta/\mu$, and $E = \varepsilon/\mu$; moreover $dt = -d\psi_0$.

In order to calculate $\Delta\psi_1 = \psi_1(2\pi)$ (the variation of ψ_1 during 2π) we integrate (5.8) between 0 and 2π which yields:

$$\psi_1(2\pi) = \tfrac{3}{8}E\rho_0\cdot 2\pi$$

therefore: $\psi(2\pi) = 2\pi\mu\frac{3}{8}E\rho_0$. We set $\psi(2\pi) = \Delta\varphi$ and $2\pi\mu = \Delta\tau$ (the element of the stroboscopic time) and obtain as usual the stroboscopic equation:

$$\frac{d\varphi}{d\tau} = \frac{3}{8}E\rho \tag{5.9}$$

It is noted that we have now the frequency correction already in the first approximation and, moreover, this correction contains the unknown ρ.

As we are interested in the stationary state, it is useful to designate the stationary amplitude ρ^* instead of ρ. We have thus

$$\frac{d\psi}{d\tau} = -1 - \frac{3}{8} E\rho^* = -1 + \gamma; \qquad \gamma = \frac{3}{8} E\rho^* \tag{5.10}$$

which means that the period is now $2\pi/(1 - \gamma) = 2\pi\eta = \lambda; \eta = 1/(1 - \gamma)$ and not 2π.

We have to consider now the first d.e. in ρ which is formed, as usual, from the combination $x\dot{x} + y\dot{y} = \frac{1}{2}(d\rho/dt)$. Using (5.3) and (5.6), one obtains

$$\frac{d\rho}{dt} = -\rho[(a + \beta) - (a + \beta)\cos 2\psi + 2\varepsilon\rho \sin\psi \cos^3\psi] \simeq 0(\mu) \tag{5.11}$$

It is clear that $\rho_0(t) = \rho_0 = \text{const}$ (the initial value of ρ) and for the first-approximation corrective term $\rho_1 = \rho_1(t)$ we have the d.e.

$$\frac{d\rho_1}{d\psi_0} = \rho_0[(A + B) - (A + B)\cos 2\psi_0 + 2E\rho_0 \sin\psi_0 \cos^3\psi_0] \tag{5.12}$$

This equation would be useless for the determination of $\rho_0 = \rho^*$ if the integration had to be performed between 0 and 2π because the term with $\int^{2\pi} \sin\psi_0 \cos^3\psi_0 d\psi_0$ would vanish and it would be impossible to determine $\rho_0 = \rho^*$. This, however, is not our case now as the period is λ and not 2π. Designating as before $\rho_1[\lambda]$ the variation of $\rho_1(t)$ during the time λ, we obtain from (5.12), by integration between 0 and λ, the expression

$$\rho_1[\lambda] = \rho_0[(A + B)(\lambda - \tfrac{1}{2}\sin 2\lambda) + \tfrac{1}{2}E(1 - \cos^4\lambda)\rho_0] \tag{5.13}$$

but λ contains ρ_0, since $\lambda = 2\pi/(1 - \gamma) = 16\pi/(8 - 3E\rho_0)$.

It is clear that one has to introduce the necessary condition $\rho_0 = \rho^* \neq 8/3E$. If one introduces this value of λ and carries out the usual passage from the difference equation to the stroboscopic d.e., one obtains

$$\frac{d\rho}{d\tau} = \rho[(A + B)(2\lambda - \sin 2\lambda)] + E\rho(1 - \cos^4\lambda) \tag{5.14}$$

The stationary amplitude $\rho = \rho^*$ is given by the formula

$$\rho^* = -\frac{(A + B)(2\lambda - \sin 2\lambda)}{E(1 - \cos^4\lambda)} \tag{5.15}$$

All depends now on the order of magnitude of $\lambda = 2\pi/(1 - \gamma)$; if the frequency correction γ is small in comparison to the fundamental frequency -1 the period will be somewhat greater than 2π (we assume $E > 0$ and, hence, $\gamma > 0$). In this case $2\lambda - \sin 2\lambda$ is positive and, as the denominator in (5.15) is also positive, one obtains $\rho^* < 0$ which is impossible as $\rho = r^2$ is essentially positive.

If, however, $E < 0$ and, hence, $\gamma < 0$ the nonlinear period $2\pi/(1 + |\gamma|)$ will be less than 2π and (5.15) yields $\rho^* > 0$ as it should be. It is seen that the sign of the nonlinear frequency correction determines the existence (or nonexistence) of ρ^*.

In the linear case it was also shown that the sign of the parameter variation $\Delta\lambda$ (around the harmonic value $\lambda = \lambda_1$) determines the sign of the decrement $\Delta\alpha$, but this does not lead to any conclusion in accordance with experimental data, since $\Delta\alpha > 0$ means that the amplitude increases indefinitely, whereas $\Delta\alpha < 0$ means that the oscillation disappears. From this point of view the nonlinear extension is in agreement with the experiment and shows under which conditions the stationary state is possible.

The problem becomes rather complicated if one tries to obtain the quantitative results. In fact, in addition to the case $E < 0$ (or $\varepsilon < 0$) with $A > 0$ and $B > 0$, one can obviously obtain the stationary amplitude with $E > 0$ but with $B < 0$ sufficiently large in absolute value so that $(A - |B|) < 0$. In this case a complication appears in the approximation procedure since it impossible to take $\rho_0(t) = \rho_0 = \text{const}$ as we did.

The question of stability is also somewhat complicated, the stroboscopic system in this case being

$$\frac{d\rho}{d\tau} = \rho[(A + B)(2\lambda - \sin 2\lambda) + E\rho(1 - \cos^4 \lambda)] = R(\rho,\varphi)$$

$$\frac{d\varphi}{d\tau} = \frac{3}{8}E\rho = \Phi(\rho,\varphi) \tag{5.16}$$

The second equation is simple but not the first one, since $\lambda = 16\pi/(8 - 3\rho)$ is a function of ρ; moreover this dependence on ρ enters also under the sign of the trigonometric functions.

Finally, even if one assumes that ρ^* has been determined and is stable, the quantitative procedure does not stop at this point since in (5.9) one has to replace ρ_0 by its stationary value ρ^* and carry out integrations again using the corrected period of integration λ' instead of the old one, λ.

In other words, in addition to the usual process of the first approximation one has to apply also the method of successive approximations, since the two equations are interdependent. In fact, the frequency correction (5.9) depends on the variable ρ yielded by (5.11), but the latter can yield a correct value of ρ if one knows the frequency correction; the latter, however, depends on this unknown ρ—whence a typical vicious circle requiring successive approximations.

We shall not go beyond these remarks which show the difficulty of the problem if one uses the standard procedure with which we were concerned in all problems of Part III.

It is noted that, although the standard theory leads to correct conclusions from the standpoint of experimental results, the procedure is cumbersome if one tries to obtain quantitative data even in the first approximation. The real reason for this difficulty lies in the linear part of the theory of the d.e. (Sections 3 and 4), that is, the problem remains transcendental, in spite of simplifications of the asymptotic case ($h \approx 0(\mu)$) which we assumed.

There appeared recently attempts to approach this subject from the standpoint of the algebraico-transcendental characteristics, but a discussion of this subject does not come within the scope of the elementary aims of this text; a reader interested in the quantitative part of these studies should consult Pinney.[4]

A remark is noteworthy: there seems to be no doubt that these oscillations exhibit all familiar features of nonlinear oscillations, namely, they appear at certain bifurcation values of the parameter; they exist within a certain *finite* region of the parameter variation (and not for only one isolated "harmonic value" of the latter, as predicted by the linear theory).

Their nature, however, remains essentially transcendental, since their frequencies are determined by zeros of the corresponding entire functions (see Section 3). For that reason they seem to be utterly unpredictable if viewed from the standpoint of the classical nonlinear theory of oscillations.

Very often an oscillation suddenly appears with a period having nothing to do with the period of the system (in which $h = 0$); moreover this oscillation may suddenly give way to another oscillation of entirely different frequency not standing in any rational ratio with respect to the one which has just disappeared. It may happen also that under seemingly identical conditions these oscillations do not appear at all.

One of the possible explanations of this erratic behavior is the lack of constancy of the time lag h. In fact, if one drops this hypothesis, nothing remains of the preceding theory and the erratic behavior of these phenomena is nothing but the result of the corresponding erratic behavior of the parameter h.

Conversely, if one secures conditions for a definite constant h the phenomena become very regular, as will be seen from Section 7.

[4] See footnote [4], page 516.

6. On the physical nature of retarded actions

The fact that in all problems governed by d.d.e. appear (generally parasitic) self-excited oscillations indicates that, under special conditions which we have investigated, certain *absorptions* of energy also appear that are converted into an oscillatory form with frequencies ω which are connected with the increments α; therefore the complex number $z = \alpha + i\omega$ corresponds to the root of a certain entire function—the characteristic equation.

It may be useful to give a more direct illustration of these retarded phenomena on the basis of energy relations. We consider the harmonic oscillator whose d.e. is

$$\ddot{x} + x = 0 \tag{6.1}$$

with its corresponding equivalent system

$$\dot{x} = y; \qquad \dot{y} = -x \tag{6.2}$$

The application of the stroboscopic transformation yields

$$\frac{d\rho}{d\tau} = 0; \qquad \frac{d\varphi}{d\tau} = -1 \tag{6.3}$$

Since $\rho = r^2 = x^2 + \dot{x}^2$ may be regarded as the total energy stored in oscillation (up to a certain constant factor and with a proper normalization), the first equation (6.3) shows that the system is *conservative*, and the second equation indicates that the frequency is *one*, that is, the period is 2π. The minus sign results merely from the existing convention to count the angles in the counterclockwise direction while the motion of the representative point takes place clockwise.

Let us consider now the same oscillator but with a retarded restoring force; this yields the simple d.d.e.

$$\ddot{x} + x_h = 0 \tag{6.4}$$

With the same assumptions as in the preceding section (the effect of the retardation is felt mostly in the phase); we have to replace x in (6.1) (and also in (6.2)) by

$$x_h = r \cos(\psi t - \beta)$$

Forming the combination $x\dot{x} + y\dot{y} = \frac{1}{2}\frac{d\rho}{dt}$, we obtain (under the assumption that β is small), after simple transformations,

$$\frac{d\rho}{dt} = -2\beta\rho \sin^2 \psi \tag{6.5}$$

The second combination $\left(x\dot{y} - y\dot{x} = \rho \dfrac{d\psi}{dt}\right)$ shows that the period remains 2π in the first approximation.

Replacing $\sin^2 \psi = \frac{1}{2} - \frac{1}{2}\cos 2\psi$ in (6.5) one obtains

$$\frac{d\rho}{dt} = -\rho\beta(1 - \cos 2\psi)$$

so that for the first-order corrective form $\rho_1(t)$ we have the d.e.

$$\frac{d\rho_1}{dt} = -\rho B(1 - \cos 2\psi); \qquad B = \beta/\mu$$

The zero-order term for the second equation is: $\psi_0(t) = \varphi_0 - t$, φ_0 being the initial phase.

We have to integrate $\rho_1(t)$ over the period 2π; if we are able to show that this integral vanishes, one would have $\rho(t) = \rho_0 = \text{const}$ and this would still be the case of the harmonic oscillator. We have, however,

$$\rho_1(2\pi) = \rho B \int_0^{2\pi} d\psi_0 - \tfrac{1}{2}B \int_0^{4\pi} \cos 2\psi_0 d2\psi_0$$

(instead of integrating with respect to t, we have done that with respect to ψ_0, since $dt = -d\psi_0$).

Hence, for $\rho(2\pi)$, we have

$$\rho(2\pi) = \Delta\rho = -\rho B \cdot 2\pi\mu = -\rho B \Delta\tau \tag{6.6}$$

where $\Delta\tau$, as usual, is the element of the stroboscopic time.

One obtains thus the stroboscopic equation (at the limit $\Delta\tau \to d\tau$)

$$\frac{d\rho}{d\tau} = -\rho B \tag{6.7}$$

whose solution is

$$\rho = \rho_0 e^{-B\tau} \tag{6.8}$$

Comparing (6.8) with (6.3) it is seen that the conservative system of the harmonic oscillator (6.1) has been transformed into a nonconservative one owing to the replacement of the nonretarded (natural) restoring force x by the retarded one x_h. If the sign of β were reversed, the sign of the exponential would be changed.

One cannot fail to notice a certain analogy with the phenomenon of parametric excitation (Chapter 20), where the natural (stable) phase $3\pi/4$ leads to an indefinite increase of energy while the unstable phase $\pi/4$ (which can be produced only artificially), leads, on the contrary, to the draining of energy away from the system.

This is, however, only a remote physical analogy, since the phenomenon of the parametric excitation is governed by an ordinary (nonlinear) d.e., while here a similar manifestation occurs in an entirely different (transcendental) system where conditions are more complicated, as was explained in the beginning of this chapter.

A remark is noteworthy: since the range in which self-excited oscillations persist is *finite* on the basis of the nonlinear theory, it may happen that two (and occasionally several) such oscillations may exist at the same time. This, as we saw, is impossible according to the linear theory where only one oscillation exists for the corresponding (isolated) value $\lambda = \lambda_1$ of the parameter.

Experiment corroborates the nonlinear theory as, very often, two such oscillations are recorded simultaneously, but as their frequencies do not stand to each other in any commensurate ratio, this gives rise to interesting phenomena which are mentioned in the following section.

7. Experimental evidence; electronic analogue

An interesting illustration of the nature of these oscillations can be given in connection with experiments with antirolling stabilization of ships by the so-called *activated tanks method.* As these experiments do not relate to this subject and are described elsewhere,[11] we shall limit ourselves only to those points that are of interest here.

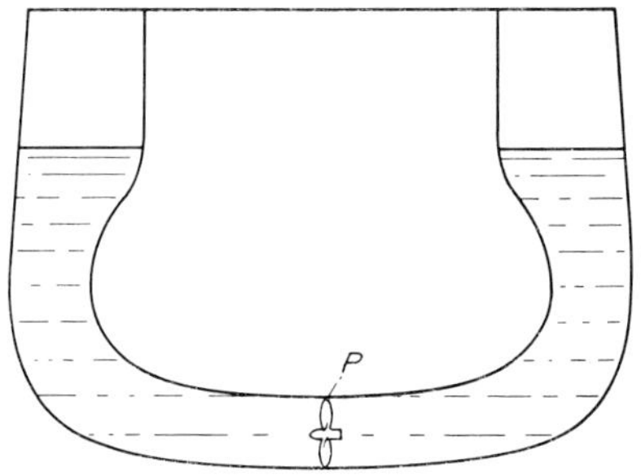

FIGURE 21.4

In the problem of stabilization by this method, the ballast is displaced between the tanks (forming a kind of a U tube) by means of an axial propeller pump P whose angle of blades is controlled by instruments responsive to ship's angular motion which the system is purported to counteract (Fig. 21.4).

We have seen (Section 7, Chapter 2) that the phase with which the ballast should move must be such that the ballast concentration must be maximum in the tank which *rises* in space (on account of rolling) at the instant when the angular velocity of rolling is also maximum. It is clear that the action of the tanks produces an artificial damping since, in this case, the negative work of gravity reduces continuously the kinetic energy produced by rolling.

[11] N. Minorsky, *Proc. Seventh Intern. Congress Appl. Mech.*, London, 1948.

In the absence of stabilization and under somewhat idealized conditions, the angular motion of rolling may be represented by a simple d.e.

$$\ddot{x} + k\dot{x} + \omega_0^2 x = a \sin \omega t \tag{7.1}$$

Under the action of the stabilizing equipment, the quenched rolling is governed by the d.e.

$$\ddot{x} + (k + K)x + \omega_0^2 x = a \sin \omega t \tag{7.2}$$

where K is the equivalent coefficient of damping produced by tanks. This is corroborated by tests on a model, the record of which is shown in Fig.

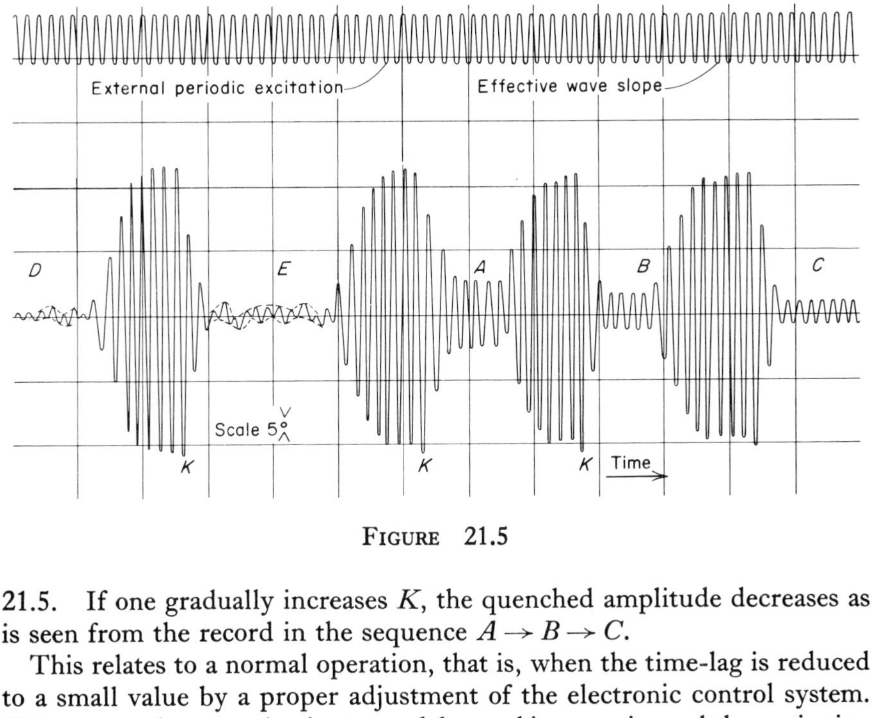

FIGURE 21.5

21.5. If one gradually increases K, the quenched amplitude decreases as is seen from the record in the sequence $A \rightarrow B \rightarrow C$.

This relates to a normal operation, that is, when the time-lag is reduced to a small value by a proper adjustment of the electronic control system. If, however, the pump begins to work beyond its capacity and the cavitation sets in, the presence of the time-lag so introduced accounts for a change in operation which, instead of being governed by the d.e. (7.2), begins to be governed by the d.d.e.

$$\ddot{x} + k\ddot{x} + K\dot{x}_h + \omega_0^2 x = a \sin \omega t \tag{7.3}$$

It is to be noted that the oscillation persists (although with a smaller amplitude) if $a = 0$. The presence of a transcendental harmonic arising from the d.d.e. is seen at the point E of the record where a certain interference pattern can be noticed between the forced oscillation and this harmonic.

If one increases the value of the parameter λ and, hence, the self-excited oscillation (Fig. 21.5), the interference pattern exhibits a distinctly non-periodic character in view of a lack of any commensurate ratio between the two frequencies.

At a later date these phenomena of retarded (or advanced) actions were reproduced electronically by means of a circuit shown in Fig. 21.6, where L and C are constants of an oscillating circuit having in series two resistors R_1 and R. The voltage across R_1 is applied to a linear amplifier A in the output of which is provided a special phase-shifting network P whose purpose is to introduce a constant time-lag between the input and the output of P. Under these conditions, the voltage across R in series with the output of P is: $e(t) = \lambda R_1 i(t-h) = \lambda R_1 i_h$, the current i_h being thus a *retarded* current in the previous terminology, and λ being the gain of the amplifier to R measured through the network P. In this manner, instead of the usual d.e. of oscillating circuit

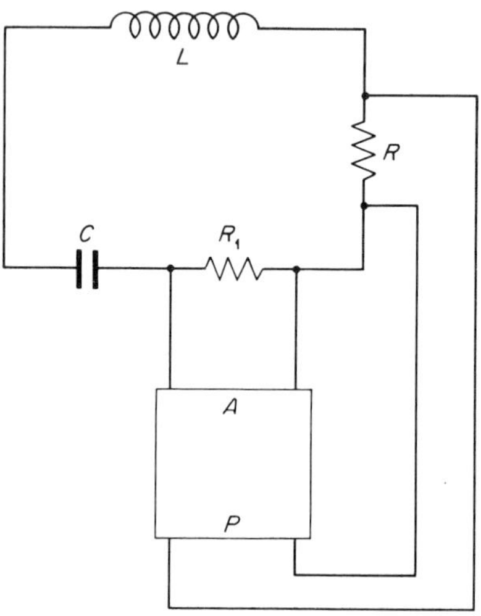

FIGURE 21.6

$$L\frac{di}{dt} + (R + R_1)i - \frac{1}{c}\int idt = 0$$

one has now the equation

$$L\frac{di}{dt} + (R + R_1)i + S(\lambda)i_h + \frac{i}{c}\int idt = 0 \tag{7.4}$$

where $S(\lambda) = \lambda R_1$. Differentiating this equation with respect to t, dividing by L, and setting $(R + R_1)/L = p$; $S(\lambda)/L = q(\lambda)$; $1/LC = \omega_0^2$, one has

$$\frac{d^2i}{dt^2} + p\frac{di}{dt} + q(\lambda)\frac{di_h}{dt} + \omega_0^2 i = 0 \tag{7.5}$$

which is a d.d.e. of a mixed type with a nonretarded $p(di/dt)$ and a retarded $q(\lambda)(di_h/dt)$ velocity terms.

This circuit gave means for investigating the transcendental spectrum of frequencies in a very simple manner and the agreement between the theoretical and experimental results is generally sufficiently good.[12] Since all these phenomena are nonlinear and are characterized by *finite ranges* in which oscillation may exist, it happens often that, for a given condition, one observes *two* transcendental frequencies of this kind. If one arranges the cathode-ray oscilloscope for the phase-plane representation and adjusts its sweep circuit so as to make one of them, say f_1 to stand still on the screen, Fig. 21.7, the wave of the other frequency f_2 cannot be maintained stationary in view of the incommensurate ratio of f_2 and f_1 and this gives rise to a pattern of a kind of a "luminous rotating bracelet." This phenomenon gives at the same time an illustration of the theory of Poincaré-Denjoy regarding the solutions on a torus (Chapter 8).

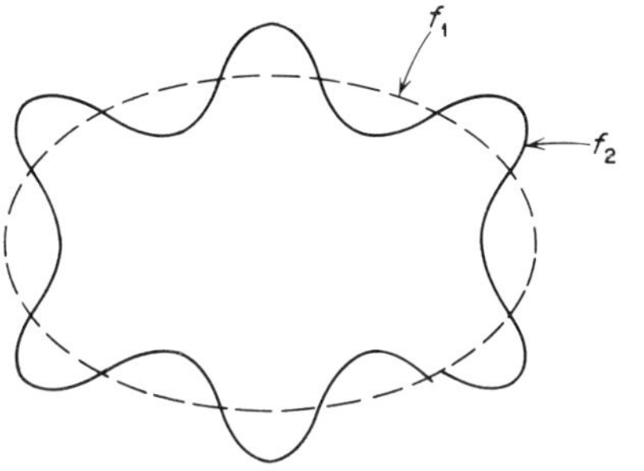

FIGURE 21.7

8. Econometric and other problems

Recent attempts have been made to give a mathematical formulation to certain econometric fluctuations ("business cycles") which were more frequent in the past than now, owing to some governmental regulations tending to prevent these fluctuations. Similar oscillatory phenomena are sometimes observed in biology (Section 9, Chapter 2), but their study is still less advanced.

As far as "econometric fluctuations" are concerned, the most difficult part, at least in the present state of these studies, is to formulate correctly the problem. There are so many unknown factors relative, for instance, to the psychology of masses (or of individuals), etc., that it is impossible to assert that a certain formulation of a given problem really takes into account all these factors with a proper mathematical interpretation.

It is necessary, therefore, to assume some simplified "econometric models" and try to form at least a rough idea on this basis. In forming such "models," it is unavoidable that a number of factors are to be neglected and, it is quite possible also that, if these factors were taken into account, the situation would be different.

As a simple example, consider a population of a certain country, more or

[12] N. Minorsky, *Trans. A.S.M.E.*, 1947.

less isolated from other countries (for example, by means of custom barriers) and having no government regulations concerning the credit control, etc. The population can subsist on a low level by raising its own food but there exists an industry capable of supplying the manufactured goods. It is clear that, if the population having made some savings, begins to spend them in buying manufactured goods, the industry begins to flourish up to a point when the demand for goods is satisfied, after which an economic crisis begins; the standard of living goes down but, as the country can still subsist on a rather primitive level, some savings are still being made and another cycle begins at a later time.

This rather crude "model" does not take into consideration many other factors. In fact, it is logical to admit that people who saved some money are not going to spend it entirely on automobiles, refrigerators, etc., but will be more or less reasonable in these expenditures. Those who do not think about their own future very likely will be still protected by certain measures on the part of the government which will try to prevent the development of a periodic phenomenon of this nature. As soon as one tries to take all this into account, the problem becomes far more complicated, but it is still uncertain as to whether some other factors have been overlooked and whether those taken into account have been formulated correctly.

These attempts to form "econometric models" are being made from time to time and it is interesting to note that most of these "models" result in d.d.e. One such attempt due to Kalecki[13] attracted particular attention and we give its brief outline in Bellman's version.[4]

If $I(t)$ is the rate of investment and $U(t)$ is the rate of depreciation of capital goods, $W(t) = I(t) - U$ may be considered as the state of economy. If $W(t)$ undergoes fluctuations, this characterizes precisely a "business cycle."

A number of other assumptions are made: for instance, it is assumed that the producers control the production just in accordance with the unfilled orders but, as a certain time interval elapses between the time when an order is received and a corresponding product is available, the question of "time-lag" appears at this point of the argument; one can normalize the problem by assuming this time-lag as $h = 1$.

We note in passing the difference between this problem and those problems considered in sections 3, 4, and 5. In the latter cases, we had physical problems (mechanical or electrical) and, on this basis, it was possible to introduce some plausible hypotheses regarding the smallness of time-lag h which enabled us to simplify the problem in advance and thus

[13] M. Kalecki, *Econometrica* **3**, 1935.

[4] See footnote [4], page 516.

obtain in the first approximation results consistent with observed facts, at least qualitatively.

In this problem we cannot introduce any such simplifications because the time-lag is *finite* and, for that reason, one has to try to obtain some conclusions from a linear d.d.e. which will appear later.

We continue the argument of Kalecki. If $A(t)$ is the production rate and as manufacturers control it only on the basis of unfilled orders, one has:

$$A(t) = \int_{t-1}^{t} I(\tau)d\tau \tag{8.1}$$

but the actual production P is obviously $P(t) = \int_{t-1}^{t} A(\tau)d\tau$. One can begin the process at $t = 0$, the integrals considered being between 0 and 1. This means that both the rate of investment and the rate of production are determined only by the unfilled orders existing during the time interval $h = 1$. Moreover, if $k(t)$ is the capital at time t and $L(t)$ is the rate of delivery of finished product, one has

$$\dot{k}(g) = L(t) - U \tag{8.2}$$

but as $L(t) = I(t - 1)$, (8.2) becomes

$$\dot{k}(t) = I(t - 1) - U \tag{8.3}$$

There is yet a relation to be assumed between the investment rate, the amount of capital, and the production rate.

The simplest assumption is that these three quantities are connected by a linear relation

$$I(t) = m[c + A(t) - nk(t)] \tag{8.4}$$

the constants m, n, and c being available for fitting the theory into the observed facts.

Differentiating (8.1) and (8.4), replacing $A(t)$ and $k(t)$ by their expressions in terms of $I(t)$, and introducing a new variable $u(t) = I(t) - U$, one has the following d.d.e.

$$\dot{u} - pu - qu_1 = 0 \tag{8.5}$$

where $u_1 = u(t - 1)$. The characteristic equation here is

$$f(z) = pe^z + q - ze^z = 0 \tag{8.6}$$

Since no simplification (resulting from the assumption $h \ll 1$) can be made here, one has to proceed by the general method of determining zeros if the analytic function $f(z)$. For more details we refer to the original work of Pontriagin[6] or to the summary of this work by R. Bellman.[4]

[6] See footnote [6], page 519.

[4] See footnote [4], page 516.

It is to be mentioned, however, that many assumptions of this theory can be criticized as being oversimplified and, therefore, not being sufficiently realistic. We mentioned this, for instance, in connection with a lack of imagination (or foresight) on the part of producers who control the production only on the basis of unfilled orders, etc. But, on the other hand, the only way to proceed in these extremely ill-defined problems is to take a very crude model and try to improve it little by little, keeping an eye on the actual econometric occurrences.

A somewhat different problem leading also to d.d.e. was considered by Y. Rocard[14] in his private publication "Coordination," in which this author attempted to formulate what may be called the "red tape" problem. As an example one may consider a production of the aeronautical material hampered by the action of some bureaucratic agencies (the "coordinators"), a lack of prompt answers to letters lying too long on the desks, etc. As the result of this, time-lags appear between the original (more or less a priori) decisions to go ahead with building a certain type of aircraft and the later "counterdecisions" based on the actual facts after a certain model has been built and tested. Oscillations resulting from such retarded decisions obviously hamper a steady production rate in which no such oscillations should occur. Here again, the author had to introduce a number of assumptions in order to be able to form his d.d.e.

In all these attempts the essence of the method of attack is always the same, viz.: to consider an effect of a certain cause only *after* a certain *finite* time interval so that what determines the *present* is not an infinitely near past (as in a d.e.) but a kind of a *finite past* resulting from time-lags.

At present all these attempts appear more or less as a matter of curiosity rather than as definite problems on account of the impossibility of a correct formulation of such problems without a full knowledge of all facts, but it is possible that, when all this information becomes more definite (particularly in biology), further applications of d.d.e. will acquire also a more definite character. In such a case it will be also necessary to develop further the theory of nonlinear d.d.e. with a view to obtaining their solutions by approximations in the manner in which we tried to accomplish this in a particular case of a very small time-lag.

[14] Y. Rocard, *C. R. Ac. Sc.* (Paris), 1948.

Chapter 22

TOPOLOGY OF LIENARD'S EQUATION IN A PARAMETER SPACE

1. Introductory remarks

In Chapter 7 was indicated an algebraic approach to the theory of bifurcations resulting from the application of the stroboscopic method. In this chapter we shall investigate this matter in more detail by introducing the concept of a *parameter space*; the topological aspect of solutions will be studied in the various regions of this space.[1]

The existing theory of Liénard's d.e. in its general form:

$$\ddot{x} + f(x)\dot{x} + g(x) = 0 \tag{1.1}$$

concerns mainly the existence of periodic solutions (Chapter 4) but in applied problems one is often interested also in the topological aspect of these solutions as well as in the evolution of the "phase portrait" when a parameter in the d.e. varies.

The stroboscopic method permits investigating these questions in a simple manner by reducing the problem ultimately to an algebraic discussion.

We note in passing that the form of the function $g(x)$ does not influence the topology in which we are interested here but merely modifies the period as will be mentioned in Section 7. For that reason it is convenient to assume first $g(x) = x$ as was done originally by Liénard.

The form of the function $f(x)$, on the other hand, has a direct bearing on the topological aspect of solutions, and in this chapter we shall be mostly concerned with this question. As in all applications $f(x)$ is represented by a polynomial, it is convenient to consider the coefficients of this polynomial as coordinates of a certain *parameter space* and to investigate the behavior of solutions in the various regions of this space.

[1] N. Minorsky, *C. R. Ac. Sc.* (Paris) **240**, 1955; *J. Phys. et radium* **18**, 1957.

In applications the polynomial representing $f(x)$ is generally of the form

$$f(x) = a + cx^2 + ex^4 \tag{1.2}$$

as we shall see from an example given in the following section. As the stroboscopic method uses generally the variable $\rho = r^2$, it will be shown that the discussion reduces to a simple quadratic equation in ρ, and it is useful to conduct it in the three-dimensional parameter space a, c, e (or, which is the same, A, C, E since $A = a/\mu$; $C = c/\mu$, and $E = e/\mu$) by considering the eight octants of this space. In this manner one obtains the "phase portraits" of the d.e. in the various octants. As these "portraits" vary from one octant to the other, one can also obtain a simple insight into the theory of bifurcations, the "coordinates" A, C, and E playing the role of the parameter in the d.e. (Chapter 7).

It is recalled that we have defined two kinds of bifurcations. In the bifurcation of the *first* kind, the stability of the singular point changes with the incidental appearance (or disappearance) of a limit cycle according to the scheme.

Stable singular point ↗ Unstable singular point
Stable singular point ↘ Stable limit cycle (1.3)

or inverse schemes in which either the word "stable" is replaced by the word "unstable" or directions of arrows are reversed in (1.3). With the convention introduced in Chapter 7, this can be written symbolically as

$$S \rightarrow US \quad (\text{or as} \quad U \rightarrow SU)$$

the *first* letter (or the letter, if it is single) relating always to the stability of the state of rest (singular point), and the subsequent letters to the stability of the corresponding limit cycles (from the singular point outward).

Likewise, by a bifurcation of the *second* kind we defined the coalescence of two neighboring limit cycles (internal within the structure) which gives rise to a semi-stable cycle that disappears with a further variation of the parameter. The following scheme illustrates the bifurcation of the *second* kind:

$$SUSUS \rightarrow S(US)US \rightarrow SUS \tag{1.4}$$

where (US) designates a semi-stable cycle which is always a critical structure.

In this chapter we shall encounter also bifurcations of the *third kind* which consists in adding (or in removing) an *external* cycle. Thus, for instance,

$$S \rightarrow SU; \qquad SU \rightarrow SUS; \qquad SU \rightarrow S \tag{1.5}$$

are examples of bifurcations of the *third* kind.

The bifurcations of the first and of the third kinds cause modifications of the structure at its ends (at the singular point and at the external cycle); the *cyclicity* (the number of limit cycles) in this case changes always by *one* unit; in the bifurcations of the second kind, it changes by *two* units and this change is *internal* in the structure.

Problems of this nature present a definite applied interest but their full significance will probably appear only when it will be possible to produce nonlinear characteristics given in advance with a sufficiently great accuracy. For the time being one is yet far from these possibilities but it is thought that they are of a sufficient importance to be mentioned here.

In the following section we establish the derivation of the stroboscopic system for the d.e. of an electron-tube oscillator which we studied already in Chapter 7. In Section 3 is outlined a study of topology of solutions of this d.e. in various regions of its parameter space.

Section 4 gives additional information concerning bifurcations of the third kind which we encounter here for the first time.

Section 5 deals with certain special cases of Liénard's equation. The last section 9 concerns relations between the form of nonlinear characteristics and the phase portraits of solutions.

2. Formation of the stroboscopic equation

In Section 4, Chapter 7 the d.e. (4.5) as an example of the bifurcation theory was discussed; we consider it again here with a view to investigating its stroboscopic system. Designating the grid voltage as v and the "saturation voltage" as V, we have the relations

$$\frac{di}{dt} = \frac{v}{M} = \frac{V}{M}x; \qquad \frac{d^2i}{dt^2} = \frac{V}{M}\dot{x}; \qquad i = \frac{V}{M}\int x\,d\tau \tag{2.1}$$

x being a "dimensionless grid voltage" which will be used as a new variable. In these notations and with the independent variable $\bar{\tau} = \omega_0 t$, $\omega_0 = 1/\sqrt{LC}$, the d.e. (4.5) can be written as

$$\frac{d^2x}{d\bar{\tau}^2} + x = \omega_0 M\left[\left(\alpha_0 - \frac{RC}{M}\right) + 2\beta_0 Vx + 3\gamma_0 V^2x^2 + 4\delta_0 V^3x^3 + 5\varepsilon_0 V^4x^4\right]\frac{dx}{d\bar{\tau}} \tag{2.2}$$

where α_0, β_0, γ_0, δ_0, and ε_0 are the coefficients of the nonlinear function

$$I = I(v) = \alpha_0 v + \beta_0 v^2 + \gamma_0 v^3 + \delta_0 v^4 + \varepsilon_0 v^5 \tag{2.3}$$

Setting

$$\omega_0 M\left(\alpha_0 - \frac{RC}{M}\right) = a; \qquad \omega_0 M2\beta_0 V = b; \qquad \omega_0 M3\gamma_0 V^2 = c; \qquad \omega_0 M4\delta_0 V^3 = d \quad \text{and} \quad \omega_0 M5\varepsilon_0 V^4 = e \tag{2.4}$$

equation (2.2) can be written as

$$\ddot{x} + x = (a + bx + cx^2 + dx^3 + ex^4)\dot{x} \tag{2.5}$$

where we write $\ddot{x}$ and $\dot{x}$ instead of $d^2x/d\bar{\tau}^2$ and $dx/d\bar{\tau}$ since no confusion can be feared from now on.

We proceed now with obtaining a stroboscopic system corresponding to (2.5) assuming that the problem is nearly linear, that is, the coefficients a, b, c, d, and e are small quantities of the first order.

If one uses the series solutions of the form $\rho(t) = \rho_0(\bar{\tau}) + \mu\rho_1(\bar{\tau}) + \ldots$; $\psi(\bar{\tau}) = \psi_0(\bar{\tau}) + \mu\psi_1(\bar{\tau}) + \ldots$ after the variables $\rho = x^2 + \dot{x}^2 = x^2 + y^2$ and $\psi = \arctan(y/x)$; $(x = r\cos\psi, y = r\sin\psi; \rho = r^2)$ have been introduced, the zero-order solutions are, as usually

$$\rho_0(t) = \rho_0, \qquad \psi_0(t) = \varphi_0 - \bar{\tau} \tag{2.6}$$

ρ_0 and φ_0 being the initial conditions.

One finds that the first-order correction term $\psi_1(2\pi)$ is zero as the integrals between 0 and 2π vanish in the first approximation. As to the first-order correction term for $\rho_1(\bar{\tau})$, it is given by the d.e.

$$\frac{d\rho_1}{d\bar{\tau}} = 2A\rho_0 \sin^2\psi_0 + 2C\rho_0^2 \sin^2\psi_0 \cos^2\psi_0 + 2E\rho_0^3 \cos^4\psi_0 \sin^2\psi_0 \tag{2.7}$$

where $A = a/\mu$; $C = c/\mu$, and $E = e/\mu$, the terms with b and d vanishing in integrations between 0 and 2π.

The variation of $\rho_1(\bar{\tau})$ during one period 2π is therefore

$$\rho_1(2\pi) = \Delta\rho_1 = 2\pi\rho_0[\tfrac{5}{8}E\rho_0^2 + \tfrac{1}{4}C\rho_0 + A] \tag{2.8}$$

and, since $\rho(2\pi) = \mu\rho_1(2\pi)$, one can set as usual, $2\pi\mu = \Delta\tau$† and, passing to the limit $\Delta\tau \to d\tau$, we obtain the stroboscopic d.e.

$$\frac{d\rho}{d\tau} = -\sigma\rho[\rho^2 + p\rho + q] = \Phi(\rho) \tag{2.9}$$

with $\sigma = \frac{5}{8}E$, $p = \frac{2}{5}C/E$, and $q = \frac{8}{5}A/E$.

As the d.e. (2.5) has no frequency correction in the first approximation,

† The variable τ is not to be confused with $\bar{\tau}$ used in the preceding formulas.

the second stroboscopic d.e. is simply $d\varphi/d\tau \equiv 0$, and we do not have to consider it here.

Before proceeding further with the discussion of (2.9) it is useful to establish a correct sign relation between the original Liénard equation (2.5) which can be written (omitting terms with b and d, since they cancel anyhow in integrations) as

$$\ddot{x} - (a + cx^2 + ex^4)\dot{x} + x = 0 \tag{2.10}$$

and its stroboscopic counterpart (2.9). There is always some uncertainty about the sign of the right-hand term of the initial d.e. which requires a careful consideration of positive directions on the circuits.

It is simpler to be guided by physical considerations in comparing (2.10) and (2.9). If one assumes that a, c, and e are positive, so are also p, q, and σ. In this case (2.9) gives $(d\rho/d\tau) < 0$, which means that the energy stored in the oscillating system decreases with time, but in such a case, for the obvious reason, one should take the plus sign in (2.10), viz.:

$$\ddot{x} + (a + cx^2 + ex^4)\dot{x} + x = 0 \tag{2.11}$$

Since the nonlinear damping term being essentially positive, the energy is also drained away from the systems for any value of x.

This gives a correct correspondence of signs between (2.11) and the stroboscopic equation (2.9) which will serve as a starting point for the analysis in the following section.

3. Phase portraits of Liénard's equation

We consider Liénard's equation in the form

$$\ddot{x} + (a + cx^2 + ex^4)\dot{x} + x = 0 \tag{3.1}$$

to which corresponds the stroboscopic equation (2.9) and consider the parameters A, C, and E as coordinate axes of the three-dimensional space in which we shall study the solutions of (2.9) in the eight octants; four for $E > 0$ and the other four for $E < 0$.

In each of the eight octants there will be a certain combination of signs of A, C, and E (or of a, c, and e, since $A = a/\mu$, $C = c/\mu$, and $E = e/\mu$, $\mu > 0$).

To determine the topological configuration, we shall need the following information: (a) stability of the state of rest (that is, of the singular point); (b) existence of a stationary amplitude (or amplitudes); and (c) its stability.

The point (a) of these data is obtained by assuming $\rho \simeq 0$ in (2.9) which permits writing it approximately as

$$\left(\frac{d\rho}{d\tau}\right)_{\rho \simeq 0} \simeq (-\sigma q)\rho \tag{3.2}$$

The sign of $(-\sigma q)$ determines the stability of the state of rest. In the upper half-space ($E > 0$, $\sigma > 0$), the state of rest is stable if $q > 0$ and is unstable if $q < 0$. For the lower half-space ($E < 0$, $\sigma < 0$) the criterion is reversed.

As regards the point (b), it is equivalent here to the existence of positive roots of the polynomial $\rho^2 + p\rho + q$ in (2.9). As the number of such roots is either 0 or 1 or 2, we call the corresponding cases as *acyclic*, *monocyclic*, or *bicyclic*, since to each of these roots ρ_0 corresponds a stationary periodic solution (that is, a limit cycle); in the first approximation, it is a circle of radius ρ_0.

The last question (c) in this program, the stability of the stationary state, is obtained in a simple manner because, in this case, we have one single stroboscopic d.e. (since the second equation is $d\varphi/d\tau = 0$) of the form:

$$\frac{d\rho}{d\tau} = -\sigma[\rho^3 + p\rho^2 + q\rho] = \Phi(\rho) \tag{3.3}$$

and the criterion of stability (equation 7.6, Chapter 7) is

$$\Phi_\rho(\rho_0) < 0 \tag{3.4}$$

where $\Phi_\rho(\rho_0) = -\sigma[3\rho_0{}^2 + 2p\rho_0 + q] = -\sigma\rho_0(2\rho_0 + p)$ since $\rho^2 + p\rho + q = 0$ for $\rho = \rho_0$.

We have to distinguish again between the upper half-space ($E > 0$, $\sigma > 0$) and the lower one ($E < 0$, $\sigma < 0$). For the former the criterion of stability of the stationary state is

$$\rho_0 + \frac{p}{2} > 0 \tag{3.5}$$

and for the latter it is

$$\rho_0 + \frac{p}{2} < 0 \tag{3.6}$$

Taking into account the expressions for p and q, one has for the eight octants the following signs of A, C, E (and, therefore, for p and q, Fig. 22.1):

For $E > 0$, $\sigma > 0$, one has for the upper half-space

(1) $A > 0, C > 0;\ (p > 0, q > 0)$
(2) $A > 0, C < 0;\ (p < 0, q > 0)$
(3) $A < 0, C < 0;\ (p < 0, q < 0)$
(4) $A < 0, C > 0;\ (p > 0, q < 0)$

and for the lower half-space ($E < 0$, $\sigma < 0$)

(5) $A > 0, C > 0;\ (p < 0, q < 0)$
(6) $A > 0, C < 0;\ (p > 0, q < 0)$
(7) $A < 0, C < 0;\ (p > 0, q > 0)$
(8) $A < 0, C > 0;\ (p < 0, q > 0)$

In the first octant the polynomial $\rho^2 + p\rho + q$ has no positive roots so that $(d\rho/d\tau) < 0$; the only stationary state is the state of rest. In accordance with our classification we may call it: S (stable state of rest); hence, the octant (1) is *acyclic.*

In the octant (2) the state of rest is stable (by 3.2). The polynomial $\rho^2 - |p|\rho + q = 0$ has two positive roots, viz.:

$$\rho_{01} = \frac{|p|}{2} - \sqrt{\frac{p^2}{4} - q} \text{ and } \rho_{02} = \frac{|p|}{2} + \sqrt{\frac{p^2}{4} - q} \text{ assuming that } \frac{p^2}{4} > q$$

Of these two roots the smaller one ρ_{01} is unstable and the larger one ρ_{02} is stable as one verifies this by means of (3.4). The topological configuration in this case is: SUS (or bicyclic).

If $(p^2/4) < q$, the polynomial has no positive roots and this clearly means an acyclic case: S.

Thus, in the octant (2) there are two configurations possible: SUS and S, and the surface Σ separating these configurations is given by the equation: $p^2 = 4q$ or, in A, C, E variables: $E = C^2/40A$.

If the parametric point is *above* the surface Σ, one has the configuration SUS; if it is *below* it, then it is simply S.

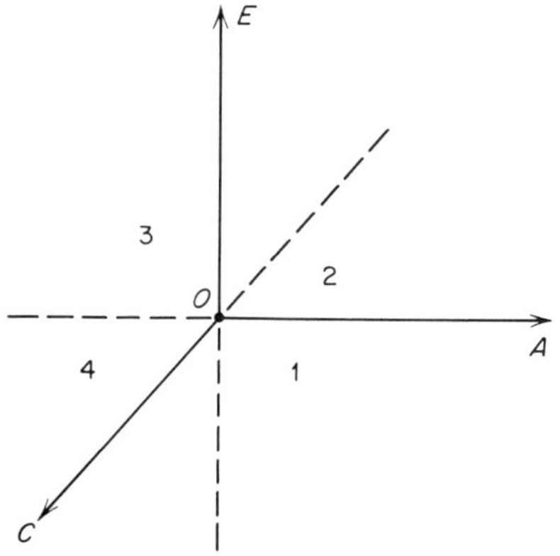

FIGURE 22.1

The surface Σ is thus the locus of the bifurcation points of the second kind, viz., $SUS \to S(US) \to S$. Algebraically, this is a locus of double roots $\rho_0 = |p|/2$. Topologically this is also a locus of semi-stable cycles (US).

Thus a d.e.:

$$\ddot{x} + (a - cx^2 + ex^4)\dot{x} + x = 0; \qquad a > 0, \quad c > 0$$

exhibits a bicyclic configuration: SUS if $c^2 > 40ae$; and an acyclic one: S if $c^2 < 40ae$. As negative coefficients of $\dot{x}$ physically mean *absorption* of energy (negative resistance), it is seen that the above condition for a bicyclic state may be regarded as a relation between the nonlinear terms of which $-cx^2\dot{x}$ is an energy absorbing term and $a\dot{x}$ and $ex^4\dot{x}$ are, on the contrary, dissipative terms; we shall return to this question in the last section of this chapter.

In octants (3) and (4) the state of rest is unstable and the polynomial has only one positive root. In octant (3) this root is

$$\rho_0 = \frac{|p|}{2} + \sqrt{\frac{p^2}{4} + q}$$

and in (4) it is

$$\rho_0 = \frac{|p|}{2} + \sqrt{\frac{p^2}{4} + |q|}$$

The criterion (3.4) shows that both these roots are stable; therefore in these octants the configuration is: *US* (that is, monocyclic).

Figure 22.2 represents the results obtained for the four octants of the

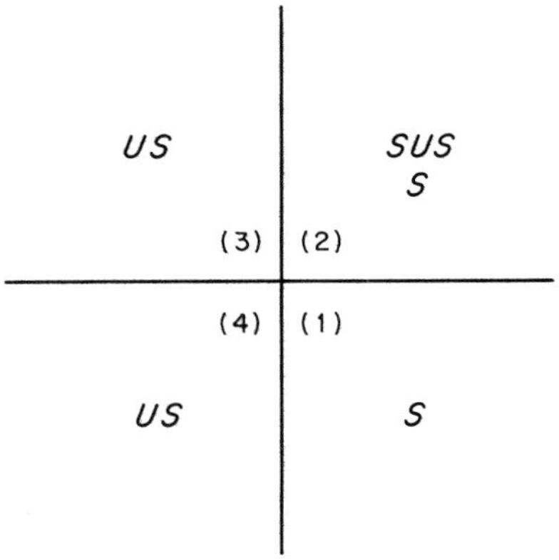

FIGURE 22.2

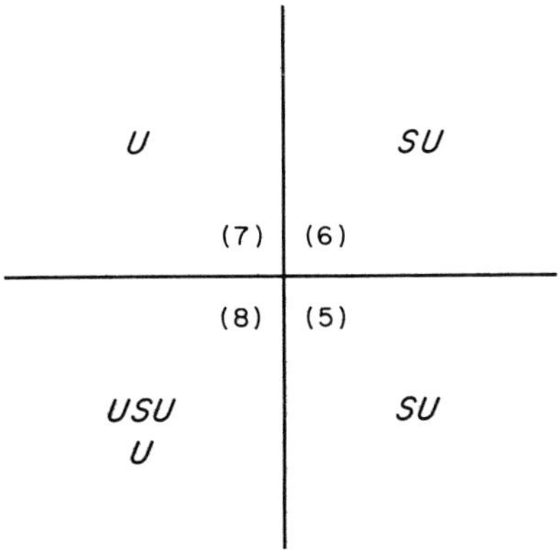

FIGURE 22.3

upper half-space ($E > 0$, $\sigma > 0$) and one readily sees the nature of bifurcations. The octant (2) is characterized by bifurcations of the *second* kind: $SUS \to S(US) \to S$, the bifurcation points (double roots) being located on the surface Σ, as was just mentioned. There are also bifurcations of the *first* kind whose locus is the plane $A = 0$; in fact in this plane there are bifurcations: $US \to SUS$; $US \to S$ for the inverse ones.

There are no bifurcations between the octants (3) and (4) since the monocyclic configuration *US* goes smoothly from one octant to the other, as one can easily verify.

For the octants (5), (6), (7), and (8) of the lower half-space ($E < 0, \sigma < 0$), Fig. 22.3, the conditions of stability are reversed both for the state of rest and for the stationary motion, whereas the polynomial remains the same; and this leads to the *inverted topological* configurations, as one sees easily by reproducing these simple calculations, which we indicate briefly.

In the octant (5): $p < 0$, $q < 0$; the polynomial is the same as in (3); therefore there is only one positive root

$$\rho_0 = \frac{p}{2} + \sqrt{\frac{p^2}{4} + q}$$

which is, however, unstable, the state of rest being stable; hence the configuration is: *SU* (compare with (3) where it is *US*). It is clear that such an "inverted" configuration is *acyclic* but of a special kind, viz.: within the area of the unstable limit cycle *U*, it is "acyclic *S*," and outside that area it is "acyclic *U*".

In the octant (6): $p > 0$, $q < 0$ the polynomial is the same as in (4). There exists thus only one positive root

$$\rho_0 = \frac{|p|}{2} + \sqrt{\frac{p^2}{4} + |q|}$$

and, again, this root is unstable, the state of rest being stable. One has again the same "inverted configuration": *SU* as in (5), leading to a "mixed acyclicity" (viz.: *S*, if the representative point is inside the area of the unstable cycle; and *U*, if it is outside that area).

The octant (7): $p > 0$, $q > 0$ is simply acyclic *U*, since the polynomial has no positive roots. This case is obvious physically; in fact, all three coefficients a, c, and e in (3.1) being negative, the amplitude increases indefinitely, the nonlinearities being turned "in a wrong manner" so to speak (that is, without limiting the increase of the amplitude).

In the octant (8) the polynomial is the same as in (2) but, as the stabilities are now reversed, one has also an "inverted configuration" *USU* and *U* (instead of *SUS* and *S* as in octant (2)). These two configurations are separated again by the surface Σ: $p^2 - 4q = 0$ as in the case (2).

It must be noted that the inverted configuration *USU* in the case of self-excitation from rest is the same as an ordinary monocyclic configuration: *US*. However, in view of the presence of an external unstable cycle, the behavior *outside* that cycle is different; instead of approaching the stable cycle from the outside, the trajectories are repelled from it if the initial conditions happen to be outside the external unstable cycle so that in this region the system is acyclic *U*. Such a situation arises in the case of the d.e.

$$\ddot{x} - (a - cx^2 + dx^4)\dot{x} + x = 0; \qquad a > 0, \quad c > 0, \quad d > 0$$

in the case when the condition $c^2 > 40ad$ is fulfilled.

Figure 22.3 summarizes the results which we obtained for the four octants of the lower half-space ($E < 0$).

The bifurcations of the first kind occur again in the plane $A = 0$ and those of the second kind on the surface Σ in the octant (8).

4. Bifurcations of the third kind; more general cases

In the preceding we encountered bifurcations of the first kind in the plane $A = 0$ and those of the second kind on the surface Σ separating bicyclic and acyclic regions.

A new type of bifurcations, those of the *third kind*, appears in the plane $E = 0$ as will be now shown. This form of bifurcation changes the cyclicity of the configuration by *one* unit just as this accurs in a bifurcation of the first kind, and it consists in adding or in removing the *external* cycle.

It is sufficient to superimpose the Figs. 22.2 and 22.3 so that the octant (1) is above the octant (5), etc., and consider the change in configuration when E varies from $E < 0$ to $E > 0$ or vice versa. One has thus the following transitions

$$(1) \to (5): S \to SU; \qquad (2) \to (6): \left\{ \begin{matrix} SUS \searrow \\ \qquad\qquad SU; \\ S \nearrow \end{matrix} \right.$$

$$(3) \to (7): US \to U \quad \text{and} \quad (4) \to (8): US \begin{matrix} \nearrow USU \\ \\ \searrow U \end{matrix}$$

These transitions are obviously reversible (that is, hold also for directions opposite to those shown by arrows).

It is seen, for instance, that the transition $(1) \to (5)$ adds an unstable cycle to the previously existing state of rest S; and the inverse transition $(5) \to (1)$, on the contrary, removes an unstable cycle previously existing, and so on for other transitions. In the transition $(2) \to (6)$ there is a situation similar to that which was previously studied, viz.: a surface Σ: $(p^2/4) - q = 0$ separates the regions in octant (2) and, according to whether $(p^2/4) - q > 0$ or < 0, the bifurcation is either $SUS \to SU$ or $S \to SU$ or an inverse one.

If the polynomial in the stroboscopic equation (2.9) is of a higher degree, for example, a cubic one, by reducing it to the standard form $\rho^3 + p\rho + q$ (where p and q have different expressions as compared to those in equation (2.9)), one can still use a tridimensional parameter space, but the relations are now also different. Since the number of real roots in this case is either one or three, according to the sign of the discriminant $\Delta = (p^3/27)$

$+ (q^2/4)$, there are several possibilities, viz.: (a) one single positive root ($\Delta > 0$); (b) if $\Delta < 0$ one may have either one, or two, or three positive roots. The conditions of stability of roots as well as of the state of rest are determined in the same manner as previously. There are thus configurations: $SUSU$ or $USUS$ with possible bifurcations of the first kind, viz.: $SUSU \rightarrow (SU)SU \rightarrow USU$; $USUS \rightarrow (US)US \rightarrow SUS$; or of the second kind: $SUSU \rightarrow S(US)U \rightarrow SU$; $USUS \rightarrow U(SU)S \rightarrow US$; or, finally, of the third kind: $SUSU \rightarrow SUS(U) \rightarrow SUS$ or $USUS \rightarrow USU(S) \rightarrow USU$.

The bifurcations of the second kind occur clearly when there appear double-roots disappearing thereafter, that is, when Δ vanishes. Those of the first and of the third kinds are related to the existence of zero roots and so on.

Summing up, the theory of bifurcations is connected with the investigation of the evolution of real positive roots of an algebraic polynomial when its coefficients vary.

5. Special cases of Liénard's equation

One obtains six special cases: (1) $A = 0$; (2) $C = 0$; (3) $E = 0$; (4) $A = C = 0$; (5) $A = E = 0$; (6) $C = E = 0$. In such cases appear simplifications.

Case (1): $A = 0$; ($q = 0$) and (2.9) becomes

$$\frac{d\rho}{d\tau} = -\sigma\rho^2(\rho + p) = \Phi(\rho) \tag{5.1}$$

The state of rest: $\left(\frac{d\rho}{d\tau}\right)_{\rho \simeq 0} \simeq -\sigma\rho^2 p$. The stationary amplitude $\rho_0 = -p$ is possible only if $p < 0$, that is, $\rho_0 = |p|$ so that $\Phi(\rho) = -\sigma\rho^3 + \sigma|p|\rho^2$; $\Phi_\rho(\rho_0) = -3\sigma\rho_0{}^2 + 2\sigma|p|\rho_0 = -\sigma|p|$. Hence, if $E > 0$, ($\sigma > 0$), the limit cycle $\rho_0 = |p|$ is stable; on the other hand, $(d\rho/d\tau)_{\rho\simeq 0} = \sigma\rho^2|p| > 0$ is unstable and therefore the configuration is US.

Thus a d.e. of the form

$$\ddot{x} - (cx^2 - ex^4)\dot{x} + x = 0; \qquad c > 0, \quad e > 0 \tag{5.2}$$

has a stable stationary solution with the "radius" of the limit cycle (always in the first approximation) given by $\rho_0 = |p| = 2|C|/5E$. In the same case but for $E < 0$, one has

$$\frac{d\rho}{d\tau} = +|\sigma|\rho^2(\rho + p) = \Phi(\rho) \tag{5.3}$$

The stationary amplitude can exist only if $p < 0$, in which case $\rho_0 = |p|$ but as $p = 2c/5e$ and, since $e < 0$, it can be negative only if $c > 0$. In such a case the state of rest is obviously stable. As to the limit cycle $\rho_0 = |p|$, one finds that it is unstable, thus leading to the configuration SU.

This corresponds to the d.e.

$$\ddot{x} + (cx^2 - ex^4)\dot{x} + x = 0; \qquad c > 0, \quad e > 0 \tag{5.4}$$

One can readily see the physical meaning of these configurations. In (5.2) and (5.4) we have written the signs explicitly: $c > 0$ and $e > 0$. In (5.2) the quadratic term is negative $(-cx^2\dot{x})$ and the fourth degree term is positive $(+ex^4\dot{x})$; hence initially (for small x) there is a *negative damping*; as x increases, the term $ex^4\dot{x}$ begins to dominate the quadratic term and for larger deviations one has a positive damping. Topologically the configuration is clearly US as we have found by using the formal argument. A similar intuitive reasoning permits ascertaining the significance of the d.e. (5.4) leading to the configuration SU.

Case (2): $C = 0$; $(p = 0)$. In this case (2.9) is

$$\frac{d\rho}{d\tau} = -\sigma\rho[\rho^2 + q] \tag{5.5}$$

For a stationary state $q = 8a/5e$ must be negative, in which case $\rho = +\sqrt{|q|}$. This requires that a and e must be of opposite signs. Assume, for example, $e > 0$, $a < 0$. In the first place, the state of rest is obviously unstable. As to the stability of the limit cycle $\rho_0 = +\sqrt{|q|}$, it is given by the formula

$$\Phi_\rho(\rho_0) = -3\sigma\rho_0{}^2 + \sigma|q| = -3\sigma|q| + \sigma|q| < 0$$

which shows that ρ_0 is stable. Thus the d.e.

$$\ddot{x} - (a - ex^4)\dot{x} + x = 0; \qquad a > 0, \quad e > 0 \tag{5.6}$$

has a configuration: US, with the "radius" of limit cycle $\rho_0 = \sqrt{\frac{8}{5}\frac{|A|}{E}}$.

One ascertains that the other possibility: $e < 0$, $a > 0$ leads to the "inverted" configuration: SU.

Case (3): $E = 0$. In this case one cannot use the d.e. (2.9) and one has to derive the stroboscopic d.e. from (2.8), setting $E = 0$. We have thus

$$\rho_1(2\pi) = \Delta\rho_1 = -2\pi\rho_0[\tfrac{1}{4}C\rho_0 + A]; \qquad \rho(2\pi) = \mu\rho_1(2\pi)$$

so that the stroboscopic d.e. is:

$$\frac{d\rho}{d\tau} = -\rho[\tfrac{1}{4}C\rho + A] = \Phi(\rho) \tag{5.7}$$

which gives the stationary amplitude $\rho_0 = -4A/C$. As $\rho_0 > 0$, clearly A and C must be of opposite signs, that is, either $A > 0$, $C < 0$ or $A < 0$, $C > 0$. One ascertains that in the first case one has an "inverted" configuration SU; but the second case is quite familiar resulting in the configuration US and corresponding to the d.e.

$$\ddot{x} - (|a| - cx^2)\dot{x} + x = 0 \tag{5.8}$$

If $|a| = c = 1$ this is the van der Pol equation with $\rho_0 = 4$; $r_0 = 2$. Thus in this general representation in the parameter space the van der Pol equation appears in the fourth quadrant ($A < 0$, $C > 0$) of the plane $E = 0$.

Case (4): $A = C = 0(p = q = 0)$. In this case (2.9) becomes

$$\frac{d\rho}{d\tau} = -\sigma\rho^3$$

For $E > 0(\sigma > 0)$ the configuration is simply S; and for $E < 0$, $(\sigma < 0)$ it is U.

Case (5): $A = E = 0$. In this case from (2.8), the stroboscopic d.e. is

$$\frac{d\rho}{d\tau} = -\tfrac{1}{4}C\rho^2$$

Hence, for $C > 0$ the configuration reduces to S, and for $C < 0$ to U.

Case (6): $C = E = 0$, by the same argument, results in the stroboscopic d.e.

$$\frac{d\rho}{d\tau} = -A\rho$$

which is the simple linear case of the d.e. $\ddot{x} + a\dot{x} + x = 0$. Hence, if $A > 0$, the configuration is S; and if $A < 0$, it is U, which is a well known result.

It is useful to summarize the results which we obtained in this section. In the following tabulation we write for the sake of simplicity $-a$ instead of $-|a|$ etc., so that a, c, and e below are positive, the proper signs taking care of different octants.

In line with each d.e. we indicate the corresponding topological (configuration of the stationary state as well as the stationary amplitude ρ_0 (or amplitude ρ_{01} and ρ_{02}). We have thus for the eight octants the following cases:

(1) $\ddot{x} + (a + cx^2 + ex^4)\dot{x} + x = 0$; S; $\rho_0 = 0$

(2) $\ddot{x} + (a - cx^2 + ex^4)\dot{x} + x = 0$

$$\begin{cases} SUS \quad \text{if} \quad c^2 > \frac{5}{2}e; \quad \rho_{01,2} = \frac{|p|}{2} \pm \sqrt{\frac{p^2}{4} - q} \\ S \quad \text{if} \quad c^2 < \frac{5}{2}e; \quad \rho_0 = 0 \end{cases}$$

(3) $\ddot{x} - (a + cx^2 - ex^4)\dot{x} + x = 0; \quad US; \quad \rho_0 = \frac{|p|}{2} + \sqrt{\frac{p^2}{4} + |q|}$

(4) $\ddot{x} - (a - cx^2 - ex^4)\dot{x} + x = 0; \quad US; \quad \rho_0 = -\frac{p}{2} + \sqrt{\frac{p^2}{4} + |q|}$

(5) $\ddot{x} + (a + cx^2 - ex^4)\dot{x} + x = 0; \quad SU; \quad \rho_0 = \frac{|p|}{2} + \sqrt{\frac{p^2}{4} + |q|}$

(6) $\ddot{x} + (a - cx^2 - ex^4)\dot{x} + x = 0; \quad SU; \quad \rho_0 = -\frac{p}{2} + \sqrt{\frac{p^2}{4} + |q|}$

(7) $\ddot{x} - (a + cx^2 + ex^4)\dot{x} + x = 0; \quad U; \quad \rho_0 \to \infty$

(8) $\ddot{x} - (a - cx^2 + ex^4)\dot{x} + x = 0$

$$\begin{cases} USU \quad \text{if} \quad c^2 > \frac{5}{2}e; \quad \rho_{01,2} \text{ the same as in (2)} \\ U \quad \text{if} \quad c^2 < \frac{5}{2}e; \quad \rho_0 \to \infty \end{cases}$$

The special cases are:

(1) Plane $A = 0$; $\ddot{x} - (cx^2 - ex^4)\dot{x} + x = 0; \quad US; \quad \rho_0 = \frac{2}{5}\frac{c}{e}$

$\ddot{x} + (cx^2 - ex^4)\dot{x} + x = 0; \quad SU; \quad \rho_0 = \frac{2}{5}\frac{e}{c}$

(2) Plane $C = 0$; $\ddot{x} - (a - ex^4)\dot{x} + x = 0; \quad US; \quad \rho_0 = \sqrt{\frac{8}{5}\frac{a}{e}}$

$\ddot{x} + (a - ex^4)\dot{x} + x = 0; \quad SU; \quad \rho_0 = \sqrt{\frac{8}{5}\frac{a}{e}}$

(3) Plane $E = 0$; $\ddot{x} - (a - cx^2)\dot{x} + x = 0; \quad US; \quad \rho_0 = \sqrt{4\frac{a}{c}}$ (the van der Pol equation)

$\ddot{x} + (a - cx^2)\dot{x} + x = 0; \quad SU; \quad \rho_0 = \sqrt{4\frac{a}{c}}$

(4) Axis E $\quad \dot{x} + ex^4\dot{x} + x = 0;\quad S$ for $E > 0$ and U for $E < 0$

(5) Axis C $\quad \ddot{x} + cx^2\dot{x} + x = 0;\quad S$ for $C > 0$ and U for $C < 0$

(6) Axis A $\quad \ddot{x} + a\dot{x} + x = 0;\quad S$ for $A > 0$ and U for $A < 0$.

In the configuration US the limit cycle is stable, whereas in SU it is unstable, but the "radius" (in the first approximation) is the same in both cases.

6. Phase portraits of Rayleigh's and mixed equations

In his "Theory of Sound" Lord Rayleigh obtained a d.e. of the form

$$\ddot{x} + f(\dot{x}) + x = 0 \tag{6.1}$$

where $f(\dot{x})$ is a certain odd function of $\dot{x}$; for instance, $a\dot{x} + c\dot{x}^3 + e\dot{x}^5$.

When, at a later date, the van der Pol equation was established, it was ascertained that the Rayleigh equation is reducible to that of van der Pol by a differentiation and by setting $\dot{x} = y$, assuming that $f(\dot{x})$ is differentiable.

With the stroboscopic method which we are using here, these results can be obtained directly from Rayleigh's equation:

$$\ddot{x} + (a + c\dot{x}^2 + e\dot{x}^4)\dot{x} + x = 0 \tag{6.2}$$

in which case the stroboscopic d.e. is:

$$\frac{d\rho}{d\tau} = -\sigma\rho(\rho^2 + p\rho + q) = \Phi(\rho) \tag{6.3}$$

with $\sigma = 5E/8$; $p = 6C/5E$ and $q = 8A/5E$.

It is noted that this is exactly the d.e. (2.9) which appears in Liénard's case, but the coefficients σ, p, and q have now different numerical values, so that our previous analysis holds also in this case.

In a later extension of Liénard's equation, N. Levinson and O. K. Smith indicated a somewhat broader form of Liénard's function $f(x)$ by introducing $f(x,\dot{x})$, which thus combines to some extent Liénard's and Rayleigh's equations and may be, for that reason, called a "mixed" (Liénard–Rayleigh) equation.

The investigation of mixed equations by the stroboscopic method does not present any difficulty and we indicate here a simple example.

Consider a mixed equation of the form:

$$\ddot{x} + (a + cx^2 + b\dot{x}^2)\dot{x} + x = 0 \tag{6.4}$$

and its corresponding stroboscopic d.e.

$$\frac{d\rho}{d\tau} = -\rho[A + \tfrac{1}{4}(C + 3B)\rho] = \Phi(\rho) \tag{6.5}$$

where $A = a/\mu$; $C = a/\mu$, and $B = b/\mu$. Setting $\frac{1}{4}(C + 3B) = m$, one has $\Phi(\rho) = -(A\rho + m\rho^2)$; $\Phi 1(\rho_0) = -(A + 2m\rho_0)$; $\rho_0 = -A/m$; with this value of ρ_0, the condition of stability is: $A < 0$.

Thus, a stable stationary state is possible if $A < 0$ and $C + 3B > 0$. In the parameter space the stable stationary state is possible in the half-space $A < 0$ provided the parameter point is above the plane $C = -3B$ passing through the A axis.

7. Frequency correction

In the derivation of the Liénard equation in Section 2 we have chosen a simple case $g(x) = x$ inasmuch as in the example of an electric circuit the capacity was assumed to be constant. With certain types of capacitors, the capacity may be, however, a function of x, and in this case one would have $g(x)$ instead of x, thus resulting in the d.e.:

$$\ddot{x} + f(x)\dot{x} + g(x) = 0 \tag{7.1}$$

In this case there appears a nonlinear frequency (or period) correction already in the first approximation, as we propose to show now.

In applied problems one has to conserve only the odd terms in $g(x)$ so that we can set:

$$g(x) = x + mx^3 + nx^5 \tag{7.2}$$

where $m > 0$ and $n > 0$ are small coefficients.

The reduction to the variables ρ and ψ results in the system

$$\begin{aligned} \frac{1}{2}\frac{d\rho}{d\tau} &= -f(x)y^2 + xy - g(x)y \\ \rho\frac{d\psi}{d\tau} &= -f(x)xy - g(x)x - y^2 \end{aligned} \tag{7.3}$$

In the calculation of $\rho_1(2\pi)$, the second and the third terms on the right-hand side of the first d.e. (7.3) cancel out so that the first stroboscopic d.e. (for $d\rho/d\tau$) remains the same as before but the difference now appears in the second d.e. (7.3). The term $-f(x)xy$ cancels out when forming $\psi_1(2\pi)$ so it is sufficient to consider only two last terms. One has

$$\rho\frac{d\psi}{d\tau} = -g(x)x - y^2 = -(x + mx^3 + nx^5)x - y^2 = -\rho - mx^4 - nx^6$$

and

$$\frac{d\psi_1}{d\tau} = -M\rho\cos^4\psi - N\rho^2\cos^6\psi; \qquad M = \frac{m}{\mu}, \quad N = \frac{n}{\mu} \tag{7.4}$$

therefore

$$\frac{d\varphi}{d\tau} = \frac{1}{16}\rho_0(6M + 5N\rho_0) \tag{7.5}$$

after the integrals $\int_0^{2\pi}\cos^4\psi_0 d\psi_0$ and $\int_0^{2\pi}\cos^6\psi_0 d\psi_0$ are evaluated taking into account that $d\psi_0 = -d\tau$ and passing through the usual procedure of the stroboscopic transformation ($2\pi\mu = \Delta\tau$; $\Delta\tau \to d\tau$; etc.).

Thus, the existence of an odd function $g(x) \neq x$ introduces generally a nonlinear frequency correction $\Delta\omega = d\varphi/d\tau$ in the first approximation which depends on the stationary amplitude ρ_0.

This correction may occasionally vanish for special values of coefficients m and n; this occurs if $\rho_0 = -6m/5n$ which requires that m and n must be of opposite signs but this is a case too special to scarcely be of any interest.

Summing up, the role of the functions $f(x)$ and $g(x)$ in (1.1) on the solution is quite different, viz.: the form of $f(x)$ accounts for the topological aspect of the solution whereas that of $g(x)$ merely determines the frequency (or period) correction. As this correction depends also on ρ_0 and the latter depends on $f(x)$, one can say that *indirectly* this correction depends also on $f(x)$. The converse, however, is not true, inasmuch as the topological aspect of the solution does not depend on the form of $g(x)$.

8. Special forms of the van der Pol equation

In Section 5 we have seen that the d.e. (5.2) and (5.4) have the same topological configuration: *US* as the van der Pol equation and may be regarded as *special forms* of that equation.

One can also consider the van der Pol equation by itself in a more general form, for instance:

$$\ddot{x} - \mu(1 - x^{2n})\dot{x} + x = 0 \tag{8.1}$$

where $n > 0$.

The well known van der Pol equation corresponds to $n = 1$. The essential point is that for a periodic solution of a d.e. of the form

$$\ddot{x} - \mu(1 - \varphi(x))\dot{x} + x = 0 \tag{8.2}$$

$\varphi(x)$ must be a positive function monotonically increasing with x beginning with $\varphi(0) = 0$.

One can even dispense with the analyticity of this function and consider it as made up of several arcs as long as the above requirement is fulfilled. It may be an odd function of x, in which case one has to consider its *absolute value* $|\varphi(x)|$. In all cases it is necessary that the initial negative damping $-\mu\dot{x}$ should become a positive damping for a sufficiently large value of x.

If one considers, therefore, that $\varphi(x) = |x^n|$, $n > 0$, the form of $\varphi(x)$ in a square: $0 < x < 1$ is shown in Fig. 22.4. For $n = 1$, it is a straight line; for $n > 1$ and increasing $\varphi(x)$ rises slowly in the begining of the interval (0,1) and, on the contrary, rapidly at its end (curves A); for $0 < n < 1$ with a decreasing n, the function $\varphi(x)$, on the contrary, rises rapidly in the beginning of (0,1) and increases slowly at the end of the interval (curves B). Thus, in terms of the "negative resistance" the curve (A) in (8.2) accounts for the existence of larger values of negative resistance in the interval (0,1) than the curves (B).

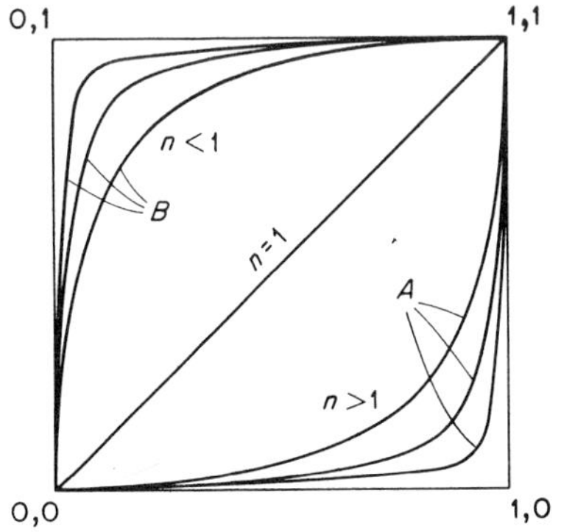

FIGURE 22.4

Consider, therefore, a van der Pol equation of the form

$$\ddot{x} + \mu(x^{2n} - 1)\dot{x} + x = 0; \qquad n = 1, 2, \ldots \tag{8.3}$$

involving an even function $\varphi(x) = x^{2n}$; for $n = 1$, it is an ordinary van der Pol equation whose stationary amplitude is $r_0 = 2$ (or $\rho_0 = 4$).

Applying the stroboscopic transformation to (8.3) we obtain

$$\frac{d\rho_1}{d\tau} = 2\rho \sin^2 \psi - 2\rho^{n+1} \sin^2 \psi \cos^{2n} \psi$$

so that

$$\rho_1(2\pi) = -2\rho_0\left(\int_0^{2\pi} \sin^2 \psi_0 \cdot d\psi_0 - \rho_0{}^n \int_0^{2\pi} \sin^2 \psi_0 \cos^{2n} \psi_0 d\psi_0\right)$$

The stroboscopic d.e. in this case is

$$\frac{d\rho}{d\tau} = -\frac{1}{\pi}\rho\left(\int_0^{2\pi} \sin^2 \psi_0 \cdot d\psi_0 - \rho^n \int_0^{2\pi} \sin^2 \psi_0 \cos^{2n} \psi_0 d\psi_0\right); \quad \psi_0 = \varphi_0 - t$$

and the stationary amplitude ρ^* is given by the expression

$$\rho^* = \sqrt[n]{\int_0^{2\pi} \sin^2 \psi_0 d\psi_0 \Big/ \int_0^{2\pi} \sin^2 \psi_0 \cos^{2n}\psi_0 d\psi_0} \tag{8.4}$$

For $n = 1$, one obtains $\rho^* = 4$ which is the well known result for the usual van der Pol equation.

For $n = 2$ one has $\rho^* = \sqrt{8} \simeq 2.83$; for $n = 3$, $\rho^* = \sqrt[3]{192/15} \simeq 2.34$ and so on; thus the stationary amplitudes decrease for increasing n.

One can also investigate the cases when n is a rational function, but in this case one has to consider only the absolute values of x^{2n}. Thus, for instance, for $n = \frac{1}{2}$, (8.3) is of the form:

$$\ddot{x} + \mu(|x| - 1)\dot{x} + x = 0 \tag{8.5}$$

In this case, instead of the variable ρ, it is preferable to use $r(\rho = r^2)$.

The d.e. of the first-order corrective term is

$$\frac{dr_1}{d\tau} = r_0 \sin^2 \psi_0 - r_0{}^2|\cos \psi| \sin^2 \psi$$

and, following the usual procedure ($dr/d\tau = \mu(dr_1/d\tau)$; $2\pi\mu = d\tau$, etc.) one obtains the stationary amplitude

$$r_0{}^* = \frac{\displaystyle\int_0^{2\pi} \sin^2 \psi_0 d\psi_0}{\displaystyle\int_0^{2\pi} |\cos \psi_0| \sin^2 \psi_0 d\psi_0} = \frac{3\pi}{4} \simeq 2.36; \qquad \rho_0{}^* \simeq 5.58$$

Summing up, for the functions of the form (A) (Fig. 121), approaching more and more the axis $y = 0$ and the straight line $x = +1$, the stationary amplitude decreases; for the functions of the form (B), approaching more and more the axis $x = 0$ and the straight line $y = 1$, they, on the contrary, increase.

9. Certain physical considerations

Results obtained in Section 3 can be given a more familiar physical interpretation. In fact, the form of the function $f(x)$ may be regarded as characterizing the damping coefficient if (1.1) represents an oscillator. As is well known, a positive damping coefficient characterizes a dissipation of energy with the incident decrease of amplitudes, while a negative one, on the contrary, means the energy absorption and, therefore, increasing amplitudes. One can consider, therefore, the Liénard equation as representing an oscillator with a *variable damping*.

On this basis it is possible to give an *interpretation* of the results obtained in Section 3 but it is impossible to *establish them*.

Without any loss of generality we may assume $A = C = E = 1$ and investigate the form of the function $f(x)$ in the first four octants ($E = +1$).

In a bicyclic case (second octant) it is necessary to have a different choice of parameters; for instance: $A = E = 1$ and $C > \sqrt{40AE} = \sqrt{40}$.

The matter reduces thus to the investigation of the form of the function $f(x)$ for the first four octants, viz.:

$$(1)\ f(x) = 1 + x^2 + x^4; \qquad (2)\ f(x) = 1 - x^2 + x^4;$$
$$f(x) = 1 - 7x^2 + x^4;$$
$$(3)\ f(x) = -1 - x^2 + x^4; \quad \text{and} \quad (4)\ f(x) = -1 + x^2 + x^4$$

It is to be mentioned also that we are using here the finite numbers A, C, and E instead of the small numbers a, c, and e which appear in (1.2), inasmuch as $A = a/\mu$, $C = c/\mu$, and $E = e/m$ but, in view of the homogeneity of $f(x)$ in these coefficients, the form of $f(x)$ remains the same.

The function $f(x)$ in (1) is positive for all values of x, which means a

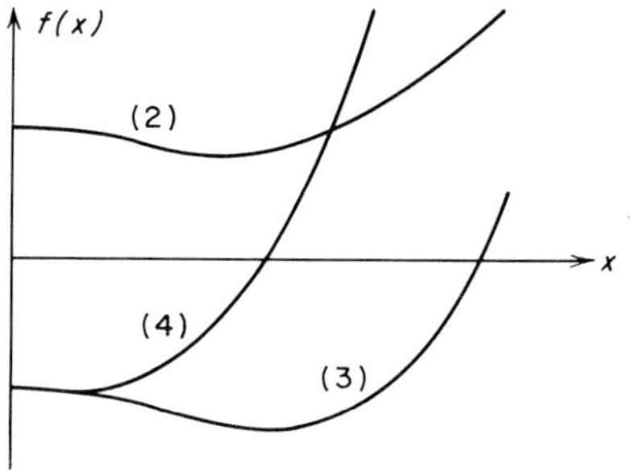

FIGURE 22.5

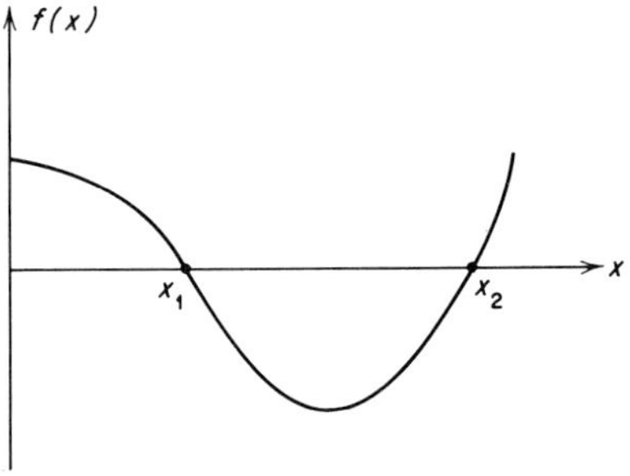

FIGURE 22.6

positive damping for any x; this case does not represent thus any interest and obviously is acyclic: S in our terminology.

For the first function: $f(x) = 1 - x^2 + x^4$ in the second octant and also for $f(x)$ in (3) and (4), the form of $f(x)$ is shown in Fig. 22.5. It is seen that the curve (2) is everywhere positive; therefore that is also an acyclic case S since the damping remains positive everywhere, although it is variable for different x. The curves (3) and (4) exhibit, however, a typical feature of the van der Pol equation, viz.: for small values of x the damping is negative (the amplitude increases), and for larger x it becomes positive which limits a further increase of the amplitude. In these two octants one has thus a typical monocyclic condition US, as we have investigated previously in connection with the van der Pol equation.

These four cases are thus obvious on physical grounds. There remains the only case to be given a further attention, namely, the bicyclic case corresponding to $f(x) = 1 - 7x^2 + x^4$ in (2).

In this case (Fig. 22.6), the curve $f(x)$ intersects twice the abscissa axis; in the first point of intersection $x = x_1$, the curve changes from positive to negative values and, in the second ($x = x_2$), on the contrary, from negative to positive values.

Thus for $x < x_1$ the damping is positive; for $x_1 < x < x_2$ it is negative, and for $x > x_2$ it is again positive.

One sees thus intuitively that a physical oscillatory system provided with a damping of this nature exhibits well the features of the configuration *SUS*. In fact, if the initial value $x_0 < x_1$, the system is in the region of a positive damping and the only stationary state is the state of rest *S*. If, however, $x_0 > x_1$, the damping is negative, amplitude increases, and the only factor that prevents their further increase is the existence of a positive damping for $x > x_2$.

Such intuitive physical arguments are helpful *in understanding* the nature of these phenomena associated with energy exchanges, but they are not sufficient for the establishment of quantitative results like this such as those in Section 3.

Thus, for instance, as regards the van der Pol equation

$$\ddot{x} + \mu(x^2 - 1)\dot{x} + x = 0$$

it is readily seen that the region of the negative damping exists in the interval $0 < x < 1$, but nothing a priori shows that the stationary amplitude is $x_0 = 2$, which is already in the region of the positive damping.

Chapter 23

INTERACTION OF NONLINEAR OSCILLATIONS

1. Introductory remarks

The fundamental property of linear d.e. (or of linear systems) of the nth order is that its general solution is obtained as a linear combination of n linearly independent particular solutions multiplied by the corresponding number of arbitrary constants.

This general property of linear systems finds a simple interpretation in the theory of oscillations in the form of the so-called *superposition principle*, which states that an oscillation in a system with several degrees of freedom consists of a number of *component* oscillations, each behaving as if it were alone. This means that the component oscillations of a linear system *do not interact* among themselves or, which is the same, that the *principle of superposition* holds for such systems.

The essential property of nonlinear oscillatory systems, on the contrary, is that they do not exhibit any superposition of component oscillations or, more specifically, that they exhibit always an *interaction* of some kind between these component oscillations. Perhaps the whole theory of nonlinear oscillations could be formed on the basis of interactions. This approach is particularly useful when a nonlinear oscillatory system consists of two nonlinear oscillators, say, (A) and (B) each of which considered separately has a definite self-sustained oscillation. The question arises as to what will occur if these two oscillators are coupled in some manner so as to form an oscillator ($A + B$) formed by (A) and (B). In such a case, if one wishes to speak in terms of an "interaction," the only way of ascertaining this is to try to compare ($A + B$) either with (A) or with (B) since any combination of the "components" (A) and (B) has no meaning whatever in the nonlinear cases.

What kind of relation exists between the "resultant" oscillation ($A + B$)

and its "components" (A) and (B) is impossible to say in advance without going first through a somewhat lengthy investigation of two preliminary questions, viz.: (1) Does oscillation ($A + B$) exist? (2) In case it exists, is it stable?

Even if these questions are answered in a positive manner, one is not always certain that such a "resultant" oscillation will actually appear. This is due to the fact that, in the case when several such oscillations are potentially possible, the problem cannot be definitely specified unless the initial conditions are prescribed, as we mentioned on a number of occasions previously in connection with problems of "hard self-excitation."

All these complications render the question of "interaction" frequently ill-defined and, for that reason, of a limited use in applications. On the other hand, it is probable that, when these seemingly complicated phenomena are better understood, a *controlled interaction* may be advantageous in a number of applied problems but, at present, one is yet far from these possibilities.

It is to be noted that the problems involving an external periodic excitation of nonlinear systems may also be regarded as problems of interaction effected by the heteroperiodic oscillation on the autoperiodic oscillation, but the usual procedure consists in looking for a periodic solution directly in such a case, as was shown, for instance, in Chapter 19, in connection with the Mandelstam–Papalexi theory.

A particularly typical form of interaction appears, however, when both oscillations are analogous with respect to the same system; for example, when they are both autoperiodic, or even nonperiodic. In such cases it is easier to see the real nature of interaction which generally consists in a modification of conditions of stability with respect to the "component" oscillations, although the term "component" here must be used with a great deal of caution because its significance in the nonlinear domain does not correspond to anything physically definite as it does in the linear field.

2. The van der Pol theory of interaction

The early work of van der Pol touched a number of subjects in the field of nonlinear oscillations and, in particular, the question of interaction of nonlinear oscillations.

In Chapter 18 we outlined the theory of synchronization by the van der Pol method but, once the d.e. 2.6 have been established, we left the argument of van der Pol and continued the exposition of the theory of synchronization on the basis of the topological method of Andronov and Witt. In fact, since the system (2.6), Chapter 18, is of the autonomous type (2.7), it is easy to show that the condition of synchronization is nothing but the singular point of (2.7) which justifies the matter here on a purely formal basis.

If, however, one continues the argument of van der Pol, the same phenomenon of synchronization is explained on the basis of the theory of interaction and there is no necessity for reducing the problem to the existence of the singular point. In fact, if one follows this argument, it is found that the autoperiodic oscillation becomes *unstable* under the effect of the heteroperiodic oscillation and is, therefore, *suppressed*, but this is precisely the phenomenon of synchronization approached from the theory of interaction rather than from a formal concept of a singular point as a condition for the stationary state.

This merely emphasizes what has been said in the introduction to Part III—that the same (idealized) physical phenomenon may occasionally appear as one or the other *aspect* of the global "mathematical phenomenon" (that is, solution of the d.e.) considered in the various regions of the parameter space. If in a certain "region" there are two possible aspects, two different physical theories are possible and generally they are both correct from their respective points of view.

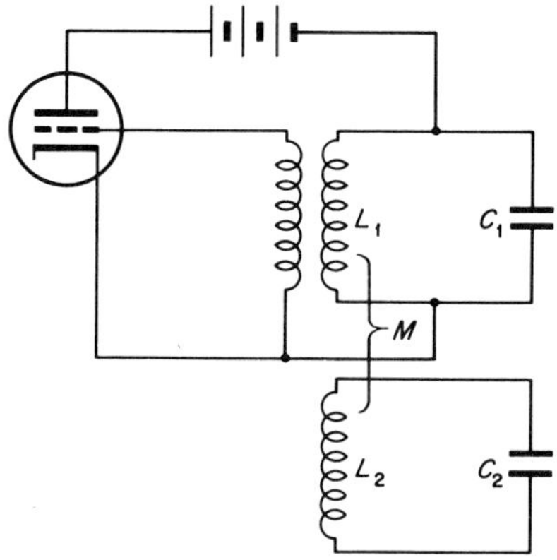

FIGURE 23.1

In another paper van der Pol[1] considers the problem of interaction more explicitly in connection with the electric circuit shown in Fig. 23.1.

In this case L_1C_1 is the normal oscillating circuit associated with the electron tube, and L_2C_2 is a second circuit coupled to L_1C_1 through an inductive coupling M. It is clear that this scheme is a particular case of the circuit which was analyzed in Section 6, Chapter 18, in connection with the problem of mutual synchronization if the second electron tube is omitted and only the oscillating circuit is left. Van der Pol gives first the result of the experiment, which consists in varying ω_2^2 at a constant value of ω_1^2 and recording i_1^2 as shown in Fig. 23.2. Thus, for instance, if one starts with $\omega_2^2 < \omega_1^2$ and increases ω_2^2, the path EFB is followed, but at B the oscillation jumps suddenly to the point C of the curve ACD which is followed if ω_2^2 is increasing. It cannot be followed, however, to the left of A, if ω_2^2 decreases, in which case the current jumps suddenly to the point F of the first curve EFB.

[1] B. van der Pol, *Phil. Mag.* (6), **43**, 1922.

It is seen, thus, that in the interval AB the function i_1^2 is not a single-valued function of ω_2^2 inasmuch as to the same value of ω_2^2 correspond two values of i_1^2, but physically only one of these two values can exist depending *on the direction* in which the parameter ω_2^2 varies. If it increases from small values, the path $EFBCD$ is followed but, if it decreases from large values, the path is $DCAFE$.

Van der Pol analyzes the condition of stability by his method which leads to rather long calculations resulting from the replacement of the solution $v = a \sin \omega_1 t + b \sin (\omega_2 t + \gamma)$ into the d.c. of the fourth order

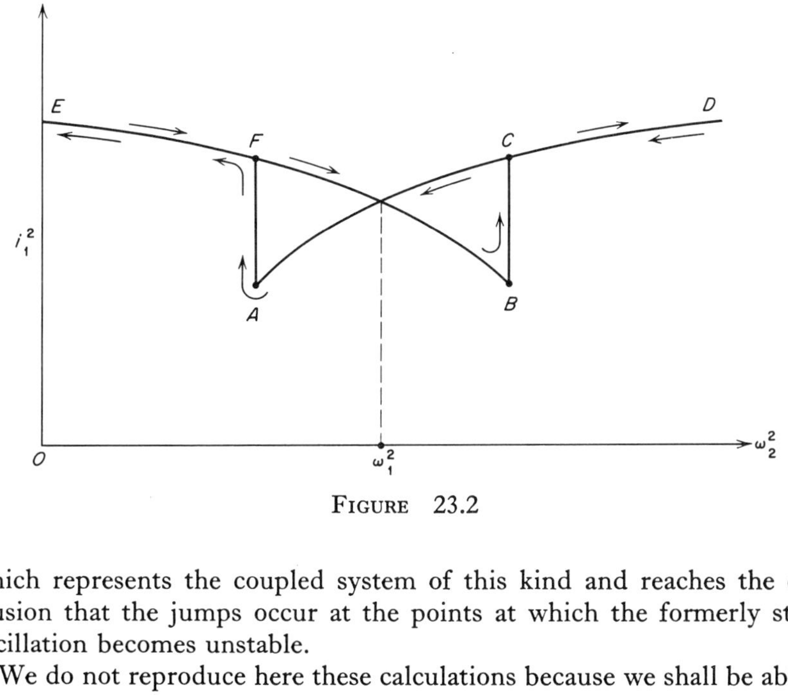

FIGURE 23.2

which represents the coupled system of this kind and reaches the conclusion that the jumps occur at the points at which the formerly stable oscillation becomes unstable.

We do not reproduce here these calculations because we shall be able to obtain the same result later by a simpler argument resulting from the use of the stroboscopic transformation.

The theory of van der Pol is, however, very valuable in that it points out in a definite manner how the question of interaction is linked to the question of stability, which explains the phenomenon of the "oscillation hysteresis," that is, a nonsingle valued determination of the amplitude.

3. Interaction of two autoperiodic oscillations

In order to give an example of analysis of interaction of nonlinear oscillations, we shall investigate the following d.e.[2]

$$\ddot{x} + e(x^2 - 1)\dot{x} + [1 + (a - cx^2)\cos 2t]x = 0 \qquad (3.1)$$

If $a = c = 0$ this equation becomes the van der Pol equation and, if $e = 0$, it becomes an equation of the parametric type which was studied in Chapter 20. In both cases the d.e. have periodic solutions (the second, at least under certain conditions), and it is natural to inquire as to what happens when these two simple "component equations" are coupled together in the form of the d.e. (3.1).

This problem is somewhat complicated and, for that reason, we introduce certain restrictions that will permit reducing the number of cases to be investigated without any loss of generality.

We will consider the case when the coefficients a, c, and e are small and positive. Clearly, if one considers the parameter space (a, c, e), this restriction amounts to the investigation of conditions only in the *first* octant of this space, but the argument developed below is applicable to the remaining octants. Besides this, we shall be interested only in the self-excited systems but, again, the nonself-excited systems can be treated in a similar manner. In other words, our investigation will bear only one of the sixteen possible cases. However, even in this case there are several ramifications, as will be shown. The complicated form of these interactions is, probably, the principal reason why so little is known on this subject.

We shall use again the stroboscopic method and, omitting intermediate calculations, write down the stroboscopic system corresponding to (3.1):

$$\frac{d\rho}{d\tau} = \frac{1}{4}\rho[E(4 - \rho) + (C\rho - 2A)\sin 2\varphi] = R(\rho,\varphi);$$

$$\frac{d\varphi}{d\tau} = \frac{1}{4}(C\rho - A)\cos 2\varphi = \Phi(\rho,\varphi) \qquad (3.2)$$

where $A = a/\mu$; $C = c/\mu$; $E = e/\mu$, μ being the small parameter in the solutions:

$$\rho(t) = \rho_0(t) + \mu\rho_1(t) + \dots$$

where the nonwritten terms contain $\mu^2, \mu^3, \dots$; we limit ourselves only to the first approximation (see the Introduction to Part II); finally φ is the variation of the angular variable ψ during the period 2π; other notations remain the same as previously.

[2] N. Minorsky, *C. R. Ac. Sc.* (Paris) **234**, 1952; *J. Franklin Inst.* **256**, 1953.

We proceed with the discussion of the stroboscopic system (3.2). The singular points of (3.2) are:

$$\begin{aligned} &(1)\ \rho_{01} = 2(A - 2E)/(C - E); \qquad \sin 2\varphi_{01} = +1 \\ &(2)\ \rho_{02} = 2(A + 2E)/(C + E); \qquad \sin 2\varphi_{01} = -1 \end{aligned} \tag{3.3}$$

There is also a third singular point $\rho_{03} = A/C$ with $\sin 2\varphi_{03}$ calculated so as to annul the bracket in the first equation (3.2); we shall not carry out the calculation for the third singular point because the procedure is already clear from the investigation of the first two singular points.

We omit again a somewhat long calculation in connection with the variational equations corresponding to (3.2) and merely indicate the characteristic equation:

$$S^2 + [\tfrac{1}{2}E(\rho - 2)]S - Q = 0 \tag{3.4}$$

where

$$Q = \tfrac{1}{4}[(C\rho - A)\sin 2\varphi + E(2 - \rho)](C\rho - A)\sin 2\varphi \tag{3.5}$$

It is recalled that the general form of the characteristic equation is:

$$S^2 - (R_r + \Phi_\varphi)S + [R_r\Phi_\varphi - R_\varphi\Phi_r] = 0 \tag{3.6}$$

where R_r, R_φ, Φ_ρ, and Φ_φ are the partial derivatives of $R(\rho,\varphi)$ and $\Phi(\rho,\varphi)$ at the singular points (ρ_0,φ_0) in question. For the first two points (3.3), the matter is simplified since $R_\varphi = \Phi_r = 0$ and this accounts for a relatively simple form of (3.6).

As we are interested only in stable singular points, one has the conditions

$$\rho_0 > 2; \qquad Q < 0 \tag{3.7}$$

The first condition means that the energy stored in the oscillation (since $\rho = r^2 = x^2 + y^2 = x^2 + \dot{x}^2$ is a measure of the total energy stored in oscillation) must be greater than the $\rho_0 = 2$; the second condition merely rules out the existence of a saddle point as it is not admitted in the stroboscopic transformation.

We proceed now with the detailed analysis of singular points (3.3).

4. Analysis of stability of singular points

First singular point:

$$\rho_{01} = \frac{2(A - 2E)}{C - E}; \qquad \sin 2\varphi_{01} = +1 \tag{4.1}$$

Inasmuch as $\rho = r^2 > 0$, one has two subcases: (1) $A > 2E$; $C > E$; and

(2) $A < 2E$; $C < E$. We shall first investigate subcase (1). We have thus the following condition:

$$A > 2E; \qquad C > E \tag{4.2}$$

To this we have to add the first condition (3.5): $\rho_{01} > 2$, which gives

$$A > C + E \tag{4.3}$$

We have to express also the second condition (3.5), namely:

$$Q = \tfrac{1}{4}[(C\rho_{01} - A) - E(\rho_{01} - 2)](C\rho_{01} - A) < 0 \tag{4.4}$$

As this inequality is given by the product of two brackets, they must be of opposite signs; on the other hand, as $E(\rho_{01} - 2) > 0$, it is clear that $(C\rho_{01} - A)$ must be positive and the bracket in (4.4) must be negative.

The condition $C\rho_{01} - A > 0$ yields

$$A > \frac{4EC}{C + E} \tag{4.5}$$

If one tries to verify that the bracket in (4.4) is negative, one finds that this is impossible. One concludes, therefore, that (4.2) does not lead to any periodic solution.

If one reverses the signs of inequalities (4.2), one obtains conditions:

$$A < C + E \tag{4.6}$$

$$A < \frac{4EC}{C + E} \tag{4.7}$$

One verifies that (4.7) is the dominating inequality (that is, one can omit (4.6)) and that the bracket in (4.4) is negative. We have thus the conditions:

$$A < 2E; \qquad A < \frac{4EC}{C + E}; \qquad C < E$$

here, again, for the same reason we can keep only $A < 4EC/(C + E)$ and omit $A < 2E$, finally obtaining:

$$C < E; \qquad A < \frac{4EC}{C + E} \tag{4.8}$$

These conditions considered in the parametric space (A, C, E) define a certain region G_1. If the parametric point is in this region, a stable stationary oscillation exists with amplitude ρ_{01} and phase φ_{01} given by (4.1). Outside G_1 there is no stable periodic solution.

It is useful to complete this study of existence of a stable periodic solution (oscillation) by the condition of self-excitation, that is, the spontaneous starting of oscillation from the state of rest.

For this one has to assume $\rho \approx 0$ and neglect terms with ρ (as compared to those without ρ) in (3.2). This gives

$$\left(\frac{d\rho}{d\tau}\right)_{\rho\sim 0} \simeq \tfrac{1}{4}[4E - 2A \sin 2\varphi^*]; \qquad \left(\frac{d\varphi}{d\tau}\right)_{\rho\sim 0} \simeq -\tfrac{1}{4}A \cos 2\varphi^* \tag{4.9}$$

As $\cos 2\varphi^* = 0$ is clearly the equilibrium position for the phase, the variational equation (replacing φ^* by $\varphi^* + \delta\varphi$, where $\delta\varphi$ is a small perturbation), yields:

$$(d\delta\varphi/d\tau) = \tfrac{1}{2}A \sin 2\varphi^* \cdot \delta\varphi$$

and it is clear for the *stable* equilibrium that one must have $\sin 2\varphi^* = -1$. With this value of $\sin 2\varphi^*$, the first equation yields

$$\left(\frac{d\rho}{d\tau}\right)_{\rho\sim 0} = \tfrac{1}{2}(2E + A) > 0 \tag{4.10}$$

which shows that the system is always self-starting. We have thus completed the study of the first singular point (4.1) yielding the stable periodic solution (in the ψ plane) if conditions (4.8) are fulfilled.

Second singular point:

$$\rho_{02} = \frac{2(A + 2E)}{C + E}; \qquad \sin 2\varphi_0 = -1 \tag{4.11}$$

Calculation develops in the same way as in the first case but here the first condition (4.2) does not exist. One finds the condition

$$E > C; \qquad A > \frac{4EC}{E - C} \tag{4.12}$$

which determine another region G_2 in the (A, C, E) space. If the parametric point is in G_2, a stable periodic oscillation in the ψ plane exists in accordance with the coordinates of the singular point (4.11).

Third singular point:

$$\rho_{03} = A/C \tag{4.13}$$

In this case the phase φ_{03} is not given directly as in the first two cases but has to be calculated by annulling the bracket in the first equation (3.2) after the substitution of (4.13). However, here there appear additional conditions: $|\sin 2\varphi_{03}| \leq 1$ which are to be imposed on the coefficients A, C, and E. Moreover, the calculation of the characteristic equation becomes more complicated since the partial derivatives R_φ and Φ_r are not zero in this case. We shall not carry out this calculation because the procedure is sufficiently clear from the preceding.

5. Special cases

A special case arises when $A = 2E$ and $C = E$. The expression for ρ_{01} gives 0/0 but, if one replaces $A = 2C$, both ρ_{01} and ρ_{02} become $\rho_{01} = \rho_{02} = 4$, which means that the energy content is the same as in the van der Pol oscillator; this case, however, is unstable as one verifies by the above procedure. Particular cases are obtained from the general case by setting one of the three parameters equal to zero.

If $A = 0$, the stable singular points of the first two kinds do not exist; hence no stable periodic oscillations are possible. The same result is obtained if $E = 0$.

If $C = 0$, the first stable singular point does not exist, but the second exists under the condition $A > C - E$.

It is to be noted that the case when $A = C = 0$ cannot be analyzed on the basis of the preceding formulas because the d.e. (3.1) in this case becomes the van der Pol equation and the second stroboscopic d.e. (3.2) disappears. However, expressions (3.3) for ρ_0 hold also in this case and give $\rho_0 = 4$. This autonomous case ($A = C = 0$) should not be confused, however, with the above investigated nonautonomous case ($A = 2C$, $C = E$) in which one has also $\rho_{01} = \rho_{02} = 4$ but which is unstable. Summing up, the above analysis shows the following peculiarities of interaction:

(1) Stable periodic oscillations exist only in certain regions defined by inequalities (like 4.8) expressing conditions of consistency of certain requirements to be fulfilled. Outside these regions there are no periodic solutions (oscillations).

(2) Condition of stability $\rho_0 > 2$ indicates the lower energy level below which oscillations cannot maintain themselves in a stable manner. As, on the other hand, $\rho_0 = 4$ is the energy level of the van der Pol oscillator, it follows that, under normal conditions, the oscillator represented by (3.1) develops *less* energy than a simple van der Pol oscillator. Everything happens as if the "parametric element" of the oscillator (that is, terms with a and c) were *absorbing* the energy developed by the van der Pol oscillator.

(3) Only at the critical threshold when $A = 2E$ and $C = E$ (that is, also when $A = 2C$) the oscillator (3.1) reaches the energy level $\rho_0 = 4$ (corresponding to that of the van der Pol oscillator), but at this point it becomes unstable.

(4) The energy content ρ of the oscillator represented by (3.1) depends

on all three parameters in a somewhat complicated manner, as is seen from (3.3). For the first oscillation (3.3), the equi-energy surface is given by the equation, $2A - KC + E(K - 4) = 0$, and for the second, $2A - KC - E(K - 4) = 0$, where K is the parameter of the family. In both cases these surfaces are certain planes in the (A, C, E) space.

6. Remarks

If one takes into account that the above analysis represents only one-sixteenth of the complete study to be made in connection with (3.1), one can readily see a considerable complexity of this d.e. from the standpoint of existence of its stable periodic solutions (oscillations); the latter exist only in some isolated regions G_i of the parameter space (A, C, E); outside these regions (as well as on their boundary surfaces) no stable periodic oscillations are possible.

One can surmise the existence of additional complications when two (or several) sets of points G_i have an intersection. In such cases several singular points (and therefore several periodic solutions) exist for points situated in these common regions but we shall not attempt to investigate such "pathological" cases here.

Chapter 24

ASYNCHRONOUS ACTIONS

1. Introductory remarks

In Chapters 18 and 19 we investigated the phenomena of synchronization and of subharmonic resonance, respectively. It was mentioned in the Introduction to Part III that these two phenomena in reality are merely two different "aspects" of a more general situation that arises in nearly linear systems acted on by an external periodic excitation ("the forcing term") when the frequency of the latter passes through the integer values $n = 2, 3, \ldots$, assuming that the autoperiodic frequency of the system is *one*.

The studies of these phenomena are usually carried out in connection with the d.e. of the form

$$\ddot{x} + x + \mu f(x,\dot{x}) = e \sin \omega t \tag{1.1}$$

and the various cases arise whether the left-hand side of (1.1) corresponds to the existence of a stable periodic solution (the autoperiodic oscillation) or not. In the first case, the left-hand side of (1.1) is generally a Liénard (or a van der Pol) equation, and the problem is to investigate the interaction of an autoperiodic oscillation (with frequency 1) and the heteroperiodic one with frequency ω. In the theory of resonance, as we saw, a stable subharmonic oscillation may exist when $\omega = 2, 3, \ldots$; this does not happen for all integers n but only for some of them for which the conditions of existence and stability of the periodic solution are fulfilled. It was also shown that these resonance oscillations exist not only for integer values of ω but for some "detuning" of ω from these integer values. In such cases, the phenomenon of synchronization appears and brings the actual frequency to the *exact* integer value inasmuch the autoperiodic frequency is "entrained" by the heteroperiodic frequency (Chapter 18).

The combined effects of subharmonic resonance and synchronization are thus distributed into certain "zones" of the heteroperiodic frequency ω around these integer values. In the intervals between these zones, there are regions in which the subharmonic resonance cannot maintain itself and, in these "asynchronous regions" or zones, the phenomena are less interesting and are rather complicated from the mathematical point of view, as we shall see below. In general, nothing of a special interest happens in these regions; there may exist a small heteroperiodic oscillation (which may be considered as "forced" oscillation, but this term is not very adequate for nonlinear systems), or there is no oscillation at all if the condition of stability of the heteroperiodic oscillation is not fulfilled. There are, however, two special cases—*asynchronous quenching* and *asynchronous excitation*—which present some interest, although these two phenomena are yet relatively little explored and at present their existence has been detected only in electrical circuits[1,2]; in mechanical systems the corresponding effects have not been ascertained as yet.

As regards the quenching, it manifests itself in the following manner: if one impresses on the grid of an electron-tube oscillator with frequency ω_0, an extraneous frequency ω (for instance, through an appropriate inductive coupling) and if ω is sufficiently high, everything happens as if this heteroperiodic frequency ω were destroying or "quenching" the previously existing autoperiodic oscillation with frequency ω_0.

In the phenomenon of asynchronous excitation the effect is in some respects opposite and can be described as follows: suppose we have a nonlinear system with the topological configuration SUS using the previously defined term (Chapters 7 and 22); in some publications on this subject such systems are also called "potentially oscillatory systems." Under certain conditions, if one applies to such a system an external periodic excitation with frequency ω, the configuration changes according to the scheme $SUS \rightarrow US$ (Chapter 22) and the system begins to oscillate with its own (autoperiodic) frequency ω_0.

In reality, these two phenomena are of entirely different natures and their apparent resemblance is misleading. In the quenching phenomenon, the principal condition is that the external (heteroperiodic) oscillation must have a sufficiently high frequency but that the *form* of the nonlinear characteristic is immaterial. In the excitation phenomenon, on the contrary, the condition that ω is large is not necessary but, instead, the necessary condition is that the nonlinear characteristic must be of a "hard" type, that is, must be represented by a polynomial of at least fifth degree. One sees readily this point from what was explained in Chapter

[1] E. A. Appleton, *Proc. Cambridge Phil. Soc.* **21**, 1922; N. Minorsky, *C. R. Ac. Sc.* (Paris), **237**, 1953; **248**, 1959; *J. Franklin Inst.* **259**, 1955.

[2] Kobsarev, *J. Tech. Phys.* (USSR) **3**, 1933.

22 dealing with bifurcations of the first kind $SUS \to US$ which is precisely the question involved here. This seems to be corroborated by the experimental evidence which, unfortunately, is rather scarce, although these asynchronous phenomena seem to be of a sufficient interest to warrant a more detailed study.

It is clear that these two phenomena, the quenching and the excitation, are possible only in the case when the system is either self-excited (or "potentially self-excited") which means that the system possesses a topological configuration US or SUS.

In the following section we shall enter into the theory of these asynchronous actions, but in the general case the problem is too difficult and is hardly of interest as was just mentioned. However, the general case will enable us to obtain some conclusions in a particular case of asynchronous quenching, which, on the contrary, is an important phenomenon. The discussion in Sections 2 and 3 is conducted under the assumption of a small external periodic excitation which introduces some simplifications. In Section 4 we take up the same subject without the above restriction; we have noticed already the difference between these two approaches in the theory of subharmonic resonance; we encounter here a similar situation, although our aim is now different inasmuch as we are interested now not in the resonance oscillation, but rather in the asymptotic range when the frequency ω of the external periodic excitation is large.

In the last section of this chapter we give an account of the phenonemon of asynchronous excitation, but instead of following the stroboscopic method which leads to somewhat long calculations, we use the method of equivalent linearization (Chapter 14) of Krylov and Bogoliubov which gives a convenient way of reaching conclusions in this case.

2. Small external periodic excitation

We shall investigate first the case when the amplitude of the forcing term is small; this will enable us to obtain the d.e. of the first approximation in a relatively simple manner. As an example we consider the simplest possible case of the d.e.

$$\ddot{x} + x = \mu[(\alpha - \gamma x^2)\dot{x} + E \sin \omega t] \qquad (2.1)$$

which is an ordinary van der Pol equation with a small forcing term $e \sin \omega t = \mu E \sin \omega t$.

Applying the usual stroboscopic transformation and using the variables $r = \sqrt{x^2 + \dot{x}^2}$; $\psi = \arctan (\dot{x}/x)$; $\dot{x} = y$, one obtains a system of two d.e. of the first order which are integrated by the series

$$r(t) = r_0(t) + \mu r_1(t) + \mu^2 r_2(t) + \ldots;$$
$$\psi(t) = \psi_0(t) + \mu\psi_1(t) + \mu^2\psi_2(t) + \ldots$$

and a simple calculation shows that the zero-order solution here is

$$r_0(t) = r_0 = \text{const}; \qquad \psi_0(t) = \varphi_0 - t \tag{2.2}$$

where r_0 and φ_0 are arbitrary constants.

The d.e. for the first-order corrective terms $r_1(t)$ and $\psi_1(t)$ are then

$$\begin{aligned} \dot{r}_1 &= \alpha r_0 \sin^2 \psi_0 - \tfrac{1}{4}\gamma r_0^3 \sin^2 2\psi_0 + E \sin \psi_0 \sin \omega t \\ \dot{\psi}_1 &= \frac{1}{2}\alpha \sin 2\psi_0 - \gamma r_0^2 \sin \psi_0 \cos^3 \psi_0 + \frac{E}{r_0} \cos \psi_0 \sin \omega t \end{aligned} \tag{2.3}$$

We shall limit ourselves to the first approximation only.

As we are interested here in a possibility of obtaining an oscillation *with the frequency of the forcing term*, our procedure will be different from that which we followed in Chapter 19 dealing with the subharmonic resonance. In fact, in the case of resonance, we were looking for a possibility of a periodic solution with period 2π under the effect of an external periodic excitation with period $2\pi/\omega$, where ω was a small integer ($\omega = 2, 3, \ldots$). Here we are interested in the existence of a periodic solution with the same period $2\pi/\omega$ as that of the external excitation, at least in the beginning of this calculation.

It follows therefore, that in carrying out the passage to the difference equations, the integrations are to be performed between 0 and $2\pi/\omega$ instead of 0 and 2π as in the case of the subharmonic resonance.

We omit a series of simple intermediate calculations resulting from the substitution of $\psi = \psi_0(t)$ in (2.3) by its value (2.2) and the subsequent integrations.

In the passage to the stroboscopic d.e., we define the stroboscopic time by its element $\Delta\tau = 2\pi\mu/\omega$. Setting $\varphi - (\pi/\omega) = \varphi'$, the stroboscopic system has the following form

$$\begin{aligned} \frac{dr}{d\tau} &= \frac{\gamma}{8}\left[\frac{\omega}{4\pi}\cos 4\varphi' \sin\frac{4\pi}{\omega} - 1\right] r^3 \\ &\qquad - \frac{\alpha}{2}\left[\frac{\omega}{2\pi}\cos\varphi' \sin\frac{2\pi}{\omega} - 1\right] r + \frac{E}{\pi}\frac{\omega^2}{\omega^2 - 1}\cos\varphi' \sin\frac{\pi}{\omega} \\ \frac{d\varphi}{d\tau} &= -\frac{1}{r\pi}\left[\frac{\gamma\omega}{8}\left(\sin 4\varphi' \sin\frac{4\pi}{\omega} + \sin 2\varphi' \sin\frac{2\pi}{\omega}\right) r^3 \right. \\ &\qquad \left. - \left(\frac{\alpha\omega}{2}\sin 2\varphi' \sin\frac{2\pi}{\omega}\right) r + \frac{E\omega^2}{\omega^2 - 1}\sin\varphi' \sin\frac{\pi}{\omega}\right] \end{aligned} \tag{2.4}$$

In the general case the discussion of these d.e. is difficult. In fact, one has to determine the singular point of this system and, for that purpose, it is necessary to determine a real positive root r_0 common to two cubic equations in r, but the difficulty is in that the coefficients of these equations contain φ. The problem is thus too complicated to be attempted and, as was mentioned, does not present any special interest.†

In the asymptotic case when ω is large, the matter is considerably simpler inasmuch as one can replace the trigonometric functions by the first terms of their series development and reduce the system (2.4) to the usual Poincaré form in terms of r and φ.

There is, however, one point to be mentioned: in the phenomenon of asynchronous quenching, a relatively high heteroperiodic frequency ω is applied to an oscillatory system oscillating with an (autoperiodic) frequency $\omega_0 = 1$. If ω were a large integer number, say $[\omega]$ and if we are able to prove the periodicity of the heteroperiodic oscillation with period $2\pi/[\omega]$, by this fact we would be able to prove also the periodicity of the oscillation with period 2π, since in this case, if $2\pi/[\omega]$ is a period, 2π is also a period. However, in our case ω is some large number, not necessarily an integer; hence if $[\omega]$ is the nearest integer to ω, the exact argument valid for $[\omega]$ will be *approximately valid* for ω, because in equation (2.4) ω appears in the denominators of arguments of the trigonometric functions and, in the approximations of these functions by polynomials, the error will be small *if ω is large enough.*

This means that, following this argument, instead of the exact singular point of the stroboscopic system, we shall be able to determine only a very small orbit around this point and that, instead of the exact periodicity, we shall have a neighboring almost periodicity; the error will be the smaller, the larger is ω. This approximation will not affect in any way our conclusions.

3. Asynchronous quenching

We can now attack the problem of quenching in the case when $\omega \to \infty$. The expressions of the form: $\omega \cos 4\varphi' \sin (4\pi/\omega)$, $\omega \cos 2\varphi' \sin (2\pi/\omega), \ldots$ have the limits $4\pi \cos 4\varphi$, $2\pi \cos 3\varphi, \ldots$ and the limits of the terms with E are obviously zero.

In the asymptotic case in which we are interested, the system (2.4) becomes

$$\begin{aligned} \frac{dr}{d\tau} &= \frac{\gamma}{8}(\cos 4\varphi - 1)r^3 - \frac{\alpha}{2}(\cos 2\varphi - 1)r = R(r,\varphi) \\ \frac{d\varphi}{d\tau} &= -\frac{1}{r\pi}\left[\left(\frac{1}{8}\sin 4\varphi + \frac{\gamma}{4}\sin 2\varphi\right)r^3 - \alpha(\sin 2\varphi)r\right] = \Phi(r,\varphi) \end{aligned} \tag{3.1}$$

† One notes that this procedure is the same as that which we used in Chapter 18 in connection with the problem of synchronization; the only difference here is the value of the parameter ω.

and it is seen that this system has singular points for arbitrary values of r if φ is either $\varphi = 0$ or $\varphi = \pi$.

In order to investigate stability conditions, we can assume that φ is small and develop the trigonometric functions around the value of $\varphi = 0$; we get thus:

$$\frac{dr}{d\tau} = r\varphi^2(\alpha - \gamma r^2); \qquad \frac{d\varphi}{d\tau} = \frac{1}{\pi}(2\alpha - \gamma r^2)\varphi \tag{3.2}$$

If $2\alpha - \gamma r^2 < 0$, $\dfrac{d \log \varphi}{d\tau} < 0$ and the orbit is stable.

If α and γ are negative, say $\alpha = -a$; $\gamma = -g$; $a > 0$, $g > 0$, we have

$$\frac{1}{r}\frac{dr}{d\tau} = \varphi^2(gr^2 - a); \qquad \frac{1}{\varphi}\frac{d\varphi}{d\tau} = \frac{1}{\pi}(gr^2 - 2a) \tag{3.3}$$

Here $\varphi = 0$ is a singular solution for every r. Suppose we perturb φ a little. In order that φ should come to its original value, we need $gr^2 - 2a < 0$; thus the condition $r < \sqrt{2a/g}$ is necessary for the stability of the phase. In such a case $(1/r)(dr/d\tau) < 0$ so that r decreases. Thus $r < \sqrt{2a/g}$ is the condition of stability for the phase and the stronger condition $r < \sqrt{a/g}$ is the condition of stability of the phase and of the amplitude as well.

These conclusions are facilitated by the fact that the amplitude of the external periodic excitation is small, which is not an important case because in applications one considers always this excitation with a finite amplitude.

For that purpose we have to modify the method, and the corresponding conclusions will acquire a more definite form.

4. Finite external periodic excitation

If the amplitude e (1.1) of the external periodic excitation is not small, the procedure indicated in Section 2 does not apply. One can, however, follow the method used in Chapter 19. We start with the d.e.

$$\ddot{\xi} - \mu(\alpha - \beta\xi^2)\dot{\xi} + \xi = e \sin \omega t \tag{4.1}$$

where we do not consider e as small; as regards ω we still assume that it is some irrational number as previously.

The change of the variable: $\xi = ex$ transforms (4.1) into the form

$$\ddot{x} - \mu(\alpha - \gamma x^2)\dot{x} + x = \sin \omega t; \qquad \gamma = \beta e^2 \tag{4.2}$$

For $\mu = 0$, the generating solution is

$$\begin{aligned} x_0(t) &= A \sin t + B \cos t + M \sin \omega t \\ y_0(t) &= \dot{x}_0(t) = A \cos t - B \sin t + M\omega \cos \omega t \end{aligned} \tag{4.3}$$

where $M = 1/(1 - \omega^2)$.

We assume that μ is small enough to justify the first approximation only, namely: $x(t) = x_0(t) + \mu x_1(t)$; $y(t) = y_0(t) + \mu y_1(t)$.

From the perturbation method one obtains

$$x_1(t) = \int_0^t \sin (t - \tau) f(x_0, \dot{x}_0) d\tau; \qquad y_1(t) = \int_0^t \cos (t - \tau) f(x_0, \dot{x}_0) d\tau \tag{4.4}$$

with $f(x_0, \dot{x}_0) = (\alpha - \gamma x_0^2)\dot{x}_0$ in this case.

From (4.3) the initial conditions are:

$$x_0(0) = B; \qquad y_0(0) = A + M\omega; \qquad x_1(0) = 0; \qquad y_1(0) = 0 \tag{4.5}$$

If a heteroperiodic oscillation with period $2\pi/\omega$ exists, one has the following conditions of periodicity (Chapter 10)

$$\begin{aligned} x\left(\frac{2\pi}{\omega}\right) - x(0) &= x_0\left(\frac{2\pi}{\omega}\right) - x_0(0) + \mu x_1\left(\frac{2\pi}{\omega}\right) = 0 \\ y\left(\frac{2\pi}{\omega}\right) - y(0) &= y_0\left(\frac{2\pi}{\omega}\right) - y_0(0) + \mu y_1\left(\frac{2\pi}{\omega}\right) = 0 \end{aligned} \tag{4.6}$$

On the other hand, one has

$$\begin{aligned} x_0\left(\frac{2\pi}{\omega}\right) - x_0(0) &= A \sin \frac{2\pi}{\omega} + B\left(\cos \frac{2\pi}{\omega} - 1\right) \\ y_0\left(\frac{2\pi}{\omega}\right) - y_0(0) &= A\left(\cos \frac{2\pi}{\omega} - 1\right) - B \sin \frac{2\pi}{\omega} \end{aligned} \tag{4.7}$$

Clearly, (4.7) is a special case of (4.6) when $\mu = 0$; in this case the only values of A and B that satisfy (4.7) are $A = B = 0$, since the determinant of the system is different from zero.

We consider now the case when $\mu \neq 0$ and determine the variations Δx and Δy during one period $2\pi/\omega$, replacing A and B by their expressions (4.5). We have thus:

$$\begin{aligned} \Delta x &= (y - M\omega) \sin \frac{2\pi}{\omega} + x\left(\cos \frac{2\pi}{\omega} - 1\right) + \mu \int_0^{2\pi/\omega} \sin \left(\frac{2\pi}{\omega} - \tau\right) f d\tau \\ \Delta y &= (y - M\omega)\left(\cos \frac{2\pi}{\omega} - 1\right) - x \sin \frac{2\pi}{\omega} + \mu \int_0^{2\pi/\omega} \cos \left(\frac{2\pi}{\omega} - \tau\right) f d\tau \end{aligned} \tag{4.8}$$

where $f = f(x_0, \dot{x}_0)$ with x_0 and $\dot{x}_0$ given by (4.3).

From these difference equations formed for one period $2\pi/\omega$ we pass now to the stroboscopic d.e. in which we define the stroboscopic time $\tau = 2\pi\mu/\omega$, which gives

$$\begin{aligned} \frac{dx}{d\tau} &= \frac{y - M\omega}{\mu}\sin\frac{2\pi}{\omega} + \frac{x}{\mu}\left(\cos\frac{2\pi}{\omega} - 1\right) + \int_0^{2\pi/\omega}\sin\left(\frac{2\pi}{\omega} - \sigma\right)f d\sigma \\ \frac{dy}{d\tau} &= \frac{y - M\omega}{\mu}\left(\cos\frac{2\pi}{\omega} - 1\right) - \frac{x}{\mu}\sin\frac{2\pi}{\omega} + \int_0^{2\pi/\omega}\cos\left(\frac{2\pi}{\omega} - \sigma\right)f d\sigma \end{aligned} \tag{4.9}$$

As our aim is to find what happens to the autoperiodic oscillation which has the period 2π, we have to change the stroboscopic time-scale from τ to $T = \tau\omega$, in which case we have relations

$$\frac{dx}{dT} = \frac{1}{\omega}\frac{dx}{d\tau}; \qquad \frac{dy}{dT} = \frac{1}{\omega}\frac{dy}{dT} \tag{4.10}$$

With this change of the stroboscopic time-scale, equations (4.9) become

$$\begin{aligned} \frac{dx}{dT} &= \left(\frac{y - M\omega}{k}\right)\omega\sin\frac{2\pi}{\omega} + \frac{x}{k}\,\omega\left(\cos\frac{2\pi}{\omega} - 1\right) \\ &\qquad + \frac{1}{\omega}\int_0^{2\pi/\omega}\sin\left(\frac{2\pi}{\omega} - \sigma\right)f d\sigma \\ \frac{dy}{dT} &= \left(\frac{y - M\omega}{k}\right)\omega\left(\cos\frac{2\pi}{\omega} - 1\right) - \frac{x}{k}\,\omega\sin\frac{2\pi}{\omega} \\ &\qquad + \frac{1}{\omega}\int_0^{2\pi/\omega}\cos\left(\frac{2\pi}{\omega} - \sigma\right)f d\sigma \end{aligned} \tag{4.11}$$

where $k = \mu\omega^2$ and $M = 1/(1 - \omega^2)$.

As ω is large, we replace the trigonometric functions by the first terms of their series expansions; one can obviously neglect the terms containing the integrals as being of a higher order of smallness. The d.e. (4.11) can then be replaced by the following approximate ones

$$\begin{aligned} \frac{dx}{dT} &= \left(-\frac{2\pi^2}{k\omega}\right)x + \left(\frac{2\pi}{k}\right)y + \left(\frac{2\pi}{k\omega}\right) = ax + by + c \\ \frac{dy}{dT} &= \left(-\frac{2\pi}{k}\right)x + \left(-\frac{2\pi^2}{k\omega}\right)y + \left(-\frac{2\pi^2}{k\omega^2}\right) = a'x + b'y + c' \end{aligned} \tag{4.12}$$

and, for the singular point of the system, one has to solve the system of two linear nonhomogeneous equations in terms of x and y, which is possible since the determinant:

$$\begin{vmatrix} a & b \\ a' & b' \end{vmatrix} \neq 0$$

With the values of the coefficients a, b,. . ., c' given by (4.12), one finds:

$$x_0 = 0; \qquad y_0 = \frac{1}{\omega} \tag{4.13}$$

where the subscript 0 refers now to the stationary values of x and y.

As the energy stored in the oscillation is $\rho = x_0^2 + y_0^2$, in this case this energy is $0(\omega^{-2})$, and, as we assumed that ω is large, this shows the effect of quenching of the autoperiodic oscillation by the heteroperiodic frequency.

5. Asynchronous excitation

As was mentioned in Section 1, the phenomenon of asynchronous excitation is of an entirely different nature as compared to quenching and consists in a change of the topological configuration from SUS to US (in the terminology of Chapter 7), under the effect of the forcing term.

One could proceed in the manner outlined in Sections 2 and 4 by considering a differential equation of the form

$$\ddot{x} + x = \mu[(\alpha + \gamma x^2 + \delta x^4)\dot{x} + E \sin \omega t] \tag{5.1}$$

and by establishing the conditions under which the bifurcation of the first kind (Chapter 22): $SUS \rightarrow US$ is possible; but this leads to rather long calculations and we prefer to outline here a simpler way of treating this problem on the basis of the method of equivalent linearization (Section 7, Chapter 14) as was done by Kobsarev.[2]

It is recalled that the method of equivalent linearization consists in the introduction of certain *equivalent parameters* owing to which the original nearly linear d.e. is replaced by an *equivalent linear* d.e., the term *equivalent* meaning that the solutions of these two d.e. differ by a quantity $0(\mu^2)$ if one limits the calculation to the first approximation.

If one considers an electrical problem, the nonlinear function is: $i_a = f(e)$, the anode current i_a of an electron tube being considered as a function of the grid voltage e, the equivalent linear problem is

$$i_a = Se \tag{5.2}$$

where S is a certain integral as was explained in Section 7, Chapter 14.

Here the problem is, however, more complicated because, instead of one single voltage e (the feedback reaction applied to the grid) we have an additional voltage representing the external periodic excitation. Krylov and Bogoliubov[3] show that in such a case the expression for the equivalent

[2] See footnote [2], page 573.

[3] N. Krylov and N. Bogoliubov, *Introduction to Nonlinear Mechanics* (in Russian), 1937.

parameter S (the linearized mutual conductance) is given by the expression of the form:

$$S = \frac{1}{2a\pi^2}\int_0^{2\pi}\int_0^{2\pi} f(E_0 + F\cos\tau_1 + a\cos\tau_2)\cos\tau_2 d\tau_1 d\tau_2 \quad (5.3)$$

where E_0 is some constant (biasing voltage) and $F\cos\tau_1$ and $a\cos\tau_2$ are the heteroperiodic and autoperiodic voltages, respectively. The integrations are carried out in angular variables and with respect to the autoperiodic oscillation, since we are interested in the behavior of this oscillation under the effect of the heteroperiodic (forcing) frequency. If one takes for the nonlinear functions, the usual polynomial approximation

$$i_a = f(x) = \alpha x + \beta x^2 + \gamma x^3 + \delta x^4 + \varepsilon x^5 \quad (5.4)$$

where $x = a\cos\varphi + b\cos\psi$ is the voltage resulting from the superposition of the autoperiodic and heteroperiodic components (reduced to angular variables), the expression for the equivalent parameter S in this case is

$$S(a,b) = \frac{1}{2a\pi^2}\int_0^{2\pi}\int_0^{2\pi} f(a\cos\varphi + b\cos\psi)\cos\varphi d\varphi d\psi \quad (5.5)$$

where a and b are the amplitudes of the autoperiodic and heteroperiodic components.

Carrying out these somewhat long but simple calculations, one obtains the expression

$$S(a,b) = \alpha + \tfrac{3}{4}\gamma a^2 + \tfrac{3}{2}\gamma b^2 + \tfrac{5}{8}\varepsilon a^4 + \tfrac{15}{4}\varepsilon a^2 b^2 + \tfrac{15}{8}\varepsilon b^4 \quad (5.6)$$

from which one can discuss the various cases.

If the heteroperiodic oscillation is absent, $b = 0$, one has

$$S(a,0) = \alpha + \tfrac{3}{4}\gamma a^2 + \tfrac{5}{8}\varepsilon a^4 \quad (5.7)$$

which gives the equivalent mutual conductance for the autoperiodic oscillation. In particular, for a "soft" characteristic, $\delta = \varepsilon = 0$; $\gamma < 0$ one has:

$$S(a,0) = \alpha - \tfrac{3}{4}|\gamma| a^2 \quad (5.8)$$

For a stationary condition, $S(a,0) = 0$ which gives the stationary amplitude $a = a_0$.

In the absence of heteroperiodic frequency, the condition of self-excitation is: $S(0,0) = \alpha$ but, with the heteroperiodic oscillation, this condition is

$$S(0,b) = \alpha + \tfrac{3}{2}\gamma b^2 = \alpha - \tfrac{3}{2}|\gamma| b^2 \quad (5.9)$$

It is seen that the presence of the heteroperiodic oscillation may prevent the self-excitation from rest ($a = 0$). In fact, the coefficient $\alpha > 0$ measures normally (that is, in an ordinary van der Pol oscillator) the energy input into a system at rest which makes the self-excitation possible. Hence, if $S(0,b) < 0$, the oscillation cannot start from rest.

In the case of a "hard characteristic," α, β, γ, and δ in (5.4) are positive, and $\varepsilon < 0$. In such a case and in the absence of the heteroperiodic oscillation ($b = 0$), the equivalent parameter S is

$$S(a,0) = \alpha + \tfrac{3}{4}\gamma a^2 - \tfrac{5}{8}|\varepsilon|a^4 \tag{5.10}$$

The maximum of $S(a,0)$ considered as a function of a^2 occurs for $a^2 = \dfrac{3\gamma}{5|\varepsilon|}$, in which case one has

$$S_{\max}(a,0) = \alpha + \tfrac{9}{40}(\gamma/|\varepsilon|) \tag{5.11}$$

The asynchronous excitation ($b \neq 0$) occurs only if

$$S(0,0) - S_{\max}(a,0) = \tfrac{3}{2}\gamma b^2 + \tfrac{15}{8}\varepsilon b^4 + \tfrac{9}{40}(\gamma^2/\varepsilon) > 0$$

that is, if

$$\tfrac{3}{2}\gamma b^2 > \tfrac{15}{8}|\varepsilon|b^4 + \tfrac{9}{40}(\gamma^2/|\varepsilon|) \tag{5.12}$$

This inequality can be fulfilled if b^2 lies in the interval

$$\tfrac{1}{5}(\gamma/|\varepsilon|) < b^2 < \tfrac{3}{5}(\gamma/|\varepsilon|) \tag{5.13}$$

Summing up, in this application of the method of equivalent linearization, the following results are obtained:

1. If the characteristic is "soft," a heteroperiodic oscillation can only quench the existing autoperiodic oscillation.
2. If the characteristic is "hard," a heteroperiodic oscillation *may* also release the autoperiodic oscillation if condition (5.13) is fulfilled.

6. Concluding remarks

The asynchronous effects are different from the manifestations of the subharmonic resonance, where the existence of a certain rational ratio between the two frequencies (or at least, of a certain neighborhood of such a ratio) is a necessary condition.

In the frequency spectrum of the heteroperiodic excitation, these asynchronous actions are located in the gaps between the subsequent zones of subharmonic resonances and associated synchronization effects.

The stroboscopic method permits approaching the study of these phenomena directly on the basis of the d.e.; in the transformation leading to the establishment of difference equations, the integrations are to be carried out between the limits 0 and $2\pi/\omega$. Since, however, the ultimate analysis concerns the behavior of the autoperiodic oscillation with period 2π, it is necessary to consider this period 2π as made up of small periods $2\pi/\omega$. If $[\omega]$ is large and integer, it is clear that the condition of periodicity for a period $2\pi/[\omega]$ remains valid (exactly) for the period 2π. Since, however, ω is large and *irrational*, the approximation remains good for the period 2π, as $2\pi/\omega \simeq 2\pi/[\omega]$ assuming that $[\omega]$ is the nearest integer to a large irrational number ω.

In view of this, the replacement of the trigonometric functions, by the first terms of their series developments, leads to a relatively simple asymptotic form of the stroboscopic system, from which it follows that the presence of a high heteroperiodic frequency destroys (or "quenches") the previously existing (in the absence of the heteroperiodic frequency) autoperiodic oscillation.

The stroboscopic method could be used also for the purpose of the establishment of conditions of the asynchronous excitation, as defined in Section 1, but this requires rather long calculations intended to express the transition $SUS \to US$ (Chapter 22).

We preferred to use a "short cut" offered by the algorithm of the method of equivalent linearization (Chapter 14), which leads to the corresponding conclusions in a simpler manner.

In general, the study of asynchronous actions is yet in a relatively early stage, and the experimental evidence available is also rather limited as compared to what is known in connection with other nonlinear phenomena.

These phenomena have been ascertained so far only in nonlinear electrical circuits, and nothing is known about their existence in mechanical oscillatory systems, where they probably also exist if one could approach their investigation with appropriate means. As we saw from the contents of this chapter, the existence of these asynchronous actions depends on certain interactions between the autoperiodic and heteroperiodic oscillations in nonlinear systems. If one leaves out the general case and concentrates attention only on two special cases which were investigated in Sections 3, 4, and 5, the matter is sufficiently clear. In fact, the phenomenon of asynchronous quenching is a relatively simple phenomenon as compared to the phenomenon of asynchronous excitation. One can visualize this effect as follows: a nonlinear system is oscillating with a certain frequency ω_0 and we impress on it an oscillation of a much higher frequency ω which ultimately extinguishes the original oscillation with frequency ω_0. Both experiment and theory show that this is so in the case of nonlinear electronic circuits, and one may expect that if one succeeds in obtaining conditions governed by the same d.e. for a mechanical system, something similar will be observed. For instance, a certain mechanical vibration existing in a steady state with frequency ω_0 will be extinguished if we impress on the system a much higher frequency ω of another vibration.

The problem has a definite applied interest and hinges obviously on the possibility of producing a mechanical system having certain nonlinear characteristics with which a corresponding electronic system is known to exhibit this quenching effect. It is also clear that, in order to be able to accomplish such a result, it is necessary to produce a mechanical system with nonlinear characteristics corresponding to the electronic system. For the time being one is yet far from such a *synthetic predetermination* of parameters in mechanical systems where these parameters are always far less definite than in electronic circuits and are subject to all kinds of couplings generally difficult to ascertain. Moreover, these parameters are often *distributed*; change in one parameter very often causes changes in the others, etc.

Summing up, all these difficulties are by no means of a basic character, but are inherent in the *practical* impossibility to produce a mechanical system amenable to the same form of d.e. as in an electronic circuit in which these phenomena exist, and where the determination of parameters and nonlinear characteristics is relatively a simple matter.

If these practical difficulties are eventually overcome and corresponding nonlinear coupling effects are obtained also in mechanical systems, very likely the whole question of asynchronous actions will acquire a greater applied interest instead of being considered as, more or less, a matter of curiosity, as it is at present.

Chapter 25

SYSTEMS WITH INERTIAL NONLINEARITIES

1. Introductory remarks

In addition to the usual nonlinear problems in which nonlinear characteristics are represented by polynomials with constant coefficients, one encounters from time to time special problems that are not quite definitely formulated so far. Whether such problems may become of importance later is difficult to say at present, but very often such new questions raise a certain amount of interest, at least theoretically.

One such question was raised a few years ago by Theodorchik[1] who investigated the behavior of an electron-tube circuit into which is inserted the so-called "thermistor"—a nonlinear conductor whose resistance is a function of the temperature, the latter being, in turn, a function of the current which flows in the conductor.

The behavior of such a circuit depends to some extent on the frequency of the oscillating current and, in this manner, is *not characterized by constant parameters* only, as in all problems with which we were concerned previously.

In fact, if the frequency of oscillating current is sufficiently high, it is clear that the temperature variations of the conductor cannot follow rapidly enough the current variations for a sufficiently high value of the *thermal time constant.* One has, therefore, a system in which the nonlinearity, instead of being fixed, is variable as a function of the root mean square value of the current. One has thus a somewhat particular form of d.e. in which, in addition to the instantaneous values of the dependent variable, the *average values* of that variable also appear. As the latter reduce ultimately to the square of the amplitude, one has a d.e. in which, besides the dependent variable $x(t)$, appears also $x_0^2(t)$. In a transient

[1] K. F. Theodorchik, *Auto-oscillatory systems* (in Russian), Moscow, 1948.

state both x and x_0^2 are functions of t but, for a stationary state while x is still $x(t)$, the amplitude x_0 becomes constant.

It seems likely that problems of this kind belong rather to the so-called integro-differential equations but as they are also nonlinear as a rule, there are too many difficulties at present to attack these problems following this argument.

If, however, in the same problem the thermal time constant is reduced, and a conductor of this kind is connected to an oscillating circuit whose frequency is within the range of the thermal time constant, it has been reported on some occasions that the circuit may become self-excited and oscillates in a stationary state without any electron tubes.† This circumstance in itself is not very surprising because it was known long before the advent of the electron tubes that self-sustained oscillations can be produced by nonlinear conductors of electricity, such as electric arcs and gaseous discharges.

There appear two distinct problems corresponding to these two extreme ranges of the thermal time constant. In one case the thermal inertia does not permit the oscillatory system to follow the instantaneous values of the variable but follows only its root mean square value proportional to x_0^2—the square of the amplitude. Since, on the other hand, there may be some other variables which vary instantaneously, say $x(t)$, one obtains a special kind of a d.e. in which, in addition to $\ddot{x}(t)$, $\dot{x}(t)$, and $x(t)$, as usual, there appear terms with $x_0^2(t)$, $x_0^2(t)\cdot\dot{x}(t)$, etc., the terms $x_0^2(t)$ tending to become constant with the approach to the stationary state, while the terms $x(t)$, $\dot{x}(t)$, and $\ddot{x}(t)$ remain variable as in an ordinary d.e.

In the following section we shall outline the results formulated by Theodorchik, and in Section 3 we shall go further into this matter by assuming a certain explicit form of the nonlinear function and applying the stroboscopic method which is particularly useful in this case, in that, for a stationary state, one can replace $x_0^2(t)$ by ρ_0 in the preceding notations, while the instantaneous terms like $x(t)$ will be still $x = r\cos\psi$ of the general theory (Chapter 16).

Once this is done, one can ascertain how the instantaneous and the average variables coexist in the ultimate stationary state, since in the latter everything appears as a relation between average values.

One can consider also the problem corresponding to the other end of the thermal time constant, that is, when the thermal time constant is so small that the instantaneous adjustments of temperature (and, hence, of nonlinearity) can occur within one period of oscillation imposed by the constants of an oscillating circuit branched across such a conductor. This is essentially a problem of self-sustained oscillations in a circuit with such a

† This holds for nonlinear conductors whose nonlinearity remains constant.

conductor but without any electron tube. This problem is more difficult, as will appear from Section 5, than the usual problem of self-excited oscillations in an electron-tube circuit. In fact, in the latter the roles of negative and positive resistances are localized in the electron tube and the rest of the circuit, respectively; moreover, the electron tube itself gives the well known transition from the energy absorption to the energy dissipation in view of its nonlinear characteristic.

As regards the nonlinear conductors, the matter is less definite since their nonlinear characteristics are hardly known as compared to those of electron tubes.

It is necessary, therefore, to make some arbitrary assumptions and try to obtain corresponding conclusions. In this manner one encounters many theoretical possibilities, some of which are probably useless whenever these questions are better studied experimentally. This matter may be of some interest as an example of a certain elasticity in the use of the stroboscopic method when it is necessary to establish the conditions of a stationary state, even if the form of the original d.e. departs from that usually encountered in applications.

In this particular case, the difficulty of the problem containing at the same time $x(t)$, $\dot{x}(t)$, and $\ddot{x}(t)$ (the instantaneous values) and the average values associated with x_0^2 is ultimately smoothed out because, in the final result (the stationary state), everything is to be averaged out so that the initial difficulty in the original d.e. disappears in the stroboscopic d.e.

2. Inertial nonlinearity

We outline first certain conclusions obtained by Theodorchik. If an alternating current of angular frequency ω and amplitude x_0 flows through a conductor having the mass m, resistance R, thermal specific capacity c, and Newton's coefficient k of cooling, the thermal equilibrium is given by the d.e.:

$$mc\dot{\theta} + k\theta = Rx_0^2 \sin^2 \omega t \tag{2.1}$$

θ being the temperature. The integration of this d.e. yields:

$$\theta = \theta_0 + \theta_1 \cos (2\omega t - \gamma) \tag{2.2}$$

where $\theta_0 = \alpha\delta$; $\alpha = Rx_0^2/2mc$; $\delta = mc/k$; $\theta_1 = \alpha\delta\eta$; $\eta = \cos \gamma$. The constant δ is called the *thermal time constant*. Expression (2.2) shows that the temperature of the conductor consists of a constant part θ_0 and a variable part of amplitude θ_1 and (angular) frequency 2ω. One has thus the ratio:

$$\theta_1/\theta_0 = \eta = 1/\sqrt{1 + (2\omega\delta)^2} \tag{2.3}$$

which shows that if the time constant δ is large enough (which will be assumed in the following sections), the amplitude θ_1 of the variable component is small, so that:

$$\theta \simeq \theta_0 = (R/2k)x_0^2 = \beta x_0^2 \tag{2.4}$$

which is obvious from the definition of the effective values.

We note that everything happens as if the conductor had a kind of *thermal inertia* which prevented its temperature from responding to the instantaneous value of the heating alternating current but allows it to follow the root-mean-square value of the latter which depends on x_0^2.

If the conductor in question is nonlinear, its nonlinearity can always be approximated by a polynomial which is a purely formal procedure consisting in an approximation of a piece of arc of a smooth (that is, continuously differentiable) experimental curve by a certain polynomial whose coefficients are determined by the interpolation procedure so as to give the best approximation.

In this manner, a nonlinear resistor of the resistance $R(\theta)$ can be represented by a polynomial:

$$R(\theta) = R_0 + c_1\theta + c_3\theta^3 \tag{2.5}$$

where θ is the *departure* of the temperature from its average value for which this resistance has the value R_0.

In accordance with the nearly linear theory, it will be assumed that c_1 and c_3 are small as compared to R_0.

Since the temperature depends on x_0^2—the square of the amplitude—the preceding expression can also be written in the form:

$$R(x_0^2) = R_0 + b_1x_0^2 + b_3x_0^6 \tag{2.6}$$

3. Van der Pol oscillator with a conductor $R(x_0^2)$

If a conductor of this kind is inserted into an oscillating circuit of a van der Pol oscillator (which can be assumed to be, for instance, of a usual inductive coupling type) it is easy to establish its d.e. First, without such conductor, the d.e. is:

$$L\frac{di}{dt} + r'i + \frac{1}{C}\int i\,dt = M\frac{dI}{dt} = M\frac{dI}{dV}\frac{dV}{dt} \tag{3.1}$$

where L, r', and C are the usual constants of the oscillating circuit, M the coefficient of mutual inductance between the anode and the grid circuits, V the voltage on the grid, and i and I the currents in the oscillating and in the anode circuits, respectively.

One has: $i = C(dV/dt)$, $dV/dt = i/C$. As to $dI/dV = S'$, this is the usual nonlinear element of the electron tube, representing the anode current considered as a function of the grid voltage.

Setting $i = x$ and changing the independent variable from t to $\tau = \omega t$, $\omega = 1/\sqrt{LC}$, (3.1) becomes after a differentiation and some simplifications:

$$\ddot{x} - \sqrt{\frac{C}{L}}\left(\frac{M}{C}\cdot S' - r'\right)\dot{x} + x = 0 \tag{3.2}$$

which can be written as:

$$\ddot{x} - \mu(nS - r)\dot{x} + x = 0 \tag{3.3}$$

where μ is a small parameter absorbing some constants of (3.2). If a nonlinear conductor $R = R(\theta)$ is inserted in the oscillating circuit, one has:

$$\ddot{x} - \mu(nS - R)\dot{x} + x = 0 \tag{3.4}$$

It is clear that, in general, the energy is added to the system through nS and dissipated owing to R.

Taking for S the usual approximation $S = S_0 - S_2x^2$, S_0 and S_2 being certain positive constants and setting $nS_0 - R_0 = A$, we obtain

$$\ddot{x} - \mu[A - (nS_2x^2 + b_1{x_0}^2 + b_3{x_0}^6)]\dot{x} + x = 0 \tag{3.5}$$

If the system is to be self-excited from rest ($x \simeq 0$), clearly one must have $A > 0$ which is obvious because the energy added to the system through the electron tube (term nS_0) in this case is greater than the energy dissipated in the conductor (term R_0).

If one applies now the stroboscopic transformation recalling that $x^2 = \rho_0 \cos^2 \psi_0$; ${x_0}^2 = \rho_0$; ${x_0}^6 = {\rho_0}^3$, one obtains the stroboscopic d.e.

$$\frac{d\rho}{d\tau} = \rho b_3\left[\rho^3 + \left(\frac{nS_2 + 4b_1}{4b_3}\right)\rho - \frac{A}{b_3}\right] = \Phi(\rho) \tag{3.6}$$

Aside the solution $\rho = 0$ which is clearly the position of equilibrium, this d.e. may have a stationary solution ρ_0 represented by the positive root of the cubic equation

$$\rho^3 + a_1\rho - a_0 = 0 \tag{3.7}$$

where $a_1 = (nS_2 + 4b_1)/4b_3$; $a_0 = A/b_3$.

Two cases are possible according to the sign of the discriminant ${a_1}^3/27 + {a_0}^2/4 = \Delta$; we shall limit ourselves here to the more important case when $\Delta > 0$ which characterizes the existence of one single real root. As we wish to have $\rho_0 > 0$, one must have $a_0 > 0$, but as we started with $A > 0$, one must also have $b_3 > 0$.

We have on the other hand

$$\Phi(\rho) = b_3[\rho^4 + a_1\rho^2 - a_0\rho]$$

whence

$$\Phi_\rho(\rho_0) = b_3[4\rho^3 + 2a_1\rho - a_0] \tag{3.8}$$

but as $\rho_0{}^3 + a_1\rho_0 - a_0 = 0$, (3.8) becomes

$$\Phi(\rho_0) = b_3[3\rho_0{}^3 + a_1\rho_0] \tag{3.9}$$

For stability this expression must be negative and, as $b_3 > 0$, we have the condition

$$3\rho_0{}^2 + a_1 < 0$$

that is

$$\rho_0{}^2 < -\frac{a_1}{3} \tag{3.10}$$

Clearly, if $a_1 > 0$, this criterion shows that there exists no stationary amplitude but, if $a_1 < 0$, it shows that the condition of stability is

$$\rho_0{}^2 < +\frac{|a_1|}{3} \tag{3.11}$$

Referring to the expression for $a_1 = (nS_2 + 4b_1)/4b_3$ in which nS_2 and b_3 are positive, the only case when a_1 can be negative is when $b_1 < 0$ and sufficiently large to render the numerator in a_1 negative.

The case which we have just considered is when $A > 0$. When $A < 0$, the existence of one positive root of the cubic polynomial, all other things being equal, requires, on the contrary, $b_3 < 0$. We have to re-examine the condition of stability, taking into account the new sign for b_3.

From (3.9) it follows that now the criterion of stability is

$$3\rho_0{}^2 + a_1 > 0 \tag{3.12}$$

Clearly, if $a_1 > 0$ the condition

$$\rho_0{}^2 > -\frac{a_1}{3} \tag{3.13}$$

is trivial and merely means $\rho_0{}^2 > 0$; that is, any positive root of the polynomial is stable. The condition $a_1 > 0$, on the other hand, means that $a_1 = (nS_2 + 4b_1)/4b_3 = -(nS_2 + 4b_1)/4|b_3| > 0$ and this may happen only if $b_1 < 0$ and sufficiently great to make $(nS_2 - 4|b_1|) < 0$, in which case, in fact, $a_1 > 0$.

There may be, however, another condition, viz.: $a_1 < 0$ in which case, instead of (3.13), one has

$$\rho_0{}^2 > +\frac{|a_1|}{3} \tag{3.14}$$

Clearly, if $a_1 < 0$ (with $b_3 < 0$), one must have $nS_2 + 4b_1 > 0$, that is, either $b_1 > 0$ or $b_1 < 0$ but not sufficiently great to render $nS_2 + 4b_1 = nS_2 - 4|b_1|$ negative.

This means that with this condition ($b_1 > 0$, $b_3 < 0$), the condition of stability requires that the stationary amplitude ρ_0 given by the positive root of the cubic polynomial must be greater than $+\sqrt{|a_1|/3}$ in order to be stable.

One can easily investigate a particular case when $S_2 \simeq 0$, in which case the electron tube works on the substantially linear part of its characteristic and the nonlinearity is brought into play by the nonlinear conductor.

This discussion merely shows that a nonlinear oscillation based on the property of *one* nonlinear characteristic may become entirely different if another nonliner characteristic is brought into play, *in addition* to the first one.

4. Oscillations produced by nonlinear conductors; physical considerations

It has been known for a long time that nonlinar conductors of electricity, such as an electric arc, for instance, can give rise to self-sustained oscillations in a circuit connected across them.

During this early period (the end of the last century and up to, possibly, 1914), the nonlinear theory had not been formed as yet and the experimental material had to be coordinated on the basis of certain graphical constructions, in conjunction with some conclusions derived from the method of small oscillations of the linear theory. It was during that early period that the concept of the "negative resistance" was introduced in connection with the characteristic $v = f(i)$, the voltage across a nonliner conductor considered as an empirical function of the current through it. According to this definition the resistance is "negative" if the characteristic exhibits a negative slope, that is, when the voltage decreases for the increasing current; the expression "negative resistance" physically means *an input* of energy into the oscillatory system. *Natural* conductors of electricity, such as arcs, gaseous discharges and the like, presented no problem, because they usually exhibit this particular feature, and the graphical constructions just mentioned are sufficient.

However, nonlinear conductors discovered relatively recently are different because they represent a synthetic product, generally an alloy of some metals or other chemical substances subjected to a certain treatment. Very often, a slight change in a chemical composition or in the treatment changes radically the form of the characteristic. For these reasons, the situation here is far less definite, particularly because there exists no well established connection between the *form* of an optimum characteristic and the properties of a corresponding oscillation.

For that reason, it is useful to attempt to establish at least some theoretical conclusions regarding these relations in general.

5. General problem

On the basis of the preceding conclusions one can formulate a more general problem[2] referring to a somewhat idealized scheme shown in Fig. 25.1, in which R is a nonlinear conductor of some kind, which for the sake of connection with the preceding argument, may be considered as a "thermal" type.† Here, however, we shall suppose that its thermal time constant is sufficiently small to permit the temperature to follow relatively rapid oscillations, for instance, those of an oscillating circuit (r, C, L) connected across the terminals of R. We assume that R carries a constant current X. The problem is to investigate whether, under some special conditions of nonlinearity of R, there may arise self-sustained oscillations in the oscillating circuit.

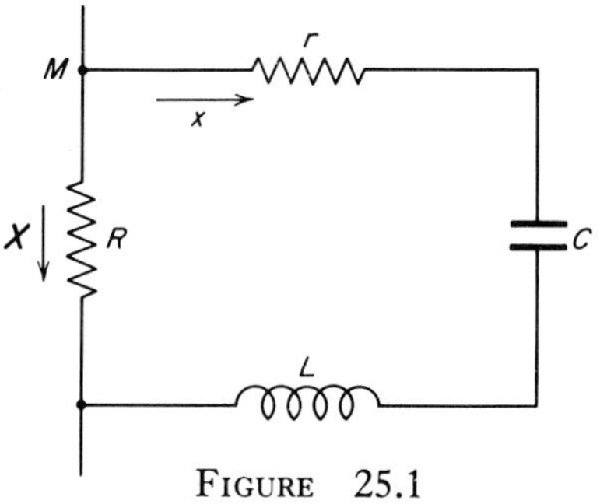

FIGURE 25.1

When R is an electric arc, it is definitely known that such oscillations occur. In the case of heat-responsive conductors of a nonlinear type, such oscillations have sometimes been reported if the thermal time constant is small enough to permit the manifestations of the nonlinearity, which we assume here.

Without attempting a complete investigation, we indicate merely an approach along the line of the preceding argument.

If there appears an oscillating current x in the circuit rCL, the question arises whether one can determine the nonlinear characteristic of R so as to obtain a stable stationary oscillation.

We may again consider the nonlinear resistance to be of the form

$$R = R_0 + b_1(X + x)^2 + b_3(X + x)^6 \tag{5.1}$$

[2] N. Minorsky, *C. R. Ac. Sc.* (Paris) **235**, 1952; *Ann. Fac. Sc. Un. of Marseille*, 1953; *Symposium on Nonlinear Circuit Theory*, Polytechnic Institute of Brooklyn, 1953.

† The assumption of the *thermal* origin is clearly not necessary here.

as we did previously, having in mind its thermal origin; in these notations $x = x_0 \cos \omega t$ is the oscillating current; by a change of time scale one can always assume that $\omega = 1$.

If $x = 0$, one has $R = R_0 + b_1 X^2 + b_3 X^6$ which is a fixed value of R. If $x = x_0 \cos t$ exists, the nonlinear resistance undergoes fluctuations and the problem is to investigate the values of the parameters b_1 and b_3 for which a stationary condition can be obtained.

Assuming that a stationary oscillation $x = x_0 \cos t$ exists, the fluctuations in resistance of the nonlinear conductor R are given by the expression

$$R = R_0 + b_1(X + x_0 \cos t)^2 + b_3(X + x_0 \cos t)^6 \tag{5.2}$$

If one develops this expression one has

$$R(x_0) = R = \sum_{n=0}^{6} a_n(x_0 \cos t)^n \tag{5.3}$$

where the coefficients $a_0, \ldots, a_6$ are calculated by identifying (5.2) and (5.3). One has thus:

$$a_0 = R_0 + b_1 X^2 + b_3 X^6; \qquad a_1 = 2b_1 X + 6b_3 X^5;$$
$$a_2 = b_1 + 15 b_3 X^4; \qquad a_3 = 20 b_3 X^3; \qquad a_4 = 15 b_3 X^2;$$
$$a_5 = 6 b_3 X; \qquad a_6 = b_3 \tag{5.4}$$

The coefficient a_0 is obviously the value of R for $x_0 = 0$.

It is more convenient to transform (5.3) into the form of a Fourier polynomial:

$$R = \sum_{n=0}^{6} m_n \cos nt \tag{5.5}$$

where the coefficients m_n contain a_n given by (5.4) and powers of x_0. Carrying out this calculation one has

$$\begin{aligned}
m_0 &= \tfrac{5}{16} a_6 x_0^6 + \tfrac{3}{8} a_4 x_0^4 + \tfrac{1}{2} a_2 x_0^2 + a_0 \\
m_1 &= \tfrac{5}{8} a_5 x_0^5 + \tfrac{3}{4} a_3 x_0^3 + a_1 x_0 \\
m_2 &= \tfrac{15}{32} a_6 x_0^6 + \tfrac{1}{2} a_4 x_0^4 + \tfrac{1}{2} a_2 x_0^2 \\
m_3 &= \tfrac{15}{16} a_5 x_0^5 + \tfrac{1}{4} a_3 x_0^3 \\
m_4 &= \tfrac{3}{16} a_6 x_0^6 + \tfrac{1}{8} a_4 x_0^4 \\
m_5 &= \tfrac{1}{16} a_5 x_0^5 \\
m_6 &= \tfrac{1}{32} a_6 x_0^6
\end{aligned} \tag{5.6}$$

In these formulas the various numerical fractions result from the trigonometric transformations from (5.3) to (5.5).

The d.e. of the circuit *MrCLNM* can be written as

$$\ddot{x} + \mu R'\dot{x} + x = 0 \tag{5.7}$$

with a proper change of the independent variable (so as to have frequency *one*), where R' is the value of R in which the constant term $R(0) = a_0$ is omitted, since it obviously does not contribute anything to the maintenance of the oscillation.

The equivalent system of (5.7) is

$$\dot{x} = y; \qquad \dot{y} = -\mu R'y - x$$

and introducing, as usual, the variables $\rho = x^2 + \dot{x}^2 = x^2 + y^2 = r^2$ and $\psi = \arctan(y/x)$; $x = r\cos\psi$, $y = r\sin\psi$ by means of two combinations: $x\dot{x} + y\dot{y} = \frac{1}{2}d\rho/dt$; $x\dot{y} - y\dot{x} = \rho(d\psi/dt)$ we obtain

$$\frac{d\rho}{dt} = -2\mu R'y^2; \quad \frac{d\psi}{dt} = -\mu R'\sin\psi\cos\psi - 1 \tag{5.8}$$

Hence, in the integration by the series of the form

$$\rho(t) = \rho_0(t) + \mu\rho_1(t) + \ldots; \qquad \psi(t) = \psi_0(t) + \mu\psi_1(t) + \ldots$$

the terms of the zero order are: $\rho_0(t) = \rho_0$; $\psi_0(t) = \varphi_0 - t$, where ρ_0 and φ_0 are the initial conditions.

Consider the first d.e. (5.8); the corresponding d.e. for the first-order corrective term $\rho_1(t)$ is

$$\frac{d\rho_1}{dt} = -2R'\rho_0\sin^2\psi_0 = -R'\rho_0(1 - \cos 2\psi_0) \tag{5.9}$$

Since $R' = m_0' + m_1\cos t + m_2\cos 2t + \ldots + m_6\cos 6t$ and $1 - \cos 2\psi_0 = 1 - \cos(2\varphi_0 - 2t) = 1 - \cos 2\varphi_0\cos 2t - \sin 2\varphi_0\sin 2t$, in the product $R'(1 - \cos 2\psi_0)$ the only terms which remain after the integration between 0 and 2π are: $m_0' - m_2\cos 2\varphi_0\cos^2 2t$, so that the difference $\rho_1(2\pi) - \rho_1(0) = \Delta\rho_1$ is

$$\Delta\rho_1 = (m_0'2\pi - \tfrac{1}{2}m_2\cos 2\varphi_0\cdot 2\pi)\rho$$

and $\Delta\rho = \mu\Delta\rho_1 = 2\pi\mu\rho(m_0' - \frac{1}{2}m_2\cos 2\varphi_0)$. Setting as usual $2\pi\mu = \Delta\tau$ and passing to the continuous variable ($\Delta\tau \rightarrow d\tau$) one has the stroboscopic d.e.

$$\frac{d\rho}{d\tau} = -\tfrac{1}{2}\rho(m_2\cos 2\varphi - 2m_0') \tag{5.10}$$

For the second equation (5.8), we have similarly

$$\frac{d\psi_1}{dt} = -\tfrac{1}{2}R' \sin 2\psi_0 \tag{5.11}$$

Likewise, the only term on the right side of (5.11) which remains after the integration between 0 and 2π in the product $R' \sin 2\psi_0$ is $\frac{1}{2}m_2 \sin 2\varphi_0 2\pi$, so that the difference equation (setting $2\pi\mu = \Delta\tau$ and $\Delta\varphi = \psi_1(2\pi) - \psi_1(0)$) is

$$\frac{\Delta\varphi}{\Delta\tau} = -\tfrac{1}{4}m_2 \sin 2\varphi_0$$

and the corresponding stroboscopic d.e. is

$$\frac{d\varphi}{d\tau} = -\tfrac{1}{4}m_2 \sin 2\varphi \tag{5.12}$$

The existence of a stable stationary oscillation reduces now to the existence of a stable singular point of the stroboscopic system (5.10) and (5.12). In the first place one observes that there is a position of equilibrium for the phase if $\sin 2\varphi_0 = 0$ (we change the notations and now by the subscript zero we designate the *stationary values* since no confusion is to be feared from the approximation procedure).

6. Stability; concluding remarks

The stability of the phase is obtained from the variational equation which is:

$$\frac{d\delta\varphi}{d\tau} = -\tfrac{1}{2}m_2 \cos 2\varphi_0 \delta\varphi$$

If $m_2 > 0$, the phase is stable for $\cos 2\varphi_0 = +1$; for $m_2 < 0$, it is stable for $\cos 2\varphi_0 = -1$.

Assume, for the sake of an example, the first case: $\cos 2\varphi_0 = +1$; then the first stroboscopic d.e. (5.10) becomes

$$\frac{d\rho}{d\tau} = -\tfrac{1}{2}\rho(m_2 - 2m_0') \tag{6.13}$$

There are obviously two positions of equilibrium, viz.: (1) $\rho = 0$, the state of rest; and (2) $m_2 = 2m_0'$, which corresponds to the stationary oscillation. We have (after replacing x_0^2 by ρ_0 and taking the values (5.6) for m_0' and m_2):

$$\frac{d\rho}{d\tau} = -\frac{1}{2}\rho^2\left(\frac{5}{32}a_6\rho^2 + \frac{1}{4}a_4\rho + \frac{1}{2}a_2\right) \tag{6.14}$$

For the stability of the state of rest we have

$$\left(\frac{d\rho}{d\tau}\right)_{\rho \simeq 0} = -\tfrac{1}{4}a_2\rho^2 \tag{6.15}$$

Hence, the state of rest is unstable (that is, the condition of self-excitation is fulfilled) if $a_2 = b_1 + 15b_3X^4 < 0$, which shows that b_1 and b_3 cannot be positive at the same time. The polynomial in (6.14) can be written as

$$\frac{5}{32}b_3\left[\rho^2 + \left(\frac{8}{5}\frac{a_4}{b_3}\right)\rho - \frac{16}{5}\frac{|a_2|}{b_3}\right] \tag{6.16}$$

since $a_6 = b_3$.

So far we have not made any hypothesis regarding the signs of the coefficients b_1 and b_3 in the nonlinear function R. We shall be guided here by physical considerations. In fact, in arcs and gaseous discharges where self-sustained oscillations are observed there is always present the so-called "negative resistance," mentioned in Section 4. With regard to energy exchanges, a negative resistance is equivalent to the *absorption* of energy in contrast with positive resistance characterized by energy *dissipation.*

If a nonlinear conductor is to exhibit an analogous property, the coefficient b_1 must be negative in the initial stage of the development of oscillation when the amplitude increases and similarly b_3 must be positive, which corresponds to the ultimate limitation of amplitude by a positive resistance.

If one makes this plausible assumption, the problem becomes more definite and a number of possibilities which do not have any physical meaning are eliminated. Under this assumption the stroboscopic d.e. (6.14), taking into account the form (6.16), can be written as:

$$\frac{d\rho}{d\tau} = -k\rho^2[\rho^2 + (k_1a_4)\rho - k_2|a_2|] = \Phi(\rho) \tag{6.17}$$

where $k = \dfrac{5}{64}b_3 > 0$, $k_1 = \dfrac{8}{5b_3} > 0$, and $k_2 = \dfrac{16}{5b_3} > 0$.

If we consider the simplest and the most important case, the one in which there is only one stationary oscillation, the polynomial in (6.17) has only one positive root

$$\rho_0 = -\frac{k_1a_4}{2} + \sqrt{\frac{k_1a_4{}^2}{2} + k_2|a_2|} \tag{6.18}$$

where a_4 and a_2 are given by (5.4). Since, however, for the self-excitation from rest a_2 must be negative, this means that in the expression for $a_2 = b_1 + 15b_3X^4$, $b_1 < 0$ must be sufficiently large in absolute value, while $b_3 > 0$ and X^4 should be relatively small.

In order to complete the problem, it is necessary to show that we can dispose of the parameters to produce a *stable* stationary amplitude ρ_0, for which we use the criterion

$$\Phi_\rho(\rho_0) < 0 \tag{6.19}$$

as we did in Chapter 22.

We have $\Phi(\rho) = -k\rho^4 - kk_1a_4\rho^2 + kk_2|a_2|\rho^2$ and, therefore, $\Phi_\rho(\rho) = -\rho(4k\rho^2 - 3kk_1a_4\rho + 2kk_2|a_2|)$. Taking into account that the polynomial in (6.17) is zero for $\rho = \rho_0$, we obtain the condition of stability in the form

$$\rho_0 > \tfrac{4}{15}X^2 \tag{6.20}$$

Hence, if $b_1 < 0$, $b_3 > 0$, and X are given, a_2 and a_4 are known and ρ_0 can be computed; if this value of ρ_0 happens to satisfy the inequality (6.20), then it is stable. If, however, it is unstable, a simple decrease of X does not settle the matter, since ρ depends also on X through the coefficients a_2 and a_4.

It is seen thus that the problem, although sufficiently simple theoretically, is rather complicated if one tries to obtain the condition of stability which requires inevitably some kind of an approximation procedure by which the stability criterion (6.20) can be computed.

There is yet another complication worthy of note. In analyzing the stroboscopic system (5.10) and (5.12) we sought first the stability of the phase ($\sin 2\varphi_0 = 0$; $\cos 2\varphi_0 = +1$) and established the existence and stability of the singular point on this basis.

It is equally possible to formulate the existence of a singular point of (5.10) and (5.12) by setting $m_2 = m_0' = 0$ from which, after substituting values for m_2 and m_0', one can determine the stationary amplitude, although the phase remains arbitrary; this means a certain parasitic oscillation without any definite frequency. It is possible, however, that in view of the stability of the phase for $\cos 2\varphi_0 = +1$, this parasitic oscillation will ultimately transform itself into an oscillation with definite amplitude and phase which we have investigated previously.

We merely mention these difficulties because, in contrast to those arising from electron-tube oscillators in which the energy absorbing and dissipating elements of characteristics are distinct, these various features are to be derived from the same characteristic that leads to more complicated conditions just mentioned.

It must be mentioned also that, in cases when the nonlinear resistance is not a function of a thermal effect but merely depends on some (empirical) function of the current, the treatment is simpler and follows the previously studied cases when the coefficients of nonlinear polynomials are constant.

PART IV

RELAXATION OSCILLATIONS

INTRODUCTION

In all quantitative methods of Part II the solution of a nonlinear d.e. is sought in the form of power series arranged in ascending powers of a certain parameter (or parameters, if there are several), and it is clear that this parameter must be sufficiently small to guarantee the convergence of such a series solution. In fact, only under this condition can the *formal solution* yielded by these methods be *the actual solution.* This is why all methods used in the theory of oscillations fall within the scope of the generic term: *methods of small parameters.*

The van der Pol equation which initiated these studies has a parameter associated with the first derivative term, and this was why the investigation of this and similar d.e. resulted in remarkable advances in the theory of oscillations once the contact with the theory of Poincaré had been established; this was possible only by assuming that the parameter was small.

However, van der Pol, having discovered his d.e., attempted to discover the nature of its solution for the larger values of μ by means of the graphical procedure of isoclines; we have touched upon this question on several occasions, particularly in Chapter 4, and it was shown that for $\mu = 10$ the trajectory departs considerably from that of the harmonic oscillator as happens for small values of μ. Indeed, for normal relaxation circuits the values of the parameter are far greater than $\mu = 10$. In one of his early publications van der Pol mentions that for a standard multivibrator circuit this value is of the order 10^5. It is obvious that for such a large value of μ the isocline procedure becomes impossible; however, a simple calculation in polar coordinates shows that at this value of μ even a very small rotation of the radius vector (of the order of $1''$ of arc) in the neighborhood of the x axis results in a change of direction of integral curve of nearly $90°$ and produces an incidental change in velocity of the representative point from a high value to almost zero.

There are thus two points on the integral curve at which the analyticity is practically lost and, for that reason, the power series solution becomes impossible. The last circumstance is seen also directly from the fact that the usual power series solution: $x(t) = x_0(t) + \mu x_1(t) + \mu^2 x_2(t) + \ldots$, which we used previously, is impossible here since it diverges.

In the above analysis of the integral curve, the fact that $y = \dot{x}$ suddenly changes from a high value to almost zero is reminiscent of the condition of a shock and, as we shall see later, this consideration gives rise to an idealization leading to a purely discontinuous treatment of relaxation oscillations (Chapters 26 and 29). These difficulties produced two different lines of approach to the subject of relaxation oscillations: (a) the discontinuous theory, and (b) the continuous theory, the first preceding the latter chronologically.

Since the oscillation phenomenon exhibits some features of quasi-discontinuities at certain points, it was natural that the effort of physicists should be directed toward a purely discontinuous treatment inspired to some extent by the analogous treatment of shocks in mechanics. This effort resulted in two distinct theories (one outlined in Chapter 26 and the other in Chapter 29); the starting points in these theories are somewhat different but, beginning with a certain intermediate point in the argument, both become almost identical. The theory outlined in Chapter 26 is based on certain physical concepts, whereas that of Chapter 29 has a more geometrical basis, although later on it, too, introduces a physical argument. Both theories use certain idealizations approximating the quasi-discontinuous phenomena by mathematical discontinuities.

The second approach arose from a series of important papers by Cartwright and Littlewood and concerns the van der Pol equation for large values of parameter; most of this work relates to the nonhomogeneous equation but there is one paper (by Cartwright) devoted to the homogeneous equation (Section 6, Chapter 30). We outline here only the latter because the study of the nonhomogeneous case would be beyond the scope of this Introduction.

The work of Cartwright–Littlewood has a somewhat limited objective, namely, to justify analytically (*without involving any a priori idealizations*) the graphical solution obtained by van der Pol by the isocline method.

The fundamental idea of this approach (Chapter 30) is simple, although the calculation process is not and is essentially as follows: the graphical (or the experimental) curve representing the solution is split into a number of the *characteristic stretches*, each of which has definite features; for example, on some of them $\ddot{x}$ is negligible, on some others $\dot{x}$ or x are negligible, etc. This permits using easily integrable abridged or "truncated" d.e. for each stretch, the difficulty being in the analysis of the order of magnitude of different quantities and in the ultimate "joining" of all these solutions of "truncated" equations; calculations are somewhat facilitated by the use of additional equations (the energy equation, the integrated equation, etc.).

Summing up, there is no definite *overall idealization* as in the discontinuous theories, but there is a precise analysis of possible simplifications in each characteristic stretch so that the problem is solved with a prescribed approximation. It should be noted that the procedure hinges on the existence of a graphical or experimental curve representing the solution, and the analysis merely *confirms* it.

At a later date Dorodnitzin showed that these difficulties can be overcome to some extent by the use of the so-called *asymptotic expansions* which by their nature do not require analyticity (Chapter 30, Sections 4 and 5); the difficulty in this case is however much the same as in the Cartwright–Littlewood (C.L.) theory (namely, the *junctions* of these expansions at certain points); otherwise the basic idea is the same; that is, to proceed from one region to the other.

In both these *asymptotic* methods it would be impossible to proceed without a preliminary knowledge of the integral curve (obtained either graphically or experimentally); in fact, as was mentioned, they merely *explain these curves analytically*. These methods have a rather limited objective, in contrast with the small parameter theory which deals uniformly with a great variety of d.e. (provided they belong to the nearly linear class) the shape of integral curves being immaterial.

There are certain advantages and disadvantages in the use of the discontinuous and asymptotic theories. The former are purely qualitative while the latter are quantitative. In the former one uses extensively the phase-plane representation, but this representation differs from the classical phase plane with which we have been dealing so far.

Thus, for instance, a quasi-discontinuous (idealized by discontinuities) motion of the representative point may describe a closed orbit containing a saddle point in its interior, etc., which is absurd according to the classical theory.

Although this and similar conclusions may appear to contradict what has been learned in the analytical theory, it has to be expected that, by dropping the essential points of that theory, its conclusions will not hold either.

In spite of this, the discontinuous theory acquires a gradually increasing importance due to the ease with which it handles the relaxation problems even of complicated types; very often even new phenomena are predicted on this basis. When a theory reaches such a state, it cannot be easily discarded only because one is more accustomed to using analytical theories.

After all, the theory of shocks in mechanics is also a kind of local "discontinuous extension" in an otherwise analytical science and the advent of discontinuous theories in studies of relaxation oscillations is not more surprising or "revolutionary" than the classical treatment of shocks.

It must be noted that difficulties exist also in the discontinuous theories but they are of a somewhat different kind.

In the asymptotic theory the idea in itself is simple; the difficult part is in applying the procedure as is shown in Section 5, Chapter 30. In the discontinuous theory, on the contrary, the application of the procedure is very simple and reduces generally to simple topological constructions in the phase plane. The difficult part lies in the justification of the theory, and it is useful to mention this briefly.

Suppose we have a d.e. of the form

$$\mu\ddot{x} + f(x,p)\dot{x} + x = 0 \tag{IV.1}$$

where μ is a small parameter and p is some finite parameter. Clearly if $\ddot{x}$ remains bounded for all t, one can neglect the term $\mu\ddot{x}$ and deal with the "abridged" d.e.

$$\dot{x} = -x/f(x,p) \tag{IV.2}$$

If $f(x,p) \neq 0$ for all t, one can integrate the preceding d.e. and obtain thus

$$t = -\int f\frac{dx}{x} \tag{IV.3}$$

Should, however, $f(x,p)$ vanish for $t = t_0$, $\dot{x}$ and $\ddot{x}$ become infinite, the use of the abridged equation (IV.2) ceases to be legitimate since then the term $\mu\ddot{x}$ cannot be neglected.

Thus it is seen that the use of a simpler equation (IV.2) instead of a more complicated equation (IV.1) is subject to certain limitations.

This raises a rather delicate question which the mathematicians designate as *singular perturbation theory* and the physicists often call *theory of degeneration*. The first designation is obvious if one writes (IV.1) as

$$\ddot{x} = -\frac{[f(x,p)\dot{x} + x]}{\mu}$$

in which case the right-hand term has a singularity for $\mu = 0$; and the second one is justified by the fact that, for $\mu \to 0$, the d.e. (IV.1) *degenerates* from the second to the first order. We shall use preferably the second term as more accessible to an intuitive understanding of the asymptotic process, the physical nature of which is simple.

In fact, it often happens in electricity and in mechanics that the coefficient of the term with the second derivative is very small (negligible inductance, negligible mass, etc.). In such a case it is *simpler* to deal with the *degenerate* equation of the first order; but this simplification involves some difficulties as will be shown later (Section 2, Chapter 26), namely, it becomes necessary to adopt an essentially discontinuous treatment.

In the discontinuous theory based on certain idealizations one assumes $\mu\ddot{x} = 0$ (in IV.1) and deals with the *degenerate* d.e. of the first order which leads to a procedure similar to that used in the theory of shocks in mechanics. In such a case one has to supplement the *intentional ignoration* of what happens during a very short time (which one wishes to idealize by a mathematical discontinuity) by certain *additional information.* In the theory of shocks this "additional information" is supplied by the theorems of momentum and kinetic energy; in problems of relaxation oscillations it appears in the form of certain *physical invariants* which permit carrying out a *physical continuation* of the solution at the points at which its analytical continuation is impossible.

The idealized (discontinuous) treatment of relaxation oscillations is more convenient for a qualitative appraisal of what may be expected in a given problem. Moreover, it permits reducing the investigations of a system amenable to two d.e. of the first order to a phase plane representation, whereas the same system on the basis of the asymptotic theory would require the study of the differential system of the fourth order which would be a very difficult problem. A simple symmetrical multivibrator scheme is, however, a problem of this kind.

Hence, having in mind primarily applications, we prefer to begin the exposition with an outline of discontinuous theories which are relatively simple and lead to the establishment of qualitative conclusions in all known cases of relaxation oscillations. The asymptotic theory is outlined briefly in Chapter 30.

Besides these two principal approaches—(1) the discontinuous and (2) the asymptotic—there is also a third one: (3) Chapter 31, in which one replaces the actual nonlinear characteristic by pieces of straight lines (this procedure is sometimes called the method of *broken characteristics*). As the result, in the interval corresponding to each, such "piece" holds its own (generally simple) d.e. and the problem of establishing the condition of periodicity reduces to the investigation of conditions under which the polygon of arcs of linear trajectories becomes closed.

As far as the physical side of these phenomena is concerned, methods (1) and (2) are applicable to the relaxation oscillations, that is, oscillations which exhibit widely different velocities of the representative point; the idealization (1) merely assumes that when these velocities are very great, they are considered *infinitely great.* The theory of degeneration (Chapter 26) aims precisely at this idealization.

The idealization (3) does not deal necessarily with relaxation oscillations and holds whenever there appears a loss of analyticity at some points of a cycle. Thus, for example, a certain dynamical system (electrical or mechanical) may acquire this feature if a relay closes or opens the circuit; there appears thus a loss of analyticity in the process without any "relaxation features" however. The idealization (3) is often called *piecewise linear method* in view of the above mentioned approximation by pieces of linear trajectories.

Although the above classification may appear somewhat complicated, in reality it merely corresponds to the existing situation. One may think that, in contrast to the "method of small parameters," one could use here the term of "large parameters"; this, however, seems to be meaningless, since most of the relaxation phenomena are governed by the d.e. in which there is no parameter and, besides, there are numerous cases in which the relaxation features are not involved either, although the system is not analytic. The only common feature between (2) and (3) is that in both one cannot use the analytical argument over the whole cycle.

Referring more specifically to the various chapters of this Part IV, Chapter 26 gives an outline of the discontinuous theory of Mandelstam–Chaikin (M.C. for short) with a number of selected examples given in Chapters 27 and 28. All this material can be found in greater detail in Andronov and Chaikin's book.[1] The approach to the theory of degeneration is also taken from this work because a more detailed presentation of this theory by Friedrichs, Wasow, and Levinson would involve a considerable amount of mathematical development in the singular perturbation theory, and is hardly needed in this elementary presentation.

Chapter 29 concerns an alternative discontinuous theory by Vogel. Because this theory is probably less known than the M.C. theory, it was thought useful to mention it also.

Both discontinuous theories were apparently inspired by the classical theory of shocks, as we have just mentioned, and were eminently successful not only in explaining all known experimental facts, but sometimes even in predicting new phenomena.

In Chapter 30 we have attempted to give only a brief review of the modern asymptotic theory, because a more or less complete survey of this difficult field would require extensive mathematical material out of keeping with the applied character of this text.

Chapter 31 gives an outline of what we have called the idealization (3) in the above classification.

[1] A. Andronov and S. Chaikin, *Theory of Oscillations* (original text in Russian), Moscow, 1937.

Chapter 26

DISCONTINUOUS THEORY OF RELAXATION OSCILLATIONS

1. Piecewise analytic phenomena

Before attempting to outline the discontinuous theory it is useful to explain certain idealizations on which it is based.

In the first place we consider the following theoretical example.[1] A perfectly elastic ball is rolling on a frictionless horizontal plane between two equally elastic walls. It is assumed that the mass of the walls is

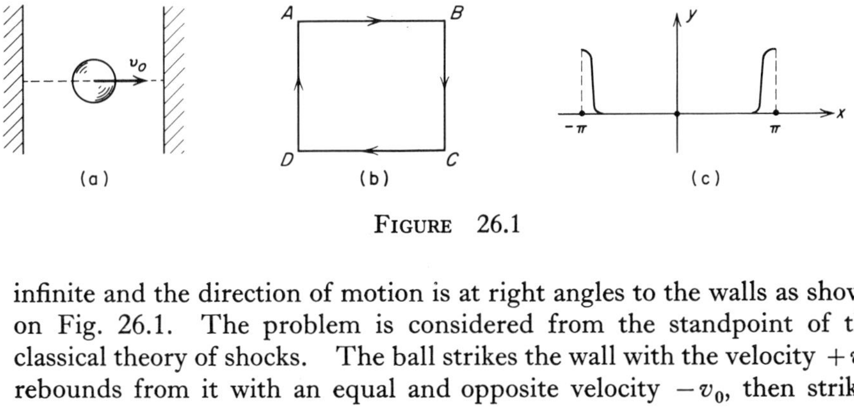

FIGURE 26.1

infinite and the direction of motion is at right angles to the walls as shown on Fig. 26.1. The problem is considered from the standpoint of the classical theory of shocks. The ball strikes the wall with the velocity $+v_0$, rebounds from it with an equal and opposite velocity $-v_0$, then strikes

[1] A. Andronov and S. Chaikin, *Theory of Oscillations* (original text in Russian), Moscow, 1937; English translation by S. Lefschetz of A. Andronov and S. Chaikin, *Theory of Oscillations*, Princeton University Press, Princeton, N.J., 1949; A. Andronov, A. Witt, S. Chaikin, *Theory of Oscillations* (in Russian); this book is the second edition (1959) of A. Andronov and S. Chaikin, *Theory of Oscillations* (original text in Russian), Moscow, 1937.

again the other wall and so on. There exists thus a periodic process represented on Fig. 26.1b by a usual phase-plane diagram. On the stretches AB and CD the motion of the representative point R occurs with a finite velocity $\pm v_0$ in accordance with a simple d.e. $\ddot{x} = 0$. The stretches BC and DA are traversed by R with an infinite velocity since, in accordance with the theory, the velocity changes discontinuously at a constant value of the coordinate x. As on stretches AB and CD the d.e. governs the motion of R, one may call these stretches as *analytic* trajectories (in general these may be some *analytic arcs*), but on stretches BC and DA the situation is different. On the basis of this theory the motion of R on these stretches is not governed by the d.e. because what happens *during* the short period of time when the ball is in contact with the wall is *intentionally ignored*, which is merely another way of saying that the process of the shock itself is not governed by the d.e. The shock is merely specified by a relation between the conditions that exist immediately *before* it and immediately *after* it; this, as is well known, results from the existence of certain *invariants* (theorems of momentum and kinetic energy), which are not *directly* related to the d.e.† One can say thus that, although the d.e. governs the motion of R on the analytic arcs, it does not govern it on the discontinuous stretches connecting these arcs; on these stretches the discontinuous motion of R is determined by means of the *invariants* of the problem (which may be regarded as a kind of an "additional information" not contained in the d.e. directly) and *not by the d.e. itself.*

What is interesting is the fact that the motion is still periodic in the sense that it is represented by a closed integral curve traversed in a finite time in spite of the fact that the d.e. $\ddot{x} = 0$ has no periodic solution. This simple model can be improved if one considers the motion of the ball not on a plane but on a curved surface (Fig. 26.1c) with steep slopes toward its limits $\pm\pi$.‡ An equation of the form $y(x) = [\sin(x/2)]^{2n}$, where $2n$ is sufficiently large, would answer this requirement. Neglecting the friction the d.e. of motion will be

$$\ddot{x} + 2\frac{g}{l}\frac{d}{dx}\left(\sin\frac{x}{2}\right)^{2n} = \ddot{x} + \frac{g}{l}n\left(\sin\frac{x}{2}\right)^{2n-2}\cdot\sin x = 0$$

In this form the motion of the ball between two walls is described completely by the d.e. of the form: $\ddot{x} + f(x) = 0$ and there is no necessity for considering the external reactions. Moreover, the simple equation $\ddot{x} = 0$ holds only for small intervals of x but the approach to $x = \pm\pi$ introduces the change in the d.e.

† By this we mean: *not* by the d.e. governing the motion of R on the analytic arcs; it does not include, however, the presence of some other (unknown) d.e. taking charge of the phenomenon during a very short time.

‡ This remark is by the late Professor B. van der Pol.

We thus reach an intuitive definition of a *piecewise analytic* periodic phenomenon having some features of a relaxation oscillation. This piecewise analytic process does not however have the property of a *limit process* in that the motion establishes itself directly on the cycle $ABCD$ without approaching it gradually.

In the following example this limit property exists. The mechanism of a clock gives a simple illustration. The clock consists of an essentially damped oscillatory system (torsional pendulum) whose integral curve is a logarithmic spiral converging to a stable focal point 0 as shown in Fig. 26.2. In addition to this, there is provided an escapement mechanism which releases substantially constant impulses applied to the pendulum at a definite point of its cycle, thus compensating for the loss of energy caused by dissipation. Therefore, while the dissipation of energy is a continuous process, its replenishing is a quasi-discontinuous one (treated as discontinuous).

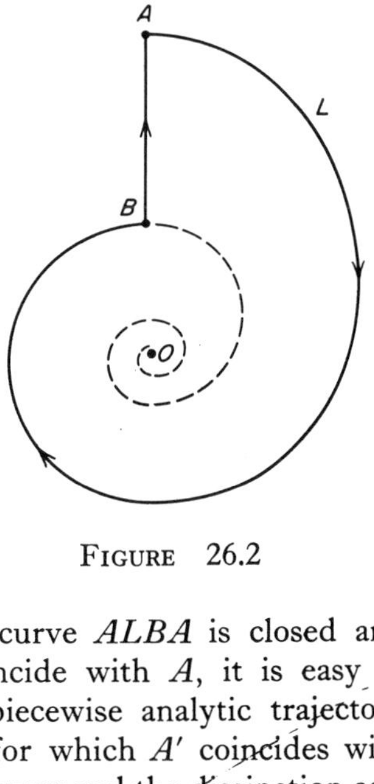

FIGURE 26.2

If one considers the motion immediately after the impact delivered by the escapement mechanism (point A), the subsequent motion takes place on the analytic arc of a logarithmic spiral L representing the integral curve in the phase plane. On L the point R moves with a finite velocity in accordance with a (substantially linear) d.e. $m\ddot{x} + k\dot{x} + cx = 0$. At the point B of the cycle, the escapement delivers an impulse increasing the velocity (at zero coordinate) in a quasi-discontinuous manner which transfers R also in a same manner from B to A'. If the point A' coincides with A (as shown in Fig. 26.2), the piecewise analytic curve $ALBA$ is closed and the motion is periodic. If A' does not coincide with A, it is easy to ascertain by a simple argument that the piecewise analytic trajectory has a tendency to approach the condition for which A' coincides with A. In fact, if A' is originally above A, L is larger and the dissipation acts for a longer time so that the trajectory has the tendency to shrink more before the next impact is delivered. Thus, inasmuch as larger arcs have a tendency to shrink, and the smaller ones are less affected by the dissipation, the motion, ultimately, will approach a trajectory for which the stretch AB caused by the dissipation is just compensated for by the stretch BA caused by the impulse, after which a periodic process is established. Here again the stretch BA is traversed by R with an infinite velocity (in reality, very high but finite). Likewise, on the analytic arc it is the d.e. that governs the phenomenon but not on the stretch BA. The phenomenon in this case is not only piecewise analytic but also of a "discontinuous limit cycle" type. It is not, however, of a relaxation type because the system cannot be considered as autonomous, if one considers the escapement mechanism as being outside the system of the pendulum proper. If, however, one considers the whole mechanism of the clock as a physical system, then obviously the clock may be considered as a relaxation oscillation mechanism.†

As the last example we consider the following phenomenon defined by an alternate sequence of two linear d.e.

$$\ddot{x} + 2h\dot{x} + \omega^2 x = 0 \quad \text{and} \quad \dot{x} + 2h\dot{x} + \omega^2 x = \omega^2 a \tag{1.1}$$

the first of these two equations taking charge of the phenomenon when $\dot{x} > 0$, and the second, when $\dot{x} < 0$. For the moment we consider the mathematical part of the problem and shall indicate later how such a scheme can be produced experimentally.

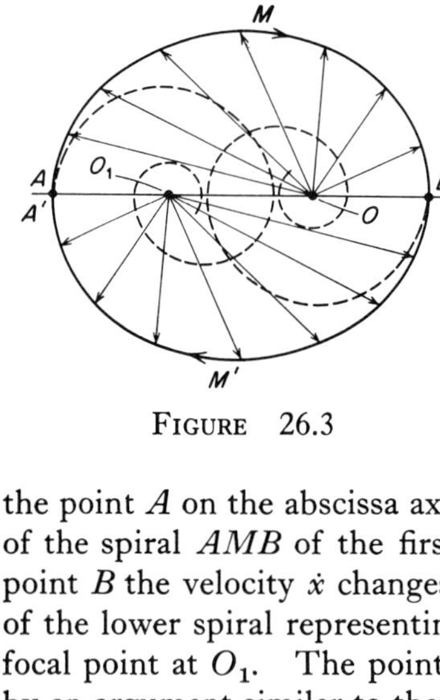

FIGURE 26.3

If one sets $x - a = x_1$, the second d.e. (1.1) can be written as

$$\ddot{x}_1 + 2h\dot{x}_1 + \omega^2 x_1 = 0 \tag{1.2}$$

so that it is sufficient to consider the first d.e. (1.1) for $\dot{x} > 0$ and (1.2) for $\dot{x} < 0$. The integral curves of these d.e. are again the logarithmic spirals but are referred to two different focal points O and O_1 as shown on Fig. 26.3. If one starts from some arbitrary initial condition represented, say, by the point A on the abscissa axis, the motion of R will take place on the arc of the spiral AMB of the first d.e. having O as its focal point. At the point B the velocity $\dot{x}$ changes its sign and R follows now the arc $BM'A'$ of the lower spiral representing the solution of the second d.e. having its focal point at O_1. The point A' generally does not coincide with A but by an argument similar to that used in connection with the clock, one can show that, after a number of cycles, the motion approaches the periodic motion for which the path is re-entrant at A. In fact, in polar coordinates at the point A, one has: $r_1 = OA$, $\varphi = 0$. After one half turn (when (1.1)

† This question is treated more fully in Section 5, Chapter 28.

governs the motion) at point B we have $r_2 = OB = r_1 \exp(-\pi h/\omega)$. After the change of the d.e. at B, R moves on the lower arc specified by the radius $r_3 = \bar{r}_2 \exp(-\pi h/\omega)$, where $\bar{r}_2$ means the radius vector at B but referred to the lower curve (with the focal point at O_1), so that $\bar{r} = r_2 + a_2$ and, therefore, $r_3 = r_2 \exp(-2\pi h/\omega) + a \exp(-\pi h/\omega)$ and $\bar{r}_3 = a + r_3$, and so on.

For the general term (after n rotations):

$$r_n = a[\exp(-\pi h/\omega) + \exp(-2\pi h/\omega)] \\ + \ldots + \exp[-(n-2)\pi h/\omega] + r_1 \exp[-(n-1)\pi h/\omega]$$

$$\bar{r}_n = a[1 + \exp(-\pi h/\omega)] \\ + \ldots + \exp[-(n-2)\pi h/\omega] + r_1 \exp[-(n-1)\pi h/\omega]$$

where n is even. Similar expressions are obtained for n odd. One finds in this manner that, for $n \to \infty$,

$$\lim r_n = r_0 = a/[1 - \exp(-\pi h/\omega)] \tag{1.3}$$

Thus, the ultimate closed curve to which such a motion approaches consists of two identical arcs of logarithmic spirals joined at the points A and B. The curve at the points of junction is not analytic.

The experimental arrangement for obtaining this particular operation consists of an ordinary electron-tube circuit (for example, a simple inductively coupled feedback system) if one provides a sufficiently high voltage on the grid. It is clear that under this condition the electron tube works practically as a switch between the blocked condition and the full value of its saturation current. In such a case the oscillating circuit receives periodically timed impulses in opposite directions and, between them, the circuit behaves itself in accordance with the alternate sequence of two linear d.e., as was just explained.

It is noted that in this example one does not have any discontinuous stretches as in the first two examples. This is due to the fact that for this choice of variables of the phase plane, these variables do not admit any discontinuities. Hence, the phenomenon is analytic everywhere *except at two points A and B*, but this is already sufficient to obtain a closed integral curve (that is, a periodic solution) although the d.e., *considered individually*, have no such solution.†

It is necessary now to ascertain which variables are capable of varying discontinuously so as to produce a piecewise analytic representation.

† This subject will be studied more fully in Chapter 31.

2. Degeneration theory and its physical significance

The discontinuous theory of relaxation oscillations is based on the use of the so-called *degenerate* d.e. and it is important to ascertain the fundamental properties of these d.e. before attempting to outline this theory.

We note in passing that this theory has recently expanded considerably[1] but for our immediate purpose we shall need only a few essential properties of these degenerate equations.

We shall call a d.e. *degenerescent* when the coefficient of the highest order derivative in the d.e. is small in comparison with the coefficients of other terms in the d.e. In what follows we shall limit ourselves to the d.e. of the second order with constant coefficients so that a degenerescent d.e. will be

$$a\ddot{x} + b\dot{x} + kx = 0 \tag{2.1}$$

provided the coefficient a is much smaller than b and k.

We shall call a d.e. a *degenerate* one, if the problem is idealized by assuming that the small coefficient a in (2.1) is *zero*, in which case (2.1) becomes

$$b\dot{x} + kx = 0 \tag{2.2}$$

Physicists became interested in the problem of degeneration because such idealizations occur frequently in physics. For example, in the case of the so-called *RC* circuits, it is generally assumed that the inductance of the circuit is so small that it is sufficient to deal with the d.e. of the first order instead of the second order. A similar situation arises often in mechanical problems when the mass (or the moment of inertia) is small as compared with damping and spring constant.

The problem is, however, far from simple, inasmuch as neglecting a term in the d.e. merely because its coefficient is small does not mean that this simplification of the d.e. applies equally to its solution.

In order to investigate this important point we determine separately the solutions of (2.1) and (2.2) under the assumption that the coefficient a in (2.1) is very small.

For the degenerescent d.e. (2.1) the solution is

$$x(t) = x_0[e^{-kt/b} - (ak/b^2)e^{-bt/a}] + (a\dot{x}_0/b)[e^{-kt/b} - e^{-bt/a}] \tag{2.3}$$

and for the degenerate d.e. (2.2) it is

$$\bar{x}(t) = x_0 e^{-kt/b} \tag{2.4}$$

In this form the dependence on the constants of integration x_0 and $\dot{x}_0$ is obvious. In the case of (2.1), the solution depends on two constants of

[1] See footnote [1], page 605.

integration, whereas (2.4) depends only on one such constant. This circumstance will be of fundamental importance for what follows.

If one forms the difference of (2.3) and (2.4), one gets

$$\alpha(t) = x(t) - \bar{x}(t) = -x_0(ak/b^2)e^{-bt/a} + \dot{x}_0(a/b)(e^{-kt/b} - e^{-bt/a}) \quad (2.5)$$

and it is seen that $\alpha(t)$ approaches zero *uniformly* for all values of $t > 0$ when $a \to 0$.

As to $\dot{\alpha}(t)$, the matter is different. In fact, in this case one has

$$\dot{\alpha}(t) = \dot{x}(t) - \dot{\bar{x}}(t) = x_0(k/b)e^{-bt/a} - \dot{x}_0(ak/b^2)e^{-kt/b} + \dot{x}_0 e^{-bt/a} \quad (2.6)$$

For a sufficiently large t, $\dot{\alpha}(t)$ approaches zero uniformly, as does $\alpha(t)$, but, if t is small enough, one has $\dot{\alpha}(t)_{t \simeq 0} = kx_0/b + \dot{x}_0$ and this expression, being independent of a, is finite, except, possibly, for a very special choice of initial conditions which, clearly, is of no interest for the general case.

This circumstance is of great importance in the discontinuous theory of relaxation oscillations for the following reason: In any physical system of the second order there are two arbitrary constants which appear as two initial conditions. More specifically, the state of rest, for instance, is specified by $x_0 = \dot{x}_0 = 0$. If, however, one adopts a degenerate d.e. for the description of the system, here there is only one constant. There appears an obvious difficulty—the state of rest is specified by two arbitrary constants and the degenerate d.e. admits only one which raised the question: What happens to the second constant when the state of rest is suddenly disturbed by the appearance of an impulse on the right side of the d.e.?

The answer to this is that the variable $\dot{x}$, whose convergence is not uniform on the basis of the theory of degeneration, will suddenly jump to its final value beginning with which the process is determined in terms of one single constant of integration as it should be. Thus, the "conflict between the constants of integration," so to speak, is removed, owing to the discontinuity of the variable which *can* vary discontinuously on the basis of the degeneration theory.

The following well known examples can illustrate what has just been explained. Consider the d.e. of a simple circuit

$$L\frac{di}{dt} + Ri + \frac{1}{C}\int idt = 0 \quad (2.7)$$

which was originally "dead" and to which is suddenly applied a constant e.m.f. E. We shall investigate the conditions that exist immediately before and after the application of this e.m.f. It is more convenient to introduce the charge x as the dependent variable, which amounts to replacing the variable i by dx/dt. Immediately after the application of E, the d.e. can be written in the form

$$LC\ddot{x} + RC\dot{x} + x = EC \tag{2.8}$$

Immediately before the application of E, when the circuit was "dead," the conditions were obviously $x_0 = \dot{x}_0 = 0$. Since in the case of the problem specified by the d.e. (2.8) of the second order the situation is well known, we omit it here and pass directly to the degenerate case.

We consider first the RC degeneration, that is, when the inductance L is so small that we use the degenerate d.e. of the first order

$$RC\dot{x} + x = EC \tag{2.9}$$

There is only one constant of integration here and it is determined by the initial condition: for $t = 0$, $A = -EC$, where A is the integration constant. The solution is then $x = EC(1 - \exp(-t/RC))$. If one differentiates this expression, one has $\dot{x} = (E/R) \exp(-t/RC)$; and for $t = 0$, this gives $\dot{x}_0 = E/R$, whereas immediately before the application of E this initial condition was obviously $\dot{x}_0 = 0$. This means that the variable $\dot{x}_0$ has to change discontinuously if the degenerate d.e. is to be used to represent a phenomenon whose state of rest is specified by two initial conditions.

Another conclusion is noteworthy: we have just found that in the case of the degenerate d.e., $x = EC[1 - \exp(-t/RC)]$ and $\dot{x} = (E/R) \exp(-t/RC)$. The ratio $x/\dot{x}$ in this case is a definite function of t and is not arbitrary as it is in the corresponding degenerescent equation. In other words, instead of the two-dimensional representation (phase plane) in the latter case, we now have a unidimensional representation, *the phase line*, because there is only one arbitrary constant of integration instead of two.

In the LR degeneration the situation can be analyzed similarly, but it is more convenient here to operate directly with the variable i (the current) instead of x (the charge) as previously. The d.e. here is

$$L\frac{di}{dt} + Ri = E \tag{2.10}$$

and, under the same assumed conditions, the solution is $i = (E/R)(1 - \exp(-Rt/L))$.

Differentiating this expression and setting $t = 0$, we find $(di/dt)_0 = -E/R$, but at the instant immediately preceding the application of E we had $(di/dt) = 0$; the conclusion is exactly the same as before: the second initial condition has to jump discontinuously if the physical existence of two initial conditions just *before* the application of E is to be reconciled with the existence of only one initial condition imposed by the degenerate d.e. of the first order, which admits only one constant of integration.

Summing up, in both cases the situation remains the same, namely: the variables in the d.e. (x, the charge, in the case of an RC degeneration, and i, the current, in the case of a LR degeneration) cannot vary discontinuously and are determined directly by the degenerate d.e. of the first order. However, the *derivatives of these variables* $dx/dt = i$ (current in the condenser circuit) and di/dt (or $L di/dt$, the voltage across the inductance) can and, in fact, *must* vary discontinuously in order to reconcile the physical existence of two initial conditions *before* the application of E with the requirement of one single constant of integration if the d.e. has to be used in the degenerate form to describe the phenomenon *after* the application of E.

It is clear that what has been said about a sudden *application* of the external impulse E holds equally well when E is suddenly removed or, generally, changed. The essential point is that the variables which appear in the degenerate d.e. vary continuously in accordance with these equations, but their derivatives jump discontinuously into the values which they must have throughout the subsequent process.

Exactly similar considerations apply to degenerate mechanical systems, the only difference being that, instead of the electrical parameters L, R, and $1/C$, one has to consider the analogous mechanical parameters m (the mass or the moment of inertia), k (the coefficient of the velocity damping) and c (the spring constant).

Thus, for instance, in a (k,c) degeneration (a system with negligible inertia but with finite damping and spring constant) the argument develops as for an electrical RC degenerate system. The vibrator system of an ordinary string loop oscillograph is a mechanical system of this kind. Because the velocity $\dot{x}$ can change discontinuously and the coordinate x cannot, the system is adequate for following rapid changes of external impulses with negligible inertial disturbance.

In the case of an (m,k) degeneration (a system with finite inertia and damping but with a very small spring constant), the accleration $\ddot{x}$ can change discontinuously but not the velocity $\dot{x}$. Ballistic instruments are typical examples. In a ballistic galvanometer, for instance, a discontinuous impulse of very short duration is corresponded to by an equally discontinuous (of course, in the idealized sense) acceleration which results in a relatively slow building up of velocity and deflection *after* the disappearance of acceleration surge.

3. Conditions imposed by invariants

The use of degenerate d.e. in the discontinuous theory is supplemented by another important condition which results from the introduction of certain physical invariants and which completes the theory. This condition is commonly known as the *condition of Mandelstam.*

From the preceding examples illustrating the application of the theory of degeneration we notice that the variables which *cannot* change discontinuously in response to discontinuous changes in forcing term are those which enter into the expression of *stored energy*. For example, in the case of the RC degeneration we saw that the charge x cannot change discontinuously and, at the same time, we note that the stored energy in this case is purely electrostatic $E = \frac{1}{2}CV^2 = \frac{1}{2}xV$, where $x = CV$. In the case of the LR degeneration, the stored energy is $E = \frac{1}{2}Li^2$ and, again, it was found that the variable i cannot change discontinuously. On the other hand, dV/dt can change discontinuously and, therefore, also $i_c = CdV/dt$, where i_c is the current flowing in the condenser circuit. Likewise, di/dt can change discontinuously, which means that the voltage Ldi/dt across the inductance can also undergo discontinuous changes.

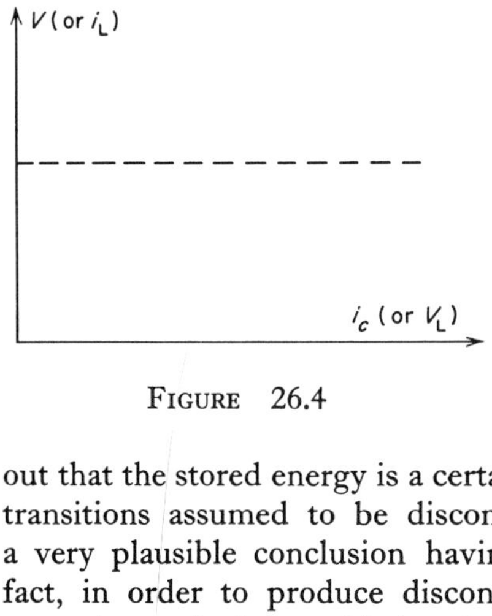

FIGURE 26.4

The fact that the variables entering into the expression of stored energy cannot change discontinuously, points out that the stored energy is a certain *invariant* in the quasi-discontinuous transitions assumed to be discontinuous in this idealization. This is a very plausible conclusion having an obvious physical meaning. In fact, in order to produce discontinuous changes in energy, an infinite *power* is required, but this is ruled out on obvious physical grounds.

It is thus clear that, if one wishes to represent a piecewise analytic phenomenon in the phase plane in the form of, say, two analytic arcs joined by discontinuous stretches, the latter must correspond to variables which can vary discontinuously. Thus, for instance, in the case of the RC degeneration, if one takes the variable $i_c = CdV/dt$ on the abscissa axis and V on the axis of ordinates, as shown in Fig. 26.4, the discontinuous stretches are possible in this representation along lines parallel to the abscissa axis, inasmuch as in this direction discontinuities are possible because the current in the condenser circuit can changes discontinuously and because the condition of Mandelstam regarding the invariant of the stored energy is fulfilled.

Similarly, in the case of the LR degeneration one can take on the abscissa axis the variable $V_L = Ldi/dt$ (the voltage across the inductance) and on the axis of ordinates the current i through the inductance, and the representation is the same as in the previous case.

Similar results can be obtained in degenerate problems in mechanics, as one ascertains easily following this argument.

If one overlooks the appropriate choice of variables, the piecewise analytic character of the problem, as defined in connection with the preceding examples, may escape notice. In the following chapter dealing with applications of the discontinuous theory this circumstance will be more definitely specified.

It is seen thus that the degeneration theory supplemented by the condition of Mandelstam permits the discovery not only that certain discontinuities are *possible* in certain variables, but also in *which direction* of the phase plane the discontinuity takes place.

Summing up, the considerations just outlined represent nothing but a further generalization of the classical theory of shocks in mechanics, the invariant of Mandelstam playing the role of the theorems of momentum and kinetic energy in the theory of shocks. In both cases certain "additional information" (not contained in the d.e. itself) permits connecting what exists before and after the discontinuity, intentionally ignoring what happens *during* the rapid transition period which is idealized by the mathematical concept of a discontinuity.

4. Discontinuous theory of Mandelstam–Chaikin

On this basis the discontinuous theory of relaxation oscillations presents itself in the following manner.

Assume that we have the d.e. of the form

$$dx/dt = X(x,y); \qquad dy/dt = Y(x,y) \tag{4.1}$$

where the functions X and Y have the form

$$X(x,y) = P(x,y)/T(x,y); \qquad Y(x,y) = Q(x,y)/T(x,y) \tag{4.2}$$

This form of functions X and Y has been ascertained also independently by T. Vogel whose theory is outlined in Chapter 29.

It is interesting to note that in practically all schemes known so far, one is led to the form (4.2) that will appear in a series of examples treated in the following chapter, whereas the van der Pol equation with its subsequent generalizations by Liénard, N. Levinson, and O. K. Smith does not enter directly into this class.†

† This is due, naturally, not to the *form* of these d.e. but to the fact that, in the usual (analytic) theory, the question of degeneration does not arise.

The essential point of the discontinuous theory in this particular formulation is that the discontinuity occurs at a point x_c, y_c, for which $T(x_c,y_c) = 0$. It is obvious that for such a point, which we shall call the *critical point*, the system (4.1) becomes meaningless.

It is also noteworthy that the existence of a critical point is in no way connected with any special values of the parameter because the latter is not involved in this case; this emphasizes once more the fact that the d.e. (4.1) have nothing to do with the van der Pol equation, once the theory of degeneration is used.

In some problems, instead of isolated critical points, there are certain *critical lines*. The description of the process in terms of the discontinuous theory proceeds then in the following way. A certain analytic arc is followed until it meets a critical line at some point. At this point the d.e. ceases to govern the phenomenon and a discontinuous stretch begins, being determined by the condition of Mandelstam. It ends at a point at which another analytic arc is encountered. Beginning with this point another analytic arc begins and the d.e. again takes charge of the phenomenon until another critical point is reached which results in another discontinuity which brings R to the first arc, etc. If this point is that at which the process started, the periodic process establishes itself at once. One has thus a piecewise analytic cycle which has no *limit* cycle feature. If, however, the process approaches the ultimate piecewise analytic cycle only after a series of rotations of the radius vector, one has a kind of a *piecewise analytic limit cycle*.

The discontinuous theory of relaxation oscillations has been checked experimentally, which adds a strong point in its favor. Once the appropriate variables are chosen, it is a very simple matter to establish the connections of a cathode ray oscilloscope so as to have the corresponding phase-plane diagram directly on the screen of the oscilloscope.

The formation of the d.e. follows directly from the Kirchhoff laws of the circuit used so that the property of the function $T(x,y)$ can be ascertained at once; if the function $T(x,y)$ vanishes at certain points of the phase plane, this gives always a reliable criterion for the existence of a piecewise analytic phenomenon; if this function does not vanish, at least in a finite domain of the phase plane, one is also certain that no piecewise analytic phenomenon is to be expected. In some cases, by introducing a parameter into the function T, it is possible either to produce the presence of its roots or, on the contrary, to remove these roots so that this function does not vanish anywhere in the phase plane. In such a case it is also possible to pass continuously from the piecewise analytic phenomena to the continuous oscillations owing to such continuous variation of parameter.

We mention in passing that in recent years it became a practice to use the term: *piecewise linear* idealization instead of: *piecewise analytic* idealization. The underlying concept is, of course, the same although the two terms do not relate to the same thing. The latter term relates to the integral curve (or trajectory) and should be understood in a geometrical sense (a continuously differentiable curve). The former concerns *the form* of the d.e. valid between the points at which the analyticity (in the above stated case) is lost. In other words, the essence of a piecewise linear method is in replacing an overall (generally complicated) d.e. by a number of *simple linear* d.e., the solutions of which are fitted by *continuity* without attempting to secure analyticity at the junction points.

Chapter 27

APPLICATION OF THE DISCONTINUOUS THEORY TO ELECTRICAL PROBLEMS

In this chapter we shall review a series of examples which the reader can find in the Andronov and Chaikin book. Most of these examples were worked out by Chaikin and Lochakov[1] on the basis of the discontinuous theory outlined in the preceding chapter. In this review these problems are somewhat abridged with a view to establishing connections with the discontinuous theory.

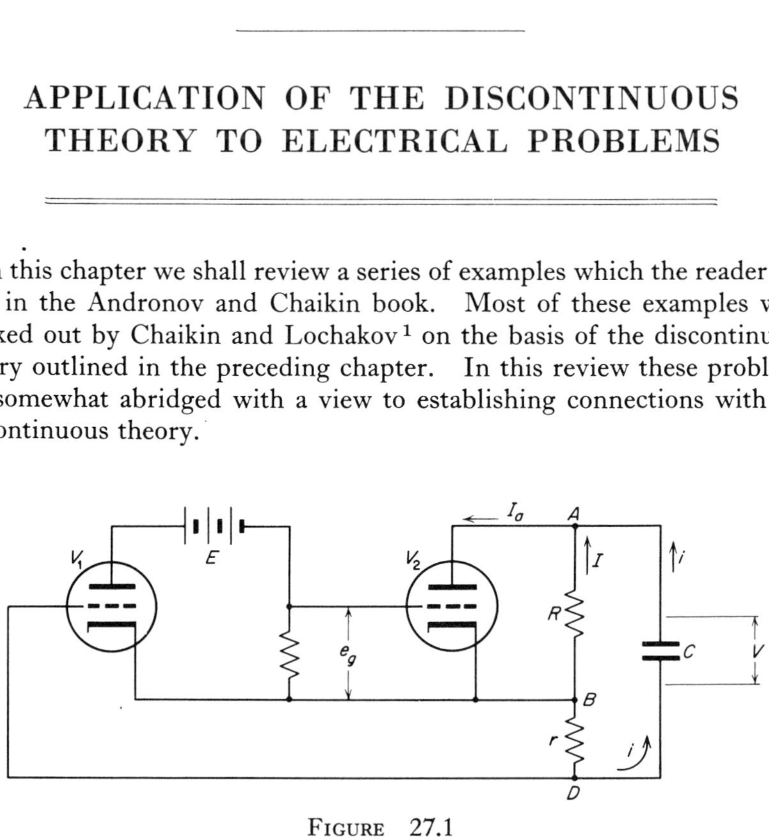

FIGURE 27.1

1. Degenerate *RC* oscillator[2]

Referring to Fig. 27.1, the electron tube V_2 may be regarded, as usual, as a nonlinear conductor of the circuit characterized by the equation $I_a = \varphi(e_g)$ where I_a is the anode current and e_g is the grid voltage. The

[1] S. Chaikin and L. Lochakov, *J. Tech. Phys.* (USSR) **11**, 1935.

[2] E. Friedlander, *Arch. f. Electrot.* **17**, 1926; **18**, 1926; B. van der Pol, *Phil. Mag.* (7) 2, 1926; *Zeitschr. f. Hochfr. Tech.* **28**, 1927.

tube V_1 appears here merely as a linear amplifier amplifying the potential difference ri between the points B and D and applying the amplified voltage $e_g = kri$ to the grid of V_2, the factor of amplification (or "gain") due to V_1 being k. The circuit is idealized in a customary manner, that is, the effects of the grid current and the anode reaction are neglected.

The fundamental assumption here is that the effect of the small inductance is neglected. This means that from the very beginning one places oneself under the conditions of the RC degenerations using the terminology of the preceding chapter.

The equations of the circuit here are obviously

$$(R + r)i + V = R\varphi(kri); \qquad i = C \cdot dV/dt \tag{1.1}$$

where the positive directions on the circuit, as well as the significance of the various constants, seen from Fig. 27.1. The quantity V is the voltage across the condenser.

The equations (1.1) reduce to one single d.e. of the first order

$$[krR\varphi'(kri) - (R + r)]di/dt = i/C \tag{1.2}$$

where φ' designates the derivative of the function φ with respect to i. The quantity in the bracket is the function T of the preceding theory.

The critical point (if any) is given by the root of $T = 0$; that is,

$$T(i_1) = [krR\varphi'(kri_1) - (R + r)] = 0 \tag{1.3}$$

The simplest way to determine this root is by a graphical procedure shown in Fig. 27.2. Part (a) of this figure shows the plot of RI_a for the tube V_2 assumed to be biased at its inflection point O taken as the origin of coordinates. If one subtracts from the ordinates of this curve the corresponding ordinates of the straight line $(R + r)i$, one has the curve $V(i)$ representing $R\varphi(kri) - (R + r)i$. The slope curve of the function $R\varphi(kri)$ multiplied by kr and referred to $M'N'$ axis is shown in Fig. 27.2b. If, finally, one subtracts from this slope curve the quantity $(R + r)$ (which amounts to referring this curve to MN axis), one obtains the function $T(i)$ and it is seen that, in view of symmetry, there are two roots $i = i_1$ and $i = -i_1$. The points P and Q are thus the critical points.

Transferring these points to the (a) portion of the figure on the curve $V(i)$, one obtains the location of the critical points at B and D.

We note in passing that this graphical construction could be eliminated if one uses the polynomial approximation and carries out the corresponding calculations.

It is noted that the function $T(i)$ is positive inside the interval $(i_1, -i_1)$

and negative outside this interval. From (1.2) it follows that the phenomenon is unstable in this interval and is stable outside it. This is indicated by the arrows on the curve $V(i)$ (Fig. 27.2a).

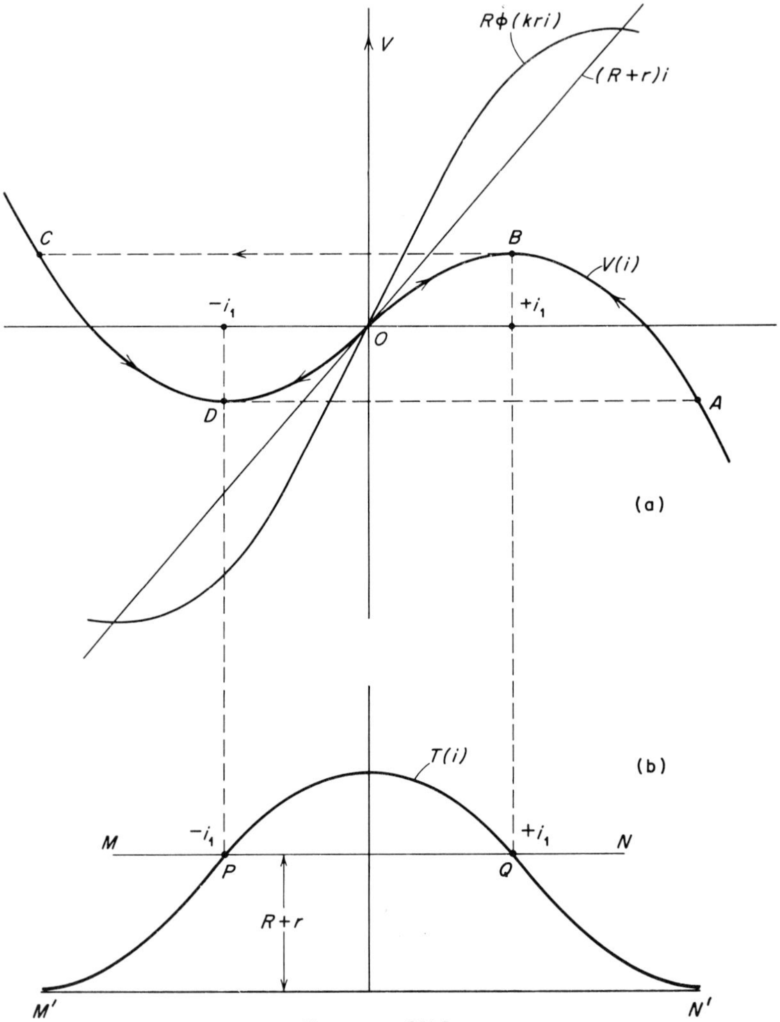

FIGURE 27.2

The discussion follows now the familiar argument. If R follows the analytic arc CD (from C to D), at D one has $di/dt = \infty$, which indicates the discontinuity at this point. The d.e. ceases to govern the motion of R

at this point and one has to use the condition of Mandelstam. Since the stored energy is electrostatic in this case (we are dealing with a degenerate problem neglecting the small inductance of the circuit), the jump occurs parallel to the abscissa axis so as to leave $V(i)$ invariant. At the point A the discontinuous stretch meets the analytic arc and the motion of R is governed again by the d.e. The analytic arc is followed up to the point B which is again a critical point which results in another jump BC.

This determines the piecewise analytic cycle $ABCD$ consisting of two analytic arcs CD and AB joined by the discontinuous stretches BC and DA.

2. Oscillator with two degrees of freedom with degeneration in each degree

If one replaces in Fig. 27.1 the resistor R by an inductance L, the oscillatory properties of the circuit are entirely different. Figure 27.3

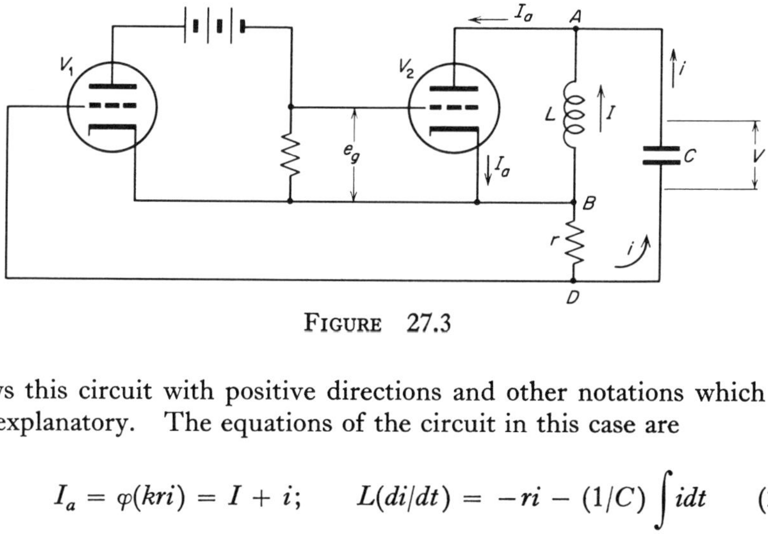

FIGURE 27.3

shows this circuit with positive directions and other notations which are self-explanatory. The equations of the circuit in this case are

$$I_a = \varphi(kri) = I + i; \qquad L(di/dt) = -ri - (1/C)\int idt \qquad (2.1)$$

In this case there are both forms of the stored energy, one in the inductance L and the other in the capacity C, but here the relations are more complicated because L and C do not belong to the same circuit.

The condition of Mandelstam holds, however, in both cases and this permits to assert that the variables I and V cannot change discontinuously; but dI/dt and i can undergo such changes, which suggests the choice of the discontinuous variables $x = kri$ and $y = dI/dt$. With these variables the system (2.1) takes the form

$$dx/dt = y/\psi(x); \qquad dy/dt = [x/krLC] + [(1/kL)y/\psi(x)] \qquad (2.2)$$

where $\psi(x) = \varphi'(x) - 1/kr$; $\varphi'(x) = (d/dx)\varphi(x)$.

The critical point obviously exists here and is given by the root of $\psi(x_1) = 0$; this occurs when $\varphi'(x_1) = 1/kr$. If the nonlinear characteristic is approximated by a polynomial, one can continue the calculation analytically from this point but we shall follow the exposition of the original paper which involves a graphical procedure. In fact, $\varphi'(x)$ is the slope of the characteristic and the condition for the existence of the critical point merely means a point x_1 for which the slope of $\varphi(x)$ has the value $1/kr$. Since the slope of the characteristic of a triode is maximum at its inflexion point and decreases monotonically on both sides of this point with increasing $|x|$, approaching zero for $|x| \to \infty$, it is clear that $\varphi'(x)$ is contained in the interval (a,O), a being the maximum value of the slope (at the inflexion point). Hence, if the constant $1/kr$ is in that interval, $\psi(x)$ has a root and, therefore, the critical point exists. If it is outside this interval, there is no critical point. In the first case the relaxation phenomenon exists and, in the second, it does not.

We assume, therefore, that the constant $1/kr$ is in the interval $(a,0)$. As soon as the critical point $x = x_1$ is reached, dx/dt and dy/dt are infinite, which indicates the discontinuity. Since both x and y are discontinuous variables, the jump can occur in any direction in the phase plane and not necessarily in one direction only, as this was the case in the preceding section where only one variable of the phase plane was discontinuous.

In order to determine the direction of the discontinuity one has to apply the condition of Mandelstam to the variables which cannot vary discontinuously, that is, I and V. We have thus

$$\Delta I\Big|_{t_0-0}^{t_0+0} = \int_{t_0-0}^{t_0+0} y\,dt = 0; \qquad \Delta V\Big|_{t_0-0}^{t_0+0} = \frac{1}{krC}\int_{t_0-0}^{t_0+0} x\,dt = 0 \qquad (2.3)$$

where t_0 is the instant at which $\psi(x)$ passes through its root. If one applies the conditions (2.3) to (2.1), one has

$$\varphi(x_1) - x_1/kr = \varphi(x_2) - x_2/kr; \qquad y_1 - y_2 = (x_1 - x_2)/kL \qquad (2.4)$$

where x_1 and x_2 are the initial and the terminal points of the discontinuous stretch and, likewise, y_1 and y_2 are the corresponding ordinates of these two points. Since one knows the initial point (x_1,y_1) which is the critical point, the terminal point is determined by these two equations. In this manner the condition of Mandelstam permits determining the point into which R jumps, once it has reached the critical point.

The direction of the jump is obtained if one eliminates t between the two d.e. (2.2) which gives the d.e. of the integral curves

$$dy/dx = x\psi(x)/krLCy + 1/kL \tag{2.5}$$

and it is seen that at the critical point $x = x_1$

$$dy/dx = 1/kL \tag{2.6}$$

which determines the direction of the jump, once R has reached the critical point.

The fact that the constant $1/kr$ is contained in the interval (a,O) suggests a further simplification which is often called "the method of broken characteristic."† In this case it amounts to replacing the actual characteristic of the electron tube (Fig. 27.2) by two straight lines with two limiting slopes a and O. For the sake of continuity one can join these straight lines by a small curve, as shown in Fig. 27.4, which thus determines a narrow shaded region around each critical point ($x = x_1$ and $x = -x_1$). The analysis is confined to the inner interval limited by these shaded strips and to the outside intervals. Since in the shaded strips one does not attempt to determine what happens, one can reduce their width as much as one wishes and the argument is confined to the investigation of what happens in the inner and in the outer intervals, respectively. Under this assumption, in the inner interval $\psi(x) = a - 1/kr$, and in the outer intervals $\psi(x) = -1/kr$. With this artificial simplification of the problem the system (2.2) can easily be discussed and one finds that, inasmuch as there is only one singular point $x = y = 0$ in this case, it is a saddle point if x is in the inner interval, and a stable singularity (either a focal or a nodal point according to the relative values of the parameters) if x is in the outer intervals.

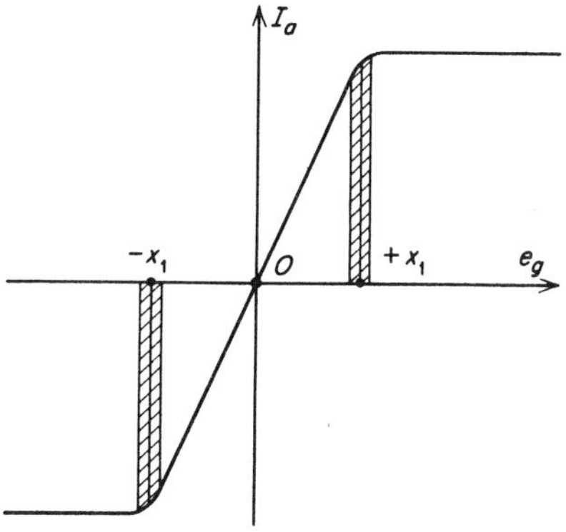

FIGURE 27.4

One is thus led to ascertain that the differential system (2.2) varies continuously *as far as its form is concerned as a function of its own dependent variable* x, but the method of the broken characteristic simplifies this

† This method can be developed along a different line as will be seen in Chapter 31.

situation by retaining out of the infinite number of possible forms only two forms that correspond to the maximum and minimum values of the function $\psi(x)$ and proceeds on this basis only.

It is to be noted that the problem in this case is two-dimensional (that is, capable of being represented in the phase plane), inasmuch as there exists no relation between the two variables x and y, as was the case in the preceding section where the differential system was reducible to one d.e. of the first order; this results in a possibility of a limit process.

We proceed now with the representation of the phenomenon in the phase plane, with reference to Fig. 27.5. In the outside intervals the

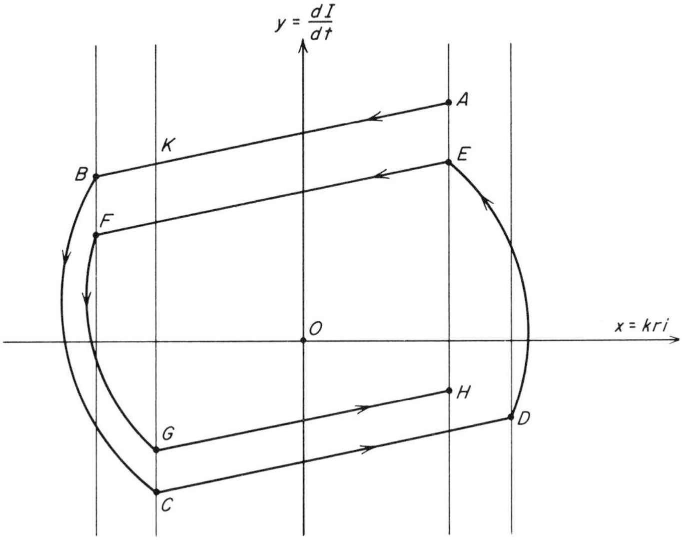

FIGURE 27.5

integral curves are either logarithmic spirals (if the origin is a focal point) or distorted parabolas, if it is a nodal point. We assume the first case, for the sake of an example.

Assume that R follows a spiral trajectory and reaches a critical point A on the critical line $x = x_1$. From this point it executes a jump specified by (2.6) to the point B, which is determined by the condition of Mandelstam. At this point B the representative point R is again on the integral curve which is followed until R reaches the point C on the critical line $x = -x_1$ resulting in another jump CD, etc. It can be shown by a detailed argument based on the analysis of the intercepts of the piecewise analytic trajectory of this kind that, if one starts with a point A having a sufficiently large value of the ordinate, the piecewise analytic curves shrink with the subsequent rotations of the radius vector. If, however, one starts with A sufficiently near to the abscissa axis, the curves, on the contrary, grow in size, which shows that there exists a unique closed trajectory with a re-entrant path on which the ultimate stationary state is established.†

The experiment corroborates these conclusions. In fact, if one connects the deflecting plates of a cathode-ray oscilloscope so as to represent the variables x and y, one observes two arcs of spirals with an empty interval between them, as is seen from Fig. 27.6, which indicates that in this inner interval the motion of the electron beam is so fast that the fluorescent material of the screen has no time to respond to the passage of the beam. It is interesting to note that, although the inner interval corresponds to the existence of the saddle point in the d.e., the hyperbolic trajectories of this point have nothing to do with the actual motion of R which is governed in this region by the condition of Mandelstam and not by the d.e.

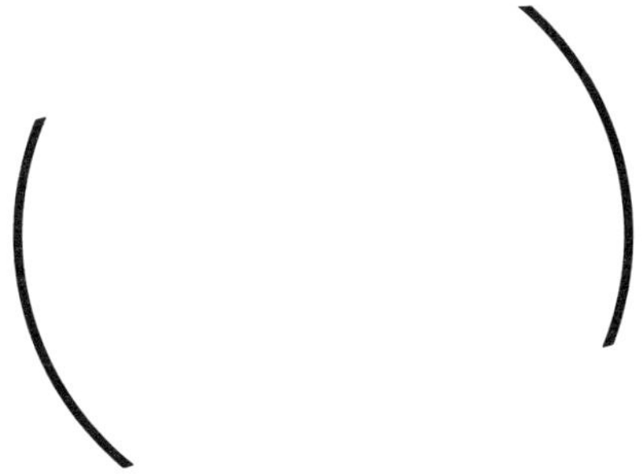

FIGURE 27.6

Another remark is noteworthy: in this case the system of two d.e. of the first order is given *directly* by the Kirchhoff laws of the circuit and cannot be reduced to the d.e. of the second order.

In the case of the van der Pol equation as well as its more recent generalizations by Liénard, Levinson-Smith and others, the matter is different, inasmuch as the d.e. of the second order *is known* and the equivalent system is obtained by the usual procedure by setting $\dot{x} = y$. Moreover, the passage from continuous oscillations to relaxation oscillations is effected by varying the parameter in the d.e., whereas in the present case the existence of relaxation oscillations depends on the existence of critical points (or lines) and these are related rather to the form of the d.e. and do not depend on any parameter. This circumstance arises from the use of the degeneration theory which reduces a d.e. of the second order to that of the first order. Since in this case the system has two degrees of freedom, there are two such equations. It is clear that, without the degeneration procedure, the oscillatory system in this case would be amenable to a differential system of the fourth order and a representation in the phase plane would be impossible.

† The procedure outlined in connection with Fig. 27.5 is frequently called the *piecewise linear* procedure. We shall enter more fully into this subject in Chapter 31 but without involving any discontinuous idealization.

3. Connection between critical points and piecewise analytic phenomena

In the preceding sections it was shown that the zeros of the function T result in the appearance of what is called here "critical points," and the latter, in turn, appear as a criterion for the existence of discontinuous stretches in the piecewise analytic representation.

Conversely, it can be shown that if $T(x,y)$ (in 4.2, Chapter 26), does not go through zero, the piecewise analytic character of oscillation disappears. An interesting experimental confirmation of this conclusion appears in the so-called Heegner circuit [3] which represents a slight modification of the RC oscillator studied in Section 1. This modification consists in shunting the resistor R by an additional condenser C_1. As the rest of the circuit remains the same, we indicate in Fig. 27.7 only the modified part of the circuit to the right of ABD in Fig. 27.1. The Kirchhoff laws for the circuit shown give

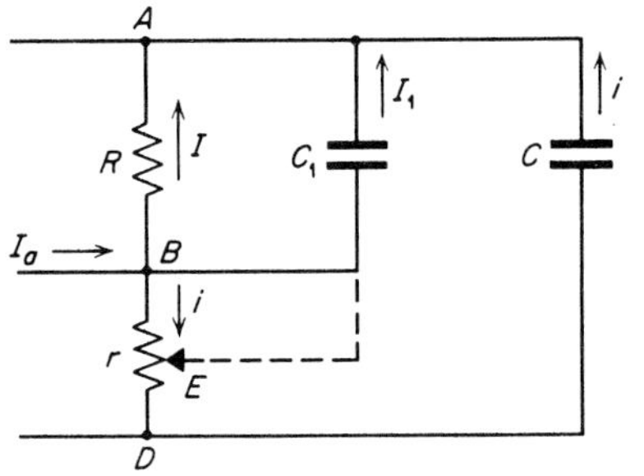

FIGURE 27.7

$$I_a = I + I_1 + i;$$
$$i = C \cdot d/dt(RI - ri);$$
$$I_1 = C_1 R dI/dt$$

Setting $I_a = \varphi(kri)$ as previously and eliminating I, one has the following system of the second order

$$\begin{aligned} \frac{di}{dt} &= -\frac{1}{rC} i + \frac{1}{rC_1} I_1; \\ \frac{dI_1}{dt} &= \frac{1 - rk\varphi'(kri)}{rC} \cdot i - \frac{R + r - Rrk\varphi'(kri)}{RrC_1} \cdot I_1 \end{aligned} \tag{3.1}$$

and it is seen that this system has no critical points and, therefore, piecewise analytic oscillations are impossible, which is also corroborated experimentally.

A slight modification of this circuit permits obtaining a critical threshold at which an analytic periodic solution of the system (3.1) gives way to a piecewise analytic one. It is sufficient for this purpose to connect the circuit of the condenser C_1, instead of to point B, to a potentiometer across the resistor r, as shown in the dotted line. If one writes Kirchhoff's laws for this circuit, one finds that in the denominators of the coefficients of I_1

[3] K. Heegner, *Zeitschr. f. Hochfr. Tech.* **29**, 1927.

and i appears the factor $(1 - \beta)$, where $\beta = r_1/r$, r_1 being the resistance between B and E.

It is clear that, for $\beta = 0$, one has the Heegner circuit whose solutions are analytic; and for $\beta = 1$, one has an ordinary RC multivibrator circuit (Section 1) which has only piecewise analytic solutions. Hence, in the interval (0,1) of β there is a point at which one type of solutions gives way to the other one. This is easily ascertained by a cathode-ray oscilloscope which shows that the continuous closed curve begins to be interrupted by a small discontinuity that gradually grows as β approaches the value 1.

4. Symmetrical multivibrator circuits

The classical circuit of this kind is the so-called Abraham–Bloch multivibrator[4] shown in Fig. 27.8. It consists of two identical RC multi-

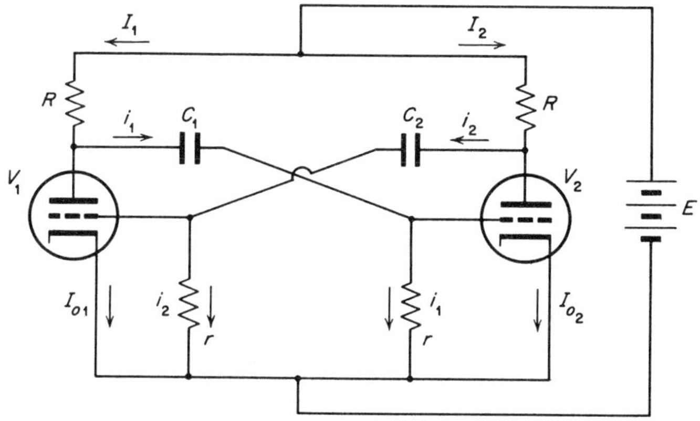

FIGURE 27.8

vibrators cross-connected as shown. Kirchhoff's laws give here

$$I_1 = I_{01} + i_1; \qquad I_2 = I_{02} + i_2;$$

$$RI_1 + \frac{1}{C}\int i_1 dt + ri_1 = E; \qquad RI_2 + \frac{1}{C}\int i_2 dt + ri_2 = E \tag{4.1}$$

and, as usual,

$$I_{01} = \varphi(e_{g1}) = \varphi(ri_2); \qquad I_{02}(e_{g2}) = \varphi(ri_1)$$

[4] N. Abraham and E. Bloch, *Ann. de Physique* **12,** 1919.

From these circuit equations one obtains the following system of d.e.

$$(R + r)\frac{di_1}{dt} + \frac{1}{C}i_1 + Rr\varphi'(ri_2)\frac{di_2}{dt} = 0$$
$$Rr\varphi'(ri_1)\frac{di_1}{dt} + \frac{1}{C}i_2 + (R + r)\frac{di_2}{dt} = 0 \qquad (4.2)$$

which is reduced to the form

$$\frac{di_1}{dt} = \frac{(R + r)i_1 - Rr\varphi'(ri_2)i_2}{C[\gamma(i_1,i_2) - (R + r)^2]}; \qquad \frac{di_2}{dt} = \frac{(R + r)i_2 - Rr\varphi'(ri_1)i_1}{C[\gamma(i_1,i_2) - (R + r)^2]} \qquad (4.3)$$

where

$$\gamma(i_1,i_2) = R^2r^2\varphi'(ri_1)\varphi'(ri_2)$$

The system (4.3) is again of the form

$$di_1/dt = P(i_1,i_2)/T(i_1,i_2); \qquad di_2/dt = Q(i_1,i_2)/T(i_1,i_2) \qquad (4.4)$$

The only singular point at a finite distance is the origin ($i_1 = i_2 = 0$). Setting $\varphi'(ri_1) = \varphi'(ri_2) = S$, one has $\gamma(0,0) = (RrS)^2$, which gives two possibilities: either $(RrS)^2 - (R + r)^2 = k^2 > 0$, or $(RrS)^2 - (R + r)^2 = -k^2 < 0$. In the first case the origin is a saddle point and, in the second, it is a stable nodal point which is of no interest here. If we consider, therefore, the case when $RrS > R + r$, both i_1 and i_2 increase initially, the position of equilibrium being unstable. Since for $i_1 = i_2 = 0$, the function $\gamma(i_1,i_2)$ has a maximum and decreases monotonically thereafter, it is clear that a moment will be reached when $\gamma(i_1,i_2) = (R + r)^2$ to which corresponds the zero of the function $T(i_1,i_2)$ and, therefore, a critical point of the system (4.4).

From that moment the phenomenon acquires a piecewise analytic character and can be investigated by the discontinuous method.

The locus of critical points i_1', i_2' is given by the equation

$$F_1(i_1',i_2') = \gamma(i_1',i_2') - (R + r)^2 = 0 \qquad (4.5)$$

Inasmuch as $\varphi'(0)rR = SrR > R + r$ and $\varphi'(ri)$ decreases monotonically with i increasing, the curve F_1 is a closed curve symmetrical with respect to the origin and containing it in its interior, as shown in Fig. 27.9.

The point (i_1'',i_2'') into which the representative point R jumps once it has reached the critical point (i_1',i_2'), is determined again by the condition of Mandelstam. As the only form of stored energy here is electrostatic, the voltage V across condensers remains invariant during the jump, which results in the relations

$$V_1 = E - R\varphi(ri_2) - (R + r)i_1; \qquad V_2 = E - R\varphi(ri_1) - (R + r)i_2 \qquad (4.6)$$

The conditions of the invariance of V during the discontinuity are thus

$$\begin{aligned} R\varphi(ri_2') + (R + r)i_1' &= R\varphi(ri_2'') + (R + r)i_1'' \\ R\varphi(ri_1') + (R + r)i_2' &= R\varphi(ri_1'') + (R + r)i_2'' \end{aligned} \tag{4.7}$$

There exists thus a (1,1) correspondence between (i_1', i_2') before the discontinuity and (i_1'', i_2'') after it. This results in another locus F_2 of points (i_1'', i_2'') which is also symmetrical with respect to the origin.

The piecewise analytic phenomenon takes place then in the following

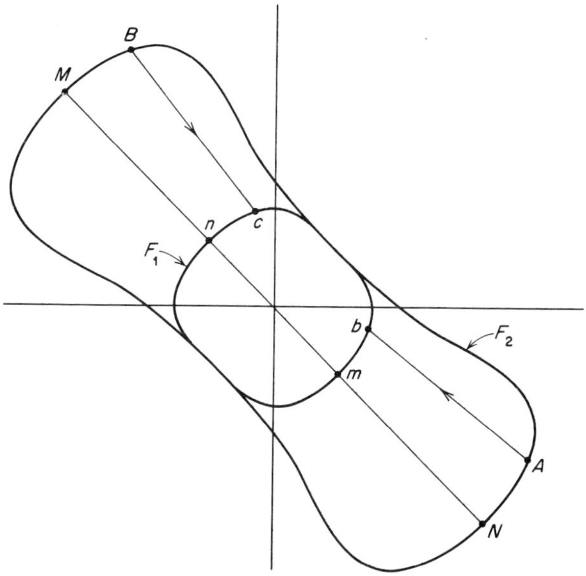

FIGURE 27.9

manner. From some point a on F_1, the point R jumps into the corresponding point A on F_2. From this point begins a continuous motion on the stretch Ab. At b begins another jump which transfers R to the point B on F_2, from where begins another continuous stretch Bc, and so on.

It can be shown that, after a series of jumps, the motion approaches the bisector line MN so that ultimately the stationary state consists of a continuous motion Mn followed by a jump nN followed again by a continuous motion Nm, etc.

The proof that the motion approaches ultimately the symmetrical motion along the bisecting line MN is relatively complicated but, if one *assumes* that, on account of the symmetry, one can expect that the motion should be symmetrical, the matter becomes very simple. In fact, it is sufficient to set $i_1 = -i_2$ and $\varphi'(kr_1) = -\varphi'(kr_2)$ in the preceding equations, which results in one single d.e.

$$\frac{di}{dt} = \frac{(R + r) + rR\varphi'(ri)}{[r^2R^2\varphi'^2(ri) - (R + r)^2]C} \cdot i \tag{4.8}$$

which has been investigated in Section 1.

Summing up, the degeneration theory permits in this case reducing a physical system describable by two d.e. of the second order to that of two d.e. of the first order. The condition of symmetry permits a further reduction of the last mentioned system to one single d.e. of the first order.

The advantage of the use of the degenerate d.e. can be seen clearly at this point. In fact, if one had to proceed along the line of the asymptotic theory (outlined in Chapter 30) which does not use the simplification afforded by the degeneration theory, one would have in this case a differential nonlinear system of the 4th order and the problem would be extremely complicated and impossible to be represented in the phase plane, as we mentioned previously.

Another remark is noteworthy. If one follows this procedure, one obtains directly the system (2.2) of two d.e. of the first order which becomes meaningless for $\psi(x) = 0$; with this point the degeneration theory associates the appearance of a discontinuity.

If, however, one considers the d.e. (2.5) of the corresponding integral curve, nothing whatever occurs for this *critical point* inasmuch as the value of dy/dx remains finite at this point. In other words, in a system with two degrees of freedom, such as the one considered here, discontinuities occur in such a manner as to preserve not only the continuity of the integral curve but even its analyticity; this is another consequence of the physical postulate of Mandelstam supplying the "additional information" without which the discontinuous theory would not be complete.

5. Concluding remarks

In this chapter we have reviewed a few typical examples of relaxation problems treated on the basis of the discontinuous theory. It is useful to analyze closer these results inasmuch as they may appear at variance with what we have learned in Chapter 3 in connection with the topological aspects of the analytical theory.

We recall briefly these aspects:

(1) In physical problems a closed integral curve (or trajectory) always represents a periodic phenomenon.

(2) Singular points are identified with positions of equilibria.

(3) A closed curve (in the phase plane) can be a trajectory only in the case when the algebraic sum of indices of singular points situated in its interior is $+1$.

These conclusions result from the analysis of d.e. of the second order which yields a simple representation in the phase plane. The assertion (2) was particularly valuable, for applied problems treated in Part I. As regards (3), it gives a necessary condition for the existence of a periodic motion. Thus, for instance, if one is able to show that the closed curve contains in its interior only one singular point with index -1, one can assert that such a closed curve cannot be a trajectory.

These topological conclusions cease to hold in the case of d.e. of the first order because such d.e. have no singular points; moreover, a system of such d.e. in general cannot be reduced to a d.e. of a higher order, and one has to deal directly with the system so that the question of singular points does not arise. The assertion (3) then becomes meaningless.

In this manner the advantage of treating quasi-discontinuous phenomena as discontinuous ones results in giving up the topological foundations of the theory with which we became familiar in Chapter 3. Only the concept (1) remains, as we have mentioned in connection with the piecewise analytic cycles.

Thus, for instance, in the case of a ball striking two walls (Fig. 26.1), the "closed trajectory" is a rectangle with two analytic sides (that is, the trajectories of the d.e.: $\ddot{x} = 0$) closed by two discontinuous stretches. In this case there is no singular point inside the rectangle and the motion is still periodic. Here one can argue that this is not an example of an autonomous system, since the reactions of the walls may be regarded as external forces. However, in the case of an asymmetrical multivibrator (Section 1 of this chapter) the closed trajectory contains in its interior only a point of an unstable equilibrium which is not a singular point, since in this case one has only one d.e. of the first order.

In a system with two degrees of freedom with the degenerate d.e. in each degree the situation is still more complicated *if viewed from the standpoint of the classical theory*; here the origin behaves as a stable focus if x is large enough, and as a saddle point if x is less than a certain critical threshold x_1.

All this does not mean that the discontinuous theory is "wrong" but merely that the change in the fundamentals of the theory results in a series of changes in its details.

On the other hand, if the criterion of validity of a physical theory is its ability to explain the observed facts and even, in some cases, to predict the new ones, then the discontinuous theory satisfies these requirements practically in all cases and, from that point of view, it must be regarded as a correct theory within its own scope.

Chapter 28

APPLICATION OF THE DISCONTINUOUS THEORY TO MECHANICAL PROBLEMS

1. Introductory remarks

The mechanical piecewise analytic phenomena are less explored at present than the electrical ones and, moreover, have no practical applications.

Most of these effects are due to the various manifestations of friction but, as the latter is still neither a predictable nor a controllable phenomenon, the corresponding studies are limited to rather qualitative aspects of mechanical relaxation oscillations.

The discontinuous theory is still a useful guide in these studies, and we propose to give here a brief review of an interesting investigation by Kaidanovsky[1] reported also in Andronov and Chaikin's book.

The theory of degeneration plays again an essential role in these studies. We have already touched this subject in Section 2, Chapter 26, where we mentioned (k,c) and (m,k) degenerations analogous to RC and LR degenerations in electrical problems. In these two forms of degeneration the quantity k, the damping coefficient, appears generally in the form of a nonlinear friction of some kind and it is generally in connection with this term that the relaxation features appear.

There is another circumstance worth mentioning. In the outline of the theory of degeneration we were primarily interested in specifying certain variables which admit a discontinuous response to an equally discontinuous stimulus. Here we are interested in a somewhat different aspect of the same problem, namely: What kind of motions may be expected in a degenerate mechanical system without any external action? This means that the interest of this study lies chiefly in the exploration of a possibility of relaxation oscillations of an essentially autonomous mechanical system.

[1] N. L. Kaidanovsky and S. E. Chaikin, *J. Tech. Phys.* (USSR) **3,** 1933.

It will be shown that this possibility presents itself owing to the nonlinear character of friction.

2. Mechanical relaxation oscillations

We consider a nonlinear d.e. of the form

$$m\ddot{x} + \mu f(\dot{x}) + cx = 0 \tag{2.1}$$

and assume that the inertia term is small compared to the damping and the restoring force terms. In order to bring the notations in accordance with previous discussion of degenerate equations, we write the d.e. (2.1) in the degenerate form as

$$F(\dot{x}) + cx = 0 \tag{2.2}$$

It is noted that the damping term is considered here as a certain function of velocity $\dot{x}$. Differentiating this equation and setting $\dot{x} = y$, one has

$$\dot{y} = dy/dt = -cy/F_y(y) \tag{2.3}$$

where $F_y = dF/dy$. If $F_y < 0$, $\dot{y}$ has the same sign as y (since $c > 0$) and this shows that the motion is unstable, because both the velocity and acceleration act in the same direction. For the same reason, if $F_y > 0$, the motion is stable. If for some value $y = y_1$, $F_y(y_1) = 0$, $dy/dt = \infty$, which, according to the discontinuous theory, shows that $y = y_1$ is a critical point; in the present case it is a point at which the acceleration is infinite. To some extent the situation resembles that considered previously of an elastic ball striking an equally elastic wall. There is, however, a difference between the two. In the case of the ball, the discontinuity in acceleration is due to the contact with the wall, that is, to an external impulsive action; whereas here it is obviously due to the loss of the dynamical equilibrium *inside* the system.

Since in this case the energy is purely potential, ($c \neq 0$, $m = 0$), the condition of Mandelstam indicates that the discontinuity can appear only in such variables which do not enter into the expression of potential energy $E = \frac{1}{2}cx^2$. Thus, the coordinate x cannot change discontinuously, although this restriction does not apply to the velocity $\dot{x}$, which can undergo a discontinuous change.

3. Relaxation oscillations of a Prony brake

The following experimental arrangement due to Chaikin and Kaidanovsky [1] illustrates the application of the discontinuous theory.

[1] See footnote [1], page 632.

A Prony brake a of a small mass (Fig. 28.1) engages frictionally a shaft K rotating with a constant angular velocity Ω. The brake is constrained by the spring S; the proper amount of friction is secured by means of an additional radial constraint not shown in Fig. 28.1.

Since the experimental arrangement is such that the mass of the brake is negligible in comparison with damping and restoring force, the case in question may be regarded as (k,c) degeneration according to our previous terminology (Chapter 26), and the degenerate d.e. here is

$$rF(v) = c\varphi \tag{3.1}$$

where $rF(v)$ is the moment due to the frictional drag applied to the brake, and $c\varphi$ is the restoring moment due to the spring S and considered as a simple proportional law with the angle φ through which the brake is

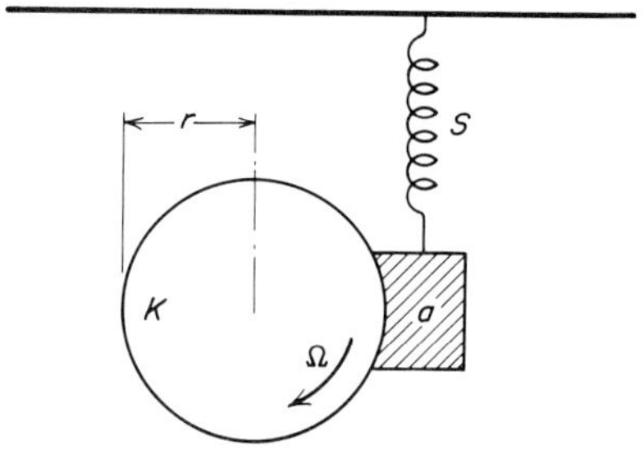

FIGURE 28.1

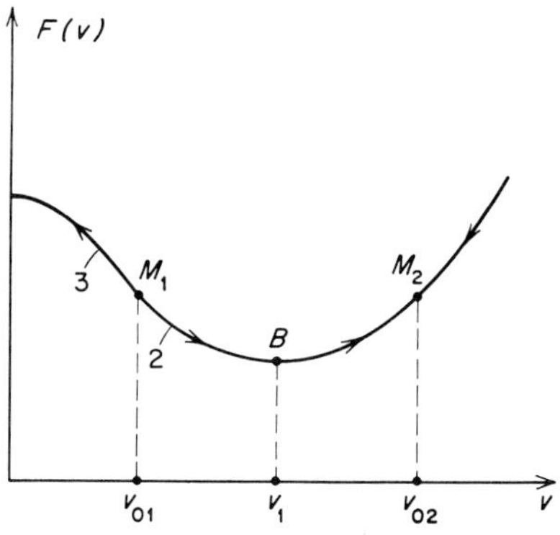

FIGURE 28.2

displaced by the friction drag. Since $v = (\Omega - \dot{\varphi})r$, the preceding equation can be written as

$$rF[(\Omega - \dot{\varphi})r] = c\varphi \tag{3.2}$$

and the question centers now on the form of the nonlinear function $F(v)$.

According to Sommerfeld, the friction force in this case (the presence of oil lubrication is assumed) passes through a minimum for a certain value $v = v_1$ which can be expressed in terms of certain parameters, such as thickness of the oil film, coefficient of viscosity, etc. The essential point here is the existence of such a minimum which accounts for the form of the function $F(v)$ shown in Fig. 28.2.

Differentiating (3.2), one has

$$\ddot{\varphi} = -e\dot{\varphi}/F'(\Omega - \dot{\varphi}) \tag{3.3}$$

where F' designates the derivative of F with respect to φ, and the constant $e > 0$ absorbs the factors r and c.

One has to ascertain first what kinds of motions are possible. It is clear that two principal types of motion are of interest here:

(1) If $v = 0$, $\Omega = \dot{\varphi}$. In this case the brake is displaced by the rotating shaft; this is the phase of a static friction. During this phase the representative point R moves along the $F(v)$ axis (Fig. 28.2) and the spring S is gradually stretched.

(2) If $v = v_0 = r\Omega$, $\dot{\varphi} = 0$. This corresponds to the standing still of the brake. In this case the point v_0 can be either to the right or to the left of the point $v = v_1$ corresponding to the minimum of the friction force $F(v)$. These two possibilities are shown by points M_1 and M_2 in Fig. 28.2.

Consider first the point M_1. For $v < v_{01}$ one has $r\Omega > r(\Omega - \dot{\varphi})$, $\dot{\varphi} > 0$. The motion of the brake a in this case is in the same direction as that of the shaft K. If $v > v_{01}$, the brake moves in the direction opposite to K. On the other hand, since the point M_1 is on the descending branch of the curve and taking into account (3.3), it follows that $\ddot{\varphi} > 0$. Thus, in this case the motion of the brake is unstable since both the velocity and acceleration have the same sign. The same argument for $v < v_{01}$ shows that again the velocity and acceleration have the same sign, so that the motion is unstable. For the point M_2 on the ascending branch of $F(v)$ the same argument shows, on the contrary, that the motion is stable. We indicate the conditions of stability by arrows on the curve $F(v)$. It is recalled that the curve $F(v)$ is actually the integral curve since, the d.e. being of the first order, there is a definite relation between the variable and its derivative. In other words, the characteristic $F(v)$ being the integral curve at the same time, the phase-plane representation is not involved and we have here a unidimensional representation by the *phase-line* (or phase-curve) $F(v)$.

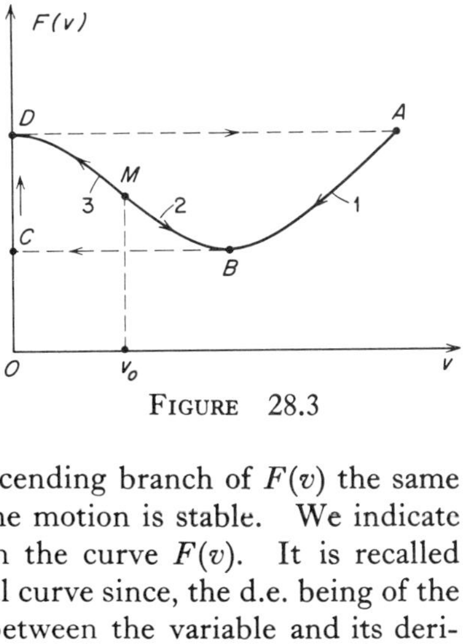

FIGURE 28.3

In considering the *production* of relaxation oscillations, only the location of M on the descending part of the curve is of interest, and in considering their *elimination*, its location on the ascending part of the curve is necessary.

Figure 28.3 shows what happens when M is on the descending branch.

Assume that initially R is somewhere on the curve $F(v)$ where v is large (point A). Since v is large and positive, φ is negative and large.

From (3.3) it follows that $\ddot{\varphi} > 0$ so that $|\dot{\varphi}|$ decreases initially as well as v which is indicated by the arrow 1. The point R follows the curve F in view of the assumed degeneration. At the point B the acceleration is infinite since F' is zero at this point. As R cannot follow the curve $F(v)$ in view of the opposite direction prescribed by the d.e. (arrow 2), it has to jump under the condition of Mandelstam. The only form of stored energy here is the potential energy of the stretched spring so that the jump occurs horizontally from B to C since the axis of ordinates is also a part of the characteristic corresponding to the static friction when the brake and the shaft have no relative velocity.

Once R is on the axis $F(\bar{v})$, the rotation of the shaft drags the brake with it, thus stretching the spring. During this static phase of the process, r moves along the $F(v)$ axis and the spring is stretched. At the point D the limit of the static friction is reached and $\ddot{\varphi}$ becomes again infinite. Since the motion along the integral line (the characteristic) is impossible (arrow 3), the system undergoes the discontinuity and the condition of Mandelstam prescribes again the horizontal direction for the jump for the same reason as previously. At the point A the representative point R finds itself again on the integral curve which thus establishes a piecewise analytic cycle $ABCD$ consisting of two analytic arcs CD and AB joined by two discontinuous stretches DA and BC on which v and, therefore, $\dot{\varphi}$, vary discontinuously.

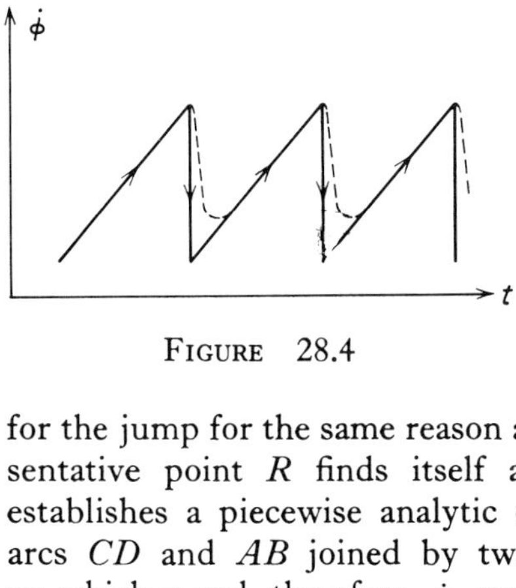

FIGURE 28.4

The amplitude of these piecewise analytic oscillations is obtained from the difference $F_{\max} - F_{\min}$ of the function F and is

$$\varphi_0 = (r/cR)(F_{\max} - F_{\min}) \tag{3.4}$$

The period is obtained by taking into account time for traversing the analytic stretches since the discontinuous stretches are traversed in no time.

On the stretch CD one has $\dot{\varphi} = \Omega$ so that the time for traversing it is

$$T_1 = \varphi_0/\Omega = (r/cR\Omega)(F_{\max} - F_{\min})$$

The time for traversing the arc AB is obtained by integrating (3.3). This gives

$$T_2 = -(r^2/c) \int_{\varphi_1}^{\varphi_2} (F'/\dot{\varphi})d\dot{\varphi}$$

which can be calculated by a graphic integration.

If one plots the piecewise analytic motion so obtained in the (t,φ) plane, one obtains a sawtooth curve shown in Fig. 28.4. In reality, on account of a finite mass of the brake, the actual curve is slightly different, as shown in the broken line.

4. Analogy between mechanical and electrical relaxation oscillations

Although the preceding analysis renders the comparison between mechanical and electrical oscillations quite clear, it is interesting to mention a case of electrical oscillations in which the phenomenon is practically the same as in the preceding section.

In the case of a simple circuit consisting of a neon tube N shunted by a

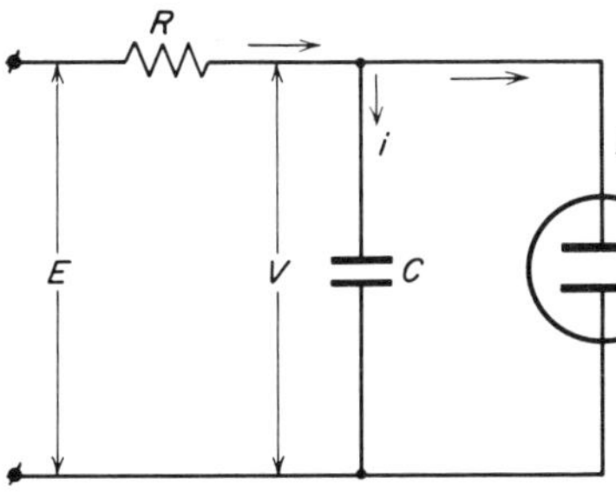

FIGURE 28.5

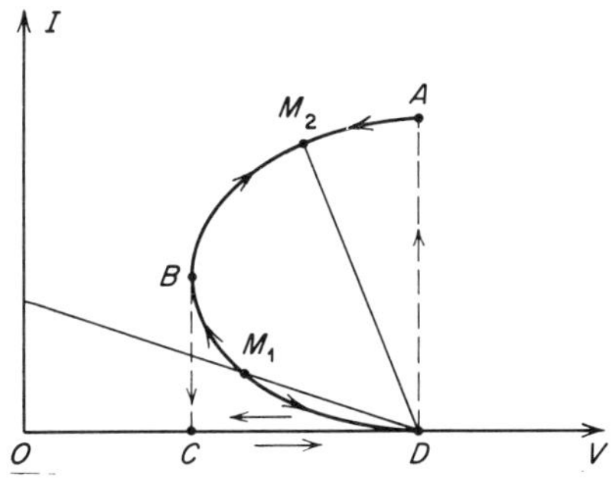

FIGURE 28.6

condenser C according to the scheme shown in Fig. 28.5, the d.e. are $i = C\dfrac{dV}{dt}$; $R(I + i) + V = E$. These equations reduce to one single d.e.

$$dV/dt = (E - V - RI)/RC = [E - V - R\varphi(V)]/RC \qquad (4.1)$$

This d.e. exists only when the tube N conducts. During the time of its extinction one has a simple d.e. of the charging condenser, viz.:

$$dV/dt = (E - V)/RC \qquad (4.2)$$

Since the oscillatory phenomenon is characterized by the alternate sequence of "on" and "off" of the discharge, it is clear that it is governed by the alternate sequence of the d.e. (4.1) and (4.2).

As the problem is a degenerate one (we neglect the effect of the residual inductance), there is only one constant of integration so that there exists a relation between the variables $I = \varphi(V)$ and V, which means that the characteristic is actually the integral curve as in all problems of this nature.

During the period of extinction (when (4.2) is in force), the integral curve is the axis V itself, Fig. 28.6 (compare this with the period of static friction in the preceding section when R moves along the axis $F(v)$). When the point D is reached, the discharge strikes and R jumps suddenly to the point A so that the d.e. (4.1) is now in force.

The position of equilibrium in the circuit when the discharge is on is obviously given by the intersection of the characteristic with the straight line given by equation $I = (E - V)/R$. If this line intersects the characteristic (Fig. 28.6) in its lower portion (point M_1), the equilibrium may be unstable, whereas at a similar point M_2 on the upper branch it is always stable. In fact, if one sets $V = V_0$ for the equilibrium point, the variational equation of (4.1) yields

$$\frac{d\delta V}{dt} = -\frac{1}{RC}(1 + R\varphi'(V_0))\delta V \tag{4.3}$$

where δV means a small departure of V from its equilibrium value V_0 and $\varphi'(V_0)$ is the slope of the characteristic at the point $V = V_0$. On the upper branch of the curve $\varphi'(V_0) > 0$ and from (4.3) it is seen that V_0 is a position of a stable equilibrium. On the lower part of the curve the equilibrium may be unstable if $|\varphi'(V_0)| > 1/R$.

Assume that R has been so chosen that the equilibrium point is on the unstable branch which begins at B and which is indicated by arrows. Since the point M_2 is absent in this case, once R has reached the point B, it cannot move along the characteristic since the d.e. prescribes on it an opposite direction. At the same time B is not a position of equilibrium. Hence, R has to jump downward in accordance with the condition of Mandelstam since V is invariant during the jump. At the point C the representative point R is again on the integral curve (the abscissa axis) of the d.e. (4.2). The piecewise analytic cycle $ABCDA$ consists thus of two analytic branches on which either one of the d.e. takes charge alternatively, "closed" by two discontinuities DA (flashing) and BC (extinction).

In this case the analogy with the mechanical relaxation phenomenon investigated in the preceding section is complete. In fact, in both cases there is a period during which the potential energy is being built up on the stretch CD. In the mechanical case this corresponds to the period of static friction and, in the electrical case, to the period of extinction of the discharge. This static period is followed by a discontinuity (flashing in the electrical case, discontinuity in velocity of the brake in the mechanical case). After this begins again the analytic stretch during which the d.e. governs again the phenomenon (when the discharge is "on" in the electrical case and when the motion of the brake approaches the minimum friction point, in the mechanical case). At the point B begins again the discontinuous stretch BC (extinction of the discharge; transition of the brake to the static friction point).

In both cases the possibility of a piecewise analytic cycle depends on the existence of two critical points D and B, but their physical nature is different. In the electrical case these points are characterized by flashing and by extinction of the discharge; in the mechanical case they are characterized by infinite accelerations. These two aspects of two different phenomena characterize, however, the same thing: the loss of internal equilibrium in the system.

In both cases the relaxation phenomenon disappears if the equilibrium point is adjusted to be on the stable branch of the characteristic. As oscillation of a neon-tube circuit is a desirable phenomenon and oscillation of the brake is, on the contrary, an undesirable one, it is of advantage to fix the equilibrium point on the unstable branch in the first case, and on the stable branch in the second case.

Remaining conclusions regarding the form of the sawtooth curve are identical in both cases.

5. Clocks

It was mentioned previously (Section 11, Chapter 3) that a clock trajectory is a kind of a nonanalytic limit cycle. It is useful to return to this question since we are now acquainted with the concept of piecewise analytic trajectories which characterize relaxation oscillations.

There is an obvious difference between the two cases because a clock is not a "phenomenon" but is a man-made mechanism; however, there is a close analogy regarding the character of periodic motions in both cases.

In fact, in relaxation oscillations there is a point in the cycle at which a quasi-discontinuous stretch begins; we have called this point the *critical point*. A somewhat similar situation occurs in a clock at the instant when the escapement mechanism delivers its impact and changes the momentum of the oscillating system in a quasi-discontinuous manner. As the result, the trajectory of a clock assumes a piecewise analytic character. If such a piecewise analytic trajectory is closed, the motion becomes periodic. So far as energy relations are concerned, a nonanalytic cycle of this kind is characterized by a continuous dissipation of energy compensated for by a quasi-discontinuous (idealized as discontinuous) energy input by impulse.

Andronov[2] was apparently the first to call attention to this fact and developed the theory of the clock on this basis. We shall follow here the

[2] A. Andronov and S. Chaikin, *Theory of Oscillations* (original text in Russian), Moscow, 1937; English translation by S. Lefschetz, Princeton University Press, Princeton, N.J., 1949; A. Andronov, A. Witt, S. Chaikin, *Theory of Oscillations* (in Russian); this book is the second edition (1959) of A. Andronov and S. Chaikin, *Theory of Oscillations* (original text in Russian), Moscow, 1937.

presentation by Bulgakov[3] who developed this subject further. This theory assumes that the dissipation of energy is not of a linear type (that is, proportional to velocity) but is of a Coulomb friction type. This is the more correct assumption, because most of the dissipation of energy here occurs in pivots where it is likely to be of the static friction (Coulomb) type.

If one assumes that the spring of a clock follows Hook's law, the potential energy is: $V(q) = \sigma q^2/2$, q being the coordinate and σ a constant.

The resisting force is then either $-\sigma B$ (for $p > 0$) or $+\sigma B$ (for $p < 0$), where p is momentum and B is a positive constant.

For a small displacement dq (not including the instant of impact) one has obviously:

$$d(p^2/2n + \sigma q^2/2) = \mp \sigma B dq \tag{5.1}$$

Integrating and rearranging the terms, one has

$$\tfrac{1}{2}\sigma(q \pm B)^2 + p^2/2m = \text{const} \tag{5.2}$$

One ascertains that for the initial condition $q_0 = a, p_0 = 0$, it is necessary to have $a > B$; otherwise the static friction would block the motion, which will be assumed. Equation (5.2) becomes

$$\tfrac{1}{2}\sigma(q - B)^2 + p^2/2m = \tfrac{1}{2}\sigma(a - B)^2 \tag{5.3}$$

where the right-hand term is a constant. From this equation one gets

$$p = -\sqrt{\sigma m[(a - B)^2 - (q - B)^2]} \tag{5.4}$$

One has to express also the condition that the torsional pendulum (part A in terms of Section 11, Chapter 3) can reach the angular coordinate $q = f$ at which the energy replenishing impact is delivered; this imposes condition

$$a > f; \qquad a > 2B - f \tag{5.5}$$

Additional conditions (the quantity under the square root must be positive) impose requirements that the first condition (5.5) should take place for $B < f$ and the second for $B > f$.

Assuming further that the impact ΔH is constant, for the quasi-discontinuous stretch one has the relation

$$p_1'^2 - p_1^2 = 2m\Delta H$$

[3] B. V. Bulgakov, *Oscillations* (in Russian), Moscow, 1954; A. Andronov and S. Chaikin, *Theory of Oscillations* (original text in Russian), Moscow, 1937; English translation by S. Lefschetz of A. Andronov and S. Chaikin, *Theory of Oscillations*, Princeton University Press, Princeton, N.J., 1949; A. Andronov, A. Witt, S. Chaikin, *Theory of Oscillations* (in Russian); this book is the second edition (1959) of A. Andronov and S. Chaikin, *Theory of Oscillations* (original text in Russian), Moscow, 1937.

where p_1' and p_1 are the values of momentum after and before the impact. This gives

$$p_1' = -\sqrt{p_1^2 + 2m\Delta H} = -\sqrt{\sigma m[(a - B)^2 - (f - B)^2 + h]} \quad (5.6)$$

where $h = 2\Delta H/\sigma$. After the impact a new motion begins and, by a similar argument, one finds that p is

$$p = -\sqrt{\sigma m[(a - B)^2 + h - (q - B)^2]} \quad (5.7)$$

One can also calculate the time of oscillation between the instants when $q = f$ and $q = -b$ from the formula

$$t = \int_f^{-\sigma} (m/p)dq \quad (5.8)$$

which leads to Airy's formula, but for our purpose here this is of no special interest.

We limit our investigation to finding conditions under which the amplitudes either grow or decrease.

The curve $p = p(q)$ cuts the abscissa axis in two points; the abscissa of the left point, say, A_2 is

$$q = -b = B - \sqrt{(a - B)^2 + h} \quad (5.9)$$

Clearly it is necessary that

$$b = \sqrt{(a - B)^2 + h} - B > 0 \quad (5.10)$$

Using the preceding equations one finds that amplitudes increase if $a < A$ and decrease if $a > A$. If $a = b = A$, one has obviously a periodic motion.

The last step in this argument is to show that the limit for the decreasing amplitudes is the same as for the increasing ones. This is done in calculating the difference $b^{b-a} - a$ by the above formulas and, as a and b approach the same limit, $b - a \to 0$; that is, $a \to A$. This shows that the motion with stationary amplitude is orbitally stable, that is, in all respects exhibits the property of a limit cycle.

Inequalities (5.5) show that the amplitude a must be greater than a certain limit value in order to start the clock going; this feature may be considered as "hard" self-excitation. In ordinary watches this limit value is very small so that the watch starts as soon as it is wound; one may call this feature a "soft" self-excitation.

Equation (5.9) can be written as

$$(b + B)^2/h - (a - B)^2/h = 1 \quad (5.11)$$

In the (a,b) plane this is a hyperbola with center at the point $(B,-B)$ and asymptote $b = a - 2B$. A bisector line, $a = b$ together with the hyperbola, permits investigating the increase (decrease) of amplitudes as they approach the stationary amplitude. The latter is given by the point of the intersection of the hyperbola and the straight line: $a = b$.

This theory can be further generalized if one considers that the impact is distributed over a finite time; likewise one can consider a more general form of the potential energy $V(q)$; moreover, the mass m which was considered as constant, in reality is a certain function of q in view of constructional details of a clock. Such generalizations introduce further complications in calculations without, however, adding anything essentially new.

Summing up, the aim of the method so far is to determine the amplitude b if one knows the preceding amplitude a. This ultimately results in relation

$$f(a,b) = 0 \tag{5.12}$$

Since the amplitude to the right is generally equal to that to the left and, for stationary state, $a = b = A$, the preceding relation becomes for the stationary state:

$$f(A,A) = 0 \tag{5.13}$$

It is noted that up to this point the procedure was elementary, namely, from the knowledge of one amplitude, the next one was determined by theoretical mechanics merely adding the condition of impact; the terminal amplitude of the first half-cycle was used as the initial amplitude for the next half-period, and so on.

Ultimately the problem consists in the investigation of what happens to these "initial conditions" (amplitudes) in the long run; if they approach the limit $a = b = A$, then the stationary state is reached (equation 5.13).

It is observed that this method has a certain analogy with the stroboscopic method (Chapter 16). In the latter we were not interested in the motion *during* the period 2π (here we have only one half-period) but the attention was focused on the isolated instants $0, 2\pi, 4\pi, \ldots$ separated in time by 2π; in this way the investigation was directed toward establishing the law of variation of the initial conditions (in each period); if this sequence of (discrete) initial conditions tends to a limit, this was sufficient for asserting that in the long run the oscillation approaches the stationary state.

From this standpoint the situation here is similar. On the other hand, as far as the details are concerned, the procedure here is more complicated inasmuch as one is obliged to interpose on the continuous (analytic) variation of the amplitude a quasi-discontinuous variation caused by impulses. As the result of this, in order to determine the terminal amplitude b given the initial one a, both the analytic and nonanalytic intervals are to be taken into account.

Once, however, the condition (5.13) of stationary state has been obtained, one can apply the standard procedure for investigating stability since now one has a situation similar to that which was encountered already in the stroboscopic method, namely: to ascertain the approach of initial conditions to the stationary state if the latter has been perturbed.

We assume, therefore, that the stationary amplitude A is perturbed by perturbations δa and δb and we have

$$a = A + \delta a; \qquad b = B + \delta b \tag{5.14}$$

Replacing these values into (5.12) and expanding this equation into Taylor's series around the stationary point keeping only the terms of the first order, one gets

$$f(A,A) + (\partial f/\partial a)\delta a + (\partial f/\partial b)\delta b = 0 \tag{5.15}$$

where the partial derivatives are taken, as usually, at the point (A,A). In view of (5.13), one has

$$\delta b = -\left[\left(\frac{\partial f}{\partial a}\right)\Big/\left(\frac{\partial f}{\partial b}\right)\right]\delta a \tag{5.16}$$

It is seen that the perturbation dies out if

$$\left(\frac{\partial f}{\partial a}\Big/\frac{\partial f}{\partial b}\right) < 1 \tag{5.17}$$

If (5.17) is considered as a curve in the (a,b) plane, the stationary amplitude is given by the coordinates of the point of intersection of this curve with the bisector line: $a = b$. The condition (5.17) means then that the slope of the tangent to the curve at the point of its intersection with the bisector line must be in the interval $(-\pi/4, +\pi/4)$.

We note in passing that this graphical condition of stability reminds us of what was outlined in Section 2, Chapter 7, in connection with the geometrical analysis of the bifurcation phenomena of limit cycles.

It is easy to show that, in view of what has been said previously, condition (5.17) is fulfilled. In fact, we have $f(a,b) = \sqrt{(a - B)^2 + h} - (b + B)$; $A = h/4B$, so that

$$\partial f/\partial a = (a - B)/\sqrt{(a - B)^2 + h};$$

$$(\partial f/\partial a)_A = (A - B)/(A + B); \qquad (\partial f/\partial b)_A = -1$$

which gives

$$\left(\frac{\partial f}{\partial a}\Big/\frac{\partial f}{\partial b}\right)_A = (A - B)/(A + B) \tag{5.18}$$

In view of (5.5) this quantity is positive and less than one.

The theory of clocks has been considerably extended by J. Haag[4] who established conditions for a subharmonic as well as harmonic synchronizations; he showed that, in general, the synchronizing force depends not only on time but also on the amplitude and velocity of the oscillating member of the clock. The method is based also on the perturbation theory. In fact, if the motion is periodic, $\Delta a = 0$ (a being the amplitude). The variation of the amplitude per one half-period is given, as previously explained, by a certain integral representing the work of driving and resisting forces.

If the dissipation increases with velocity, the excessive initial amplitude will decrease and vice versa. These intuitive considerations are not sufficient, however, and Haag establishes a series of theorems based on the properties of the variational equations which we cannot review here; we mention only the conclusions: synchronization can be obtained by means of a sinusoidal (harmonic) force; it can be produced also by quasi-discontinuous impacts; in all cases there exists a certain *zone* of synchronization as in the analytic cases which we studied in Chapter 18; the problems here are more complicated since in addition to the analytic components of forces, the quasi-discontinuous forces due to impacts of the escapement mechanism are to be taken into account also; moreover, the nonlinearity in the restoring force accounts for a possibility of synchronization on subharmonics.

We merely mention these various conclusions as their even superficial review here would be outside the limit of this brief outline of the theory of clocks.

6. Froude's pendulum

Another case of self-excited mechanical oscillations is the so-called Froude pendulum; it consists essentially of the following arrangement: if one mounts a pendulum on a rotating shaft with a certain amount of friction, it is observed that the pendulum begins to oscillate under certain conditions and the amplitude reaches a stationary state.

The origin of this phenomenon is due to the friction which acts as a coupling between the shaft rotating with a constant angular velocity Ω and the pendulum. The conditions under which these self-excited oscillations appear are very similar to those which we have investigated already in Sections 2 and 3.

Here again all depends on the nature of friction but, as the latter is

[4] J. Haag, *C. R. Ac. Sc.* (Paris) **202,** 1936; **204,** 1937; **206,** 1938; *Ann. Ec. Norm. Sup. Series,* 3, **60,** 1943; 3, **61,** 1944.

erratic and uncontrollable, it is useful to formulate first the theoretical conditions under which this phenomenon is bound to appear.

We start with the following d.e.

$$I\ddot{x} + b\dot{x} + k\dot{x}^3 + cx = F(\Omega - \dot{x}) \tag{6.1}$$

where x is the angle of the pendulum, and I, b, k, and c are constant parameters which have the obvious meaning. We have added a nonlinear term $k\dot{x}^3$ for reasons which will appear later. The right-hand term is the moment transferred to the pendulum by the rotating shaft. This moment is due to friction and is generally a certain nonlinear function of $\dot{x}$. The usual procedure is to develop the function F in Taylor's series, assuming that $\dot{x}$ is small in comparison with the constant Ω. This gives

$$F(\Omega - \dot{x}) = F(\Omega) - \dot{x}F'(\Omega) + \frac{\ddot{x}}{2!}F''(\Omega) - \dots \tag{6.2}$$

where $F'(\Omega) = (d/dt)F(\Omega)$, etc. Setting $F(\Omega) = m$; $F'(\Omega) = n$, etc., and assuming that the series converges rapidly enough to justify the use of the first two terms only, the insertion of (6.2) into (6.1) gives

$$I\ddot{x} + (b + n)\dot{x} + k\dot{x}^3 + cx - m = 0 \tag{6.3}$$

The last constant term, m, can be obviously dropped; it merely means that the oscillation occurs not around the position $x = 0$ but around some other position owing to a constant frictional drag exerted by the rotating shaft.

If the coefficient n is negative and such that $|n| > b$, after a division by I, (6.3) becomes

$$\ddot{x} - \alpha'\dot{x} + \beta'\dot{x}^3 + \omega_0^2 x = 0 \tag{6.4}$$

where $\alpha' = (|n| - b)/I$; $\beta' = k/I$ and $\omega_0^2 = c/I$.

Introducing the new independent variable: $\tau = \omega_0 t$, one obtains the d.e.

$$\ddot{x} - \alpha\dot{x} + \beta\dot{x}^3 + x = 0 \tag{6.5}$$

in which the differentiations are with respect to τ but, since no confusion is to be feared, we can still use the previous symbols: $\ddot{x}$ and $\dot{x}$. In this equation $\alpha = \alpha'/\omega_0$ and $\beta = \beta'\omega_0$.

Equation (6.5) is the standard Rayleigh equation which is known to possess a periodic solution. We reproduce briefly the calculation under the assumption of a near-linearity, that is, when α and β are small.

With the equivalent system: $\dot{x} = y$; $\dot{y} = \alpha y - \beta y^3 - x$ and with the new variables: $\rho = x^2 + y^2$ and $\psi = \arctan(y/x)$ the system becomes

$$\begin{aligned} \frac{d\rho}{dt} &= 2\rho(\alpha \sin^2 \psi - \beta\rho \sin^4 \psi) \\ \frac{d\psi}{dt} &= -1 + \alpha \sin \psi \cos \psi - \beta\rho \cos \psi \sin^3 \psi \end{aligned} \tag{6.6}$$

and we attempt to satisfy it by series solutions of the form

$$\rho(t) = \rho_0(t) + \mu\rho_1(t) + \ldots; \qquad \psi(t) = \psi_0(t) + \mu\psi_1(t) + \ldots \tag{6.7}$$

where μ is a small parameter. For the first approximation we need only two terms as written in (6.7).

As α and β are small, the zero-order solution, as usual, is:

$$\rho_0(t) = \rho_0; \qquad \psi_0(t) = \varphi_0 - t \tag{6.8}$$

where ρ_0 and φ_0 are the initial conditions. The d.e. for the first order corrective terms are then:

$$\begin{aligned} \frac{d\rho_1}{dt} &= 2\rho_0(A \sin^2 \psi_0 - B\rho_0 \sin^3 \psi_0) \\ \frac{d\psi_1}{dt} &= A \sin \psi_0 \cos \psi_0 - B\rho_0 \cos \psi_0 \sin^3 \psi_0 \end{aligned} \tag{6.9}$$

where $A = \alpha/\mu$ and $B = \beta/\mu$.

We can use now the stroboscopic procedure in which we evaluate the differences $\rho_1(2\pi)$ and $\psi_1(2\pi)$ of $\rho_1(t)$ and $\psi_1(t)$ for one period 2π. It is noted that the integration of the second equation between 0 and 2π gives identically $\psi_1(2\pi) \equiv 0$ so that we need to consider only the first equation (6.9). It should be mentioned in passing that this circumstance is due to the fact that the d.e. (6.5) has only a linear term in x; if we had also a nonlinear term, say, γx^3, the result would be different and the second equation (6.9) would not be identically zero, as was explained in Section 7, Chapter 22.

The integration yields: $\rho_1(2\pi) = 2\pi\rho_0(A - \frac{3}{4}B\rho_0)$ and, therefore,

$$\rho(2\pi) = (2\pi\mu)\rho_0(A - \tfrac{3}{4}B\rho_0)$$

Setting $2\pi\mu = \Delta\tau$; $\rho(2\pi) = \Delta\rho$ and passing to the limit: $\Delta\tau \to d\tau$; $\Delta\rho \to d\rho$ one obtains the stroboscopic d.e.:

$$d\rho/d\tau = \rho(A - \tfrac{3}{4}B\rho) = \Phi(\rho) \tag{6.10}$$

whence the stationary amplitude

$$\rho^* = \frac{4A}{3B} = \frac{4}{3}\frac{[|n| - b]}{k\omega_0^{\,2}} \tag{6.11}$$

It is seen now why the term $k\dot{x}^3$ was necessary in (6.1); in fact, without it it would be impossible to determine ρ^*. The question of stability is determined by the usual criterion:

$$\Phi_\rho(\rho^*) < 0 \tag{6.12}$$

One ascertains easily that this criterion is fulfilled.

Thus, under the assumptions made, the Froude phenomenon is quite clear. These assumptions are, however, too special and we have to examine them. In fact, we assumed: (1) that $F'(\Omega)$ is negative and that $|F'(\Omega)|$ is just slightly greater than b so that the difference $b - |F'(\Omega)|$ is a small quantity of the first order, as is the coefficient k in (6.3); and (2) that the higher-order terms $F''(\Omega)$, $F'''(\Omega)$, etc., in the expansion of $F(\Omega - \dot{x})$ are negligible.

It is clear that if the assumption (1) is not fulfilled, the oscillation will either not appear at all or, if α is not small, we cannot use the nearly-linear procedure and the calculations would be much more complicated since the first-zero-order solution (6.8) would not hold then.

Still more difficult would be the case when the development of $F(\Omega - \dot{x})$ would have an important second-order term: $\frac{1}{2}F''(\Omega) = p$. In such a case, instead of (6.3), one would have the d.e.

$$(I - p)\ddot{x} + (b + n)\dot{x} + cx = 0 \tag{6.13}$$

and it is seen that, if $p > 0$ happens to be of the same order as I, at least in a certain interval of Ω, one would have a degenerate problem (Section 2, Chapter 26) with a possibility of a quasi-discontinuous variation of $\dot{x}$, and this brings the problem within the scope of relaxation oscillations.

All these uncertainties arise because the frictional coupling function $F(\Omega - \dot{x})$ is generally known only approximately and, except that occasionally the friction may become "negative," practically nothing is known about the quantitative end of the problem.

On the other hand, if, with a better technique, this experimental function could be produced with a greater certainty and, better still, could be modified at will in a predetermined manner, then an arrangement of Froude's type could become an interesting tool in the exploration of these complicated phenomena, particularly as regards transitions between the analytic and piecewise analytic forms of these oscillations which is still an entirely unknown field.

Chapter 29

DISCONTINUOUS THEORY OF VOGEL

1. Introductory remarks

In this chapter we propose to give a brief outline of another discontinuous theory developed independently by Vogel.[1]

If the M.C. theory (outlined in Chapter 26) may be regarded as a purely physical theory, Vogel's approach appears to be of a somewhat postulational nature based more on a geometrical ground. If one regards this theory without any reference to applications, it appears as a somewhat abstract essay along the line of concepts introduced by Whitney[2] and Birkhoff.[3]

If, however, this abstract (and to some extent heuristical) approach is applied to dynamics, the familiar aspects of the M.C. theory appear; in such a case both theories lead to the same conclusions although the argument is different.

As the purpose of this text is primarily an applied one, we shall limit the theoretical part to a minimum, trying to follow Vogel's own exposition. However, its connection with the analytical argument will be of a greater interest as it will enable us to reach physical applications.

It is useful to introduce first the concept of regularity in connection with trajectories filling the phase plane.

[1] Th. Vogel, *C. R. Ac. Sc.* (Paris) **231**, 1950, *Ann. des Telecomm.* **6**, 1951; *Bull. Soc. Math.* (France) **81**, 1953; *Rend. Semin. Padova* (22), 1953; *Ann. des Telecomm.* **8**, 1953, *Colloque Int. Porquerolles* (France), 1951.

[2] H. Whitney, *Ann. of Math.* **34**, 1933.

[3] G. D. Birkhoff, *Dynamical Systems*, Am. Math. Soc., 1927; A. Andronov, A. Witt, S. Chaikin, *Theory of Oscillations* (in Russian); this book is the second edition (1959) of A. Andronov and S. Chaikin, *Theory of Oscillations* (original text in Russian), Moscow, 1937.

We use the definition of Nemitzky.[4]

(1) We call the family F of curves *regular* if for any point p on a curve C and for every arc $\widehat{pq}$ of C, for every $\varepsilon > 0$, one can find $\delta > 0$ such that, from the condition $p' \subset C'$, where C' is some curve of F, and $\rho(p,p') < \delta$ (ρ being the "distance"), follows the existence of an arc $\widehat{p'q'}$ on C' such that

(a) $\widehat{p'q'} \subset S(\widehat{pq}, \varepsilon)$; $\rho(q',q) < \varepsilon$, where $S(\widehat{pq}, \varepsilon)$ is the ε-neighborhood of $\widehat{pq}$.

(b) If r' and s' are points of the arc $\widehat{p'q'}$ and $\rho(r',s') < \delta$, then the diameter of the arc connecting r' with s' is less than ε.

The first property may be regarded as the *geometric continuity*, and the second, *as equal local connectedness.*

In what follows the expression: the "arc of the curve C" will mean a one-to-one image of either a segment or a simple closed line; we will exclude singular points at first.

(2) We shall say that the family F is *orientable* if on all arcs it is possible to establish the direction consistently, that is, for any arc $\widehat{pq}$ from the condition that the sequence $\widehat{p_iq_i}$ converges to $\widehat{pq}$ it follows that, for a sufficiently large i, $\widehat{p_iq_i}$ is positively oriented if $\widehat{pq}$ is positively oriented.

With these definitions the basic theorem of Whitney[2] can be formulated as follows:

If a family of curves covering (a certain region) R is regular and orientable, it can act as a family of trajectories of some dynamical system.

For dynamical systems of the second order with which we are generally concerned, the theorem of Whitney permits treating trajectories as a flow of a planar irrotational "fluid of trajectories," a concept which has been introduced intuitively in Chapter 3.

The theory of Vogel starts with this concept and considers what happens when this "flow" is suddenly changed at a certain critical threshold, thus giving rise to "reflections" or "refractions" of the "fluid of trajectories."

2. Fundamentals of the theory

Consider an autonomous differential system of the second order whose representative point follows a certain trajectory of a regular family up to a point where the law governing its motion changes suddenly in such a manner that the further path of the point is along another regular trajectory;

[4] V. V. Nemitzky, *Topological Problems*, Am. Math. Soc. translation, 103, 1954.

[2] See footnote [2], page 648.

such behavior is exemplified by a mechanical system in which a shock occurs at some moment. If the trajectories corresponding to all possible initial conditions are thus made up of portions belonging to two or more regular and oriented families of curves, the systems will be said to possess *piecewise regularity*. The motions of such systems can be nonperiodic (as, for instance, the motions of a mass moving on a ground formed by two levels separated by a cliff), or periodic (as in the case of the "sawtooth" oscillator, where a condenser is being charged until the voltage at its terminals is sufficient to produce a discharge).

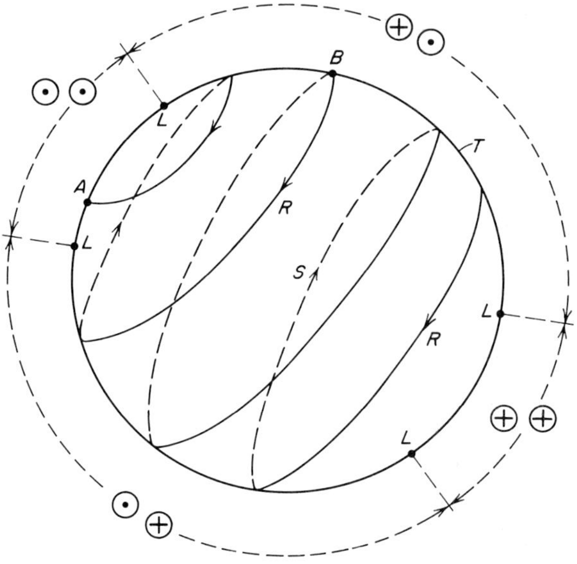

FIGURE 29.1

With a view to a periodic case (which is obviously the more interesting of the two) one can investigate the problem of an abstract system possessing two regular oriented families of trajectories, say R and S, and a "boundary curve" T on reaching which the representative point must switch over to an S (or R) curve if it was moving along an R (or S) one. Looking at the boundary from one side, every R or S curve is either inwardly or outwardly bound (denoted respectively as ⊕ or ⊙), unless it is tangential to the limit curve. The limiting points L on T, where a trajectory of one of the families does not cut the boundary T, divide it into regions, all points of one region having the same character; we can thus have regions ⊙⊙, ⊙⊕, ⊕⊕, or ⊕⊙ (Fig. 29.1).

At a $\odot\odot$ or $\oplus\oplus$ point, such as A in Fig. 29.1, the path of the representative point of the system will cross the boundary; making use of an optical analogy, we may say that it has a "refractive" effect on the motion. At a $\oplus\odot$ or $\odot\oplus$ point, on the contrary (such as B in Fig. 29.1) there will occur a "reflection." This case is obviously the more interesting of the two, since the representative point turns back, and there may thus be a chance that its path will be closed in the long run, which would mean that the motion is periodic. Any closed path will be composed of a certain number, n, of R arcs alternating with an equal number of S arcs; the

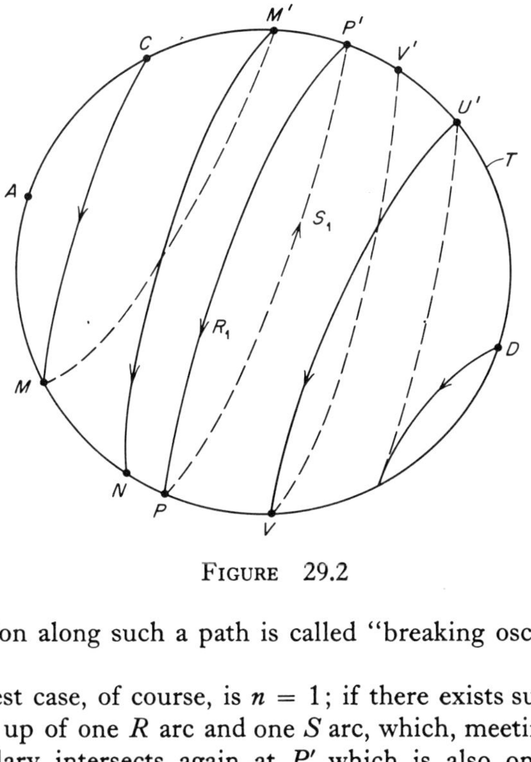

FIGURE 29.2

periodic motion along such a path is called "breaking oscillation" of the nth kind.

The simplest case, of course, is $n = 1$; if there exists such a motion its path is made up of one R arc and one S arc, which, meeting at a point P of the boundary intersects again at P' which is also on the boundary (Fig. 29.2). The motion is, symbolically, $R_1S_1R_1S_1R_1\ldots$.

Let us inquire into the necessary and sufficient condition for such a situation: both families of trajectories form a net which can be thought of as determining curvilinear coordinates R, S in the phase plane. If we map that net onto a plane with cartesian coordinates R, S, each R curve will appear as a straight line parallel to the S axis, and each S curve as a parallel to the R axis; a piecewise regular motion will be mapped as a ladder with its steps parallel to the axes. The image of boundary T is a curve $\overline{T}$ such that all points of T will be mapped 1:1 except for such points as P, P', whose common image will be a double point $\overline{P}$ of $\overline{T}$ (Fig. 29.3). Conversely, let there exist a double point $\overline{P}$ on $\overline{T}$, at the intersection of arcs $\overline{N}\overline{V}$ and $\overline{M}'\overline{U}'$: This means that on the original arcs NV and MU there are two points $P'P'$ having the same R, S-coordinates: that is, the situation shown in Fig. 29.2. There exists then a periodic motion of the first kind.

The stability of such a solution is easily discussed by means of the above

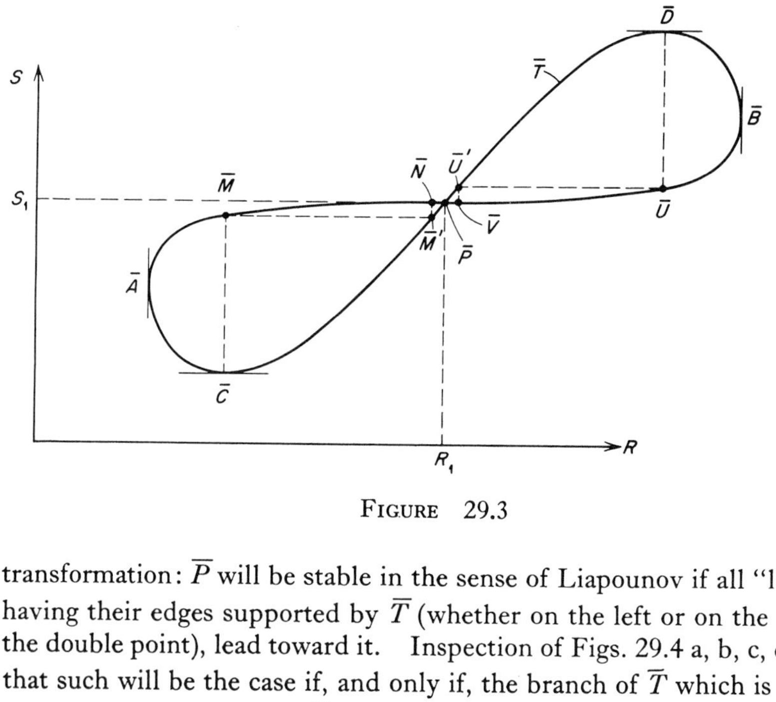

FIGURE 29.3

transformation: $\overline{P}$ will be stable in the sense of Liapounov if all "ladders" having their edges supported by $\overline{T}$ (whether on the left or on the right of the double point), lead toward it. Inspection of Figs. 29.4 a, b, c, d shows that such will be the case if, and only if, the branch of $\overline{T}$ which is $\oplus$ with respect to R (or S) has at $\overline{P}$ a greater slope than the $\odot$ branch.

A periodic solution of the first kind is possible for the case shown in Fig. 29.2 in which there are no singular points.

We shall not investigate oscillations of higher orders ($n > 1$) but will merely mention the conditions for their occurrence.

Consider the oscillation of the second kind; the closed path $R_1S_1R_2S_2$

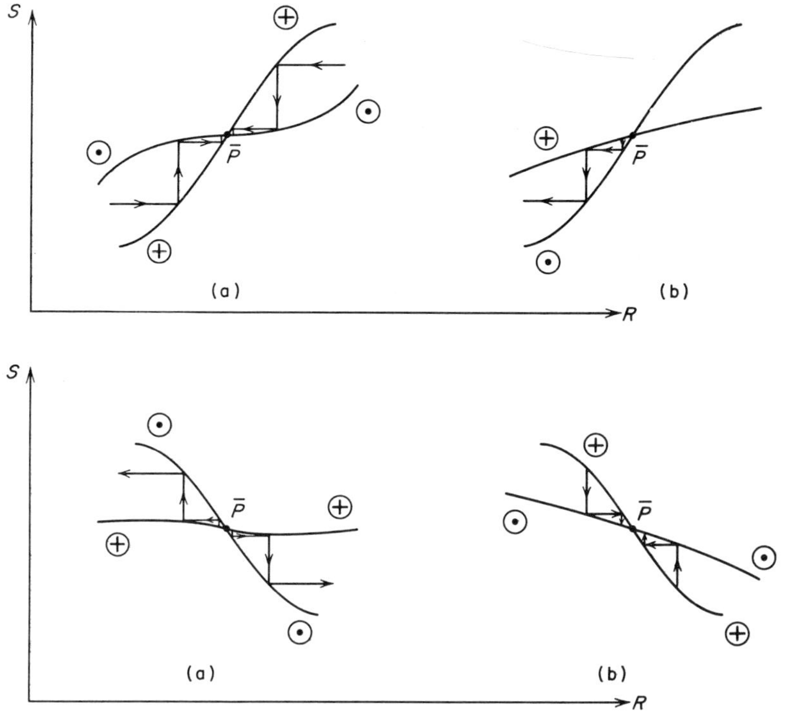

FIGURE 29.4

$R_1S_1R_2 \ldots$ is made up of two R arcs and two S arcs and can be described in the indicated order only if the direction of motion along the R_1 stretch is opposite to that along the R_2 stretch and likewise for S_1 and S_2 (Fig. 29.5).

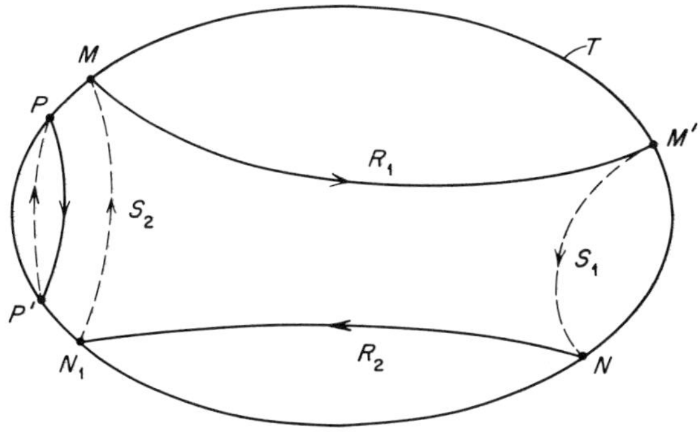

FIGURE 29.5

This implies the presence of a saddle point or center of each family inside the curve T. If such a saddle point exists, both periodic solutions, one of the first kind (such as PP') and the other of the second kind (such as $MM'NN'$) may exist. However no higher order ($n > 2$) solutions are possible; they become possible only when there are singularities of higher order inside T.

One can also discuss the limit cases when singular points (either for one or for both families) appear *on the boundary* T.

A particular interesting case arises when one of the families has a node on T.

Assume that N is a node for the R family (Fig. 29.6) and R_0 is the R

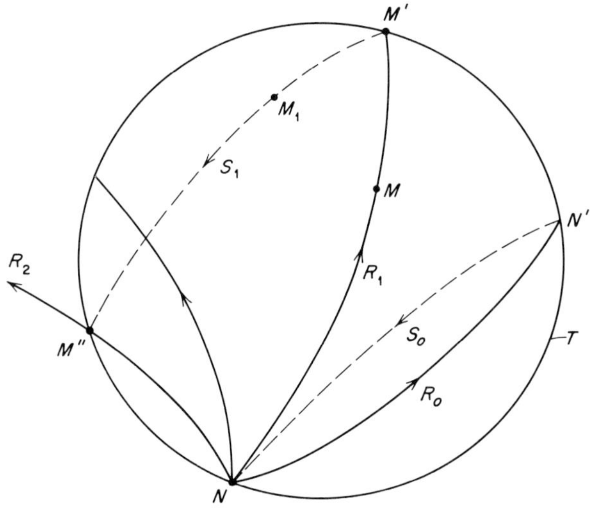

FIGURE 29.6

trajectory; this trajectory intersects T again at N'. The cycle $R_0N'S_0N$ is the periodic solution of the first kind. However, if N is $\oplus$ for R and $\odot$ for S, this solution will be unstable. In fact, if the representative point ρ is initially at M (on R_1), it will move to M' (on T); after that it will move along on S curve to M'' (on T) and, finally, outward along the R_2 curve through M'' since this point is $\odot$ (because its other end at N is $\oplus$). Likewise the initial position M_1 on S_1 will bring ρ to M'' and from there outward.

On the contrary, if N is $\odot$ for R (and $\oplus$ for S) as in Fig. 29.7, one obtains a stable behavior; an initial point at M (on R_1) will lead to N and from there it will move along the cycle S_0R_0.

Likewise, an initial point M_1 on S_1 will move to M' on the boundary T and from there to N along the R_1 stretch and again to the S_0R_0 cycle. Thus the periodic motion will be established at once and not after an infinite number of successive steps as in the case of the oscillator illustrated in Fig. 29.2. This will occur after meeting *one* discontinuity at N or, at the most, two discontinuities (one at M' and the other at N). It will be shown that this permits explaining the transient behavior of multivibrators.

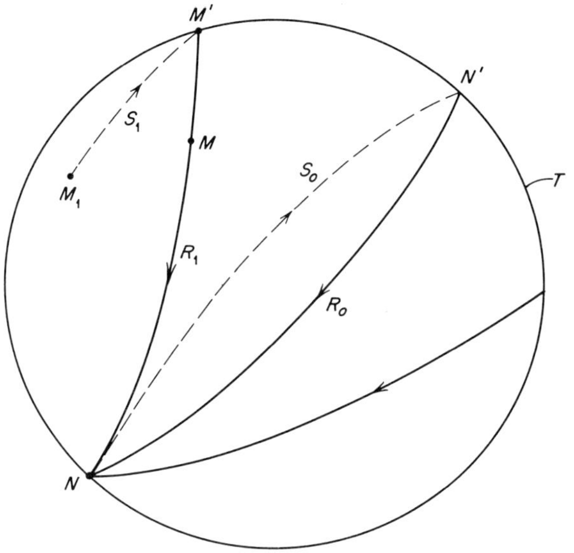

FIGURE 29.7

3. Analytical formulation; a special case

An objection can be made that a node is reached after an infinite time so that the above stated periodicity is not reached in a finite time. This objection would be valid if the system were described by two d.e. of the first order with the analytic right-hand sides.

It is necessary, therefore, to supplement the preceding theory by considering a system of d.e.

$$dx/X(x,y) = dy/Y(x,y) = dt/T(x,y) \tag{3.1}$$

(compare to equation (4.2), Chapter 26) where X, Y, T are analytic functions of x and y.

It is clear that this system breaks down at certain critical points x_0, y_0

for which $T(x_0,y_0) = 0$ while everywhere else the solution of (3.1) is analytic, that is, knowing the solution at $t = t_1$ it is known also for $t = t_1 + \Delta t$.

It is noted that at this point the present theory becomes identical with the M.C. theory outlined in Chapter 26, although the arguments leading to this point are different in both theories.

We continue, however, the argument of Vogel; it is stated that, in spite of the fact that for $x = x_0, y = y_0$ the system (3.1) breaks down, the physical system "continues to exist" and that some other "continuation" should exist, although there is no *analytical* continuation.

The theory does not specify the nature of this continuation but merely postulates that it is represented by the equation:

$$S(x,y) = s \tag{3.2}$$

while the regular one-parameter family (when $x \neq x_0, y \neq y_0$) is given by: $dx/X = dy/Y$, for which

$$R(x,y) = r \tag{3.3}$$

It is seen thus that the regular (3.3) and the "irregular" (3.2) situations appear to some extent on equal basis without involving so far any physical considerations as in the M.C. theory.

4. Physical interpretation

The particular features of the problem are in that the boundary points on the curve T are reached with a finite velocity in accordance with the d.e. (3.1) while the motion along the S arcs occurs with an infinite velocity if we assume for $S(x,y)$ the same condition as in the M.C. theory. Moreover motions along the R curves are reversed on the interior and the exterior portions so that no crossings of the boundary are possible.

One can question: What is the meaning of the ⊕⊕ and ⊙⊙ boundary points? In fact, the first can never be reached from the inside and the second from the outside so that, ultimately, one has the same situation, namely the point ρ arriving at the boundary from some definite side with an infinite velocity is unable to proceed any further. This may be interpreted as a breakdown of the physical system; this means, as we saw, that ρ crosses T and no further breaking oscillation takes place (for example, point M'' in Fig. 29.6).

If, on the other hand, a boundary point M is reached for which one has either ⊕⊙ or ⊙⊕ there appears a jump governed by (3.2); in such a case the next position of ρ is M' determined by the point of the intersection of

T with the S curve through M; hence ρ will proceed along the R_1 curve through M'.

As was said, the theory does not specify the principle from which (3.2) is derived; this depends on the physical meaning of each problem. If this problem involves the invariance of energy, then it becomes identical to that investigated in the M.C. theory.

5. A numerical example

Assume that we have the system

$$dx/y = dy/(2 - x) = dt/[(x - 1)^2 + (y - 1)^2 - 1]; \qquad y/x = s \quad (5.1)$$

The R curves are circles

$$(2 - x)^2 + y^2 = r^2 \quad (5.2)$$

and the S curves are straight lines through the origin. Equation of the curve T is:

$$(x - 1)^2 + (y - 1)^2 - 1 = 0 \quad (5.3)$$

so that its (r,s) transform $\overline{T}_0$ is

$$(r^2 - 3)^2(1 + s^2) - 4(r^2 - 3)(s^2 - 1) + 4(s - 1)^2 = 0 \quad (5.4)$$

The conditions for the existence of a double point $\bar{P}$ of $\overline{T}_0$ are:

$$(\overline{T}_0)_r = (\overline{T})_s = 0 \quad (5.5)$$

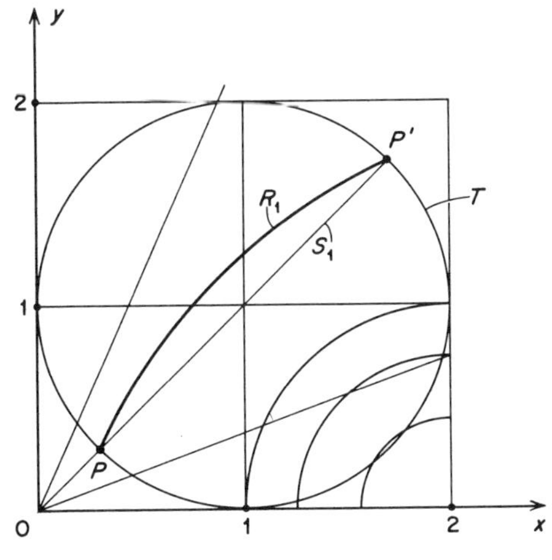

FIGURE 29.8

that is, here

$$2(r^2 - 3)(1 + s^2) - 4(s^2 - 1) = 0$$
$$4s(r^2 - 3) - 8s(r^2 - 3) + 8(s - 1) = 0 \qquad (5.6)$$

It is seen that the piecewise periodic motion $PS_1P'R_1P$ (Fig. 29.8), admits a unique solution: $s = 1$, $r^2 = 3$. One verifies that this point of intersection of the R and the S curves is actually on $\overline{T}_0$.

The interesting point is that the double point $\bar{P}$ of $\overline{T}_0$ corresponds to two points PP' on $\overline{T}_0$; all other points of T_0 have a unique image on $\overline{T}_0$ the relation between the x,y and the r,s coordinates being

$$x = (r^2 - 3)/2(1 - s); \qquad y = s(r^2 - 3)/2(1 - s) \qquad (5.7)$$

These expressions become indeterminate (0/0) for $r^2 = 3$, $s = 1$. Moreover the points of T_0 which are $\oplus$ have also $\oplus$ images on T_0 but the sequence of $\oplus$ and $\odot$ points on the transform is different from that on the original boundary. In particular, the branches of the transform through the double point keep their $\oplus$ or $\odot$ character when this point is crossed.

6. Multivibrator

As an example illustrating this theory, we consider a multivibrator scheme shown in Fig. 29.9. If one neglects the effects of the grid current and of the anode reaction, the d.e. of the currents are

$$r_1 i_1 - \frac{1}{C_1}\int_0^t I_1 dt - R_1 I_1 = r_2 i_2 - \frac{1}{C_2}\int_0^t I_2 dt - R_2 I_2 = 0 \qquad (6.1)$$

$$i_1 = I_{a1} - I_1 = \varphi_1(V_1 - v_1) - I_1; \qquad i_2 = \varphi_2(V_2 - v_2) - I_2$$
$$v_1 = R_2 I_2; \qquad v_2 = R_1 I_1 \qquad (6.2)$$

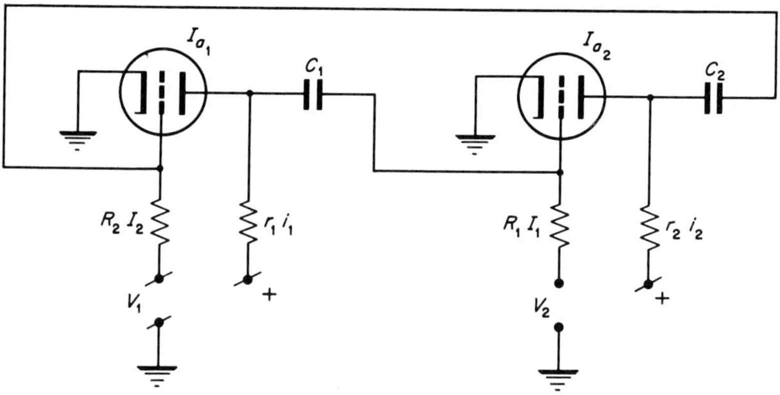

FIGURE 29.9

where φ are the nonlinear characteristics of the tubes, and V are fixed biasing voltages.

Introducing the variables $u_j = V_j - v_j$; $j = 1, 2$, equations (6.1) and (6.2) become

$$\begin{aligned} u_1 - V_1 &= a_2(u_2)\dot{u}_2 + b_2\dot{u}_1 \\ u_2 - V_2 &= b_1\dot{u}_2 + a_1(u_1)\dot{u}_1 \end{aligned} \tag{6.3}$$

where $a_j(u_j) = C_jR_jr_j\varphi_j'(u_j)$; $b_j = C_j(R_j + r_j)$; $j = 1, 2$.

The d.e. (6.3) reduce to the system

$$\begin{aligned} \dot{u}_1 &= [a_2(u_2)u_2 - b_1u_1 + V_1b_1 - V_2a_2(u_2)]/T_0 \\ \dot{u}_2 &= [a_1(u_1)u_1 - b_2u_2 + V_2b_2 - V_1a_1(u_1)]/T_0 \end{aligned} \tag{6.4}$$

where

$$T_0 = a_1(u_1)a_2(u_2) - b_1b_2 \tag{6.5}$$

The critical points appear for the value of u_1 and u_2 for which $T_0 = 0$ which is of the same type as the d.e. (4.1), Chapter 26.

It is noted that in the (u_1,u_2) plane, the integral curve du_1/du_2 behaves normally since the discontinuity occurs in $\dot{u}_1$ and $\dot{u}_2$ and not in the integral curve itself. This result follows also from the form of the d.e. (4.1), Chapter 26.

Inasmuch as the variables u_1 and u_2 remain bounded, although the d.e. (5.4) lose their meaning for the values $u_1 = u_{10}$, $u_2 = u_{20}$ for which $T_0 = 0$ the analytic continuation is impossible (since $\dot{u}_1 = \infty$ and $\dot{u}_2 = \infty$), the physical continuation of the solution is still possible, because $\int u_j dt$, $j = 1, 2$, remain continuous at the point (u_{10},u_{20}). If t_0 is the instant when $u_1 = u_{10}$, $u_2 = u_{20}$, the condition of continuity is

$$U_1(t_0 - 0) = U_1(t_0 + 0); \qquad U_2(t_0 - 0) = U_2(t_0 + 0) \tag{6.6}$$

where $U = \int u dt$. In terms of the function S, this leads to the expressions

$$\frac{U_1(t_0)}{U_2(t_0)} = \frac{U_1(t_0 \pm 0)}{U_2(t_0 \pm 0)} = \frac{u_1(t_0)}{u_2(t_0)} = \frac{u_1(t_0 \pm 0)}{u_2(t_0 \pm 0)} \tag{6.7}$$

Approximating the characteristics of the electron tubes by straight lines (Fig. 27.4), we can assume that the slope of this characteristic is S_j in the interval (O,V) of the grid voltage and zero outside that interval. In such a case, the curve T_0 of equation $a_1(u_1)a_2(u_2) - b_1b_2 = 0$ is a rectangle $u_1 = 0$, $u_1 = V_1$, $u_2 = 0$, $u_2 = V_2$ as shown in Fig. 29.10.

If $T > 0$, inside this rectangle, this reduces to the condition

$$S_1S_2 > \left(\frac{1}{R_1} + \frac{1}{r_1}\right)\left(\frac{1}{R_2} + \frac{1}{r_2}\right) \tag{6.8}$$

The multivibrator is capable of oscillating since $a = CRr\varphi'$, $b = C(R + r)$, φ' being the slope of φ; the trajectories R (inside the rectangle) are then given by equation:

$$(a_1u_1 - b_2u_2 + g_2)du_1 = (a_2u_2 - b_1u_1 + g_1)du_2 \tag{6.9}$$

where $g_j = V_jb_j - V_ka_k$ are constants.

It can be shown that these curves in the (V_1V_2) plane have the appearance indicated in Fig. 29.10. Moreover, it is also shown that the expression $T_xX + T_yY$ is positive for any point inside the rectangle T except at the point A which is a node. Thus any point of T is a repulsive point.

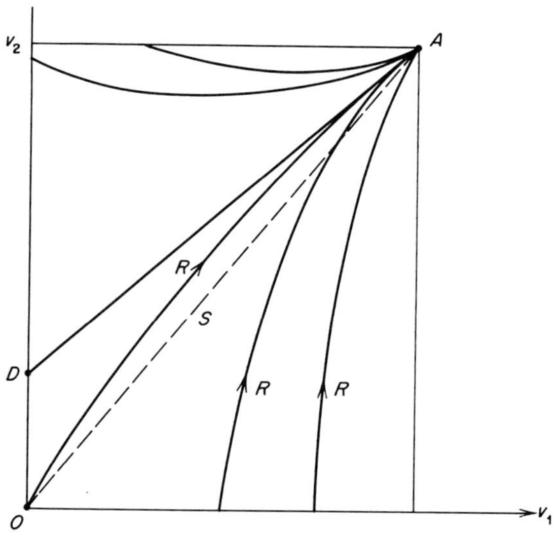

FIGURE 29.10

Inside the rectangle T the trajectories are given by equation (6.9). These R curves are shown in Fig. 29.10. The establishment of the piecewise analytic oscillation will depend on the slope of the initial solution DA. If DA goes through O, the variables u_1 and u_2 will grow up simultaneously, so that the curves representing these variables will have the same form up to a certain factor of proportionality. If DA approaches the sides of the rectangle T_0 (that is, the ratio a_1/a_2 deviates from unity), there will be an asymmetrical oscillation; one of the voltages will increase rapidly and the other, on the contrary, will increase slowly.

A great deal of experimental work was done to check these conclusions but we refer to Vogel's publications[1] for further details.

7. A further extension of the theory

The preceding discussion applies to systems that exhibit a piecewise analytic operation interrupted by discontinuous stretches as is also the case of the M.C. theory (Chapter 26).

It is possible, however, to generalize this theory also for cases when two different laws of motion replace each other during one cycle without involving any discontinuities like stretches S.

Since each law of motion requires its own system of d.e., we reach a different formulation which is usually termed *piecewise linear* theory, as it is customary to take for each of these laws the simplest possible form, for instance, amenable to a simple linear d.e. We shall postpone the investigation of piecewise linear phenomena to Chapter 31, but will give a brief outline of Vogel's approach to this question.

If one follows the preceding argument, it is clear that for such a representation we shall need two systems of d.e. of the first order which will alternatively replace each other when the representative point reaches the boundary of the curve T. Since both systems are analytic, the trajectories will be analytic throughout the cycle except at the points at which one system of d.e. replaces the other.

We have thus a familiar situation, but in this theory the approach to this question is more general as it also takes into account the so-called *hereditary actions* with which we have been partially concerned in Chapter 21.

Consider a system of d.e. of the form

$$\dot{x} = X(x,y) + U(x,y,H); \qquad \dot{y} = Y(x,y) + V(x,y,H) \tag{7.1}$$

where U and V are functions (we can assume them to be analytic) of the present state $[x(t), y(t)]$ and also of the functional H of all the past states $[x(t'), y(t')]$; $0 \leq t' < t$ where t begins to be counted from the moment when the physical system begins to operate.

It is usually assumed that H is a linear functional appearing as a curvilinear integral taken along the trajectory C from $t = 0$ up to the present moment t, that is,

$$H = \int_C h(x,y)ds(x,y) \tag{7.2}$$

[1] See footnote [1], page 648.

We need a similar formulation for this problem in which $\dot{x}$ and $\dot{y}$, respectively, are either X_1, Y_1, or X_2, Y_2 with the change of functions each time the trajectory reaches the boundary T.

One has to represent the integral as the sum of discrete quantities; every point not on the boundary does not contribute to the value of the integral.

This is what is known as a Stieltjes integral and is written as $\int dm(x,y)$ with zero "mass density" except on T. Moreover, the value of the integral must alternate between two fixed numbers each time the boundary is reached. This means that the mass densities must be of opposite signs for the $\oplus$ points and for the $\odot$ points. In such a case, in fact, we shall have $H = 0$ for the start; for the first boundary point $\odot$ we have $H = +1$, and for the second boundary point $\oplus$ the value of H will be -1, and so on.

Finally the values of $X + U$ and $Y + V$ must alternate between X_1, X_2 and Y_1, Y_2; this can be obtained by taking

$$X = X_1; \qquad Y = Y_1; \qquad U = (X_2 - X_1)H; \qquad V = (Y_2 - Y_1)H \tag{7.3}$$

An adequate set of integro-differential equations for such a system will be of the form

$$dx/[X_1 + (X_2 - X_1)\int dm(x,y)] = dy/[Y_1 + (Y_2 - Y_1)\int dm(x,y)] = dt \tag{7.4}$$

with the mass densities:

0 if (x,y) is not on the boundary
$+1$ if (x,y) is on the boundary and $\oplus$
-1 if (x,y) is on the boundary and $\odot$

8. Concluding remarks

In the general form which has been outlined in Section 7 this theory has not found yet any definite applications. In the more restricted form explained in the beginning of this chapter, on the contrary, this theory gives the same results as the M.C. theory (Chapter 26) as may be expected, of course. There is, however, a definite advantage of Vogel's theory in that the question of stability is reduced to a simple formulation owing to the transformation $(x,y) \to (r,s)$ by which the problem of stability is reduced to the form used in the theory of analytic limit cycles (Chapter 7).

It is recalled that this criterion consists in comparing the slopes of the characteristics at the point of their intersection.

In Chapter 31 we shall outline a similar theory but based on the existence of a fixed point of the point transformation theory.

In conclusion it may be worth mentioning that, although the generalized theory (Section 7) has not found any applications so far, the principal reason is likely to be that all existing experimental evidence relates to electronic circuits in which hereditary actions are, practically, absent. The situation might be different if one were to undertake the analysis of oscillating discharges or similar plasma-phenomena where these actions are likely to be felt.

On the other hand, the more elementary theory (Sections 2 and 3) has been confirmed by numerous experiments by Vogel[1] himself and Sideriades.[5] A formal justification of this theory is also discussed.[6]

[1] See footnote [1], page 648.

[5] L. Sideriades, *Methodes Topologiques appliquées à l'electronique, Ministère de l'Air*, No. N.T. 84, Paris, 1959.

[6] A. D. Myshkis and A. J. Hochriakov, *Mat. Sbornik* (in Russian) **45**, 1, 1958.

Chapter 30

ASYMPTOTIC METHODS

1. Introductory remarks

As we mentioned in the Introduction to Part IV, in the asymptotic methods no physical idealization is made and a degenerescent d.e. of the second order is considered as such. The situation is thus much clearer to begin with since no idealizations are needed.

The impossibility of using power solutions directed attention to other methods. Liénard was probably the first to draw attention to the asymptotic procedure (Chapter 4). Later on Haag[1] suggested the use of the asymptotic expansions, but the procedure was not explained sufficiently and the matter was more or less ignored by physicists until Flanders and Stoker[2] presented this subject in a simple manner in connection with the van der Pol equation. Their outline, however, touched on a purely asymptotic case ($\mu \to \infty$).

Beginning in 1945 there appeared a series of papers by M. L. Cartwright and J. E. Littlewood concerning the nonhomogeneous van der Pol equation in which μ is large.

This is obviously the most general case since the simplifications resulting from the asymptotic assumption ($\mu \to \infty$) are given up and, besides the right-hand term (containing t explicitly) makes the analysis still more general.

In what follows, we have restricted this outline only to the homogeneous case investigated by Cartwright in one of her papers,[3] omitting many

[1] J. Haag, *C. R. Ac. Sc.* (Paris) **202**, 1936; **204**, 1937; **206**, 1938; *Ann. Ec. Norm. Sup. Series* **3**, 60, 1943; **3**, 61, 1944; L. Cesari, *Asymptotic Behavior and Stability Problems*, Springer, Berlin, 1959.

[2] D. A. Flanders and J. J. Stoker, *Inst. of Math. and Mech.*, New York University, 1946.

[3] M. L. Cartwright, *Contributions to the Theory of Nonlinear Oscillations*, ed. by S. Lefschetz, II, 1955.

details and trying to present only the essence of the method (Section 4). The great simplification of the problem is due to the use of the (t,x) plane (instead of the conventional $(x,\dot{x})$ plane) which permits simplifying the procedure in the different regions of the *curve supposed to be known.* It is to be noted that the procedure does not require any special mathematical device and proceeds along the line of the standard theory of d.e. The difficult point, as we mentioned in the Introduction, is not in the basic idea of the procedure but in a rather delicate way of carrying it through. In this connection it becomes necessary to use, in addition to the d.e. itself, the corresponding integrated, differentiated and energy equations. We give an abridged version of this method in Section 4.

Finally, still later, Dorodnitzin[4] took up this problem from the point of view of the theory of asymptotic expansions (Sections 5 and 6), thus completing the work of Haag.

As was mentioned previously, the asymptotic methods now constitute a considerable advance in the theory by *justifying* the existence of the solution (known either from a graphical procedure of isoclines or from the experimental data).

It is obvious that their applications are less important. In fact from the preceding survey of the relaxation problems (Chapters 27 and 28) it is observed that most of them are not reducible to the d.e. of van der Pol and, moreover, very often no parameter is involved.

However, as some of these problems have two or more degrees of freedom, they are amenable to differential systems of at least fourth order, and these extensions are yet to be made.

It is recalled that one of the most frequently encountered circuits, the multivibrator, belongs to the case which is beyond the reach of a simple van der Pol equation, particularly in its asymmetrical form.

All this seems to indicate that there is still much work to be done on the asymptotic methods before they can become a working tool in hands of physicists and engineers. This, however, has nothing to do with the intrinsic value of these methods but merely concerns their applications.

2. Asymptotic theory versus discontinuous theory

A particularly simple approach to this problem was indicated by Flanders and Stoker[2] in connection with Rayleigh's d.e.

$$\ddot{x} + \mu F(\dot{x}) + x = 0 \tag{2.1}$$

[4] A. Dorodnitzin, *Prikl. Math. i Meh* (in Russian) **2**, 1947; English translation by Am. Math. Soc., No. 81, 1953.

[2] See footnote [2], page 664.

Since this d.e. reduces to the van der Pol equation by a change of the dependent variable, conclusions hold also for the latter equation, although the discussion is simpler for the d.e. (2.1).

The equivalent system in this case is

$$\dot{x} = y; \qquad \dot{y} = -\mu F(y) - x \tag{2.2}$$

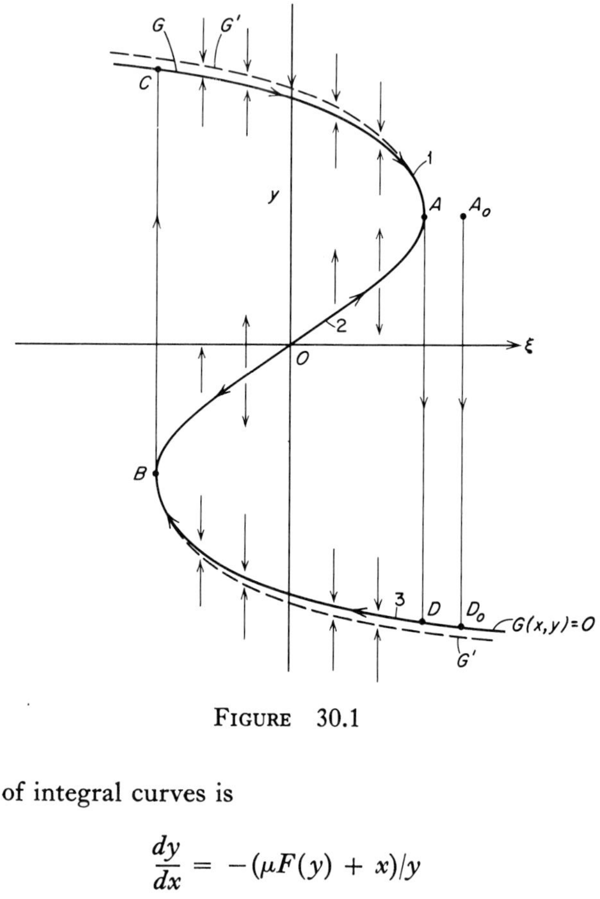

FIGURE 30.1

and the d.e. of integral curves is

$$\frac{dy}{dx} = -(\mu F(y) + x)/y \tag{2.3}$$

Introducing the new variable ξ defined by the relation: $x = \mu\xi$, these equations become

$$\xi = y/\mu; \qquad \dot{y} = -\mu[F(y) + \xi] \tag{2.4}$$

$$\frac{dy}{d\xi} = -\mu^2\left[\frac{F(y) + \xi}{y}\right] \tag{2.5}$$

We consider the asymptotic case when μ is a large number. Assuming a "soft" characteristic, one has

$$F(y) = -y + \tfrac{1}{3}y^3 \tag{2.6}$$

in which case (2.5) becomes

$$\frac{dy}{d\xi} = \mu^2[y - \tfrac{1}{3}y^3 - \xi] = \mu^2 G(\xi,y) \tag{2.7}$$

As μ is assumed to be large, $dy/d\xi$ is very large, that is, the integral curve $y = y(\xi)$ is practically vertical everywhere in the (ξ,y) plane except on the curve $G(\xi,y) = 0$, where its direction is horizontal.

Figure 30.1 shows the curve $G(\xi,y) = 0$ consisting of three branches separated by points A and B of coordinates $(\frac{2}{3},1)$ and $(\frac{2}{3},-1)$, respectively; at these points the curve $G(\xi,y) = 0$ has vertical tangents.

The direction field is obtained by investigating the sign of $G(\xi,y)$ for the points above and below the branches 1, 2, and 3 of the curve $G(\xi,y) = 0$. This direction field is indicated by the arrows in Fig. 30.1. It is seen that near to the branches 1 and 3 the direction field is directed *toward* these branches on both sides, while on 2 it is, on the contrary, directed *away* from that branch. This shows that the branches 1 and 3 are stable and the branch 2 unstable, since any small deviation of the representative point R moving on G is corrected on branches 1 and 3 so as to turn it back toward the curve; on branch 2 any such deviation is still more emphasized by the direction field. This statement needs, however, a more accurate formulation. In fact, on the curve $G(\xi,y)$ itself $dy/d\xi = 0$, so that R cannot follow G curve being *exactly* on it. But, as we have assumed that μ is very large, $dy/d\xi \simeq \infty$ very near to the curve G on both sides, as is shown in Fig. 30.2. This means that R can still follow the general direction of the curve G being slightly above the branch 1 and slightly below the branch 3, as shown in Fig. 30.1 in broken lines G'. These curves G' are the nearer to the branches 1 and 3 the greater is the value of μ. In order to determine the direction of motion of the representative point R along the curve G', one has to investigate, as usual, the sign of $\dot{\xi}$ and $\dot{y}$ along the curve with the assumed form (2.6) for $F(y)$. One ascertains that these directions are

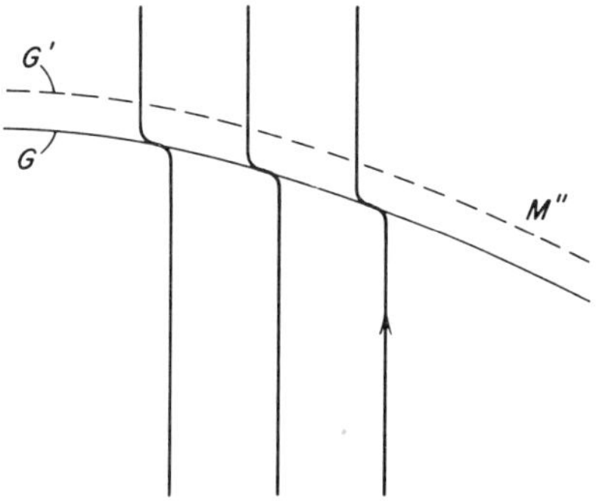

FIGURE 30.2

those which are indicated by arrows *along* the curve $G(\xi,y) = 0$ (Fig. 30.1). It is easy to see what happens if one assumes some arbitrary initial condition represented by a point A_0. As at this point the direction field is vertical downward, R will drop vertically downwards until it meets the curve G (or, more correctly, a curve G' slightly below G as was just explained) which it follows in the prescribed direction (from right to left) until it reaches the point B. At this point a special situation arises: in fact, B is not a position of equilibrium; moreover R cannot follow the branch 2 because the allowed direction on this branch is from 0 to B. Hence the only (nonanalytical) issue is to follow the vertical direction (from B to C), since it is available at point B. Arriving at point C on branch 1, the representative point begins to follow again the analytic branch 1 (from left to right) until it reaches point A, where the situation is exactly the same as it was at point B. This results again in another downwardly directed vertical jump from A to D.

This determines the motion on the piecewise analytic cycle $ADBCA$ consisting of two analytic branches CA and DB "closed" by discontinuous stretches AD and BC.

The familiar concept of a piecewise analytic cycle is reached here in a very simple manner by investigating the direction field, although there are some tacit assumptions which must be taken into account.

In the first place, as R cannot follow *exactly* the curve $G(\xi,y) = 0$ as long as μ is finite, as is assumed here, it follows a neighboring curve G', which is a plausible assumption since, $dy/d\xi$ being zero on G and, practically, infinite near to G, there is always an appropriate value of $dy/d\xi$ in the narrow band around the curve $G(\xi,y) = 0$ which permits following the average direction closely enough to the G curve.

There is another resemblance between the two theories, namely: as long as R is on the analytic branch (1 or 3) the motion is governed by the d.e. but, on arriving either at A or at B, the representative point "has no other issue" but to follow the vertical field. In the discontinuous theory we had a similar conclusion owing to the introduction of the condition of Mandelstam; here we reach it owing to the availability of the vertical field which permits the representative point to escape from a kind of an *analytical deadlock* just mentioned.

In the discontinuous theory these stretches AD and BC are traversed in no time; here they are traversed with a large but finite velocity as long as μ is finite.

There is thus a close resemblance between the two theories but what is "discontinuous" and happens "in no time" in the discontinuous theory here, in the asymptotic theory is "quasi-discontinuous" and happens in a very short time.

One can reach similar conclusions in a slightly different manner. Suppose we have a d.e.

$$\mu\ddot{x} + f(x,p)\dot{x} + x = 0 \tag{2.8}$$

where μ is a small parameter and p is a finite parameter. The term $\mu\ddot{x}$ is small if $\ddot{x}$ is bounded; in such a case (2.8) can be replaced by the degenerate d.e. of the first order

$$\dot{x} = -x/f(x,p) \tag{2.9}$$

and if $f(x,p) \neq 0$ for all t, this d.e. gives a good approximation.

To be more specific, assume that (2.8) has the form:

$$\mu\ddot{x} + (p - x^2)\dot{x} + x = 0 \tag{2.10}$$

If one neglects the term $\mu\ddot{x}$, it becomes

$$[(p - x^2)/x]\dot{x} = -1 \tag{2.11}$$

and the integration yields

$$\int_{x(0)}^{x(t)} \left(\frac{p}{x} - x\right)dx = p \log \frac{x}{x_0} - \frac{x^2 - x_0^2}{2} = -t$$

Thus

$$t = h(x) = \frac{x^2 - x_0^2}{2} - p \log \frac{x}{x_0} \tag{2.12}$$

whence

$$\frac{dt}{dx} = h'(x) = x - \frac{p}{x} = \frac{x^2 - p}{x} \tag{2.13}$$

Assume that $h'(x) = (dt/dx) < 0$ for $x < \sqrt{p}$. As $(dt/dx) < 0$, clearly $(dx/dt) < 0$ so that, if $x_0 < \sqrt{p}$, x decreases and for all t one has $x < \sqrt{p}$. Likewise, if $x_0 > \sqrt{p}$, one has $(dx/dt) > 0$ so that $x > \sqrt{p}$ for all t. As in this case x and x remain bounded, we can use the degenerate equation (2.9) instead of the full equation (2.8).

Suppose now that the sign of the second term in (2.10) is changed. If one proceeds as previously, one obtains

$$t = h(x) = p \log \frac{x}{x_0} - \frac{x^2 - x_0^2}{2} \qquad \text{and} \qquad \frac{dt}{dx} = \frac{p - x^2}{x}$$

If now $x_0 < \sqrt{p}$, $(dt/dx) > 0$ and, therefore, $(dx/dt) > 0$. When x reaches the value $\sqrt{p}$, one has $\dot{x} = \infty$ and thus $\ddot{x} = \infty$.

If one uses the (t,x) plane representation (Fig. 30.3), the fact that $\dot{x} \to \infty$ means that the trajectory approaches the verticality and $\ddot{x} \to \infty$ means infinite curvature at this point (that is, an infinitely small radius of curvature). In this case the trajectory will have the appearance shown in Fig. 30.3. One could use a similar elementary argument in connection with $a(x,v)$ plane where $v = \dot{x}$. Here again if $\ddot{x}$ is bounded for all t, one can neglect the term $\mu\ddot{x}$ and one has simply $v = x/(p - x^2)$ which can be discussed directly (Fig. 30.4) without integration. However, if $x = \sqrt{p}$ or is close to this value, $v = \dot{x}$ becomes large as well as $\ddot{x}$ and one has to use the full equation (with $\mu\ddot{x}$) as the result of which, instead of the theoretical curve $v(x)$ for the degenerate equation having $x = \sqrt{p}$ as asymptote, the real curve $v(x)$ will have the appearance shown in broken line. This shows that the effect of the term $\mu\ddot{x}$ in this range limits the instantaneous jump in v from $+\infty$ to $-\infty$ to a finite change occurring rapidly but not instantaneously.

These simple examples do not introduce anything essentially new but merely show the difference between the results obtained on the basis of the idealized theory and those yielded by the complete theory in which the effect of the term $\mu\ddot{x}$ is taken into account.

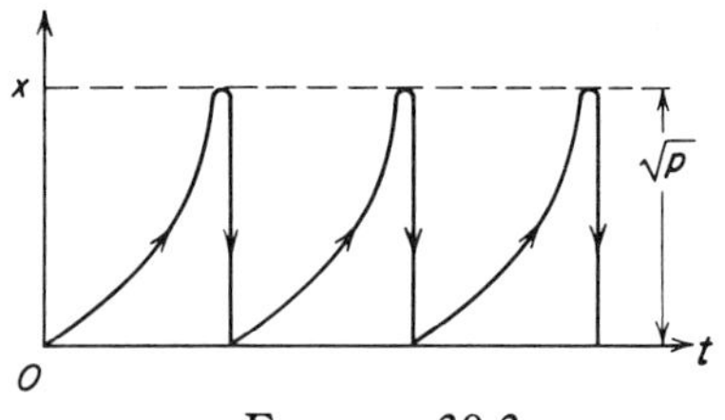

FIGURE 30.3

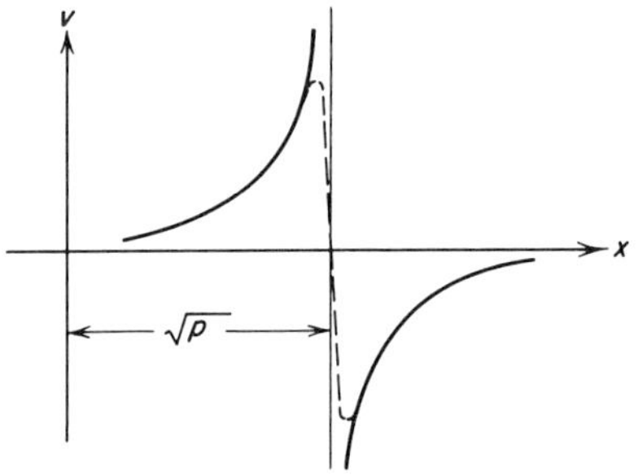

FIGURE 30.4

3. Asymptotic theory applied to an asymmetrical multivibrator

As a second example we consider the case of a multivibrator with one degree of freedom which has been investigated in Section 1, Chapter 27, by the discontinuous theory which resulted in a d.e. of the form

$$-\varphi(x)\dot{x} + x = 0 \tag{3.1}$$

Since in the asymptotic theory, the degeneration theory is not used, we have to use the d.e. of the second order

$$\mu\ddot{x} - \varphi(x)\dot{x} + x = 0 \tag{3.2}$$

where μ is a small number. In physical terms this means that we do not neglect the effect of a small inductance as we did in the discontinuous theory dealing with the degenerate d.e. of the first order.

It is noted that this d.e. is not exactly of the van der Pol type since dividing it by μ one obtains two large terms, one associated with $\dot{x}$, as in the van der Pol case, and the other x/μ which is also large. The latter *does not* correspond to the van der Pol equation. In view of this one may expect a result different from that of the preceding section.

Written as an equivalent system, the d.e. (3.2) becomes

$$\dot{x} = y; \qquad \dot{y} = \lambda[\varphi(x)y - x]; \qquad \lambda = \frac{1}{\mu} \tag{3.3}$$

and the d.e. of integral curves is

$$\frac{dy}{dx} = \lambda[\varphi(x) - x/y] \tag{3.4}$$

If one assumes, for instance, $\varphi(x) = 1 - x^2$, the preceding equation becomes

$$\frac{dy}{dx} = \lambda(1 - x^2 - x/y) = \lambda G(x,y) \tag{3.5}$$

where λ is a large number.

The argument remains the same as in the preceding section, namely: dy/dx is large everywhere in the (x,y) plane except on the curve

$$G(x,y) = 1 - x^2 - x/y = 0 \tag{3.6}$$

where it is zero (this presupposes that λ is large but not infinity). The curve $G(x,y) = 0$ is shown in Fig. 30.5; it consists again of three branches; the branches 1 and 3 are stable and 2 unstable which one ascertains by the same argument as previously. Likewise, as on the curve $G(x,y) = 0$ the direction field is horizontal, R cannot follow this curve but it can follow curves G' slightly above the branch 3 and slightly below the branch 1 as shown in broken lines.

As y increases, $\dot{x}$ also increases so that the line $x = -1$ is crossed with a high velocity from left to right. In the interval $(-1, +1)$ this velocity continues to increase until the line $x = +1$ is crossed, after which the velocity begins to decrease; but this decrease is not discontinuous since the d.e. is of the second order.

Once R reaches the branch 3, it follows G' toward the line $x = +1$ which is crossed from right to left with an ever-increasing velocity in the interval $(+1, -1)$. As soon as $x = -1$ is crossed, the velocity decreases and, later, reverses, which brings R on some point of the curve G' on the branch 1 which continues the process.

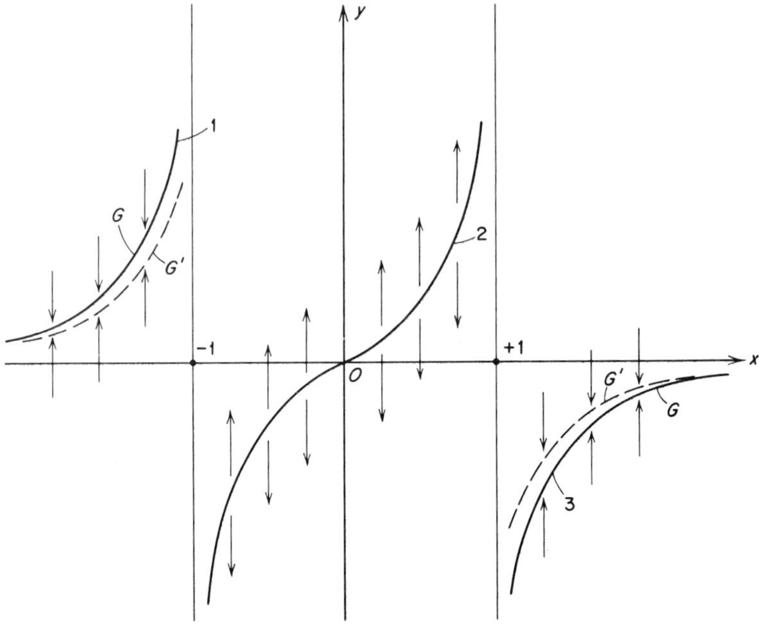

FIGURE 30.5

Everything happens as if the branches 1 and 3 were sending R toward the unstable interval $(-1, +1)$ through which R is conveyed with an ever-increasing velocity and, on crossing the limits of this interval, the stable branches (1 or 3) direct it again into the interval $(-1, +1)$ which maintains the oscillation. It is noted that in this representation there are no discontinuous stretches since the d.e. is of the second order, but the actual calculation of trajectories requires a step-by-step determination which ultimately reduces to a tedious graphical construction.

In the remaining sections of this chapter we outline briefly the exact asymptotic methods needed for these calculations.

4. Method of Cartwright-Littlewood

Since it is impossible to give here a full account of this work [5] we mention briefly its scope and the conclusions in a particular case which interests us here.

[5] M. L. Cartwright and J. E. Littlewood, *J. London Math. Soc.*, 1945; *Ann. of Math.* 1947; *Proc. Cambridge Math. Soc.*, Vol. 45; J. E. Littlewood, *Acta Math.* **97** and **98**; L. Cesari, *Asymptotic Behavior and Stability Problems*, Springer, Berlin, 1959.

The d.e. is considered in the form

$$\ddot{x} - \mu f(x)\cdot\dot{x} + g(x) = b\mu p(t) \tag{4.1}$$

where f, g and p are independent of μ; μ large. This d.e. has been discussed in a number of papers[5] but we shall confine our attention to the autonomous case only ($p = 0$) which is given in one of the papers of M. L. Cartwright.[3] The investigation is conducted in the (t,x) plane.

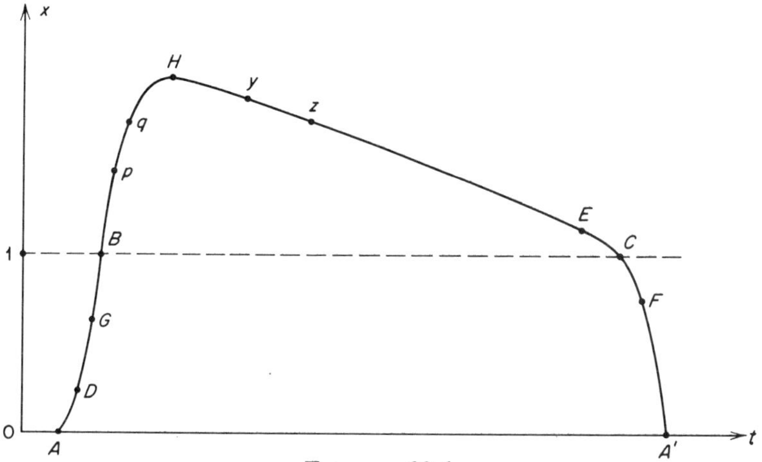

FIGURE 30.6

The purpose of this analysis is to show that for μ large the d.e.

$$\ddot{x} + \mu(x^2 - 1)\dot{x} + x = 0 \tag{4.2}$$

has the solution: $x(t)$ of the form obtained experimentally by van der Pol and shown in Fig. 30.6.

In addition to (4.2) two other equivalent equations are used: the *integrated equation*

$$\Delta\dot{x} = \dot{x} - \dot{x}_0 = \mu\left(x - \frac{x^3}{3} - x_0 + \frac{x_0^3}{3}\right) - \int_{t0}^{t} x\,dt \tag{4.3}$$

and the *energy equation*:

$$\Delta\dot{x}^2 = \dot{x}^2 - \dot{x}_0^2 = 2\mu\int_{t0}^{t}(1 - x^2)\dot{x}^2dt - x^2 + x_0^2 \tag{4.4}$$

The use of the (t,x) plane is particularly convenient here as the analysis

[5] See footnote [5], page 672.
[3] See footnote [3], page 664.

of the curve in its different regions gives directly information regarding the relative order of magnitude of x, $\dot{x}$, and $\ddot{x}$. As the quantity $\dot{x}$ is measured by the *slope* of the curve and $\ddot{x}$ by its *curvature*, the possible simplifications in equations (4.2), (4.3), and (4.4) become obvious. For instance, on the stretch Aq the velocity $\dot{x}$ is high and nearly constant (so that $\ddot{x} \approx 0$); on the contrary, in the region around H the acceleration $\ddot{x}$ is large (since $\dot{x}$ varies rapidly) but, on the other hand, x changes very little; on the stretch HE the velocity is negative and not very large; besides this, it is also nearly constant (so that $\ddot{x} \approx 0$ again), and so on. One can simplify the d.e. on this basis proceeding from one region to the other. One integrates these simplified d.e. and joins the integral curves by taking the terminal conditions of one stretch as the initial conditions of the next one.

In fact this approach remains practically the same as that outlined in Section 5 but the procedure does not require the use of asymptotic expansions. One can readily see that the estimate of the order of magnitude becomes very important here because the procedure hinges largely on a judicious choice of what has to be retained and what can be neglected.

The first thing is to ascertain how rapidly a solution starting from various points converges to the periodic solution. In what follows we shall designate the solution† passing through the point H (Fig. 30.6) as $x = h$; $\dot{x} = \dot{h}$ and, likewise, for other points.

Clearly for H we have: $x = h$, $\dot{x} = 0$; for C: $x = 1$, $\dot{x} = \dot{c}$; and for A: $x = 0$, $\dot{x} = \dot{a}$, where h, $\dot{c}$, and $\dot{a}$ may be any numbers except that, obviously $h > 1$, $\dot{a} > 0$, $\dot{c} < 0$.

One establishes first a number of lemmas necessary for the proof of the final theorems. We shall give only a short outline here and refer the reader to Cartwright[3] for details. In lemma 1, we use the energy equation for the stretch HC; in lemma 7 the same equation is used for the stretch CA'. A more detailed study shows that a similar procedure holds for other parts of the curve, taking into account the corresponding orders of magnitude of x, $\dot{x}$, and $\ddot{x}$ on each particular stretch. In addition to using the integrated equation (4.3), one also uses the differentiated equation

$$\dddot{x} + \mu(x^2 - 1)\ddot{x} + 2\mu x\dot{x}^2 + \dot{x} = 0 \tag{4.5}$$

All these expressions are strictly equivalent to the original equation (4.2) and there is no point in using one in preference to another. The energy equation (4.4) is particularly useful in showing that the energy $\rho = x^2 + \dot{x}^2$ decreases for $x^2 > 1$ and increases for $x^2 < 1$. In a similar way one proceeds to more detailed approximations. Use of the above criteria:

† The term "solution" used in this section is equivalent to the term "trajectory" used previously.

[3] See footnote [3], page 664.

(1) one shows that on the stretch HE the last two terms: $\mu(x^2 - 1)\dot{x} + x$ are the most important; (2) for the stretch EF the significant terms reduce to $\ddot{x} + x$; (3) for the stretch FA' the first two terms (as the interval near $x = -1$ is very short in time); and so on. In all cases one is more interested in x and $\dot{x}$ than in $\ddot{x}$. On the stretch HE the time is long and, as μ is large, in the integrated equation the terms

$$\mu\left(\frac{e^3}{3} - e\right) - \mu\left(\frac{h^3}{3} - h\right) - \int_0^t x dt$$

are large compared with $\int_0^t \ddot{x} dt = \dot{e} - \dot{h} = \dot{e}$. On EF the term $\mu \int_0^t (x^2 - 1)\dot{x}^2 dt$ is small so that the change of energy is small provided E and F are sufficiently near C. In addition to the form of integral curves in different regions, the time intervals for crossing them can also be calculated; this results in the calculation of the period.

The problem studied by Cartwright-Littlewood (μ large but finite) occupies the intermediate position between the two asymptotic cases; (1) μ small, and (2) $\mu \to \infty$ in which conditions are simpler. We have seen that for μ small there is no necessity for using the additional equations (4.3) and (4.4) since in this range the homogeneous van der Pol equation can be integrated directly by approximations.

As to the other asymptotic case ($\mu \to \infty$), the discontinuous theory (Chapter 26) gives also a simple qualitative approach in which, on the contrary, the d.e. (at the points of discontinuities) gives way to the invariant (analogous to equation (4.3) and (4.4)) which governs the idealized discontinuous transition. In the difficult intermediate range with which the Cartwright-Littlewood theory is dealing, it becomes necessary to make use of all three equations: the van der Pol equation as well as equations: (4.3) and (4.4).

From this short survey it is noted that the difficult problem of van der Pol's equation when μ is large is solved "piece by piece" by splitting the integral curves into certain characteristic regions in each of which the problem can be simplified to the extent that the integration becomes possible. The remaining part of the work is merely to "fit the pieces" together by imposing some plausible conditions.

The work of Dorodnitzin (Section 6) appeared after the first papers of Cartwright-Littlewood were published; it introduced the use of asymptotic expansions which are convenient. Both approaches lead ultimately to the same result as one could expect from the "piecewise" analytic treatment of the initial d.e.

It must be noted that in this and in the preceding sections it was possible to *explain* by analysis the form of the integral curve obtained either graphically (by the method of isoclines) or experimentally; this analysis becomes possible if one splits the whole periodic process into "pieces" which can be joined. Idealization of this nature is partly the same as that used in Chapter 26, but merely arranged in a different manner: instead of treating points of very bad analyticity as definitely nonanalytic (for example, having two distinct tangents), the same situation here is treated by introducing a small "joining arc" with a very large curvature (a very small radius of curvature) so that the local idealization proceeds, so to speak, systematically through all regions without introducing any special "critical" points or lines at which features suddenly appear that did not exist on other points of otherwise analytic arcs (compare Chapter 26).

5. Asymptotic expansions

As was mentioned in the Introduction to Part III, the power series solution cannot be used in connection with the van der Pol equations when the parameter μ is not sufficiently small, because in this case the series diverges and cannot be used.

There exists, however, a possibility of obtaining the desired result even in this case if, instead of a power series, one uses a special *asymptotic expansion.*

This question was known already to Euler, but the modern developments in this field began with the work of Poincaré [6] and it is useful to say a few words about this series.

Consider the function

$$f(x) = \int_x^\infty t^{-1}e^{x-t}dt \tag{5.1}$$

where x is real and positive and the integration is taken along the real axis.

By repeated integration by parts one gets

$$f(x) = x - x^{-2} + 2!x^{-3} + \dots + (n-1)!(-1)^{n-1}x^{-n} + n!(-1)^n \int_0^x t^{-(n+1)}e^{x-t}dt \tag{5.2}$$

We set

$$u_{n-1} = (n-1)!(-1)^{n-1}x^{-n}$$

and

$$s_n(x) = \sum_{m=0}^{n} u_m = x^{-1} - x^{-2} + 2!x^{-3} - \dots + (-1)^n n!x^{-(n+1)} \tag{5.3}$$

As $u_m/u_{m-1} = mx^{-1} \to \infty$ for $m \to \infty$, the series diverges for all values of x.

[6] H. Poincaré, *Les méthodes nouvelles de la mécanique céleste* **T.3**, Gauthier-Villars, Paris, 1892; E. Goursat, *Cours d'Analyse* **T.2**; Gauthier-Villars, Paris, 1918.

In spite of this the series can be still used for the calculation of $f(x)$. In fact, suppose we fix the number n and calculate the value s_n, which gives

$$f(x) - s_n(x) = (-1)^{n+1}(n+1)! \int_x^\infty t^{-(n+2)} e^{x-t} dt$$

Since $e^{x-t} \leq 1$, one has

$$|f(x) - S_n(x)| = (n+1)! \int_x^\infty t^{-(n+2)} e^{x-t} dt < (n+1)! \times \int_x^\infty t^{-(n+2)} dt = \frac{n!}{x^{n+1}} \tag{5.4}$$

For a sufficiently large value of x the right-hand side of (5.4) is very small and the function $f(x)$ can be approximated sufficiently well by taking the sum of sufficient number of terms.

Thus, for instance, for $n = 5$ and $x = 10$, one has $|f(10) - S_5(10)| < 0.00012$. This series is called the *asymptotic expansion* of the function $f(x)$. The following definition of an asymptotic expansion was given by Poincaré:

A series:

$$A_0 + A_1 z^{-1} + A_2 z^{-2} + \ldots + A_n z^{-n} + \ldots \tag{5.5}$$

in which the sum of the first $(n + 1)$ terms is $S_n(z)$ is said to be an asymptotic expansion of a function $f(z)$ (for a given argument of z) if the expression $R_n(z) = z^n|f(z) - S_n(z)|$ fulfils the condition

$$\lim_{|z|\to\infty} R_n(z) = 0; \qquad (n \text{ fixed but otherwise arbitrary}) \tag{5.6}$$

This holds even if

$$\lim_{n\to\infty} |R_n(z)| = \infty; \qquad (z \text{ fixed})$$

In such a case one can always make

$$|z^n[f(z) - S_n(z)]| < \varepsilon \tag{5.7}$$

where ε is arbitrarily small if $|z|$ is sufficiently large.

The question whether a function $f(z)$ possesses an asymptotic expansion and what are the values of its coefficients is answered by expressing that the successive limiting values for $z \to \infty$ along a radius must exist, namely:

$$f(z) \to A_0; \qquad f(z) - A_0 \to A_1; \qquad f(z) - A_0 - A_1 z^{-1} \to M_2 \ldots$$

where the limit is taken either along the real axis or in a region of the complex plane.

On the other hand, for

$$f(z) = e^{-z}; \qquad -\frac{\pi}{2} < \arg z < \frac{\pi}{2}$$

all A_n are zero since $z^m e^{-z} \to 0$ for every integer $m > 0$, if $z \to \infty$ in $-\pi/2 < \arg z < \pi/2$; this result shows that different functions may have the same asymptotic expansion. Thus, if $f(z)$ has an asymptotic expansion for the above argument of z, $f(z) + ae^{-bz}$; $a > 0$, $b > 0$ has the same asymptotic expansion.

If one writes

$$f(z) = A_0 + A_1 z^{-1} + A_2 z^{-2} + \dots + A_n z^{-n} + W_n(z) z^{-n}$$

with $W_n(z) \to 0$ for $z \to \infty$ in some interval $\theta_1 < \arg z < \theta_2$, the following theorems can be proved:

(1) If $f_1(z) = f_2(z)$, then $A_n = B_n$ (uniqueness theorem)

(2) If $F(z) = f_1(z) \times f_2(z) = C_0 + C_1 z^{-1} + C_2 z^{-2} + \dots + C_n z^{-n} + W_n(z) z^{-n}$, then

$$C_n = \sum_0^n A_1 B_{n-1} \quad \text{(the product theorem)}$$

(3) If $F(z) = f_1(z)/f_2(z) = C_0 + C_1 z^{-1} + C_2 z^{-2} + \dots + C_n z^{-n} + W_n(z) z^{-n}$, then

$$B_0 C_0 = A_0; \quad B_1 C_0 + B_0 C_1 = A_1; \quad B_2 C_0 + B_1 C_1 + B_0 C_2 = A_2$$

(the quotient theorem)

(4) If $F(z) = A_2 z^{-2} + A_3 z^{-3} + \dots + A_n z^{-n} + W_n(z) z^{-n}$, then

$$\int_0^\infty F(z)dz = -A_2 z^{-1} - (A_3/2) z^{-2} - \dots + [A_n/(n-1)] z^{-n+1} + \varepsilon_n(z) z^{-n+1} \quad \text{(the integration theorem)}$$

(5) If $F(z) = A_1 z^{-1} + A_2 z^{-2} + \dots + A_n z^{-n} + W_n z^{-n}$, then

$$F'(z) = -A_1 z^{-2} - 2A_2 z^{-3} - \dots - nA_n z^{-n-1} - \varepsilon_n(z) z^{-n-1} \quad \text{(the differentiation theorem)}$$

In this case it is assumed that $F'(z)$ can be developed asymptotically. We shall limit ourselves to this short outline as this subject can be found easily in textbooks.

In the following section we shall see the application of these expansions to relaxation problems.

6. Method of Dorodnitzin

A "method of junctions" of asymptotic expansions was developed by Dorodnitzin[4] in connection with the van der Pol equation with large

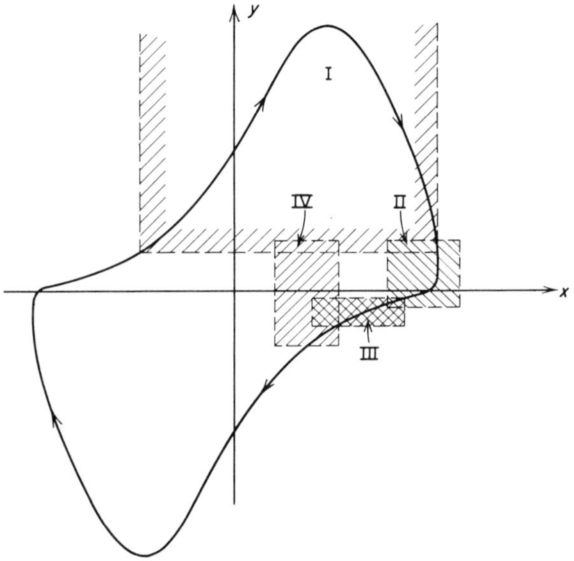

FIGURE 30.7

values of μ. We indicate this procedure following the exposition of Bogoliubov and Mitropolsky.[7] The idea of the method is to introduce certain intermediate regions in the phase-plane (Fig. 30.7) in which the asymptotic expansions representing the solution of the van der Pol equation can be "joined"; this amounts to a special form of *continuation.*

The van der Pol equation can be written as

$$y\frac{dy}{dx} - \mu(1 - x^2)y + x = 0 \tag{6.1}$$

[4] See footnote [4], page 665.

[7] N. Bogoliubov and J. Mitropolsky, *Asymptotic Methods in the Theory of Nonlinear Oscillations* (in Russian), Moscow, 1958.

In regions I and III of Fig. 30.6 this d.e. may be approximated by two abridged d.e.†:

$$y\frac{dy}{dx} - \mu(1 - x^2)y = 0 \tag{6.2}$$

$$-\mu(1 - x^2)y + x = 0 \tag{6.3}$$

These two regions cannot be joined because it is impossible to determine the integration constant of (6.2) so as to obtain the analytic continuation into region III for which (6.3) is valid. In order to obtain a junction of integral curves between these two regions, one introduces two *joining regions* II and IV in which one constructs asymptotic expansion with a view to joining them together, since in these regions II and IV the abridged equations (6.2) and (6.3) do not hold.

In view of the symmetry with respect to the origin 0, it is sufficient to carry out the argument for one half-cycle.

We construct first the solution for region I. If a_1 and a_2 are two values of x for which $dy/dx = 0$ (which in the case of a limit cycle is $a_1 = a_2 = a$, a being the amplitude of the oscillation), the region I is specified by inequalities.

$$\begin{aligned} &-1 + \varepsilon < x < a_1 - \varepsilon; \quad y > 0; \quad \varepsilon > 0 \\ &-a_2 + \varepsilon < 1 - \varepsilon; \quad y < 0; \quad \varepsilon > 0 \end{aligned} \tag{6.4}$$

The solution is sought in the form of a series

$$y = \mu \sum_{n=0}^{\infty} \mu^{-2n} f_n(x) \tag{6.5}$$

where μ is a variable increasing indefinitely.

If one substitutes this series in (6.2) and equates the coefficients of like powers of μ, a system of equations is obtained from which the functions $f_i(x)$ can be determined recursively.

† The idea of approximating (6.1) by two abridged d.e. is due to van der Pol who introduced two independent variables τ and τ' defined by relations (a) $t = \mu\tau$ and (b) $t = \tau'/\mu$. This results in the two d.e. (after dropping terms with $1/\mu^2$):

$$(x^2 - 1)\frac{dx}{d\tau} + x = 0; \qquad \frac{d^2x}{d\tau'^2} + (x^2 - 1)\frac{dx}{d\tau'} = 0$$

which can be both integrated. By this procedure one ascertains qualitatively what happens around the "sharp corners" of the (x,t) integral curve but the problem of "fitting" (which constitutes the essence of this section) has not been carried out by van der Pol.

For the first two functions the following expressions are obtained

$$f_0(x) = c + x - \tfrac{1}{3}x^3 \tag{6.6}$$

$$f_1(x) = \frac{x_1}{x_1^2 - 1}\left[\log\left(1 - \frac{x}{x_1}\right) - \frac{1}{2}\log\frac{(2x + x_1)^2 + 3(x_1^2 - 4)}{4(x_1^2 - 3)}\right]$$

$$+ \frac{x_1^2 - 2}{x_1^2 - 1}\sqrt{\frac{3}{x_1^2 - 4}}\left[\arctan\frac{2x + x_1}{\sqrt{3(x_1^2 - 4)}} - \arctan\frac{x_1}{\sqrt{3(x_1^2 - 4)}}\right] \tag{6.7}$$

where x_1 is a real positive root of $f_0(x) = 0$. It is assumed that $c > \frac{2}{3}$ as required for a limit cycle.

The procedure is the same as that which we encountered previously only, instead of the power series, we use the asymptotic series (6.5).

The functions $f_1(x)$ and also the others have special features for $x = x_1$ but the series (6.5) preserves its asymptotic character up to the value of x satisfying the inequality

$$O(x_1 - x) > O\left(\frac{\log \mu}{\mu^2}\right)\dagger$$

In particular, the series is an asymptotic series for $x = x_1 - O(1/\mu)$, in which case $y \simeq O(1)$.

Region II corresponds to the neighborhood of the points; $(y = 0, x = a_1)$ and $(y = 0, x = -a_2)$. Consider, for instance, the first point: $(y = 0, x = a_1)$ and introduce a new variable $z = -\mu y$. If one expresses x as a function of z, (6.1) becomes

$$\frac{dx}{dz} = \frac{1}{\mu^2}\frac{z}{z(x^2 - 1) - x} \tag{6.8}$$

We look again for a solution of the form:

$$x = \sum_{n=0}^{\infty} \varphi_n(z)\mu^{-2n} \tag{6.9}$$

Proceeding as previously, we determine recursively the functions $\varphi_n(z)$. The first two functions have the form

$$\varphi_1(z) = \frac{1}{a_1^2 - 1}\left[z + \frac{a_1}{a_1^2 - 1}\log\left(1 - \frac{a_1^2 - 1}{a_1}\cdot z\right)\right] \tag{6.10}$$

† The logarithms below are Naperian logarithms.

$$\varphi_2(z) = \frac{a_1}{(a_1^2 - 1)^4}\left\{(a_1^2 - 1)z\left(z + \frac{a_1^2 + 1}{a_1(a_1^2 - 1)}\right)\right.$$

$$+ \left[\frac{a_1^2 + 1}{a_1^2 - 1} + 2a_1 z - 2(a_1^2 - 1)z^2\right] \frac{\log\left[1 - \dfrac{z(a_1^2 - 1)}{a_1}\right]}{\left[1 - z\dfrac{a_1^2 - 1}{a_1}\right]}$$

$$\left. + \frac{3a_1^2 + 1}{2(a_1^2 - 1)} \log^2\left(1 - z\frac{a_1^2 - 1}{a_1}\right)\right\} \tag{6.11}$$

The functions $\varphi_n(z)$ have again special features when $z \to \dfrac{a_1}{a_1^2 - 1}$ and when $z = -\infty$, but the series (6.9) maintains its asymptotic character for all z satisfying the condition

$$O\left(\frac{a_1}{a_1^2 - 1} - z\right) > O\left(\frac{\log \mu}{\mu^2}\right)$$

For $z < 0$ the condition is: $O(z) < O(\mu^2)$. The asymptotic convergence of the series (6.9) occurs for $z = -\mu$, that is, for $y = 1$.

As the series (6.5) and (6.9) converge asymptotically for the same value of x, they form a system. For this purpose it is necessary to determine the constant a_1 from the given value of the constant c.

If one sets in both series (6.5) and (6.9) $y = 1$, one obtains two equations with two unknowns: x^* and a_1, namely:

$$1 = \mu \sum_{n=0}^{\infty} f_n(x^*)\mu^{-2n}; \qquad x^* = \sum_{n=0}^{\infty} \varphi_n(-\mu)\mu^{-2n} \tag{6.12}$$

From the first equation one finds x^* and, from the second, a_1, expressed in terms of x_1 and c.

Region III is specified by the intervals:

$$a_1 - \varepsilon > x > 1 + \varepsilon; \qquad y < 0; \qquad \varepsilon > 0$$

$$-a_2 + \varepsilon < x < -1 - \varepsilon; \qquad y > 0; \qquad \varepsilon > 0 \tag{6.13}$$

This region is important because as soon as x enters into this region, oscillation acquires a stationary character.

We consider the region III for $y < 0$. Omitting a series of intermediate calculations, the solution of (6.1) appears in the form of a series

$$y = -\frac{1}{\mu} \sum_{n=0}^{\infty} P_n(x)\mu^{-2n} \tag{6.14}$$

where $P_n(x)$, as previously, are given by a recursive procedure, viz.:

$$P_0(x) = \frac{x}{x^2 - 1}; \qquad P_1(x) = -\frac{x(x^2+1)}{(x^2-1)^4} \tag{6.15}$$

The series (6.14) preserves the asymptotic character as long as

$$O(x - 1) > O(\mu^{-2/3})$$

Near the limit of convergence $y = O(\mu^{-1/3})$.

Region IV is specified by the intervals

$$\begin{aligned} 1 - \varepsilon < x < 1 + \varepsilon; \qquad & y < 0; \qquad \varepsilon > 0 \\ -1 - \varepsilon < x < -1 + \varepsilon; \qquad & y > 0; \qquad \varepsilon > 0 \end{aligned} \tag{6.16}$$

As, at the approach to region III, $y \to O(\mu^{-1/3})$, it is useful to introduce a change of the variables:

$$y = -\mu^{-1/3}Q(u); \qquad u = \mu^{2/3}(x - 1) \tag{6.17}$$

With these variables (6.1) becomes

$$Q\frac{dQ}{du} - 2uQ + 1 = \mu^{-2/3}(u^2Q - u) \tag{6.18}$$

We look for the solution of this d.e. again in the form of an asymptotic series

$$Q(u) = \sum_{n=0}^{\infty} Q_n(u)\mu^{-2n/3} \tag{6.19}$$

This gives a recursive determination of the functions Q_n, viz.:

$$Q_0(u) = u^2 + \alpha + \frac{1}{u} - \frac{\alpha}{3u^2} - \frac{1}{u^3} - \frac{1}{4u^4} + \frac{\alpha}{5u^5} + \cdots$$

$$Q_1(u) = \frac{1}{A(u)}\left[C + \int_0^u A(u)\left(u^2 - \frac{u}{Q_0}\right)du\right]; \qquad A = \exp\left(-\int_0^u \frac{du}{Q_0{}^2}\right) \tag{6.20}$$

where α is the smallest root of the equation

$$J_{1/3}(\tfrac{2}{3}\tau^{3/2}) + J_{-1/3}(\tfrac{2}{3}\tau^{3/2}) = 0$$

where $J_{1/3}$ and $J_{-1/3}$ are Bessel functions of orders $\frac{1}{3}$ and $-\frac{1}{3}$. For the junction with the solution (5.14) of region III, it is necessary that $\mu^{-2/3}Q_1(u)$ be bounded for $u = Q(\mu^\epsilon)$. From the form of expressions $Q_n(u)$, one observes that the series (5.19) maintains its asymptotic character up to the values of u satisfying the conditions $Q(u) < Q(\mu^{-2/3})$, that is, for x satisfying the condition $O(x - 1) < O(1)$. It follows that the regions in which the solutions (6.19) and (6.14) hold, do actually overlap.

It remains now to join the solutions of regions I and IV. For this purpose it is necessary to join the solutions (6.5) and (6.19) taking into account the new variables (6.17). For $x = -1$, $y > 0$, the constant c must be greater than $\frac{2}{3}$. Let $c = \frac{2}{3} + \gamma$; $\gamma > 0$. As $y(-1) = O(\mu^{-1/3})$, $\mu\gamma$ is of the order $\mu^{-1/3}$ and, thus, $\gamma = O(\mu^{-4/3})$.

It can be shown that (6.5) conserves its asymptotic character up to values of x satisfying the condition $O(x + 1) > O(\mu^{-1/3})$. Thus the regions in which solutions (6.19) and (6.5) hold, do overlap, the asymptotic approximation being guaranteed for $x = -1 + \mu^{-1/3}$.

One can then determine the constant c by equating the values of y from (6.19) and (6.5) corresponding to $x = -1 + \mu^{-1/3}$, viz.:

$$\mu^{-1/3} \sum_{n=0}^{\infty} \mu^{-2/3} Q_n(-\mu^{-1/3}) = \mu \sum_{n=0}^{\infty} \mu^{-2n} f_n(-1 + \mu^{-1/3}) \tag{6.21}$$

From this relation one determines γ with the accuracy $O(\mu^{-8/3})$ and hence $c = \frac{2}{3} + \gamma$. This permits determining x_1 as the root of $f_0(x_1) = 0$.

One obtains for the amplitude of oscillation the expression:

$$\alpha = 2 + \frac{\alpha}{3}\mu^{-4/3} - \frac{16 \log \mu}{27\mu^2} + \frac{1}{9}(3b_0 - 1 + 2 \log 2 - 8 \log 3)\mu^{-2} + O(\mu^{-8/3}) \tag{6.22}$$

As to the period, it is obtained from the expression

$$T = 2 \int_{-a}^{a} \frac{dx}{y(x)} \tag{6.23}$$

For the actual calculation the interval of integration is in five parts according to the various regions, namely:

(1) From $-a$ to $-x_2$, region II where x_2 is given by (6.9) with

$$z = (1 - \mu^{-4/3})a/(a^2 - 1)$$

(2) From $-x_2$ to $-(1 + \mu^{-1/3})$, region III.
(3) From $-(1 + \mu^{-1/3})$ to $-(-\mu^{-1/3})$, region IV.
(4) From $-(1 - \mu^{-1/3})$ to x^*, region I, where x^* is given by

$$x^* = x_1 - \frac{1}{\mu}\frac{1}{x_1^2 - 1} - \frac{\log \mu}{\mu^2}\frac{x_1}{(x_1^2 - 1)^2} - \frac{1}{\mu^2} \left[\frac{x_1}{(x_1^2 - 1)^2} \log x_1(x_1^2 - 1) - \frac{g}{x_1^2 - 1} + \frac{x_1}{(x_1^2 - 1)^3}\right] - \frac{\log \mu}{\mu^3}\frac{2x_1}{(x_1^2 - 1)^4} + O(\mu^{-3}) \tag{6.24}$$

(5) From x^* to a, region II.

The period T is given by

$$T = 2(T_1 + T_2 + T_3 + T_4 + T_5) \tag{6.25}$$

If one carries out the integration with the numerical values of the coefficients, one finds

$$T \simeq 1.613706\mu + 7.01432\mu^{-1/3} - \frac{22}{g}\frac{\log\mu}{\mu} + 0.0087\mu^{-1} + O(\mu^{-4/3}) \tag{6.26}$$

For a sufficiently large μ one can neglect all terms except the first one.

It is observed that for large values of μ the expansions become very complicated. In fact for $\mu \ll 1$ it was seen (Chapters 14 and 15) that the series solutions are given in the form of power series. Here the series solution contains fractional powers and logarithmic terms. The *form* of the d.e. for large values of μ exerts a greater influence on the analytical expression of the solution than happens in the other asymptotic case when $\mu \to 0$.

7. Concluding remarks

From the preceding review it appears that, in spite of the theoretical simplicity of the asymptotic procedure, calculations are long and tedious. If one considers that all this relates to the simplest possible case of an autonomous one-degree-of-freedom system, one has to admit that there is yet much to be done before these methods can be extended to the formulation of numerous phenomena which were analyzed in Part III in connection with nearly linear systems.

The difficulty seems to be inherent in the very nature of these phenomena of a quasi-discontinuous type, when one tries to reduce their formulation to terms of the analytic theory. On the other hand, the idealization of these effects by mathematical discontinuities leads to much simpler results as happened also in theoretical mechanics in connection with the theory of shocks.

It is interesting to note that many years before the advent of modern theories of relaxation oscillations, Boussinesq[8] specified this general subject in a very clear manner which we quote:

"Si la continuité simplifie les choses quand elle en relie plusieurs qui suivent la même loi, elle les complique au contraire, le plus souvent, lorsqu'elle établit la

[8] Boussinesq, *Applications des potentiels à l'etude d'équilibre*, Gauthier-Villars, Paris, 1885.

transition entre deux catégories d'objets ou de faits régis par deux lois simples différentes et c'est alors une discontinuité fictive, un passage brusque de la première catégorie à la seconde qui rend les questions abordables."

In fact, in the above analysis of discontinuous theories there are two distinct "categories of objects"—the analytic arcs and the discontinuous stretches with a definite passage from one type of phenomena to the other; the continuous (and even analytic) part is distinctly separated from the quasi-discontinuous (idealized as discontinuous) part and, on this basis the problem becomes simple.

On the other hand, if one does not wish to introduce this idealization and prefers to proceed in the orthodox analytical manner, this also is possible (Sections 5 and 6) but at the cost of a more complicated procedure.

Chapter 31

PIECEWISE LINEAR IDEALIZATION

1. Introductory remarks

In Chapters 26 and 29 we outlined the salient points of the discontinuous theory. This theory is based essentially on the theory of degeneration which introduces certain discontinuities in the phase plane, namely: parts of trajectories are traversed in no time, whereas on other parts of the closed trajectory the motion of the representative point is continuous and takes place, as usual, in accordance wth the d.e. of the process. There is another way of idealizing the problem; namely: by means of the so-called *piecewise linear idealization.* It consists in replacing the actual nonlinear characteristic by "pieces" of straight lines. Thus, for instance, the usual characteristic of an electron tube (plate current against grid voltage) may be idealized by two straight lines: one corresponding to the "rectilinear part" of it (which is generally used in applications) and the other, the saturation part, being merely a straight line parallel to the abscissa axis (Fig. 31.1).

The phenomenon represented by this idealization will differ slightly from the actual phenomenon since it represents the performance by means of two straight lines disregarding the curvilinear part between them (shown in broken lines).

If one looks mainly for the qualitative behavior of an oscillator, such idealized representation is, generally, sufficient and a slight quantitative difference between the two modes of representation is justified by the gain in simplicity inherent in the idealized "piecewise linear" representation.

This approach was given considerable attention in recent years owing to the availability of the so-called *point transformation method.* This permits using a graphical procedure resulting from certain transcendental equations. The condition of periodicity in this case reduces to that of the existence of the *fixed point* under the repeated transformations. This

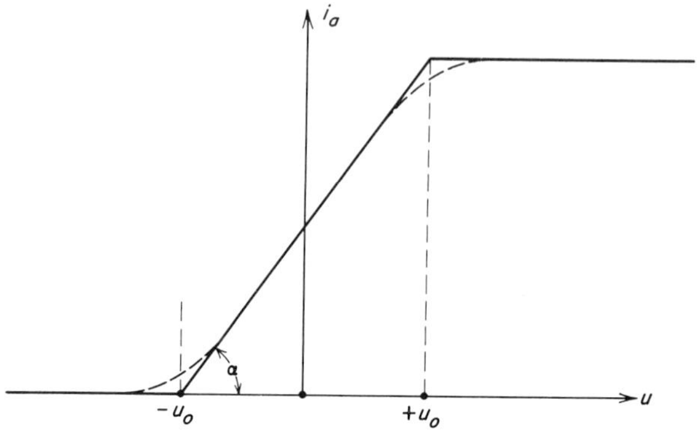

FIGURE 31.1

method is related to some extent to the theory of successor functions (Chapter 7).

2. Point transformation method

It is recalled that the theory of Poincaré considers a "segment without contact" in this field of spiral trajectories which determines two sequences of points of intersection of the segment with trajectories: $A_1, A_2, \ldots, A_n$ (outside) and $B_1, B_2, \ldots, B_n$ (inside). The existence of the limit cycle is proved when these two sequences tend to a common limit for $n \to \infty$ (see beginning of Chapter 16). In this manner every point of a limit cycle may be regarded as a fixed point. The theory of Poincaré presupposes the analytic case. Here, on the contrary, we are interested, in a nonanalytic or, more precisely, in a piecewise analytic (or, piecewise linear) case in the sense which was defined in the preceding section. It is recalled also that equations (1.1) characterize this approach. Each of these d.e. is linear and, hence, has no periodic solution but one of them is replaced by the other (and vice versa) when $\dot{x} = 0$. Although each of these d.e. has no closed trajectory, their alternative operation with a "change over" for $\dot{x} = 0$, as we saw, leads to a closed (although nonanalytic) trajectory. This may be also regarded as a representation of a trajectory with two different phase planes replacing each other discontinuously along the x axis with a displacement OO'. Thus, the impossibility of obtaining a closed trajectory for a linear d.e. in a *fixed phase plane* becomes a possibility if the same d.e. refers to two different phase planes displaced discontinuously along the x axis by the segment OO'.

A closed trajectory of this kind ceases to be analytic at the points where it cuts the x axis, and this loss of analyticity (at the two points on the trajectory) introduces a change in the whole argument.

This particular case can be generalized if we consider it in relation to the point transformation theory. We consider the phase plane yox (Fig. 31.2) and three regions I, II, III with boundaries S, S_3 and S_1, S_2. Regions I and III are limited on one side (either by S, S_3 or S_1, S_2). Region II is between S, S_3 and S_1, S_2. In each of the three regions governs a simple linear d.e. Assume that the representative point starts from the point s somewhat on S (in the following, small letters indicate the ordinates on the ordinate axes shown by the capital letters). If we assume that the

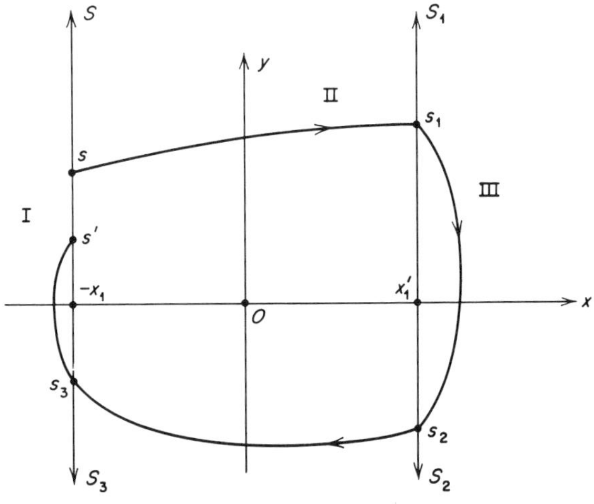

FIGURE 31.2

point s is already in region II, the representative point will traverse the arc ss_1.

As on this stretch (in region II) the d.e. is linear and we know the initial conditions $(-x_1, s)$, we can easily calculate the terminal conditions (x_1', s_1) for this stretch. As we are now on the border of region III, the terminal conditions of the preceding stretch become the initial conditions of the following stretch s, s_2 taking place in III. Arriving at s_2 we continue the argument for the following two stretches s_2s_3 and s_3s'. The important point in the argument is the position of s' with respect to s: we cannot form any conclusion after the first turn of the representative point through the sequence of regions II $\rightarrow$ III $\rightarrow$ II $\rightarrow$ I, but for the subsequent turns we can see whether s' comes nearer to s or not. If s' approaches s and coalesces with it for the repeated transformations I → II → III → I, etc., then we obtain the fixed point of the transformation which leads to the following obvious theorem: *The existence of a fixed point of a piecewise linear transformation is the criterion of periodicity of a piecewise linear phenomenon.*

In this argument we deal with the transformations of a semi-line S into another semi-line S_1; of S_1 into S_2, etc.; these transformations are continuous and (1:1), we designate them respectively by T_1, T_2, T_3 and T_4.

More precisely we have:

$$\begin{aligned} &\text{For } T_1\colon s_1 = \varphi_1(\tau_1); && s = \psi_1(\tau_1) \\ &\text{For } T_2\colon s_2 = \varphi_2(\tau_2); && s_1 = \psi_2(\tau_2) \\ &\text{For } T_3\colon s_3 = \varphi_3(\tau_3); && s_2 = \psi_3(\tau_3) \\ &\text{For } T_4\colon s' = \varphi_4(\tau_4); && s_3 = \psi_4(\tau_4) \end{aligned} \tag{2.1}$$

where τ_1, τ_2, τ_3, and τ_4 are the times of traversing the arcs in question. It is clear that

$$s' = f(s) \tag{2.2}$$

is the successor function (Chapter 7) but this time the question of analyticity does not arise since all points of junction of arcs (on SS_3 and S_1S_2) are nonanalytic. As $T = T_1T_2T_3T_4$, one can carry out the calculation of the successor function (2.2). The principal difficulty lies in the necessity of solving systems of transcendental equations which appear in this case, namely

$$\begin{aligned} \varphi_1(\tau_1) &= \psi_2(\tau_2); & \varphi_2(\tau_2) &= \psi_3(\tau_3) \\ \varphi_3(\tau_3) &= \psi_4(\tau_4); & \varphi_4(\tau_4) &= \psi_1(\tau_1) \end{aligned} \tag{2.3}$$

The stability of the fixed point and, hence, of the (nonanalytic) limit cycle can be determined by noting that at a stable fixed point (bars above τ) one must have:

$$\frac{ds'}{ds} = \frac{\varphi_1'(\bar{\tau}_1)\,\varphi_2'(\bar{\tau}_2)\,\varphi_3'(\bar{\tau}_3)\,\varphi_4'(\bar{\tau}_4)}{\psi_1'(\bar{\tau}_1)\,\psi_2'(\bar{\tau}_2)\,\psi_3'(\bar{\tau}_3)\,\psi_4'(\bar{\tau}_4)} < 1 \tag{2.4}$$

where ds is the perturbation of the initial conditions ("the cause") and ds' is the corresponding perturbation of the terminal condition ("the effect") after one turn of the radius vector.

3. Calculation of a piecewise linear limit cycle

As an example we shall investigate the stationary self-excited oscillation of an electron-tube oscillator. As we have met this question previously on several occasions, we simply indicate the d.e.

$$LC\ddot{u} + [RC - MG(u)]\dot{u} + u = 0 \tag{3.1}$$

where u is the grid voltage and $G(u)$ is the mutual conductance. In general, the nonlinear function is: $i_a = i_a(u)$ (the anode current considered as a nonlinear function of u), but we shall introduce the piecewise linear idealization by approximating the nonlinear characteristic by a "broken" one defined as follows (see Fig. 31.1):

$$i_a = \begin{cases} 0 \quad \text{for } u < -u_0 \\ G(u + u_0) \quad \text{for } |u| < u_0 \\ 2Gu_0 \quad \text{for } u > u_0 \end{cases} \tag{3.2}$$

The usual idealizations (negligible grid current, anode reaction, internal capacity of the electron tube, etc.) will be assumed. Changing the

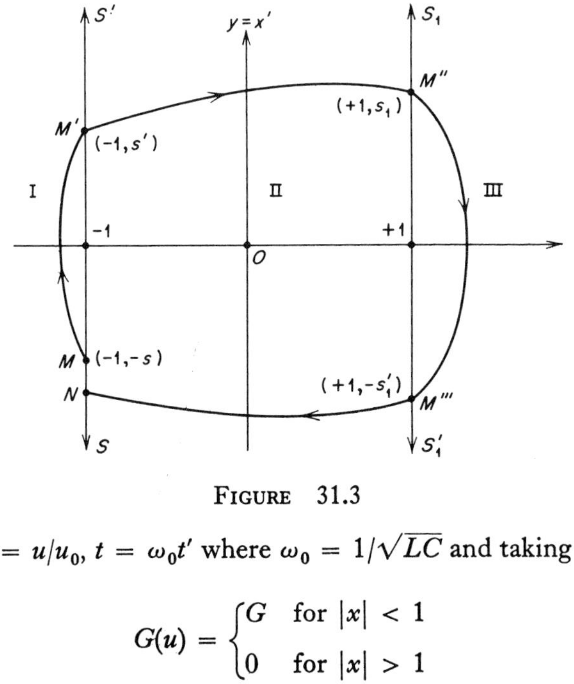

FIGURE 31.3

variables: $x = u/u_0$, $t = \omega_0 t'$ where $\omega_0 = 1/\sqrt{LC}$ and taking into account that

$$G(u) = \begin{cases} G \quad \text{for } |x| < 1 \\ 0 \quad \text{for } |x| > 1 \end{cases}$$

(3.1) results finally in the d.e.

$$\begin{aligned} \ddot{x} + 2h_1\dot{x} + x &= 0 \qquad \text{for } |x| > 1 \\ \ddot{x} - 2h_2\dot{x} + x &= 0 \qquad \text{for } |x| < 1 \end{aligned} \tag{3.3}$$

where

$$h_1 = RC\omega_0/2; \qquad h_2 = (MG - RC)\omega_0/2 \tag{3.4}$$

Instead of one single phase plane, we consider now a phase plane separated by straight lines $S'S$ and S_1S_1' into three regions: I, II, and III (Fig. 31.3).

In each of these three regions its own linear d.e. is valid. In view of the point transformation it is more convenient to consider the lines $S'S$ and S_1S_1' as consisting each of two semi-lines S', S and S_1, S_1', respectively, which is indicated by the arrows in Fig. 31.3. Since the d.e. do not change when x and y are replaced by $-x$, $-y$, the stationary trajectories in I and III are symmetrical with respect to the origin, and a similar symmetry exists for the upper and the lower parts of the middle region II. The point of rest O may be either a focus or a node, stable if $h_2 < 0$ (that is, $MG < RC$) or unstable if $h_2 > 0$. The case when $h_2 > 0$ is more interesting because it leads to the phenomenon of self-excitation studied below.

The procedure is sufficiently clear but we will discuss a specific case in detail. In different regions of the phase plane will be fitted "pieces" of trajectories of simple d.e. One can expect that if the points M and N must ultimately coincide and, if the stability condition is fulfilled, there will be a piecewise linear cycle. The trajectory will be continuous but not analytic at the points at which it cuts the lines $x = +1$ and $x = -1$; this cycle will be symmetrical with respect to the origin as just explained. The overall point transformation T can be represented by the product $T = T_1T_2T_3T_4$ where the component transformations transform one semi-line into the following one; thus T_1 will transform the points of S' into those of S_1, etc.

In view of the symmetry one has $T_3 \equiv T_1$ and $T_4 \equiv T_2$. Hence, the transformation T transforming S' into S_1' will be: $T' = T_1T_2$; and $T = (T')^2$. In other words: the total (overall) transformation T is obtained by applying twice the transformation T'. It will be sufficient, therefore, to study only the transformation T', since everything is symmetrical with respect to the origin.

4. Successor function and the fixed point

The subsequent calculations are simple but long; we shall indicate the salient points of the procedure referring to the cited reference for details. One starts with the point: $M(x_0, y_0)$ (Fig. 31.3) where $x_0 = -1$; $y_0 = -s$; $s > 0$, and write the solution of the d.e. corresponding to the region I; this gives:

$$\begin{aligned} x &= \exp(-h_1 t)\left(x_0 \cos \omega_1 t + \frac{y_0 + h_1 x_0}{\omega_1} \sin \omega_1 t\right) \\ y = \dot{x} &= \exp(-h_1 t)\left(y_0 \cos \omega_1 t - \frac{x_0 + h_1 y_0}{\omega_1} \sin \omega_1 t\right) \end{aligned} \tag{4.1}$$

where $\omega_1 = \sqrt{1 - h_1^2}$. One has to replace in these equations x_0 and y_0 by their values and then express the terminal condition, namely: the representative point R following the trajectory (4.1) will reach the point M' on the semi-line S'. This permits a gradual calculation of s and s', etc. The successor function of the transformation T_1 is given by equations:

$$\begin{aligned} s &= [\exp(\gamma_1\tau_1) - \cos\tau_1 - \gamma_1 \sin\tau_1]/\sqrt{1 + \gamma_1^2}\sin\tau_1 \\ s' &= [\exp(-\gamma_1\tau_1) - \cos\tau_1 + \gamma_1 \sin\tau_1]/\sqrt{1 + \gamma_1^2}\sin\tau \end{aligned} \tag{4.2}$$

where τ_1 is the parameter of the transformation. Thus the representative point, having been initially on the semi-line S', after τ_2 comes on the semi-line S_1 (point M'') and this constitutes the transformation T_2. If one substitutes here $t_2 = \tau_2/\omega_2 > 0$; $x = +1$; $y = s_1 > 0$ and solves again for s' and s, one obtains again the parametric equations

$$\begin{aligned} s_1 &= [\exp(\gamma_2\tau_2) + \cos\tau_2 + \gamma_2 \sin\tau_2]/\sqrt{1 + \gamma_2^2}\sin\tau_2 \\ s' &= [\exp(-\gamma_2\tau_2) + \cos\tau_2 - \gamma_2 \sin\tau_2]/\sqrt{1 + \gamma_2^2}\sin\tau_2 \end{aligned} \tag{4.3}$$

where $\gamma_2 = h_2/\omega_2 = h_2/\sqrt{1 - h_2^2}$.

At this point we omit some details of calculations concerning the general appearance of functions in some special cases (for example, $\tau_2 \to +0$, s' and $s_1 \to \infty$, etc.), which can be found in (1) and indicate only the argument leading to the establishment of the fixed point of the transformation: $T = T_1T_2$. If one builds the curves (4.2) and (4.3), their intersection Q (Fig. 31.4) will be the fixed point of the transformation: $T = T_1T_2$ transforming points of S into those of S_1. In the case when $0 < h_1 < 1$ and $0 < h_2 < 1$ the existence of a fixed point is expressed by equations

$$\begin{aligned} \frac{\exp(\gamma_1\tau_1) - \cos\tau_1 - \gamma_1\sin\tau_1}{\sqrt{1 + \gamma_1^2}\sin\tau_1} &= \frac{\exp(\gamma_2\tau_2) + \cos\tau_2 + \gamma_2\sin\tau_2}{\sqrt{1 + \gamma_2^2}\sin\tau_2} \\ \frac{\exp(-\gamma_1\tau_1) - \cos\tau_1 + \gamma_1\sin\tau_1}{\sqrt{1 + \gamma_1^2}\sin\tau_1} &= \frac{\exp(-\gamma_2\tau_2) + \cos\tau_2 - \gamma_2\sin\tau_2}{\sqrt{1 + \gamma_2^2}\sin\tau_2} \end{aligned} \tag{4.4}$$

which is obtained from (4.2) and (4.3) if one equates s' and sets $s_1 = s$. It can be shown that there exists only *one* point of intersection of curves (4.2) and (4.3); the argument is based on a detailed study of these curves which we omit here and only indicate the form of the curves on Fig. 31.4. It is recalled that the stability of the fixed point Q is determined by the approach of the representative point R toward Q in case of a perturbation. If one takes a phase trajectory other than the closed one (corresponding to the point Q), it is easy to ascertain the sequence of its intersections with the semi-axis $y = 0$, $x > 0$; let this sequence be: $x_1, x_2, \ldots$. In this sequence each following point is determined by the preceding one by means of the successor function. Thus, if one starts with a point x_1 on the s' axis, one obtains the point Q_1 on the curve D to which corresponds the ordinate x_1'; if one puts on the abscissa axis the value x_1', to this value will correspond a point Q_2 on D, so that the points Q_i approach the point Q *from the left* as the procedure continues.

If the initial point x_1 were to the right from the point $\bar{x}$ (corresponding

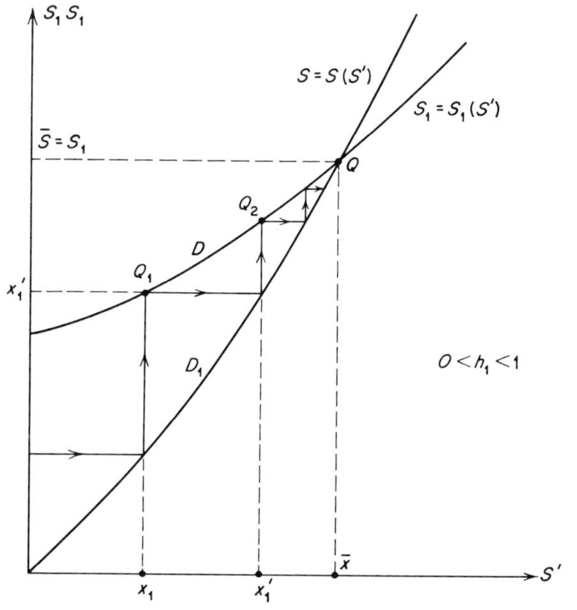

FIGURE 31.4

to the fixed point Q), the same procedure would result in a gradual approach of points Q_i to the fixed point Q *from the right*. In such a case the fixed point and, therefore, the nonanalytic limit cycle is stable.†

It is easy to show that in the case when the slope of the curve D is greater than that of D_1, the fixed point Q is unstable (see Chapter 7). One has thus a graphical criterion (sometimes called "the stairway of Lemeray") for establishing the existence as well as stability of nonanalytic limit cycles.

† It is noted that this graphical procedure is the same as that encountered in Chapters 7 and 29.

5. Successor function and the fixed point (hard self-excitation)

As the second example we consider the case when the plate current i_a can be approximated by relations

$$i_a = \begin{cases} I_s & \text{for } u_g > 0 \\ 0 & \text{for } u_g < 0 \end{cases} \tag{5.1}$$

assuming that in the condition of equilibrium ($u_g = 0$) the electron tube does not conduct owing to a constant biasing voltage: $-E_g$. The d.e. of the oscillator in this case is

$$LC\frac{d^2i}{dt^2} + RC\frac{di}{dt} + i = \begin{cases} I_s & \text{for } u_g > 0 \\ 0 & \text{for } u_g < 0 \end{cases} \tag{5.2}$$

where $u_g = -E_g - M(di/dt)$ and I_s is the saturation current (that is, i_a in the region of the saturation). With the change of the variables

$$x = i/I_s; \qquad t' = \omega_0 t; \qquad \omega_0 = 1/\sqrt{LC} \tag{5.3}$$

equation (5.2) is brought to the form

$$\ddot{x} + 2h\dot{x} + x = \begin{cases} 1 & \text{for } \dot{x} > b \\ 0 & \text{for } \dot{x} < b \end{cases} \tag{5.4}$$

where the differentiations are with respect to t'; $2h = \omega_0 RC$ and $b = E_g/\omega_0|M|I_s$.

The phase plane x, y ($y = \dot{x}$) is divided by the line $y = b$ into two regions: I (shaded) for $y > b$ and II (unshaded) for $y < b$ (Fig. 31.5). The nonanalytic junction of trajectories occurs on the line $y = b$. We define on the line $y = b$ two semi-lines: S' (to the right of the y axis) and S (to the left), and consider the segment $-s > 2bh - 1$ on the semi-line S and the segment $s' > -2hb$ on the semi-line S'. From the first segment trajectories move away into region I and from the second, they move into II. From the segment $y = b$, $-2hb < x < 1 - 2hb$ belonging to both semi-lines, trajectories move away into I for $y = b + 0$ or into II for $b - 0$. This segment may be called the *segment of repulsion*. It can be shown that the system (8.3) has the origin as its only singular point (that is, position of equilibrium) which is a focus if $h < 1$ or a node if $h > 1$. As in the latter case there are no limit cycles, oscillations are impossible. We shall consider only the case when $0 < h < 1$.

It is clear that if a (nonanalytic) limit cycle exists, it must pass through both regions I and II so as to have the origin in its interior. Hence, the trajectories must necessarily intersect the line: $y = b$. It will be sufficient,

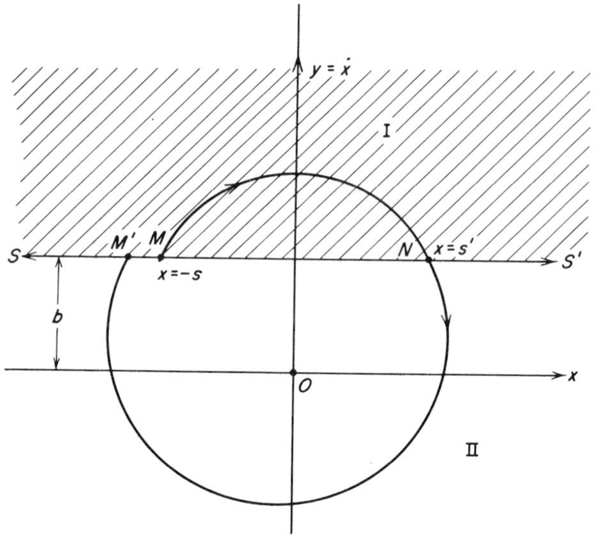

FIGURE 31.5

therefore, to study the point transformation T of the semi-line S into itself, which it is convenient to split into two partial transformations: T_1 transforming S into S', and T_2 transforming S' into S; thus $T = T_1T_2$. One obtains these transformations in a manner similar to that which we used in the preceding section. We start from a point $M\,(-s,b)$ on the semi-line S and investigate the trajectory starting from this point for $t = 0$. This trajectory in I is given by the equations:

$$
\begin{aligned}
x &= 1 + \exp(-ht)\left[-(1+s)\cos\omega t + \frac{b - h(1+s)}{\omega}\sin\omega t\right] \\
y &= \exp(-ht)\left[b\cos\omega t + \frac{1+s-hb}{\omega}\sin\omega t\right]
\end{aligned}
\tag{5.5}
$$

where $\omega = \sqrt{1-h^2}$.

The representative point moving along this trajectory will reach after some time t_1 the point $N(s',b)$ on the semi-line S'. This point is given by equations

$$
\begin{aligned}
s' &= 1 + \exp(-ht_1)\left[-(1+s)\cos\omega_1 t + \frac{b - h(1+s)}{\omega}\sin\omega t_1\right] \\
b &= \exp(-ht_1)\left[b\cos\omega t_1 + \frac{1+s-hb}{\omega}\sin\omega t_1\right]
\end{aligned}
\tag{5.6}
$$

Solving these equations for s and s' one obtains

$$\begin{aligned} S &= b\omega[\exp(\gamma\tau_1) - \cos\tau_1 + \gamma\sin\tau_1]/\sin\tau_1 - 1 \\ S' &= b\omega[\exp(-\gamma\tau_1) - \cos\tau_1 - \gamma\sin\tau_1]/\sin\tau_1 + 1 \end{aligned} \tag{5.7}$$

where $\gamma = h/\omega = h/\sqrt{1-h^2}$; $\tau_1 = \omega t_1$. Setting: $u = s/b\omega$; $v = s'/b\omega$, one obtains the successor function for the transformation T_1:

$$\begin{aligned} u &= [\exp(\gamma\tau_1) - \cos\tau_1 + \gamma\sin\tau_1]/\sin\tau_1 - a \\ v &= [\exp(-\gamma\tau_1) - \cos\tau_1 - \gamma\sin\tau_1]/\sin\tau_1 + a \end{aligned} \tag{5.8}$$

where $a = 1/b\omega = (\omega_0|M|I_s)/E_g\sqrt{1-h^2}$.

In a similar manner one obtains the successor function for the transformation T_2 transforming the point $N(b\omega v, b)$ of the semi-line S' into the points $M'(-b\omega u_1, b)$ of the semi-line S, namely:

$$\begin{aligned} u_1 &= -[\exp(-\gamma\tau_2) - \cos\tau_2 - \gamma\sin\tau_2]/\sin\tau_2 \\ v &= -[\exp(\gamma\tau_2) - \cos\tau_2 + \gamma\sin\tau_2]/\sin\tau_2 \end{aligned} \tag{5.9}$$

where $\tau_2 = \omega t_2$ is the dimensionless time of travel of the representative point in region II. The analysis of these successor functions is similar to that indicated previously. We mentioned only the conclusion; for details, see footnote 1. When the parameter of the transformation T_1 varies in: $0 < \tau_1 < \pi$, u increases monotonically from $u_0 = 2\gamma - a$ to $+\infty$ and v varies from $v_0 = a - 2\gamma$ also to $+\infty$. Moreover $(du/dv) > 0$ and $(d^2u/dv^2) > 0$ (we omit the explicit form of these derivatives). The curve (u,v) has an asymptote (when $\tau_1 \to \pi$) given by equation: $u = \exp(\gamma\pi)v - a(1 + \exp(\gamma\pi))$. This (together with some other details) permits tracing the curves $u = u(v)$ for (5.8).

A similar analysis of equations (5.9) shows that $(du_1/dv) > 0$ and $(d^2u_1/dv^2) < 0$ and that the asymptote for the (u_1,v) curve is: $u_1 = \exp(-\gamma\pi)v$. If we call the curves corresponding to equations (5.8) and (5.9) C and C_1, respectively, a more detailed analysis shows that, for $a \leq 2\gamma$, the curves C and C_1 do not intersect. Moreover, the curve C_1 does not depend on the parameter a, whereas C does; when this parameter varies the curve C is displaced in the $(u,v;\, u_1,v)$ diagram of Fig. 31.6 and occupies different positions: C, C', C'',.... It follows that beginning with a certain critical value $a = a^*$ of the parameter a, the intersection of curves C with curve C_1 becomes possible and, according to the general theory, the points of intersection determine the fixed points of the transformation and, hence, express the existence of closed trajectories (limit cycles). In this case, in view of the difference of signs of the second derivatives: $(d^2u/dv^2) > 0$ and $(d^2u_1/dv^2) < 0$ of the curves C and the

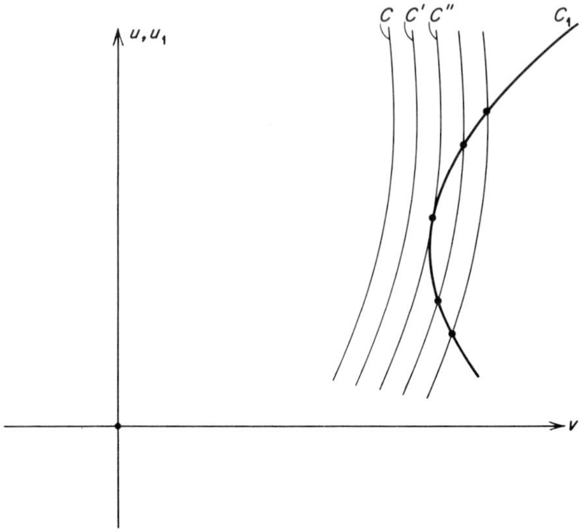

FIGURE 31.6

curve C_1, respectively, there are two points of intersection of C_1 with the family C so that there are two limit cycles, as we know already from Chapter 3. One can expect, therefore (Chapter 7), that we will come across the two possible topological configurations: *USU* or *SUS*. We consider the latter as corresponding to what is called the *hard self-excitation*. In this case for the first fixed point (smaller $\bar{u} = \bar{u}^{(1)}$, $\bar{v} = \bar{v}_1$, $\bar{\tau}_1$ and larger $\bar{\tau}_2$) one has

$$\left(\frac{du_1}{du}\right)_{v=\bar{v}_1} = \left(\frac{du_1}{dv}\right)_{v=\bar{v}_1} : \left(\frac{du}{dv}\right)_{v=\bar{v}} > 1 \tag{5.10}$$

and for the second ($\bar{u} = \bar{u}^{(2)} > \bar{u}^{(1)}$; $\bar{v} = \bar{v}_2 > \bar{v}_1$)

$$0 < \left(\frac{du_1}{dv}\right)_{v=\bar{v}_2} = \left(\frac{du_1}{dv}\right)_{v=\bar{v}_2} : \left(\frac{du}{dv}\right)_{v=\bar{v}_2} < 1 \tag{5.11}$$

that is, the first limit cycle is unstable and the second is stable, which is well the configuration: *SUS* using the notations of Chapter 7. If there is only one fixed point, it is always stable.

These considerations can be further amplified by constructing the regions in which the trajectories have the same orientation (either approaching the external cycle or the singular point) according to the position of the intermediate unstable cycle.

For the graphical construction of these piecewise linear trajectories, see footnote 1. It is sufficient to mention that the points of nonanalyticity of these closed trajectories are situated on the line: $y = b$ which may be expected, of course, from an offhand consideration. Summing up, this direct study of nonanalytic trajectories permits obtaining qualitatively the same results which we know already from the analytic theory (Chapters 3 and 7) but through a more complicated argument of the successor functions and the resulting fixed point of the point transformation.

6. Topology of certain relay systems

In many problems of the theory of automatic regulation and control involving relays, the splitting of the phase plane into certain *regions* in the above specified sense becomes convenient.

A few words about the relay mechanism itself may be useful. A relay is an electromechanical mechanism, which, depending on the magnitude of a signal ξ releases the control action u in one or the other direction. As the control action (for example, the current) is generally constant: $u = u_0$, a relay is merely an "on" or "off" mechanism. In a polarized relay the "on" phase changes its sign with ξ as is shown in Fig. 31.7. There is another peculiarity of the relay arising from the *hysteresis* effect; this manifests itself as follows: assume that the positive signal ($\xi > 0$) is gradually applied from zero; the relay closes the circuit only when $\xi = \sigma_0$; if the signal increases, nothing is changed in the control action which remains $u = u_0$. If, however, ξ begins to decrease, the relay does not open the circuit when $\xi = \sigma_0$ but opens it only for $\xi = \sigma_1$, where $\sigma_1 < \sigma_0$; there is thus a hysteresis loop as indicated by the arrows at the points $\xi = \sigma_0$ and $\xi = b_1 = \lambda\sigma_0$; $\lambda < 1$. For the negative values of ξ everything is symmetrical.

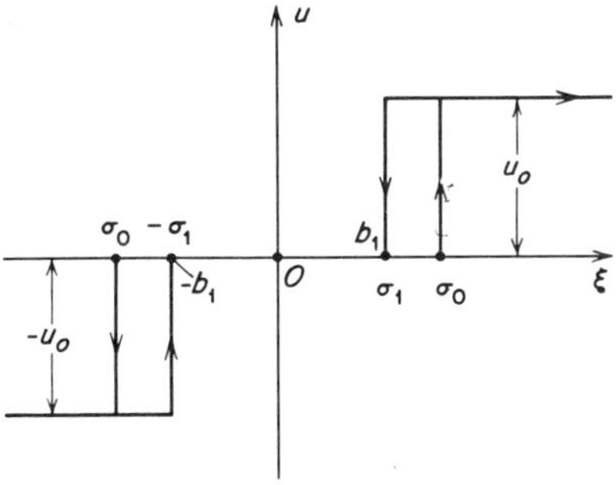

FIGURE 31.7

The response of a relay to the signal manifests itself thus in the three different ways, namely: u_0, 0 and $-u_0$ but the relation between these values of the response and the signal ξ is not a single valued one as, in addition to the direct dependence on ξ, it depends also (somewhat implicitly) on $\dot{\xi}$ at least in certain regions, as explained. By comparison with the previously studied piecewise linear manifestations in which an explicit dependence on $\dot{x}$ appears at certain thresholds (see equation (1.1), Chapter

26), one may expect a relay mechanism to be capable of producing similar manifestations in a system of which it is a component.

As an example we shall investigate the problem of a follow-up mechanism which keeps two shafts in alignment with each other. If θ_1 is the angle of the leading shaft (counted from an arbitrary origin of angles) and θ_2 the angle of the control led shaft, the control action is derived from the error function $\varphi = \theta_1 - \theta_2$. In the control scheme we assume the presence of a relay which reverses a servomotor so as to maintain the alignment between the two shafts. Assuming that the control action at the relay input is of the form $\sigma = \varphi + B\dot{\varphi}$, I is the moment of inertia of the servomotor, $k\dot{\theta}_2$ the counter e.m.f. in the rotor of the servomotor, Ai the moment developed by the rotor, and R the ohmic resistance of the armature, equations of motion can be written as

$$I\ddot{\theta}_2 = Ai; \qquad Ri = V - k\theta_2 \tag{6.1}$$

or simply

$$I\ddot{\theta}_2 + \frac{Ak}{R}\dot{\theta}_2 = \frac{A}{R}V \tag{6.2}$$

To simplify the matter further we assume that $\theta_1 = \text{const}$. In this case we consider the follow action *per se*, disregarding the variation of the angle θ_1 to be followed. In such a case one can write (6.2) in the form

$$I\ddot{\varphi} + \frac{Ak}{R}\dot{\varphi} = -\frac{A}{R}V \tag{6.3}$$

to this one has to add the law of the control

$$\sigma = \varphi + B\dot{\varphi} \tag{6.4}$$

Equations (6.3) and (6.4) can be reduced to the form

$$\ddot{x} + \dot{x} = -u(\xi); \qquad \xi = x + \beta\dot{x} \tag{6.5}$$

where

$$x = \left(\frac{Ak^2}{IRU_0}\right)\varphi; \qquad \xi = \left(\frac{Ak^2}{IRV_0}\right)\sigma; \qquad u = \frac{V}{V_0} \tag{6.6}$$

and where the differentiations are performed with respect to the dimensionless time t' defined by the formula

$$t' = \frac{Ak}{IRV_0}$$

Finally, $\beta = (Ak/IR)B$ is the dimensionless coefficient corresponding to B. The system (6.5) is sufficiently general to warrant its further discussion; setting $y = \dot{x}$, it can be written as equivalent system of the form

$$\dot{y} = -y - u(\xi); \qquad \xi = (1 - \beta)y - \beta u(\xi) \tag{6.7}$$

As we have two regions of nonsingle-valuedness of the relay characteristic: $u = u(\xi)$, it is convenient to use a *three-sheet phase plane* shown in Fig. 31.8 containing three regions, region I (for $|\xi| < \varepsilon$) corresponding to $u = 0$ (relay is open). On sheet I are superimposed (with certain overlappings) two other sheets: II with $\xi > \lambda\varepsilon$, and III with $\xi < -\lambda\varepsilon$, to which correspond the closed position of the relay (for II, one has $u = +1$, and for III, $u = -1$).

The passage of the representative point from I either to II or to III takes place only on its borderlines, that is, for $\xi = \pm\varepsilon$; whereas the inverse passage (that is, from either II or III to the middle region I) occurs either on $\xi = +\lambda\varepsilon$ or on $\xi = -\lambda\varepsilon$. This takes into account the hysteresis action as previously explained. In any case, all passages take place *continuously*, which is essential for what follows. It is noted that, resulting from the symmetry of the relay element, the trajectories are symmetrical with respect to the origin. In region I the relay remains open and (6.7) has the form

$$\dot{x} = -y; \qquad \dot{\xi} = (1 - \beta)y \tag{6.8}$$

which yields $dy/d\xi = -1/(1 - \beta)$.

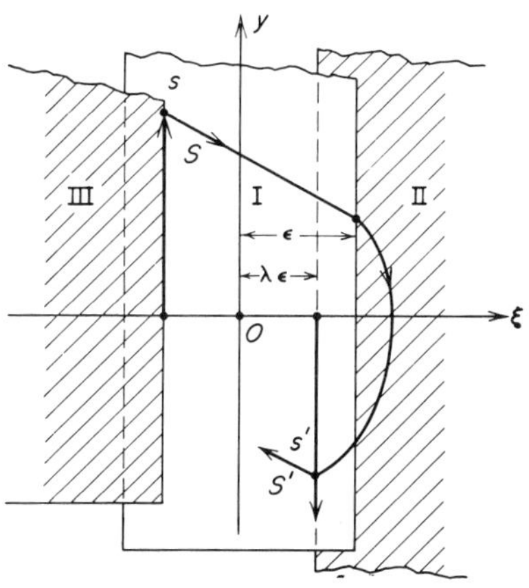

FIGURE 31.8

This shows that all points of the abscissa in region I (that is, $|\xi| < \varepsilon$) are positions of stable equilibria as the representative point R moves in I on the rectilinear trajectories: $\xi + (1 - \beta)y = \text{const}$ *toward* the abscissa axis. To these positions of equilibria tend also ($t \to +\infty$) all trajectories for which

$$|\xi + (1 - \beta)y| < \varepsilon$$

In region II the d.e. (6.7) become

$$\dot{y} + y = -1; \qquad \dot{\xi} = (1 - \beta)y - \beta \tag{6.9}$$

In this region there are no positions of equilibria and all trajectories approach asymptotically the line: $y = -1$; $\xi = -t + \text{const}$. Hence once R enters into region II it is bound to return to I. The ultimate analysis depends on whether $\beta < 1$ or $\beta > 1$.

7. Point transformation for $\beta < 1$

We leave out the case when $|\xi + (1 - \beta)y| < \varepsilon$ as it corresponds to the state of rest with the relay open. As previously, we consider the transformation of the semi-line S into the semi-line S' (Fig. 31.8) as the transformation is symmetrical. We have for the semi-line S (without contact):

$$y = -\lambda\varepsilon; \qquad y > \frac{\beta}{1-\beta}$$

and for the semi-line S':

$$\xi = +\lambda\varepsilon; \qquad y < \frac{\beta}{1-\beta}$$

Across these semi-lines trajectories pass from regions III or II to region I. As there is a symmetry about the origin, trajectories passing through symmetrical points are also symmetrical so that there is no necessity to distinguish S and S'. In other words, one can use one single point transformation $s' = T(s)$ independently with whichever of the two semi-lines (S or S') the trajectory in question intersects.

For the calculation of the successor function of this transformation we consider an arbitrary trajectory passing from region III to region I at some point s of the semi-line S. Within region I this trajectory will be a straight line

$$\xi + (1-\beta)y = -\lambda\varepsilon + (1-\beta)s \tag{7.1}$$

Hence for $|-\lambda\varepsilon + (1-\beta)s| < \varepsilon$, that is, for

$$-\frac{1-\lambda}{1-\beta}\varepsilon < s \frac{1+\lambda}{1-\beta}\varepsilon \tag{7.2}$$

the point R will approach the position of equilibrium without leaving region I so that there will be no further successor points on S and S'. For $s > [(1+\lambda)/(1-\beta)]\varepsilon$ the representative point will emergy from I through its right-hand border at the point $\xi = \varepsilon; y = y_0 = s - [(1+\lambda)/(1-\beta)]\varepsilon \geq 0$ and will continue its motion in II in accordance with the d.e. (6.9) with the initial conditions: $\xi = +\varepsilon$, $y = y_0$ along the trajectory

$$y = -1 + (1 + y_0)\exp(-t);$$
$$\xi = \varepsilon - t + (1-\beta)(1+y_0)(1-\exp(-t)) \tag{7.3}$$

and will return to the frontier S' of this region.

Let τ be the time of traversing region II. Then for $\tau > 0$, $\xi = +\lambda\varepsilon$, $y = -s'$ one obtains the following expression for the successor function $\left(\text{taking into account that } s = y_0 + \frac{1+\lambda}{1-\beta}\varepsilon\right)$; for $s \geq \frac{1+\lambda}{1-\beta}\varepsilon$:

$$\begin{aligned} s &= -1 + \frac{1+\lambda}{1-\beta}\varepsilon + \frac{\tau - (1-\lambda)\varepsilon}{(1-\beta)[1-\exp(-\tau)]} \\ s' &= +1 + \frac{\tau - (1-\lambda)\varepsilon}{(1-\beta)[\exp(\tau) - 1]} \end{aligned} \tag{7.4}$$

for $s \leq -[(1 + \lambda)/(1 - \beta)]\varepsilon$ (which is possible only for $\beta > (1 - \lambda)s$), the point R moving along the trajectory (7.1) will leave II from its left borderline at the point: $y_0{}' = s + [(1 - \lambda)/(1 - \beta)]\varepsilon \leq 0$ and will move thereafter in region I; after a time τ it will come back to the semi-line S. It can be shown that the successor function in this case will be, for $s \leq -[(1 - \lambda)/(1 - \beta)]\varepsilon$,

$$\begin{aligned} s &= +1 - \frac{1 - \lambda}{1 - \beta}\,\varepsilon - \frac{\tau - (1 - \lambda)\varepsilon}{(1 - \beta)[1 - \exp(-\tau)]} \\ s' &= +1 - \frac{\tau - (1 - \lambda)\varepsilon}{(1 - \beta)[\exp(\tau) - 1]} \end{aligned} \tag{7.5}$$

Formulas (7.4) and (7.5) determine the point transformation $s' = T(s)$ of the semi-lines S and S' into themselves.

From the coordinate s outside the interval (7.4) one determines the parameter τ of the transformation; the latter, in turn, determines the coordinate of the subsequent point s'. The relations of τ as functions of s and s' are yielded by single valued continuous functions which we shall designate as $\tau = f(s)$ and $s' = g(\tau)$.

8. Nonanalytic cycles and their stability

It is convenient to introduce auxiliary functions $\psi_1(\tau)$ and $\psi_2(\tau)$ defined by the equations

$$\psi_1(\tau) = \frac{\tau - \alpha}{1 - \exp(-\tau)}; \qquad \psi_2(\tau) = \frac{\tau - \alpha}{\exp(\tau) - 1} = \psi_1(\tau)\exp(-\tau) \tag{8.1}$$

Where we shall set $\alpha = (1 - \lambda)\varepsilon \geq 0$; it is recalled that for a relay:

$$-1 \leq \lambda \leq +1$$

The expression for the successor function can be written then

$$s = \begin{cases} -1 + \dfrac{1 + \lambda}{1 - \beta}\,\varepsilon + \psi_1(\tau)/(1 - \beta) & \text{for } s \geq \dfrac{1 + \lambda}{1 - \beta}\,\varepsilon \\[2ex] +1 - \dfrac{1 - \lambda}{1 - \beta}\,\varepsilon - \psi_1(\tau)/(1 - \beta) & \text{for } s \leq \dfrac{1 - \lambda}{1 - \beta}\,\varepsilon \end{cases} \tag{8.2}$$

$$s' = 1 - \psi_2(\tau)/(1 - \beta)$$

Setting $\tau = \tau_0$ the value of τ corresponding to $s = [(1 + \lambda)/(1 - \beta)]\varepsilon$ and

$s = -[(1-\lambda)/(1-\beta)]\varepsilon$, this value is determined from $\psi_1(\tau_0) = 1-\beta$ in a single valued manner. An intermediate calculation (which we omit) shows that

$$(s')_{\tau=\tau_0} = s_0' = 1 - \exp(-\tau_0) = [\tau_0 - (1-\lambda)\varepsilon]/(1-\beta)$$

It is noted that ψ_1 is a monotonically increasing function of τ while ψ_2 goes through a maximum (minimum). In view of this for the values $s \geq [(1+\lambda)/(1-\beta)]\varepsilon$ or for $s < -[(1-\lambda)/(1-\beta)]\varepsilon$, the parameter τ of the transformation should be varied in the interval: $\tau_0 \leq \tau < +\infty$.

With these details the diagram is obtained by plotting the values of s and s' against τ; it is recalled that s is the preceding and s' is the following points

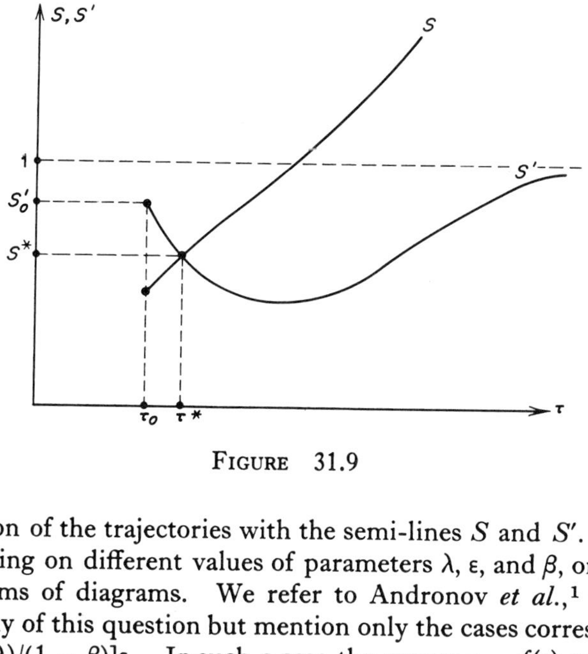

FIGURE 31.9

of intersection of the trajectories with the semi-lines S and S'. It is clear that, depending on different values of parameters λ, ε, and β, one can have different forms of diagrams. We refer to Andronov *et al.*,[1] for a more detailed study of this question but mention only the cases corresponding to $s_0' > [(1+\lambda)/(1-\beta)]\varepsilon$. In such a case the curves $\tau = f(s)$ and $s' = g(\tau)$ have a point of intersection, and the point transformation $s' = T(s)$ has only one stable fixed point $s^* > [(1+\lambda)/(1-\beta)]\varepsilon$ to which corresponds a symmetrical limit cycle (Fig. 31.9).

[1] A. Andronov, A. Witt, S. Chaikin, *Theory of Oscillations* (in Russian); this book is the second edition (1959) of A. Andronov and S. Chaikin, *Theory of Oscillations* (original text in Russian), Moscow, 1937; N. A. Geleszov, *J. Tech. Phys.* (USSR) **13**, 1948; **20**, 1950; *Radio-Physics* **2**, 1958.

The value of the parameter τ^* corresponding to the fixed points s^* is given by equation $s' = s$, that is:

$$[\tau^* - (1 - \lambda)\varepsilon] \cot \frac{\tau^*}{2} = 2(1 - \beta) - (1 + \lambda)\varepsilon$$

It is to be noted that the existence of the fixed point $s^* > [(1 + \lambda)/(1 - \lambda)]\varepsilon$ results from the continuity of curves $\tau = f(s)$ and $s' = g(\tau)$ and also from the fact that the difference $s' - s$ has different signs for $\tau = \tau_0$ and for $\tau \to +\infty$; the stability of s^* depends on: $\left|\frac{ds'}{d\tau}\right| < \left|\frac{ds}{d\tau}\right|$. The appearance of the limit cycle is shown in Fig. 31.10 in heavy line.

It is to be noted that for different points of the parameter space conditions may be entirely different; for instance, there may be either no fixed point of the transformation or the latter may not be stable. Finally in order to complete this study one has to split the family of trajectories into subfamilies separated from each other by separatrices. This determines the strips of attraction of trajectories for the ultimate nonanalytic cycle.

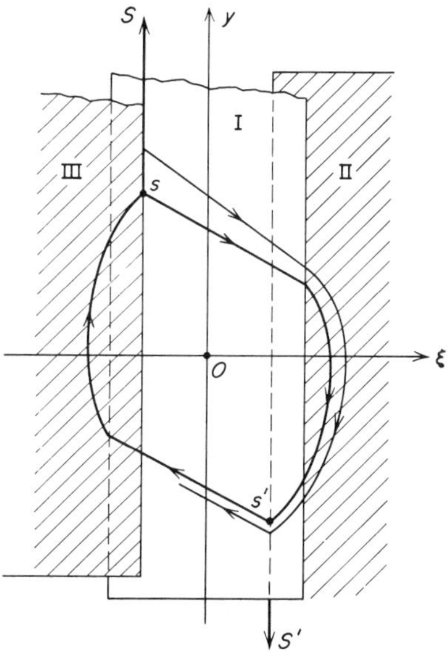

FIGURE 31.10

9. Remarks

In this approach we have outlined only the principal points of the point transformation theory as applied to these problems.

It is noted that the starting point is always somewhat arbitrary and depends on the way in which a nonlinear characteristic is approximated by "broken lines." However, as the parameter does not appear in the d.e. and the procedure is ultimately graphical, there is no restriction concerning small parameter values as in the analytical methods. Moreover, discontinuities are not involved in this idealization, as was the case of Chapters 26, 27, 28, and 29 dealing with discontinuous theories.

From this point of view the piecewise linear idealization appears to be sufficiently general to be able to account for a considerable number of phenomena encountered in the theory of oscillations.

Unfortunately the complexity of calculations involved in the determination of successor functions, fixed points and their stability, is a serious handicap to the practical use of the method in its present form. In fact, at present it is possible to apply this method only to the d.e. of second order, and even here the difficulties are considerable, particularly in solving systems of transcendental equations and in the graphical construction of the successor functions. In spite of these difficulties the method has been successfully applied to the theory of nonlinear and nonanalytic mechanisms of second order,[2] particularly those involving relays.

Another difficulty which deserves mention is the somewhat critical behavior of trajectories for a given combination of parameters λ, ε, β of a relay system. In other words, although one can sometimes carry out calculations and establish conditions of periodicity, it may happen that one cannot do so for a somewhat different parametric point λ, ε, β. The topological considerations in this nonanalytic case are much less certain (to guide the analysis) than in analytic cases. In other words a purely analytical procedure (applied, for instance, to the investigation of an electron-tube oscillator) leads always to a definite result if one knows the parameters of the problem. Here in nonanalytic cases this is less certain; it is also well known from the practice of systems involving quasi-discontinuous elements, such as relays. As has been seen, relations here are so complicated that it is impossible to be guided by any offhand considerations, and very often a little can account for much if one happens to be in the neighborhood of a critical threshold which it is difficult to foresee.

It is possible that with a more extended use of tables, calculating machines, etc., these long and tedious calculations can be simplified and the method can be made of greater practical use. At present, despite its great theoretical interest, the method has not yet reached a stage when it can be advantageously compared with the existing analytical and discontinuous theories.

10. Concluding remarks

It is useful to summarize the different methods which we attempted to outline in Part IV. In contrast to the first three parts, Part IV appears to be more complicated and less definite. This reflects to some extent the inherent difficulties of the subject, which have not been mastered completely so far.

We have considered three different approaches resulting from a very bad analyticity of the solution at least at some points; besides this, these

[2] N. S. Gorskaya *et al.*, *Dynamics of Nonlinear Servomechanisms* (in Russian), Moscow, 1959.

approaches do not have the same objective and so are not quite comparable.

We have classified these methods in three principal categories:

(1) Discontinuous methods (Chapters 26 and 29).
(2) Asymptotic methods (Chapter 30).
(3) Piecewise linear method (Chapter 31).

Approaches (1) and (3) use certain idealizations of a basic character, whereas in (2) these idealizations are of rather minor importance (for example, replacing the full d.e. by "truncated" equations on some stretches of the integral curve) so that the method (2) may be regarded as the *exact method* since the solution can be calculated with any prescribed accuracy.

The approach (1) splits into two different methods: (1a) the Mandelstam-Chaikin theory (Chapter 26) and (1b) the Vogel theory (Chapter 29); although the starting points in these theories are different, they merge into one single theory when applied to relaxation oscillations. Both discontinuous theories (1) are based on the concept of *degeneration* (Chapter 26). This concept has been elaborated by mathematicians[3] to a much greater extent than is indicated in Chapter 26. However, for the elementary presentation, aiming mostly at applications, a simplified theory of degeneration (Chapter 26) is probably sufficient.

The essential feature resulting from the degeneration theory is the introduction of discontinuous variables and this, in turn, brings about a typical aspect of these theories, namely: the cycle is composed of analytic arcs on which the d.e. governs the motion of the representative point and of the discontinuous stretches traversed in no time; on the latter stretches the d.e. recedes, so to speak, into the background and the discontinuous jump is governed by a certain physical invariant. One has a kind of a "hybrid theory" to which both mathematics and physics contribute.

This theory had useful applications (Chapters 27 and 28), but it must be recalled that it is a purely *qualitative* theory.

The asymptotic theory (2) (Chapter 30) pursues a purely analytic approach in problems in which the analyticity is practically evanescent, at least at some points of the cycle, and this results in considerable difficulties. Here again there are two possible methods: (2a) Cartwright-Littlewood approach, Section 4 (Chapter 30) and (2b), Haag-Dorodnitzin approach, Section 6 (Chapter 30).

The essential point of (2) is that *one must know the integral curve in advance* in order to apply this analysis. There is a difference of a formal nature between (2a) and (2b) as to what kind of analysis is to be used; in

[3] K. O. Friedrichs and W. Wasow, *Duke Math. J.* **13**, 1946; N. Levinson, *Acta Math.* **82**, 1951.

(2a) one proceeds by the general theory (taking into account *local* conditions in different parts of the integral curve); in (2b) use is made of the asymptotic expansions; this, however, is not a really important difference. The important point is that in order to use these asymptotic theories in their present form, one has to know the solution at least qualitatively which is not always possible.

For one single van der Pol equation, this does not yet constitute any difficulty. For more complicated systems (like those of multivibrators, etc.) this prerequisite may be a serious handicap; it is precisely in this connection that the discontinuous theories have a definite advantage since they yield the solutions.

Summing up, the asymptotic methods (2), at least at present, appear rather as *existence theorems* permitting us to *justify analytically* the existence of a *known* solution, but they do not give the means for determining this solution. However, if the solution is known, the asymptotic methods permit its determination *quantitatively* with a prescribed accuracy. The latter feature is outside the reach of the discontinuous theories which are essentially qualitative.

Considering now the piecewise linear method (3), it must be said that it is not concerned with relaxation oscillations proper, and hence should not be included in Part IV. We have considered it here along with the relaxation problems only because of the nonanalytical treatment which is partly similar to the discontinuous theories (1). In fact, in both (1) and (3) the classical theory does not apply; there are some "closed" (non-analytic) trajectories not obeying the theorem of the index and some other seemingly "absurd" situations if considered from the point of view of the classical theory.

The idealization used in (3) is more drastic than that in (1); in fact, in (1) one "simplifies" the d.e. by means of the theory of degeneration (if this is possible). In the idealization (3) one replaces the *actual nonlinear characteristic* by a polygon of straight lines. To this rectilinear polygon in *the characteristic* corresponds a curvilinear polygon in *the solution* and the principal aim of the problem is to determine the conditions under which this polygon becomes closed, that is, when the piecewise linear idealization leads to a periodic phenomenon.

Here, again, the procedure actually *yields a solution* but this solution is continuous without being analytic at the point of the junctions of analytic *arcs*. Contrary to (1) there are no discontinuous stretches. There is however a feature common to (1) in that the theorem of the index does not hold here either.

Summing up, there are two groups of different theories: the continuous ones (1) and the asymptotic one (2); in each of these groups there are two subgroups: (1a) the Mandelstam-Chaikin theory and (1b) Vogel's theory; (2a) Cartwright-Littlewood theory and (2b) Haag-Dorodnitzin theory. To these two major groups is to be added also the piecewise linear approach (3).

To these various approaches one has to add the most recent method initiated by Pontriagin[4] and Mishchenko[5] concerning d.e. of the form

$$\mu \dot{x} = f(x,y) \tag{10.1}$$

In this theory the relaxation oscillations are considered as systems characterized by rapid motions (quasi-discontinuities) at some points of a cycle and having also equilibrium positions resulting from stationary solutions. Starting from this point, it can be shown that it is possible to obtain a number of formulas for the determination of periods and similar details without any necessity for the actual integration of the system

$$\mu \dot{x} = f(x,y); \qquad \dot{y} = g(x,y) \tag{10.2}$$

In particular, Pontriagin also shows that it is possible to use this procedure in cases when there are no positions of equilibrium having an asymptotic stability and when there are no isolated limit cycles.

This method can be used in two principal cases:

(1) When it is desired to ascertain whether the difference between the exact solution and its asymptotic approximation can be made as small as desired during a finite interval of time.

(2) To investigate the correspondence between the exact solution and its asymptotic approximation during an unlimited time interval (III).

For instance, it is sometimes possible to obtain an invariant set† of a toroidal type; in such a case the problem presents itself in the following manner: it is desired to show that there exists an invariant set for the exact equation which is in the neighborhood of corresponding sets of approximate equations; here it is necessary to supplement the investigation by conditions of stability.

These generalizations are somewhat reminiscent of the theory of Poincaré concerning the existence of a periodic solution of a *local* type (that is, in the neighborhood of the generating solution). In this new approach one tries to determine certain functions characterizing the integral set in question.

[4] L. Pontriagin, *Bull. Acad. Sci.* (USSR), Ser. math. **21**, 1957.

[5] E. Mishchenko, *ibid.*

† An invariant set (or manifold) is a hypersurface Σ having the property that if a point of the solution of the system of d.e. is on the hypersurface Σ, the whole solution belongs to Σ.

Another important point of this theory is that it permits determining these integral sets without any additional hypotheses regarding the structure of the right-hand term (that is, independently of whether the right-hand term contains t explicitly or not).[6] For the time being these developments are not yet completed and connections with physical facts are not yet completely established.

In general, in these new developments, the tendency is to avoid as much as possible the somewhat artificial arguments like the "junction" of solutions in different regions of the phase space, theory of degeneration, etc., or the introduction of physical concepts of some nature. For the time being this new theory has not yet reached the state of codification, and connections with applied problems are not yet in sight. It is impossible, therefore, to advance any opinion on this subject as far as applications are concerned.

Summing up, all attempts outlined in Part IV, one has to admit that there exists no general theory (similar to the theory of Poincaré in the small parameters domain) capable of yielding solutions with a prescribed accuracy uniformly in all possible cases.

Aside the discontinuous methods which are able to yield only qualitative results, the quantitative methods are either too cumbersome or insufficiently developed; most of them concern theorems of existence which are not yet sufficient as a working tool in hands of physicists or engineers.

[6] N. Bogoliubov and I. Mitropolsky, *Intern. Congress, I.F.A.C.*, Moscow, 1960.

Index

Abraham, 627
Abridged equations, 19
Aizerman, 147
Almost periodic functions, 283
Almost periodic solutions, 283
Amplitude, external force, 343, 444
  of heteroperiodic oscillation, 224
  modulation, 380, 458
Andronov, 23, 48, 51, 79, 82, 95, 169, 639
Andronov-Chaikin, 610
Andronov-Witt, 441
Appleton, 573
Approximation methods, 212
Asymptotic expansions, 676
  methods, 329
Asynchronous excitation, 580
  quenching, 576
Autonomous systems, 243
  change of time scale, 247
  translation, 244
Autoperiodic oscillations, 566
Averaging, 273

Beats, 438
Bellman, 516, 539
Bendixson, 82
Besicovich, 282
Bethenod, 490
Bifurcation theory, 31
  applications, 173
  diagrams, 182
  value, 47
Birkhoff, 648
Blaquière, 418
Bloch, 627
Bogoliubov-Mitropolsky, 103, 330, 356, 509, 679, 710
Bohr, 282
Boussinesq, 685
Brillouin, 488
Brouwer, 80
Brownell, 526
Bulgakov, 640

Canonical transformation, 19
Cartwright, 672
Cauchy, 4
Center, 32
Cesari, 3, 672
Chaikin, 633
Characteristic, equation, 20
  exponents, 125
  hard, 75
  roots, 21
  soft, 74
Characteristics (see integral curves)
Coddington, 78
Combination tones, 463
Configuration, polycyclic, 75
  topological, 98
Conservation of energy, 41
Contact curves, 85
Conti, 103
Coulomb damping (see damping)
Critical points, 616
Cycle (see limit cycles)
Cylindrical phase space, 190

Damping, negative, 559
  nonlinear, 338
  positive, 559
Degeneration, 609
Denjoy, 191
Detuning, 449
Difference equations, 398
  differential equations, 514
Differential equations, 4
Direction field, 667
Dissipative systems, 480
Dorodnitzin, 665
Duffing equation, 480

Energy criterion, 116
fluctuations, 219
integral, 41
Equilibrium, indifferent, 133
stable, 118
unstable, 140
Equivalent damping, 350
linearization, 348
restoring force, 350
Excitation, asynchronous, 580
parametric, 488
self, 74

Favard, 285
Fixed point, existence of, 688
stability of, 411
Flanders, 665
Floquet theory, 127
Focus (spiral point), 13
weak, 38
Forced oscillation, 301
Free oscillation (see autoperiodic)
Frequency, autoperiodic, 438
correction, 244
heteroperiodic, 438
stability of, 421
transcendental, 522
Friedrichs, 215, 617, 707
Fundamental (set of solution), 125

Geleszov, 704
Generating solution, 233
Gliding flight, 207
Gomory, 412
Gorelik, 506
Gorskaya, *et al.*, 706
Goursat, 118, 163, 228
Graphical construction of Lienard's curves, 103

Haag, 644, 665
Hamiltonian Principle, 53
Harmonic oscillator, 9
Hayashi, 448
Heegner, 626
Helmholtz, 462
Hill's equation, 503
Hill-Meissner equation, 503
Huygens, 438
Hysteresis, 379
nonlinear, 379

Index theorem, 77
of a closed curve, 77
Inertial nonlinearity, 587
Initial conditions, 74, 396
Integral curves, 6
invariant, 53
Invariant, physical, 614
Isolated cycles (see limit cycles)

Jacobian, 79, 238
Joukovsky, 207
Jump phenomenon, 375
Junctions, method of, 679

Kaden, 451
Kaidanovsky, 633
Kalechi, 438
Kobsarev, 573
Kolmogorov, 69
Krylov (Bogoliubov), 286, 580

Lagrangian equations, 54
La Salle, 84
Le Corbeiller, 109
Lefshetz, 81, 104
Letov, 154, 162
Levinson, 78, 103, 707
Liapounov, 118, 155, 315
Liénard, 101, 108
Limit cycles, 71
definition, 71
existence, 3
nonanalytic, 96, 703
piecewise linear, 690
physical significance, 74
semistable, 73
stable, 72
unstable, 73
Lindstedt, 224
Linear oscillations, 10
Linear systems with constant coefficients, 124
with periodic coefficients, 127
Linearization, equivalent, 348
Lipshitz (see Cauchy)
Littlewood (see Cartwright), 672
Lochakov, 618
Lourj'e, 159

Ludeke, 484

Malkin, 118, 216, 309
Mandelstam, 461, 464, 488
Mathieu oscillator, linear, 401
  nonlinear, 404
Meissner (see Hill), 503
Minorsky, 390, 444, 496, 523, 534, 541, 566, 592
Mischenko, 709
Mitropolsky, 380, also see Bogoliubov-Mitropolsky
Mono-rail car oscillations, 301
Motion in the large, 45
Myshkis, 516, 663

Nemitzky, 115, 649
Node (nodal point), 12, 15
Nonautonomous systems, 7, 234, 260, 356
Noncritical systems, 185
Nonlinear damping, 338
Nonstationary processes, 380
Nyquist criterion, extension of, 416

Orbital stability, 119
Ordinary point, 8
Oscillations, interaction of, 562
  nonresonance, 234
  resonance, 236
  self-excited, 270
Oscillator, parametric, 489
  several degrees of freedom, 270

Papalexi, 461, 464, 488
Parameter, bifurcation value, 47
  variable, 47
Parametric excitation, 488
Period correction, 244
Periodic motion, 234
  condition for, 234
  solution, 234
Perturbation method, 219
Phase plane, 8
  portrait, 545
Piecewise analyticity, 606
  linearity, 625, 688
Pinney, 516
Poincaré, 3, 41, 71, 163, 206, 228, 232, 461, 488
Poincaré-Bendixson theorem, 84
Point transformation method, 688
Poisson, 212
Pol (see van der Pol)
Pontriagin, 519, 709
Potential energy, 42
Power series method of approximation, 217, 228

Rayleigh, 438
Relaxation oscillations, 599
Resonance, nonlinear, 236
  passage through zones of, 357
  subharmonic, 464, 473
Restoring force, nonlinear, 42
Rocard, 495, 540

Sansone, 103
Schiffer, 216
Schmidt, 516
Secular terms, 212
Sekerska, 508
Separatrix, 45
Sideriades, 663
Singular points, 10, 45
  classification, 10, 14
  at infinity, 91
  in plane (φ), 9
Slow time, 383
Smith, 103
Solution, perturbed, 122
  unperturbed, 122
Spiral point (see focus)
Stable equilibrium, 137
Stability
  asymptotic, 119
  definition of, 137
  critical cases, 150
  in the large, 161
  Liapounov's second method, 134
  orbital, 119, 130
  of periodic solutions, 313
  Poincaré's criterion of, 47
  zones of, 319
Stationary oscillations (see limit cycles)
Stepanov, 113
Stoker, 87, 438, 664
Stroboscopic method, 390
  differential equations, 395
  plane, 391

Subharmonic resonance (see resonance)
Successor function, 165
Superposition principle, 562
Symmetry (principle of), 113
Synchronization, 438
  equation, 440
  mutual, 448
  by stroboscopic method, 444
  van der Pol theory, 439
Synchronous motor, asynchronous behavior, 197
  oscillations of, 202

Theodorchik, 419, 586
Topographic system, 86
Topological configurations, 98
Topology of relay systems, 699
Toroidal phase space, 204
Trajectory, definition of, 99
Transformations, 390
Tricomi, 197

Undamped oscillations (see limit cycles)
Unstable, equilibrium, 11
Urabe, 413

Van der Pol equation, 219
  relaxation oscillations, 599
  solution, 219, 222
  theory of interaction, 563
  theory of synchronization, 439
  topology of, 557
Variational equations, 118, 121, 124, 127
Vlasov, 202
Vogel, 649
Volterra, 65

Wasov, 617, 707
Whitney, 648
Witt (see Andronov)

Yorinaga, 413